Critical Reviews of

OXIDATIVE STRESS
AND AGING

Advances in Basic Science, Diagnostics and Intervention

Volume I

Critical Reviews of

OXIDATIVE STRESS AND AGING

Advances in Basic Science, Diagnostics and Intervention

Volume I

Editors

Richard G. Cutler

Kronos Longevity Research Institute, Arizona, USA

Henry Rodriguez

National Institute of Standards and Technology, Gaithersburg, Maryland, USA

 **World Scientific**
New Jersey • London • Singapore • Hong Kong

Published by

World Scientific Publishing Co. Pte. Ltd.

5 Toh Tuck Link, Singapore 596224

USA office: Suite 202, 1060 Main Street, River Edge, NJ 07661

UK office: 57 Shelton Street, Covent Garden, London WC2H 9HE

British Library Cataloguing-in-Publication Data
A catalogue record for this book is available from the British Library.

ISBN 981-02-4636-6 (Set)
ISBN 981-238-996-2 (Vol. 1)
ISBN 981-238-997-0 (Vol. 2)

Printed in Singapore by Uto-Print

I dedicate this book to my wife Jessica and sons Ronald and Roman that have not only been so encouraging, but also understanding of why science is so important to me particularly when it deals with the possibility of significantly extending the healthy and productive years of human life span.

Richard G. Cutler

This book is dedicated to my parents Ricardo and Micaela Rodriguez, who taught me to believe in myself and to my wife Joy and daughter Dana who gives me the strength to take on and conquer the many challenges that life has to offer.

Henry Rodriguez

Acknowledgments

The editors of *Critical Reviews of Oxidative Stress and Aging: Advances in Basic Science, Diagnostics, and Intervention* and The Oxidative Stress and Aging Association (O2SA) would like to thank the following associations for their outstanding patronage and support of the Second International Conference on Oxidative Stress and Aging and their continuing support towards advancing the field of Oxidative Stress and Aging in research, diagnostics and therapeutics worldwide.

American Aging Association
American Federation For Aging Research
Cayman Research
Ciphergen
Galileo Laboratories
Great Smokies Diagnostic Laboratories
HealthSpan Sciences
InterHealth
Karger
Kronos
Medinox
MitoKor
National Institute of Standards and Technology
NitroMed
Oxford Biomedical Research
Polyphenols Laboratory
The Kronos Longevity Research Institute
Zenith Technology

Preface

The fastest-growing segment of the United States population is age 65 years and older. At current rates, demographers forecast that by the year 2025, 65 year-olds will outnumber teenagers by almost 2:1. Fifty percent or more of the "baby boomer" population are projected to live to be Centenarians. Outside the United States, demographic trends are similar. As longer life span becomes a reality, there is a new emphasis on diagnostic and therapeutic applications that not only prolong life, but also increase the quality of life and productivity of the aging population.

Aging is associated with an increased incidence and prevalence of numerous diseases, many of them chronic. It is now known that free radicals cause extensive damage to cellular components that can lead to serious dysfunctions and death. Recently, free radicals have been associated with the aging process, several clinical disorders and a range of age-related diseases including atherosclerosis, cancer, and neurodegenerative diseases such as Parkinson's and Alzheimer's.

As a result, a new market is emerging: Oxidative Stress Diagnostics and Therapeutics. This new discipline in the field of Clinical Age Management and Longevity Medicine is revolutionizing the economics, science and practice of health care in new ways. Virtually every medical discipline will be reshaped by longevity research and its social implications.

The purpose of *Critical Reviews of Oxidative Stress and Aging: Advances in Basic Science, Diagnostics and Intervention* is to provide a comprehensive review of the most up-to-date information pertaining to the translational research field of oxidative stress and aging. **The book focuses on understanding the molecular basis of oxidative stress and its associated age-related diseases with the goal being the development of new and novel methods in treating human aging processes.** Over 100 of the leading experts in this field whose specialty includes biogerontology, geriatric medicine, free radical chemistry and biology, oncology, cardiology, neurobiology, dermatology, pharmacology, nutrition, and molecular medicine, have contributed information to this book. Accordingly, this reference book is essential reading to a broad range of individuals including researchers, physicians, corporate industry leaders, graduate and medical school students, as well as the many health conscious individuals who wish to learn more about the emerging field of oxidative stress and aging with an emphasis on diagnostics and intervention.

This is an exciting time to be involved in oxidative stress and longevity research. Answers to some of the most intractable problems in the field now appear to be within reach.

Richard G. Cutler   Henry Rodriguez

List of Contributors

In writing this book, the following authors have provided invaluable information and data. Special thanks are due to the following contributors (in alphabetical order), but the responsibility for any errors or omissions in the text is entirely that of the authors.

Acworth, Ian N.
Adinolfi, Christy
Akman, Steven A.
Alessio, Helaine M.
Allen, R.G.
Alleva, Renata
Ames, Bruce N.
Arik, Tali
Arking, Robert
Arteel, Gavin E.
Ayala, Antonio
Azzi, Angelo
Bagchi, Debasis
Balin, Arthur K.
Banks, Dwayne A.
Baudry, Michel
Beckman, Joseph S.
Block, Gladys
Blumberg, Jeffrey B.
Bogdanov, Mikhail B.
Bohr, Vilhelm
Bonkovsky, Herbert L.
Bourdat, A.-G.
Bray, Tammy M.
Bronkovsky, Herbert L.
Brown-Borg, Holly M.
Buettner, Garry R.
Cadet, Jean
Cai, Jiyang

Cao, Guohua
Carrillo, Maria-Cristina
Carroll, A.K.
Chen, Hsiu-Hua
Chen, XinLian
Christen, Stephan
Clarke, Charlotte H.
Clarke, Mark S.F.
Coles, L. Stephen
Collins, Andrew R.
Cooke, Marcus S.
Cutler, Richard G.
Cutler, Roy G.
Dai, Shu-Mei
Davies, Kelvin J. A.
de Grey, Aubrey D. N. J.
Dietrich, Marion
Dizdaroglu, Miral
Doctrow, Susan
Douki, T.
Duthie, Susan J.
Dykens, James A.
Elbirt, Kimberly K.
Flanagan, Steven D.
Fleck, B.
Fossel, Michael
Frelon, S.
Giorgio, Marco
Goto, Sataro

Griffiths, Helen R.
Harman, S. Mitchell
Hattori, Itaro
Heward, Christopher B.
Hiai, Hiroshi
Hirano, Takeshi
Hoberg, A-M.
Holmgren, Arne
Holmquist, Gerald P.
Huffman, Karl
Ivy, Gwen O.
Janero, David R.
Jaruga, Pawel
Jazwinski, S. Michal
Jensen, B.R.
Jiang, Qing
Jones, Dean P.
Jones, George D.D.
Joseph, James A.
Kanai, Setsuko
Kang, Y. James
Kasai, Hiroshi
Kelley, Eric E.
Khanna, Savita
Kirkwood, Thomas B.L.
Kitani, Kenichi
Kregel, Kevin C.
Krinsky, Norman I.
Lai, Ching-San (Monte)
Li, Jun
Loft, Stephen
Luccy, Britt
Lunec, Joseph
Machado, Alberto
Malfroy, Bernard
Marcotte, Richard
Marcus, Catherine Bucay
Maruyama, Wakako
Mason, Ronald P.
Masutani, Hiroshi
Matson, Wayne R.
Mattson, Mark P.

McCord, Joe M.
Mele, James
Melov, Simon
Meydani, Mohsen
Milbury, Paul
Minami, Chiyoko
Mitsui, Akira
Monnier, Vincent M.
Montine, Thomas J.
Morrow, Jason D.
Nakamura, Hajime
Niki, Etsuo
Nishinaka, Yumiko
Noguchi, Noriko
Norkus, Edward
O'Connor, Timothy R.
Okamoto, Takashi
Orida, Norman K.
Osawa, Toshihiko
Ozeki, Munetaka
Packer, Lester
Pelicci, Pier G.
Pong, Kevin
Pouget, J.-P.
Poulsen, Henrik E.
Preuss, Harry G.
Radák, Zsolt
Ravanat, J.-L.
Rayment, S. J.
Reaven, Peter
Reich, Erin E.
Remmen, Holly Van
Rice-Evans, Catherine
Richardson, Arlan
Riggs, Arthur D.
Roberts, L. Jackson
Rodell, Timothy C.
Rodriguez, Henry
Rong, Yongqi
Routledge, Michael N.
Roy, Sashwati
Saxena, Amit

Saxena, Poonam
Schafer, Freya Q.
Schroeter, Hagen
Seidel, Chris
Sell, David R.
Sen, Chandan K.
Sevy, Alexander
Shigenaga, Mark K.
Shmookler Reis, Robert J.
Shukitt-Hale, Barbara
Sies, Helmut
Singh, Keshav K.
Sinibaldi, Ralph M.
Smart, Janet L.
Sorenson, M.
Souza-Pinto, Nadja C.
Spencer, Jeremy P. E.
Spickett, Corinne M.
Sternberg, Jr., Paul
Subramaniam, Ram
Takahashi, Ryoya
Tanaka, Tomoyuki
Teoh, Cheryl Y.

Terai, Hidetomi
Termini, John
Tessier, Frederic
Tetsuka, Toshifumi
Tocco, Georges
Toyokuni, Shinya
Vacanti, Joseph P.
Villeponteau, Bryant
Visarius, Theresa
Wang, Eugenia
Warner, Huber R.
Weimann, A.
Weinberger, Scot R.
Weiss, Miriam F.
Wilson, Rhoda
Wise, Bradley
Wright, A.C.
Yarosh, Daniel
Yodoi, Junji
Youdim, Kuresh A.
Zubik, Ligia
Zullo, Steven J.

Contents

Volume 1

VOLUME 1

SECTION 1

Historical Perspectives

Chapter 1

Metabolic Rate, Free Radicals and Aging

R.G. Allen and Arthur K. Balin*

R.G. Allen and **Arthur K. Balin** • Sally Balin Medical Center for Dematology, Cosmetic Surgery and Longevity Medicine, 110 Chesley Drive, Media, PA 19063

*Corresponding Author.
Tel: (610) 565-3300, E-mail: akbalin@aol.com

1. Introduction

Aging is a progressive, time-dependent deterioration in the capacity of an organism to respond adaptively to environmental change that results in increased vulnerability to death. The process is irreversible and occurs in all members of a population. Aging increases vulnerability to certain diseases; however, it is distinct from any known disease pathology. Aging crosses virtually all species barriers and unlike any known disease, aging affects all members of a species.[1] The precise cause of the phenomenon remains unknown. Numerous aging theories exist but all of these can be broadly categorized as either genetic or stochastic (probabilistic) in nature. In this discussion, we examine oxidative stress-associated mechanisms that may affect longevity.

2. Metabolic Rate and Lifespan (Rubner's Metabolic Potential)

The fact that each species exhibits a characteristic lifespan suggests that lifespan is controlled through inheritable mechanisms; however, the existing experimental data also suggest that environmental factors can strongly influence lifespan. One of the earliest attempts to explain aging on the basis of a single underlying mechanism, was presented in 1908 by Rubner.[2] He observed that the total lifetime energy expenditure per unit weight was similar in 5 different domesticated species of animals with dissimilar lifespans. On the basis of this observation, Rubner inferred that the number of molecular rearrangements possible in living material was limited and that the total amount of metabolic work possible during life was thus a fixed constant. This limit became known as the metabolic potential. A number of studies have challenged the existence of the universal metabolic potential originally proposed by Rubner.[3-5] Indeed, Rubner's own work would discount a universal fixed metabolic potential by showing that humans exert more metabolic work per unit weight than other animals that he examined.[6] In mammals alone, there are multiple metabolic potentials.[4, 5, 7] Furthermore, birds have unusually high metabolic potentials and long lifespans as compared to mammals.[8] Although it is true that unrelated species with different lifespans sometimes exhibit similar total energy expenditures during their life, it is also true that some closely related species can have vastly different metabolic potentials. For example, humans, which are genetically very similar to other primates have twice the metabolic potential of other primate species.[4] These observations suggests that metabolic potential and possibly lifespan are influenced by a relatively small number of genes.

Attempts have been made to relate surface area/body size[9] as well as brain/body size[10] ratios to species metabolic potential and lifespan. Although many correlations have been presented subsequently, the existence of such relationships remains a subject of intense controversy.[7, 9, 11] It is also noteworthy that Rubner's contemporaries questioned his hypothesis because it seemed to account only for effects measured at the level of the intact organism but was inadequate to explain variations in the rates of metabolism or the lifespans of different types of cells within an organism.[12] Indeed, although aging is believed to occur at a cellular level, no relationship between lifespan and metabolism at a cellular level has ever been described.

3. The Rate of Living Theory

While there is no universal constant that defines the relationship between metabolism and lifespan, there remains strong evidence that metabolic processes influence longevity especially in poikilothermic (cold-blooded) animals. The clearest description of this relationship was encapsulated by Pearl[13] in "The Rate of Living Theory", which stated that all species exhibit a genetically determined (inherited) metabolic potential and that species lifespan was therefore dependent on the rate of metabolism. In fact, subsequent studies have supported the hypothesis that poikilotherms (cold-blooded animals) exhibit a maximum energy expenditure that is species specific.[14–16] Furthermore, the lifespan of poikilothermic animals can be predictably altered by experimentally varying their metabolic rate[17–20] Although the metabolic potential of different phylogenetic groups is variable, within each group the rate of metabolism and lifespan are inversely correlated[4, 16, 21, 22] and the effects can be profound. For example, milkweed bugs, *Oncopeltus fasciatus*, raised at 18°C live 4-fold longer and consume oxygen at only about one-fourth the rate observed in insects reared at 30°C;[23] yet, the total amount of oxygen consumed during life by both groups of insects is similar. Similar results have been reported in a number of other insect species.[16] Restricting the flight activity of houseflies by confining them to small bottles more than doubles their lifespan as compared to high activity controls. Trout and Kaplan[24] observed that mutant *Drosophila*, which made characteristic twitching motions, exhibited a higher rate of metabolism and shorter life span than normal controls. However, the total volume of oxygen consumed during life was similar in the "shaker" mutants and normal controls. The total number of heartbeats is relatively constant in water fleas maintained under different temperatures regimes even though these conditions alter their heart rate and lifespan by more than 100%.[25] Many of the mutations that increase lifespan in nematodes also diminish aerobic metabolism.[26] These results suggest that if there is a genetically controlled clock that governs longevity, it runs in relationship to energy expenditure rather than time.

Regrettably few studies that have used molecular manipulations to increase lifespan in poikilotherms have made any serious attempt to determine metabolic potential. Failure to account for changes in total metabolic work may result in skewed or simply incorrect interpretations of experimental data. In some cases, effects arising entirely from stochastic mechanisms have been interpreted to support genetic regulation. For example, it is well known that exposure of various insect species to ionizing radiations can sometimes increase their lifespan. For decades this was explained as a result of stimulation of repair mechanisms; albeit, the factors involved in the repair could not be identified. A closer examination revealed that the effect was completely lost when the metabolic rate of controls and exposed organisms was normalized by restriction of physical activity.[27, 28] Measuring metabolic potential in houseflies revealed that it was decreased by radiation exposure whether or not lifespan is increased. This clearly indicated that the exposure resulted in unrepaired damage that was proportional to dose. Decreases in the rate of metabolism were able to increase lifespan even though the total amount of metabolic work possible had decreased. Hence, the "beneficial" effects of radiation were observed only under conditions that permitted manifestation of activity differences between groups. The repair mechanism postulated to account for radiation induced life-lengthening either does not exist or is insignificant to post-irradiation changes in longevity. Without an accurate determination of metabolic potential, based on oxygen consumption, interpretations of effects on lifespan, whether as a result of radiation exposure or genetic manipulation, must be viewed with great caution.

Only scant evidence exists for metabolic effects in determining mammalian longevity. It is important to note that comparisons of mammals maintained under different activity regimes are extremely difficult to design. Mammals exhibit hypertrophy in response to activity increases, whereas insects and other short-lived poikilotherms do not.[14] Furthermore, maintenance of mammals under sedentary conditions for prolonged periods results in atrophy and is generally deleterious, while similar conditions extend the life of cold-blooded animals. It has been observed that hibernating Turkish hamsters live significantly longer than hamsters prevented from hibernation.[29] Additionally, it has been reported that while moderate exercise is beneficial, high levels of physical activity can be deleterious in humans.[30] However, it would be extremely difficult to conclude that either of these studies establishes the validity of the Rate of Living Theory in mammals. A number of studies have tried to test the validity of the Rate of Living Theory by comparing lifespans as well as other biochemical parameters in different species that exhibit different basal rates of metabolism[3–5, 8] Unfortunately, because such comparisons cross species boundaries they actually test the Rubner hypothesis and not the Rate of Living Theory. Probably the most compelling studies are those that have maintained mammals at reduced temperatures to stimulate their metabolism. Exposure of rats to cold temperatures throughout their adult life was reported to cause a marked reduction in lifespan;[31–33] however, these studies were also confounded by upper respiratory infections. A more recent study used pathogen free animals and observed no effects on lifespan.[34]

Unfortunately, controls and experimental groups appear to have been administered similar numbers of calories in their diet. Because of this, the greater metabolic demands on the animals maintained in cold[35] would be expected to produced a dietary restriction effect that should have increased longevity; yet, no increase was observed. Evidence of a dietary restriction effect in this study is seen in the fact that the pathologies associated with death differed between control and experimental groups.[34] Nevertheless, this has been a promising area of investigation; albeit, further research is needed before conclusions can be drawn about the effects of metabolism on mammalian longevity.

4. The Free Radical Theory of Aging

One seemingly plausible explanation for the influence of metabolic rate on lifespan was that oxygen radicals generated in metabolic pathways damaged cells and ultimately killed them. Many of the effects of ionizing radiations in aqueous systems are mediated through free radical generation. A number of studies of ionizing radiation-induced life-shortening led some researchers to postulate that radiation exposure caused accelerated aging or a precocious aging effect (for review see Refs. 20 and 36). However, there was no immediate association between metabolic rate and free radical generation. Originally free radicals were believed to result from background radiation and to therefore play a very limited role, if any, in biological processes. There was no clear consensus that free radicals actually existed in living material at all. Indeed, as late as 1962, all of the data supporting the existence of free radicals in cells could be summarized in less than one page.[37] In 1954, an Electron Spin Resonance (ESR) spectrometer was first used to demonstrate that free radicals existed in cells[38] and due to artifacts associated with this type of procedure there was no widespread acceptance of appreciable free radical generation in cells until the 1969 discovery of superoxide dismutase activity by McCord and Fridovich.[39] Yet, even at this early stage, Harman suggested that free radical reactions could account for the progressive deterioration of biological systems with time because of their inherent ability to produce random change. He proposed that environmentally and metabolically generated free radicals were the underlying cause of the aging process.[40, 41] Since the "Free Radical Theory of Aging" was first presented, many changes have occurred in the perceived importance of free radicals in biological systems; nevertheless, the basic tenets of the Free Radical Theory of Aging remain largely unaltered. The theory can still be distilled to the simple premise that free radicals and peroxide generated intracellularly cause damage to cells and ultimately limits their ability to adapt to environmental change, thus, making them more vulnerable to death.[42] Tests of the Free Radical Theory have generally involved attempts to extend life with antioxidants or to show that aging-related changes result from oxidative damage.[43–46]

In contrast to the relatively minor role to which they were originally relegated, oxygen free radicals are now believed to play a fundamental role in a wide variety of pathologies and other biological phenomena as well as aging.[37, 47-50] As predicted by the free radical theory of aging, evidence of free radical damage can be found in most aerobic cells. Free radicals and oxidants modify proteins and inactivate enzymes,[51-59] damage DNA,[60-65] the cellular transcriptional machinery[66, 67] and initiate the chain reactions that peroxidize lipids.[68-70] Damage inflicted by reactive oxygen species is believed to be the underlying cause of accumulation of cellular lipofuscin,[71] a cause of ischemic damage,[49, 72, 73] to increase the incidence of neoplastic transformation[37, 74-77] and to promote metastasis.[78] Furthermore, it has been demonstrated that species longevity is inversely correlated with the rate of free radical generation.[79, 80]

5. Free Radicals, Damage and Aging

Although it has been possible to find evidence of free radical damage in cells and to demonstrate that such damage accumulates with age, the postulate that damage alone was the basis for aging left many unresolved questions. The free radical theory of aging has sometimes been criticized because most investigations have failed to identify the target molecules that are damaged during aging.[81] While it is relatively simple to damage molecules by free radical bombardment *in vitro*, the identification of damage *in vivo* that can account for any age-associated changes has remained elusive. An examination of age-associated changes reveals that neither gross structural damage to cellular components, nor decreased repair capacity can account for cellular dysfunction and death.[48, 61, 81, 82] A second problem stems from the fact that antioxidant administration decreases the rate of accumulation of some types of cellular damage, such as lipofuscin pigment,[83] but fails to increase maximum lifespan.[81, 84] Several studies indicate that antioxidants can increase mean lifespan,[85, 86] but their failure to influence maximum lifespan probably indicates that their effects are actually due to a decrease in vulnerability to aging-unrelated causes rather than an alteration of the rate of aging.[46, 81] These inconsistencies do not negate the free radical theory of aging but do highlight some of the limitations of using chemical antioxidants to control levels of cellular oxidation.

Many antioxidant compounds were actually developed for industrial uses such as stabilizing rubber or petroleum products. These agents tend to be distributed differently and metabolized much more rapidly than normal dietary antioxidants.[87] Furthermore, for some vitamins such as tocopherol, increased intake tends to decrease absorption,[88] and chemical antioxidants can themselves become toxic after they react with oxidants.[89] In order to avoid these problems, there have been several attempts to extend lifespan through overexpression of various antioxidant defense enzymes. *Drosophila,* overexpressing bovine Cu/Zn SOD (SOD-1), were reported to exhibit a statistically significant increase in lifespan,[90] but high levels of overexpression appeared to be toxic. The toxicity of SOD-1 overexpression was postulated to increase intracellular peroxide levels and therefore limit its effectiveness. No increase in lifespan was observed when the insect SOD-1 was used.[91] Similarly transfection of animals with catalase alone has no effect on lifespan.[92] When both SOD-1 and catalase were overexpressed, both mean and maximum lifespans were greatly extended.[93] In a subsequent study, it was also determined that simultaneous overexpression of SOD-1 and catalase increased the metabolic potential of flies exhibiting a greater maximum lifespan.[94] Extension of life was also reported in flies that overexpressed human SOD-1 only in their neurons.[95] These results support the view that when the location and by-products of antioxidant reactions are controlled, they exert very positive effects on maximum lifespan. Recently, it was reported that treatment of nematodes with SOD/catalase mimetics extended lifespan by up to 44%. Furthermore, these treatments were able to extend life without the decrease in fecundity or altered growth rate often seen in mutants with extended life;[96] however, the effects on aerobic metabolism were not reported. These studies clearly demonstrate the importance of free radicals in regulating longevity.

6. The Oxidant/Antioxidant Equilibrium

One of the more puzzling aspects of free radical biology is the apparent existence of an equilibrium between oxidants and antioxidants.[4] Organisms challenged by an oxidative stress often decrease their rate of metabolism, which presumably would lead to a corresponding decrease in their rate of free radical generation.[28, 97, 98] This type of compensatory response has been implicated as the underlying cause of the radiation resistance observed in post-mitotic organisms[27, 28] and may also account for the fact that some chemical oxidants fail to decrease longevity in poikilothermic organisms.[21, 22] Chemical oxidants also stimulate endogenous antioxidant defenses, which can prevent at least some damage.[99] A second aspect of the equilibrium is that administration of exogenous antioxidants tends to depress endogenous antioxidant levels.[22, 100, 101] This suggests that cells respond to either excessive oxidation or anti-oxidation. The net effect of these responses is to maintain a tightly controlled cellular balance between oxidizing and reducing equivalents. Given the deleterious nature of free radical reactions, the reason for maintaining such a balance is not immediately clear. More recently it has been determined that oxidants play an important role in certain types of cellular signaling, which may partly account for cellular resistance to a complete elimination of oxidants. Of course, this also indicates that free radicals may also influence lifespan by mechanisms that are independent of structural damage.

7. Active Oxygen and Signal Transduction

Reactive oxygen reactions influence molecular and biochemical processes and directly cause some of the changes observed in cells during differentiation, aging and transformation.[48, 102–107] On the basis of these observations, it has been suggested that cells evolved strategies to utilize ROS as biological stimuli.[48, 107–110] A possible link between genetic and stochastic mechanisms of aging is the fact that oxidants and reductants influence the expression of certain controlling genes and signaling pathways.[111] Oxidants are also thought to act as subcellular messengers for certain growth factors.[110, 112–114]

Many studies that have examined redox effects on gene expression have employed chemical oxidants and antioxidants, which raises some of the same concerns associated with the use of these compounds in aging studies. However, it seems improbable that secondary effects of chemical treatments rather than their oxidant/antioxidant properties are responsible for the effects on signal transduction and gene expression because oxidants have been used to block the effects of antioxidant compounds[115–118] and *vice versa*[66, 110, 119–137] in a number of independent studies. Choi and Moore[138] examined a series of structural isomers of BHT and observed that only those compounds with relatively high antioxidant potential are capable of inducing *c-fos*. Those structural analogs that lacked antioxidant properties fail to induce *c-fos*.[138] Furthermore, in several cases antioxidant enzymes, which presumably exert specific effects, have been employed to modulate signal transduction pathways and gene expression.[113, 119, 139–149] Taken together, these observations suggest that the redox potential of oxidant/antioxidant treatments rather than other secondary characteristics of these chemicals is, partly, if not totally responsible for their effects on gene expression.

According to Lander,[150] cellular responses stimulated by active oxygen and nitrogen species can be divided into five categories including (1) modulation of cytokine, growth factor or hormone action and secretion; (2) ion transport; (3) transcription; (4) neuromodulation; and (5) apoptosis. At a molecular level, the mechanisms of redox regulation of transcription factors and signal transduction are not fully understood; however, in many cases the effects of redox changes appear to be mediated through changes in the redox state of protein sulfhydryls.[107, 134, 151–154] Changes in the local redox state of protein sulfhydryls results in conformational changes that, depending on the protein, can either diminish or augment DNA binding activity,[107, 116, 124, 125, 155, 156] release inhibitory subunits[110, 157, 158] or promote protein complex formations[159, 160] that are necessary for signal transduction or transcription to proceed. In fact, the presence of multiple conserved cysteine residues in some PKCs[161] that could potentially serve as targets for redox regulation may explain redox effects on pathways influenced by PKC activation.[162, 163] It is also important to note that not all redox effects are direct. Oxidants induce surface down-modulation of the transferrin receptor through redistribution (internalization) rather than direct oxidation and inactivation or other modification of the receptor.[164] Also, antioxidants decrease the mRNA abundance of macrophage scavenger receptor-1 (MSR-1) by suppression of NF-κB and decreasing MSR-1 mRNA stability rather than through direct effects on transcription.[165]

The impact of redox regulation of certain genes on the aging process may not be apparent until it is considered that while the levels of most antioxidant defenses tend to remain stable with age,[166] the relative rates of oxidant generation greatly increase during aging. For example, the rates of superoxide ($\cdot O_2^-$)[79, 80, 167–171] and H_2O_2 generation[167, 171–175] increase in the cells of aging organisms while glutathione concentration declines progressively with advancing age.[176–179]

The importance of age-associated changes in redox status, and subsequent effects on regulatory pathways, in the aging process has not been determined. The search for this type of "damage" will be somewhat more challenging than previous demonstrations of structural damage to various cellular components. Deleterious changes in redox regulation of gene expression that result from unfavorable redox changes in the surrounding milieu will not destroy a physical target nor will damage of this "target" result in an accumulation of debris. Of course, changes in signal transduction can be measured but demonstrating which point of a pathway is affected and that the change results from oxidation is more difficult than the simple demonstrations of chemical alterations to various classes of cellular macromolecules that has typified much of free radical research in aging. Nevertheless such demonstrations are possible.[66, 67, 107]

8. Summary

A relationship between metabolic work and lifespan has long been suspected although the basis of the relationship is still unknown. One link between metabolic rate and lifespan is the production of deleterious oxidants in aerobic metabolic pathways. Oxidants and antioxidants influence a number of cellular processes including aging. What remains to be determined is the extent to which free radicals and other oxidants control lifespan. Although genetic influences impact on longevity as well as the levels of oxidant production and removal, it is known that cellular oxidation/reduction potential influences genetic controls over at least some cellular pathways. It is also possible that at least some redox sensitive pathways are important to the aging process.

References

1. Haflick, L. (1999). Aging is not a disease. *Aging Clin. Exp. Res.* **10**: 146.
2. Rubner, M. (1908). *Das Problem der Lebensdauer und seine Beziehunger zum Wachstum und Ernaibrung*, Oldenburg, Munich.

3. Cutler, R. G. (1985). Antioxidants and longevity in mammalian species. *In* "Molecular Biology of Aging" (A. D. Woodhead, A. D. Blackett, and A. Hollaender, Eds.), pp. 15–73, Plenum Press, New York.

4. Cutler, R. G. (1984). Antioxidants, aging and longevity. *In* "Free Radicals in Biology" (W. A. Pryor, Ed.), pp. 371–428, Academic Press, New York.

5. Austad, S. N. and Fischer, K. E. (1991). Mammalian aging, metabolism, and ecology: evidence from bats and marsupials. *J. Gerontol.* **46**: B47–B53.

6. Rubner, M. (1909). *Kraft und Stoff im Haushalte der Natur*, Liepzig.

7. West, G. B., Brown, J. H. and Enquist, B. J. (1999). The fourth dimension of life: fractal geometry and allometric scaling of organisms. *Science* **284**: 1677–1679.

8. Holmes, D. J. and Austad, S. N. (1995). Birds as animal models for the comparative biology of aging: a prospectus. *J. Gerontol. A Biol. Sci. Med. Sci.* **50**: B59–B66.

9. McMahon, T. (1973). Size and shape in biology. *Science* **179**: 1201–1204.

10. Friedenthal, H. (1910). Über die gültigkeit der massenwirkung für der energieumsatz der lebendigen substanz. *Zentralbl. Physiologie* **24**: 321–327.

11. Miller, R. and Austad, S. (1999). Large animals in the fast lane. *Science* **285**: 199.

12. Child, C. M. (1915). *Senescence and Rejuvenescence*, Chicago University Press, Chicago.

13. Pearl, R. (1928). *The Rate of Living*, A.A. Knopf, New York.

14. Sohal, R. S. (1976). Metabolic rate and life span. *Interdiscipl. Top. Gerontol.* **9**: 25–40.

15. Sohal, R. S. (1982). Oxygen consumption and adult life span in the adult housefly, *Musca domestica. Age* **5**: 21–24.

16. Sohal, R. S. (1986). The rate of living theory: a contemporary interpretation. *In* "Comparative Biology of Insect Aging: Strategies and Mechanisms" (K. G. Collatz, and R. S. Sohal, Eds.), pp. 23–44, Springer-Verlag, Heidelberg.

17. Liu, R. K. and Walford, R. L. (1975). Mid-life temperature transfer effects on life-span of the annual fish. *J. Gerontol* **30**: 129–131.

18. Ragland, S. S. and Sohal, R. S. (1973). Mating behavior, physical activity and aging in the housefly, *Musca domestica. Exp. Gerontol.* **8**: 135–145.

19. Miquel, J., Lungren, P. R., Bensch, K. J. and Atlan, H. (1976). Effect of temperature on the life span vitality and fine structure of *Drosophila melanogaster. Mech. Ageing Dev.* **5**: 347–370.

20. Sacher, G. A. (1977). Life table modification and life prolongation. *In* "The Handbook of the Biology of Aging" (C. E. Finch, and L. Hayflick, Eds.), pp. 582–638, Van Nostrand Reinhold, New York.

21. Sohal, R. S. and Allen, R. G. (1985). Relationship between metabolic rate, free radicals, differentiation and aging: a unified theory. *In* "The Molecular Biology of Aging (Brookhaven Symposium)" (A. Woodhead, A. D. Blackett, and A. Hollaender, Eds.), pp. 75–104, Plenum Press, New York.

22. Sohal, R. S. and Allen, R. G. (1986). Relationship between oxygen metabolism, aging and development. *Adv. Free Radic. Biol. Med.* **2**: 117–160.

23. McArthur, M. C. and Sohal, R. S. (1982). Relationship between metabolic rate, aging, lipid peroxidation and fluorescent age pigment in milkweed bug, *Oncopeltus fasciatus* (Hemiptera). *J. Gerontol.* **37**: 268–274.

24. Trout, W. E. and Kaplan, W. D. (1970). A relationship between longevity and activity in shaker mutants of *Drosophila melanogaster*. *Exp. Gerontol.* **5**: 83–92.

25. MacArthur, J. W. and Bailie, W. H. T. (1929). Metabolic rate and the duration of life. II. Metabolic rates and their relation to longevity in *Daphnia magna*. *J. Exp. Zool.* **53**: 243–286.

26. Van Voorhies, W. A. and Ward, S. (1999). Genetic and environmental conditions that increase longevity in *Caenorhabditis elegans* decrease metabolic rate. *Proc. Natl. Acad. Sci. USA* **96**: 11 399–11 403.

27. Allen, R. G. (1985). Relationship between γ-irradiation, life span, metabolic rate and accumulation of fluorescent age pigment in the adult male housefly *Musca domestica*. *Arch. Gerontol. Geriatr.* **4**: 169–178.

28. Allen, R. G. and Sohal, R. S. (1982). Life-lengthening effects of γ-radiation on the adult male housefly, *Musca domestica*. *Mech. Ageing Dev.* **20**: 369–375.

29. Lyman, C. P., O'Brian, R. C., Green, G. C. and Papafrangos, E. D. (1981). Hibernation and longevity in the Turkish hamster *Mesocritus brandti*. *Science* **212**: 668–670.

30. Paffenbarger, R. S., Hyde, R. T., Wing, A. L. and Hsieh, C. C. (1986). Physical activity, all-cause mortality, and longevity of college alumni. *New England J. Med.* **314**: 605–613.

31. Johnson, H. D., Kinter, I. D. and Kibler, H. H. (1963). Effects of 48°F. (8.9°C) and 83°F. (28.4°C) on the longevity and pathology of male rats. *J. Gerontol.* **18**: 235–239.

32. Kibler, H. H., Silsby, H. D. and Johnson, H. D. (1963). Metabolic trends and life spans of rats living at 9°C and 28°C. *J. Gerontol.* **18**: 235–239.

33. Kibler, H. H. and Johnson, D. (1966). Temperature and longevity in rats. *J. Gerontol.* **21**: 52–56.

34. Holloszy, J. O. and Smith, E. K. (1986). Longevity of cold-exposed rats: a reevaluation of the "rate-of-living theory". *J. Appl. Physiol.* **61**: 1656–1660.

35. Kiang-Ulrich, M. and Horvath, S. M. (1985). Age-related differences in food intake, body weight, and survival of male F344 rats in 5°C cold. *Exp. Gerontol.* **20**: 107–117.

36. Lamb, M. J. (1977). *Biology of Aging*, Halsted Press/John Wiley and Sons, New York.

37. Harman, D. (1962). Role of free radicals in mutation, cancer aging and the maintenance of life. *Rad. Res.* **16**: 753–763.

38. Commoner, B., Townsend, J. and Pake, G. E. (1954). Free radicals in biological materials. *Nature* **174**: 689–691.

39. McCord, J. M. and Fridovich, I. (1969). Superoxide dismutase. *J. Biol. Chem.* **244**: 6049–6055.

40. Harman, D. (1955). Aging a theory based on free radical and radiation chemistry. Radiation Laboratory Report 3078. University of California.

41. Harman, D. (1956). Aging: a theory based on free radical and radiation biology. *J. Gerontol.* **11**: 298–300.

42. Harman, D. (1991). The aging process: major risk factor for disease and death. *Proc. Natl. Acad. Sci. USA* **88**: 5360–5363.

43. Harman, D. (1968). Free radical theory of aging: effect of free radical reaction inhibitors on the mortality rate of male LAF mice. *J. Gerontol.* **23**: 476–482.

44. Harman, D. (1979). Free radical theory of aging: beneficial effects of adding antioxidants to the maternal mouse diet on the lifespan of offspring; possible explanation of the sex difference in longevity. *Age* **2**: 109–122.

45. Harman, D. (1980). Free radical theory of aging: beneficial effects of antioxidants on the lifespan of male NZB mice: role of free radical reactions in the deterioration of the immune system and in the pathogenesis of systemic lupus erythematosus. *Age* **3**: 64 073.

46. Balin, A. K. (1982). Testing the free radical theory of aging. *In* "Testing the Theories of Aging" (R. C. Adelman, and G. C. Roth, Eds.), pp. 137–182, CRC Press, Boca Raton, FA.

47. Oberley, T. D., Allen, R. G., Schultz, J. L. and Lauchner, L. J. (1991). Antioxidant enzymes and steroid-induced proliferation of kidney tubular cells. *Free Radic. Biol. Med.* **10**: 79–83.

48. Sohal, R. S. and Allen, R. G. (1990). Oxidative stress as a causal factor in differentiation and aging: a unifying hypothesis. *Exp. Gerontol.* **25**: 499–522.

49. McCord, J. M. (1985). Oxygen-derived free radicals in postischemic tissue injury. *New England J. Med.* **312**: 159–163.

50. Oberley, L. W. (1988). Free radicals and diabetes. *Free Radic. Biol. Med.* **5**: 113–124.

51. Harman, D. (1960). Athrosclerosis: oxidation of serum lipoproteins and its relationship to pathogenesis. *Clin. Res.* **8**: 108.

52. Friguet, B., Stadtman, E. R. and Szweda, L. I. (1994). Modification of glucose-6-phosphate dehydrogenase by 4-hydroxy-2-nonenal. Formation of cross-linked protein that inhibits the multicatalytic protease. *J. Biol. Chem.* **269**: 21 639–21 643.

53. Stadtman, E. R., Oliver, C. N., Starke-Reed, P. E. and Rhee, S. G. (1993). Age-related oxidation reaction in proteins. *Toxicol. Ind. Heath* **9**: 187–196.

54. Kong, S. and Davidson, A. J. (1980). The role of the interactions between O_2, H_2O_2, $OH^\cdot$, e^-, and $^1O_2$ in the free radical damage to biological systems. *J. Biol. Chem.* **204**: 18–29.

55. Stadtman, E. R. (1988). Protein modification in aging. *J. Gerontol.* **43**: B112–B120.

56. Stadtman, E. R. (1990). Covalent modification reactions are marking steps in protein turnover. *Biochemistry* **29**: 6323–6331.

57. Szweda, L. I., Uchida, K., Tsai, L. and Stadtman, E. R. (1993). Inactivation of glucose-6-phosphate dehydrogenase by 4-hydroxy-2-nonenal. Selective modification of an active-site lysine. *J. Biol. Chem.* **268**: 3342–3347.

58. Agarwal, S. and Sohal, R. S. (1993). Relationship between aging and susceptability to protein oxidative damage. *Biochem. Biophys. Res. Commun.* **194**: 1203–1206.

59. Stadtman, E. R. (1993). Oxidation of free amino acids and amino acid residues in proteins by radiolysis and by metal-catalyzed reactions. *Ann. Rev. Biochem.* **62**: 797–821.

60. Brawn, K. and Fridovich, I. (1980). Superoxide radical and superoxide dismutases: threat and defense. *Acta Physiol. Scand. Suppl.* **492**: 9–18.

61. Newton, R. K., Ducore, J. M. and Sohal, R. S. (1989). Effect of age on endogenous DNA single-strand breakage, strand break induction and repair in the adult housefly, *Musca domestica*. *Mut. Res.* **219**: 113–120.

62. Agarwal, S. and Sohal, R. S. (1994). DNA oxidative damage and life expectancy in houseflies. *Proc. Natl. Acad. Sci. USA* **91**: 12 332–12 335.

63. von Zglinicki, T., Saretzki, G., Döcke, W. and Lotze, C. (1995). Mild hyperoxia shortens telomeres and inhibits proliferation of fibroblasts: a model for senescence? *Exp. Cell Res.* **220**: 186–192.

64. Yakes, F. M. and Houten, B. V. (1997). Mitochondrial DNA damage is more extensive and persists longer than nuclear DNA damage in human cells following oxidative stress. *Proc. Natl. Acad. Sci. USA* **94**: 514–519.

65. Reid, T. M. and Loeb, L. A. (1993). Tandem double CC → TT mutations are produced by reactive oxygen species. *Proc. Natl. Acad. Sci. USA* **90**: 3904–3907.

66. Ammendola, R., Mesuraca, M., Russo, T. and Cimino, F. (1994). The DNA binding efficiency of Sp1 is affected by redox changes. *Eur. J. Biochem.* **225**: 483–489.

67. Ammendola, R., Mesuraca, M., Russo, T. and Cimino, F. (1992). Sp1 DNA binding efficiency is highly reduced in nuclear extracts from aged rat tissues. *J. Biol. Chem.* **267**: 17 944–17 948.

68. Uchida, K., Toyokuni, S., Nishikawa, K., Kawakishi, S., Oda, H., Hiai, H. and Stadtman, E. R. (1994). Michael addition-type 4-hydroxy-2-nonenal adducts in modified low-density lipoproteins: markers for atherosclerosis. *Biochemistry* **33**: 12 487–12 494.

69. Pryor, W. A. (1976). The role of free radical reactions in biological systems. *In* "Free Radicals in Biology" (W. A. Pryor, Ed.), pp. 1–49, Academic Press, New York.

70. Rosen, G. M., Barber, M. J. and Rauckman, E. J. (1983). Disruption of erythrocyte membrane organization by superoxide. *J. Biol. Chem.* **258**: 2225–2228.

71. Harman, D. (1989). Lipofuscin and ceroid formation: the cellular recycling system. *Adv Exp. Med. Biol.* **266**: 3–15.
72. Kramer, J. H., Arroyo, C. M., Dickens, B. F. and Weglicki, W. B. (1987). Spin-trapping evidence that graded myocardial ishemia alters post-ischemic superoxide production. *Free Radic. Biol. Med.* **3**: 153–159.
73. Darley-Usmar, V. M., Smith, D. R., O'Leary, V. J., Hardy, D. L., Stone, D. and Clark, J. B. (1990). Hyperoxia-reoxygenation induced damage in the myocardium: the role of mitochondria. *Biochem. Soc. Trans.* **18**: 526–528.
74. Harman, D. (1961). Mutation, aging and cancer. *Lancet* **I**: 200–201.
75. Weitzman, S., Schmeichel, C., Turk, P., Stevens, C., Tolsma, S. and Bouck, N. (1988). Phagocyte-mediated carcinogenesis: DNA from phagocyte transformed C3H T10 1/2 cells can transform NIH/3T3 cells. *In* "Membrane in Cancer Cells" (T. Galeotti, A. Cittadini, G. Neri, and A. Scarpa, Eds.), pp. 103–110, *Ann. NY Acad. Sci.*, New York.
76. Oberley, L. W. and Oberley, T. D. (1984). The role of superoxide dismutase and gene amplification in carcinogenesis. *J. Theor. Biol.* **106**: 403–422.
77. Oberley, L. W. (1983). Superoxide dismutases in cancer. *In* "Superoxide Dismutase" (L. W. Oberley, Ed.), pp. 127–165, CRC Press, Boca Raton, FL.
78. Safford, S. E., Oberley, T. D., Urano, M. and St. Clair, D. K. (1994). Suppression of fibrosarcoma metastasis by elevated expression of manganese superoxide dismutase. *Cancer Res.* **54**: 4261–4265.
79. Sohal, R. S., Arnold, L. A. and Sohal, B. H. (1990). Age-related changes in antioxidant enzymes and prooxidant generation in tissues of the rat with special reference to parameters in two insect species. *Free Radic. Biol. Med.* **9**: 495–500.
80. Sohal, R. S., Svensson, I., Sohal, B. H. and Brunk, U. T. (1989). Superoxide radical production in different animal species. *Mech. Ageing Dev.* **49**: 129–135.
81. Mehlhorn, R. J. and Cole, G. (1985). The free radical theory of aging: a critical review. *Adv. Free Radic. Biol. Med.* **1**: 165–223.
82. Newton, R. K., Ducore, J. M. and Sohal, R. S. (1989). Relationship between life expectancy and endogenous DNA single-strand breakage, strand break induction and DNA repair capacity in the adult housefly, *Musca domestica*. *Mech. Ageing Dev.* **49**: 259–270.
83. Nandy, K. (1984). Effects of antioxidants on neuronal lipofuscin pigment. *In* "Free Radicals in Molecular Biology, Aging and Disease" (D. Armstrong, R. S. Sohal, R. G. Cutler, and T. F. Slater, Eds.), pp. 223–233, Raven Press, New York.
84. Kohn, R. R. (1971). Effects of antioxidants on the lifespan of C57BL mice. *J. Gerontol.* **26**: 378–380.
85. Harman, D. (1982). The free radical theory of aging. *In* "Free Radicals in Biology" (W. A. Pryor, Ed.), pp. 255–275, Academic Press, New York.

86. Harman, D. (1984). Free radicals in aging. *Mol. Cell. Biol.* **84**: 155–161.
87. Whitting, L. A. (1980). Vitamin E and lipid antioxidants in free radical initiated reactions. *In* "Free Radicals in Biology" (W. A. Pryor, Ed.), pp. 295–319, Academic Press, New York.
88. Losowsky, M. S., Kelleher, J., Walker, B. E., Davies, T. and Smith, C. L. (1972). Intake and absorption of tocopherol. *Ann. NY Acad. Sci.* **203**: 212–222.
89. Burton, G. W. and Ingold, K. U. (1984). β-carotene: an unusual type of lipid antioxidant. *Science* **224**: 569–573.
90. Reveillaud, I., Neidzwiecki, A., Bensch, K. G. and Fleming, J. E. (1991). Expression of bovine superoxide dismutase in *Drosophila melanogaster* augments resistance to oxidative stress. *Mol. Cell. Biol.* **11**: 632–640.
91. Orr, W. C. and Sohal, R. S. (1993). Effects of Cu-Zn superoxide dismutase overexpression of life span and resistance to oxidative stress in transgenic *Drosophila melanogaster*. *Arch. Biochem. Biophys.* **301**: 34–40.
92. Orr, W. C. and Sohal, R. S. (1992). The effects of catalase gene overexpression on life span and resistance to oxidative stress in transgenic *Drosophila melanogaster*. *Arch. Biochem. Biophys.* **297**: 35–41.
93. Orr, W. C. and Sohal, R. S. (1994). Extension of lifespan by overexpression of superoxide dismutase and catalase in *Drosophila melanogaster*. *Science* **263**: 1128–1130.
94. Sohal, R. S., Agarwal, A., Agarwal, S. and Orr, W. C. (1995). Simultaneous overexpression of copper- and zinc-containing superoxide dismutase and catalase retards age-related oxidative damage and increases metabolic potential in *Drosophila melanogaster*. *J. Biol. Chem.* **270**: 15 671–15 674.
95. Parkes, T. L., Elia, A. J., Dickinson, D., Hilliker, A. J., Phillips, J. P. and Boulianne, G. L. (1998). Extension of Drosophila lifespan by overexpression of human SOD1 in motorneurons. *Nature Genetics* **19**: 171–174.
96. Melov, S., Ravenscroft, J., Malik, S., Gill, M. S., Walker, D. W., Clayton, P. E., Wallace, D. C., Malfroy, B., Doctrow, S. R. and Lithgow, G. J. (2000). Extension of life-span with superoxide dismutase/catalase mimetics. *Science* **289**: 1567–1569.
97. Allen, R. G., Farmer, K. J., Newton, R. K. and Sohal, R. S. (1984). Effects of paraquat administration on longevity, oxygen consumption, lipid peroxidation, superoxide dismutase, catalase, glutathione reductase, inorganic peroxides and glutathione in the adult housefly. *Comp. Biochem. Physiol.* **78C**: 283–288.
98. Allen, R. G., Farmer, K. J. and Sohal, R. S. (1984). Effect of diamide administration on longevity, oxygen consumption, superoxide dismutase, catalase, inorganic peroxides and glutathione in the adult housefly, *Musca domestica*. *Comp. Biochem. Physiol.* **78C**: 31–33.
99. Halliwell, B. (1981). Free radicals, oxygen toxicity and aging. *In* "Age Pigments" (R. S. Sohal, Ed.), pp. 1–62, Elsevier/North Holland, Amsterdam.

100. Blakely, S. R., Slaughter, L., Adkins, J. and Knight, E. V. (1988). Effect of β-carotene and retinyl palmitate on corn oil-induced superoxide dismutase and catalase in rats. *J. Nutri.* **118**: 152–158.

101. Sohal, R. S., Allen, R. G., Farmer, K. J., Newton, R. K. and Toy, P. L. (1985). Effects of exogenous antioxidants on the levels of endogenous antioxidants, lipid-soluble fluorescent material and life span in the housefly, *Musca domestica*. *Mech. Ageing Dev.* **31**: 329–336.

102. Allen, R. G., Farmer, K. J., Toy, P. L., Newton, R. K., Sohal, R. S. and Nations, C. (1985). Involvement of glutathione in the differentiation of the slime mold, *Physarum polycephalum*. *Dev. Growth Differ.* **27**: 615–620.

103. Allen, R. G., Newton, R. K., Sohal, R. S., Shipley, G. L. and Nations, C. (1985). Alterations in superoxide dismutase, glutathione, and peroxides in the plasmodial slime mold *Physarum polycephalum* during differentiation. *J. Cell. Physiol.* **125**: 413–419.

104. Allen, R. G., Balin, A. K., Reimer, R. J., Sohal, R. S. and Nations, C. (1988). Superoxide dismutase induces differentiation in the slime mold, *Physarum polycephalum*. *Arch. Biochem. Biophys.* **261**: 205–211.

105. Beckman, B. S., Balin, A. K. and Allen, R. G. (1989). Superoxide dismutase induces differentiation in Friend erythroleukemia cells. *J. Cell. Physiol.* **139**: 370–376.

106. Allen, R. G. and Balin, A. K. (1989). Oxidative influence on development and differentiation: an overview of a free radical theory of development. *Free Radic. Biol. Med.* **6**: 631–661.

107. Abate, C., Patel, L., Rauscher, F. J. and Curran, T. (1990). Redox regulation of FOS and JUN DNA-binding activity in vitro. *Science* **249**: 1157–1161.

108. Allen, R. G. (1993). Free radicals and differentiation: the interrelationship of development and aging. *In* "Free Radicals in Aging" (B. P. Yu, Ed.), pp. 11–37, CRC Press, Boca Raton, FL.

109. Allen, R. G. (1991). Oxygen-reactive species and antioxidant responses during development: the metabolic paradox of cellular differentiation. *Proc. Soc. Exp. Biol. Med.* **196**: 117–129.

110. Schreck, R., Rieber, P. and Baeuerle, P. A. (1991). Reactive oxygen intermediates as apparently widely used messengers in the activation of the NF-κB transcription factor and HIV-1. *EMBO J.* **10**: 2247–2258.

111. Allen, R. G. and Tresini, M. (2000). Oxidative stress and gene regulation. *Free Radic. Biol. Med.* **28**: 463–499.

112. Wagner, A. M. (1995). A role for active oxygen species as second messengers in the induction of alternate oxidase gene expression in *Petunia hybrida* cells. *FEBS Lett.* **368**: 339–342.

113. Shibanuma, M., Kuroki, T. and Nose, K. (1991). Release of H_2O_2 and phosphorylation of 30 kilodalton proteins as early responses of cell cycle-dependent inhibition of DNA synthesis by transforming growth factor beta 1. *Cell Growth Diff.* **2**: 583–591.

114. Sundaresan, M., Yu, Z., Ferrans, K. I. and Finkel, T. (1995). Requirement for generation of H_2O_2 for platlet-derived growth factor signal transduction. *Science* **270**: 296–299.

115. Bannister, A. J., Cook, A. and Kouzarides, T. (1991). In vitro DNA binding of *Fos/Jun* and BZLF1 but not C/EBP is affected by redox changes. *Oncogene* **6**: 1243–1250.

116. Huang, R.-P. and Adamson, E. D. (1993). Characterization of the DNA-binding properties of the early growth response-1 (EGR-1) transcription factor: evidence for modulation by a redox mechanism. *DNA Cell Biol.* **12**: 265–273.

117. Galang, C. K. and Hauser, C. A. (1993). Cooperative DNA binding of the Human HoxB5 (Hox-2.1) protein is under redox regulation in vitro. *Mol. Cell. Biol.* **13**: 4609–4617.

118. Adler, V., Schaffer, A., Kim, J., Dolan, L. and Ronai, Z. (1995). UV irradiation and heat shock mediate JNK activation via alternate pathways. *J. Biol. Chem.* **270**: 26 071–26 077.

119. Maki, A., Berezesky, I. K., Fargnoli, J., Holbrook, N. J. and Trump, B. F. (1992). Role of $[Ca^{2+}]_i$ in induction of *c-fos*, *c-jun*, and *c-myc* mRNA in rat PTE after oxidative stress. *FASEB J.* **6**: 919–924.

120. Jimenez, L. A., Zanella, C., Fung, H., Janssen, Y. M., Vacek, P., Charland, C., Goldberg, J. and Mossman, B. T. (1997). Role of extracellular signal-regulated protein kinases in apoptosis by asbestos and H_2O_2. *Am. J. Physiol.* **273**: L1029–L1035.

121. DeForge, L. E., Preston, A. M., Takeuchi, E., Boxer, L. A. and Remick, D. G. (1993). Regulation of interleukin 8 gene expression by oxidant stress. *J. Biol. Chem.* **268**: 25 568–25 576.

122. Datta, R., Hallahan, D. E., Kharbanda, S. M., Rubin, E., Sherman, M. L., Huberman, E., Weichselbaum, R. R. and Kufe, D. W. (1992). Involvement of reactive oxygen intermediates in the induction of *c-jun* gene transcription by ionizing radiation. *Biochemistry* **31**: 8300–8306.

123. Kurata, S., Matsumoto, M. and Nakajima, H. (1996). Transcriptional control of the heme oxygenase gene in mouse M1 cells during their TPA-induced differentiation into macrophages. *J. Cell. Biochem.* **62**: 314–324.

124. Toledano, M. B. and Leonard, W. J. (1991). Modulation of transcription factor NF-κB binding activity by oxidation-reduction in vitro. *Proc. Natl. Acad. Sci. USA* **88**: 4328–4332.

125. Molitor, J. A., Ballard, D. W. and Greene, W. C. (1991). κB-specific DNA binding proteins are differentially inhibited by enhancer mutations and biological oxidation. *The New Biologist* **3**: 987–996.

126. Kurata, S. I., Matsumoto, M., Tsuji, Y. and Nakajima, H. (1996). Lipopolysaccharide activates transcription of the heme oxygenase gene in mouse M1 cells through oxidative activation of nuclear factor kappa-B. *Eur. J. Biochem.* **239**: 566–571.

127. Guyton, K. Z., Liu, Y., Gorospe, M., Xu, Q. and Holbrook, N. J. (1996). Activation of mitogen-activated protein kinase by H$_2$O$_2$. *J. Biol. Chem.* **271**: 4138–4142.

128. Rao, G. N., Glasgow, W. C., Eling, T. E. and Runge, M. S. (1996). Role of hydroperoxyeicosatetraenoic acids in oxidative stress-induced activating protein 1 (AP-1) activity. *J. Biol. Chem.* **271**: 27 760–27 764.

129. Das, K. C., Lewis-Molock, Y. and White, C. W. (1995). Activation of NF-κB and elevation of MnSOD gene expression by thiol reducing agents in lung adenocarcinoma (A549) cells. *Am. J. Physiol.* **269**: L588–L602.

130. Stevenson, M. A., Pollock, S. S., Coleman, C. N. and Calderwood, S. K. (1994). X-irradiation, phorbol esters, and H$_2$O$_2$ stimulate mitogen-activated protein kinase activity in NIH-3T3 cells through the formation of reactive oxygen intermediates. *Cancer Res.* **54**: 12–15.

131. Tate, D. J., Miceli, M. V. and Newsome, D. A. (1995). Phagocytosis and H$_2$O$_2$ induced catalase and metallothionein gene expression in human retinal pigment epithelial cells. *Invest. Ophthalmol. Vis. Sci.* **36**: 1271–1279.

132. Barker, C. W., Fagan, J. B. and Pasco, D. S. (1994). Down-regulation of P4501A1 and P4501A2 mRNA expression in isolated hepatocytes by oxidative stress. *J. Biol. Chem.* **269**: 3985–3990.

133. Miyazaki, Y., Shinomura, Y., Tsutsui, S., Yasunaga, Y., Zushi, S., Higashiyama, S., Taniguchi, N. and Matsuzawa, Y. (1996). Oxidative stress increases gene expression of heparin-binding EGF-like growth factor and amphiregulin in cultured rat gastric epithelial cells. *Biochem. Biophys. Res. Commun.* **226**: 542–546.

134. Hecht, D. and Zick, Y. (1992). Selective inhibition of protein tyrosine phosphatase activities by H$_2$O$_2$ and vanadate in vitro. *Biochem. Biophys. Res. Commun.* **188**: 773–779.

135. Savino, G., Briat, J. F. and Lobréaux, S. (1997). Inhibition of the iron-induced ZmFer1 maize ferritin gene expression by antioxidants and serine/threonine phosphatase inhibitors. *J. Biol. Chem.* **272**: 33 319–33 326.

136. Barchowsky, A., Munro, S. R., Morana, S. J., Vincenti, M. P. and Treadwell, M. (1995). Oxidant-sensitive and phosphorylation-dependent activation of NF-κB and AP-1 in endothelial cells. *Am. J. Physiol.* **269**: L829–L836.

137. Uchida, K., Shiraishi, M., Naito, Y., Torii, Y., Nakamura, Y. and Osawa, T. (1999). Activation of stress signaling pathways by the end product of lipid peroxidation — 4-hydroxy-2-nonenal is a potential inducer of intracellular peroxide production. *J. Biol. Chem.* **274**: 2234–2242.

138. Choi, H.-S. and Moore, D. D. (1993). Induction of *c-fos* and *c-jun* gene expression by phenolic antioxidants. *Mol. Endocrin.* **7**: 1596–1602.

139. Kiningham, K. K. and St. Clair, D. K. (1997). Overexpression of manganese superoxide dismutase selectively modulates the activity of *Jun*-associated transcription factors in fibrosarcoma cells. *Cancer Res.* **57**: 5265–5271.

140. Schmidt, K. N., Amstad, P., Cerutti, P. and Baeuerle, P. A. (1995). The roles of hydrogen peroxide and superoxide as messengers in the activation of transcription factor NF-κB. *Chem. Biol.* **2**: 13–22.

141. Kretz-Remy, C., Mehlen, P., Mirault, M. E. and Arrigo, A. P. (1996). Inhibition of IκB-α phosphorylation and degradation and subsequent NF-κB activation by glutathione peroxidase overexpression. *J. Cell Biol.* **133**: 1083–1093.

142. Larsson, R. and Cerutti, P. (1988). Oxidants induce phosphorylation of ribosomal protein S6. *J. Biol. Chem.* **263**: 17 452–17 458.

143. Li, Y. B. and Chan, P. H. (1998). Identification of the pro-oncogene *stathmin/op18* mRNA in the brain of mitochondrial Mn-superoxide dismutase-deficient mMice by a modified differential display PCR. *Mol. Brain Res.* **55**: 277–284.

144. Manna, S. K., Zhang, H. J., Yan, T., Oberley, L. W. and Aggarwal, B. B. (1998). Overexpression of manganese superoxide dismutase suppresses tumor necrosis factor-induced apoptosis and activation of nuclear transcription factor-κB and activated protein-1. *J. Biol. Chem.* **273**: 13 245–13 254.

145. Mikawa, S., Sharp, F. R., Kamii, H., Kinouchi, H., Epstein, C. J. and Chan, P. H. (1995). Expression of *c-fos* and *hsp70* mRNA after tramatic brain injury in transgenic mice overexpressing Cu/Zn-superoxide dismutase. *Mol. Brain Res.* **33**: 288–294.

146. Kamii, H., Kinouchi, H., Sharp, F. R., Koistinaho, J., Epstein, C. J. and Chan, P. H. (1994). Prolonged expression of *hsp70* mRNA following transient focal cerebral ischemia in transgenic mice overexpressing CuZn-superoxide dismutase. *J. Cereb. Blood Flow Metab.* **14**: 478–486.

147. Kamii, H., Kinouchi, H., Sharp, F. R., Epstein, C. J., Sagar, S. M. and Chan, P. H. (1994). Expression of *c-fos* mRNA after a mild focal cerebral ischemia in SOD-1 transgenic mice. *Brain Res.* **662**: 240–244.

148. Kondo, T., Sharp, F. R., Honkaniemi, J., Mikawa, S., Epstein, C. J. and Chan, P. H. (1997). DNA fragmentation and prolonged expression of *c-fos, c-jun,* and *hsp70* in kainic acid-induced neuronal cell death in transgenic mice overexpressing human CuZn-superoxide dismutase. *J. Cereb. Blood Flow Metab.* **17**: 241–256.

149. Li, J. J., Oberley, L. W., Fan, M. and Colburn, N. H. (1998). Inhibition of AP-1 and NF-κB by manganese-containing superoxide dismutase in human breast cancer cells. *FASEB J.* **12**: 1713–1723.

150. Lander, H. M. (1997). An essential role for free radicals and derived species in signal transduction. *FASEB J.* **11**: 118–124.

151. Walters, D. W. and Gilbert, H. F. (1986). Thiol/disulfide exchange between rabbit muscle phosphofructokinase and glutathione. *J. Biol. Chem.* **261**: 15 372–15 377.

152. Okuno, H., Akahori, A., Sato, H., Xanthoudakis, S., Curran, T. and Iba, H. (1993). Escape from redox regulation enhances the transforming activity of *fos*. *Oncogene* **8**: 695–701.

153. Meyer, M., Pahl, H. L. and Baeuerle, P. A. (1994). Regulation of the transcription factors NF-κB and AP-1 by redox changes. *Chem. Biol. Interact.* **91**: 91–100.

154. Cappel, R. E. and Gilbert, H. F. (1988). Thiol/dissulfide exchange between 3-hydroxy-3-methylglutaryl-CoA reductase and glutathione. A thermodynamically facile dithiol oxidation. *J. Biol. Chem.* **263**: 12 204–12 212.

155. Knoepfel, L., Steinkühler, C., Carrì, M.-T. and Rotilio, G. (1994). Role of zinc-coordination and of glutathione redox couple in the redox susceptibility of human transcription factor Sp1. *Biochem. Biophys. Res. Commun.* **201**: 871–877.

156. Iwata, E., Asanuma, M., Nishibayashi, S., Kondo, Y. and Ogawa, N. (1997). Different effects of oxidative stress on activation of transcription factors in primary cultured rat neuronal and glial cells. *Brain Res. Mol. Brain Res.* **50**: 213–220.

157. Pantopoulos, K. and Hentze, M. W. (1998). Activation of iron regulatory protein-1 by oxidative stress in vitro. *Proc. Natl. Acad. Sci. USA* **95**: 10 559–10 563.

158. Hentze, M. W., Rouault, T. A., Harford, J. B. and Klausner, R. D. (1989). Oxidative-reduction and the molecular mechanism of a regulatory RNA-protein interaction. *Science* **244**: 357–359.

159. Arnone, M. I., Zannini, M. and Di Lauro, R. (1995). The DNA binding activity and dimerization ability of the thyroid transcription factor I are redox regulated. *J. Biol. Chem.* **270**: 12 048–12 055.

160. Rao, G. N. (1996). Hydrogen peroxide induces complex formation of SHC-Grb2-SOS with receptor tyrosine kinase and activates Ras and extracellular signal-regulated protein kinases group of mitogen-activated protein kinases. *Oncogene* **13**: 713–719.

161. Ohno, S., Akita, Y., Konno, Y., Imajoh, S. and Suzuki, K. (1988). A novel phorbol ester receptor/protein kinase, nPKC, distantly related to the protein kinase C family. *Cell* **53**: 731–741.

162. Keogh, B. P., Tresini, M., Cristofalo, V. J. and Allen, R. G. (1995). Effects of cellular aging on the induction of *c-fos* by antioxidant treatments. *Mech. Ageing Dev.* **86**: 151–160.

163. Keogh, B. P., Allen, R. G., Tresini, M., Furth, J. J. and Cristofalo, V. J. (1998). Antioxidants stimulate transcriptional activation of the *c-fos* gene by multiple pathways in human fetal lung fibroblasts (WI-38). *J. Cell. Physiol.* **176**: 624–633.

164. Malorni, W., Testa, U., Rainaldi, G., Tritarelli, E. and Peschle, C. (1998). Oxidative stress leads to a rapid alteration of transferrin receptor intra-vesicular trafficking. *Exp. Cell Res.* **241**: 102–116.

165. De Kimpe, S. J., Änggård, E. E. and Carrier, M. J. (1998). Reactive oxygen species regulate macrophage scavenger receptor Type I, but not Type II, in the human monocytic cell line THP-1. *Mol. Pharmacol.* **53**: 1076–1082.

166. Matsuo, M. (1993). Age-related alterations in antioxidant defense. *In* "Free Radicals in Aging" (B. P. Yu, Ed.), pp. 143–181, CRC Press, Boca Raton, FL.

167. Farmer, K. J. and Sohal, R. S. (1989). Relationship between superoxide anion generation and aging in the housefly, *Musca domestica*. *Free Radic. Biol. Med.* **7**: 23–29.

168. Nohl, H. and Hegner, D. (1978). Do mitochondria produce oxygen radicals in vivo? *Eur. J. Biochem.* **82**: 563-567.

169. Nohl, H. (1986). Oxygen release in mitochondria: influence of age. *In* "Free Radicals, Aging, and Degenerative Disease" (J. E. Johnson, R. Walford, D. Harman, and J. Miquel, Eds.), pp. 79–97, Alan Liss, New York.

170. Hegner, D. (1980). Age-dependence of molecular and functional changes in biological membranes. *Mech. Ageing Dev.* **14**: 101–118.

171. Ku, H.-H., Brunk, U. T. and Sohal, R. S. (1993). Relationship between mitochondrial superoxide and hydrogen peroxide production and longevity of mammalian species. *Free Radic. Biol. Med.* **15**: 621–627.

172. Sohal, R. S., Svensson, I. and Brunk, U. T. (1990). Hydrogen peroxide production by liver mitochondria in different species. *Mech. Ageing Dev.* **53**: 209–215.

173. Allen, R. G., Farmer, K. J. and Sohal, R. S. (1983). Effect of catalase inactivation on the levels of inorganic peroxides, superoxide dismutase, glutathione, oxygen consumption and life span in adult houseflies, *Musca domestica*. *Biochem J.* **216**: 503–506.

174. Sohal, R. S. (1993). Aging, cytochrome oxidase activity, and hydrogen peroxide release by mitochondria. *Free Radic. Biol. Med.* **14**: 583–588.

175. Sohal, R. S. and Sohal, B. H. (1991). Hydrogen peroxide release by mitochondria increases during aging. *Mech. Ageing Dev.* **57**: 187–202.

176. Lang, C. A., Naryshkin, S., Schneider, D. L., Mills, B. J. and Linderman, R. D. (1992). Low blood glutathione levels in healthy aging adults. *J. Lab. Clin. Med.* **120**: 720–725.

177. Noy, N., Schwartz, H. and Gafni, A. (1985). Age-related changes in the redox status of rat muscle cells and their role in enzyme aging. *Mech. Ageing Dev.* **29**: 63–69.

178. Rikans, L. E. and Moore, D. R. (1988). Effects of aging on aqueous-phase antioxidants in tissues of male Fischer rats. *Biochem. Biophys. Acta* **966**: 269.

179. Sohal, R. S., Farmer, K. J., Allen, R. G. and Cohen, N. R. (1984). Effects of age on oxygen consumption, superoxide dismutase, catalase, glutathione, inorganic peroxides, and chloroform-soluble antioxidants in the adult male housefly, *Musca domestica*. *Mech. Ageing Dev.* **24**: 185–195.

SECTION 2

Free Radical Chemistry and Biology

Part II

Basic Radical Chemistry and Biology

Chapter 2

Spin-Trapping Methods for Detecting Superoxide and Hydroxyl Free Radicals *In Vitro* and *In Vivo*

Garry R. Buettner and Ronald P. Mason*

Garry R. Buettner • Free Radical Research Institute, The University of Iowa, Iowa City, IA 52242
Ronald P. Mason • Laboratory of Pharmacology and Chemistry, NIEHS, MD: F0-01, P.O. Box 12233, Research Triangle Park, NC 27709
*Corresponding Author.
Tel: (919) 541-3910, E-mail: mason4@niehs.nih.gov

1. Summary

Spin trapping has become a valuable tool in the study of transient free radicals as evidenced by the many investigations in which it has been employed.[1] Oxygen-centered radicals are of particular interest because they have been implicated in many adverse reactions *in vivo*. Their short lifetimes and broad line widths make many of these radicals difficult, if not impossible, to detect by direct electron spin resonance (ESR) in room temperature aqueous solutions. Spin trapping provides a means, in principle, to overcome these problems, but it is not without its pitfalls and limitations. We discuss some of these problems in this chapter.

2. Choice of Spin Trap

Two types of spin traps have been developed, nitrone and nitroso compounds. In aqueous solutions, however, oxygen-centered spin adducts in nitroso spin traps such as 2-methyl-2-nitrosopropane are, in general, quite unstable. Thus, the nitrone spin traps are by far the most popular. The most used radical trap for the study of oxygen-centered free radicals is 5,5-dimethyl-1-pyrroline N-oxide (DMPO), which has been used extensively to study superoxide[1, 2] and hydroxyl radicals[1, 3] in biochemical and biological systems.

3. Superoxide

The spin trapping of superoxide has been of such interest because of the involvement of superoxide in many physiological processes. DMPO/superoxide (the superoxide radical adduct of DMPO) has a distinctive spectrum ($a^N = 14.2$ G, $a_\beta^H = 11.3$ G, and $a_\gamma^H = 1.25$ G)[1] that is easily recognizable. However, other oxygen-centered adducts of DMPO such as alkoxyl have a similar appearance.[4] Thus, the real proof that the spectrum observed is indeed due to DMPO/superoxide is gained by using superoxide dismutase (SOD) to inhibit the signal.[5]

Although the DMPO/superoxide spectrum is distinctive, the spin trapping of superoxide is not without its problems. The actual reaction of superoxide with DMPO is very slow (k_{obs} is 60 M^{-1} sec^{-1} at pH 7 and only 30 M^{-1} sec^{-1} at pH 7.4).[6] Thus, in most superoxide-generating systems, the spin-trap concentration must be quite high (~0.1 M) in order to outcompete the self-decay, namely, spontaneous dismutation of superoxide. In addition, the DMPO/superoxide adduct is unstable, decaying by a first-order process with a half-life of about 60 s at pH 7.[7] Therefore, one must always be prepared to deal with a relatively weak signal; that is, the concentration of DMPO/superoxide will, under most circumstances, be much less than 10 μM.

4. Hydroxyl Radical

The DMPO/hydroxyl adduct is the most often reported radical adduct of DMPO ($a^N \cong a_\beta{}^H = 14.9$ G). Much of the interest in the spin trapping of ·OH is due to its formation in the superoxide-dependent Fenton reaction:

$$O_2{}^{\cdot -} + HO_2{}^\cdot \xrightarrow{\ H+\ } H_2O_2 + O_2 \tag{1}$$

$$O_2{}^{\cdot -} + Fe(III) \longrightarrow Fe(II) + O_2 \tag{2}$$

$$Fe(II) + H_2O_2 \longrightarrow {}^\cdot OH + OH^- + Fe(II). \tag{3}$$

Thus, SOD will inhibit DMPO/superoxide and/or DMPO/hydroxyl formation if this reaction sequence is operative. However, catalase will always inhibit the formation of hydroxyl in reaction (3) above. A failure of catalase to inhibit the formation of DMPO/hydroxyl when the superoxide-driven Fenton reaction is suspected indicates that something artifactual is occurring or that another mechanism must be sought.

Two additional SOD-inhibitable routes to DMPO/hydroxyl from DMPO/superoxide itself should be considered: the reduction of DMPO/superoxide[5] (a hydroperoxide) to the alcohol DMPO/hydroxyl, for example, by glutathione peroxidase;[8] and the possible homolytic cleavage of the oxygen–oxygen bond of DMPO/superoxide to produce free ·OH,[2] which is subsequently trapped by unreacted DMPO.[9] Finkelstein *et al.* indicate that approximately 3% of DMPO/superoxide decomposes to produce hydroxyl. Unfortunately, no experimental data or details are given to indicate how this estimate was made; thus, it is difficult to assess how this number should be used. Therefore, weak DMPO/hydroxyl signals that are not catalase inhibitable should always be viewed cautiously because they quite often are artifactual rather than the result of the spin trapping of free ·OH generated by the system under study. Possible sources of artifactual DMPO/hydroxyl signals are (1) hydrolysis of DMPO to produce DMPO/hydroxyl as an impurity signal;[10] (2) the one-electron oxidation of DMPO followed by hydration of DMPO$^+$;[12] (3) the apparently concerted hydrolysis-oxidation reaction by photochemically excited molecules;[12] and (4) the presence of a strong oxidant such as hypochlorous acid.[13]

To establish the existence of free hydroxyl radical in spin-trapping experiments, it is necessary to perform kinetic competition experiments with hydroxyl radical scavengers.[14] For example, ethanol, formate, and dimethyl sulfoxide can be used in these competition experiments because, upon hydroxyl radical attack, they form carbon-centered radicals that can subsequently be trapped by DMPO:

$$^\cdot OH + DMPO \xrightarrow{\ k_1\ } DMPO/hydroxyl \tag{4}$$

$$^\cdot OH + HCO_2{}^- \xrightarrow{\ k_2\ } CO_2{}^{\cdot -} + H_2O \tag{5}$$

$$CO_2^{\bullet -} + DMPO \xrightarrow{\;k_3\;} DMPO/{}^{\bullet}CO_2^{-} \tag{6}$$

$$^{\bullet}OH + CH_3CH_2OH \xrightarrow{\;k_4\;} {}^{\bullet}CH(OH)CH_3 + H_2O \tag{7}$$

$$^{\bullet}CH(OH)CH_3 + DMPO \xrightarrow{\;k_5\;} DMPO/{}^{\bullet}CH(OH)CH_3. \tag{8}$$

Most artifacts leading to DMPO/hydroxyl radical adduct formation will be excluded by the use of hydroxyl radical scavengers if the scavenger-derived radical adduct is detected, if a corresponding decrease in the DMPO/hydroxyl radical adduct concentration is found, and if quantitative kinetic criteria are used.[14]

Measurement of the initial rates of formation of the DMPO/hydroxyl and DMPO scavenger radical adducts removes the effects of the differential radical adduct decay rates.[14] Using this approach, the relative efficiency of two hydroxyl radical scavengers, if quantitatively predictable from the known rate constants, can be calculated for the reactions of the hydroxyl radical with these scavengers. For example, using formate (k_2) and ethanol (k_4) we can calculate k_2/k_4 from the ratio of the rates of formation of these two radical adducts:

$$\frac{k_2}{k_4} = \frac{d[DMPO/{}^{\bullet}CO_2^{-}]/dt}{d[DMPO/{}^{\bullet}CH(OH)CH_3]/dt} \times \frac{[CH_3CH_2OH]}{[HCO_2^{-}]}. \tag{9}$$

In Eq. (9), the ratio k_2/k_4 from spin trapping should agree with the ratio of rate constants for the reaction of the hydroxyl radical with these scavengers as determined from pulse radiolysis. It should be kept in mind that to arrive at this expression, it is assumed that the predominant route of scavenger radical decay is via the trapping reaction. This kinetic approach has been successfully applied to an enzyme-dependent hydroxyl radical-generating system.[15]

A similar approach has been presented by Buettner *et al.*[16] In this approach a ${}^{\bullet}OH$ scavenger is included in the spin-trapping mixture at a concentration calculated to reduce the intensity of the DMPO/hydroxyl signal by 50%. In other words, the rate of the reaction of ${}^{\bullet}OH$ with scavenger (Scav) is equal to its rate of reaction with DMPO:

$$k_{Scav}[Scav][{}^{\bullet}OH] = k_{DMPO}[DMPO][{}^{\bullet}OH] \tag{10}$$

$$[Scav] = k_{DMPO}[DMPO]/k_{scav}. \tag{11}$$

The spin trapping of hydroxyl radical scavenger-derived radicals is the most reliable method of detecting hydroxyl radical in complex biological systems. Examples of the success of this approach are the detection of spontaneous peroxynitrite homolysis to form the hydroxyl radical,[17] which was initially very controversial, and the detection of hydroxyl radical from the inactivation of mitochondrial aconitase by superoxide.[18]

Samuni *et al.*[19] have demonstrated that $O_2^{\cdot-}$ reacts very efficiently with DMPO/hydroxyl and DMPO/$\cdot$CH$_3$ radical adducts, destroying the nitroxide and thus producing an ESR-silent species. If the flux of superoxide is high enough, some DMPO radical adduct may not be observed because of its rapid removal. Thus, a high flux of superoxide would not be desirable if additional free radical reactions are expected in a superoxide spin-trapping system.

5. *In Vivo* Hydroxyl Radical Detection

The DMPO/hydroxyl adduct has also been detected from cell organelles, intact cells, and organs, and the DEPMPO/hydroxyl adduct has been detected in a living mouse, but the numerous hydroxyl radical-independent pathways to DMPO/hydroxyl make the interpretation of this data problematic.[20] In addition, the determination of free hydroxyl radical using the hydroxyl radical scavenger approach is also somewhat problematic because the presence of classic hydroxyl radical scavengers such as ethanol, dimethyl sulfoxide, or formate can have a severely perturbing influence on the system, especially at the high concentrations that are required to outcompete the reaction of any $\cdot$OH formed with the numerous biochemicals present at millimolar concentrations. Thus, we believe that the unambiguous determination that free hydroxyl has been spin-trapped requires very careful experimental design and interpretation, especially when the goal is to examine free radical production in cells, organs, or whole animals.

In the study of hydroxyl radical formation *in vivo*, we used the scavenging reaction in which the hydroxyl radical is converted into the methyl radical via its reaction with dimethyl sulfoxide, the most inert of the classical hydroxyl radical scavengers.[21] The methyl radical is then detected as its long-lived phenyl *N-tert*-butylnitrone (PBN) adduct. Alone, DMSO is relatively nontoxic with a 24 h LD$_{50}$ in the rat (*ip*) of 13.7 g/kg and is, therefore, an ideal reagent for the *in vivo* detection of the hydroxyl radical.

We usually assay untreated bile for radical adducts. Experiments are initiated by *ip* injection of DMSO containing PBN, followed by intragastric injection of ferrous sulfate or another hydroxyl radical-generating agent. The resulting PBN/methyl radical adduct (PBN/$\cdot$CH$_3$) is detected by ESR, and the DMSO- and iron-dependence of *in vivo* adduct formation is demonstrated using collection of bile into dipyridyl, which inhibits *ex vivo* hydroxyl radical generation. Collection directly into dipyridyl is necessary to stop *ex vivo* iron chemistry due to the iron excreted into the bile along with the radical adducts. Having unambiguously demonstrated *in vivo* iron-dependent free radical formation, we next determined the effect of Desferal in this system. Desferal is a ferric iron chelator used to treat iron overload. After the treatment of rats with ferrous sulfate and an *ip* injection of Desferal, the six-line signal from the PBN/$\cdot$CH$_3$ adduct was almost abolished, suggesting that Desferal can inhibit hydroxyl radical generation during iron overload, presumably by binding iron in the ferric state.[22]

Since the Fenton reaction requires hydrogen peroxide, we thought that a substance that catalyzes hydrogen peroxide formation would increase the signal. The activity of the herbicide paraquat (PQ^{2+}) is attributed to its ability to catalyze the formation of superoxide and, subsequently, hydrogen peroxide. The herbicide undergoes an enzymatic one-electron reduction to form the paraquat radical-cation, $PQ^{\cdot +}$, which is then oxidized by molecular oxygen to form the superoxide radical, $O_2^{\cdot -}$. Through its participation in repeated cycles of reduction and oxidation, PQ^{2+} catalyzes superoxide radical formation. The formation of superoxide radical and the resulting hydrogen peroxide during the "futile cycling" of PQ^{2+} is thought responsible for its pulmonary toxicity to man. The paraquat radical has been detected using direct ESR in microsomal,[23] hepatocyte,[24] alveolar type II, and Clara cell incubations.[25] Unexpectedly, when we administered paraquat to our iron-poisoned rat model, only a modest increase of radical adduct formation occurred.

In contrast to iron, radical adducts were detected in the bile of copper-poisoned rats only after they had been given paraquat.[26] Apparently hydrogen peroxide was limiting *in vivo* in the copper analog of the Fenton reaction.

$$Cu^{1+} + H_2O_2 \rightarrow Cu^{2+} + OH^- + {}^{\cdot}OH. \qquad (12)$$

When the experiment was repeated in the absence of copper or PQ^{2+}, no radical adducts were detected, thereby confirming the dependence of radical formation on the co-administration of both copper and PQ^{2+}. The fact that copper or PQ^{2+} alone caused little detectable radical adduct formation may be attributed to their inability to form hydroxyl radicals at detectable concentrations due to strong defense systems against oxidative stress in living organisms. For instance, GSH binds Cu^{1+} as a stable complex that reacts slowly, if at all, with hydrogen peroxide to form the hydroxyl radical.[27]

In an attempt to demonstrate PBN/${}^{\cdot}CH_3$ formation as detected in the bile of animals treated with Cu(II) and PQ^{2+} or with Fe^{2+}, ^{13}C-labeled DMSO was used.[21] The presence of hyperfine splittings in the ESR spectrum from ^{13}C ($I = 1/2$) allows an unambiguous assignment of the PBN/${}^{\cdot}CH_3$ radical adduct formed *in vivo*. The appearance of ^{13}C-hyperfine splittings is unambiguous proof that the PBN/${}^{13}CH_3$ radical adduct was formed. ESR detection of PBN/${}^{13}CH_3$ from DMSO has also been used to investigate hydroxyl radical generation in rats with chronic dietary iron loading.[28] Desferal completely inhibited *in vivo* hydroxyl radical generation stimulated by high dietary iron intake. No radical adducts were detected in rats which were fed the control diet for the same period of time. This was the first ESR evidence of hydroxyl radical generation in chronic iron-loaded rats.[28]

6. *In Vivo* and *In Vitro* Superoxide

Many studies are pursuing the possible production of superoxide or hydroxyl radicals by cell organelles, intact cells, and organs. The detection of superoxide by spin trapping with DMPO has been achieved in all of the above. For success, however, experimental protocols must allow for the relatively short lifetime of DMPO/superoxide[7] and the possible interference by metal catalysts such as iron.[29] For example, in studying free radicals produced in myocardial ischemia/reperfusion, Arroyo *et al.*[30] immediately froze the coronary effluents in liquid nitrogen to prevent spin adduct decay. By monitoring the ESR spectra of the effluents immediately after thawing, they were successful in observing DMPO/superoxide. The DMPO/superoxide adduct has also been detected in perfusate from isolated perfused rat livers subjected to ischemia/reperfusion.[31]

7. DTPA, EDTA, and Desferal

The presence of transition metals (particularly iron) and various chelating agents can significantly alter the results of spin-trapping experiments.[29] Although contaminating catalytic metals can be removed from buffer and biochemical systems,[32] this would be a difficult and uncertain (perhaps impossible) process for cells and organs. Thus, chelating agents are much needed tools. When studying a superoxide-generating system, EDTA will, in general, enhance the catalytic activity of iron in the reaction sequence,[1-3, 29] thereby increasing the yield of DMPO/hydroxyl while decreasing or eliminating the appearance of DMPO/superoxide. DTPA (diethylenetriaminepentaacetic acid) reduces or eliminates many of the problems generated by catalytic iron in superoxide-generating systems,[29] but under circumstances where a reducing agent stronger than superoxide is responsible for iron reduction, DTPA can increase DMPO/hydroxyl formation.[15] In studying stimulated neutrophils, Britigan *et al.*[33] found DTPA (1–100 μM) to be a very useful tool; it had no effect on neutrophil superoxide production or oxygen consumption, whereas it enhanced the detection of superoxide by DMPO in their cellular experiments.

The iron chelator Desferal (deferrioxamine mesylate) renders iron essentially catalytically inactive in reactions (2)–(3) above. Unfortunately, the hydroxamic acid moieties of Desferal can undergo one-electron oxidation by superoxide (mostly likely ˙OOH), hydroxyl radical, and horseradish peroxidase.[34-36] The nitroxide radical so formed is stable, for a free radical; nevertheless, it reacts rapidly with cysteine, methionine, glutathione, ascorbate, and a water-soluble form of vitamin E.[35] This radical may also deactivate enzymes, as demonstrated for alcohol dehydrogenase.[35] If Desferal is present at a relatively high concentration (compared to spin trap), it can effectively compete for superoxide and hydroxyl radical.[36] Since adventitious transition metals are present at only micromolar concentrations and spin traps are used at millimolar concentrations, scavenging by Desferal is perhaps less of a problem than the interference caused by the detection of the Desferal nitroxide radical itself. In any case, the Desferal concentration should be kept as low as possible to minimize scavenging.

Although DMPO is at least 20-fold more sensitive than the reduction of cytochrome c for the measurement of superoxide,[37] recent work has focused on the development of spin traps even more sensitive than DMPO. The first of these spin traps is 5-diethyloxy-phosphoryl-5-methyl-1-pyrroline $\underline{N}$-oxide (DEPMPO). Although the rate of superoxide trapping by DEPMPO is still relatively slow, the DEPMPO/superoxide adduct is 15-fold more persistent than the DMPO/superoxide adduct.[38] In fact, the use of DEPMPO has improved the detection of superoxide during reperfusion of ischemic rat hearts[38, 39] over that which was possible with DMPO,[30] although the experiment is still difficult. The DEPMPO/superoxide radical adduct is detectable from phorbal ester-activated polymorphornuclear leukocytes with as few as 2×10^3 cells.[40] Recently, a nitrone derivative, 5-ethoxycarboxyl-5-methyl-1-pyrroline $\underline{N}$-oxide (EMPO), has been synthesized and found to have a superoxide adduct that is 5-fold more persistent than the DMPO analogue without the considerable spectral complexity of the DEPMPO/superoxide adduct.[41] In fact, the ESR spectrum of EMPO/superoxide adduct is virtually identical to that of the DMPO/superoxide adduct. The ^{15}N-labeled EMPO increases the sensitivity of this radical adduct by 50% as a result of having one-third fewer lines.[42]

Since the introduction of nitrone spin traps as a tool for the detection superoxide and hydroxyl radical, thousands of publications have used this technique. Not a single artifact has been reported for the detection of superoxide with DMPO in all this time. In contrast, many hydroxyl radical-independent pathways to DMPO/hydroxyl have been discovered. Fortunately, the hydroxyl radical scavenger-derived trapping approach as exemplified by the trapping of methyl radical from DMSO also appears to be artifact free, although it may not be possible to distinguish the hydroxyl radical from hydroxyl-like species without careful kinetic studies. In retrospect, the nitrone spin-trapping approach to the detection of superoxide and hydroxyl radical has been an outstanding success.

In summary, success in biological spin trapping requires:

(1) the appropriate choice of spin trap for detecting the radicals of interest;
(2) an experimental design that considers the kinetics of the reaction of the radicals of interest with the spin trap and potential competing reactants;
(3) careful attention to possible artifacts;
(4) consideration of the chemistry that adventitious catalytic metals can introduce; and
(5) use of appropriate spectrometer parameters to obtain the best spectra possible in the lifetime of the free radical-generating system.

References

1. Buettner, G. R. (1987). Spin trapping: ESR parameters of spin adducts. *Free Radic. Biol. Med.* **3**: 259–303.
2. Thornalley, P. J. and Bannister, J. V. (1985). The spin trapping of superoxide radicals. *In* "Handbook of Methods for Oxygen Radical Research" (R. A. Greenwald, Ed.), pp. 133–136, CRC Press, Boca Raton, FL.
3. Buettner, G. R. (1985). Spin trapping of the hydroxyl radical. *In* "Handbook of Methods for Oxygen Radical Research" (R. A. Greenwald, Ed.), pp. 151–156, CRC Press, Boca Raton, FL.
4. Dikalov, S. I. and Mason, R. P. (1999). Reassignment of organic peroxyl radical adducts. *Free Radic. Biol. Med.* **27**: 864–872.
5. Finkelstein, E., Rosen, G. M., Rauckman, E. J. and Paxton, J. (1979). Spin trapping of superoxide. *Mol. Pharmacol.* **16**: 676–685.
6. Finkelstein, E., Rosen, G. M. and Rauckman, E. J. (1980). Spin trapping: kinetics of the reaction of superoxide and hydroxyl radicals with nitrones. *J. Am. Chem. Soc.* **102**: 4994–4999.
7. Buettner, G. R. and Oberley, L. W. (1978). Considerations in the spin trapping of superoxide and hydroxyl radical in aqueous systems using 5,5-dimethyl-1-pyrroline-1-oxide. *Biochem. Biophys. Res. Commun.* **83**: 69–74.
8. Rosen, G. M. and Freeman, B. A. (1984). Detection of superoxide generated by endothelial cells. *Proc. Natl. Acad. Sci. USA* **81**: 7269–7273.
9. Finkelstein, E., Rosen, G. M. and Rauckman, E. J. (1982). Production of hydroxyl radical by decomposition of superoxide spin-trapped adducts. *Mol. Pharmacol.* **21**: 262–265.
10. Floyd, R. A. and Wiseman, B. B. (1979). Spin-trapping free radicals in the autooxidation of 6-hydroxydopamine. *Biochim. Biophys. Acta* **586**: 196–207.
11. Chandra, H. and Symons, M. C. R. (1986). Hydration of spin-trap cations as a source of hydroxyl adducts. *J. Chem. Soc., Chem. Commun.* **16**: 1301–1302.
12. Chew, V. S. F. and Bolton, J. R. (1980). Photochemistry of 5-methylphenazinium salts in aqueous solution. Products and quantum yield of the reaction. *J. Phys. Chem.* **84**: 1903–1908.
13. Janzen, E. G., Jandrisits, L. T. and Barber, D. L. (1987). Studies on the origin of the hydroxyl spin adduct of DMPO produced from the stimulation of neutrophils by phorbol-12-myristate-13-acetate. *Free Radic. Res. Commun.* **4**: 115–123.
14. Castelhano, A. L., Perkins, M. J. and Griller, D. (1983). Spin trapping of hydroxyl in water: decay kinetics for the ·OH and $CO_2^{·-}$ adducts to 5,5-dimethyl-1-pyrroline-N-oxide. *Can. J. Chem.* **61**: 298–299.
15. Morehouse, K. M. and Mason, R. P. (1988). The transition metal-mediated formation of the hydroxyl free radical during the reduction of molecular oxygen by ferredoxin-ferredoxin: NADP + oxidoreductase. *J. Biol. Chem.* **263**: 1204–1211.

16. Buettner, G. R., Motten, A. G., Hall, R. D. and Chignell, C. F. (1986). Free radical production by chlorpromazine sulfoxide, an ESR spin-trapping and flash photolysis study. *Photochem. Photobiol.* **44**: 5–10.

17. Gatti, R. M., Alvarez, B., Vasquez-Vivar, J., Radi, R. and Augusto, O. (1998). Formation of spin trap adducts during the decomposition of peroxynitrite. *Arch. Biochem. Biophys.* **349**: 36–46.

18. Vasquez-Vivar, J., Kalyanaraman, B. and Kennedy, M. C. (2000). Mitochondrial aconitase is a source of hydroxyl radical. *J. Biol. Chem.* **275**: 14064–14069.

19. Samuni, A., Black, C. D. V., Krishna, C. M., Malech, H. L., Bernstein, E. F. and Russo, A. (1988). Hydroxyl radical production by stimulated neutrophils reappraised. *J. Biol. Chem.* **263**: 13 797–13 801.

20. Timmins, G. S., Liu, K. J., Bechara, E. J. H., Kotake, Y. and Swartz, H. M. (1999). Trapping of free radicals with direct in vivo EPR detection: a comparison of 5,5-dimethyl-1-pyrroline-*N*-oxide and 5-diethoxyphorphoryl-5-methyl-1-pyrroline-*N*-oxide as spin traps for HO$^{\bullet}$ and SO$_4^{\bullet-}$. *Free Radic. Biol. Med.* **27**: 329–333.

21. Burkitt, M. J. and Mason, R. P. (1991). Direct evidence for *in vivo* hydroxyl-radical generation in experimental iron overload: an ESR spin-trapping investigation. *Proc. Natl. Acad. Sci. USA* **88**: 8440–8444.

22. Burkitt, M. J., Kadiiska, M. B., Hanna, P. M., Jordan, S. J. and Mason R. P. (1993). Electron spin resonance spin-trapping investigation into the effects of paraquat and desferrioxamine on hydroxyl radical generation during acute iron poisoning. *Mol. Pharmacol.* **43**: 257–263.

23. Mason, R. P. and Holtzman, J. L. (1975). The role of catalytic superoxide formation in the O$_2$ inhibition of nitroreductase. *Biochem. Biophys. Res. Commun.* **67**: 1267–1274.

24. DeGray, J. A., Rao, D. N. R. and Mason, R. P. (1991). Reduction of paraquat and related bipyridylium compounds to free radical metabolites by rat hepatocytes. *Arch. Biochem. Biophys.* **289**: 145–152.

25. Horton, J. K., Brigelius, R., Mason, R. P. and Bend, J. R. (1986). Paraquat uptake into freshly isolated rabbit lung epithelial cells and its reduction to the paraquat radical under anaerobic conditions. *Mol. Pharmacol.* **29**: 484–488.

26. Kadiiska, M. B., Hanna, P. M. and Mason, R. P. (1993). In vivo ESR spin trapping evidence for hydroxyl radical-mediated toxicity of paraquat and copper in rats. *Toxicol. Appl. Pharmacol.* **123**: 187–192.

27. Hanna, P. M. and Mason. R. P. (1992). Direct evidence for inhibition of free radical formation from Cu(I) and hydrogen peroxide by glutathione and other potential ligands using the EPR spin-trapping technique. *Arch. Biochem. Biophys.* **295**: 205-213.

28. Kadiiska, M. B., Burkitt, M. J., Xiang, Q.-H. and Mason, R. P. (1995). Iron supplementation generates hydroxyl radical in vivo. *J. Clin. Invest.* **96**: 1653–1657.

29. Buettner, G. R., Oberley, L. W. and Leuthauser, S. W. H. C. (1978). The effect of iron on the distribution of superoxide and hydroxyl radicals as seen by spin trapping and on the superoxide dismutase assay. *Photochem. Photobiol.* **28**: 693–695.

30. Arroyo, C. M., Kramer, J. H., Dickens, B. F. and Weglicki, W. B. (1987). Identification of free radicals in myocardial ischemia/reperfusion by spin trapping with nitrone DMPO. *FEBS Lett.* **221**: 101–104.

31. Togashi, H., Shinzawa, H., Yong, H., Takahashi, T., Noda, H., Oikawa, K. and Kamada, H. (1994). Ascorbic acid radical, superoxide, and hydroxyl radical are detected in reperfusion injury of rat liver using electron spin resonance spectroscopy. *Arch. Biochem. Biophys.* **308**: 1–7.

32. Buettner, G. R. (1988). In the absence of catalytic metals ascorbate does not autoxidize at pH 7: ascorbate as a test for catalytic metals. *J. Biochem. Biophys. Meth.* **16**: 27–40.

33. Britigan, B. E., Cohen, M. S. and Rosen, G. M. (1987). Detection of the production of oxygen-centered free radicals by human neutrophils using spin trapping techniques: a critical perspective. *J. Leuk. Biol.* **41**: 349–362.

34. Morehouse, K. M., Flitter, W. D. and Mason, R. P. (1987). The enzymatic oxidation of Desferal to a nitroxide free radical. *FEBS Lett.* **222**: 246–250.

35. Davies, M. J., Donkor, R., Dunster, C. A., Gee, C. A., Jonas, S. and Willson, R. L. (1987). Desferrioxamine (Desferal) and superoxide free radicals. Formation of an enzyme-damaging nitroxide. *Biochem. J.* **246**: 725–729.

36. Hinojosa, O. and Jacks, T. J. (1986). Interference by desferrioxamine of spin trapping oxyradicals for electron spin resonance analysis. *Anal. Lett.* **19**: 725–733.

37. Sanders, S. P., Harrison, S. J., Kuppusamy, P., Sylvester, J. T. and Zweier, J. L. (1994). A comparative study of EPR spin trapping and cytochrome C reduction techniques for the measurement of superoxide anions. *Free Radic. Biol. Med.* **16**: 753–761.

38. Frejaville, C., Karoui, H., Tuccio, B., Le Moigne, F., Culcasi, M., Pietri, S. Lauricella, R. and Tordo, P. (1995). 5-(diethoxyphosphoryl)-5-methyl-1-pyrroline *N*-oxide: a new efficient phosphorylated nitrone for the in vitro and in vivo spin-trapping of oxygen-centered radicals. *J. Med. Chem.* **38**: 258–265.

39. Pietri, S., Liebgott, T., Frejaville, C., Tordo, P. and Culcasi, M. (1998). Nitrone spin traps and their pyrrolidine analogs in myocardial reperfusion injury: hemodynamic and ESR implications. Evidence for a cardioprotective phosphonate effect for 5-(diethoxyphosphoryl)-5-methyl-1-pyrroline *N*-oxide in rat hearts. *Eur. J. Biochem.* **254**: 256–265.

40. Roubaud, V., Sankarapandi, S., Kuppusamy, P., Tordo, P. and Zweier, J. L. (1998). Quantitative measurement of superoxide generation and oxygen consumption from leukocytes using electron paramagnetic resonance spectroscopy. *Anal. Biochem.* **257**: 210–217.

41. Olive, G., Mercier, A., Le Moigne, F., Rockenbauer, A. and Tordo, P. (2000). 2-ethoxycarbonyl-2-methyl-3,4-dihydro-2*H*-pyrrole-1-oxide: evaluation of the spin trapping properties. *Free Radic. Biol. Med.* **28**: 403–408.
42. Zhang, H., Joseph, J., Vasquez-Vivar, J., Karoui, H., Nsanzumuhire, C., Martasek, P., Tordo, P. and Kalyanaraman, B. (2000). Detection of superoxide anion using an isotopically labeled nitrone spin trap: potential biological applications. *FEBS Lett.* **473**: 58–62.

Chapter 3

Peroxyl and Alkoxyl Radical Mediated DNA Damage

John Termini

John Termini • Department of Molecular Biology, Beckman Research Institute of the City of Hope, 1450 East Duarte Road, Duarte, California 91010
Tel: (626) 301-8169, E-mail: termini@coh.org

1. Introduction: Alkyl Radicals to Oxyradicals — The Oxygen Effect

The vast majority of studies on oxidative DNA damage have concentrated on the chemical reactions and biochemical and biological consequences of primary reactive oxygen species. These may be loosely defined as the initial oxygen derived intermediates from the radiolysis of water, or radicals resulting from the metal ion induced decomposition of H_2O_2. Typically these are relatively short-lived, highly reactive intermediates with relatively poor chemical selectivity such as hydroxyl radicals ($\cdot$OH), hydrogen radicals ($\cdot$H) and solvated electrons (e^-_{aq}). Solvated electrons, like hydrogen radicals, are produced only during water radiolysis, thus their potential for involvement in DNA damage is limited to situations of radiation exposure. Hydroxyl radicals are produced during this process and also via endogenous Fenton reaction-related processes, thus damage to DNA by $\cdot$OH would be expected to be more ubiquitous. Much effort has therefore been expended towards the analyses of the reactions of hydroxyl radicals with nucleic acid bases and carbohydrate moieties in an effort to understand the kinds of damage incurred. Some theoretical considerations for reactions of $\cdot$OH with DNA have been considered in detail by Chatterjee and Holley.[1]

Hydroxyl radicals may add as nucleophiles to double bonds of the DNA bases or abstract hydrogen atoms from either the bases or the carbohydrate backbone to generate DNA radicals. Unless rapidly repaired, these DNA radicals may further react leading to oxidative base damage and/or strand breaks. Radicals on DNA may be repaired by donation of an H atom from a suitable donor. Intracellular thiols are believed to play an important role in the chemical repair of free radical damage via the reduction shown in Fig. 1.

This reaction, although possessing a rate constant far from the diffusion controlled limit of $\sim 10^{10} M^{-1} s^{-1}$ is thought to be adequate for repair since thiols such as glutathione are found in relatively high intracellular concentrations. The DNA repair reaction shown in Fig. 1 generates a thiyl radical as a result of H atom donation to the carbon centered radical. One might well wonder as to the potential for the newly generated thiyl radical to carry out additional oxidations of the type shown in Fig. 2.

Fortunately, the extremely slow rate constant for reaction of thiyl radicals with DNA, at least two orders of magnitude slower than the repair reaction, precludes

$$\text{DNA}\bullet \xrightarrow{\text{RSH}} \text{DNAH} + \text{RS}\bullet$$

$$k \sim 8 \times 10^4 M^{-1} s^{-1}$$

Fig. 1. Repair of carbon radicals by thiols.

$$RS\bullet + DNA \longrightarrow RSH + DNA\bullet$$

$$k \sim 1 \times 10^{2-3} M^{-1} s^{-1}$$

Fig. 2. Hypothetical oxidation reaction of DNA by a thiyl radical.

$$DNA\bullet \xrightarrow{O_2} DNAOO\bullet$$

$$k \sim 10^8 M^{-1} s^{-1}$$

Fig. 3. The fixation of free radical damage by oxygen to form a DNA peroxyl radical.

its occurrence to any significant extent under physiological conditions.[2] For this reason, thiols are excellent supressors of alkyl radical formation in DNA.

DNA radicals which escape repair may go on to form secondary radical products by "fixation" of molecular oxygen. This reaction is depicted in Fig. 3. The damage to DNA is said to be "fixed" in the sense that oxygen addition is not readily reversible. The well documented enhancement of radiation damage in biological systems in the presence of oxygen ("the oxygen effect") is due to this reaction.[3] The reaction with oxygen possesses a rate constant which is close to the diffusion limit, far faster than the repair reaction shown in Fig. 1. At high physiological concentrations of oxygen, the reaction shown in Fig. 3 likely dominates the distribution of free radical intermediates. The product of the reaction in Fig. 3 is a DNA peroxyl radical. It is important to emphasize that peroxyl radicals (ROO$\bullet$) may be formed from the reaction of any alkyl radical with O_2, and thus they are of special importance in aerobic free radical chemistry. These are the same intermediates which propagate the radical chain reactions which take place during lipid peroxidation. Peroxyl radicals generated from alkyl radicals of many different types would be expected to display similar reactivities. This is because the R group in ROO$\bullet$ is two bonds removed from the unpaired electron, and would not be expected to influence the redox properties of the radical to a significant degree. This is not the case for other radicals such as alkyl and alkoxyl.

The issue of whether thiols may participate in the quenching of peroxyl radicals in DNA and thus inhibit the propagation of free radical reactions has been examined in some detail. Since this reaction would result in the formation of a hydroperoxide (ROOH), technically this could not be considered repair, however, further free radical chain propagation via peroxyl radical intermediates would be prevented. This reaction is shown in Fig. 4.

The extremely slow rate constant observed for quenching of DNA and other peroxyl radicals by thiols such as glutathione explains their inefficiency at protecting against DNA oxidation during aerobic irradiation.[4] However, other

$$\text{DNAOO} \bullet \xrightarrow{\text{RSH}} \text{DNAOOH} + \text{RS} \bullet$$

$$k \sim 2 \times 10^2 M^{-1} s^{-1}$$

Fig. 4. Quenching of DNA peroxyl radicals by thiols.

$$2\text{ROO} \bullet \longrightarrow [\text{ROOOOR}] \longrightarrow 2\text{RO} \bullet + O_2$$

Fig. 5. Self-reaction of peroxyl radicals leading to production of alkoxyl radicals.

antioxidants such as tocopherols (vitamin E) are extremely efficient at quenching peroxyl radicals, and their use as "chain breaking" antioxidants to inhibit lipid peroxidation is well known. The intracellular concentration of tocopherols is lower than thiols however, and is strongly dependent on dietary intake.

Peroxyl radicals can dimerize in solution to form tetraoxide intermediates which rapidly decompose according to the scheme in Fig. 5 to yield alkoxyl radicals and oxygen.[5] Thus peroxyl radicals can serve as precursors for alkoxyl radicals, and it is likely that under physiological conditions both are present simultaneously. In addition to this pathway, tetraoxide intermediates may decompose to yield non-radical products and singlet oxygen when the R group possesses a readily abstractable α hydrogen.[6] The predominant pathway for tetraoxide decomposition (alkoxyl versus non-radical) depends to a large extent on the nature of R, but both pathways likely contribute to damage. Technically speaking, singlet oxygen produced in the non-radical pathway is not a free radical, however its role as a biological oxidant is well known.

2. Lifetimes and Diffusion of Oxyradicals

In order to assess the relative importance of each of these radical species as potential biological damaging agents, some examination of their relative reactivities is required. The chemical lifetimes of reactive intermediates are an indicator of potential reactivity. Intermediates with extremely short half lives ($\leq 10^{-9}$s) are expected to be highly reactive and to display little chemical selectivity in their reactions. Essentially they must encounter a substrate close to the site of generation in order to form products. Radicals with longer half-lives are predicted to be more selective in their reactions, and are capable of significant intracellular diffusion. Table 1 gives the half-lives of some reactive intermediates of biological interest.

The extremely short half-life of the hydroxyl radical, shown in Table 1, suggests that it is a highly reactive oxidant with little chemical selectivity. The extraordinary

Table 1. Half Lives of Some Reactive Intermediates of Biological Interest*

HO$\cdot$	10^{-9} s	R$\cdot$	10^{-8} s
RO$\cdot$	10^{-6} s	$^1O_2$	10^{-6} s
ROO$\cdot$	0.5–7 s		

*Data adapted from Refs. 4 and 31.

half life of the peroxyl radical suggests the opposite characteristics, while alkoxyl radicals would be predicted to be intermediate in reactivity. The capacity for diffusion over large intracellular distances prior to reaction is a predicted property of peroxyl, but not alkoxyl or hydroxyl radicals. Thus peroxyl radicals may promote oxidation reactions far from their site of generation. In order to dramatize this important difference, let us consider the expression for the distance-time relationship for diffusion.

$$x = \sqrt{2Dt_{1/2}}.$$

The variable x is the average distance of diffusion, D is the diffusion coefficient, which is different in viscous and non-viscous media, and $t_{1/2}$ is the half-life of the intermediate. In non-viscous media (e.g. non-ionic organic solvent), the value of D is 5×10^{-5} cm^{2}s^{-1}, whereas in viscous media (possibly closer to the cytosol) an approximate value of 1×10^{-10} cm^{2}s^{-1} is used.[7] Using the values from Table 1, the diffusion distance (x) that $\cdot$OH travels during one half life in non-viscous media is 32 Å, a distance equal to about one turn of a DNA B helix. However in viscous media, this value is computed to be 0.04 Å. These values may be compared to those calculated for peroxyl radicals, where the half-life diffusion distance in non-viscous media can be be shown to be 1.4×10^6 Å, approximately 1 mm. Even in viscous media the diffusion distance is substantial, 2000 Å. Values for alkoxyl radicals are estimated to be 1 and 1000 Å in viscous and non-viscous media, respectively. Assuming the *in vivo* situation is intermediate between a viscous and non-viscous media, it is apparent that peroxyl radicals can traverse large intracellular distances, which would appear to magnify their potential for harm. Alkoxyl radicals would also appear to possess some capacity for diffusion. Their actual potential for carrying out oxidation of biological macromolecules is also dependent upon their efficiency in carrying out oxidation reactions, either by H abstraction reactions or via addition to double bonds.

From the standpoint of mutation induction via free radical induced base damage, one would expect that the more stable radical species would give rise to a smaller number of damage products and hence a characteristic mutation pattern, whereas highly reactive intermediates might generate a wide spectrum of damage products and a broader range of mutations. Peroxyl radicals would be predicted to fall into the former category, alkoxyl radicals in the latter. This hypothesis is supported by the peroxyl radical induced mutation spectrum in *E. coli*.[28]

3. Relative Reactivities of Peroxyl and Alkoxyl Radicals

As implied above in the discussion of the half-lives of reactive oxygen species, alkoxyl radicals are more reactive (better oxidants) than peroxyl radicals. The values for the reduction potentials for ROO$\cdot$/ROOH (E° versus NHE, pH 7) have been estimated at ~ 1.0 V, whereas the RO$\cdot$/ROH value is ~ 1.6 V, where R is an aliphatic alkyl group. These values may be related to the bond dissociation energies (BDE) for breaking a C–H or O–H bond. BDE values are a convenient way to evaluate the potential for hydrogen abstraction reactions by a particular radical. The higher the energy required to break a bond, the more reactive the radical intermediate. For example, the BDE required to remove a hydrogen atom to generate an ethyl radical ($\cdot$CH$_2$CH$_3$/C$_2$H$_6$) is ~ 104 kcal/mol. The BDE for formation of an alkoxyl radical from ethanol ($\cdot$OCH$_2$CH$_3$/HOC$_2$H$_5$) is nearly isoenergetic, with value of 104 ± 1.[8] The BDE value for formation of peroxyl radicals (ROO$\cdot$/ROOH) is ~ 90 kcal/mol. Thus alkoxyl radicals are energetic enough to abstract hydrogen from ethane to form alkyl radicals, whereas peroxyl radicals are not sufficiently reactive in this regard. In general, alkoxyl radicals are better at H abstraction.

Consideration of the BDE required for removal of a bis-allylic hydrogen reveals a value of ~ 80 kcal/mol. Peroxyl radicals should easily be able to carry out this reaction to form a bis-allylic alkyl radical. This is consistent with the important role of peroxyl radicals in the chain propagation reactions which take place during autooxidation of poly unsaturated fatty acids (PUFA).[9] Published BDE values for the functional groups of the DNA bases exist only for the 5-Me group of thymidine.[10] It would be of great interest to determine relevant BDE values for functional groups of the remaining bases in order to determine the likelihood of reaction with various oxyradicals. This could be helpful in identifying reactive sites along DNA susceptible to reaction by peroxyl and other radicals.

The higher reactivity of alkoxyl radicals may limit their ability to carry out biological oxidations *in vivo*. They may be quenched intermolecularly by thiols, for example, or they may react intramolecularly to yield other radical species. An example of the latter reaction is well known for alkoxyl radicals of PUFA as shown in Fig. 6.

The PUFA alkoxyl radical rapidly undergoes addition to the vicinal double bond to form an epoxy allylic radical. This reaction is so efficient that intermolecular reactions of the PUFA alkoxyl radical intermediate are not observed.[11] Under aerobic conditions, this intermediate alkyl radical rapidly reacts with oxygen to form a peroxyl radical. The contribution of lipid alkoxyl radicals to the process

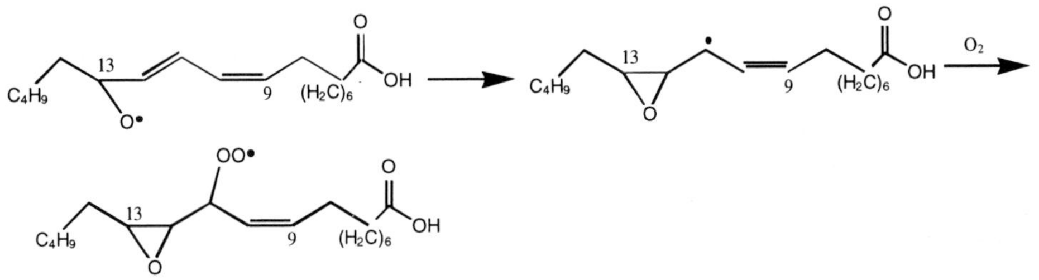

Fig. 6. Intramolecular reaction of PUFA alkoxyl radicals and subsequent trapping by O_2 to yield peroxyl radicals.

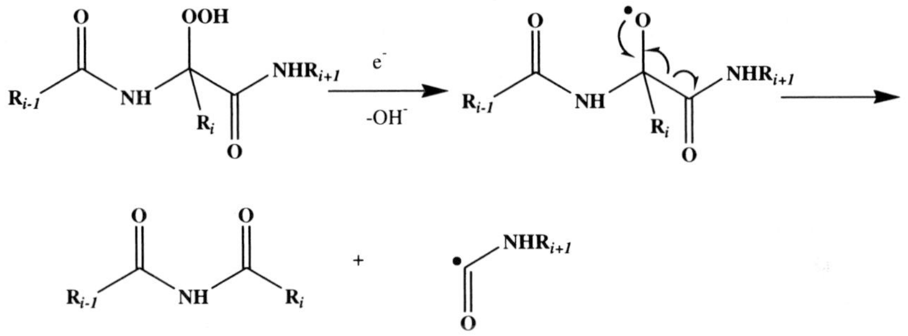

Fig. 7. Polypeptide backbone cleavage mediated by α-hydroperoxide decomposition.

of lipid peroxidation is thought to be small for the reason that their free concentration at any point in time is low.

However, alkoxyl radicals have been detected in tissues and organelles such as mouse epidermis[12] and liver microsomes,[13] suggesting they likely play some role in biological oxidations. Alkoxyl radicals can be formed from the low valent transition metal ion mediated decomposition of hydroperoxides (see Sec. 4). This reaction for amino acid hydroperoxides has been studied in some detail. The metal ion catalyzed decomposition of polypeptide α hydroperoxides results not only in protein backbone cleavage but also generates diffusible reactive radical intermediates according to the reactions shown in Fig. 7. These intermediates are capable of oxidizing other biological molecules.

Analogous to the *bis*-allylic hydrogen of polyunsaturated lipids, the hydrogen occupying the α position of the polypeptide backbone is readily abstracted by oxidants. The resulting carbon radical reacts with O_2 to yield a peroxyl radical, which in turn abstracts a hydrogen atom from a suitable donor to generate the hydroperoxide shown in Fig. 7. Hydroperoxide reduction generates alkoxyl radical intermediates which rapidly ($\sim 10^7 s^{-1}$) fragment to yield a stable terminal carbonyl peptide fragment and a free radical intermediate.[14] This radical intermediate has been detected as a dominant species by EPR in the metal ion induced decomposition of polypeptide hydroperoxides. This intermediate would be expected to be relatively stable owing to amide delocalization of the unpaired electron, and thus demonstrate interesting chemical selectivity. Free radical oxidation of proteins has the potential therefore to generate other radical species with the potential to carry out further biological damage. Additional studies of the reactions of this radical are warranted in order to assess its potential for reaction with DNA.

4. Methods for the Generation of Peroxyl and Alkoxyl Radicals

Reagents used for modeling biologically relevant reactions of peroxyl and alkoxyl radicals must optimally be capable of generating radicals in a steady, quantifiable manner so that kinetic reproducibility may be obtained. This criteria also implies that other competing radical intermediates are not formed to any appreciable extent during reaction. These criteria are satisfied when thermolysis of diazo initiators in the presence of oxygen are used to generate peroxyl radicals according to Fig. 8.

When R is a water soluble functional group it then becomes a relatively simple matter to generate peroxyl radicals in the presence of DNA or other biological macromolecules for oxidation model studies. A reagent used by us and others for model studies of biological peroxyl radical oxidations is 2,2'-azobis(2-amidino-propane) dihydrochloride (ABAP). The kinetics of radical formation by this reagent have been well studied, and peroxyl radicals are produced in a linear, kinetically reproducible fashion.[15] The low, steady concentration of peroxyl radicals produced largely prevents self-condensation reactions and thus the formation of alkoxyl radicals as in Fig. 5.

Peroxyl radicals may also be generated from alkyl hydroperoxides (ROOH) via reaction with high valent iron-oxo porphyrin complexes. Two equivalents of ROOH are consumed in this reaction, yielding a reduced alcohol (ROH) and $ROO^{\bullet}$ according to the mechanism in Fig. 9(a).[16] One equivalent of hydroperoxide is consumed in Fig. 9(a) to effect the oxidation of Fe^{3+}(porphyrin) to the high valent iron-oxo species, which can then oxidize an additional equivalent of hydroperoxide to $ROO^{\bullet}$, with reduction of the porphyrin radical cation. It is important to point out that non-heme iron chelates react with hydroperoxides by

$$ R-N{=}N-R \xrightarrow{\Delta} 2R^{\bullet} $$

$$ R^{\bullet} + O_2 \longrightarrow ROO^{\bullet} $$

Fig. 8. Aerobic thermolysis of diazo compounds to generate peroxyl radicals.

(a)

   i) $Fe^{3+}(porphyrin) + ROOH \longrightarrow Fe^{4+}=O(porphyrin)^{+ \bullet} + ROH$

   ii) $Fe^{4+}=O(porphyrin)^{+ \bullet} + ROOH \longrightarrow Fe^{4+}=O(porphyrin) + ROO^\bullet + H^+$

(b) $Fe^{3+}(bipyridyl) + ROOH \longrightarrow Fe^{2+}(bipyridyl) + RO^\bullet + {}^-OH$

Fig. 9. Reactions of alkyl hydroperoxides with iron chelates.

$$ROOR \xrightarrow{\quad h\nu \quad} 2RO\bullet$$

Fig. 10. Photodissociation of alkyl endoperoxides to yield alkoxyl radicals.

reduction to yield alkoxyl radicals, as shown in Fig. 9(b).[17] It is likely that reduction of organic hydroperoxides by adventitious low-valent transition metal ions is an important biological route to alkoxyl radicals.

The production of alkoxyl radicals via the reaction in Fig. 9(b) should be useful for studying the nature of this oxidant if carried out under anaerobic conditions. In the presence of O_2, alkyl radicals generated as a result of H atom abstraction by $RO^\bullet$ would rapidly form peroxyl radicals. A more reliable method of generating this species independently which has proven to be of considerable utility is the photochemical decomposition of endoperoxides, shown in Fig. 10.

The research group of Prof. W. Adam has introduced the use of the peroxy ester 4-(*tert*-butyldioxycarbonyl)benzyl triethylammonium chloride as a model compound for generating tert-butoxyl radicals in water for biochemical studies of the reactions of alkoxyl radicals.[18] The cleavage of *tert*-butyl hydroperoxide by submitochondrial particles to yield alkoxyl radicals has also been described.[19]

5. Peroxyl and Alkoxyl Radical Mediated DNA Strand Breaks

The ability to induce strand breaks in genomic DNA by free radicals is predicated upon the ability of the reactive intermediates to abstract H atoms from the carbohydrate moiety. Peroxyl radicals of DNA were first proposed to explain the dependence of oxygen on the yield of strand breaks caused by irradiation. It had long been known that irradiation of DNA under anoxic conditions yielded little or no strand breaks. Irradiation in the presence of oxygen yielded extensive strand breaks, whose rate of formation appeared to coincide with decay of a long-lived spectrophotometric transient assigned to peroxyl radicals.[20, 21] These experiments, along with those of others, led to the proposed involvement of base peroxyl radicals, formed in the manner outlined in Fig. 3, in the initiation of strand breaks induced by oxic irradiation. The abstraction of a C4′ or C1′ H atom of a neighboring carbohydrate residue by the base peroxyl radical was proposed to be the rate determining step in strand break formation. A direct test of this model was accomplished by generating a peroxyl radical intermediate *in situ* from a photochemically generated 5,6 dihydrothymid-5-yl radical within a oligonucleotide fragment. Strand breaks in both the 5′ and 3′ directions were clearly observed only in the presence of O_2.[22] The induction of strand breaks in genomic DNA using the $ABAP/O_2$ peroxyl radical generating system has also been studied in detail using denaturing agarose electrophoresis.[23] This reagent may cause formation of base peroxyl radicals or abstract H atoms directly from the carbohydrate backbone. Strand breaks displayed a logarithmic dependence on peroxyl radical concentration. At a concentration of $1\,\mu M$ $ROO^\bullet$, ~ 1 single strand break (ssb)/2 kb was observed.

Strand break induction by alkoxyl radicals has also been studied using supercoiled pBR322 DNA as a substrate. The photochemical system described by Adam and co-workers which generates *tert*-butoxyl radicals from a water soluble perester was used. A concentration of 1mM perester yielded 1.2 ssb/4 kb after irradiation for 12 hours. Although this appears to be less efficient than the ssb yield produced by peroxyl radicals, it is difficult to compare these results directly, since the quantum yield for alkoxyl radicals produced by photolysis of the perester was not reported. Alkoxyl radicals, owing to their greater reactivity, would be predicted to be more efficient than peroxyl radicals at promoting DNA strand breaks. However, their significantly shorter lifetime may limit their diffusion, as discussed above, and unlike peroxyl radicals, alkoxyl radicals are efficiently quenched by thiols.

6. DNA Base Modifications by Peroxyl and Alkoxyl Radicals

An interesting example of the effect of radical reactivities on base oxidation product distributions can be illustrated for the case of thymidine. Figure 11 illustrates the distribution of thymidine oxidation products resulting from reaction with $^\bullet OH$, $ROO^\bullet$ and $RSO_2OO^\bullet$. This latter radical, the sulphonyl peroxyl radical, is formed from the consecutive reactions of thiyl radicals with oxygen, and is similar in redox capacity to the peroxyl radical.[10]

The reactions of the hydroxyl radical with thymidine can be seen to consist of nucleophilic attack at the 5,6 double bond to form 6-and 5-yl radicals respectively, and H atom abstraction at the 5-Me group to form an allylic radical. The major products result from radical addition at the 5,6 double bond, accounting for ~ 90% of initial radical products. The allylic radical resulting from H atom abstraction from the methyl group is a minor product when $^\bullet OH$ is the attacking species.[24–26] This range of radical intermediates can subsequently give rise to a variety of unsaturated and 5-methyl modified thymidine oxidation products, each

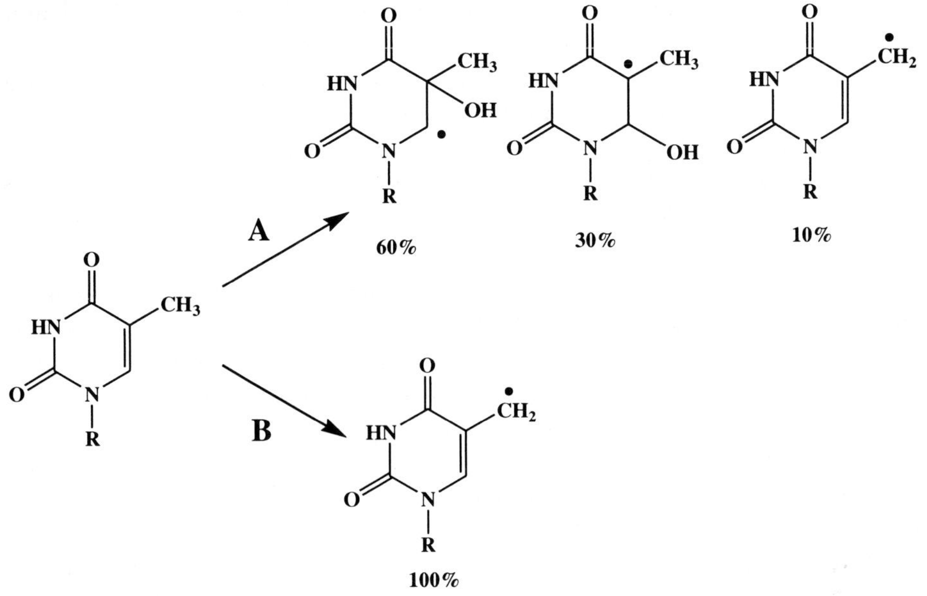

Fig. 11. Oxidation of thymine by various oxyradicals. (A) Hydroxyl radical (·OH) and (B) Peroxyl (ROO·) or sulphonyl peroxyl (SO₂OO·) radicals.

with a unique mutagenic potential. In contrast to this situation, the reaction of peroxyl radical with thymidine gives rise to products arising solely from oxidation of the 5-methyl group. No products resulting from addition of ROO· to the 5,6 double bond have been detected in model reactions.[27] This is in accord with the prediction that peroxyl radicals should demonstrate greater chemical selectivity than other reactive oxygen radicals. Sulfonyl peroxyl radicals have redox potentials similar to ROO·, thus it is not suprising that they also react exclusively via H atom abstraction from thymidine.[10] In view of the narrow range of products formed, one would expect a limited set of mutations resulting from peroxyl radical oxidation of *dT* or other bases. This suggests the possibility that exposure of DNA to peroxyl radicals, either through the oxidation of lipid membranes or proteins, may yield a specific pattern of DNA damage which gives rise to a "signature" pattern of mutations characteristic of this type of damage.

Reactions of *dG* with peroxyl radicals are also found to yield a limited number of detectable products, only two detected by HPLC/diode array analysis.[28] In fact, *dG* is ~10 times more oxidizable than *dT* by ROO·. Interestingly, the oxidation products do not include 8-oxo *dG*, but the peroxyl radical induced lesions in DNA are substrates for the formamidopyrimidine glycosylase (FPG protein). It has also been demonstrated that these lesions at *dG* are not alkali labile.[28] Preliminary mass spectrometric evidence indicates that one of the products may be an oxidized ring-opened purine derivative, but further stuctural clarification is required.

Deoxyadenosine as a model compound is completely unreactive towards oxidation by peroxyl radicals, whereas deoxycytosine gives a minor yield of oxidized products. The relative reactivities towards oxidation by peroxyl radicals of free nucleosides in solution has recently been shown to reflect trends of reactivity of the bases in DNA.[23] Maps of peroxyl radical induced DNA damage were mapped on genomic DNA using DNA glycosylases/AP lyases to create strand breaks at oxidized bases; then these breaks were mapped using the ligation mediated PCR method (LMPCR). The order of reactivity of DNA bases towards oxidation by peroxyl radicals was found to be $G \gg C > T$. Interesting nearest neighbor effects were observed on the oxidizability of DNA bases. For example, every guanine base flanked on the 3′ side by cytosine was detected as oxidized by peroxyl radicals, regardless of the identity of the 5′ neighbor, whereas 5′ and 3′ purine neighbors drastically reduced the frequency of G oxidation by this radical.

The reactions of chemically generated alkoxyl radicals with DNA have been recently reported.[18] These workers detected oxazolone and oxoimidazolidine upon oxidation of either calf-thymus DNA or 2′deoxyguanosine model compounds. These products are normally associated with the oxidation of 8-oxo guanine, however; alkoxyl radicals did not appear to generate this product when reactions were monitored using the extremely sensitive electrochemical detection method. A 4-hydroperoxy guanine intermediate is proposed to account for the formation of these products. Other base products were not characterized. Abstraction of hydrogen by alkoxyl radicals appears to be the main reaction pathway based upon product analysis. However, other workers have presented EPR evidence that suggests alkoxyl radical addition to the C_5–C_6 double bond of thymidine and uridine (in DNA and RNA, respectively) to give a 5-yl radical.[29] Both groups used the same alkoxyl radical (tBuO·) although different chemical methods were used to generate the radicals from stable precursors. Some theoretical and experimental results would suggest that alkoxyl radicals are poor nucleophiles compared to hydroxyl radicals, thus favoring H atom abstraction as the preferred pathway for reactions with DNA.[30] Additional work must be done in order to clarify this issue.

7. Conclusions

The secondary radicals produced in the presence of oxygen following H atom abstraction or 1e- oxidation of biological substrates are likely peroxyl radicals. These in turn may oxidize other substrates by abstracting a hydrogen atom to form a hydroperoxide (ROOH) and a new radical center, thus propagating free radical chain oxidation. This autooxidation is well studied in the case of polyunsaturated lipids, but autooxidation of proteins and DNA is likely a major source of endogenous biological damage. Whenever peroxyl radicals and hydroperoxides are produced, alkoxyl radicals are also likely produced. This occurs via self-reaction of peroxyl radicals, and also via the reduction of hydroperoxides by metal ions. Thus the two radical intermediates are likely present simultaneously under physiological conditions. The challenge is to determine the patterns of damage produced by each reactive intermediate, in order to evaluate their relative importance in the generation of endogenous DNA damage. One possibility is that peroxyl radicals, which possess unusually long chemical lifetimes and enhanced chemical selectivity, may be the principal propagators of DNA autooxidation, analogous to their role in lipid oxidation. Alkoxyl radicals, although more chemically reactive, may possess greater importance through the products generated by their decomposition reactions, as we have discussed for the case of polypeptide backbone fragmentation. Analogous fragmentation of DNA alkoxyl radicals (e.g. decomposition of DNA hydroperoxides) may lead to novel base lesions and strand breakage. The mechanistic elucidation of these oxidative/degradative pathways remains an important challenge in free radical chemistry with profound implications for mutagenesis.

Acknowledgment

Work in the authors laboratory on oxidative DNA damage is supported by PHS grant GM59219.

References

1. Chatterjee, A. and Holley, W. R. (1993). Computer simulation of initial events in the biochemical mechanisms of DNA damage. *Adv. Rad. Biol.* **17**: 180–226.
2. Liphard, M., Bothe, E. and Schulte-Frohlinde, D. (1990). The influence of glutathione on single-strand breakage in single-stranded DNA irradiated in aqueous solution in the absence and presence of oxygen. *Int. J. Rad. Biol.* **58**: 589–602.
3. Quintiliani, M. (1986). The oxygen effect in radiation inactivation of DNA and enzymes. *Int. J. Rad. Biol.* **50**: 573–594.
4. Hildebrand., K. and Schulte-Frohlinde, D. (1997). Time resolved EPR studies on the reaction rates of peroxyl radicals of poly(acrylic acid) and of calf thymus DNA with glutathione. Re-examination of the rate constant for DNA. *Int. J. Rad. Biol.* **71**: 377–385.
5. Howard, J. A. (1978). Self-reactions of alkylperoxyl radicals in solution. *Am. Chem. Soc. Symp. Ser.* **69**: 413–432.

6. Nakano, M., Takayama, K., Shimizu, Y., Tsuji, Y., Inaba, H. and Migita, T. (1976). Spectroscopic evidence for the generation of singlet oxygen in self-reaction of sec-peroxy radicals. *J. Am. Chem. Soc.* **98**: 1974–1975.

7. Turro, N. (1978). *Modern Molecular Photochemistry*, pp. 316–317. Benjamin/ Cummings Publishing Co., Menlo Park CA.

8. Koppenol, W. H. (1990). Oxyradical reactions: from bond-dissociation energies to reduction potential. *FEBS Lett.* **264**: 165–167.

9. Wagner, B.A., Buettner, G. R. and Burns, C. P. (1994). Free radical-mediated lipid peroxidation in cells: oxidizability is a function of cell lipid bis-allylic hydrogen content. *Biochemistry* **33**: 4449–4453.

10. Razskazovskii, Y. and Sevilla, M. D. (1996). Reactions of sulphonyl peroxyl radicals with DNA and its components: hydrogen abstraction from the sugar backbone versus addition to pyrimidine double bonds. *Int. J. Rad. Biol.* **69**: 75–87.

11. Marnett, L. J. and Wilcox, A. L. (1995). The chemistry of lipid alkoxyl radicals and their role in metal-amplified lipid peroxidation. *Biochem. Soc. Symp.* **61**: 65–72.

12. Timmins, G. S. and Davies, M. J. (1993). Free radical formation in murine skin treated with tumor promoting organic peroxides. *Carcinogenesis* **14**: 1499–1503.

13. Davies, M. J. (1989). Detection of peroxyl and alkoxyl radicals produced by reaction of hydroperoxides with rat live microsomal fractions. *Biochem. J.* **257**: 603–606.

14. Davies, M. J. (1996). Protein and peptide alkoxyl radicals can give rise to C-terminal decarboxylation and backbone cleavage. *Arch. Biochem. Biophys.* **336**: 163–172.

15. Niki, E. (1990). Free radical initiators as source of water or lipid soluble peroxyl radicals. *Meth. Enzymol.* **186**: 100–108.

16. Davies, M. J. (1988). Detection of peroxyl and alkoxyl radicals produced by reaction of hydroperoxides with heme proteins by electron spin resonance spectroscopy. *Biochim. Biophys. Acta.* **964**: 28–35.

17. Winston, G. W., Harvey, W., Berl, L. and Cederbaum, A. I. (1983). The generation of hydroxyl and alkoxyl radicals from the interaction of ferrous bipyridyl with peroxides. *Biochem. J.* **216**: 415–421.

18. Adam, W., Grimm, G. N., Saha-Moller, C. R., Dall'Acqua, F., Miolo, G. and Vedaldi, D. (1998). DNA damage by *tert*-butoxyl radicals generated in the photolysis of a water soluble, DNA binding peroxyester acting as a radical source. *Chem. Res. Toxicol.* **11**: 1089–1097.

19. Massa, E. M. and Giulivi, C. (1993). Alkoxyl and methyl radical formation during cleavage of tert-butyl hydroperoxide by a mitochondrial membrane-bound, redox active copper pool: an EPR study. *Free Radic. Biol. Med.* **14**: 559–565.

20. Jones, G. D. D. and O'Neill, P. (1990). The kinetics of radiation induced strand breakage in polynucleotides in the presence of oxygen: a time resolved light-scattering study. *Int. J. Rad. Biol.* **57**: 1123–1139.
21. Bothe, E., Behrens, G., Bohm, E., Sethuram, B. and Schulte-Frohlinde, D. (1986). Hydroxyl radical induced strand break formation of poly(U) in the presence of oxygen: comparison of rates as determined by conductivity, esr and rapid-mix experiments with a thiol. *Int. J. Rad. Biol.* **49**: 57–66.
22. Barvian, M. R. and Greenberg, M. M. (1995). Independent generation of 5,6-dihydrothymid-5-yl in single stranded polythymidylate. O_2 is necessary for strand scission. *J. Am. Chem. Soc.* **117**: 8291–8292.
23. Rodriguez, H., Valentine, M. R., Holmquist, G. P., Akman, S. A. and Termini, J. (1999). Mapping of peroxyl radical induced damage on genomic DNA. *Biochemistry* **38**: 16578–16588.
24. Rustgi, S. and Riesz, P. (1978). ESR of free radicals in aqueous solutions of substituted pyrimidines. *Int. J. Rad. Biol.* **33**: 21–39.
25. Fujita, S. and Steenken, S. (1981). Pattern of OH radical addition to uracil and methyl- and carboxyl-subsituted uracils. Electron transfer of OH adducts with N,N,N'N''-tetramethyl-p-phenylenediamine and tetranitromethane. *J. Am. Chem. Soc.* **103**: 2540–2545.
26. Von Sonntag, C. and Schuchmann, H.-P. (1986). The radiolysis of pyrimidines in aqueous solutions: an updating review. *Int. J. Rad. Biol.* **49**: 1—34.
27. Martini M., and Termini, J. (1997). Peroxyl radical oxidation of thymidine. *Chem. Res. Toxicol.* **10**: 231–241.
28. Valentine, M. J., Rodriguez, H. and Termini, J. (1998). Mutagenesis by peroxyl radicals is dominated by transversions at deoxyguanosine: evidence for the lack of involvement of 8-oxo-dG and/or abasic site formation. *Biochemistry* **37**: 7030–7038.
29. Hazlewood, C. and Davies, M. J. (1995). Damage to DNA and RNA by tumor promoter-derived alkoxyl radicals: an EPR spin trapping study. *Biochem. Soc. Trans.* **23**: 259.
30. Betrand, M. P. and Surzur, J.-M. (1976). Reactivite des radicaux alkoxy vis a vis des olefines une interpretation basee sur les effets orbitalaires. *Tet. Lett.* **38**: 3451–3454.
31. Pryor, W. A. (1986). Oxy-radicals and related species: their formation, lifetimes and reactions. *Ann. Rev. Physiol.* **48**: 657–667.

Chapter 4

Nitric Oxide, Peroxynitrite and Ageing

Joseph S. Beckman

Joseph S. Beckman • Linus Pauling Institute, Department of Biochemistry and Biophysics, Oregon State University, Corvallis, Oregon 97331
E-mail: Joe.Beckman@orst.edu

1. Introduction

Since the initial radical hypothesis of Harmon in the 1950s, a significant cause of ageing has been postulated to result from the stochastic generation of oxygen radicals as an unavoidable byproduct of oxidative metabolism. The continuous flux of oxidants is postulated to randomly damage biological molecules and eventually degrade tissues beyond the point of effective repair. Although oxidative stress has been well-accepted to contribute to ageing, the specific role of oxidants remains perplexing. Can differences in metabolic rate really explain why mice survive for two years and humans for over eighty? Why do antioxidants not substantially increase life span? Although dietary antioxidants have clear benefits in malnourished individuals, ageing is not stopped by taking massive dosages of tocopherol, ascorbate and other antioxidants. Oxidants may have a subtler and less random role in the ageing process.

Some oxygen radicals such as hydroxyl radical (HO$^{\bullet}$) are so indiscriminately reactive that every part of every biological molecule is about equally susceptible to attack. The toxicity and perhaps the biological significance of the most highly reactive oxidants has been overemphasized[1] while the more significant biological effects of less reactive oxidants like superoxide (O$^{2\bullet-}$) have largely been overlooked.[2] It has become clear that many biologically relevant oxidants can selectively react with specific functional moieties important for signal transduction and electron transport. These targets can have far more profound biological effects than the random damage inflicted by hydroxyl radical.

Furthermore, the flux of oxidants produced by metabolic accidents may be small compared to the flux produced by specific enzymes induced in many tissues that can produce large quantities of superoxide and nitric oxide (NO). Oxidants serve critical functions in host-defense, wound healing and signaling functions. Growing evidence suggests that some oxidants activate specific signaling cascades that can promote mitosis, favor differentiation or activate apoptotic cascades.

Although not a strong oxidant itself, the free radical nitric oxide is synthesized in virtually all tissues as a local modulator of biological activity. The repertoire of potentially biologically relevant oxidants grew enormously in the past decade with the realization that virtually all cells in the body are capable of producing nitric oxide. Here, we will consider how the production of oxidants derived from nitric oxide such as peroxynitrite (ONOO$^-$) might influence the ageing process. Our understanding of the reactive chemistry of peroxynitrite is still rudimentary, but growing evidence suggests it may participate in many diseases associated with ageing. Both the vasculature and skeletal muscles appear to be significant targets of peroxynitrite in ageing. However, the central nervous system may be most vulnerable to oxidative attack because it lacks many of the regenerative capacities of other tissues.

Oxidative death of one particular type of neuron in the spinal cord could have a surprisingly important role in ageing. In the human spinal cord, approximately one million motor neurons control all muscle contraction and are therefore necessary for all voluntary movement. Nitric oxide-dependent oxidative stress may play an important role in the loss of motor neurons that characterizes the neurodegenerative disease, amyotrophic lateral sclerosis (ALS).[3] Mutations to the antioxidant enzyme, superoxide dismutase (SOD) remain the only established cause of this disease.[4] Exciting experiments conducted with *Drosophila* implicate oxidative stress in motor neurons having a major contribution to ageing of the whole organism.[5]

2. Why Do Tissues Produce Oxidants?

In many ways, ageing is a luxury that few species in the wild survive long enough to experience. Death from disease, starvation or predation is a far more likely fate, even for humans until recently. From an evolutionary perspective, it is difficult to rationalize how selective forces could favor genes that promote ageing unless those genes conferred an early adaptive advantage. The genes encoding enzymes that produce superoxide and nitric oxide are important examples of such genes.[6]

These radical-producing enzymes are crucial components of the nonspecific immune responses that ward off infection.[6] Oxidative stress is generally viewed as deleterious in the medical literature because it amplifies tissue injury. However, there has also been strong evolutionary pressure for cells to produce oxygen radicals as antimicrobial defenses. Because these molecules are broadly reactive, energetically cheap to make and difficult to defend against, they are produced by a variety of inflammatory cells including neutrophils and macrophages to injure or kill invading microorganisms and parasites. Surviving infection imposes huge selective pressures during early childhood. Consequently, the body produces large amounts of oxidants as antimicrobial defenses at the expense of suffering greater collateral damage in degenerative diseases occurring well past the age of reproduction. Oxidant production related to the ability to defend against infection may be a far more significant determinant of ageing than radical production resulting from aerobic metabolism.

Both macrophages and neutrophils contain highly active NADPH oxidases that reduce molecular oxygen to superoxide:

$$\text{·O–O· } (O_2) + 1 \text{ e}^- \rightarrow \text{·O–O:}^- (O_2\text{·}^-). \tag{1}$$

The structure of molecular oxygen is drawn above to show that it normally has two unpaired electrons occupying separate orbitals. Consequently, molecular oxygen can readily accept a single electron to fill one orbital and leaving one unpaired electron that makes superoxide a free radical.

Activation of NADPH oxidases in the phagolysosomes surrounding invading microorganisms produces huge fluxes of superoxide and hydrogen peroxide within a confined space that will eventually kill most microorganisms.[7] Neutrophils in addition release the enzyme myeloperoxidase into phagolysosomes. Myeloperoxidase utilizes the hydrogen peroxide and chloride to produce hypochlorous acid (HOCl), the principal ingredient in Clorox bleach. Even more surprising, expression of NADPH oxidase is not restricted to inflammatory cells but can also be induced in endothelium, vascular muscle and even neurons.[8, 9]

3. Nitric Oxide

Macrophages.[10–14] and neutrophils.[15, 16] have been shown to produce nitric oxide. Nitric oxide itself is not toxic, but adds a whole new layer of secondary reactive species that can be generated within these inflammatory cells. The broadened range of oxidants produced from a few simple progenitors greatly expands the toxic potential against all types of microorganisms including viruses.

Since the initial discovery of nitric oxide almost two decades ago, nitric oxide has been implicated in an enormous array of physiological and pathological processes and recognized by the awarding of the 1998 Nobel Prize in Physiology to Drs. Furchgott, Murad and Ignarro. Nitric oxide is extremely useful as an intercellular signaling agent because it can freely diffuse between cells and serves to modulate local cellular responses.[17] It is produced by three distinct nitric oxide synthases (NOS) through the oxidation of a nitrogen in the side chain of arginine. NADPH and oxygen are also essential substrates used in a stoichiometry of 1.5 NADPH and 2 O_2 per nitric oxide produced. These isozymes are generally referred to by the tissues where they were first described: the endothelial NOS, the neuronal NOS and the inducible NOS expressed in cytokine-activated macrophages. However, their distribution is wider than first thought, with the endothelial NOS being found in neurons and the neuronal NOS being found in skeletal muscle and throughout the peripheral nervous system. Cytokines and related inflammatory stimuli can induce the expression of inducible NOS in most all tissues. It also can be constitutively expressed in several tissues under noninflammatory conditions.

Nitric oxide was initially described in biological systems as the endothelium derived relaxing factor, which was found to activate guanylate cyclase in vascular smooth muscle to promote the vasorelaxation of blood vessels.[18, 19] Nitric oxide also modulates neuronal transmission and can be produced in much greater concentrations by the central nervous system.[20–22] The major physiological action of nitric oxide is activation of guanylate cyclase to increase cGMP synthesis. Guanylate cyclase contains the same heme protoporphyrin IX as hemoglobin with the iron in the ferrous form. Activation depends upon nitric oxide binding reversibly to the ferrous heme iron of guanylate cyclase.[23]

$$\text{Heme–Fe}^{2+} + {}^{\bullet}\text{NO} \rightleftarrows \text{Heme–Fe}^{2+}\text{–NO}. \tag{2}$$

The binding of NO to ferrous iron in heme proteins is strong and rapid.[24] An important feature of guanylate cyclase for signal transduction is how it can be rapidly inactivated, but the mechanisms that promote the release of nitric oxide from guanylate cyclase are still not well understood. This depends first upon the displacement of nitric oxide from the heme of guanylate cyclase and second upon the efficient removal of nitric oxide from tissues so it cannot rebind to guanylate cyclase.[25]

4. Why is Nitric Oxide Used as an Intercellular Messenger?

Most messenger molecules encode information within their shape, which is recognized by a specific receptor. Nitric oxide is the smallest of biological messenger molecules with the possible exception of carbon monoxide. Because of its chemical simplicity, nitric oxide conveys information by its concentration, which is interpretable by the spatial proximity of the source and target cells and its short duration *in vivo*.

As a small hydrophobic gas, nitric oxide crosses cell membranes as readily as molecular oxygen and carbon dioxide without mediation of channels or receptors. Thus, nitric oxide will diffuse isotropically to surrounding tissue. The diffusion coefficient of nitric oxide in water is higher than oxygen, carbon dioxide or carbon monoxide at 37°C,[26] which is ideal for propagating at signal. The signal is inherently short lived and localized because nitric oxide is consumed by hemoglobin in the vasculature.[27]

Release of nitric oxide from endothelium causes a local relaxation only of the underlying vascular smooth muscle. The arterial bed consists of a series of smaller and smaller branching blood vessels, which constrict and relax rapidly according to local tissue demand for oxygen. If a blood vessel supplying one region should suddenly constrict, the upstream pressure rises and can cause turbulent flow in other branches should the Reynolds number for laminar flow be exceeded.[17] Turbulent flow will further disrupt perfusion to other vascular branches by increasing shear stress, which will constrict upstream blood vessels by triggering myogenic contraction. This constriction will further amplify turbulent flow, eventually causing a catastrophic collapse of blood flow to distal vascular beds. However, shear stress induced by turbulence induces endothelium to synthesize nitric oxide to relax the neighboring smooth muscle by stimulating cGMP synthesis. The local relaxation of the underlying smooth muscle counterbalances myogenic responses to turbulent flow and thereby insures a laminar distribution of blood between vessels.[17] In effect, nitric oxide acts as a shock absorber to dampen local responses to stress. Rather than being a direct signaling molecule, nitric oxide is better viewed as a local modulator of other more rapid signaling mechanisms.

5. The Limited Reactivity of Nitric Oxide *In Vivo*

Nitric oxide is known to be a toxic gas and air pollutant, so its use as a biological messenger seems maladaptive. While nitric oxide is generally described as being highly reactive and toxic, further studies showed that nitric oxide itself is surprisingly inert towards reactions with most biological molecules.[28, 29] Perhaps the clearest evidence of this is the widespread clinical use of nitric oxide to treat pulmonary hypertension. Patients have inhaled micromolar concentrations of nitric oxide in breath for months with no apparent toxicity.[30, 31]

At the low concentrations of nitric oxide produced for single transduction, nitric oxide is rapidly removed by diffusing into blood vessels where it reacts with hemoglobin to form met-hemoglobin plus nitrate and to a lesser extent nitrosyl(Fe^{2+})hemoglobin complexes.[32, 33] The major route for the destruction of nitric oxide *in vivo* is the fast and irreversible reaction with *oxy*-hemoglobin (Hb) or *oxy*-myoglobin to produce nitrate.

$$Hb\text{–}Fe^{2+}\text{–}O_2 + {}^{\cdot}NO \rightarrow Hb\text{–}Fe^{2+}OONO \rightarrow Hb\text{–}Fe^{3+} + NO_3^-. \tag{3}$$

Red blood cells typically contain 20 mM *oxy*-hemoglobin and thus will destroy nitric oxide diffusing into the vascular stream. Nitric oxide can diffuse over 50 μm, more than the average distance to a capillary, in about one second. The resulting *met*-hemoglobin will be reduced by NADPH-dependent mechanisms in the red blood cell.

6. Nitrogen Dioxide

The reputation of nitric oxide as being highly toxic is derived from the formation of secondary reactive species. One of the obvious products is the orange-brown gas, nitrogen dioxide (NO_2), which is formed when pure nitric oxide reacts with molecular oxygen. Nitrogen dioxide is an insidious air pollutant, causing severe lung and mucosal irritation at low dosages and death results from pulmonary edema after moderate exposure.

The formation of nitrogen dioxide from oxygen occurs by a trimolecular reaction that must be rare because it involves the collision between three molecules:

$$2\,NO + O_2 \rightarrow 2\,NO_2. \tag{4}$$

Because the reaction depends upon two nitric oxides colliding with oxygen, the reaction rate depends upon the square of nitric oxide concentration. When nitric oxide is diluted a thousand fold, the overall formation of nitrogen dioxide decreases a million fold. Because the concentration of nitric oxide is ten thousand to a million times more dilute that saturating concentrations, the formation of nitrogen dioxide by reaction with oxygen in biological systems is trivial compared to other reactions that consume nitric oxide *in vivo*.

Several studies have suggested that nitrogen dioxide formation may be greatly accelerated in the hydrophobic phases of lipid bilayers or within the interiors of proteins.[34, 35] The reaction rate is known to accelerate about 6–8 fold in the gas phase compared to water, but the greatest acceleration comes from nitric oxide preferentially partitioning into these phases. A five-fold increase in nitric oxide concentration accelerates the reaction by 25 fold. However, the reaction of nitric oxide with oxygen in the hydrophobic phase is unlikely to produce a significant amount of nitrogen dioxide *in vivo* because other reactions, such as with hemoglobin in red blood cells, will consume nitric oxide much faster.

It has been well established in air pollution studies that the reaction of nitric oxide with oxygen produces almost no nitrogen dioxide. The major reaction producing nitrogen dioxide is the reaction of the conjugate acid of superoxide with nitric oxide to produce peroxynitrous acid, which immediately decomposes in the gas phase to form hydroxyl radical and nitrogen dioxide:[36, 37]

$$\text{HOO}^{\bullet} + {}^{\bullet}\text{NO} \rightarrow \text{HOONO} \rightarrow \text{HO}^{\bullet} + {}^{\bullet}\text{NO}_2. \tag{5}$$

Surprisingly, this reaction is also likely to be the major source of nitrogen dioxide in biological systems.

7. Peroxynitrite

In 1990, we proposed that a similar reaction between superoxide and nitric oxide could be a biologically important source of hydroxyl radical and nitrogen dioxide. Superoxide ($O_2^{\bullet-}$) reacts at close to the diffusion limit with nitric oxide[38, 39] to form peroxynitrite anion ($ONOO^-$).

$$^{\bullet}\text{NO} + {}^{\bullet}\text{O--O}{:}^{-} \rightarrow \text{ONOO}^- + \text{H}^+ \rightarrow \text{HOONO} \rightarrow \text{HO}^{\bullet} + {}^{\bullet}\text{NO}_2. \tag{6}$$

The major difference between the gas phase reaction and the solution reaction is that peroxynitrite anion ($ONOO^-$) is a remarkably stable molecule in water that can be synthesized by a variety of methods and stored for years. In the gas phase, peroxynitrous acid decomposes too rapidly to be observed directly. In solution, peroxynitrite must be protonated to form peroxynitrous acid ($ONOOH$) before decomposing to form radicals. This means that peroxynitrite anion can survive long enough in biological systems to be selectively reactive with a variety of targets *in vivo*. Peroxynitrite is able to directly oxidize many molecules such as sulfhydryls,[40] tyrosine phosphatases,[41] zinc-thiolate fingers[42] and iron-sulfur centers.[43, 44] This means that it can be far more toxic than either hydroxyl radical or nitrogen dioxide, which are far more reactive but correspondingly less specific.

Peroxynitrite can be produced in substantial concentrations by activated macrophages and neutrophils.[45] It also appears to be a major damaging species produced after cerebral and myocardial ischemia as well as by inflammation, sepsis and many other pathological conditions.[46-49]

A major target for the reaction with superoxide became apparent with the discovery of nitric oxide as a biological molecule.[50] Because both nitric oxide and superoxide are free radicals, they react at near diffusion-limited rates to form peroxynitrite anion. The diffusion-limited limit implies that essentially every collision between nitric oxide with superoxide results in the formation of peroxynitrite. Because small molecules can diffuse much more rapidly than large proteins, nitric oxide reacts several times faster with superoxide than SOD can possibly scavenge superoxide. This allows nitric oxide to outcompete SOD for superoxide under physiological conditions. The high intracellular concentration of SOD inside of cells, estimated to be as high as $10\,\mu M$,[51] greatly reduces the formation of peroxynitrite *in vitro*. However, the formation of peroxynitrite can still occur in the presence of SOD and is even more favorable than a simple competitive analysis of the relative rate constants would indicate.[52]

Peroxynitrite is a far stronger oxidant and much more toxic than either nitric oxide or superoxide acting separately. One can consider peroxynitrite to be a binary weapon assembled from two less reactive intermediates. Peroxynitrite anion is remarkably stable for being such a strong oxidant. We have kept crystals of peroxynitrite for as long as 10 years with little decomposition. Peroxynitrite anion has a pK_a of 6.8, allowing a substantial fraction to be protonated at neutral pH. The resulting peroxynitrous acid is less stable with 30% decomposing to form hydroxyl radical and nitrogen dioxide.[7, 53] In addition, peroxynitrite can react with metal ions bound to proteins as well as carbon dioxide that catalyzes the addition of NO_2 groups to nitrate biological molecules. The most common targets include tyrosine, tryptophan and guanine.

8. Nitrative versus Nitrosative Stress

Nitration is often confused with nitrosation and nitrosylation, but all three are quite distinct chemical moieties with widely different properties (Fig. 1). Nitration is the addition of a nitro ($-NO_2$) group to tyrosine or other chemical moieties. Biological molecules with aromatic rings like tyrosine or guanine are the most susceptible to nitration. Nitro groups stable chemical adducts that are generally removed by degradation of protein and secretion in urine.[54]

Nitrosation is the addition of an NO group to an organic molecule, while nitrosylation refers to the binding of nitric oxide to a metal. Nitrosyl adducts with metals change the valence state of the metal. On the other hand, formation of nitroso adducts to organic molecules requires oxidation by one electron.

$$\cdot NO + RSH \rightarrow RSNO + H^+ + e^-. \tag{7}$$

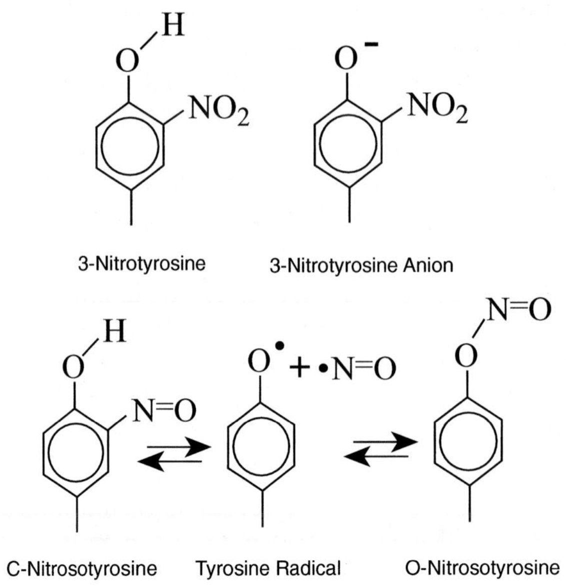

Fig. 1. Nitration versus nitrosation of tyrosine. Structure of nitrotyrosine versus nitrosotyrosine and tyrosine radical plus nitric oxide. Nitrosotyrosine is relatively unstable whereas nitrotyrosine is effectively a permanent modification.

Thiol nitrosation is also a transient modification in biological systems where an excess of thiols will reduce S-nitroso groups back to thiols. Lifetimes of nitrosothiols *in vivo* range from a few seconds to as long as several hours in certain proteins.

Nitrosation occurs most commonly on thiols but also is a well known modification of hydroxyl and amine groups. Nitrosotyrosine can also be formed by the addition of nitric oxide to tyrosine radicals, but is not formed by a direct reaction between tyrosine and nitric oxide itself (Fig. 1). Formation of nitrosotyrosine is readily reversible.[55]

9. Tyrosine Nitration

The ability of peroxynitrite to nitrate biological molecules has been an important factor for establishing a role for peroxynitrite in many pathological processes. Quantitatively, the amount of nitration found in human diseased tissues can be surprisingly high. Circulating levels of free nitrotyrosine as high as 120 μM have been found in septic patients in renal failure.[56] From 1–3% of tyrosine in spinal cords of ALS patients is nitrated.[57, 58] Using mass spectrometry, the amount of nitrotyrosine found in low density lipoprotein isolated from human atherosclerotic lesions was about 0.1% of total tyrosine.

Because the nitro group is a stable chemical modification that dramatically changes the chemical properties of tyrosine, it can have long lasting effects upon protein function. The nitro group is a large bulky addition that is strongly electron withdrawing. This decreases the pK_a of the phenol hydroxy group on tyrosine from 10 to approximately 7.5. At neutral pH, approximately 50% of nitrotyrosine is negatively charged. In the negatively charged form, nitrotyrosine is visibly yellow. In fact, we initially discovered tyrosine nitration by peroxynitrite by observing that proteins became yellow after exposure to peroxynitrite. This was particularly dramatic with bovine Cu, Zn SOD. Bovine SOD contains a single tyrosine far removed from its active site. However, we discovered that at high concentrations bovine SOD catalyzed the nitration of this tyrosine, turning the normally blue-green SOD into a yellow enzyme that still remained fully active. A careful kinetic analysis showed that one bovine SOD molecule was forming a complex with peroxynitrite that catalyzed the nitration of a tyrosine on a completely separate SOD molecule.[59] In a mixture of proteins, SOD will catalyze the nitration of many other proteins before it will catalyze nitration of itself.[60] Curiously, many of the proteins most susceptible to nitration in such a complex biological mixture appear to be subunits of structural proteins including neurofilaments and actin.[61] In 1993, we proposed that SOD-catalyzed tyrosine nitration by peroxynitrite may be the gain of function increased by mutations to SOD in ALS.[62] This proved to be too simple but led to a more interesting hypothesis.[52, 61, 63, 64]

Because antibodies to phosphotyrosine have been extremely useful, we undertook the development of antibodies to recognize nitrotyrosine.[65, 66] Both polyclonal and monoclonal antibodies were successfully generated and are now commercially available from several sources. These antibodies work particularly well for immunohistochemistry. We first described tyrosine nitration occurring in human atherosclerosis, pulmonary lesions resulting from ARDS, and myocarditis.[65, 67–69] The literature has grown rapidly to include over 600 publications showing tyrosine nitration is present in a wide range of diseases.[70] Tyrosine nitration is reported to occur in ALS, Alzheimer's disease, Huntington's disease, multiple sclerosis, and bacterial meningitis. Specific antibodies raised to nitrated α-synuclein have been shown to colocalize with Lewy bodies in Parkinson's disease and related syndromes.[71]

An important contribution by the use of nitrotyrosine antibodies has been to show that substantial amounts of reactive nitrogen species can be produced in human diseases. In the early 1990s, strong criticisms were expressed that human monocytes and other human cells did not produce significant amounts of nitric oxide and therefore nitric oxide was not playing a major role in human disease.[72–74] The identification of nitrotyrosine colocalized with active inflammatory processes provided dramatic confirmation that reactive nitrogen species were being produced in substantial quantities in a wide range of disease processes. It turns out that expression of the inducible nitric oxide synthase in humans is regulated by quite different mechanisms and found in rats and mice. Only recently have some of the complexities of the regulation of nitric oxide synthesis in humans started to be unraveled.[75, 76]

The mechanisms by which tyrosine is nitrated *in vivo* remains controversial.[77–79] Identification of nitration by antibodies can be subject to many artifacts. However, there are several excellent controls to verify specificity for nitrotyrosine. Specificity of the nitrotyrosine antibodies can be confirmed by blocking the nitrotyrosine antibodies with small peptides containing nitrotyrosine. In addition, one can treat tissue sections or Western blot membranes with the reducing agent sodium hydrosulfite (dithionite; $S_2O_6^{2-}$), which rapidly reduces nitrotyrosine to aminotyrosine. Aminotyrosine is not recognized by the antibody, so immunoreactivity is lost after the treatment. The immunoreactivity can be largely restored by reacting a dithionite-treated sample with hydrogen peroxide plus copper, which reoxidizes a substantial portion of aminotyrosine back to nitrotyrosine. The presence of nitrotyrosine in these tissue sections can be confirmed by hydrolyzing proteins followed by HPLC analysis of the individual amino acids. In some human samples, the amount of nitrotyrosine can exceed 1% of total tyrosines present in proteins isolated from diseased tissues.[80] While there have been a number of controversies over possible artifactual formation of nitrotyrosine during tissue hydrolysis, we have found that such artifacts are relatively easy to control and minimize. Multiple methods have confirmed high levels of nitrotyrosine in essentially all tissues. In addition, for some proteins in tissue samples, acid hydrolysis can destroy a substantial amount of nitrotyrosine during sample preparation. With appropriate cautions followed during sample preparation, tyrosine nitration is a useful marker for the formation of peroxynitrite and possibly other oxidants derived from nitric oxide in biological samples.

10. Alternative Mechanisms of Nitration

The production of nitric oxide itself does not result in tyrosine nitration and can substantially inhibit nitration *in vitro*. Substantial criticisms have been raised recently about other mechanisms of tyrosine nitration occurring *in vivo*. Two major culprits that might result in tyrosine nitration include nitrogen dioxide and nitrylchloride (NO_2Cl) which can be formed by the reaction of hypochlorous acid with nitrite.[81, 82] In many of the these studies, the mechanisms of tyrosine nitration were studied with free tyrosine dissolved in simple phosphate buffers, which may not be the same as nitration of tyrosine in proteins *in vivo*.[55, 83, 84] In simple solutions, radical processes can certainly be the major cause of nitration and dimerization by peroxynitrite. However, thiols and other biological molecules are efficient inhibitors of radical dependent nitration that greatly reduces their contribution to nitration of proteins in cells.[55, 60, 77]

Nitrogen dioxide is a strong one electron oxidant ($E^{o'} = +0.99$ V) and at low concentrations efficiently initiates free radical oxidation of unsaturated lipids, thiols and proteins.[85, 86]

$$R\text{–}H + {}^{\bullet}NO_2 \rightarrow R^{\bullet} + NO_2^-. \tag{8}$$

At higher concentrations, nitrogen dioxide will further react with most organic radicals at near diffusion limited rates to form a nitro-derivative ($R\text{–}NO_2$).

$$R^{\bullet} + {}^{\bullet}NO_2 \rightarrow R\text{–}NO_2. \tag{9}$$

For example, tyrosine residues on proteins are readily nitrated by a free radical mechanism,[87] where one nitrogen dioxide oxidizes the tyrosine to a phenyl radical that reacts with a second nitrogen dioxide to give nitrotyrosine.

Prütz[87] showed that either thiols or ascorbate will completely block tyrosine nitration by nitrogen dioxide. While large amounts of nitrogen dioxide, acidified nitrite or nitryl chloride can nitrate free tyrosine,[81] they are known to be extremely inefficient at nitrating tyrosine in proteins.[65, 70] For example, nitryl chloride (NO_2Cl) produced by the reaction of hypochlorous acid with nitrite does not give detectable nitration of proteins unless 2–3 millimolar concentrations of each are added.[88] Consequently, the combination of competing targets in addition to the presence of endogenous antioxidants present *in vivo* effectively prevent tyrosine nitration by nitrogen dioxide and nitryl chloride.

Under the same conditions, these factors only partially suppress nitration by peroxynitrite because a single peroxynitrite molecule can carry out a rapid two electron oxidation of tyrosine necessary to form nitrotyrosine.[77] As the number of competing targets increases, peroxynitrite becomes more likely to be the predominant nitrating agent *in vivo*. In complex biological systems, peroxynitrite also becomes increasingly selective for modifying specific tyrosine residues on certain proteins in complex biological systems.[61] Increased vulnerability of specific tyrosine groups to nitration by peroxynitrite can in part be traced to subtle differences in the local tyrosine environment imposed by the three dimensional structure of the protein.[89] These tyrosines appear to be more susceptible because the protein environment substantially decreases the phenoxyl pK_a.[89] In addition to a locally decreased tyrosine concentration, the presence of redox active metals within enzymes can also facilitate the nitration of specific tyrosines.[60, 90]

Although controversial,[91] we believe that tyrosine nitration is still an excellent marker of peroxynitrite formation *in vivo*. In biological systems, not all tyrosine nitration necessarily results from peroxynitrite, but strong evidence indicates that peroxynitrite is likely to play a major role in tyrosine nitration. Within an acute inflammatory lesion, multiple mechanisms undoubtedly could contribute to tyrosine nitration occurring in parallel. However, peroxynitrite will still be a dominant player under these conditions.

11. Proteins Susceptible to Nitration

Tyrosine nitration of selected proteins is difficult to identify by simple Western blotting using nitrotyrosine antibodies. However, immunoprecipitation with nitrotyrosine antibodies to enrich nitrated proteins has resulted in approximately 20 different proteins being identified to date that are nitrated *in vivo*.[70]

The first endogenously nitrated protein to be identified was manganese SOD,[92] which is the major enzymatic defense against superoxide in mitochondria. Peroxynitrite rapidly inactivates manganese SOD[59] and causes the selective nitration of three different tyrosines of the six present in the protein.[93] This paradoxically should amplify the formation of peroxynitrite by decreasing the ability of mitochondria to scavenge superoxide. Recently, tyrosine nitration of manganese SOD and loss of SOD activity has been shown to occur in ageing endothelium, whereas endothelial nitric oxide synthase activity is upregulated.[94] The authors propose that endothelium compensates for the loss of nitric oxide production by increasing endothelial nitric oxide synthase activity, which only accelerates the further nitration of manganese SOD in a futile cycle.

While nitration of manganese SOD correlates with the loss of activity, it is unclear whether nitration *per se* is responsible for the loss of activity. Tyrosine 34 near the active site of manganese SOD has been shown to be selectively nitrated by peroxynitrite.[95] However, mutation of this tyrosine to phenylalanine did not prevent the inactivation of manganese SOD by peroxynitrite.[96] There are at least two other tyrosines susceptible to nitration whose function has not been investigated yet. Still, nitration of manganese SOD correlates with the loss of its enzymatic activity and is commonly observed in many different disease processes.

Another important target of peroxynitrite attack is the type II isoform of the sarcoplasmic reticulum calcium ATPase.[97–100] Nitration of this protein was shown to increase greatly in ageing skeletal muscles in rats. Another protein susceptible to nitration is surfactant protein A, important for reducing surface tension of pulmonary surfactant and minimizing mycoplasma infections.[101–103] The list of nitrated proteins continues to grow.

12. Nitration of Prostacyclin Synthase

Prostacyclin synthase is a crucial enzyme in endothelium for maintaining antithrombotic activities of endothelium. This enzyme converts the product PGH_2 produced from arachidonate by cyclooxygenase into prostacyclin. Ming Zou and Volker Ullrich have identified prostacyclin synthase as being remarkably susceptible to nitration and inactivation by peroxynitrite.[104–106] Submicromolar concentrations of peroxynitrite added to endothelium almost completely inactivate prostacyclin synthase.

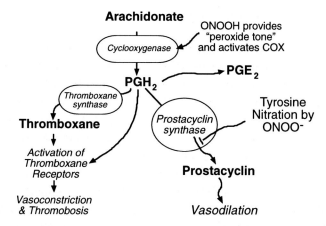

Fig. 2. Proinflammatory effects of peroxynitrite on prostaglandin production. Dual actions of peroxynitrite on prostanoid biosynthesis. Peroxynitrite can activate cyclooxygenase by providing the peroxide tone needed to put the iron in a highly oxidized state. This maximally stimulates the formation of PGH_2, which is a precursor for prostacyclin, thromboxane and PGE_2. Prostacyclin synthase is rapidly inactivated and nitrated by peroxynitrite, which results in the accumulation of PGH_2. PGH_2 has many of the same actions as thromboxane and will rapidly promote platelet aggregation and vasoconstriction. Thus, inactivation of prostacyclin synthase alone is sufficient to cause the physiological actions of thromboxane even in tissues lacking thromboxane synthase.

Peroxynitrite plays a subtle and complex role in arachidonate metabolism, because it can promote the formation of PGH_2 in addition to inactivating prostacyclin synthase. Liberation of arachadonate by phospholipases can result in the formation of prostacyclin through two steps (Fig. 2). The first step involves activation of cyclooxygenase to oxidize arachadonate to PGH_2. This enzyme contains an iron that must be converted to a higher oxidation state to oxidize arachadonate. To fully activate the enzyme, biochemists have found that a small amount of peroxide must be added to the enzyme to initiate the chemical reactions. The biological source of this peroxide tone *in vivo* has remained quite mysterious. Marnett *et al.*[107] showed that peroxynitrous acid, the protonated form of peroxynitrite, is an efficient activator of cyclooxygenase *in vitro*. More recently, they provided evidence that the formation of peroxynitrite *in vivo* is critical to the full activation of cyclooxygenase. On the other hand, peroxynitrite can rapidly inactivate prostacyclin synthase, which would take the product of cyclooxygenase and turn it into the vasodilating and antithrombogenic agent prostacyclin. Consequently, sustained production of peroxynitrite could completely inactivate prostacyclin synthase activity in seconds, while leading to the rapid formation of PGH_2. Curiously, PGH_2 is almost as effective at activating thromboxane receptors as thromboxane itself. Therefore, the build up of PGH_2 due to inactivation of prostacyclin synthase will produce the same vasoconstricting effects even in the absence of thromboxane synthase.

Consequently, endothelium maintains an antithrombotic and vasodilating state through the combined production of nitric oxide and prostacyclin. However, inflammatory stimuli can activate superoxide production by endothelium,[8, 108] thereby forcing endogenous nitric oxide production to be rapidly converted to peroxynitrite. Peroxynitrite will increase the peroxide tone that can fully activate cyclooxygenase to oxidize more arachadonate to PGH_2, while shutting down the synthesis of prostacyclin through irreversible inactivation of prostacyclin synthase. The accumulating PGH_2 can substitute for thromboxane. In this manner, endothelium in a blood vessel can become strongly pro-thrombotic and constricted in a period of a matter of seconds. While this would have negative effects in stroke and myocardial ischemia, it is an important adaptive response to limit trauma-induced bleeding and the spread of infectious agents.

Zou and Ullrich[104–106] have demonstrated nitration of prostacyclin synthase in atherosclerosis and have shown that exogenous PEG-SOD can prevent nitration of prostacyclin synthase in endothelium treated with pro-inflammatory stimuli. Given the propensity of endothelium to become positive for nitrotyrosine after cerebral ischemia, nitration of prostacyclin synthase might be an important target of peroxynitrite in brain. Protection of the endothelium may be a major contributing factor to explain how PEG-SOD was protective in stroke.[109]

13. Nitration of Structural Proteins

Other major targets for nitration *in vivo* are the disassembled subunits of structural proteins. Structural proteins are the most abundant proteins expressed in cells. For instance, actin makes about 8% of total brain protein, while neurofilaments are the predominant protein expressed in motor neurons. Structural proteins depend upon making many hydrophobic contacts between subunits to form stable macromolecular structures. Tyrosine tends to be particularly abundant in such proteins occurring in 3–5% of total amino acids in a number of structural proteins. The hydroxyl group on tyrosine can hydrogen bond to water allowing it to be relatively hydrophilic, but also stable within hydrophobic environments. When structural proteins are disassembled, many of these tyrosines become exposed to the solvent that greatly increases their susceptibility to nitration. We found that neurofilament-L (NFL), the smallest member of the neurofilament triplet family, was particularly susceptible to nitration at tyrosines located in the coiled coil domain.[61]

Nitration of structural proteins can have major functional consequences. When nitrated neurofilaments are mixed with normal neurofilament protein, they greatly disrupt the assembly into functional neurofilament protein. Structural proteins are major targets for attack by peroxynitrite and other reactive nitrogen species because they are abundant, contain tyrosines that are susceptible to nitration, and nitration may have profound affects upon their ability to assemble into functional proteins. Tyrosine nitration adds a bulky hydrophilic group and introduces a negative charge to an amino acid that must normally fit tightly with the surface of other interacting subunits. Therefore, nitration has the capability of disrupting the assembly of macromolecular structures. Only a few tyrosines on a minority of subunits need to be modified to profoundly disrupt the correct assembly of structural elements. In addition, nitrated neurofilaments can be found in the spinal cord of ALS patients as well as patients suffering from a variety of other neurological diseases.[110–112]

14. Resistance to Viral Infections

Although one tends to think of tyrosine as being crucial of signaling mediated by phosphorylation, these only account for 0.2% of all tyrosines in proteins. Tyrosines occur much more commonly in structural proteins where many residues must exist as buried residues in the assembled structures, but also be hydrophilic enough to allow unassembled monomer subunits to remain soluble in relatively high concentrations. Tyrosines fill this role nicely because they are hydrophobic enough to tolerate being in hydrophobic environments while the hydroxyl group allows hydrogen bonding in the aqueous phase.

Nitration of tyrosine is a major disruptive modification. The nitro group is bulky and will generate steric interference within the closely packed interfaces between structural subunits. The nitro group is also strongly electron withdrawing and reduces the pK_a of tyrosine to near neutral pH, effectively making nitrotyrosine a partially charged residue. When structural proteins are disassembled, the tyrosines involved in subunit contacts are exposed to solvent and become more susceptible to nitration. Nitration of tyrosine of the disassembled subunits can have major functional consequences. Because structural proteins like actin and neurofilaments depend upon the assembly of many thousands of monomeric subunits, only a few tyrosines need to be modified to have profound effects on the overall structure, because a few nitrated subunits can weaken the overall structure. We have previously shown this to be true of nitration of neurofilaments and have proposed that nitration of neurofilaments may be an important target in ALS.[61]

Could there be an adaptive advantage for cells to produce oxidants that damage structural proteins? One possibility is that peroxynitrite is a defense against viral infections. The capsids of viruses contain hundreds to thousands of structural protein subunits that must form a protective package for viral RNA or DNA. These subunits may be susceptible to nitration and other oxidative modifications as viruses replicate and result in fewer infective particles being produced. It also turns out that RNA is particularly susceptible to damage by peroxynitrite compared to DNA. The double strands of DNA apparently allow for more efficient repair. It is difficult to prove that peroxynitrite is a viral defense. However, the strongest inducers of nitric oxide synthases *in vivo* are the interferons and interleukins, which are well known to induce antiviral defenses. Inhibition of nitric oxide synthesis is also known to increase viral loads. If peroxynitrite were important in viral defense, it would explain why peroxynitrite and tyrosine nitration occur so frequently in such a wide range of disease states.

15. Ageing, Nitrative Stress and the Central Nervous System

Because neurons cannot divide, their loss over time ultimately limits the ability of the body to regenerate. Although the existence of adult stem cells in the central nervous system has recently been established, neurons have limited capacity to regenerate. Even if cardiovascular disease and cancer could be prevented, the risk rises exponentially for developing Alzheimer's disease, Parkinson's disease and amyotrophic lateral sclerosis. The gradual loss of sensory, motor memory and cognitive function limit the quality of life as we age.

The importance of motor neurons in ageing has received surprising support from recent studies in *Drosophila*. Amazingly, expression of SOD using a motor neuron specific promotor provides a substantial increase in life span.[5] Targeted deletion of superoxide dismutase and catalase in *Drosophila* results in a 60% reduction in life span. Overexpression of SOD using a general promoter restores almost a normal life span.

All voluntary movement strictly depends on the activation of motor neurons localized within the spinal cord. Each muscle cell is innervated by a single motor neuron although each motor neuron can control multiple muscle fibers. In larger muscles, the motor unit size dependent upon a single motor neuron can be as large as several thousand muscles cells. Remarkably, there are only about a million motor neurons in humans that are responsible for all muscle activity. These neurons are gradually lost with ageing and contribute to the loss of muscle mass and strength. In part, the loss of individual motor neurons is compensated by other motor neurons expanding their motor units size. This process occurs normally as part of ageing.

If large numbers of motor neurons are lost, the central nervous system loses all ability to communicate with the outside world. Unfortunately, this occurs all too often in the disease known as ALS. The loss of innervation in ALS leads to muscle degeneration resulting in progressive paralysis. Motor neurons are the only means for the nervous system to activate muscles and thereby translate thought into action. Their complete loss eliminates the ability to communicate with the outside world.

In 1993, mutations to Cu, Zn SOD were found in about 20% of familial patients.[4] Because familial patients only account for about 10% of all ALS patients, SOD mutations can only account for 2–3% of all ALS patients. Of the ALS patients carrying SOD mutations, over 70 different mutations have been identified to the SOD protein.[113] The vast majority of these mutations are missense point mutations occurring in all five exons of this small protein. The SOD mutations are dominant, suggesting they somehow confer a toxic gain of function.[113] Overexpression of certain SOD mutations in transgenic mice results in mice developing progressive paralysis.[114] In contrast, knocking out the endogenous mouse SOD gene does not result in the development of motor neuron disease, although motor neurons in these mice do become more susceptible to injury.[115]

To better characterize what the toxic gain of function might be for these mutations to SOD, we expressed several mutants in bacteria and purified the protein for biochemical studies. The majority of the SOD proteins folded to give perfectly normal functionally active Cu, Zn superoxide dismutases[63] if care was taken to incorporate one zinc and one copper atom per subunit. The ALS-SOD mutant proteins has the same superoxide-scavenging activity and catalyzed tyrosine nitration by peroxynitrite to the same extent as wild-type enzyme.[61] While the mutants appeared to be the same as wild type SOD when fully loaded with copper and zinc, it became clear that much of the ALS-SOD protein had been discarded during the purification procedures because it was missing zinc. This led us to propose that the ALS mutants may have diminished affinity for zinc.[63, 77] Similar conclusions were proposed by Valentine's group.[116]

The majority of ALS-associated SOD mutations occur at sites that structurally weaken the beta barrel that is the foundation supporting the metal binding regions in the active site of SOD.[117, 118] Because zinc is held about 7000-fold less tightly than copper by the SOD protein, structural defects favor the loss of zinc over copper.[63] One can roughly order zinc affinity of the SOD mutations with the rate of disease progression after diagnosis. There is no clear correlation between any of the mutations with the age of onset.

The loss of zinc from SOD visibly changes the color of Cu, Zn SOD, changing it from a blue-green protein with zinc to azure-blue.[52] When ascorbate is added to the blue colored zinc-deficient SOD, the protein immediately becomes colorless for a period of minutes until all the ascorbate is fully oxidized. The blue color gradually reappears as oxygen is reduced to superoxide. In the presence of a low concentration of nitric oxide, peroxynitrite can be made by the reaction of nitric oxide with superoxide produced by zinc-deficient SOD. In effect, zinc-deficient SOD can operate in reverse, stealing electrons from antioxidants present in the cell to reduce oxygen to superoxide and can even catalyze the formation of peroxynitrite (Fig. 3). Curiously, addition of wild-type Cu, Zn SOD does not slow the formation of peroxynitrite by zinc-deficient SOD. This is important because it shows how mutant SODs can confer a toxic gain of

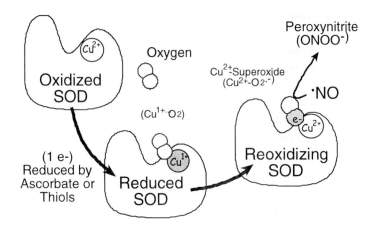

Fig. 3. Production of peroxynitrite by zinc-deficient SOD. The copper in zinc-deficient SOD rapidly oxidizes intracellular antioxidants like ascorbate while becoming reduced itself. The reduced copper slowly reacts with oxygen to form superoxide which in turn reacts with nitric oxide to form peroxynitrite, which is able to induce apoptosis in motor neurons.

function through the loss of zinc even in the presence of a large excess of wild type Cu, Zn SOD.

Zinc-deficient SOD is toxic to motor neurons in culture.[52] Motor neurons are one of the two neuronal cell cultures that can be grown in essentially pure culture without a mixture of neuronal types. When grown in the presence of brain derived neurotrophic factor, motor neurons isolated from embryonic day 15 spinal cords of rats will attach to cell culture plates and begin to spread neurites.[3] After 3 days, an axon is clearly delineated by its uniform diameter after branching. After 7 days in culture, motor neurons take on a mature phenotype of a motor neuron and no longer express the p75 neurotrophic receptor.

If motor neurons are deprived of trophic factors when initially plated, they still attached to plates and sent out neurites for the first 12–24 hours. However, the cells then apoptosed with the vast majority of neurons dying within 3 days. Blockade of nitric oxide synthesis within these trophic factor-deprived motor neurons blocked the increase in apoptosis and kept motor neurons alive for at least 6 days.[3] Generation of a constant flux of about 100 nM nitric oxide reversed the protection provided by nitric oxide synthase inhibitors to trophic factor deprived motor neurons. Nitric oxide itself was not toxic to motor neurons cultured in the presence of trophic factors. Toxicity also required the motor neurons to produce superoxide, since low molecular weight SOD mimics as well as liposomal delivery of Cu, Zn SOD to motor neurons were equally protective as blocking nitric oxide synthesis.[64] As motor neurons were dying, they became immunoreactive for nitrotyrosine. These results strongly indicated that peroxynitrite was an essential intermediate to activate apoptosis in trophic factor-deprived motor neurons.

The ability to deliver SOD entrapped in liposomes to motor neurons enabled us to test the toxicity of zinc-deficient SOD *in vitro*.[64] We expressed four different ALS mutants as well as wild-type SOD and prepared both containing Cu plus Zn as well as in the zinc-deficient state.[52] Delivery of either wild-type Cu, Zn SOD or any of the four ALS mutant SODs protected motor neurons equally well from trophic factor deprivation. These results indicate that the ALS mutant genes can produce fully active SODs that are protective to motor neurons. However, the zinc-deficient forms of ALS mutant and wild-type SOD induced motor neuron death in the presence of trophic factors. Inhibition of nitric oxide synthesis prevented death and blocked accumulation of nitrotyrosine in the motor neurons.

Copper was essential for toxicity because copper chelators such as batho-cuproine fully protected against the toxicity of zinc-deficient SOD and apoSOD itself was not toxic to the motor neurons.[52] In addition, delivering copper citrate or copper BSA at the same concentrations of the SOD protein also was not toxic to the motor neurons. These results indicate that copper bound to zinc-deficient SOD was increasing the death of motor neurons through the formation of peroxynitrite. Overall these results indicate that the loss of zinc from either wild-type or ALS mutant SOD is sufficient to cause the death of motor neurons. We propose that mutations to SOD do not directly confer a gain of function to the protein but rather increase the susceptibility to the loss of zinc and the zinc-deficient SOD is responsible for the dominant gain of function. This provides an exciting connection whereby SOD may participate in sporadic ALS as well as in familial ALS.

16. Conclusions

Nitric oxide plays an intimate role in free radical biology because it can be produced throughout the body in both physiology and pathology in concentrations that can interact with other oxidants. Oxidants are certainly produced by random autooxidation and leakage from electron transport mechanisms, which might contribute to the ageing process. However, oxidant production is important for host defense and may be produced in substantial concentrations by noninflammatory cells. Less reactive oxidants can have substantial effects on cell signaling networks that might have important roles in initiating wound healing and revascularization. While all tissues suffer from ageing, the loss of motor neurons in particular may play a crucial role determining ultimate life span, because they only number about one million in humans and their loss occurs continuously with ageing. These neurons are protected by overexpression of SOD and particularly vulnerable to mutations in SOD showing their peculiar susceptibility to oxidative stress.

Abbreviations Used

ALS — amyotrophic lateral sclerosis
NADPH — nicotine adenine dinucleotide phosphate (reduced)
NOS — nitric oxide synthase
PGH_2 — prostaglandin H_2
SOD — superoxide dismutase

References

1. Beckman, J. S. (1994). Peroxynitrite versus hydroxyl radical: the role of nitric oxide in superoxide-dependent cerebral injury. In "The Neurobiology of NO and OH" (C. C. Chiueh, D. L. Gilbert, and C. A. Colton, Eds.), Vol. 738, pp. 69–75, New York Academy of Sciences, New York.
2. Fridovich, I. (1986). Biological effects of the superoxide radical. *Arch. Biochem. Biophys.* **247**: 1–11.
3. Estévez, A. G., Spear, N., Manuel, S. M., Radi, R., Henderson, C. E., Barbeito, L. and Beckman, J. S. (1998). Nitric oxide and superoxide contribute to motor neuron apoptosis induced by trophic factor deprivation. *J. Neurosci.* **18**: 923–931.
4. Rosen, D. R., Siddique, T., Patterson, D., Figlewicz, D. A., Sapp, P., Hentati, A., Donaldson, D., Goto, J., O'Regan, J. P., Deng, H.-X., Rahmani, Z., Krizus, A., McKenna-Yasek, D., Cayabyab, A., Gaston, S. M., Berger, R., Tanszi, R. E., Halperin, J. J., Herzfeldt, B., van den Bergh, R., Hung, W.-Y., Bird, T., Deng, G., Mulder, D. W., Smyth, C., Lang, N. G., Soriana, E., Pericak-Vance, M. A., Haines, J., Rouleau, G. A., Gusella, J. S., Horvitz, H. R. and Brown, R. H., Jr. (1993). Mutations in Cu, Zn superoxide dismutase gene are associated with familial amyotrophic lateral sclerosis. *Nature* **362**: 59–62.
5. Parkes, T. L., Elia, A. J., Dickinson, D., Hilliker, A. J., Phillips, J. P. and Boulianne, G. L. (1998). Extension of Drosophila lifespan by overexpression of human SOD-1 in motorneurons. *Natural Genetics* **19**: 171–174.
6. Nathan, C. and Shiloh, M. U. (2000). Reactive oxygen and nitrogen intermediates in the relationship between mammalian hosts and microbial pathogens. *Proc. Natl. Acad. Sci. USA* **97**: 8841–8848.
7. Hurst, J. K. and Lymar, S. V. (1999). Cellularly generated inorganic oxidants as natural microbicidal agents. *Acc. Chem. Res.* **32**: 520–528.
8. Pagano, P. J., Griswold, M. C., Najibi, S., Marklund, S. L. and Cohen, R. A. (1999). Resistance of endothelium-dependent relaxation to elevation of O_2-levels in rabbit carotid artery. *Am. J. Physiol.* **277**: H2109–H2114.
9. Tammariello, S. P., Quinn, M. T. and Estus, S. (2000). NADPH oxidase contributes directly to oxidative stress and apoptosis in nerve growth factor-deprived sympathetic neurons. *J. Neurosci.* **RC53**: 51–55.

10. Drapier, J. D. and Hibbs, J. B. (1988). Differentiation of murine macrophages to express nonspecific cytotoxicity for tumor cells results in L-arginine-dependent inhibition of mitochondrial iron-sulfur enzymes in the macrophage effector cells. *J. Immunol.* **140**: 2829–2838.

11. Granger, D. L., Hibbs, J. B., Jr., Perfect, J. R. and Durack, D. T. (1988). Specific amino acid (L-arginine) requirement for the microbiostatic activity of murine macrophages. *J. Clin. Invest.* **81**: 1129–1136.

12. Green, S. J., Meltzer, M. S., Hibbs, J. B., Jr. and Nacy, C. A. (1990). Activated macrophages destroy intracellular *Leishmania major* amastigotes by an L-arginine-dependent killing mechanism.

13. Hibbs, J. B., Jr., Taintor, R. R. and Vavrin, Z. (1987). Macrophage cytotoxicity: role of L-arginine deminiase and imino nitrogen oxidation to nitrite. *Science* **235**: 473–476.

14. Hibbs, J. B., Jr., Taintor, R. R., Vavrin, Z. and Rachlin, E. M. (1988). Nitric oxide: a cytotoxic activated macrophage effector molecule. *Biochem. Biophys. Res. Commun.* **157**: 87–94.

15. McCall, T. B., Boughton-Smith, N. K., Palmer, R. M. J., Whittle, B. J. R. and Moncada, S. (1989). Synthesis of nitric oxide from L-arginine by neutrophils. *Biochem. J.* **261**: 293–296.

16. Evans, T. J., Buttery, L. D. K., Carpenter, A., Springall, D. R., Polak, J. M. and Cohen, J. (1996). Cytokine-treated human neutrophils contain inducible nitric oxide synthase that produces nitration of ingested bacteria. *Proc. Natl. Acad. Sci. USA* **93**: 9553–9558.

17. Griffith, T. M., Edwards, D. H., Davies, R. L., Harrison, T. J. and Evans, K. T. (1987). EDRF coordinates the behaviour of vascular resistance vessels. *Nature* **329**: 442–445.

18. Ignarro, L. J., Adams, J. B., Horwitz, P. M. and Wood, K. S. (1986). Activation of soluble guanylate cyclase by NO-hemoproteins involves NO-heme exchange. *J. Biol. Chem.* **261**: 4997–5002.

19. Ignarro, L. J. (1989). Heme-dependent activation of soluble guanylate cyclase by nitric oxide: regulation of enzyme activity by porphyrins and metallo-porphyrins. *Sem. Hematol.* **26**: 63–76.

20. Garthwaite, J., Charles, S. L. and Chess-Williams, R. (1988). Endothelium-derived relaxing factor release on activation of NMDA receptors suggests role as intercellular messenger in the brain. *Nature* **336**: 385–388.

21. Garthwaite, J. (1991). Glutamate, nitric oxide and cell-cell signalling in the nervous system. *Trans. Neurosci.* **14**: 75–82.

22. Garthwaite, J. and Boulton, C. L. (1995). Nitric oxide signaling in the central nervous system. *Ann. Rev. Physiol.* **57**: 683–706.

23. Traylor, T. G. and Sharma, V. S. (1992). Why NO? *Biochemistry* **31**: 2847–2849.

24. Sharma, V. S., Traylor, T. G., Gardiner, R. and Mizukami, H. (1987). Reaction of nitric oxide with heme proteins and model compounds of hemoglobin. *Biochemistry* **26**: 3837–3843.

25. Kharitonov, V. G., Russwurm, M., Magde, D., Sharma, V. S. and Koesling, D. (1997). Dissociation of nitric oxide from soluble guanylate cyclase. *Biochem. Biophys. Res. Commun.* **239**: 284–286.

26. Wise, D. L. and Houghton, G. (1968). Diffusion coefficients of neon, krypton, xenon, carbon monoxide and nitric oxide in water at 10–60°C. *Chem. Eng. Sci.* **23**: 1211–1216.

27. Furchgott, R. F. and Vanhoutte, P. M. (1989). Endothelium-derived relaxing and contracting factors. *FASEB J.* **3**: 2007–2018.

28. Beckman, J. S. (1996). The physiological and pathological chemistry of nitric oxide. *In* "Nitric Oxide: Principles and Actions" (J. R. Lancaster, Ed.), pp. 1–82, Academic Press.

29. Beckman, J. S. and Koppenol, W. H. (1996). Nitric oxide, superoxide, and peroxynitrite — the good, the bad, and the ugly. *Am. J. Physiol.* **271(Cell Physiol. 40)**: C1424–C1437.

30. Frostell, C. G., Blomqvist, H., Hedenstierna, G., Lundberg, J. and Zapol, W. M. (1993). Inhaled nitric oxide selectively reverses human hypoxic pulmonary vasoconstriction without causing systemic vasodilation. *Anesthesiology* **78**: 427–435.

31. Roberts, J. D., Jr., Fineman, J. R., Morin III, F. C., Shaul, P. W., Rimar, S., Schreiber, M. D., Polin, R. A., Zwass, M. S., Zayek, M. M., Gross, I., Heymann, M. A. and Zapol, W. M. (1997). Inhaled nitric oxide and persistent pulmonary hypertension of the newborn. *New England J. Med.* **336**: 605–610.

32. Lancaster, J. R. (1994). Simulation of the diffusion and reaction of endogenously produced nitric oxide. *Proc. Natl. Acad. Sci. USA* **91**: 8137–8141.

33. Liu, X., Miller, M. J. S., Joshi, M. S., Sadowska-Krowicka, H., Clark, D. A. and Lancaster, J. R., Jr. (1998). Diffusion-limited reaction of free nitric oxide with erythrocytes. *J. Biol. Chem.* **273**: 18709–18713.

34. Thomas, D. D., Liu, X., Kantrow, S. P. and Lancaster, J. R., Jr. (2001). The biological lifetime of nitric oxide: implications for the perivascular dynamics of NO and O_2. *Proc. Natl. Acad. Sci. USA* **98**: 355–360.

35. Nedospasov, A., Rafikov, R., Beda, N. and Nudler, E. (2000). An autocatalytic mechanism of protein nitrosylation. *Proc. Natl. Acad. Sci. USA* **97**: 13 543–13 548.

36. Payne, W. A., Stief, L. J. and Davis, D. D. (1973). A kinetics study of the reaction of HO_2 with SO_2 and NO. *J. Am. Chem. Soc.* **95**: 7614–7619.

37. Simonaltis, R. and Heicklen, J. (1974). Reaction of HO_2 with NO and NO_2. *J. Phys. Chem.* **78**: 653–657.

38. Blough, N. V. and Zafiriou, O. C. (1985). Reaction of superoxide with nitric oxide to form peroxonitrite in alkaline aqueous solution. *Inorg. Chem.* **24**: 3504–3505.
39. Huie, R. E. and Padmaja, S. (1993). The reaction rate of nitric oxide with superoxide. *Free Radic. Res. Commun.* **18**: 195–199.
40. Radi, R., Beckman, J. S., Bush, K. M. and Freeman, B. A. (1991). Peroxynitrite-mediated sulfhydryl oxidation: the cytotoxic potential of superoxide and nitric oxide. *J. Biol. Chem.* **266**: 4244–4250.
41. Takakura, K., Beckman, J. S., MacMillan-Crow, L. A. and Crow, J. P. (1999). Rapid and irreversible inactivation of protein tyrosine phosphatases PTP1B, CD45, and LAR by peroxynitrite. *Arch. Biochem. Biophys.* **369**: 197–207.
42. Crow, J. P., Beckman, J. S. and McCord, J. M. (1995). Sensitivity of the essential zinc-thiolate moiety of yeast alcohol dehydrogenase to hypochlorite and peroxynitrite. *Biochemistry* **34**: 3544–3552.
43. Castro, L., Rodriguez, M. and Radi, R. (1994). Aconitase is readily inactivated by peroxynitrite, but not by its precursor, nitric oxide. *J. Biol. Chem* **269**: 29 409–29 415.
44. Hausladen, A. and Fridovich, I. (1994). Superoxide and peroxynitrite inactivate aconitases, nitric oxide does not. *J. Biol. Chem.* **269**: 29 405–29 408.
45. Ischiropoulos, H., Zhu, L. and Beckman, J. S. (1992). Peroxynitrite formation from macrophage-derived nitric oxide. *Arch. Biochem. Biophys.* **298**: 446–451.
46. Matheis, G., Sherman, M. P., Buckberg, G. D., Haybron, D. M., Young, H. H. and Ignarro, L. J. (1992). Role of L-arginine-nitric oxide pathway in myocardial reoxygenation injury. *Am. J. Physiol.* **262**: H616–H620.
47. Mulligan, M. S., Hevel, J. M., Marletta, M. A. and Ward, P. A. (1991). Tissue injury caused by deposition of immune complexes is L-arginine dependent. *Proc. Natl. Acad. Sci. USA* **88**: 6338–6342.
48. Mulligan, M. S., Warren, J. S., Smith, C. W., Anderson, D. C., Yeh, C. G., Rudolph, A. R. and Ward, P. A. (1992). Lung injury after deposition of IgA immune complexes. Requirements for CD18 and L-arginine. *J. Immun.* **148**: 3086–3092.
49. Nowicki, J. P., Duval, D., Poignet, H. and Scatton, B. (1991). Nitric oxide mediates neuronal death after focal cerebral ischemia in the mouse. *Eur. J. Pharmacol.* **204**: 339–340.
50. Moncada, S., Palmer, R. M. J. and Higgs, E. A. (1991). Nitric oxide: physiology, pathophysiology, and pharmacology. *Pharmacol. Rev.* **43**: 109–142.
51. Rae, T. D., Schmidt, P. J., Pufahl, R. A., Culotta, V. C. and O'Halloran, T. V. (1999). Undetectable intracellular free copper: the requirement of a copper chaperone for superoxide dismutase. *Science* **284**: 805–808.
52. Estévez, A. G., Crow, J. P., Sampson, J. B., Reiter, C., Zhuang, Y.-X., Richardson, G. J., Tarpey, M. M., Barbeito, L. and Beckman, J. S. (1999). Induction of

nitric oxide-dependent apoptosis in motor neurons by zinc-deficient superoxide dismutase. *Science* **286**: 2498–2500.

53. Coddington, J. W., Hurst, J. K. and Lymar, S. V. (1999). Hydroxyl radical formation during peroxynitrous acid decomposition. *J. Am. Chem. Soc.* **121**: 2438–2443.

54. Ohshima, H., Friesen, M., Brouet, I. and Bartsch, H. (1990). Nitrotyrosine as a new marker for endogenous nitrosation and nitration of proteins. *Food Chem. Toxicol.* **28**: 647–652.

55. Goldstein, S., Czapski, G., Lind, J. and Merényi, G. (2000). Tyrosine nitration by simultaneous generation of $^\bullet$NO and $O_2^{\bullet-}$ under physiological conditions: how the radicals do the job. *J. Biol. Chem.* **275**: 3031–3036.

56. Fukuyama, N., Takebayashi, Y., Hida, M., Ishida, H., Ichimori, K. and Nakazawa, H. (1997). Clinical evidence of peroxynitrite formation in chronic renal failure patients with septic shock. *Free Radic. Biol. Med.* **22**: 771–774.

57. Ferrante, R. J., Shinobu, L. A., Schulz, J. B., Matthews, R. T., Thomas, C. E., Kowall, N. W., Gurney, M. E. and Beal, M. F. (1997). Increased 3-nitrotyrosine and oxidative damage in mice with a human copper/zinc superoxide dismutase mutation. *Ann. Neurol.* **42**: 326–334.

58. Ferrante, R. J., Browne, S. E., Shinobu, L. A., Bowling, A. C., Baik, M. J., MacGarvey, U., Kowall, N. W., Brown, R. H., Jr. and Beal, M. F. (1997). Evidence of increased oxidative damage in both sporadic and familial amyotrophic lateral sclerosis. *J. Neurochem.* **69**: 2064–2074.

59. Ischiropoulos, H., Zhu, L., Chen, J., Tsai, H. M., Martin, J. C., Smith, C. D. and Beckman, J. S. (1992). Peroxynitrite-mediated tyrosine nitration catalyzed by superoxide dismutase. *Arch. Biochem. Biophys.* **298**: 431–437.

60. Beckman, J. S., Ischiropoulos, H., Zhu, L., van der Woerd, M., Smith, C., Chen, J., Harrison, J., Martin, J. C. and Tsai, M. (1992). Kinetics of superoxide dismutase- and iron-catalyzed nitration of phenolics by peroxynitrite. *Arch. Biochem. Biophys.* **298**: 438–445.

61. Crow, J. P., Strong, M. J., Zhuang, Y., Ye, Y. and Beckman, J. S. (1997). Superoxide dismutase catalyzes nitration of tyrosines by peroxynitrite in the rod and head domains of neurofilament L. *J. Neurochem.* **69**: 1945–1953.

62. Beckman, J. S., Carson, M., Smith, C. D. and Koppenol, W. H. (1993). ALS, SOD and peroxynitrite. *Nature* **364**: 584.

63. Crow, J. P., Sampson, J. B., Zhuang, Y., Thompson, J. A. and Beckman, J. S. (1997). Decreased zinc affinity of amyotrophic lateral sclerosis-associated superoxide dismutase mutants leads to enhanced catalysis of tyrosine nitration by peroxynitrite. *J. Neurochem.* **69**: 1936–1944.

64. Estévez, A. G., Sampson, J. B., Zhuang, Y.-X., Spear, N., Richardson, G. J., Crow, J. P., Tarpey, M. M., Barbeito, L. and Beckman, J. S. (2000). Liposome-delivered superoxide dismutase prevents nitric oxide-dependent motor

neuron death induced by trophic factor withdrawal. *Free Radic. Biol. Med.* **28**: 437–446.

65. Beckman, J. S., Ye, Y. Z., Anderson, P., Chen, J., Accavetti, M. A., Tarpey, M. M. and White, C. R. (1994). Extensive nitration of protein tyrosines in human atherosclerosis detected by immunohistochemistry. *Biol. Chem. Hoppe-Seyler* **375**: 81–88.

66. Ye, Y. Z., Strong, M., Huang, Z.-Q. and Beckman, J. S. (1996). Antibodies that recognize nitrotyrosine. *In* "Methods in Enzymology" (L. Packer, Ed.), Vol. 269, pp. 201–209, Academic Press, San Diego.

67. Haddad, I., Pataki, G., Hu, P., Galliani, C., Beckman, J. S. and Matalon, S. (1994). Quantitation of nitrotyrosine levels in lung sections of patients and animals with acute lung injury. *J. Clin. Invest.* **94**: 2407–2413.

68. Kooy, N. W., Royall, J. A., Ye, Y. Z., Kelly, D. R. and Beckman, J. S. (1995). Evidence for *in vivo* peroxynitrite production in human acute lung injury. *Am. J. Respir. Crit. Care Med.* **151**: 1250–1254.

69. Kooy, N. W., Lewis, S. J., Royall, J. A., Ye, Y. Z., Kelly, D. R. and Beckman, J. S. (1997). Extensive tyrosine nitration in human myocardial inflammation: evidence for the presence of peroxynitrite. *Crit. Care Med.* **25**: 812–819.

70. Ischiropoulos, H. (1998). Biological tyrosine nitration: a pathophysiological function of nitric oxide and reactive oxygen species. *Arch. Biochem. Biophys.* **356**: 1–11.

71. Giasson, B. I., Duda, J. E., Murray, I., Chen, Q., Souza, J. M., Hurting, H. I., Ischiropoulos, H., Trojanowski, J. Q. and Lee, M.-Y. (2000). Selective alpha-synuclein nitration in synucleiopathy lesions links nitrative damage to neurodegeneration. *Science* **290**: 985–989.

72. Cameron, M. L., Granger, D. L., Weinberg, J. B., Kozumbo, W. J. and Koren, H. S. (1990). Human alveolar and peritoneal macrophages mediate fungistasis independently of L-arginine oxidation to nitrite or nitrate. *Am. Rev. Respir. Dis.* **142**: 1313–1319.

73. Padgett, E. L. and Pruett, S. B. (1992). Evaluation of nitrite production by human monocyte-derived macrophages. *Biochem. Biophys. Res. Commun.* **186**: 775–781.

74. Albina, J. E. (1995). On the expression of nitric oxide synthase by human macrophages. Why no NO? *J. Leukoc. Biol.* **58**: 643–649.

75. Sherman, M. P., Loro, M. L., Wong, V. Z. and Tashkin, D. P. (1991). Cytokine- and pneumocystis carinii-induced L-arginine oxidation by murine and human pulmonary alveolar macrophages. *J. Protozool.* **38**: 234S–236S.

76. Weinberg, J. B. (1998). Nitric oxide production and nitric oxide synthase Type-2 expression by human mononuclear phagocytes: a review. *Mol. Med.* **4**: 557–591.

77. Beckman, J. S. (1996). Oxidative damage and tyrosine nitration from peroxynitrite. *Chem. Res. Toxicol.* **9**: 836–844.

78. Eiserich, J. P., Cross, C. E., Jones, A. D., Halliwell, B. and van der Vliet, A. (1996). Formation of nitrating and chlorinating species by reaction of nitrite with hypochlorous acid: a novel mechanism for nitric oxide-mediated protein modification. *J. Biol. Chem.* **271**: 19 199–19 208.

79. Eiserich, J. P., Hristova, M., Cross, C. E., Jones, A. D., Freeman, B. A., Halliwell, B. and van der Vliet, A. (1998). Formation of nitric oxide-derived inflammatory oxidants by myeloperoxidase in neutrophils. *Nature* **391**: 393–397.

80. Banks, B. A., Ischiropoulos, H., McClelland, M., Ballard, P. L. and Ballard, R. A. (1998). Plasma 3-nitrotyrosine is elevated in premature infants who develop bronchopulmonary dysplasia. *Pediatrics* **101**: 870–874.

81. Van der Vliet, A., Eiserich, J. P., O'Neill, C. A., Halliwell, B. and Cross, C. E. (1995). Tyrosine modification by reactive nitrogen species: a closer look. *Arch. Biochem. Biophys.* **319**: 341–349.

82. Van der Vliet, A., Eiserich, J. P., Halliwell, B. and Cross, C. E. (1997). Formation of reactive nitrogen species during peroxidase-catalyzed oxidation of nitrite. *J. Biol. Chem.* **272**: 7617–7625.

83. Daiber, A., Mehl, M. and Ullrich, V. (1998). New aspects in the reaction mechanism of phenol with peroxynitrite: the role of phenoxy radicals. *Nitric Oxide: Biol. Chem.* **2**: 259–269.

84. Pfeiffer, S., Schmidt, K. and Mayer, B. (2000). Dityrosine formation out-competes tyrosine nitration at low steady-state concentrations of peroxynitrite. *J. Biol. Chem.* **275**: 6346–6352.

85. Pryor, W. A. and Lightsey, J. W. (1981). Mechanisms of nitrogen dioxide reactions: initiation of lipid peroxidation and the production of nitrous acid. *Science* **214**: 435–437.

86. Pryor, W. A., Church, D. F., Govindan, C. K. and Crank, G. (1982). Oxidation of thiols by nitric oxide and nitrogen dioxide: synthetic utility and toxicological implications. *J. Org. Chem.* **147**: 156–158.

87. Prütz, W. A., Mönig, H., Butler, J. and Land, E. J. (1985). Reactions of nitrogen dioxide in aqueous model systems: oxidation of tyrosine units in peptides and proteins. *Arch. Biochem. Biophys.* **243**: 125–134.

88. Sampson, J. B., Ye, Y. Z., Rosen, H. and Beckman, J. S. (1998). Myeloperoxidase and horseradish peroxidase catalyze tyrosine nitration in proteins from nitrite and hydrogen peroxide. *Arch. Biochem. Biophys.* **356**: 207–213.

89. Souza, J. M., Daikhin, E., Yudkoff, M., Raman, C. S. and Ischiropoulos, H. (1999). Factors determining the selectivity of protein tyrosine nitration. *Arch. Biochem. Biophys.* **371**: 169–178.

90. Mehl, M., Daiber, A., Herold, S., Shoun, H. and Ullrich, V. (1999). Peroxynitrite reaction with heme proteins. *Nitric Oxide: Biol. Chem.* **3**: 142–152.

91. Halliwell, B., Zhao, K. and Whiteman, M. (1999). Nitric oxide and peroxynitrite. The ugly, the uglier and the not so good: a personal view of recent controversies. *Free Radic. Res.* **31**: 651–669.

92. MacMillan-Crow, L. A., Crow, J. P., Kerby, J. D., Beckman, J. S. and Thompson, J. A. (1996). Nitration and inactivation of manganese superoxide dismutase in chronic rejection of human renal allografts. *Proc. Natl. Acad. Sci. USA* **93**: 11 853–11 858.

93. MacMillan-Crow, L. A., Crow, J. P. and Thompson, J. A. (1998). Peroxynitrite-mediated inactivation of manganese superoxide dismutase involves nitration and oxidation of critical tyrosine residues. *Biochemistry* **37**: 1613–1622.

94. Van der Loo, B., Labugger, R., Skepper, J. N., Bachschmid, M., Kilo, J., Powell, J. M., Palacios-Callender, M., Erusalimsky, J. D., Quaschning, T., Malinski, T., Gygi, D., Ullrich, V. and Luscher, T. F. (2000). Enhanced peroxynitrite formation is associated with vascular aging. *J. Exp. Med.* **192**: 1731–1744.

95. Yamakura, F. (1997). *In* "5th International Meeting on the Biology of Nitric Oxide" (S. Moncada, N. Toda, H. Maeda, and E. A. Higgs, Eds.), 34 pp., Portland Press, Kyoto, Japan.

96. MacMillan-Crow, L. A. and Thompson, J. A. (1999). Tyrosine modifications and inactivation of active site manganese superoxide dismutase mutant (Y34F) by peroxynitrite. *Arch. Biochem. Biophys.* **366**: 82–88.

97. Viner, R. I., Ferrington, D. A., Hühmer, A. F. R., Bigelow, D. J. and Schöneich, C. (1996). Accumulation of nitrotyrosine on the SERCA2a isoform of SR Ca-ATPase of rat skeletal muscle during aging: a peroxynitrite-mediated process? *FEBS Lett.* **379**: 286–290.

98. Klebl, B. M., Ayoub, A. T. and Pette, D. (1998). Protein oxidation, tyrosine nitration, and inactivation of sarcoplasmic reticulum CA^{2+}-ATPase in low-frequency stimulated rabbit muscle. *FEBS Lett.* **422**: 381–384.

99. Viner, R. I., Ferrington, D. A., Williams, T. D., Bigelow, D. J. and Schöneich, C. (1999). Protein modification during biological aging: selective tyrosine nitration of the SERCA2a isoform of the sarcoplasmic reticulum Ca^{2+}-ATPase in skeletal muscle. *Biochem. J.* **340**: 657–669.

100. Viner, R. I., Williams, T. D. and Schöneich, C. (1999). Peroxynitrite modification of protein thiols: oxidation, nitrosylation, and S-glutathiolation of functionally important cysteine residue(s) in the sarcoplasmic reticulum Ca-ATPase. *Biochemistry* **38**: 12 408–12 415.

101. Haddad, I. Y., Ischiropoulos, H., Holm, B. A., Beckman, J. S., Baker, J. R. and Matalon, S. (1993). Mechanisms of peroxynitrite-induced injury to pulmonary surfactants. *Am. J. Physiol.* **265**: L555–L564.

102. Greis, K. D., Zhu, S. and Matalon, S. (1996). Identification of nitration sites on surfactant protein A by tandem electrospray mass spectrometry. *Arch. Biochem. Biophys.* **335**: 396–402.

103. Hickman-Davis, J., Gibbs-Erwin, J., Lindsey, J. R. and Matalon, S. (1999). Surfactant protein A mediates mycoplasmacidal activity of alveolar macrophages by production of peroxynitrite. *Proc. Natl. Acad. Sci. USA* **96**: 4953–4958.

104. Zou, M.-H. and Ullrich, V. (1996). Peroxynitrite formed by simultaneous generation of nitric oxide and superoxide selectively inhibits bovine aortic prostacyclin synthase. *FEBS Lett.* **382**: 101–104.

105. Zou, M., Martin, C. and Ullrich, V. (1997). Tyrosine nitration as a mechanism of selective inactivation of prostacyclin synthase by peroxynitrite. *Biol. Chem. Hoppe-Seyler* **378**: 707–713.

106. Zou, M. H., Klein, T., Pasquet, J. P. and Ullrich, V. (1998). Interleukin 1-beta decreases prostacyclin synthase activity in rat mesangial cells via endogenous peroxynitrite formation. *Biochem. J.* **336**: 507–512.

107. Landino, L. M., Crews, B. C., Timmons, M. D., Morrow, J. D. and Marnett, L. J. (1996). Peroxynitrite, the coupling product of nitric oxide and superoxide, activates prostaglandin biosynthesis. *Proc. Natl. Acad. Sci. USA* **93**: 15 069–15 074.

108. Wolin, M. S., Burke-Wolin, T. M. and Mohazzab-H, K. M. (1999). Roles for NAD(P)H oxidases and reactive oxygen species in vascular oxygen sensing mechanisms. *Respir. Physiol.* **115**: 229–238.

109. Liu, T. H., Beckman, J. S., Freeman, B. A., Hogan, E. L. and Hsu, C. Y. (1989). Polyethylene glycol-conjugated superoxide dismutase and catalase reduce ischemic brain injury. *Am. J. Physiol.* **256**: H589–H593.

110. Strong, M. J., Sopper, M. M., Crow, J. P., Strong, W. L. and Beckman, J. S. (1998). Nitration of the low molecular weight neurofilament is equivalent in sporadic amyotrophic lateral sclerosis and control cervical spinal cord. *Biochem. Biophys. Res. Commun.* **248**: 157–164.

111. Chou, S. M., Wang, H. S. and Taniguchi, A. (1996). Role of SOD-1 and nitric oxide/cyclic GMP cascade on neurofilament aggregation in ALS/MND. *J. Neurol. Sci.* **139**: S16–S26.

112. Chou, S. M., Wang, H. S. and Komai, K. (1996). Colocalization of NOS and SOD-1 in neurofilament accumulation within motor neurons of amyotrophic lateral sclerosis: an immunohistochemical study. *J. Chem. Neuroanat.* **10**: 249–258.

113. Brown, R. H., Jr. (1996). Superoxide dismutase and familial amyotrophic lateral sclerosis: new insights into mechanisms and treatments. *Ann. Neurol.* **39**: 145–146.

114. Gurney, M. E., Pu, H., Chiu, A. Y., Dal Corto, M. C., Polchow, C. Y., Alexander, D. D., Caliendo, J., Hentati, A., Kwon, Y. W., Deng, H.-X., Chen, W., Zhai, P., Sufit, R. L. and Siddique, T. (1994). Motor neuron degeneration in mice that express a human Cu, Zn superoxide dismutase mutation. *Science* **264**: 1772–1775.

115. Reaume, A. G., Elliott, J. L., Hoffman, E. K., Kowall, N. W., Ferrante, R. J., Siwek, D. F., Wilcox, H. M., Flood, D. G., Beal, M. F., Brown, R. H., Jr., Scott, R. W. and Snider, W. D. (1996). Motor neurons in Cu, Zn superoxide

dismutase-deficient mice develop normally but exhibit enhanced cell death after axonal injury. *Nature Genetics* **13**: 43–47.

116. Lyons, T. J., Liu, H., Goto, J. J., Nersissian, A., Roe, J. A., Graden, J. A., Café, C., Ellerby, L. M., Bredesen, D. E., Gralla, E. B. and Valentine, J. S. (1996). Mutations in copper-zinc superoxide dismutase that cause amyotrophic lateral sclerosis alter the zinc binding site and the redox behavior of the protein. *Proc. Natl. Acad. Sci. USA* **93**: 12 240–12 244.

117. Deng, H.-X., Hentati, A., Tainer, J., Igbal, Z., Cayabyab, A., Hung, W.-Y., Getzoff, E., Hu, P., Herzfeldt, B., Roos, R., Warner, C., Deng, G., Soriano, E., Smyth, C., Parge, H., Ahmed, A., Roses, A., Hallewell, R., Pericak-Vance, M. and Siddique, T. (1993). Amyotrophic lateral sclerosis and structural defects in Cu, Zn superoxide dismutase. *Science* **261**: 1047–1051.

118. Deng, H. X. and Taylor, R. (1993). Superoxide dismutase dysfunction: a radical route to familial ALS. *J. NIH Res.* **5**: 64–67.

SECTION 3

Redox Regulation and Cell Signaling

Chapter 5

Thioredoxin-Dependent Redox Regulation — Implication in Aging and Neurological Diseases

Itaro Hattori, Hajime Nakamura, Hiroshi Masutani,
Yumiko Nishinaka, Akira Mitsui and Junji Yodoi*

Keywords: Redox regulation, thioredoxin, reactive oxygen species.

**Itaro Hattori, Hajime Nakamura, Hiroshi Masutani, Yumiko Nishinaka, Akira Mitsui
and Junji Yodoi** • Department of Biological Responses, Institute for Virus Research, Kyoto
University, 53 Shogoin-Kawaharacho, Sakyo, Kyoto 606-8507, Japan

*Corresponding Author.
Tel: +81-75-751-4024, E-mail: yodoi@virus.kyoto-u.ac.jp

1. Summary

Increasing evidence has indicated that oxidative stress mediates various cellular responses. Regulation of reduction/oxidation (redox) is important to maintain homeostasis of life.

Thioredoxin (TRX) and related molecules is one of key systems to control cellular redox status, as well as the glutathione system. TRX is a small multi-functional protein with a redox-active disulfide/dithiol in the active site. TRX physiologically has cytoprotective effects against oxidative stress by scavenging reactive oxygen species (ROS) together with peroxiredoxin and is induced by various oxidative stresses through the activation of responsive elements in its promoter.

In addition, TRX is translocated from the cytoplasm into the nucleus upon oxidative stress, physically interacting with Ref-1 (redox factor1)/APEX, an endoexonuclease located in the nucleus. Several reports showed that TRX and/or Ref-1 enhance the DNA binding activity of AP-1, polyoma enhancer binding protein-2 (PEBP2), NF-κB, p53, and other transcription factors. Based on these findings, besides a ROS-scavenging activity, TRX is considered as a regulator/modulator involved in various steps of cellular signaling against oxidative stress. We produced transgenic mice overexpressing human TRX (hTRX) gene that is under the control of human β-actin promoter. From the viewpoint of the clinical implication, ischemic brain injury was attenuated in the hTRX transgenic mice. Moreover, our preliminary study showed that these hTRX transgenic mice exhibited extended 22% in the maximum life-span and 35% in the median life-span. Further understanding of TRX-dependent redox regulation will give us a new strategy for preventing diseases related to oxidative stress.

2. Introduction

ROS are naturally generated in eukaryotic cells from oxygen during respiration for energy metabolism, or in response to various stimuli, such as UV-irradiation, X-ray, ischemia/reperfusion, inflammatory cytokines and chemical carcinogens. ROS can alter or disrupt the balance of redox potential in cells and that causes various cellular dysfunction and diseases.[1, 2] Eukaryotic cells have acquired several systems to maintain intracellular redox status by scavenging ROS in evolution. Those systems include the glutathione (GSH)[3] and the thioredoxin (TRX) systems.[4] Recently, in addition to this function, evidence has accumulated that reducing molecules such as TRX play important roles in cellular signaling through the reduction of cysteine residues of, as well as the interaction with, various important components of signal transduction pathways.

In this chapter we focus on TRX and its associated molecules and discuss the role of TRX-dependent redox regulation in oxidative stress-evoked cellular signaling.

3. TRX and Its related Molecules

TRX is a small protein having oxidoreductase activity via its redox-active disulfide/dithiol site within the conserved active sequence, -Cys-Gly-Pro-Cys-,[4] and operates together with NADPH and TRX reductase. TRX appears to be present in all living cells.[4] We identified and cloned an adult T cell leukemia (ATL)-derived factor (ADF) from HTLV-I positive cell line ATL-2 which was found to be a human homologue of TRX.[6] Several cytokine-like factors such as 3B6-IL-1, eosinophil cytotoxicity enhancing factor (ECEF), T cell hybridoma MP6 derived B-cell growth factor, and early pregnancy factor, are identical or related to TRX, indicating that TRX plays a multifunctional role.[7] TRX reductase has a selenium-containing active center in the C-terminus and there exist several isoforms of TRX reductase.[8]

In the past years, new members of TRX-related molecules in the mammalian system have also been identified. They are called TRX superfamily. Table 1 summarizes the members of human TRX superfamily.

Table 1. TRX Superfamily

	kDa	Localization	Active Site Sequence
Thioredoxin	12	Cytosol	-Cys-Gly-Pro-Cys-
Thioredoxin 2	12	Mitochondria	-Cys-Gly-Pro-Cys-
TRX related protein (TRP32)	32	Cytosol	-Cys-Gly-Pro-Cys-
Glutaredoxin (GRX)	12	Cytosol	-Cys-Pro-Tyr-Cys-
Nucleoredoxin	48	Nucleus	-Cys-Pro-Pro-Cys-
Protein disulfide isomerase (PDI)	55	Endoplasmic reticulum	-[Cys-Gly-His-Cys]$_2$-
Ca binding protein 1 (CaBP1)	49	Endoplasmic reticulum	-[Cys-Gly-His-Cys]$_2$-
Ca binding protein 2 (ERp72)	72	Endoplasmic reticulum	-[Cys-Gly-His-Cys]$_3$-
Phospholipase C gamma	61	Endoplasmic reticulum	-[Cys-Gly-His-Cys]$_2$-

Table 2. Peroxiredoxin Family in Human
(modified from Refs. 12 and 30)

Name	Length in Amino Acids	Localization
Prx 1: PAG, NKEFA	199	Cytosol and nucleus
Prx 2: TSA, NKEFB	198	Cytosol
Prx 3: AOP-1	256	Mitochondoria
Prx 4: AOE372, TRANK	271	Cytosol/secreted
Prx 5: AOEB166	214	Mitochondoria/microsome
Prx 6: ORF6	224	Cytosol

Glutaredoxin (GRX), known as thioltransferase, has GSH-disulfide oxido-reductase activity with redox-active site, -Cys-Pro-Tyr-Cys-.[9] GRX reduces low molecular weight disulfides and proteins in concert with NADPH and GSH reductase.

Mammalian thioredoxin 2 (TRX2) has high homology with TRX and has an active site Cys-Gly-Pro-Cys with thiol-reducing activity.[10] It has mitochondrial insertion signal and is specifically localized in mitochondria. It has been showed that TRX2 regulates the intra-mitochondrial redox state and plays an important role in cell proliferation in *in vitro* study using DT40 cell line (Tanaka *et al.* in preparation). The biological functions of these members have not yet been clarified. Thioredoxin-dependent peroxidases (Peroxiredoxins: Prx) are considered to be members of a family for intracellular hydrogen peroxidase[11] (Table 2). Six members of Prx family have been identified in human, all of which utilizes TRX as the electron donor except Prx VI. The features and functions of Prx family were well described in a recent review elsewhere.[12] Thus, the TRX system is composed of several related molecules forming a network of recognition and interaction through its active site cysteine residues.

4. TRX Binding Proteins

Several TRX-binding proteins have been isolated. MAP kinase is involved in one of signaling pathways activated by various oxidative stress.[13] Apoptosis signal-regulating kinase (ASK1), which mediates apoptosis signal by activating the c-Jun N-terminal kinase (JNK) and p38 MAP kinase pathways, was shown to become inactive by binding to reduced TRX.[14] When TRX is oxidized under oxidative stress, it dissociates from ASK1 and apoptosis signal is transduced (Fig. 2). Nishiyama *et al.* have reported another TRX binding protein, TBP-2, which is identical to vitamin D3 upregulated protein1 (VDUP1).[15] In 1α, 25-dihydroxyvitamin D3-treated HL-60 cells, TBP-2/VDUP1 expression was enhanced, whereas TRX expression and the reducing activity were down-regulated.

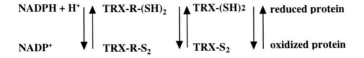

$$\text{NADPH} + \text{H}^+ \quad \uparrow \downarrow \quad \text{TRX-R-(SH)}_2 \quad \uparrow \downarrow \quad \text{TRX-(SH)2} \quad \uparrow \downarrow \quad \text{reduced protein}$$

$$\text{NADP}^+ \qquad \text{TRX-R-S}_2 \qquad \text{TRX-S}_2 \qquad \text{oxidized protein}$$

TRX-R: thioredoxin reductase
TRX: thioredoxin

Fig. 1. TRX reducing cycle.

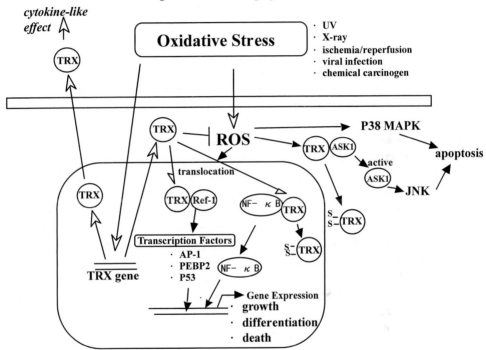

Fig. 2. TRX-dependent redox regulation.

1α, 25-dihydroxy vitamin D3 is important for regulation of calcium homeostasis and hormone secretion[17] and is a potent inducer of myeloid cell differentiation. It is likely that TRX-TBP-2/VDUP1 interaction plays an important role in the redox regulation of growth and differentiation of the cells sensitive to a variety of inducers, including 1α, 25-dihydroxy vitamin D3 responses. The redox active site in TRX is required for the binding to both TBP-2 and ASK1. The conformation of TRX adjacent to the active site may be critical for the binding between these molecules. A phagocyte oxidase component, p40 phox was also reported to have TRX binding property, although the physiological role of TRX-p40 phox interaction is still unclear.[16] These three proteins were all cloned by yeast two-hybrid system. Watson *et al.* identified TRX as a protein kinase C-interacting protein using phage display system and showed that the exogenously added TRX inhibits autophosphorylation of PKC and PKC-mediated phosphorylation of histone *in vitro*.[5] Cloning of other TRX binding proteins using cDNA libraries from different sources is in progress. The finding of new binding proteins would give us understanding of novel roles of TRX.

5. Redox Regulation of Transcription Factors by TRX

Transcription factors are important sensing and signaling components of oxidative signaling. Redox regulation appears to be involved in various steps of activation of transcription factors.

Several important transcription factors have been shown to be modulated by reducing agents.[2, 18, 19] TRX is one of key mediators to modulate activities of transcription factors (Fig. 1). Ref-1, AP endonuclease (APEX), enhances DNA-binding activities of many nuclear transcription factors such as AP-1, NF-κB, ATF/CREB, Myb, HIF-1α[20] and p53 through redox-dependent mechanism. The reduction of Ref-1 by TRX is required for its activity.[21, 22] Thus, TRX regulates many transcriptional events through Ref-1. TRX can also directly enhance the DNA binding activities of several factors. Ueno *et al.* reported that TRX facilitates p53-mediated p21 activation through both Ref-1-independent and -dependent mechanisms.[23] The enhancement of DNA binding activity by TRX was also reported in nuclear receptors such as glucocorticoid receptor[24] and estrogen receptor.[25] Regulation of NF-κB by TRX has been extensively investigated.[26-28] Treatment of NF-κB with oxidants such as diamide or hydrogen peroxide renders the protein incapable of DNA binding, while reduction of the protein by TRX as well as thiol reductants such as DTT enhances its DNA-binding activity.

This action is associated with the reduction of Cys62 in the DNA-binding loop of NF-κB p50 by TRX.[28] Interestingly, studies on targeted-overexpression of TRX either in cytosol or in nucleus have indicated that TRX has dual and opposing roles depending on its localization in the regulation of NF-κB.[29] In cytosol, TRX interferes the activities of NF-κB by blocking the dissociation of I-κB from NF-κB, whereas TRX enhances its DNA binding activities in nucleus.

TRX translocation from cytosol to nucleus was induced by a wide variety of oxidative stresses including UV irradiation,[29] hydrogen peroxide,[32] or hypoxia,[31] treatment with CDDP.[23] Therefore, it is assumed that TRX is translocated to the nuclear compartment upon oxidative stress to interact with Ref-1. Nuclear localization of TRX is often observed in pathological tissue. In cervical tissue, TRX expression is observed in human papilloma virus-infected cells and TRX is localized in the nucleus.[33] In the renal proximal tubules, TRX is induced and translocated to nuclei by oxidative damage mediated by Fe-nitrilotriacetate.[34]

Although further investigation is needed, TRX translocation may be related to the cyotoprotection and pathogenesis of oxidative stress-related disorders. Structural analysis of TRX complexed with Ref-1 or NF-κB was performed by Qin *et al.* and revealed that these complexes represent kinetically stable mixed disulfide intermediates in the same binding location of TRX, but in opposing orientations.

This indicates TRX has the potential to target a wide range of proteins by balancing versatility in substrate recognition.[35, 36]

6. Redox Regulation of Apoptosis by TRX

Apoptosis can be considered as a finely regulated mechanism to maintain genome stability by eliminating cells with severe DNA damage caused by oxidative stress. Jun kinase/ SAP kinase and p38 kinase play an important role in oxidative stress-induced apoptosis.[38] TRX has been found to bind to ASK1, a MAPKKK, and inhibits the apoptosis process in the specificity of the cell type.[37] When TRX is oxidized by oxidative stress, ASK1 is dissociated from oxidized TRX and activated to induce an apoptosis signal. Indeed, we previously showed that TRX prevents apoptosis induced by TNF,[39] or L-cystin and glutathione depletion.[40] In addition, overexpression of TRX negatively regulates p38 Map kinase activation and IL-6 production by TNF-α -stimulated cells. And TRX reduces ROS generated by NADPH oxidase including Rac-1 and inhibits p38 MAP kinase activation in response to epidermal growth factors (Hirota, K. *et al.* submitted).

These results indicate that TRX plays a critical role for p38 Map kinase activation.[41] Interestingly, Prx II is reported to be an inhibitor of apoptosis with a mechanism distinct from that of Bcl-2.[42]

7. Cytoprotective Effects of TRX

TRX has been shown to play crucial roles in cytoprotection against a variety of oxidative stress. Recombinant TRX can protect cells from anti-Fas antibody-induced apoptosis and cytotoxicity induced by TNF-α, hydrogen peroxide and activated neutrophils.[32, 39] TRX is also a potent costimulator of various cytokine expression.[43, 44] Recently, Nilsson *et al.* reported that TRX induces the secretion of TNF-α and maintains the expression of Bcl-2, whereby prolongs survival of B-LCL.[45] Overexpression of TRX has been observed in a wide variety of oxidative conditions such as viral infection, diabetes, ischemic/reperfusion and malignant tissues.[7, 46, 47] During viral infection, considerable amount of ROS is generated, causing tissue damage and DNA breaks. As TRX was first purified from HTLV-1 transformed cells,[46] TRX is induced and/or secreted from transformed cells related to infection of viruses such as HTLV-1, EBV,[48] hepatitis C virus and papilloma virus. Elevated TRX level in serum was also reported in late stage HIV patients.[49] Recently, Sono *et al.* have reported that TRX suppresses lytic replication of EBV induced by 12-O-tetradecanoylphorbol-13-acetate (TPA) and prevented the cell death evoked by the lytic induction.[50] These observations suggest that TRX is closely involved in both the process of virus infection and the prognosis of the infected patients.

In vivo study showed that recombinant TRX attenuated ischemia/reperfusion lung injury in rat.[51]

8. Aging and TRX Overexpressing Transgenic Mice

TRX overexpressing transgenic (Tg) mice were produced in which human TRX gene is systemically expressed under the control of human beta-actin promoter.[60] The Tg mice were shown to be resistant to oxidative tissue damage such as fasting-induced lipid peroxidation in liver and UV-induced cytocide of hemopoietic progenitor cells. Telomerase activity in spleen tissues in Tg mice was higher than that in wild type mice. Preliminary study showed that the Tg mice exhibited median and maximum life span. Relative to the controls, the percent increase in the Tg mice was 22% in the maximum life-span and 35% in the median life-span (Fig. 3) (Mitsui, A. *et al.* submitted). These results show that TRX gives resistance against general oxidative stress a possible extension of life span without apparent abnormality in mammals. Miyazaki and coworkers have produced another type of TRX overexpressing mice in which human TRX gene

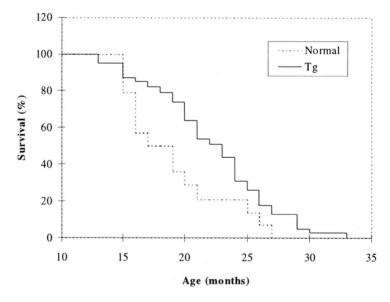

Fig. 3. Survivorship of the TRX Tg mice.

was expressed in pancreatic beta-cells.[52] TRX overexpression attenuated the onset of autoimmune and drug-induced diabetes in which oxidative stress is suggested to be involved in the pathogenesis.

On the other hand, homozygous mutant carrying a targeted disruption of TRX was lethal after implantation, suggesting that TRX expression is essential for early differentiation and morphogenesis of the mouse embryo.[53]

9. Oxidative Stress and Central Nervous System

ROS generation may also contribute to neuronal dysfunction and death in a range of disorders, including stroke, Alzheimer's disease, and Parkinson's disease.[54, 55] Redox regulation is an important issue in the central nervous system.

Recombinant TRX has a neuroprotective effect on murine primary cultured neurons obtained from the striatum and cortex.[58] TRX functions as a neurotrophic factor for central cholinergic neurons[63] and was upregulated in mechanical injury of brain[59] and peripheral motor neuron.[65] Takagi *et al.* reported that protein and mRNA levels of TRX are decreased in the ischemic core and increased in the penumbra ischemic region during permanent middle cerebral artery occlusion in rats.[64] Moreover, ischemic brain injury and excitotoxic hippocampal injury were attenuated in TRX transgenic mice.[60, 62]

Asahina *et al.* showed that TRX and TRX mRNA are expressed in glial cells in the white matter but not in neuronal cells in brain of Alzheimer's disease.[64] Lovell *et al.* reported that TRX protein levels is decreased in brain of Alzheimer's disease and neuron in the primary hippocampal cultures treated with TRX show resistance to amyloid β-peptide-induced toxicity in TRX dose-dependent manner.[57] Kojima *et al.* reported that low-dose γ-ray irradiation activates antioxidant potency, including TRX, and attenuates brain damages in experimental mice model of Parkinson's disease.[56] These findings indicate that TRX and its redox modulation have a neuroprotective function and play a role during brain injury. In this sense, TRX can be a new target molecule for prevention of diseases in central nervous system.

10. Conclusion

As we described in this chapter, TRX, a redox regulating protein, and its network play important roles in regulating cellular signaling and gene expression. Further analysis of the role of TRX in the oxidative stress response and the mechanism of the TRX gene regulation by oxidative stress should help to elucidate how cells link the oxidative stress response to gene regulation.

Because oxidative stress is involved in various diseases as well as cachexia and aging process, increasing knowledge in the redox regulation will lead us to plan a new strategy for diagnosis and treatment of diseases including pathological aspects of aging.

References

1. Packer, L. and Yodoi, J. (Eds.) (1999). *Redox Regulation of Cell Signaling and Its Clinical Application*, Marcel Dekker, Inc., New York.
2. Masutani H., Ueno, M., Ueda, S., Yodoi, J., Sen, C. K., Sies, H. and Baeuerle, P.A., Eds. (2000). *Antioxidant and Redox Regulation of Genes*, Academic Press, San Diego.
3. Hayes, J. D. and McLellan, L. I. (1999). Glutathione and glutathione-dependent enzymes represent a co-ordinately regulated defence against oxidative stress. *Free Radic. Res.* **31**: 273–300.
4. Holmgren, A. (1985). Thioredoxin. *Ann. Rev. Biochem.* **54**: 237–271.
5. Watson, J. A., Rumsby, M. G. and Wolowacz, R. G. (1999). Phage display identifies thioredoxin and superoxide dismutase as novel protein kinase C-interacting proteins: thioredoxin inhibits protein kinase C-mediated phosphorylation of histone. *Biochem. J.* **343 (Part 2)** 301–305.
6. Tagaya, Y., Maeda, Y., Mitsui, A., Kondo, N., Matsui, H. *et al.* (1989). ATL-derived factor (ADF), an IL-2 receptor/Tac inducer homologous to hioredoxin; possible involvement of dithiol-reduction in the IL-2 receptor induction. *EMBO J.* **8**: 757–764.
7. Yodoi, J. and Uchiyama, T. (1992). Diseases associated with HTLV-I: virus, IL-2 receptor dysregulation and redox regulation. *Immunol. Today* **13**: 405–411.
8. Liu, S. Y. and Stadtman, T. C. (1997). Heparin-binding properties of selenium-containing thioredoxin reductase from HeLa cells and human lung adenocarcinoma cells. *Proc. Natl. Acad. Sci. USA* **94**: 6138–6141.
9. Wells, W. W., Yang, Y., Deits, T. L. and Gan, Z. R. (1993). Thioltransferases. *Adv. Enzymol. Rel. Areas Mol. Biol.* **66**: 149–201.
10. Spyrou, G., Enmark, E., Miranda, V. A. and Gustafsson, J. (1997). Cloning and expression of a novel mammalian thioredoxin. *J. Biol. Chem.* **272**: 2936–2941.
11. Jin, D. Y., Chae, H. Z., Rhee, S. G. and Jeang, K. T. (1997). Regulatory role for a novel human thioredoxin peroxidase in NF-kappa B activation. *J. Biol. Chem.* **272**: 30 952–30 961.
12. Butterfield, L. H., Merino, A., Golub, S. H. and Shau, H. (1999). From cytoprotection to tumor suppression: the multifunctorial role of peroxiredoxins. *Antioxidants and Redox Signaling* **1**: 385–402.
13. Ichijo, H. (1999). From receptors to stress-activated MAP kinases. *Oncogene* **18**: 6087–6093.

14. Saitoh, M., Nishitoh, H., Fujii, M., Takeda, K., Tobiume, K., Sawada, Y., Kawabata, M., Miyazono, K. and Ichijo, H. (1998). Mammalian thioredoxin is a direct inhibitor of apoptosis signal-regulating kinase (ASK) 1. *EMBO J.* **17**: 2596–2606.

15. Nishiyama, A., Matsui, M., Iwata, S., Hirota, K., Masutani, H., Nakamura, H., Takagi, Y., Sono, H., Gon, Y. and Yodoi, J. (1999). Identification of thioredoxin-binding protein-2/vitamin D(3) up-regulated protein 1 as a negative regulator of thioredoxin function and expression. *J. Biol. Chem.* **274**: 21 645–21 650.

16. Nishiyama, A., Ohno, T., Iwata, S., Matsui, M., Hirota, K., Masutani, H., Nakamura, H. and Yodoi, J. (1999). Demonstration of the interaction of thioredoxin with p40phox, a phagocyte oxidase component, using a yeast two-hybrid system. *Immun. Lett.* **68**: 155–159.

17. Suda, T., Shinki, T. and Takahashi, T. (1990). The role of vitamin D in bone and intestinal cell differentiation. *Ann. Rev. Nutri.* **10**: 195–211.

18. Arrigo, A. P. (1999). Gene expression and the thiol redox state. *Free Radic. Biol. Med.* **27**: 936–944.

19. Flohe, L., Brigelius-Flohe, R., Saliou, C., Traber, M. G. and Packer, L. (1997). Redox regulation of NF-kappa B activation. *Free Radic. Biol. Med.* **22**: 1115–1126.

20. Ema, M., Hirota, K., Mimura, J., Abe, H., Yodoi, J., Sogawa, K., Poellinger, L. and Fujii-Kuriyama, Y. (1999). Molecular mechanisms of transcription activation by HLF and HIF1alpha in response to hypoxia: their stabilization and redox signal-induced interaction with CBP/p300. *EMBO J.* **18**: 1905–1914.

21. Rothwell, D. G., Barzilay, G., Gorman, M., Morera, S., Freemont, P. and Hickson, I. D. (1997). The structure and functions of the HAP1/Ref-1 protein. *Oncol. Res.* **9**: 275–280.

22. Hirota, K., Matsui, M., Iwata, S., Nishiyama, A., Moroi, K. and Yodoi, J. (1997). AP-1 transcriptional activity is regulated by a direct association between thioredoxin and Ref-1. *Proc. Natl. Acad. Sci. USA* **94**: 3633–3638.

23. Ueno, M., Masutani, H., Arai, R. J., Yamauchi, A., Hirota, K., Sakai, T., Inamoto, T., Yamaoka, Y., Yodoi, J. and Nikaido, T. (1999). Thioredoxin-dependent redox regulation of p53-mediated p21 activation. *J. Biol. Chem.* **274**: 35 809–35 815.

24. Makino, Y., Okamoto, K., Yoshikawa, N., Aoshima, M., Hirota, K., Yodoi, J., Umesono, K., Makino, I. and Tanaka, H. (1996). Thioredoxin: a redox-regulating cellular cofactor for glucocorticoid hormone action. Cross talk between endocrine control of stress response and cellular antioxidant defense system. *J. Clin. Invest.* **98**: 2469–2477.

25. Hayashi, S., Hajiro-Nakanishi, K., Makino, Y., Eguchi, H., Yodoi, J. and Tanaka, H. (1997). Functional modulation of estrogen receptor by redox state with reference to thioredoxin as a mediator. *Nucleic Acids Res.* **25**: 4035–4040.

26. Schulze-Osthoff, K., Schenk, H. and Droge, W. (1995). Effects of thioredoxin on activation of transcription factor NF-kappa B. *Meth. Enzymol.* **252**: 253–264.

27. Okamoto, T., Sakurada, S., Yang, J. P. and Merin, J. P. (1997). Regulation of NF-kappa B and disease control: identification of a novel serine kinase and thioredoxin as effectors for signal transduction pathway for NF-kappa B activation. *Current Topic Cell Regul.* **35**: 149–161.

28. Matthews, J. R., Wakasugi, N., Virelizier, J. L., Yodoi, J. and Hay, R. T. (1992). Thioredoxin regulates the DNA binding activity of NF-kappa B by reduction of a disulphide bond involving cysteine 62. *Nucleic Acids Res.* **20**: 3821–3830.

29. Hirota, K., Murata, M., Sachi, Y., Nakamura, H., Takeuchi, J., Mori, K. and Yodoi, J. (1999). Distinct roles of thioredoxin in the cytoplasm and in the nucleus. A two-step mechanism of redox regulation of transcription factor NF-kappa B. *J. Biol. Chem.* **274**: 27 891–27 897.

30. Koops B., Clippe A., Bogard C., Arsalane K., Wattiez R., Hermans C., Duconseille E., Falmagne P. and Bernard A. (1999). Cloning and characterization of AOEB166, a novel mammalian antioxidant enzyme of the peroxiredoxin family. *J Biol. Chem.* **274**: 30 451–30 458.

31. Ema, M., Hirota, K., Mimura, J., Abe, H., Yodoi, J., Sogawa, K., Poellinger, L. and Fujii-Kuriyama, Y. (1999). Molecular mechanisms of transcription activation by HLF and HIF1alpha in response to hypoxia: their stabilization and redox signal-induced interaction with CBP/p300. *EMBO J.* **18**: 1905–1914.

32. Nakamura, H., Matsuda, M., Furuke, K., Kitaoka, Y., Iwata, S., Toda, K., Inamoto, T., Yamaoka, Y., Ozawa, K. and Yodoi, J. (1994). Adult T cell leukemia-derived factor/human thioredoxin protects endothelial F-2 cell injury caused by activated neutrophils or hydrogen peroxide. *Immun. Lett.* **42**: 75–80.

33. Fujii, S., Nanbu, Y., Nonogaki, H., Konishi, I., Mori, T., Masutani, H. and Yodoi, J. (1991). Coexpression of adult T-cell leukemia-derived factor, a human thioredoxin homologue, and human papillomavirus DNA in neoplastic cervical squamous epithelium. *Cancer* **68**: 1583–1591.

34. Tanaka, T., Nishiyama, Y., Okada, K., Hirota, K., Matsui, M., Yodoi, J., Hiai, H. and Toyokuni, S. (1997). Induction and nuclear translocation of thioredoxin by oxidative damage in the mouse kidney: independence of tubular necrosis and sulfhydryl depletion. *Lab. Invest.* **77**: 145–155.

35. Qin, J., Clore, G. M., Kennedy, W. M., Huth, J. R. and Gronenborn, A. M. (1995). Solution structure of human thioredoxin in a mixed disulfide intermediate complex with its target peptide from the transcription factor NF-kappa B. *Structure* **3**: 289–297.

36. Qin, J., Clore, G. M., Kennedy, W. P., Kuszewski, J. and Gronenborn, A. M. (1996). The solution structure of human thioredoxin complexed with its target from Ref-1 reveals peptide chain reversal. *Structure* **4**: 613–620.

37. Saito, M., Nishitoh, H., Fujii, M., Takeda, K., Tobiume, K., Sawada, Y., Kawabata, M., Miyazono, K. and Ichijo, H. (1998). Mammalian thioredoxin is a direct inhibitor of apoptosis signal-regulating kinase ASK1. *EMBO J.* **17**: 2596–2606.
38. Ichijo, H., Nishida, E., Irie, K., Dijke, P. T., Saitoh, M., Moriguchi, T., Takagi, M., Matsumoto, K., Miyazono, K. and Gotoh, Y. (1997). Induction of apoptosis by ASK1, a mammalian MAPKKK that activates SAPK/JNK and p38 signaling pathways. *Science* **275**: 90–94.
39. Matsuda, M., Masutani, H., Nakamura, H., Miyajima, S., Yamauchi, A., Yonehara, S., Uchida, A., Irimajiri, K., Horiuchi, A. and Yodoi, J. (1991). Protective activity of adult T cell leukemia-derived factor (ADF) against tumor necrosis factor-dependent cytotoxicity on U937 cells. *J. Immun.* **147**: 3837–3841.
40. Iwata, S., Hori, T., Sato, N., Ueda, T. Y., Yamabe, T., Nakamura, H., Masutani, H. and Yodoi, J. (1994). Thiol-mediated redox regulation of lymphocyte proliferation. Possible involvement of adult T cell leukemia-derived factor and glutathione in transferrin receptor expression. *J. Immun.* **152**: 5633–5642.
41. Hashimoto, S., Matsumoto, K., Gon, Y., Furuichi, S., Maruoka, S., Takeshita, I., Hirota, K., Yodoi, J. and Horie, T. (1999). Thioredoxin negatively regulates p38 MAP kinase activation and IL-6 production by tumor necrosis factor-alpha. *Biochem. Biophys. Res. Commun.* **258**: 443–447.
42. Zhang, P., Liu, B., Kang, S. W., Seo, M. S., Rhee, S. G. and Obeid, L. M. (1997). Thioredoxin peroxidase is a novel inhibitor of apoptosis with a mechanism distinct from that of Bcl-2. *J. Biol. Chem.* **272**: 30 615–30 618.
43. Wakasugi, N., Tagaya, Y., Wakasugi, H., Mitsui, A., Maeda, M., Yodoi, J. and Tursz, T. (1990). Adult T-cell leukemia-derived factor/thioredoxin, produced by both human T-lymphotropic virus type I- and Epstein-Barr virus-transformed lymphocytes, acts as an autocrine growth factor and synergizes with interleukin-1 and interleukin-2. *Proc. Natl. Acad. Sci. USA* **87**: 8282–8286.
44. Schenk, H., Vogt, M., Droge, W. and Schulze-Osthoff, K. (1996). Thioredoxin as a potent costimulus of cytokine expression. *J. Immun.* **156**: 765–771.
45. Nilsson, J., Soderberg, O., Nilsson, K. and Rosen, A. (2000). Thioredoxin prolongs survival of B-type chronic lymphocytic leukemia cells. *Blood* **95**: 1420–1426.
46. Yodoi, J. and Tursz, T. (1991). ADF, a growth-promoting factor derived from adult T cell leukemia and homologous to thioredoxin: involvement in lymphocyte immortalization by HTLV-I and EBV. *Adv. Cancer Res.* **57**: 381–411.
47. Nakamura, H., Nakamura, K. and Yodoi, J. (1997). Redox regulation of cellular activation. *Ann. Rev. Immun.* **15**: 351–369.

48. Wakasugi, H., Wakasugi, N., Tursz, T., Tagaya, Y. and Yodoi, J. (1989). Production of an IL-1-like factor by 3B6 EBV-infected B cells [letter]. *J. Immun.* **142**: 2569–2570.

49. Nakamura, H., DeRosa, S., Roederer, M., Anderson, M. T., Dubs, J. G., Yodoi, J., Holmgren, A., Herzenberg, L. A. and Herzenberg, L. A. (1996). Elevation of plasma thioredoxin levels in HIV infected individuals. *Int. Immun.* **8**: 603–611.

50. Sono, H., Teshigawara, K., Sasada, T., Takagi, Y., Nishiyama, A., Ohkubo, Y., Maeda, Y., Tatsumi, E., Kanamaru, A. and Yodoi, J. (1999). Redox control of Epstein-Barr virus replication by human thioredoxin/ATL-derived factor: differential regulation of lytic and latent infection. *Antioxidants and Redox Signaling* **1**: 155–165.

51. Fukuse, T., Hirata, T., Yokomise, H., Hasegawa, S., Inui, K., Mitsui, A., Hirakawa, T., Hitomi, S., Yodoi, J. and Wada, H. (1995). Attenuation of ischaemia reperfusion injury by human thioredoxin. *Thorax* **50**: 387–391.

52. Hotta, M., Tashiro, F., Ikegami, H., Niwa, H., Ogihara, T., Yodoi, J. and Miyazaki, J. (1998). Pancreatic beta cell-specific expression of thioredoxin, an antioxidative and antiapoptotic protein, prevents autoimmune and streptozotocin-induced diabetes. *J. Exp. Med.* **188**: 1445–1451.

53. Matsui, M., Oshima, M., Oshima, H., Takaku, K., Maruyama, T., Yodoi, J. and Taketo, M. M. (1996). Early embryonic lethality caused by targeted disruption of the mouse thioredoxin gene. *Dev. Biol.* **178**: 179–185.

54. Jesberger, J. A. and Richardson, J. S. (1991). Oxygen free radicals and brain dysfunction .*Int. J. Neurosci.* **57**: 1–7.

55. Simonian, N. A. and Coyle, J. T. (1996). Oxidative stress in neurodegenerative disease. *Ann. Rev. Pharmacol. Toxicol.* **36**: 83–106.

56. Kojima, S., Matsuki, O. Nomura, T. Yamaoka, K. Takahashi, M. and Niki, E. (1999). Elevation of antioxidant potency in the brain of mice by low-dose γ-ray irradiation and its effect on 1-methyl-4-phenyl-1,2,3,6-tetrahydropyridine (MPTP)-induced brain damage. *Free Radic. Biol. Med.* **26**: 388–395.

57. Lovell, A. M., Xie, C., Gabbita, P. S. and Markesbery, R. W. (2000). Decreased thioredoxin and increased thioredxin reductase levels in Alzheimer's disease brain. *Free Radic. Biol. Med.* **28**: 418–427.

58. Hori K., Katayama M. and Sano N. *et al.* (1994). Neuroprotection by glial cells through adult T-cell leukemia-derived factor/human thioredoxin (ADF/thioredoxin). *Brain Res.* **652**: 304–310.

59. Lippoldt A., Padilla C. A., Gerst H., Andbjer B., Richer E., Holmgren A. and Fuxe K. (1995). Localization of thioredoxin in the rat brain and functional implications. *J. Neurosci.* **15**: 6747–6756.

60. Takagi Y., Mitsui A., Nishiyama A., Nozaki K., Sono H., Gon Y., Hashimoto N. and Yodoi J. (1999). Overexpression of thioredoxin transgenic mice attenuates focal ischemic brain damage. *Proc. Natl. Acad. Sci. USA* **96**: 4131–4136.

61. Takagi Y., Tokime T., Nozaki K., Gon Y., Kikuchi H. and Yodoi J. (1998). Redox control of neuronal damage during brain ischemia following middle cerebral artery occlusion in the rat: immunohistochemical and hybridization studies of thioredoxin. *J. Cereb. Blood Flow Metab.* **18**: 206–214.
62. Takagi Y., Hattori I., Nozaki K., Mitsui A., Ishikawa M., Hashimoto N. and Yodoi J. (2000). Excitotoxic hippocampal injury is attenuated in thioredoxin transgenic mice. *J. Cereb. Blood Flow Metab.* **20**: 829–833.
63. Endoh M., Kunishita T. and Tabira T. (1993). Thioredoxin from activated macrophages as a trophic factor for central cholinergic neurons in vitro. *Biochem. Biophys. Res. Commun.* **192**: 760–765.
64. Asahina, M., Yamada, T., Yoshiyama, Y. and Yodoi, J. (1998). Expression of adult T cell leukemia-derived factor in human brain and peripheral nerve tissues. *Dementia and Geriatric Cognitive Disorders* **9**: 181–185.
65. Mansur, K., Iwanishi, Y., Kiryu, S., Su, Q., Namikawa, K., Yodoi, J. and Kiyama, H. (1998). Up-regulation of thioredoxin expression in motor neurons after nerve injury. *Mol. Brain. Res.* **62**: 86–91.

Chapter 6

Redox Regulation of Genes and Cell Function

Arne Holmgren

Keywords: Thioredoxin, thioredoxin reductase, glutathione, glutaredoxin, glutathione reductase, selenium, oxidative stress, thiol redox
control, disulfide, redox potential transcription factors, enzymes,
oxidant, antioxidants, hydrogen peroxide, superoxide, thioredoxin
peroxidase, redox regulation, stress response.

Arne Holmgren • Medical Nobel Institute for Biochemistry, Department of Medical
Biochemistry and Biophysics, Karolinska Institute, SE-171 77 Stockholm, Sweden
Tel: +46 8 728 7686, E-mail: arne.holmgren@mbb.ki.se

1. Introduction

Redox regulation is now established as a major mechanism in cell signaling affecting a large and growing number of enzymes, transcription factors and receptors.[1] Particularly, effects of stresses like oxidative stress, UV irradiation, heat shock or osmotic stress are linked to changes in intracellular redox potential which is transmitted to changes in activity of numerous genes and pathways. Generation of free radicals of oxygen (ROS) is induced in all aerobic cells and are widely known to participate in deleterious reactions like in degenerative diseases and ageing.[2] However, ROS and antioxidants have also been realized to be used by cells in signaling for a growing range of biological stimuli and growth factors.[3, 4] Oxidants and antioxidants modulate gene expression and are involved in mitogenic responses and signal transduction leading to transformation, differentiation and senescence.[1, 2] As cells approach the later phases of their proliferative life span, they exhibit a progressive loss of capacity to respond to environmental stresses.[5, 6] The importance of a more oxidized cell state in senescence has been shown since chronic inhibition of glutathione (GSH) synthesis decreases proliferative life span, whereas stimulation of GSH synthesis increases the same phenomenon.[7] Treatment of early passage human diploid fibroblast cells with an oxidant like H_2O_2 causes cells to have many properties of the senescent phenotype.[8, 9] As an example of the relation between transcription factor activity and proliferation, activator protein 1 (AVP-1) activity is required for initiation of DNA synthesis and is lost during senescence.[10] This is particularly due to loss of *c-fos* expression which is regulated in a complicated may by both antioxidants and oxidants including possible influence from PKC regulated pathways. Induction of *c-fos* can be given by the antioxidant N-acetyl cysteine (NAC),[11] a known precursor of GSH-synthesis. Also the regulation of *c-jun*-NH2 terminal kinase (JNK) involves redox changes being a target for antioxidants and working with the glutathione transferase pi (GSTp).[12]

In this short review the two major disulfide reduction systems of cells will be summarized. The thioredoxin and glutaredoxin systems ultimately get electrons from NADPH and have distinct properties and target molecules in mammalian cells. They have central roles in redox signaling and are also major antioxidants in cells of particular significance for the control of the levels of ROS. The latter have a central role in the biology of ageing.[13]

2. The Thioredoxin System

The thioredoxin system consists of NADPH, thioredoxin reductase (TrxR) and thioredoxin (Trx) and is a major electron transport system in cells. The thioredoxin system is ubiquitous from Archea to man.[14, 15] Via Reactions 1 and 2, reduced thioredoxin (Trx-(SH)$_2$) operates as a main disulfide reductase and oxidized thioredoxin (Trx-S$_2$) is regenerated by TrxR:

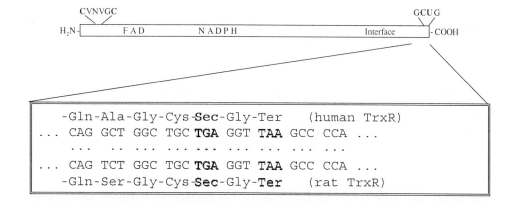

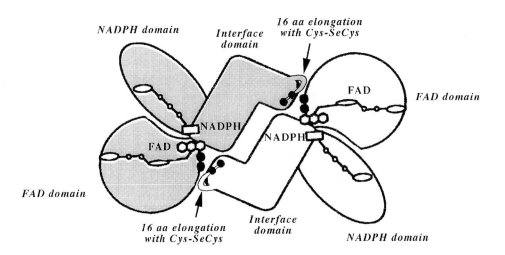

Thioredoxin Reductase

Fig. 1. Structure of mammalian thioredoxin reductases (TrxRs). Schematic subunit structure of human and rat TrxRs.[19] The N-terminal GR-like active site disulfide (CVNVGC) and the active site GCUG denoting the active site -Gly-Cys-SeCys-Gly. Enlarged the region of that part of the corresponding genes with the TGA codon for selenocysteine (Sec). Below structural model of mammalian TrxR based on the homology to glutathione reductase with the head to tail arrangement in the dimer.[21]

$$\text{Trx-S}_2 + \text{NADPH} + \text{H}^+ \xrightarrow{\text{TrxR}} \text{Trx-(SH)}_2 + \text{NADP}^+ \tag{1}$$

$$\text{Trx(SH)}_2 + \text{Protein-S}_2 \longrightarrow \text{Trx-S}_2 + \text{Protein-(SH)}_2. \tag{2}$$

Human and mammalian thioredoxin reductases are dimeric selenoenzymes and radically different from the smaller enzymes from bacteria, yeast or plants.[16–18] The TrxRs from mammalian cells including the bovine, rat or human enzymes are larger selenium-dependent enzymes with a close homology to glutathione reductase (Fig. 1). However, the polypeptide chain is elongated by 16-residues with the conserved C-terminal sequence: Gly-Cys-SeCys-Gly, where SeCys is selenocysteine.[19] This is the active site with a selenenylsulfide in the oxidized enzyme and a selenolthiol in the reduced enzyme formed from the Cys-SeCys sequence.[20, 21] A structural model of the enzyme based on the homology to glutathione reductase (Fig. 1) predicts a head to tail structure with electrons from the redox-active disulfide/dithiol of one subunit transferred to the active site in the other subunit.[21]

Thioredoxin is a 12 kDa-protein from *E. coli* to man with a redox active disulfide in the conserved active site sequence: Cys-Gly-Pro-Cys.[14] The three-dimensional structure of thioredoxins (the thioredoxin fold) shows that the active site is located on a protrusion of the molecule and that localized conformational changes a company reduction disulfide to a dithiol.[22] Trx-(SH)$_2$ is the cells major disulfide reductase reacting with protein disulfides like those of insulin or in an enzyme like ribonucleotide reductase 10^4 to 10^6 times faster than the stronger chemical reductant dithiothreitol (DTT).[14] The mechanism of thioredoxin has been well investigated involving docking of the protein to a target protein in an initial non-covalent complex. This is followed by attack of the nucleophilic N-terminal Cys-residue of the active site motif on the disulfide to make a short-lived covalent mixed disulfide intermediate. Attack by the C-terminal active site Cys-residue, generates thioredoxin disulfide and a dithiol in the target protein.[22]

Thioredoxin has a large and growing number of functions, reviewed in Ref. 15, which involves transfer of electrons to essential reactions like the biosynthesis of the deoxyribonucleotides for DNA-synthesis via ribonucleotide reductase, repair of proteins by reduction of methionine sulfoxide residues by methionine sulfoxide reductases, or defense against oxidative stress by thioredoxin peroxidases. The latter are at least six enzymes, which have a low K_m-value for hydrogen peroxide and operate by Cys residues.[23] Other roles of thioredoxin involve general reduction of disulfides. A major role for thioredoxin is redox regulation of target proteins including numerous transcription factors. This involves e.g. NF-$_\kappa$B and AP-1.[24] By its low redox potential (−270 mV), thioredoxin is the major reductant responsible for keeping the inside of the cell reduced.[14, 15] Novel functions for thioredoxin involve being a secreted co-cytokine as well as a chemokine.[15] Levels of thioredoxin in plasma reflect inflammation, or oxidative stress.[25]

The mammalian thioredoxin reductase and thioredoxin system is dependent on an adequate supply of selenium for activity.[20] The SeCys$_{498}$ to Cys mutant has a very low activity and the truncated enzyme from TGA acting as a stop codon is inactive.[20] Apart from reducing thioredoxin, the mammalian thioredoxin reductases have a large number of different substrates including selenium compounds, like selenite, dehydroascorbate or ascorbyl free radical.[15] Reduction of selenite generates selenide used in synthesis of selenocysteine. The TrxR has an inherent hydroperoxide reductase activity working with lipid hydroperoxides as well as hydrogen peroxide (review in Ref. 15). The importance of this activity of the enzyme in cell physiology or ageing is not known presently. Almost no studies have so far been published regarding the relative activity of thioredoxin reductase with its selenium dependence during ageing.

Inhibitors of thioredoxin reductase (review in Ref. 26) involve e.g. 1-chloro-2, 4-dinitrobenzene (DNCB) with irreversible covalent modification of both the active site selenocysteine residue and the adjacent cysteine residue. DNCB treatment activates the enzyme *in vitro* and *in vivo* the enzyme can still be active and produce a 35-fold increase in superoxide. Other inhibitors of the enzyme like gold thioglucose and clinically used derivatives for treating tumor cells are most likely targeted to the selenocysteine residue. Potentially, selenium deficiency may cause the synthesis of a truncated enzyme with prooxidant activity.[21] The physiological consequences of this in general or in ageing have not yet been studied. However, the known inverse relationship between selenium levels and cancer incidence[27] requires more studies of thioredoxin reductase and thioredoxin.

3. The Glutaredoxin System

The glutaredoxin system is comprised of GSH, NADPH, glutathione reductase and glutaredoxin.[28, 29] The level of GSH in cells are high (1–10 mM) and it is kept reduced by NADPH and glutathione reductase (GR) as shown in Reaction 3:

$$\text{GSSG} + \text{NADPH} + \text{H}^+ \xrightarrow{\text{GR}} 2 \text{ GSH} + \text{NADP}^+. \tag{3}$$

Via glutaredoxin (Grx) electrones from GSH are used to reduce disulfides (Reactions 4 and 5)

$$\text{Grx-S}_2 + 2 \text{ GSH} \longrightarrow \text{Grx-(SH)}_2 + \text{GSSG} \tag{4}$$

$$\text{Grx-(SH)}_2 + \text{Protein-S}_2 \longrightarrow \text{Grx-S}_2 + \text{Protein-(SH)}_2. \tag{5}$$

Grx also catalyzes formation and cleavage of glutathionylated proteins (Reaction 6):

$$\text{Protein-SH} + \text{GSSG} \underset{\longrightarrow}{\overset{\text{Grx}}{\longleftarrow}} \text{Protein-S-SG} + \text{GSH}. \tag{6}$$

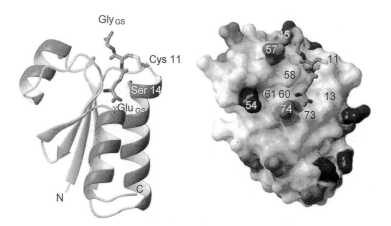

Fig. 2. Left: NMR solution structure of *E. coli* Grx1 in mixed disulfide with GSH (Glygs) via Cys 11.[31] Right: Molecular surface with residues interacting with the GSH molecule.

Glutaredoxin was discovered as a GSH-dependent hydrogen donor for ribonucleotide reductase.[30] Today glutaredoxins are a multifunctional family of GSH-disulfide oxidoreductases which belong to the thioredoxin fold super family.[31] They have a GSH-binding site[32] and a redoxactive disulfide with the concensus sequence -Cys-Pro-Tyr-Cys (Fig. 2). Only the N-terminal nucleophilic Cys-residue is required for catalyzing reversible glutathiorylation reaction (Reaction 6) whereas both Cys residue are required for disulfide reduction (Reaction 5).[29]

Glutaredoxins in mammalian cells (also called thioltransferase) have a growing list of functions such as reduction of dehydroascorbic acid,[29] cellular differentiation, or regulation of transcription factor activity.[33] A new class of monothiol glutaredoxins in yeast and many other organisms appear to be particularly important in defense against oxidative stress.[34] Glutaredoxin also protects cerebellar granulae neurons from dopamine induced oxidative stress by activating $NF_\kappa B$ via Ref-1.[35]

4. Redox Regulation by Thioredoxin and Glutaredoxin

Control of the activity of proteins by the reversible oxidation of SH-groups or thiol redox control[14, 28] is now recognized as a major mechanism for signal transduction. Oxidants generated upon cell activation or exposure to oxidative stress are converted to a disulfide signal via GSH-peroxidases or thioredoxin peroxidases (Fig. 3) and are balanced by antioxidants from the thioredoxin and glutaredoxin systems.[15] Transcription factor binding to DNA is particularly sensitive to the redox state of critical SH-groups. The outside of cells is an oxidizing environment dominated by disulfides whereas the cytosol is rich in SH-groups

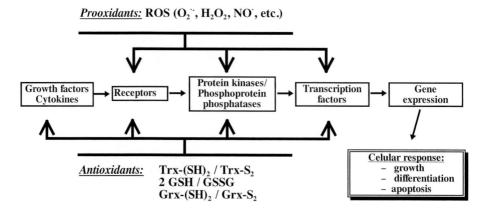

Fig. 3. Redox regulation of cellular systems via thiol redox control.[15]

and proteins in the cytosol have few or no disulfides.[36] Changes in the levels of GSH and GSSG will be an important global parameter in determining the intracellular redox potential[37] since glutathione is the major redox buffer of mammalian cells.[36] For details of effects on oxidative stress and gene regulation see.[1, 3, 13]

5. Effects of Oxidants and Antioxidants

The level of thioredoxin and thioredoxin mRNA is regulated by oxidative stress showing induction by hydrogen peroxide.[38] Also the protein levels go up.[38] Upon stimulation of lymphocytes by hydrogen peroxide thioredoxin is secreted[38] and a truncated form (Trx 80) is also secreted.[39] Trx 80 has recently been shown to be a mitogenic cytokine for normal human peripheral blood mononuclear cells (PBMC).[39] As an example of differential regulation, thioredoxin reductase and glutathione synthesis is upregulated by t-butylhydroquinone in cortical astrocytes but not in cortical neurons.[40]

6. Future Perspectives

Almost nothing is today known about the regulation and expression of thioredoxin or glutaredoxin systems during ageing. Particularly, the role of these systems in replicative senescence should be studied. Since senescent cells arrest growth and become resistant to apoptotic cell death yet unknown redox signaling mechanisms involving thioredoxins and glutaredoxins can be predicted. Ultimately, this knowledge may also be used clinically by gene therapy or by novel drug development.

Acknowledgments

Research support by the Swedish Medical Research (13X-3529) Council and the Swedish Cancer Society (961) is gratefully acknowledged.

References

1. Allen, R. G. and Tresini, M. (2000). Oxidative stress and gene regulation. *Free Radic. Biol. Med.* **28**: 463–499.
2. Beckman, K. B. and Ames, B. N. (1998). The free radical theory of ageing. *Physiol. Rev.* **78**: 547–581.
3. Abate, C., Patel, L., Rauscher, F. J. and Curran, T. (1990). Redox regulation of FOS and Jun DNA binding activity in vitro. *Science* **249**: 1157–1161.
4. Schreck, R., Rieber, P. and Bauerle, P. A. (1991). Reactive oxygen intermediates as apparently widely used messengers in the activation of the NF_kB transcription factor and HIV-1. *EMBO J.* **10**: 2247–2258.
5. Wheaton, K., Atadja, P. and Riabouel, K. (1996). Regulation of transcription factor activity during ageing. *Biochem. Cell. Biol.* **74**: 523–534.
6. Seshadri, T. and Campisi, J. (1990). Regression of *c-fos* transcription and an altered gene program in senescent human fibroblasts. *Science* **247**: 205–209.
7. Honda, S. and Matsuo, M. (1988). Relationships between the cellular glutathione level and in vitro life span of human diploid fibroblasts. *Exp. Gerontol.* **23**: 81–86.
8. Chen, Q., Fisher, A., Reagan, J. D., Yan, L. J. and Ames, B. N. (1995). Oxidative DNA damage and senescence of human diploid fibroblasts. *Proc. Natl. Acad. Sci. USA* **92**: 4337–4341.
9. Chen, Q. M., Bartholomew, J. C., Campisi, J., Acosta, M., Reagan, J. D. and Ames, B. N. (1998). Molecular analysis of H_2O_2-induced senescent-like growth arrest in normal human fibroblasts: p53 and Rb control G_1 arrest but not cell replication. *Biochem. J.* **332**: 43–50.
10. Riabowol, K., Schiff, J. and Gilman, M. Z. (1992). Transcription factor AP-1 activity is required for initiation of DNA synthesis and is lost during cellular aging. *Proc. Natl. Acad. Sci. USA* **89**: 157–161.
11. Keogh, B. P., Allen, R. G., Tresini, M. and Cristofalo, V. J. (1998). Antioxidants stimulate transcriptional activation of the *c-fos* gene by multiple pathways in human fetal lung fibroblasts (W1-38). *J. Cell. Physiol.* **176**: 624–633.
12. Adler, V., Yin, Z., Fuchs, S. Y., Benezra, M., Rosario, L., Tew, K. D., Pincus, M. R., Sardana, M., Henderson, C. J., Wolf, C. R., Davis, R. J. and Ronai, Z. (1999). Regulation of JNK singaling by GSTp. *EMBO J.* **18**: 1321–1334.
13. Finkel, T. and Holbrook, N. J. (2000). Oxidants, oxidative stress and the biology of ageing. *Nature* **408**: 239–247.

14. Holmgren, A. (1985). Thioredoxin. *Ann. Rev. Biochem.* **54**: 237–271.
15. Arnér, E. S. J. and Holmgren, A. (2000). Physiological functions of thioredoxin and thioredoxin reductase. *Eur. J. Biochem.* **267**: 6102–6109.
16. Luthman, M. and Holmgren, A. (1982). Purification and characterization of thioredoxin and thioredoxin reductase. *Biochemistry* **21**: 6628–6633.
17. Tamura, T. and Stadtman, T. C. (1996). A new selenoprotein from human lung adenocarcinoma cells: purification properties and thioredoxin reductase activity. *Proc. Natl. Acad. Sci. USA* **93**: 1006–1011.
18. Williamns, C.H., Jr., Arscott, L. D., Müller, S., Lennon, B. W., Ludwig, M. L., Wang, Pan-Fen, Veine, D. M., Becker, K. and Schirmer, H. (2000). Thioredoxin reductase. Two modes of catalysis have evolved. *Eur. J. Biochem.* **267**: 6110–6117.
19. Zhong, L., Arnér, E. S. J., Ljung, J., Åslund, F. and Holmgren, A. (1998). Rat and calf thioredoxin reductase are homologous to glutathione reductase with a carboxyterminal elongation containing a conserved catalytically active penultimate selenocysteine residue. *J. Biol. Chem.* **273**: 8581–8591.
20. Zhong, L. and Holmgren, A. (2000). Essential role of selenium in the catalytic activities of mammalian thioredoxin reductase revealed by characterization of recombinant enzymes with selenocysteine mutations. *J. Biol. Chem.* **275**: 18 121–18 128.
21. Zhong, L., Arnér, E. S. J. and Holmgren, A. (2000). Structure and mechanism of mammalian thioredoxin reductase: the active site is a redoxactive selenolthiol/selenenylsulfide formed from the conserved cysteine-selenocysteine sequence. *Proc. Natl. Acad. Sci. USA* **97**: 5854–5859.
22. Holmgren, A. (1995). Thioredoxin structure and mechanism: conformational changes on oxidation of the active site sulfhydryls to a disulfide. *Structure* **3**: 239–243.
23. Chae, H. Z., Kang, S. W. and Rhee, S. G. (1999). Isoforms of mammalian peroxiredoxin that reduce peroxides in presence of thioredoxin. *Meth. Enzymol.* **300**: 219–226.
24. Hirota, K., Murata, M., Sachi, Y., Nakamura, H., Takeuchi, J. K. M. and Yodoi, J. (1999). Distinct roles of thiroedoxin in the cytoplasm and in the nucleus. A two-step mechanism of redox regulation of transcription factor NF-kappa B. *J. Biol. Chem.* **274**: 27 891–27 897.
25. Nakamura, H., DeRosa, S., Roederer, M., Anderson, M. T., Dubs, J. G., Yodoi, J., Holmgren, A., Herzenberg, L. A. and Herzenberg, L. A. (1996). Elevation of plasma thioredoxin levels in HIV-infected individuals. *Int. Immunol.* **8**: 603–611.
26. Becker, K., Gromer, S., Schirmer, R. H. and Müller, S. (2000). Thioredoxin reductase as a pathophysiological factor and drug target. *Eur. J. Biochem.* **267**: 6118–6125.
27. Combs, G. F., Jr. and Gray, W. P. (1998). Chemopreventive agents: selenium. *Pharmacol. Ther.* **79**: 179–192.

28. Holmgren, A. (1989). Thioredoxin and glutaredoxin systems. *J. Biol. Chem.* **264**: 13 963–13 966.

29. Holmgren, A. and Åslund, F. (1995). Glutaredoxin. *Meth. Enzymol.* **252**: 283–292.

30. Holmgren A. (1976). Hydrogen donor system for *Escherichia coli* ribonucleoside-diphosphate reductase dependent upon glutathione. *Proc. Natl. Acad. Sci. USA* **73**: 2275–2279.

31. Sodano, P., Xia, T., Bushweller, J. H., Björnberg, O., Holmgren, A., Billeter, M. and Wüthrich, K. (1991). Sequence-specific ^{1}H n.m.r. assignments and determination of the three-dimensional structure of reduced *Escherichia coli* glutaredoxin. *J. Mol. Biol.* **221**: 1311–1324.

32. Bushweller, J. H., Billeter, M., Holmgren, A. and Wüthrich, K. (1994). The NMR solution structure of the mixed disulfide between *Escherichia coli* Glutaredoxin (C14S) and Glutathione. *J. Mol. Biol.* **235**: 1585–1597.

33. Nakamura, T., Ohno, T., Hirota, K., Nishiyama, A., Nakamura, H., Wada, H. and Yodoi, J. (1999). Mouse glutaredoxin — cDNA cloning, high level expression in *Escherichia coli* and its possible implication in redox regulation of the DNA binding activity in transcription factor PEBP2. *Free Radic. Res.* **4**: 357–365.

34. Rodriguez-Manzaneque, M. T., Ros, J., Cabiscol, E., Sorribas, A. and Herrero, E. (1999). Grx5 glutaredoxin plays a central role in protection against protein oxidative damage in Saccharomyces cerevisiae. *Mol. Cell Biol.* **12**: 8180–8190.

35. Daily, D., Vlamis-Gardikas, A., Offen, D., Mittelman, L., Melamed, E., Holmgren, A. and Barzilai, A. (2001). Glutaredoxin protects cerebellar granule neurons from dopamine induced apoptosis by activating NF-${}_\kappa$B via Ref-1. *J. Biol. Chem.* **276**: 1335–1344.

36. Gilbert, H. F. (1990). Molecular and cellular aspects of thiol-disulfide exchange. *Adv. Enzymol. Rel. Areas Mol. Biol.* **63**: 69–172.

37. Åslund, F., Berndt, K. D. and Holmgren, A. (1997). Redox potentials of glutaredoxins and other thiol-disulfide oxidoreductases of the thioredoxin superfamily determined by direct protein-protein redox equilibria. *J. Biol. Chem.* **272**: 30 780–30 786.

38. Nakamura, H., Nakamura, K. and Yodoi, J. (1997). Redox regulation of cellular activation. *Ann. Rev. Immun.* **15**: 351–369.

39. Pekkari, K., Gurunath, R., Arnér, E. S. J. and Holmgren, A. (2000). Truncated thioredoxin is a mitogenic cytokine for resting human peripheral blood mononuclear cells and is present in human plasma. *J. Biol. Chem.* **275**: 37 474–37 480.

40. Eftekharpour, E., Holmgren, A. and Juurlink, B. H. J. (2000) Thioredoxin reductase and glutathione synthesis is upregulated by t-Butylhydroquinone in cortical astrocytes but not in cortical neurons. *GLIA* **31**: 241–248.

Chapter 7

Oxidative Stress and Signal Transduction Pathway of Redox Sensor Proteins: Clinical Applications and Novel Therapeutics

Toshifumi Tetsuka* and Takashi Okamoto[†]

Toshifumi Tetsuka and **Takashi Okamoto** • Department of Molecular Genetics, Nagoya City University Medical School, 1 Kawasumi, Mizuho-cho, Mizuho-ku, Nagoya 467-8601, Japan

*Corresponding Author. E-mail: tetsuka@med.nagoya-cu.ac.jp
[†]Corresponding Author. E-mail: tokamoto@med.nagoya-cu.ac.jp

1. Introduction

Reactive oxygen species (ROS) are produced in the cell by various environmental stimuli such as infection of microbes (viruses, bacteria, etc.), ionizing and UV irradiation, and pollutants (i.e. oxidants), which are collectively called "oxidative stress". These ROS are highly reactive with biological macromolecules to result in producing lipid peroxides (which are often radicals), inactivating proteins and mutating DNA (by producing 8-OH-dG or breaking nucleic acid chains). Therefore, cells must have acquired the multiplicated endogenous anti-oxidant system for the maintenance of a stable form of life under such harmful

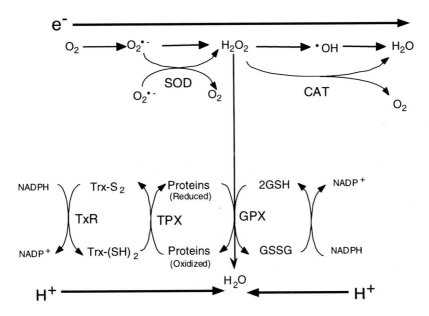

Fig. 1. Cellular redox system. Reactive oxygen species (ROS) are generated during the process of oxygen reduction. Among ROS, H_2O_2 has the longest half life and is considered to be a mediator of oxidative signal. To maintain redox homeostasis, there are multiple antioxidant defense mechanisms within the cells. Major component of the first line defense system is glutathione (GSH). GSH is present in the cell at milimolar level. GSH reduces peroxides in the presence of glutathione peroxidase (GPX) and the oxidized GSH (GSSG) is enzymatically restored by NADPH-dependent reduction of GSH reductase. Other antioxidant molecules in the cells include nutrients such as tocopherol (vitamin E) and ascorbate (vitamin C) and antioxidant enzymes that deoxify ROS such as superoxide dismutase (SOD) and catalase. The second line defense system include reducing enzyme systems such as thioredoxin (Trx) and glutaredoxin (Grx). While GSH is directly involved in scavenging ROS, Trx and Grx appear to participate in this cascade by repairing the oxidized proteins through its reducing activity. This reversible modification of functional proteins by oxidation and reduction are used as "redox signals". TxR, thioredoxin reductase; TPX, thioredoxin peroxidase; GPX, glutathione peroxidase.

conditions (Fig. 1). These defense mechanisms include reducing enzymes such as thioredoxin (Trx) and glutaredoxin (Grx).[1, 2] Oxidized protein molecules by ROS are reversibly reduced by Trx or Grx. This reversible oxidation and reduction of a functional protein could determine its activity. Thus, the term "redox regulation" has been proposed indicating the active role of oxido-reductive modifications of proteins in regulating their activities. In other words, oxidation and reduction of biomolecules is considered to be "signals" in certain instances and is utilized for the maintenance of cellular homeostasis.[1–5]

2. Source of ROS as Signaling Molecules

One of the major sources of superoxide and hydrogen peroxide generation occurs through basal cellular metabolism due to mitochondrial metabolic activity. For example, Schultz-Oshchoff *et al.*[6] showed that TNFα-induced cytotoxicity and NF-κB activation were abolished by the mitochondrial electron transport system inhibitor, rotenone. Thus, ROS generated at mitochondrial electron transport system may participate in signal transduction process.

NADPH oxidase associated with cytoplasmic membrane could be another source of ROS which is involved in signal transduction. In phagocytic cells, a variety of stimuli cause the massive release of superoxide anion through the assembly of NADPH oxidase. This multimetric protein complex is essential for neutrophil function and is composed of both membrane-bound proteins and cytosolic factors.[7] In phagocytic cells, small-GTP-binding Rac proteins are involved in the assembly of the neutrophil NADPH oxidase system. In non-phagocytic cells, it has been demonstrated that cytokine- and growth factor-stimulated ROS production occurs through a Rac-dependent pathway.[8, 9] Subsequent cellular responses, such as NF-κB activation, protein kinase activation and DNA synthesis, are also rac-dependent and are effectively blocked by ROS scavenging agents.[8, 10–12] Although many components of NADPH oxidase in phagocytic cells are ubiquitously expressed, the major cytochrome, gp91 appears unique to phagocytes. However recent reports have identified a homolog of gp91 that is expressed in non-phagocytic cells.[13, 14] Therefore, similar to neutrophils, non-phagocytic cells appears to have a Rac-regulated NADPH oxidase complex which may be responsible for ligands-induced ROS production and subsequent cellular events.

3. Mechanism of Oxidant Signaling

If oxidative status can signal cells to respond in various ways, how these signals are transduced, carried, and interpreted by the cells? ROS has been implicated in the regulation of nonreceptor- and receptor-type of protein tyrosine kinases (PTKs),[15, 16] protein tyrosine phosphatases (PTPases), ras,[17] protein kinase C,[18] mitogen-activated protein kinases.[4, 19]

ROS can alter signaling pathways by specifically altering the oxidation of reactive cysteine residues in proteins. Under physiological conditions, most cysteines in proteins are in the protonated (SH) form. However, some cysteines of intracellular proteins have their sulfhydryl group oxidized at physiological pH. A variety of proteins in mammalian cells have been demonstrated to contain reactive cysteines. Among these proteins, redox-mediated regulation of PTPases activity has been most extensively studied.[20, 21] All members of this family have a reactive cysteine in their active site, and oxidant stress can reversibly inactivate cellular PTPase activity. The mechanism for transient inactivation presumably involves oxidation of the reactive cysteine to form a sulfenic ion. In this form, the enzymatic activity of the PTPase is significantly reduced or abolished. The oxidized reactive cysteine in the sulfenic form is most likely capable of interacting with cellular glutathione to form a protein mix disulfide. This mix disulfied can be reduced probably through the actions of cellular enzymes such as Grx or Trx to restore the original molecule. In this scenario, specificity comes from the local amino acid environment surrounding the cysteine residue that leads to its reactivity. In many ways, this is analogous to amino acid motifs in other proteins surrounding specific serine, threonine or tyrosine residues that serve as recognition for phosphorylation by cellular kinases. In both cases, enzymatic activity is regulated in a reversible fashion by targeting specific amino acids using either phosphate addition or cysteine oxidation.

Another piece of evidence linking the cellular redox status to specific signaling pathways came from the interaction between the apoptosis signal regulated kinase 1 (ASK1) and Trx. ASK1 is a member of the mitogen-activated kinase kinase kinase (MAPKKK) family that activates downstream kinases including *c-Jun* NH$_2$-terminal kinase (JNK) and p38 mitogen-activated kinase. Interestingly, Trx was found to form a complex with ASK1.[22] When Trx is bound to ASK1, the activity of ASK1 is inhibited. Trx is dissociated from ASK1 by the rise in ROS levels after TNF stimulation, followed by activation of ASK1.[22, 23] Thus, it is likely that cellular reducing enzymes, such as Trx and Grx, could also function as redox sensor molecules.

4. Transcription Factor AP-1 and Its Redox Regulation

Homo- and heterodimers of the *c-Jun* and *c-Fos* transcription factor family constitute the transcription factor AP-1, which is tightly regulated by redox processes. This family of transcription factors is critically important in regulating stress-response genes that control cellular proliferation and cell growth mediators.[24]

By interaction between their leucine zipper domains, *c-Fos* and *c-Jun* form a dimeric complex. This allows their regions rich in bacic amino acids to come into a proper juxtaposition and allows the binding to DNA sequence containing the appropriate AP-1 sites. AP-1 is present in quiescent state until it is phosphorylated on specific serine residues of *c-Jun*. Transcriptional activity of the *c-Jun* protein is greatly enhanced by phosphorylation of two serine residues in its activation domain by the *c-Jun* NH$_2$-terminal kinase (JNK), a member of the mitogen-activated protein kinase family. It has been demonstrated that JNK is activated in a redox-sensitive manner.[25] Another independent mechanism of AP-1 regulation involves the reversible oxidation of a conserved cysteine residue in the DNA binding domain of *c-Fos* and *c-Jun*.[26] Reduction of this residue by reducing agents stimulates AP-1 binding activity *in vitro*. Furthermore, the nuclear redox factor, Ref-1, has been shown to mediate the redox-sensitive cysteine within *c-Jun* and *c-Fos*, allowing an increase in DNA binding.[27]

5. Transcription Factor NF-κB and Its Redox Regulation

Nuclear factor κB (NF-κB) is an inducible cellular transcription factor that regulates a wide variety of cellular and viral genes.[28-31] These genes include cytokines such as IL-2, IL-6, IL-8, GM-CSF and TNF, cell adhesion molecules such as ICAM-1 and E-selectin, inducible nitric oxidase synthase (iNOS) and viruses such as human immunodeficiency virus (HIV) and cytomegalovirus. Since NF-κB is responsible for transcriptional induction of these genes, it is considered to be involved in the currently intractable diseases such as acquired immunodeficiency syndrome (AIDS), hematogenic cancer cell metastasis and rheumatoid arthritis (RA).

The members of the NF-κB family in mammalian cells include the proto-oncogene c-Rel, Rel A (p65), Rel B, NF-κB1 (p50/105), and NF-κB2 (p52/p100). These proteins share a conserved 300 amino acids region known as the Rel homology domain, which is responsible for DNA binding, dimerization, and nuclear translocation of NF-κB. In most cells, Rel family members form hetero- and homo-dimers with distinct specificities in various combinations. A common feature of the regulation of NF-κB family is their sequestration in the cytoplasm as inactive complexes with a class of inhibitory molecules known as IκBs.[28-31] Upon stimulation of the cells such as by proinflammatory cytokines, IL-1 and TNF, IκB is dissociated and NF-κB is translocated to the nucleus and activates expression of target genes (Fig. 1). In addition to these physiological stimuli, NF-κB activation cascade is also triggered by ionizing and UV irradiation and oxidative reagent such as hydrogen peroxide.[32-34]

There are two biochemically independent steps in the NF-κB activation cascade: kinase pathways and redox-signaling pathway. However, these two distinct

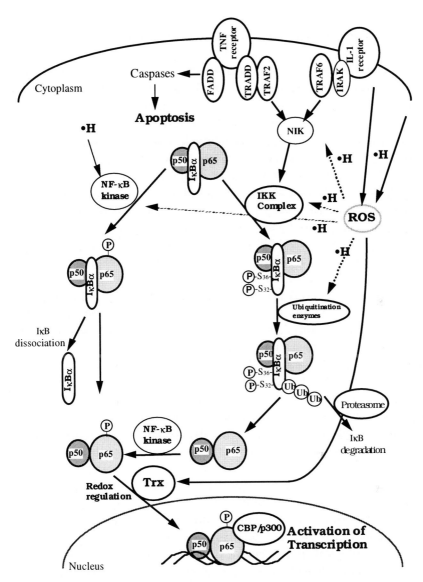

Fig. 2. Signal transduction pathways for NF-κB activation. The first step involves kinase pathways such as NF-κB and IκB kinases (IKKs). The second step involves "redox regulation" by thioredoxin (Trx). After the stimulation of the cells by TNF or IL-1, for example, radical oxygen species (ROS) are produced. ROS activates kinase cascade by direct or indirect mechanisms. ROS also induces production of thioredoxin (Trx). TRAF molecules transduces signals from TNF receptor or IL-1 receptor and stimulate the downstream NF-κB activation pathway. Phosphorylation of NF-κB or IκB will lead to dissociation of NF-κB from IκB. The phosphorylated IκBs by IKK complex will be ubiquitinated and then degraded by proteasome. After liberated from IκBs, NF-κB must go through the Trx-mediated reduction of the "redox-sensitive" cysteine to recognize the target DNA sequence (κB site).

pathways are involved in the NF-κB activation cascade in a coordinate fashion, which may contribute to a fine tune, as well as fail-safe, regulation of NF-κB activity.

At least two distinct types of kinase pathways are known to be involved in NF-κB activation: NF-κB kinase and IκB kinase (Fig. 2). TNF and IL-1 initiate their signaling through receptor-associated signal transducers, TRAF2 and TRAF6, respectively. These distinct TNF and IL-1 pathways merge at the level of the protein kinase NIK, NF-κB-inducing kinase.[35] Recently, several groups identified IκB kinases, IKK-α and IKK-β, which directly phosphorylate IκBs. IKK-α and IKK-β exist in a heterodimer complex that is able to interact with NIK. Phosphorylation on specific serine residues of IκBs leads to ubiquitination of IκBs and subsequent degradation by proteasome complex. There are other evidences suggesting the presence of kinase cascade(s) that phosphorylate NF-κB. We found a 43 kD serine kinase, named NF-κB kinase, that is associated with NF-κB.[36] This kinase phosphorylates the subunits, particularly p65, of NF-κB and dissociates it from IκB. Similar findings were reported from others.[37–39] Recently, catalytic subunit of protein kinase A (PKAc) was found to be associated with the NF-κB/IκB complex in the cytoplasm.[40] However, PKAc alone, when overexpressed in culture cells, did not stimulate NF-κB dependent gene expression.[41, 42] There is still some controversy regarding the kinase responsible for p65 phosphorylation.

After dissociation from IκB, NF-κB must go through the redox regulation by cellular reducing catalyst, thioredoxin (Trx)[43, 44] in order to recognize the target DNA sequence and induce transcription. Trx is a cellular reducing catalyst and is known to participate in redox reactions through reversible oxidation of its active center dithiol to a disulfide (Figs. 3). We and others have demonstrated *in vitro* that NF-κB cannot bind to the κB DNA sequence of the target genes until it is reduced.[43–46] Based on the estimation of high local pI value near one of the conserved cysteine residues, we have assigned the cysteine residue at the 62nd amino acid position of p50 subunit as a target of the redox regulation.[44] It was confirmed by site-directed mutagenesis study by others[47] in which the cysteine 62 substitution abolished the DNA binding activity.

Structural biological approaches have provided physical evidences supporting the molecular mechanism of the redox regulation of NF-κB by Trx. Crystalographic examination have demonstrated the 3-dimensional structure of the NF-κB subunit p50 homodimer associated with the target DNA.[48, 49] NF-κB appears to have a novel DNA-binding structure called β-barrel, a group of beta sheets stretching toward the target DNA. There is a loop in the tip of the β-barrel structure that intercalates with the target nucleotide bases and is considered to make a direct contact with the DNA. This DNA-recognition loop of p50 contains the residue corresponding to cysteine 62 that we predicted to be the target of redox regulation, although in both studies this cysteine was replaced with alanine presumably because of technical reasons for crystallization.[48, 49] Furthermore,

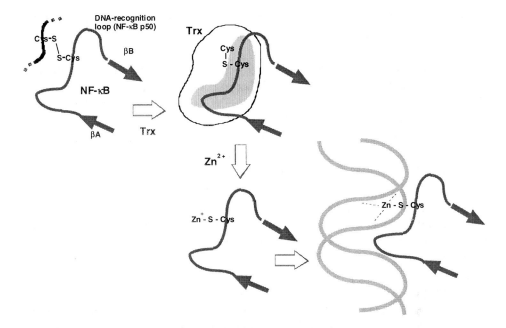

Fig. 3. Redox regulation of the DNA-binding activity of NF-κB by Trx. A boot-shaped hollow on the surface of Trx, also containing the redox-sensitive cysteines, could stably recognize the DNA-binding loop of p50 and reduce the oxidized cysteine by donating protons in a structure-dependent way [based on the 3-D structure of Trx molecule with the DNA-binding loop peptide of p50[50]]. Zinc is required to make NF-κB competent for the DNA-binding.[76] When zinc dissociates with NF-κB, Trx may be dissociated from NF-κB.

another group has solved the 3-D structure of Trx molecule that is associated with the DNA-binding loop of p50 by using NMR.[50] A boot-shaped hollow on the surface of Trx containing the redox-active cysteines could stably recognize the DNA-binding loop of p50[48, 49] and is likely to reduce the oxidized cysteine on p50 by donating protons in a structure-dependent fashion. Therefore, the reduction of NF-κB by Trx is considered to be specific and dependent on the structural compatibility between the target protein and Trx. However, the inter-molecular disulfide bridge formation between Trx and NF-κB might be transient since the binding of Trx to the NF-κB DNA-binding loop would block the recognition of DNA moiety because of the apparent competition of the same cysteine residue. In favor of this model, we have demonstrated that NF-κB and Trx concomitantly migrated to the nucleus in the rheumatoid synovial cells during the early phase of the NF-κB activation process.[51] Trx was relocated in the cytoplasm after 30 mins of stimulation, while the NF-κB was predominantly present at the nucleus for several hours.

6. Redox-Mediated Intervention in NF-κB Associated Diseases

What triggers these NF-κB activating cascades? Not much is known about what happens immediately downstream of the cell surface receptor. A hypothesis has been proposed whereby divergent agents activate NF-κB by increasing ROS, although direct evidence is still lacking to indicate where and how ROS are generated and involved in NF-κB activating pathway. This hypothesis is based on several lines of evidence. First, most of the agents activating NF-κB tend to trigger the formation of ROS.[32, 33, 52, 53] Second, H_2O_2 or organic hydroperoxide has been shown to activate NF-κB in some cell lines in the abscence of any physiological stimulus. Third, antioxidants, such as N-acetyl- L-cysteine (NAC) or α-lipoic acid, were shown to be effective in blocking NF-κB activation in response to diverse stimuli.[32–34, 54, 55] Interestingly, we found that NAC could also block the induction of Trx.[56] Therefore, anti-NF-κB actions of antioxidants are considered to be two-fold: (1) blocking the signaling immediately downstream of the signal elicitation, and (2) suppression of induction of the redox effector Trx.

Based on the understanding of redox regulation of NF-κB, novel therapeutic approaches for NF-κB associated diseases have been developed. In AIDS, NF-κB plays a pivotal role in the life cycle of HIV, especially in virus reactivation process within the latently infected cells.[31, 57] After activation through intracellular signaling pathways such as those elicited by T cell receptor antigen complex or by receptors for IL-1 or TNF, NF-κB initiates HIV gene expression by binding to the target DNA element within LTR.[57–60] Then, the virus-encoded transactivator Tat is produced and triggers explosive viral replication.[61–63] Since activation of HIV gene expression by cellular transcription factors conceptually precedes the production of Tat, NF-κB is regarded as a critical determinant of the maintenance and breakdown of the viral latency. Antioxidant compounds have been suggested to be effective in preventing the clinical development of AIDS by blocking HIV replication.[54, 64, 65]

Another situation where NF-κB plays a role is hematogenic cancer cell metastasis. NF-κB induces E-selectin on the surface of vascular endothelial cells[66–68] as well as ICAM-1 and VCAM-1. Since most of the leukocytes and some cancer cells constitutively express a ligand for E-selectin, called sialyl-Lewis[X] antigen, on their cell surface, induction of E-selectin on the endothelium is considered to be a rate determining step of cell-cell interaction.[69, 70] We have demonstrated that various compounds known to block NF-κB, such as N-acetylcysteine, aspirin and pentoxyphillin, eventually prevented these interactions by interfering the induction of these CAMs.[66]

Similarly, there have been accumulating evidences that suggest the involvement of NF-κB in the pathogenesis of rheumatoid arthritis (RA).[51, 71–73] Because of its regulatory role in gene expression of IL-1, TNF, IL-2, IL-6, IL-8, GM-CSF, chemokines such as RANTES and MIP-1 α, ICAM-1, E-selectin and iNOS that are known to be overexpressed in the rheumatoid synovium, NF-κB is considered to be a major regulator in the expansion and maintenance of chronic inflammatory response in the affected joints. For example, sustained NF-κB activation would induce production of cytokines and thus activate maturation of B lymphocytes to produce antibodies while GM-CSF and chemokine production together with overexpression of cell adhesion molecules would support recruitment of leukocytes from blood stream thus augment the local inflammatory response. Additionally, some of the anti-rheumatic drugs including corticosteroids, aspirin and gold compounds are known to block the NF-κB cascade.[74–76]

Interestingly, NF-κB could be blocked by gold ion by a redox mechanism.[76] We found that the zinc ion is a necessary component of the active NF-κB. Addition of monovalent gold ion could efficiently block its activity by oxidizing the redox-active cysteins on NF-κB. Since gold did not appear to replace zinc, it is likely that gold ion oxidizes these thiolate anions on NF-κB into disulfides. Thus, gold ion abrogates the DNA-binding activity because of its higher oxidation potential over zinc ion. Our finding could explain why gold is effective in RA and suggests that NF-κB might have a crucial role in the disease process.[77, 78] It may be that gold compound is potentially effective in other diseases where NF-κB plays a pathological role. For example, we have demonstrated that gold compounds effectively block TNF-induced HIV replication in latently infected cells.[79]

7. Conclusion

Recognition of ROS and redox-mediated protein modifications as signals has opened up a new field of cell regulation and provided a novel strategy for controlling disease processes. One such approach has been shown to be feasible for gene expression regulated by the transcription factor NF-κB. Thus, intervention of NF-κB activity by antioxidants could be useful for currently intractable diseases such as AIDS, rheumatoid arthritis, and hematologic cancer metastasis in which NF-κB plays a critical role.

Acknowledgment

This work was supported by grants in aid from the Ministry of Health, Labour and Welfare, the Ministry of Education, Culture, Sports, Science and Technology of Japan and from the Japan Health Sciences Foundation.

References

1. Holmgren, A. (1985). Thioredoxin. *Ann. Rev. Biochem.* **54**: 237–271.
2. Holmgren, A. (1989). Thioredoxin and glutaredoxin systems. *J. Biol. Chem.* **264**: 13 963–13 966.
3. Sen, C. K. and Packer, L. (1996). Antioxidant and redox regulation of gene transcription. *FASEB J.* **10**: 709–720.
4. Gabbita, S. P., Robinson, K. A., Stewart, C. A., Floyd, R. A. and Hensley, K. (2000). Redox regulatory mechanisms of cellular signal transduction. *Arch. Biochem. Biophys.* **376**: 1–13.
5. Finkel, T. (2000). Redox-dependent signal transduction. *FEBS Lett.* **476**: 52–54.
6. Schulze-Osthoff, K., Beyaert, R., Vandevoorde, V., Haegeman, G. and Fiers, W. (1993). Depletion of the mitochondrial electron transport abrogates the cytotoxic and gene-inductive effects of TNF. *EMBO J.* **12**: 3095–3104.
7. DeLeo, F. R. and Quinn, M. T. (1996). Assembly of the phagocyte NADPH oxidase: molecular interaction of oxidase proteins. *J. Leukoc. Biol.* **60**: 677–691.
8. Sulciner, D. J., Irani, K., Yu, Z. X., Ferrans, V. J., Goldschmidt-Clermont, P. and Finkel, T. (1996). Rac1 regulates a cytokine-stimulated, redox-dependent pathway necessary for NF-kappaB activation. *Mol. Cell Biol.* **16**: 7115–7121.
9. Sundaresan, M., Yu, Z. X., Ferrans, V. J., Sulciner, D. J., Gutkind, J. S., Irani, K., Goldschmidt-Clermont, P. J. and Finkel, T. (1996). Regulation of reactive-oxygen-species generation in fibroblasts by Rac1. *Biochem. J.* **318**: 379–382.
10. Irani, K., Xia, Y., Zweier, J. L., Sollott, S. J., Der, C. J., Fearon, E. R., Sundaresan, M., Finkel, T. and Goldschmidt-Clermont, P. J. (1997). Mitogenic signaling mediated by oxidants in Ras-transformed fibroblasts. *Science* **275**: 1649–1652.
11. Joneson, T. and Bar-Sagi, D. (1998). A Rac1 effector site controlling mitogenesis through superoxide production. *J. Biol. Chem.* **273**: 17 991–17 994.
12. Kheradmand, F., Werner, E., Tremble, P., Symons, M. and Werb, Z. (1998). Role of Rac1 and oxygen radicals in collagenase-1 expression induced by cell shape change. *Science* **280**: 898–902.
13. Suh, Y. A., Arnold, R. S., Lassegue, B., Shi, J., Xu, X., Sorescu, D., Chung, A. B., Griendling, K. K. and Lambeth, J. D. (1999). Cell transformation by the superoxide-generating oxidase Mox1. *Nature* **401**: 79–82.
14. Banfi, B., Maturana, A., Jaconi, S., Arnaudeau, S., Laforge, T., Sinha, B., Ligeti, E., Demaurex, N. and Krause, K. H. (2000). A mammalian H+ channel generated through alternative splicing of the NADPH oxidase homolog NOH-1. *Science* **287**: 138–142.
15. Abe, J., Okuda, M., Huang, Q., Yoshizumi, M. and Berk, B. C. (2000). Reactive oxygen species activate p90 ribosomal S6 kinase via Fyn and Ras. *J. Biol. Chem.* **275**: 1739–1748.

16. Herrlich, P. and Bohmer, F. D. (2000). Redox regulation of signal transduction in mammalian cells. *Biochem. Pharmacol.* **59**: 35–41.

17. Lander, H. M., Milbank, A. J., Tauras, J. M., Hajjar, D. P., Hempstead, B. L., Schwartz, G. D., Kraemer, R. T., Mirza, U. A., Chait, B. T., Burk, S. C. and Quilliam, L. A. (1996). Redox regulation of cell signalling. *Nature* **381**: 380–381.

18. Klann, E., Roberson, E. D., Knapp, L. T. and Sweatt, J. D. (1998). A role for superoxide in protein kinase C-activation and induction of long-term potentiation. *J. Biol. Chem.* **273**: 4516–4522.

19. Janssen-Heininger, Y. M., Macara, I. and Mossman, B. T. (1999). Cooperativity between oxidants and tumor necrosis factor in the activation of nuclear factor (NF)-kappaB: requirement of Ras/mitogen-activated protein kinases in the activation of NF-kappaB by oxidants. *Am. J. Respir. Cell. Mol. Biol.* **20**: 942–952.

20. Lee, S. R., Kwon, K. S., Kim, S. R. and Rhee, S. G. (1998). Reversible inactivation of protein-tyrosine phosphatase 1B in A431 cells stimulated with epidermal growth factor. *J. Biol. Chem.* **273**: 15 366–15 372.

21. Barrett, W. C., DeGnore, J. P., Keng, Y. F., Zhang, Z. Y., Yim, M. B. and Chock, P. B. (1999). Roles of superoxide radical anion in signal transduction mediated by reversible regulation of protein-tyrosine phosphatase 1B. *J. Biol. Chem.* **274**: 34 543–34 546.

22. Saitoh, M., Nishitoh, H., Fujii, M., Takeda, K., Tobiume, K., Sawada, Y., Kawabata, M., Miyazono, K. and Ichijo, H. (1998). Mammalian thioredoxin is a direct inhibitor of apoptosis signal-regulating kinase (ASK) 1. *EMBO J.* **17**: 2596–2606.

23. Gotoh, Y. and Cooper, J. A. (1998). Reactive oxygen species- and dimerization-induced activation of apoptosis signal-regulating kinase-1 in tumor necrosis factor-alpha signal transduction. *J. Biol. Chem.* **273**: 17 477–17 482.

24. Karin, M., Liu, Z. and Zandi, E. (1997). AP-1 function and regulation. *Current Opinion Cell Biol.* **9**: 240–246.

25. Lo, Y. Y. C., Wong, J. M. S. and Cruz, T. F. (1996). Reactive oxygen species mediate cytokine activation of c-Jun NH_2-terminal kinases. *J. Biol. Chem.* **271**: 15 703–15 707.

26. Abate, C., Patel, L., Rauscher, F. J. D. and Curran, T. (1990). Redox regulation of *fos* and *jun* DNA-binding activity in vitro. *Science* **249**: 1157–1161.

27. Xanthoudakis, S., Miao, G., Wang, F., Pan, Y. C. and Curran, T. (1992). Redox activation of *Fos-Jun* DNA binding activity is mediated by a DNA repair enzyme. *EMBO J.* **11**: 3323–3335.

28. Baldwin, A. S., Jr. (1996). The NF-κB and IκB proteins: new discoveries and insights. *Ann. Rev. Immun.* 649–683.

29. Baeuerle, P. A. and Baichwal, V. R. (1997). NF-κB as a frequent target for immunosuppressive and anti-inflammatory molecules. *Adv. Immun.* **65**: 111–137.

30. Ghosh, S., May, M. J. and Kopp, E. B. (1998). NF-kappa B and Rel proteins: evolutionarily conserved mediators of immune responses. *Ann. Rev. Immun.* **16**: 225–260.

31. Okamoto, T., Sakurada, S., Yang, J.-P. and Merin, J. P. (1997). Regulation of NF-κB and disease control: identification of a novel serine kinase and thioredoxin as effectors for signal transduction pathway for NF-κB activation. *Current Topic Cell Regul.* **35**: 149–161.

32. Schreck, R., Rieber, P. and Baeuerle, P. A. (1991). Reactive oxygen intermediates as apparently widely used messengers in the activation of the NF-κB transcription factor and HIV-1. *EMBO J.* **10**: 2247–2258.

33. Schreck, R., Albermann, K. and Baeuerle, P. A. (1992). Nuclear factor kappa-B: an oxidative stress-responsive transcription factor of eukaryotic cells (a review). *Free Radic. Res. Commun.* **17**: 221–237.

34. Meyer, M., Schreck, R. and Baeuerle, P. A. (1993). H202 and antioxidants have opposite effects on activation of NF-κB and AP-1 in intact cells: AP-1 as secondary antioxidant-responsive factor. *EMBO J.* **12**: 2005–2015.

35. Malinin, N. L., Boldin, M. P., Kovalenko, A. V. and Wallach, D. (1997). MAP3K-related kinase involved in NF-κB induction by TNF and IL-1. *Nature* **385**: 540–544.

36. Hayashi, T., Sekine, T. and Okamoto, T. (1993). Identification of a new serine kinase that activates NF-κB by direct phosphorylation. *J. Biol. Chem.* **268**: 26 790–26 795.

37. Ostrowski, J., Sims, J. E., Sibley, C. H., Valentine, M. A., Dower, S. K., Meier, K. E. and Bomsztyk, K. (1991). A serine/threonine kinase activity is closely associated with a 65-kDa phosphoprotein specifically recognized by the kappa-B enhancer element. *J. Biol. Chem.* **266**: 12 722–12 733.

38. Naumann, M. and Scheidereit, C. (1994). Activation of NF-κB in vivo is regulated by mutiple phosphorylations. *EMBO J.* **13**: 4597–4607.

39. Li, C.-C. H., Dai, R.-M., Chen, E. and Longo, D. L. (1994). Phosphorylation of NF-KB1-p50 is involved in NF-κB activation and stable DNA binding. *J. Biol. Chem.* **269**: 30 089–30 092.

40. Zhong, H., SuYang, H., Erdjument-Bromage, H., Tempst, P. and Ghosh, S. (1997). The transcriptional activity of NF-κB is regulated by the IκB-associated PKAc subunit through a cyclic-AMP independent mechanism. *Cell* **89**: 413–424.

41. Neumann, M., Grieshammer, T., Chuvpilo, S., Kneitz, B., Lohoff, M., Schimpl, A., Franza, B. R., Jr. and Serfling, E. (1995). Rel A/p65 is a molecular target for the immunosuppressive action of protein kinase A. *EMBO J.* **14**: 1991–2004.

42. Ollivier, V., Parry, G. C. N., Cobb, R. R., de Prost, D. and Mackman, N. (1996). Elevated cyclic AMP inhibits NF-κB-mediated transcription in human monocytic cells and endothelial cells. *J. Biol. Chem.* **271**: 20 828–20 835.

43. Okamoto, T., Ogiwara, H., Hayashi, T., Mitsui, A., Kawabe, T. and J, Yodoi. (1992). Human thioredoxin/adult T cell leukemia-derived factor activates the enhancer binding protein of human immunodeficiency virus Type-1 bt thiol redox control mechanism. *Int. Immun.* **4**: 811–819.
44. Hayashi, T., Ueno, Y. and Okamoto, T. (1993). Oxidoreductive regulation of nuclear factor kappa-B. Involvement of a cellular reducing catalyst thioredoxin. *J. Biol. Chem.* **268**: 11 380–11 388.
45. Molitor, J. A., Ballard, D. W. and Greene, W. C. (1991). Kappa-B-specific DNA binding proteins are differentially inhibited by enhancer mutations and biological oxidation. *New Biol.* **3**: 987–996.
46. Toledano, M. B. and Leonard, W. J. (1991). Modulation of transcription factor NF-kappa B binding activity by oxidation-reduction *in vitro*. *Proc. Natl. Acad. Sci. USA* **88**: 4328–4332.
47. Matthews, J. R., Wakasugi, N., Virelizier, J.-L., Yodoi, J. and Hay, R. T. (1992). Thioredoxin regulates the DNA binding activity of NF-kappa B by reduction of a disulfide bond involving cystein 62. *Nucleic Acids Res.* **20**: 3821–3830.
48. Ghosh, G., Van Duyne, G., Ghosh, S. and Sigler, P. B. (1995). Structure of NF-kappa B p50 homodimer bound to a kappa-B site. *Nature* **373**: 303–310.
49. Müller, C. W., Rey, F. A., Sodeoka, M., Verdine, G. L. and Harrison, S. C. (1995). Structure of the NF-kappa B p50 homodimer bound to DNA. *Nature* **373**: 311–317.
50. Qin, J., Clore, G. M., Kennedy, W. M. P., Huth, J. R. and Gronenborn, A. M. (1995). Solution structure of human thioredoxin in a mixed disulfide intermediate complex with its target peptide from the transcription factor NF-kappa B. *Structure* **3**: 289–297.
51. Sakurada, S., Kato, T. and Okamoto, T. (1996). Induction of cytokines and ICAM-1 by proinflammatory cytokines in primary rheumatoid synovial fibroblats and infibition by N-acetyl-L-cysteine and aspirin. *Int. Immun.* **8**: 1483–1493.
52. Schmidt, K. N., Traenckner, E. B., Meier, B. and Baeuerle, P. A. (1995). Induction of oxidative stress by okadaic acid is required for activation of transcription factor NF-κB. *J. Biol. Chem.* **270**: 27 136–27 142.
53. Los, M., Schenk, H., Hexel, K., Baeuerle, P. A., Droge, W. and Schulze-Osthoff, K. (1995). IL-2 gene expression and NF-κB activation through CD28 requires reactive oxygen production by 5-lipoxygenase. *EMBO J.* **14**: 3731–3740.
54. Suzuki, Y. J., Aggarwal, B. B. and Packer, L. (1992). Alpha-lipoic acid is a potent inhibitor if NF-kappa B activation in human T cells. *Biochem. Biophys. Res. Commun.* **189**: 1709–1715.
55. Suzuki, Y. J., Mizuno, M. and Packer, L. (1994). Signal transduction for nuclear factor-kappa-B activation. Proposed location of antioxidant-inhibitable step. *J. Immun.* **153**: 5008–5015.

56. Sachi, Y., Hirota, K., Masutani, H., Toda, K., Okamoto, T., Takigawa, M. and Yodoi, J. (1995). Three NF-kappa B binding sites in the human E-selectin gene required for maximal tumor necrosis factor alpha-induced expression. *Immun. Lett.* **44**: 189–193.

57. Okamoto, T., Matsuyama, T., Mori, S., Hamamoto, Y., Kobayashi, N., Yamamoto, N., Josephs, S. F., Wong-Staal, F. and Shimotohno, K. (1989). Augmentation of human immunodeficiency virus Type-1 gene expression by tumor necrosis factor alpha. *AIDS Res. Human Retrovir.* **5**: 131–138.

58. Nabel, G. and Baltimore, D. (1987). An inducible transcription factor activates expression of human immunodeficiency virus in T cells. *Nature* **326**: 711–713.

59. Bohnlein, E., Lowenthal, J. W., Siekevitz, M., Ballard, D. W., Franza, B. R. and Greene, W. C. (1988). The same inducible nuclear proteins regulates mitogen activation of both the interleukin-2 receptor-alpha gene and Type-1 HIV. *Cell* **53**: 827–836.

60. Okamoto, T., Benter, T., Josephs, S. F., Sadaie, M. R. and Wong-Staal, F. (1990). Transcriptional activation from the long-terminal repeat of human immunodeficiency virus in vitro. *Virology* **177**: 606–614.

61. Arya, S. K., Guo, C., Josephs, S. F. and Wong-Staal, F. (1985). Trans-activator gene of human T-lymphotropic virus type III (HTLV-III). *Science* **229**: 69–73.

62. Sodroski, J., Patarca, R. and Rosen, C. (1985). Location of the trans-activating region on the genome of human T-cell lymphotropic virus type III. *Science* **229**: 74–77.

63. Okamoto, T. and Wong-Staal, F. (1986). Demonstration of virus-specific transcriptional activator(s) in cells infected with HTLV-III by an in vitro cell-free system. *Cell* **47**: 29–35.

64. Roederer, M., Staal, F. J. T., Raju, P. A., Ela, S. W. and Herzenberg, L. A. (1990). Cytokine-stimulated human immunodeficiency virus replication is inhibited by N-acetyl-L-cysteine. *Proc. Natl. Acad. Sci. USA* **87**: 4884–4888.

65. Merin, J. P., Matsuyama, M., Kira, T., Baba, M. and Okamoto, T. (1996). α-Lipoic acid blocks HIV-1 LTR-dependent expression of hygromycin resistance in THP-1 stable transformants. *FEBS Lett.* **394**: 9–13.

66. Tozawa, K., Sakurada, S., Kohri, K. and Okamoto, T. (1995). Effects of anti-nuclear factor kappa-B reagents in blocking adhesion of human cancer cells to vascular endothelial cells. *Cancer Res.* **55**: 4162–4167.

67. Montgomery, K. F., Osborn, L., Hession, C., Tizard, R., Goff, D., Vassallo, C., Tarr, P. I., Bomsztyk, K., Lobb, R., Harlan, J. M. and Pohlman, T. H. (1991). Activation of endothelial-leukocyte adhesion molecule 1 (ELAM-1) gene transcription. *Proc. Natl. Acad. Sci. USA* **88**: 6523–6527.

68. Whelan, J., Ghersa, P., Huijsduijnen, R. H., Gray, J., Chandra, G., Talabot, F. and DeLamarter, J. F. (1991). An NF kappa-B like factor is essential but not sufficient for cytokine induction of endothelial leukocyte adhesion molecule 1 (ELAM-1) gene transcription. *Nucleic Acid Res.* **19**: 2645–2653.

69. Dejana, E., Bertocci, F., Bortolami, M. C., Regonesi, A., Tonta, A., Breviario, F. and Giavazzi, R. (1988). Interleukin 1 promotes tumor cell adhesion to cultured human endothelial cells. *J. Clin. Invest.* **82**: 1466–1470.

70. Takada, A., Ohmori, K., Yoneda, T., Tsuyuoka, K., Hasegawa, A., Kiso, M. and Kannagi, R. (1993). Contribution of carbohydrate antigens sialyl Lewis A and sialyl Lewis X to adhesion of human cancer cells to vascular endothelium. *Cancer Res.* **53**: 354–361.

71. Alvaro-Gracia, J. M., Zvaifler, N. J., Brown, C. B., Kaushansky, K. and Firestein, G. S. (1991). Cytokines in chronic arthritis. VI. Analysis of the synovial cells involved in granulocyte macrophage colony-stimulating factor production and gene expression in rheumatoid arthritis and its regulation by IL-1 and TNF-α. *J. Immun.* **146**: 3365–3372.

72. Ulfgren, A. K., Lindblad, S., Klareskog, L., Andersson, J. and Andersson, U. (1995). Detection of cytokine producing cells in the synovial membrane from patients with rheumatoid arthritis. *Ann. Rheum. Dis.* **54**: 654–659.

73. Handel, M. L., McMorrow, L. B. and Gravallese, E. M. (1996). Nuclear factor-κB in rheumatoid synovium. Localization of p50 and p60. *Arthritis Rheum.* **38**: 1762–1770.

74. McKay, L. I. and Cidlowski, J. A. (1999). Molecular control of immune/ inflammatory responses: interactions between nuclear factor kappa-B and steroid receptor-signaling pathways. *Endocr. Rev.* **20**: 435–459.

75. Yin, M. J., Yamamoto, Y. and Gaynor, R. B. (1998). The anti-inflammatory agents aspirin and salicylate inhibit the activity of I(kappa)B kinase-beta. *Nature* **396**: 77–80.

76. Yang, J.-P., Merin, J. P., Nakano, T., Kato, T., Kitade, Y. and Okamoto, T. (1995). Inhibition of the DNA-binding activity of NF-κB by gold compounds in vitro. *FEBS Lett.* **361**: 89–96.

77. Skosey, J. L. (1993). Treatment of rheumatoid arthritis. *In* "Arthritis and Allied Conditions" (McCarty D.J. and W. J. Koopman, W. J., Eds.), pp. 603–614, Lea and Febiger, Philadelphia.

78. Insel, P. A. (1996). Analgesic-antipyretic and anti-inflammatory agents and drugs employed in the treatment of gout. *In* "The pharmacological basis of therapeutics" (Hardman, J. G. *et al.*, Ed.), pp. 670–681, Macmillan, New York

79. Traber, K. E., Okamoto, H., Kurono, C., Baba, M., Saliou, C., Soji, T., Packer, L. and Okamoto, T. (1999). Anti-rheumatic compound aurothioglucose inhibits tumor necrosis factor-alpha-induced HIV-1 replication in latently infected OM10.1 and Ach2 cells. *Int. Immun.* **11**: 143–150.

Measurement of Oxidative Stress: Technology, Biomarkers and Applications

(Nucleic Acid Damage)

Chapter 8

Measuring Oxidative Stress and Interpreting Its Clinical Relevance for Humans

Roy G. Cutler*,† and Mark P. Mattson‡

Keywords: Cancer, cardiovascular disease, diabetes; DNA damage, folic acid, stroke; vitamin.

Roy G. Cutler and **Mark P. Mattson** • Laboratory of Neurosciences, National Institute on Aging Gerontology Research Center, 5600 Nathan Shock Drive, Baltimore, MD 21224
†Laboratory of Neurosciences
‡Department of Neuroscience, Johns Hopkins University School of Medicine, 725 North Wolfe Street, Baltimore, MD 21205
*Corresponding Author.
Tel: (410) 558-8239, E-mail: RCutler@nih.gov

1. Aging is Our #1 Health Problem

Aging is the leading cause of disability, pain and death in the United States and other developed countries. The National Vital Statistics Report shows that about 74% of all deaths are caused by age-related diseases.[1] A list of the top 10 killers in the USA is shown in Table 1.

Table 1. Major Causes of Death in the USA (1998 statistics)

Cause of Death	Number of Deaths
Heart disease	724 859
Cancer	541 532
Stroke	158 448
Pulmonary diseases	112 584
Accidents	97 835
Pneumonia and influenza	91 871
Diabetes	64 751
Suicide	30 575
Nephritis (kidney)	26 182
Liver disease	25 331
Other*	463 290
Total	2 337 258

*Neurodegenerative disorders such as Alzheimer's and Parkinson's disease are the fourth leading cause of death in persons over the age of 70.

2. Degenerative Effects of Aging are Universal

What we call aging is the noticeable loss of function or the onset of the degenerative diseases that is caused by the aging process. Most scientists or physicians can define aging as the loss of a cell/organ/organisms peak function that continues until its death. Most humans' physiological functions such as hearing, eyesight, taste, lung capacity, agility, immune response, adaptation to change, etc... reach peak ability between the ages of 11 and 20 years old. After that age there is a slow decline in performance as the degenerative process of aging begins.[2] The rate of the degenerative effects of aging is different for each cell/organ/organism due to genetics and environmental exposure. Most of us will not reach our species' maximum life span potential due to the inheritance of genes that gives us a predisposition that accelerates the onset and/or the progression of an age related disease and/or due to our environmental exposure to harmful agents that can also accelerate the onset of particular age related disease. If we are genetically and environmentally lucky, we can live to, or near, our species' maximum life span potential, which is around 122 years old. However, the degenerative effect of aging is, in general, uniform for every cell/organ/organism. Even though an elderly person has not died earlier from a genetic or environmentally induced weak link in their vital life processes, their bodies are still showing signs of the degenerative effects of aging. A good example to further illustrate that the degenerative effects of aging are universal is to look at a person that is over 100 years of age. They have not yet died of a particular disease, but they do have signs of degeneration and loss of function throughout their body such as loss of: hearing, hormone regulation, skin elasticity, taste, vision, memory, immune responsiveness, etc... .[3, 4] It has been estimated that if we were able to completely eliminate and cure all forms of cancer that the impact of increased health and life span in the U.S. would only go up an average of 2.3 years. The reason for this is that most of the people who suffer from major killer diseases are over 65 years old and would be soon dying of another failing vital function.[5]

3. Why do Humans Live So Long? It is in Our Genes

Humans live about two times longer than our nearest relatives, the chimpanzee and bonobo. Humans also have an unusually long span of healthy and productive years of life relative to most other species. So what makes humans live so long and so healthy? Research indicates it is the expression of our genes, which has

Table 2

Gene Changed	Extension in Maximum Life Span	Species
p66shc −/− (apoptosis, phenoptosis)	30%	mouse[6]
PROP1 −/− (GH, prolactin or TSH)	50%	mouse[7]
GHR/BP −/− (GH receptor binding protein)	50%	mouse[8]
Cu/ZnSOD ++/++ (in motorneurons)	40%	drosophila[9]
MnSOD ++/++ (in motorneurons)	30%	drosophila[10]
mth −/− (putative G-protein-coupled receptor)	35%	drosophila[11]
indy −/− (sodium dicarboxylate cotransporter)	45%	drosophila[12]
age-1 −/− (PI3 kinase homologue)	65%	C. elegans[13]
daf 2 −/− (insulin receptor homologue in neurons)	100%	C. elegans[14]
clk-1 −/− (CoQ biosynthesis homologue)	40%	C. elegans[15]
spe-10 −/− (unknown)	40%	C. elegans[16]
spe-26−/− (unknown)	65%	C. elegans[17]
old-1 −/− (putative receptor tyrosine kinase)	65%	C. elegans[18]

A list of some of the genetically regulated chemical and physiological differences that may affect a species maximum lifespan, as well as, an individual's health and longevity, are shown in Table 3.

Table 3

Positive Correlation	No Correlation	Negative Correlation
Uric acid	DHEA	Production of free radicals
Ceruloplasmin	Ascorbate	DNA damage (8OHdG and
		mitochondrial DNA deletions)
Carotenoids	Retinol	Auto-oxidation potential of tissue
Tocopherols	Growth hormone	Insulin or IGF-1
DHEA-S		Iron or Ferritin
Choline		P450 oxidase enzymes
Albumin		

become coined as longevity determinant genes. Many of the genes listed below (Table 2) have been shown to extend the maximum life span of transgenic animals with few side effects, thereby giving some evidence that small changes in a single gene can have a big impact on increasing the life span and health span of a species.

What most of these genes have in common is that they relate to oxidative stress and developmental growth factors such as insulin and IGF-1. Past and recent research has made it clear that oxidative damage plays a key role in the onset and progression of almost all the age related diseases and that therapies to reduce oxidative stress helps to slow down these diseases and to maintain good health for as long as possible.

4. What is Oxidative Stress?

Oxidative stress is the amount of free radical damage that is being produced in an organelle/cell/organ/organism. A free radical is a very reactive atom with an unpaired electron, which can be in a reduced or oxidized state. The majority of free radicals that damage biological systems are oxygen radicals. What determines the amount of free radicals being produced depends on the balance of many different factors, some of which are shown in Fig. 1.

5. Measuring Oxidative Damage

There is much discussion and publicity about eating foods that contain high amounts of antioxidants, but what are these antioxidants supposed to do for you? The answer is that antioxidants are to help prevent and/or reduce the amount of oxidative damage — which may be the bottom line for improving our health. One way to reduce the risk of age-related disease is to lower the amount of

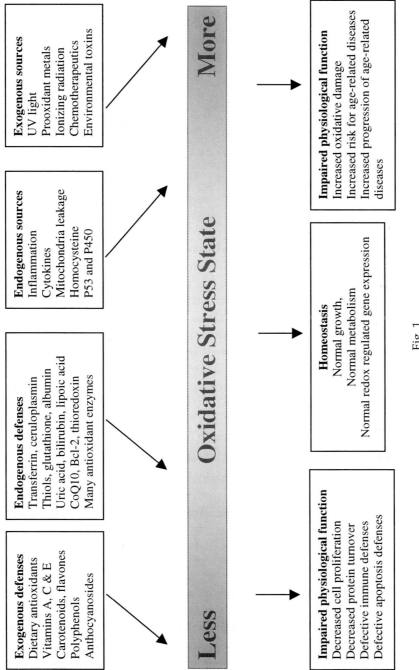

Fig. 1

oxidative damage to our cells. In order to do this it is of value to have the following information: (1) what is the current overall oxidative stress state, (2) what factors are a probable source of free radicals, (3) what can be done to maximize antioxidant protection, and (4) feed-back to see if the intervention to lower oxidative stress is working.

Below is a list of many accurate and commercially available assays that can be done for laboratory or clinical research to determine a persons oxidative stress state. Every research lab performing these assays should make a strong effort to work with the other labs in establishing and performing a proficiency testing program or round robin to help standardize the accuracy of the results and acceptance of oxidative stress biomarkers by colleagues, patients, and the medical community.

5.1. Alkenals

These are products of lipid peroxidation (e.g. lipid peroxides, malondialdehyde and 4-hydroxynonenal) that are liberated when the unsaturated bonds of membrane fatty acids are attacked by free radicals. The amount of lipid peroxides is a strong indicator of the average amount of free radical damage in the body at a given time. A high serum concentration of lipid peroxides has been correlated with an increased risk of most age-related diseases including diabetes, heart disease and cancer.[19] Increased levels of 4-hydroxynonenal-protein adducts have been documented in analyses of affected tissues in patients with cardiovascular disease and neurodegenerative disorders such as Alzheimer's and Parkinson's diseases.

Commercial availability: spectrophotometric, e.g. R&D Systems.

Human normal range: serum 0.5–4.1 µM; *urine 3.5–25.0 µM.

Human healthy average: serum 2.0 µM; *urine 14 µM.

Method reference: Ollinger, K. and Brunmark, A. (1994). Effect of different oxygen pressures and N,N'-Diphenyl-p-phenylenediamine adrimycin toxicity to cultured neonatal rat heart myocytes. *Biochem. Pharm.* **49**: 1707.

5.2. Hydroperoxides (Aqueous)

This method measures the actual amount of <u>aqueous</u> hydroperoxides (i.e. hydrogen peroxide). Hydroperoxides can react with prooxidant metals to form the very reactive hydroxyl radical, through Fenton reactions. The measure of the amount of serum hydroperoxides reflects the amount of free radicals being produced in the body at that time. A high serum concentration of lipid hydroperoxides has been correlated with a higher risk of most age-related diseases such as diabetes, heart disease and cancer.[20]

Commercial availability: spectrophotometric e.g. Pierce.

Human normal range: serum 0.3–3.2 µM; *urine 6.0–34.0 µM.

Human healthy average: serum 1.0 µM; *urine 12.1 µM.

Method reference: Wolff, S. (1994). Ferrous ion oxidation in presence of ferric ion indicator xylenol orange for measure of hydroperoxides. Methods in Enzymology (Lester Packer, Ed.) **233**, 182–189.

5.3. Hydroperoxides (Lipid)

This method measures the actual amount of lipid hydroperoxides (not hydrogen peroxide or products of lipid peroxide damage, i.e. aldehydes (MDA)). The measure of the amount of serum lipid hydroperoxides reflects the amount of free radicals being produced in the body at that time. A high serum concentration of lipid hydroperoxides has been correlated with a higher risk of most age-related diseases such as diabetes, heart disease and cancer.[21]

Commercial availability: spectrophotometric, e.g. Kaymia Biomedical.

Human normal range: serum 1.6–2.4 µM; *urine 5.1–15.1 µM.

Human healthy average: serum 2.0 µM; *urine 10.1 µM.

Method reference: Ohishi, N. *et al.* (1985). A new assay method for lipid peroxides using methylene blue derivative. *Biochem. Int.* **10**, 205–211.

5.4. 8-Hydroxydeoxyguanosine (8-OHdG)

This product of oxidative damage to DNA is considered by many researchers to be the "gold" standard for measuring oxidative damage to chromosomal and mitochondrial DNA. It is the most used and published biomarker of oxidative DNA damage. 8-OHdG reflects a person's DNA mutation potential, which strongly correlates with their risk of developing cancer.[22] 8-OHdG is a hydroxyl radical-damaged guanine nucleotide that has been excised from DNA by endonuclease repair enzymes. Since repair is known to normally occur quickly and efficiently, the amount of excised DNA adducts in serum or urine directly reflects the amount of damage within the entire body. Long-lived species have been shown to have lower levels of DNA damage than genetically similar shorter-lived species and therefore low DNA damage levels are thought to be a longevity determinant.[23]

Commercial availability: ELISA, e.g. Genox; HPLCEC, e.g. ESA Laboratory.

Human normal range: *urine 0.0–49.0 ng/ml.

Human healthy average: *urine 23.0 ng/ml.

Method reference: ESA application note #70-1181; http://www2.esainc.com

5.5. 8-Epi-Prostaglandin F2α (8-epi-PGF2α)(Isoprostane)

This marker of oxidative stress is formed *in vivo* by the free radical catalyzed non-enzymatic peroxidation of arachidonic acid in cellular membranes and lipoproteins (i.e. LDL). The damaged lipid peroxide is excised from the cell wall into the serum and then excreted in urine. Unlike the reactive aldehydes, once isoprostanes are formed they are chemically stable and can be accurately measured in serum or urine. 8-epi-PGF2α formation is unaffected by aspirin consumption and has been shown to be a sensitive measure of oxidative stress and measurement of the effectiveness of antioxidant supplementation.[24]

Commercial availability: ELISA, e.g. Oxford Biomedical; gas chromatograph mass spectrometer, e.g. Hewlett Packard.

Human normal range: *urine 0–3500 pg/ml.

Human healthy average: *urine 1750 pg/ml.

Method reference: Bachi, A. *et al.* (1996). Measurement of urinary 8-epi-prostaglandin F2α, a novel index of lipid peroxidation *in vivo*, by immuno-affinity extraction/gas chromatography-mass spectrometry. Basal levels in smokers and nonsmokers. *Free Radic. Biol. Med.* **20**: 619–624.

5.6. Urine Creatinine

Creatinine is used to normalize each sample by calculating the patient's metabolic efficiency [amount of free radical production (damage) per energy utilized (ATP synthesis)]. Creatinine is a product of the breakdown of ATP/creatine utilization which is excreted in the urine. Urine creatinine values from 12 or 24-hour samples are used to calculate the amount of damage being produced relative to the amount of energy or activity of that person (ratio). Urine creatinine values reflect a person's basal metabolic rate. Dividing creatinine by a person's lean body mass gives the amount of metabolic activity (ATP utilization) per cell over a given period of time.

Commercial availability: spectrophotometric, e.g. Sigma Diagnostics.

Human normal range: urine 58–146 mg/dl.

Human healthy average: urine 102 mg/dl.

Method reference: Bowers, L. and Wong, E. (1980). Kinetic serum creatinine assays. II. A critical evaluation and review. *Clin. Chem.* **26**: 555.

6. Measuring Prooxidants

The prooxidants listed below are compounds that reflect the potential for free radical damage. In general, they fall into three categories: (1) indicators of inflammatory responses, (2) blockers of mitochondrial electron transport that results in leakage of electrons, mainly as superoxide, and (3) Haber–Weiss and Fenton reactive metals and reducing agents. Haber–Weiss reactions can produce free radicals from peroxides (e.g. hydrogen peroxide) and superoxide and require a transitional metal (e.g. copper, iron). Fenton reactions also produce free radicals (i.e. hydroxyl radical) from a hydroperoxide (e.g. hydrogen peroxide) and a reduced transitional metal that can easily donate an electron to the splitting oxygen molecules. Strong reducing agents (e.g. ascorbate or homocysteine) can reduce the now oxidized metal to recycle its electron donating potential thereby propagating the generation of free radicals.

6.1. Antimony

Antimony is a prooxidant metal and toxin that can generate free radicals via the Fenton reaction.[25] There is no established biological function of antimony.

Commercial availability: inductive coupled plasma mass spectrometer, e.g. Hewlett Packard; flame atomic absorption, e.g. Perkin Elmer.

Human normal range: serum 0.0–0.5 µg/L.

Human healthy average: serum > 0.1 µg/L.

Method reference: Inagaki, K. and Haraguchi, H. (2000). Determination of rare earth elements in human blood serum by inductively coupled plasma mass spectrometry after chelating resin preconcentration. *Analyst* **125**(1): 191–196.

6.2. Arsenic

Arsenic is a well-known poison that inhibits respiration. It produces superoxide and hydroperoxides by blocking the electron flow in mitochondria oxidative phosphorilation.[26]

Commercial availability: inductive coupled plasma mass spectrometer, e.g. Hewlett Packard; flame atomic absorption, e.g. Perkin Elmer.

Human normal range: serum 0–11 µg/L.

Human healthy average: serum 4 µg/L.

Method reference: Inagaki, K. and Haraguchi, H. (2000). Determination of rare earth elements in human blood serum by inductively coupled plasma mass spectrometry after chelating resin preconcentration. *Analyst* **125**(1): 191–196.

6.3. C-Reactive Protein

C-reactive protein is a protein, that's released as part of acute and long-term inflammation responses. Levels of C-reactive protein reflect the patient's level of inflammation, which has been show to be strongly related to risk and progression of many forms of arthritis, cancer, heart disease and diabetes. The inflammation response has many mechanisms that produce free radicals such as TNFα and activated macrophages. Therefore, inflammation can be a major source of free radical generation, damage and consequential risk and progression of age-related diseases in humans.[27] Some of the most common and lethal causes of chronic inflammation in humans comes from low titer viral infections such as hepatitis B, hepatitis C, all sexually transmitted diseases, along with many other low grade chronic bacterial infections such as pneumonias, gingivitis, etc.

Commercial availability: spectrophotometric kit, e.g. Sigma Diagnostics.

Human normal range: serum 0.1–12.0 mg/dl.

Human healthy average: serum 0.4 mg/dl or less.

6.4. Cadmium

Cadmium is a prooxidant metal that can generate free radicals through the Fenton reaction. However, the majority of free radicals cadmium produces is from inhibiting the mitochondrial oxidative phosphorylation pathway causing the release of electrons, mostly as superoxide radicals.[28] It is a well-known toxin with similar properties as mercury and currently has no known human need.

Commercial availability: inductive coupled plasma mass spectrometer, e.g. Hewlett Packard; flame atomic absorption, e.g. Perkin Elmer.

Human normal range: serum 0.0–2.6 μg/L.

Human healthy average: serum > 0.3 μg/L.

Method reference: Inagaki, K. and Haraguchi, H. (2000). Determination of rare earth elements in human blood serum by inductively coupled plasma mass spectrometry after chelating resin preconcentration. *Analyst* **125**(1): 191–196.

6.5. Chromium

Industrial chromium (i.e. Cr^{+6}) is a very toxic prooxidant metal that can generate free radicals via Fenton reactions.[29] Small amounts of Cr^{+3} are required in glucose tolerance proteins that function to regulate blood glucose levels.

Commercial availability: inductive coupled plasma mass spectrometer, e.g. Hewlett Packard; flame atomic absorption, e.g. from Perkin Elmer.

Human normal range: serum 0.0–1.0 µg/L.

Human healthy average: serum 0.4 µg/L.

Method reference: Inagaki, K. and Haraguchi, H. (2000). Determination of rare earth elements in human blood serum by inductively coupled plasma mass spectrometry after chelating resin preconcentration. *Analyst* **125**(1): 191–196.

6.6. Chlorine

Chlorine is a halogen (like fluorine and iodine) and can readily act as a free radical generator.[30] High amounts of chlorine can be very toxic.

Commercial availability: inductive coupled plasma mass spectrometer, e.g. Hewlett Packard; flame atomic absorption, e.g. Perkin Elmer.

Human normal range: serum 0.00–1.23 µg/L.

Human healthy average: serum 0.25 µg/L.

Method reference: Inagaki, K. and Haraguchi, H. (2000). Determination of rare earth elements in human blood serum by inductively coupled plasma mass spectrometry after chelating resin preconcentration. *Analyst* **125**(1): 191–196.

6.7. Iron

Unbound iron is a prooxidant metal that can catalyze the production of the very reactive hydroxyl radical from hydrogen peroxide via Haber–Weiss or Fenton reactions. The now oxidized iron can then be reactivated in the presence of strong reducing agents such as ascorbate or homocysteine. High iron (overload) is associated with high amounts of free radical damage and a higher risk for developing most of the age-related diseases such as diabetes, heart disease and cancer.[34]

Commercial availability: spectrophotometric kit, e.g. Sigma diagnostics; inductive coupled plasma mass spectrometer, e.g. Hewlett Packard; flame atomic absorption, e.g. Perkin Elmer.

Human normal range: 35–140 µg/dl.

Human healthy average: 88 µg/dl.

Method reference: Goodwin, J., Murphy, B. and Guillemette, M. Direct measurement of serum iron and binding capacity. *Clin. Chem.* **12**: 47, 1966.

Method reference: Inagaki, K. and Haraguchi, H. (2000). Determination of rare earth elements in human blood serum by inductively coupled plasma mass spectrometry after chelating resin preconcentration. *Analyst* **125**(1): 191–196.

6.8. Ferritin

Although not a prooxidant *per se*, ferritin (an iron binding protein synthesized by the liver in response to the amount of iron in the serum) is a potential source of iron, a major prooxidant. Ferritin is an indicator of the body's iron storage and possible long-term iron overload. High ferritin levels in serum have been associated with high amounts of free radical damage and a higher risk for developing most of the age-related diseases such as diabetes, heart disease and cancer.[32]

Commercial availability: immunological aggregation performed on clinical analyzer, e.g. Abbott, Hitachi or Roche.

Human normal range: 15–160 ng/ml.

Human healthy average: 32 ng/ml.

Method reference: Borsch, B. (1995). Determination of iron status; brief review of physiological effects on iron measures. *Analyst* **120**(3): 891–893.

6.9. Copper

Unbound copper is a prooxidant metal that is known to be an even more reactive transitional metal than iron. Free radicals can be evolved via Haber–Weiss or Fenton reactions when reduced copper is allowed to oxidize hydrogen peroxide to create a hydroxyl radical. This reaction can cycle over and over again in the presence of strong reducing agents such as ascorbate (vitamin C) or homocysteine. High copper levels in serum can induce, and are associated with, high levels of oxidative damage.[31] Small amounts are required for various biological processes such as in Cu/Zn-SOD and ceruloplasmin.

Commercial availability: spectrophotometric; inductive coupled plasma mass spectrometer, e.g. from Hewlett Packard; flame atomic absorption, e.g. Perkin Elmer.

Human normal range: serum 20–130 µg/L.

Human healthy average: serum 100 µg/L.

Method reference: Abe, A. *et al.* (1989). Copper determination in serum and plasma. *Clin. Chem.* **552**: 35.

Method reference: Inagaki, K. and Haraguchi, H. (2000). Determination of rare earth elements in human blood serum by inductively coupled plasma mass spectrometry after chelating resin preconcentration. *Analyst* **125**(1): 191–196.

6.10. Homocysteine

Homocysteine is a strong reducing agent and a compound that promotes methyl donor deficiency and DNA damage. Elevated levels of homocysteine are highly associated with a high risk of cardiovascular disease, stroke and other age-related diseases.[33] In most patients, high homocysteine is caused by deficiencies in folic acid, vitamin B6 and/or vitamin B12.

Commercial availability: fluorescent polarization, e.g. Abbott; ELISA, e.g. BioRad; high-pressure liquid chromatography (HPLC).

Human normal range: 5–15 µM.

Human healthy average: 8 µM.

6.11. Lead

Lead is a prooxidant metal and well-known neurotoxin that has no known function in humans. It can act as a transitional metal to produce free radicals via Haber–Weiss or Fenton reactions. However, lead does most of its damage by inhibiting mitochondrial oxidative phosphorilation thereby causing the release of electrons, mostly as superoxide.[35]

Commercial availability: inductive coupled plasma mass spectrometer, e.g. Hewlett Packard; flame atomic absorption, e.g. Perkin Elmer.

Human normal range: serum 0–11 µg/L.

Human healthy average: serum > 4 µg/L.

Method reference: Inagaki, K. and Haraguchi, H. (2000). Determination of rare earth elements in human blood serum by inductively coupled plasma mass spectrometry after chelating resin preconcentration. *Analyst* **125**(1): 191–196.

6.12. Manganese

Unbound manganese is known to be a strong prooxidant metal and toxic at high levels. It readily produces free radicals in the presence of hydroperoxides via Haber–Weiss and Fenton reactions; these reactions can be recycled if there is also strong reducing agents such as ascorbate or homocysteine. Small amounts are required for various biological processes such as endogenous antioxidant enzyme MnSOD.

Commercial availability: inductive coupled plasma mass spectrometer, e.g. Hewlett Packard; flame atomic absorption, e.g. Perkin Elmer.

Human normal range: serum 0.44–0.76 µg/L.

Human healthy average: serum 0.60 µg/L.

Method reference: Inagaki, K. and Haraguchi, H. (2000). Determination of rare earth elements in human blood serum by inductively coupled plasma mass spectrometry after chelating resin preconcentration. *Analyst* **125**(1): 191–196.

6.13. Mercury

Mercury is a prooxidant metal that is a well-known neuro-toxin and has no known human need. It can produce free radicals via the Fenton reaction, but the majority of free radicals it produces is from inhibiting the mitochondrial oxidative phosphorylation causing a leakage of electrons, mainly as superoxide.[36]

Commercial availability: inductive coupled plasma mass spectrometer, e.g. Hewlett Packard; flame atomic absorption, e.g. Perkin Elmer.

Human normal range: serum 0.0–3.4 µg/L.

Human healthy average: serum > 1.4 µg/L.

Method reference: Inagaki, K. and Haraguchi, H. (2000). Determination of rare earth elements in human blood serum by inductively coupled plasma mass spectrometry after chelating resin preconcentration. *Analyst* **125**(1): 191–196.

6.14. Nickel

Unbound nickel is known to be a strong prooxidant metal and is toxic at high levels.[37] Nickel is a transitional metal and a strong electron donor, so consequently it can actively produce free radicals via the Haber–Weiss and Fenton reactions. Small amounts are required for various biological functions such as in red blood cells and liver functions.

Commercial availability: inductive coupled plasma mass spectrometer, e.g. from Hewlett Packard; flame atomic absorption, e.g. from Perkin Elmer.

Human normal range: serum 0.6–7.5 µg/L.

Human healthy average: serum 1.3 µg/L.

Method reference: Inagaki, K. and Haraguchi, H. (2000). Determination of rare earth elements in human blood serum by inductively coupled plasma mass spectrometry after chelating resin preconcentration. *Analyst* **125**(1): 191–196.

7. Measuring Total Antioxidant Capacity

Data from cross-species analysis comparing the biochemical differences between short- versus long-lived species strongly suggest that oxidative stress state plays a critical role in determining a species maximum health and lifespan potential. Below are some of the most clinically relevant non-invasive tests for chemistries that have likely evolved to promote a long and healthy life. Plasma antioxidants can be classified into two major types: (1) primary antioxidants such as ceruloplasmin and transferrin, which act to reduce the initiation of free radicals and lipid peroxidation by binding prooxidant metals, and (2) secondary antioxidants such as tocopherol, which help reduce the chain propagation and amplification of lipid peroxide radicals. Many antioxidants have multiple antioxidant properties such as uric acid, which can both bind many prooxidant metals, as well as, directly scavenge oxidized species.

7.1. Lipid Peroxidation Inhibition Capacity (LPIC)* Assay

LPIC assay measures the activity of both the primary and secondary antioxidant systems as they are working together in a sample. The basic principle of this assay is to take a standardized rat brain homogenate, let it autooxidize and measure the amount of lipid peroxides. A sample's ability to inhibit the autooxidation reaction relative to the blank is proportional to the antioxidant capacity in that sample. Long-lived species have been shown to have a higher LPIC value than genetically similar shorter-lived species and therefore high LPIC is thought to be a longevity determinant (method reference). In humans, low serum LPIC has been shown to strongly predict development of adult onset diabetes.[38] Actual assay results will vary from lab to lab depending on the method used.

Commercial availability: microplate reader with temperature control.

Human normal range: serum 45–85% inhibition of autooxidation of lipid substrate.

Human healthy average: serum 55%.

Method reference: Cutler, R. G. Peroxide-producing potential of tissues: correlation with the longevity of mammalian species. *Proc. Natl. Acad. Sci. USA* **82**: 4798–4802.

*LPIC name was coined by Dr Richard Cutler for Genex Corporation.

7.2. Oxygen Radical Absorption Capacity (ORAC) Assay

ORAC assay measures the total antioxidant capacity in a sample. Many researchers have shown that the primary and secondary antioxidants that help to lower oxidative stress do not just work by themselves, but they interact with each other in ways we are just learning about such as recycling.[39] It is nearly impossible to measure all of the antioxidants individually in a biological sample. Therefore, the ORAC assay was developed in order to qualitatively and quantitatively examine all of the antioxidants in a sample as they are working together. The principle of the assay is the use of the fluorescent protein β-phycoerythrin (β-PE) that loses its fluorescence when a free radical sensitive tryptophan amino acid is damaged. The free radical generator used in this assay is typically AAPH (2, 2′-azobis(2-amindinopropane) dihydrochloride), a heat-activated peroxide generator, or copper/iron Fenton reactions. The amount of antioxidants can be calculated by taking the differences of areas under the reaction curve from the sample reaction versus the blank. The quality or antioxidant activity of the sample can be determined by calculating the area or time that the sample was able to protect the β-PE protein 100%. The fast acting antioxidants are the first line of defense. In human serum these include: thiols, uric acid, lipid soluble antioxidants, ascorbate, bilirubin, biofavonoids, polyphenols, and many unknowns. Long lived species have been shown to have a higher ORAC value than similar shorter-lived species and therefore high ORAC is thought to be a longevity determinant.[40] Actual assay results will vary from lab to lab depending on the method used.

Commercial availability: fluorescent assay can be performed by a robotic chemical analyzer, e.g. Roche or fluorescent microplate reader with temperature control.

Human normal range: serum 3300–5000 μM equivalents of Trolox.

Human healthy average: serum 4386 μM equivalents of Trolox.

Method reference: Cao, G., Alessio, H. and Cutler, R. (1993). Oxygen-radical absorbance capacity assay for antioxidants. *Free Radic. Biol. Med.* **14**: 303–311.

7.3. ORAC (Aqueous Soluble)

ORAC (aqueous soluble) measures the antioxidants in a serum sample after the removal of proteins and lipids. Because of their abundance, almost half of the whole serum ORAC value is from proteins and lipids. Most proteins and lipids are not very active antioxidants (sacrificial) and decrease the sensitivity of the ORAC assay in measuring less abundant but more active antioxidants. Removing the lipids and proteins from the serum sample increases the sensitivity of the ORAC assay in measuring the other aqueous soluble antioxidants.

Commercial availability: fluorescent assay can be performed by a robotic chemical analyzer, e.g. Roche; or fluorescent microplate reader with temperature control.

Human normal range: serum 480–1000 μM equivalents of Trolox.

Human healthy average: serum 752 μM equivalents of Trolox.

Method reference: Cao, G., Giovanoni, M. and Prior, R. (1996). Antioxidant capacity in different tissues of young and old rats. *Proc. Soc. Exp. Biol. Med.* **211**(4): 359–365.

7.4. ORAC (Lipid Soluble)

ORAC (lipid soluble) measures the antioxidants in a serum sample after the removal of proteins and the aqueous phase of the sample. Because of their abundance, over half of the whole serum ORAC value is from proteins and aqueous antioxidants. Removing the proteins and aqueous antioxidants from the serum sample increases the sensitivity of the assay (i.e. focuses) on measuring the many lipid soluble antioxidants.

Commercial availability: fluorescent assay can be performed by a robotic chemical analyzer, e.g. Roche; or fluorescent microplate reader with temperature control.

Human normal range: serum 960–1910 µM equivalents of Trolox.

Human healthy average: serum 1520 µM equivalents of Trolox.

Method reference: Cao, G., Giovanoni, M. and Prior, R. (1996). Antioxidant capacity in different tissues of young and old rats. *Proc. Soc. Exp. Biol. Med.* **211**(4): 359–365.

8. Measuring Primary Antioxidants

Primary antioxidants are the body's first line of defense. Many act by keeping prooxidants like iron and copper from initiating the production of free radicals, resulting in a form of prevention. So what characteristics make-up a good antioxidant? A good antioxidant should have the following qualities:

(1) specificity for free radical quenching;
(2) metal chelating activity;
(3) interaction with other antioxidants (i.e. recycling quenching potential);
(4) good absorption and bioavalibility; and
(5) effective residing location and concentration — best if cell can control concentrations based on its need for protection.

8.1. Available Iron Binding Capacity (AIBC)

AIBC is the amount of transferrin, ferritin and albumin that is not binding iron and therefore can accept (capture) a free iron atom. Iron binding proteins are known to be a very effective prevention of iron catalyzed free radical production. High AIBC offers good protection against the initiation of oxidative damage reactions. Epidemiology studies have shown that high AIBC strongly correlates with a low risk for cardiovascular disease.[41] AIBC proteins are synthesized in the liver and are kept at a fairly constant steady state level in the serum; generally the higher the iron, the lower the AIBC. Total Iron Binding Capacity (TIBC) is the

AIBC plus the Total Iron. This value is used to reflect the liver's total capacity in making iron binding proteins.

Commercial availability: spectrophotometric assay performed on a robotic chemical analyzer, e.g. Beckman, Roche or Hitachi.

Human normal range: serum 130–375 µg/dl.

Human healthy average: serum 253 µg/dl.

Method reference: Goodwin, J., Murphy, B. and Guillemette, M. (1996). Direct measurement of serum iron and binding capacity. *Clin. Chem.* **12**: 47.

8.2. % Iron Saturation (Total Iron/TIBC)*100%

100% Iron saturation (total iron/TIBC) is the relative ratio of iron to iron binding capacity; the higher the iron saturation, the higher the risk for iron to participate in catalyzing free radical species.

8.3. Ceruloplasmin

Ceruloplasmin binds up to 95% of the copper found in serum. In normal patients the amount of ceruloplasmin is directly proportional to the amount of copper in the serum; high amounts of free (i.e. unbound) copper are usually associated with high amounts of lipid peroxidation and risk for cardiovascular disease. Ceruloplasmin not only binds free copper, but has ferroxidase (oxidizes free iron) like catalytic activity and thereby helps inhibit free iron from participating as a prooxidant and is essential in removing excess iron out of the body. Ceruloplasmin also acts as a superoxide dismutase.[42] Lower than normal amounts of ceruloplasmin are found in patients with Wilson's disease.

Commercial availability: immunoprecipitation (turbidimetric) assay performed on a robotic chemical analyzer, e.g. Roche.

Human normal range: serum 15–63 mg/dl.

Human healthy average: serum 56 mg/dl.

9. Measuring Secondary Antioxidants

Secondary antioxidants act as free radical scavengers that catch or donate an electron, which then stops the propagation of a free radical reaction. Once the antioxidant has accepted or donated one of its electrons to stabilize a free radical, it now has an unpaired electron which energy is then delocalized around the phenol ring or hydrocarbon chain such that it now has a low energy of reaction.

As discussed in the description of the ORAC assay, most of these antioxidants work together to recycle each other and overlap the areas in the body in which they are designed to protect.

10. Aqueous Soluble Antioxidants

10.1. Ascorbic Acid (vitamin C)

Ascorbic acid (vitamin C) is one of the most popular diet supplemented vitamins. Ascorbic acid can directly scavenge many different types of oxidative species, as well as, regenerate other oxidized antioxidants such as vitamin E. However, under conditions where there are free prooxidant metals around, such as iron and copper, vitamin C's strong reductive capacity can help catalyze the production of oxidative free radicals by reducing the transition metal to participate again in a Haber–Weiss or Fenton reaction. Humans have lost the ability to synthesize ascorbate; this may be due to the unwanted production of hydrogen peroxide during its synthesis and/or because of its ability to catalyze prooxidant metal free radical reactions (i.e. Fenton).[43] It is thought that ascorbate has been replaced by uric acid in humans as our body's primary aqueous antioxidant. Nonetheless, moderate levels of ascorbate are required for health and are associated with the prevention of many diseases.[44]

Commercial availability: spectrophotometric assay; HPLC method.

Human normal range: serum 7.0–26.5 µg/ml.

Human healthy average: serum 13.0 µg/ml.

Method reference: McGown, E. *et al.* (1982). Tissue ascorbic acid analysis using Ferrozine compared with the dinitrophenylhydrazine method. *Anal. Biochem.* **119**: 55–61.

10.2. Bilirubin (Conjugated and Total)

Considered a waste product of heme metabolism, bilirubin is known to be a very active lipid and aqueous soluble serum antioxidant. Hyperbilirubinemia can give unusually high antioxidant values as measured by the ORAC and LPIC assays, but usually represent another problem in the liver. Direct (conjugated) bilirubin is the form of bilirubin that can be absorbed and removed from the body in the bile. Epidemiology studies have shown that families with genetically high levels of bilirubin tend to have very long lives and low risk for many of the age-related diseases.[45]

Commercial availability: spectrophotometric assay performed on a robotic chemical analyzer, e.g. Bechman, Roche, or Hitachi.

Human normal range: serum conjugated 0.00–0.65 mg/dl, total 0.10–1.20 mg/dl.
Human healthy average: serum conjugated 0.30 mg/dl, total 0.65 mg/dl.

10.3. Thiols

Thiols are very active antioxidants and reducing agents. Most serum thiols are found in albumin followed by free cysteine and glutathione. Albumin thiols are thought to act as sacrificial antioxidants that have little biological consequences of being damaged. Because of their high antioxidant reactivity and high concentration, albumin thiols act as a major defense against free radical damage to cell membranes, LDL, etc.[46]

Commercial availability: spectrophotometric assay; or HPLC methods.

Human normal range: serum 105–230 μM.

Human healthy average: serum 200 μM.

Method reference: Motchnik, P., Frei, B. and Ames, B. (1994). Measurement of antioxidants in human blood plasma. *Meth. Enz.* **234**: 269–279.

10.4. Uric Acid

Uric acid is a multifunctional antioxidant as it can directly scavenge oxidative species and chelate prooxidant metals. Uric acid is known to be unusually high in long-lived species including birds, bats, and humans. Therefore, it is thought to be a longevity determinant. It is also interesting that uric acid is a methylxanthine (like caffeine) which acts to stimulate cAMP and other cytoprotective signaling mechanisms.[47] Our body's concentration of uric acid is also thought to be able to up regulate itself as a protection mechanism in times of high free radical insult.[48] Cross species analysis has shown a strong correlation between uric acid and maximum lifespan potential.[49, 50]

Commercial availability: spectrophotometric assay performed on a robotic chemical analyzer e.g. Bechman, Roche, or Hitachi.

Human normal range: serum 3.5–7.7 mg/dl.

Human healthy average: serum 5.6 mg/dl.

11. Lipid Soluble Antioxidants

Most biological molecules have multiple functions within a cell and throughout the body. The lipid soluble antioxidants are no exception. They have unique absorption and membrane positioning which enables them to not only act as free radical scavengers but also to initiate other life benefiting mechanisms. For instance, many of these phenolic antioxidants stimulate the synthesis and activation of phase II antioxidant enzymes, as well as suppress many of the free radical generating p450 oxidases. Many of the following lipid soluble phenolic antioxidants also stimulate peroxisome proliferator activated receptor alpha (PPARα) which stimulates peroxisomes to remove damaging lipid peroxides, steroids and lipid growth factors that are thought to promote many age-related diseases. Unlike most other antioxidants, the dietary availability and absorption of the carotenoids are actually enhanced by cooking, particularly in the presence of other fats. High serum and tissue lipid soluble antioxidants such as tocopherols and carotenoids are strongly correlated with maximum health and lifespan potential. The data showing the association of lipid soluble antioxidants and a species' lifespan potential is particularly striking because it compares primates (e.g. monkeys and apes) with very similar genetic codes.[51]

11.1. α-Carotene

α-carotene is found in pumpkin and carrot and is readily absorbed in human serum. It is a known antioxidant and precursor to vitamin A. Some researchers have reported that α-carotene may be a stronger antioxidant and cellular differentiating agent than β-carotene and therefore may be better in preventing cancer.[52] Long-lived species have high amounts of carotenoids in their serum and tissue compared to similar shorter-lived species, and therefore, high carotenoid levels are thought to be a longevity determinant.

Commercial availability: High-Pressure Liquid Chromatography (HPLC) standardized and certified by the National Institute of Standards and Technology (NIST).

Human normal range: serum 0.01–0.37 µg/ml.

Human healthy average: serum 0.05 µg/ml.

Method reference: Browne, R. W. and Armstrong, D. (1998). Simultaneous determination of serum retinol, tocopherols, and carotenoids by HPLC. *Meth. Mol. Biol.* **108**: 269–275.

11.2. β-Carotene

β-carotene is a known antioxidant and precursor to vitamin A, which is a strong cellular differentiating agent and is therefore thought to help prevent cancer. β-carotene is the most widely studied carotenoid, as well as, the most diet supplemented. Recent epidemiology data has shown some possible negative effects of heavily supplementing just one carotenoid. Therefore, it is recommended that dietary supplements contain a balance of all the carotenoids — from natural sources.[53] Long-lived species have high amounts of carotenoids in their serum and tissue compared to similar shorter-lived species, and therefore, high carotenoid levels are thought to be a longevity determinant.

Commercial availability: High-Pressure Liquid Chromatography (HPLC) standardized and certified by the National Institute of Standards and Technology (NIST).

Human normal range: serum 0.07–0.68 µg/ml.

Human healthy average: serum 0.19 µg/ml.

Method reference: Browne, R. W. and Armstrong, D. (1998). Simultaneous determination of serum retinol, tocopherols, and carotenoids by HPLC. *Meth. Mol. Biol.* **108**: 269–275.

11.3. β-Cryptoxanthin

β-cryptoxanthin is found in papaya and oranges. It is readily absorbed into the serum and is a very active lipid soluble antioxidant.[54] We have found that β-cryptoxanthin is the most active of the lipid soluble antioxidants, 3.1 times higher than Vitamin E as measured by the ORAC assay (unpublished).

Commercial availability: High-Pressure Liquid Chromatography (HPLC) standardized by using the National Institute of Standards and Technology (NIST) lipid soluble antioxidant standard reference material.

Human normal range: serum 0.007–0.18 µg/ml.

Human healthy average: serum 0.06 µg/ml.

Method reference: Browne, R. W. and Armstrong, D. (1998). Simultaneous determination of serum retinol, tocopherols, and carotenoids by HPLC. *Meth. Mol. Biol.* **108**: 269–275.

11.4. Lutein

Litein is the most abundant carotenoid in green leafy vegetables. It is readily absorbed into the serum and is a very active lipid soluble antioxidant, 2.3 times higher than vitamin E as measured by the ORAC assay (unpublished). Recent studies have shown that lutein and zeaxanthin are major factors in the prevention of macular degeneration, which is the leading cause of blindness in the elderly and represents 10% of all blindness in humans.[55]

Commercial availability: High-Pressure Liquid Chromatography (HPLC) standardized and certified by the National Institute of Standards and Technology (NIST).

Human normal range: serum 0.05–0.57 µg/ml.

Human healthy average: serum 0.17 µg/ml.

Method reference: Browne, R. W. and Armstrong, D. (1998). Simultaneous determination of serum retinol, tocopherols, and carotenoids by HPLC. *Meth. Mol. Biol.* **108**: 269–275.

11.5. Lycopene

Lycopene is found in tomatoes and watermelon and is readily absorbed in human serum. We have found lycopene to be one of the most active of the lipid soluble antioxidants, 2.8 times higher than vitamin E as measured by the ORAC assay (unpublished). Research has indicated that lycopene may be very important in the prevention of heart disease, prostate cancer and may block some of the toxic effects of testosterone and 5-dihydro-testosterone.[56, 57]

Commercial availability: High-Pressure Liquid Chromatography (HPLC) standardized and certified by the National Institute of Standards and Technology (NIST).

Human normal range: serum 0.01–0.33 µg/ml.

Human healthy average: serum 0.18 µg/ml.

Method reference: Browne, R. W. and Armstrong, D. (1998). Simultaneous determination of serum retinol, tocopherols, and carotenoids by HPLC. *Meth. Mol. Biol.* **108**: 269–275.

11.6. Retinol (vitamin A)

Retinol (vitamin A) is a known antioxidant and cellular differentiating agent. Data suggest that it may aid in the prevention of many forms of cancer. Caution: retinol is a terogen and is toxic at high levels.

Commercial availability: High-Pressure Liquid Chromatography (HPLC) standardized and certified by the National Institute of Standards and Technology (NIST).

Human normal range: serum 0.35–1.25 µg/ml.

Human healthy average: serum 0.65 µg/ml.

Method reference: Browne, R. W. and Armstrong, D. (1998). Simultaneous determination of serum retinol, tocopherols, and carotenoids by HPLC. *Meth. Mol. Biol.* **108**: 269–275.

11.7. Retinyl Palmitate

Retinyl palmitate is the form of vitamin A that is most commonly used in dietary supplements and foods as a source for vitamin A. Caution: Retinyl

Palmitate and its cleaved product retinol are terogens and are toxic at high levels.

Commercial availability: High-Pressure Liquid Chromatography (HPLC), standardized using the National Institute of Standards and Technology (NIST) lipid soluble antioxidant standard reference material.

Human normal range: serum 0.01–0.17 µg/ml.

Human healthy average: serum 0.07 µg/ml.

Method reference: Browne, R. W. and Armstrong, D. (1998). Simultaneous determination of serum retinol, tocopherols, and carotenoids by HPLC. *Meth. Mol. Biol.* **108**: 269–275.

11.8. α-Tocopherol (vitamin E)

α-tocopherol (vitamin E) is one of the best characterized and diet supplemented lipid soluble antioxidants. In addition to its antioxidant capabilities, it has cellular differentiation properties which are believed to be beneficial in preventing cancer.[58] Long-lived species have high amounts of tocopherols in their serum and tissue compared to similar shorter-lived species, and therefore, high tocopherol levels are thought to be a longevity determinant.

Commercial availability: High-Pressure Liquid Chromatography (HPLC) standardized and certified by the National Institute of Standards and Technology (NIST).

Human normal range: serum 6.5–17.2 µg/ml.

Human healthy average: serum 10.1 µg/ml.

Method reference: Browne, R. W. and Armstrong, D. (1998). Simultaneous determination of serum retinol, tocopherols, and carotenoids by HPLC. *Meth. Mol. Biol.* **108**: 269–275.

11.9. δ-Tocopherol (vitamin E)

δ-tocopherol (vitamin E) is normally found at lower amounts in foods and human serum than the alpha or gamma form of tocopherol. Scientists are just beginning to study the unique beneficial effects of δ-tocopherol. It is recommended that dietary vitamin E supplements contain a mix of all the tocopherols.

Commercial availability: High-Pressure Liquid Chromatography (HPLC) standardized and certified by the National Institute of Standards and Technology (NIST).

Human normal range: serum 0.06–0.20 µg/ml.

Human healthy average: serum 0.1 µg/ml.

Method reference: Browne, R. W. and Armstrong, D. (1998). Simultaneous determination of serum retinol, tocopherols, and carotenoids by HPLC. *Meth. Mol. Biol.* **108**: 269–275.

11.10. γ-Tocopherol (vitamin E)

γ-tocopherol (vitamin E) has been reported to be the major type of vitamin E found in heart tissue and therefore may be selected for by the body because of its unique properties either as an antioxidant and/or as a differentiation agent. Scientists are just beginning to study the unique beneficial effects of γ-tocopherol. It is recommended that vitamin E supplements contain a mix of all the tocopherols.

Commercial availability: High Pressure Liquid Chromatography (HPLC) standardized and certified by the National Institute of Standards and Technology (NIST).

Human normal range: serum 0.7–4.6 μg/ml.

Human healthy average: serum 2.0 μg/ml.

Method reference: Browne, R. W. and Armstrong, D. (1998). Simultaneous determination of serum retinol, tocopherols, and carotenoids by HPLC. *Meth. Mol. Biol.* **108**: 269–275.

11.11. Tocopherols/(Chol. + Trig.)

Tocopherols/(chol. + trig.) is the ratio of lipid antioxidants per amount of lipids that there is to be protected. Research and clinical studies have used this calculation and have found it to give a much better indication of risk for developing cardiovascular disease then by evaluating each of these biomarkers alone.[59]

Human normal range: serum 0.013–0.125 ratio (μg/ml)/(mg/dl).

Human healthy average: serum 0.04 ratio (μg/ml)/(mg/dl).

11.12. Ubiquinol (Coenzyme Q10)

Ubiquinol (coenzyme Q10) is normally synthesized in cells as part of the mitochondria oxidative phosphorilation redox system. CoQ10 can also be absorbed through the diet and is a very popular dietary supplement. Research has found that CoQ10, which is normally in lipid membranes, can act as a very active antioxidant and protect LDL from becoming oxidized.[60]

Commercial availablility: High-Pressure Liquid Chromatography (HPLC).

Human normal range: serum 0.6–1.0 μg/ml.

Human healthy average: serum 0.8 µg/ml.

Method reference: Menke, T. *et al.* (2000). Simultaneous detection of ubiquinol-10, ubiquinone-10, and tocopherols in human plasma microsamples and macrosamples as a marker of oxidative damage in neonates and infants. *Anal. Biochem.* **282**(2): 209–217.

11.13. Zeaxanthin

Zeaxanthin is a very abundant carotenoid in green leafy vegetables. It is readily absorbed into the serum and is a very active lipid soluble antioxidant, 2.8 times higher than vitamin E as measured by the ORAC assay (unpublished). Recent studies have shown that lutein and zeaxanthin are major factors in the prevention of macular degeneration, which is the leading cause of blindness in the elderly and represents 10% of all blindness in humans.[61]

Commercial availablility: High-Pressure Liquid Chromatography (HPLC) standardized and certified by the National Institute of Standards and Technology (NIST).

Human normal range: serum 0.02–0.13 µg/ml.

Human healthy average: serum 0.04 µg/ml.

Method reference: Browne, R. W. and Armstrong, D. (1998). Simultaneous determination of serum retinol, tocopherols, and carotenoids by HPLC. *Meth. Mol. Biol.* **108**: 269–275.

12. Antioxidant Supporting Factors

12.1. Albumin

Albumin is a dual antioxidant in that it contains many very active thiol groups that act as potent antioxidants and has a strong prooxidant metal binding capacity. Because of the high amount of albumin in the body, it is considered to be one of the major antioxidants. Albumin is known as a sacrificial antioxidant because it has no recycling pathway and the consequences of its damage do not directly affect cellular function. Albumin has a high turnover rate; damaged albumin is degraded by specific proteases and the body reuses the good amino acids. In contrast, most other antioxidant mechanisms use some sort of direct regeneration system (e.g. vitamin E, vitamin C, or glutathione peroxidase). Human epidemiology studies have shown serum albumin levels to be positively correlated with a low risk of cardiovascular disease.[62] There is also a strong correlation between serum albumin levels and lifespan potential of a species.

Commercial availablility: spectrophotometric assay performed on a robotic chemical, e.g. Bechman, Roche or Hitachi.

Human normal range: serum 2.2–5.0 g/dl.

Human healthy average: serum 4.5 g/dl.

12.2. Total Protein

Total protein includes albumin and the immunoglobulins. Therefore, the amount of globulins in a serum sample can be calculated by subtracting the albumin value from the amount of total proteins.

Commercial availablility: spectrophotometric assay performed on a robotic chemical, e.g. Bechman, Roche or Hitachi.

Human normal range: serum 6.0–8.2 g/dl.

Human healthy average: serum 7.3 g/dl.

12.3. Albumin/Globulins Ratio

Albumin/globulins ratio is used as a general marker of health and well being. The ideal ratio is 1.85 or higher. A low ratio usually means that you have a high amount of immunoglobulins. High serum immunoglobulins can indicate a long history of infections, which may increase the risk of developing autoimmune diseases, and indicates your history of inflammation derived exposure to free radicals. Human epidemiology studies have shown that high serum globulins are strongly associated with a person's risk of developing many of the age-related diseases such as cancer and heart disease.[63]

12.4. Magnesium

Magnesium is necessary for many biological functions such as RNA/DNA synthesis, protein synthesis, ADP synthesis and muscle contraction. Because it has a fixed outer electron valence of +2, it can inhibit many iron based enzymatic free radical generating reactions (e.g. oxidases, peroxidases) by displacing iron from its active binding site. Magnesium has been shown to be helpful in preventing heart disease.[64]

Commercial availablility: inductive coupled plasma mass spectrometer, e.g. Hewlett Packard; flame atomic absorption, e.g. Perkin Elmer.

Human normal range: serum 0.50–0.80 mmol/L.

Human healthy average: serum 0.75 mmol/L.

Method reference: Inagaki, K. and Haraguchi, H. (2000). Determination of rare earth elements in human blood serum by inductively coupled plasma mass spectrometry after chelating resin preconcentration. *Analyst* **125**(1): 191–196.

12.5. Selenium

Unbound selenium is known to be a strong prooxidant with similar outer valance electrons as oxygen. It is toxic at high levels. Small amounts of selenium are required for many biological functions such as glutathione peroxidase. Human epidemiology studies have shown that people low in selenium have a much greater risk for early development of age-related diseases.[65]

Commercial availablility: inductive coupled plasma mass spectrometer, e.g. Hewlett Packard; flame atomic absorption. e.g, Perkin Elmer.

Human normal range: serum 23–190 µg/L.

Human healthy average: serum 105 µg/L.

Method reference: Inagaki, K. and Haraguchi, H. (2000). Determination of rare earth elements in human blood serum by inductively coupled plasma mass spectrometry after chelating resin preconcentration. *Analyst* **125**(1): 191–196.

12.6. Sulfur

Sulfur is essential for protein structures and enzyme activity and is also required in many detoxification reactions. Higher than average levels have been shown to help lower your risk of developing cancer and many of the age-related diseases.[66]

Commercial availablility: inductive coupled plasma mass spectrometer, e.g. Hewlett Packard; flame atomic absorption, e.g. Perkin Elmer.

Human normal range: serum 0.8–3.5 µmol/L.

Human healthy average: serum 1.9 µmol/L.

Method reference: Inagaki, K. and Haraguchi, H. (2000). Determination of rare earth elements in human blood serum by inductively coupled plasma mass spectrometry after chelating resin preconcentration. *Analyst* **125**(1): 191–196.

12.7. Zinc

Zinc is thought to be an active part of more different enzymes than any other metal, including many proteins involved in metabolism, RNA polymerases and CuZnSOD. Because it has a fixed outer electron valence of +2, it can inhibit many iron based enzymatic free radical reactions by displacing iron from its active binding site. At normal levels zinc is essential for proper function and maintenance,[64] but at high levels, it can be toxic.

Commercial availablility: inductive coupled plasma mass spectrometer e.g. Hewlett Packard; flame atomic absorption e.g. Perkin Elmer.

Human normal range: serum 650–2910 µg/L.

Human healthy average: serum 900 µg/L.

Method reference: Inagaki, K. and Haraguchi, H. (2000). Determination of rare earth elements in human blood serum by inductively coupled plasma mass spectrometry after chelating resin preconcentration. *Analyst* **125**(1): 191–196.

13. Clinical Interpretation and Use of Oxidative Stress Biomarkers for Metabolic Optimization

Many of us take dietary antioxidant supplements like vitamin E. But how do we know if it is working and are we taking the most effective dosage? Taking dietary interventions blindly is much like thinking that your cholesterol levels are low because you have a low fat diet. As it is with lipid metabolism and risk for heart disease, there are a lot of variables that can affect your oxidative stress state. So, you do not really know how you are doing unless you are tested.

The type of damage that a free radical produces depends on its origin of evolution and the type of free radical (e.g. peroxides will react faster with lipophilic fats and amino acids than they will with hydrophilic substrates). Therefore, the most accurate and clinically relevant measurement of oxidative damage is to measure many different types of damage from many different origins within a cell and then put them all together as a profile. Each antioxidant varies in the location and types of free radicals that it can neutralize based on its structure and cellular absorption (e.g. phenol antioxidant receptors).

One of the best examples for the oxidative stress profiling approach can be seen in how physicians now use a patient's lipid profile as the most accurate determination for risk of cardiovascular disease (CVD). Cholesterol alone is actually not a very strong predictor of a persons risk of CVD with a correlation with of below 30%. However, when cholesterol values are combined with other clinical measurements, its risk predicting ability can reach up to a 87% rate of accuracy.[67] Our recommendation for the use of these oxidative stress state tests is to first look at a few of the oxidative damage biomarkers to see if there is a serious problem. Then look at the prooxidants because one of the most common sources of free radicals that we have seen in patients is iron overload, which affects about 10% of the population, particularly elderly men. Other common sources of free radicals are inflammation (i.e. C-reactive protein) and high homocysteine. We have also seen many patients that have heavily supplemented their diet with a single antioxidant like β-carotene, which displaces and causes a deficiency of many of the other beneficial carotenoids like lycopene and lutein. The blood serum profiles of patients who supplement their diet heavily with α-tocopherol also frequently show an unbalanced profile deficient in the other important tocopherols like γ-tocopherol.

Our bodies are genetically programmed to synthesize and absorb a variety of antioxidants from our diet in order to maintain a low oxidative stress state. Many of these protective factors do in fact overlap each other, but not completely as seen with the degenerative effects caused by the deficiency of protective agents like selenium and vitamin E. We believe that each species has a genetic set point for its oxidative stress state that is not easily overcome by mega doses of dietary antioxidants. Though few studies have been able to show that such mega dosing will result in an extension of maximum lifespan, it does seem to extend the average lifespan closer to the species' genetically programmed maximum lifespan potential. Therefore, the greatest clinical benefit of measuring a person's oxidative stress state would be to help them peak-out their maximum health and lifespan potential. We conclude by stressing the point that oxidative damage is part of the degenerative processes involved in age-related diseases. However, it seems possible to use oxidative stress profiling as a tool to lower a patient's oxidative stress state and to increase a patient's antioxidant reserve capacity in order to significantly increase healthy years of life.

References

1. National Vital Statistics Report Volume 48, Number 11; final mortality for 1998. http://www.cdc.gov/nchs/fastats/deaths.html
2. Bafitis, H. and Sargent, F. (1977). Human physiological adaptability through the life sequence. *J. Gerontol.* **32**: 402–410.
3. Kohn, R. R. (1982). Cause of death in very old people. *J. Am. Med. Assoc.* **247**: 2793–2797.
4. Maynard, S. J. (1966). Theories of aging. *In* "Topics in the Biology of Aging", (P. L. Krohn, Ed.), pp. 1–35, Interscience, New York.
5. National Center of Health Statistics, USPHS & US Bureau of the Census, "Some Demographic Aspects of Aging in the United States", February 1973.
6. Migliaccio, E. *et al.* (1999). The p66shc adaptor protein controls oxidative stress response and life span in mammals. *Nature* **402**(6759): 309–313.
7. Bartke, A. *et al.* (1998). Does growth hormone prevent or accelerate aging? *Exp. Gerontol.* **33**(7–8): 675–687.
8. Coschigano, K. T. *et al.* (2000). Assessment of growth parameters and life span of GHR/BP gene-disrupted mice. *Endocrinology* **141**(7): 2608–2613.
9. Parkes, T. L. *et al.* (1998). Extension of *Drosophila* lifespan by over expression of human SOD-1 in motorneurons. *Nat. Genet.* **19**(2): 171–174.

10. Phillips, J. P. *et al.* (2000). Targeted neuronal gene expression and longevity in *Drosophila. Exp. Gerontol.* **35**(9–10): 1157–1164.

11. Schmidt, P. S. *et al.* (2000). Adaptive evolution of a candidate gene for aging in *Drosophila. Proc. Natl. Acad. Sci. USA* **97**(20): 10 861–10 865.

12. Rogina, B. *et al.* (2000). Extended life-span conferred by cotransporter gene mutations in *Drosophila. Science* **290**(5499): 2137–2140.

13. Johnson, T. E. *et al.* (2000). Gerontogenes mediate health and longevity in nematodes through increasing resistance to environmental toxins and stressors. *Exp. Gerontol.* **35**(6–7): 687–694.

14. Wolkow, C. A. *et al.* (2000). Regulation of *Caenorhabditis elegans* life-span by insulin-like signaling in the nervous system. *Science* **290**(5489): 147–150.

15. Mutations in the clk-1 gene of *Caenorhabditis elegans* affect developmental and behavioral timing. *Genetics* **139**(3): 1247–1259.

16. Cypser, J. R. and Johnson, T. E. (1999). The spe-10 mutant has longer life and increased stress resistance. *Neurobiol. Aging* **20**(5): 503–512.

17. Lithgow, G. J. *et al.* (1995). Thermotolerance and extended life-span conferred by single-gene mutations and induced by thermal stress. *Proc. Natl. Acad. Sci. USA* **92**(16): 7540–7544.

18. Rikke, B. A. *et al.* (2000). Paralogy and orthology of tyrosine kinases that can extend the life span of *Caenorhabditis elegans. Mol. Biol. Evol.* **17**(5): 671–683.

19. Steinberg, R. *et al.* (1989). Beyond Cholesterol. Modifications of low-density lipoprotein that increase its atherogenicity. *New England J. Med.* **320**: 915–924.

20. Jiang, Z., Woollard, A. and Wolff, S. (1990). Hydrogen peroxide production during experimental protein glycation. *FEBS* **268**(1): 69–71.

21. Chajes, V. *et al.* (1996). Photometric evaluation of lipid peroxidation products in human plasma and copper oxidized low-density lipoproteins: correlation of different oxidation parameters. *Atherosclerosis* **121**(2): 193–203.

22. Osawa, T. *et al.* (1995). Protective role of dietary antioxidants in oxidative stress. In *"Oxidative Stress and Aging"* (R. Cutler, L. Packer, J. Bertram, and A. Mori, Eds.), pp. 367–377.

23. Shigenaga, M. and Ames, B. (1991). Assays for 8-hydroxy-2'-deoxyguanosine: a biomarker of in vivo oxidative DNA damage. *Free Radic. Biol. Med.* **10**: 211–216.

24. Obata, T. *et al.* (2000). Smoking and oxidant stress: assay of isoprostane in human urine by gas chromatography-mass spectrometry. *J. Chromat. Biomed. Sci. Appl.* **746**(1): 11–15.

25. Kurasawa, Y. *et al.* (1978). Color reaction of cholesterol with trichloracetic acid and antimony trichloride. On the reaction mechanism. *Steroids* **31**(2): 163–174.

26. Barchowsky, A. *et al.* (1999). Stimulation of reactive oxygen, but not reactive nitrogen species, in vascular endothelial cells exposed to low levels of arsenite. *Free Radic. Biol. Med.* **27**(11–12): 1405–1412.

27. Rader, D. J. (2000). Inflammatory markers of coronary risk. *New England J. Med.* **343**(16): 1179–1182.
28. Almazan, G. *et al.* (2000). Exposure of developing oligodendrocytes to cadmium causes HSP72 induction, free radical generation, reduction in glutathione levels, and cell death. *Free Radic. Biol. Med.* **29**(9): 858–869.
29. Hojo, Y. *et al.* (2000). In vivo singlet-oxygen generation in blood of chromium (VI)-treated mice: an electron spin resonance spin-trapping study. *Biol. Trace Element Res.* **76**(1): 85–93.
30. Vile, G. F. *et al.* (2000). Initiation of rapid, P53-dependent growth arrest in cultured human skin fibroblasts by reactive chlorine species. *Arch. Biochem. Biophys.* **377**(1): 122–128.
31. Salonen, J. *et al.* (1991). Serum copper and the risk of acute myocardial infarction: a prospective population study in men in Eastern Finland. *Am. J. Epidemiol.* **134**: 268–276.
32. Sempos, C. T. *et al.* (2000). Serum ferritin and death from all causes and cardiovascular disease: the NHANES II Mortality Study. National Health and Nutrition Examination Study. *Ann. Epidemiol.* **10**(7): 441–448.
33. Aronow, W. S. *et al.* (2000). Increased plasma homocysteine is an independent predictor of new atherothrombotic brain infarction in older persons. *Am. J. Cardiol.* **86**(5): 585–586, A10.
34. Salonen, J. *et al.* (1992). High stored iron levels are associated with excess risk of myocardial infarction in Eastern Finnish men. *Circulation* **86**: 803–811.
35. Ding, Y. *et al.* (2000). Lead promotes hydroxy radical generation and lipid peroxidation in cultured aortic endothelial cells. *Am. J. Hypertens.* **13(Part 1)**(5): 552–555.
36. Sarafian, T. A. (1994). bcl-2 expression decreases methyl mercury-induced free-radical generation and cell killing in neural cell line. *Toxic. Lett.* **74**(2): 149–155.
37. Novelli, E. L. *et al.* (1995). Superoxide radical and toxicity of environmental nickel exposure. *Human Exp. Toxicol.* **14**(3): 248–251.
38. Asayama, K. *et al.* (1993). Antioxidants in the serum of children with insulin-dependent diabetes mellitus. *Free Radic Biol. Med.* **15**: 597–602.
39. Kagan, V. E., Sebinova, E. A. and Packer, L. (1990). Generation and recycling of radicals from phenolic antioxidants. *Arch. Biochem. Biophys.* **280**(1): 33–39.
40. Ninfali, P. and Aluigi, G. (1998). Variability of oxygen radical absorbance capacity (ORAC) in different animal species. *Free Radic. Res.* **29**(5): 399–408.
41. Gillum, R. F. *et al.* (1996). Serum transferrin saturation, stroke incidence, and mortality in women and men. The NHANES I Epidemiologic Follow-up Study. National Health and Nutrition Examination Survey. *Am. J. Epidemiol.* **144**(1): 59–68.
42. Pogosian, G. *et al.* (1983). Inhibition of lipid peroxidation by superoxide dismutase and ceruloplasmin. *Biokhimiia* **48**: 7, 1129–1134.

43. Burkitt, M. J. and Gilbert, B. C. (1990). Model studies of the iron-catalyzed Haber–Weiss cycle and the ascorbate-driven Fenton reaction. *Free Radical Res. Commun.* **10**(4–5): 265–280.
44. Enstrom, J., Kanim, L. and Klein, M. (1992). Vitamin C intake and mortality among a sample of the United States population. *Epidemiology* **3**: 194–202.
45. Madhavan, M. *et al.* (1997). Serum bilirubin distribution and its relation to cardiovascular risk in children and young adults. *Atherosclerosis* **131**(1): 107–113.
46. Li, X., Song, L. and Jope, R. (1998). Glutathione depletion exacerbates impairment by oxidative stress of phosphoinositide hydrolysis, AP-1, and NF-kappa B activation by cholinergic stimulation. *Brain Res. Mol. Brain Res.* **53**(1–2): 196–205.
47. Neito, F. J., Iribarren, C., Gross, M. D., Comstock, G. W. and Cutler, R. C. (2000). Uric acid and serum antioxidant capacity: a reaction to atherosclerosis? *Atherosclerosis* **148**(1): 131–139.
48. Davis, J. *et al.* (1996). Observations on serum uric acid levels and the risk of idiopathic Parkinson's disease. *Am. J. Epidemiol.* **144**(5): 480–484.
49. Cutler, R. G. (1985). Antioxidants and longevity of mammalian species. *Basic Life Sci.* **35**: 15–73.
50. Cutler, R. G. (1991). Antioxidants and aging. *Am. J. Clin. Nutri.* **53(1 Suppl)**: 373S–379S.
51. Cutler, R. G. (1984). Carotenoids and retinol: their possible importance in determining longevity of primate species. *Proc. Natl. Acad. Sci. USA* **81**(23): 7627–7631.
52. Levy, J. *et al.* (1995). Lycopene is a more potent inhibitor of human cancer cell proliferation than either alpha-carotene or beta-carotene. *Nutri. Cancer* **24**(3): 257–266.
53. Woodson, K. *et al.* (1999). Association between alcohol and lung cancer in the alpha-tocopherol, beta-carotene cancer prevention study in Finland. *Cancer Causes Control* **10**(3): 219–226.
54. Dorgan, J. *et al.* (1998). Relationships of serum carotenoids, retinol, alpha-tocopherol, and selenium with breast cancer risk: results from a prospective study in Columbia, Missouri (United States). *Cancer Causes Control* **9**(1): 89–97.
55. Beatty, S. *et al.* (2000). The role of oxidative stress in the pathogenesis of age-related macular degeneration. *Survey Ophthalmol.* **45**(2): 115–134.
56. Khachik, F., Beecher, G. and Smith, J. (1995). Lutein, lycopene, and their oxidative metabolites in chemoprevention of cancer. *J. Cell Biochem. Suppl.* **22**: 236–246.
57. Clinton, S. *et al.* (1996). Cis-trans lycopene isomers, carotenoids, and retinol in the human prostate. Cancer Expedition. *Biomarkers Prev.* **5**(10): 823–833.
58. Packer, L. (1991). Protective role of vitamin E in biological systems. *Am. J. Surgery* **161**: 488–503.

59. Gey, K. *et al.* (1991). Inverse correlation between plasma vitamin E and mortality from ischemic heart disease in cross-cultural epidemiology. *Am. J. Clin. Nutri.* **53**: 326S–334S.

60. Beal, M. and Matthews, R. (1997). Coenzyme Q10 in the central nervous system and its potential usefulness in the treatment of neurodegenerative diseases. *Mol. Aspects Med.* **18(suppl)**: S169–S179.

61. Hammond, B., Wooten, B. and Snodderly, D. (1997). Density of the human lens is related to macular pigment carotenoids, lutein and zeaxanthin. *Optom. Vis. Sci.* **74**(7): 499–504.

62. Laban, A. *et al.* (1997). Iron-binding antioxidant potential of plasma albumin. *Clin. Sci. (Colch).* **93**(5): 445–451.

63. Mayr, M. *et al.* (2000). Infections, immunity, and atherosclerosis: associations of antibodies to Chlamydia pneumoniae, Helicobacter pylori, and cytomegalovirus with immune reaction to heat-shock protein 60 and carotid or femoral atherosclerosis. *Circulation* **102**(8): 833–839.

64. Chipperfield, B. and Chipperfield, J. R. (1973). Heart-muscle magnesium, potassium, and zinc concentrations after sudden death from heart disease. *Lancet* **2**(7824): 293–296.

65. Comstock, G. W. *et al.* (1997). The risk of developing lung cancer associated with antioxidants in the blood: ascorbic acid, carotenoids, alpha-tocopherol, selenium, and total peroxyl radical absorbing capacity. *Cancer Epidemiol. Biomarkers Prev.* **6**(11): 907–916.

66. Ogasawara, Y. *et al.* (1999). Antioxidant effects of albumin-bound sulfur in lipid peroxidation of rat liver microsomes. *Biol. Pharm. Bull.* **22**(5): 441–415.

67. Gey, K. F. *et al.* (1991). Inverse correlation between plasma vitamin E and mortality from ischemic heart disease in cross-culture epidemiology. *Am. J. Clin. Nutri.* **53(1 suppl)**: 326S–334S.

Chapter 9

Oxidative Damage to DNA: Mechanisms of Product Formation and Measurement by Mass Spectrometric Techniques

Miral Dizdaroglu*, Pawel Jaruga and Henry Rodriguez

Miral Dizdaroglu, Pawel Jaruga and **Henry Rodriguez** • Chemical Science and Technology Laboratory, National Institute of Standards and Technology, Bldg. 227/A243, Gaithersburg, Maryland, 20899-8311, USA

*Corresponding Author.
Tel: 301-975-2581, E-mail: miral@nist.gov

1. Introduction

Free radicals are produced in living cells by normal metabolism and by exogenous sources such as ionizing radiation, redox-cycling drugs and carcinogenic compounds [reviewed in Ref. 1]. Free radicals, most notably highly reactive hydroxyl (OH) radicals cause oxidative damage to DNA and generate multiple products including modified bases and sugars, strands breaks and DNA-protein crosslinks by a variety of mechanisms [reviewed in Refs. 2 and 3]. This type of damage is implicated to play a role in carcinogenesis and other age-related diseases.[1] Measurement of oxidative damage to DNA is essential for understanding the mechanisms, repair and biological consequences of this type of damage. There are various analytical techniques for this purpose with their own advantages and drawbacks [reviewed in Refs. 4 and 5]. Most of them measure only one product with no spectroscopic evidence. At present, there is no consensus between laboratories in terms of the measurement of oxidative damage to DNA, especially in terms of the measurement of 8-hydroxy-2′-deoxyguanosine (8-OH-dGuo) and its endogenous levels in cells.[6, 7] Techniques with mass spectrometry provide unequivocal identification and quantification of DNA products.[5] GC/MS has been used for over a decade for this purpose. DNA base and sugar lesions and DNA-protein crosslinks were identified and quantified in cells and *in vitro*.[5] Recently, LC/MS/MS emerged as a new technique for the measurement of oxidative damage to DNA. First, it was exclusively used for the measurement of 8-OH-dGuo.[8, 9] Subsequently, several other products were also measured.[10, 11] Just recently, LC/MS was applied to the measurement of 8-OH-dGuo, and 8,5′-cyclo-2′-deoxyadenosine (8,5′-cdAdo).[12, 13] This article reviews mechanistic aspects of oxidative damage to DNA and its measurement by techniques that use mass spectrometry.

2. Mechanisms of Oxidative Damage to DNA

Hydroxyl (OH) radical reacts with organic compounds at or near diffusion-controlled rates by addition to double bonds and by abstraction of H atoms from C–H bonds [reviewed in Ref. 14]. In the case of DNA, it adds to double bonds of heterocyclic bases at second-order rate constants ranging from 3×10^9 to 10×10^9 M^{-1} s^{-1} and abstracts an H atom from each of the five carbon atoms of 2′-deoxyribose at rate constants of $\approx 2 \times 10^9$ $M^{-1} s^{-1}$.[14] An H atom from the methyl group of thymine is also abstracted leading to the allyl radical. Addition reactions yield OH-adduct radicals of DNA bases. Carbon-centered sugar radicals are formed from abstraction reactions. Further reactions of these radicals lead to the formation of a variety of products. This type of free radical-induced DNA damage is also called oxidative damage to DNA.

In the presence of oxygen, OH-adduct radicals of DNA bases and sugar radicals react with oxygen at diffusion-controlled rates to give peroxyl radicals. Some of purine adduct radicals might not react with oxygen.[14, 15] There are many types of reactions that DNA base and sugar moieties undergo. It is beyond the scope of this article to present all mechanistic aspects of free radical-induced DNA damage. A few representative reactions will be discussed. Further information on this broad subject can be obtained from various original and review articles in the literature.

Addition of OH radicals to the C5– and C6–positions of cytosine gives rise to C5–OH– and C6–OH–adduct radicals, respectively. Figure1 illustrates the product formation from the C5–OH–adduct radical. Oxidation followed by addition of OH^- (or addition of water followed by deprotonation) yields cytosine glycol.[2, 3, 14] In the presence of oxygen, C5–OH–adduct radicals react with oxygen at diffusion-controlled rates to give 5-hydroxy-6-peroxyl radicals, which may eliminate $O_2^{\cdot-}$ followed by reaction with water (addition of OH^-). This reaction also leads to cytosine glycol. C6–OH–adduct radicals undergo analogous reactions. Similarly, oxidation of thymine OH-adduct radicals yields thymine glycol. Two thymine-specific products, 5-(hydroxymethyl)uracil and 5-formyluracil, result from oxidation of the allyl radical of thymine.[2, 3, 14] Reduction of 5–OH– and 6–OH–adduct radicals of pyrimidines in the absence of oxygen followed by protonation causes formation of 5-hydroxy-6-hydro- and 6-hydroxy-5-hydropyrimidines, respectively, as shown in Fig. 1. One unique reaction of cytosine products is that they are readily converted into corresponding uracil derivatives by deamination depending on reaction conditions. Thus, cytosine glycol yields 5-hydroxycytosine by dehydration, uracil glycol by deamination, and 5-hydroxyuracil by deamination and dehydration (Fig. 1).[2, 3, 16–18] Other uracil derivatives such as 5,6-dihydrouracil, isodialuric acid (or 5,6-dihydroxyuracil) and 5-hydroxy-6-hydrouracil are also formed by deamination of corresponding cytosine products.[3, 19, 20] Alloxan found in DNA using DNA repair enzymes is a product of cytosine and can decarboxylate depending on reaction conditions to give rise to 5-hydroxyhydantoin.[17, 19] Ring-reduction reactions of thymine OH-adduct radicals also occur to yield 5-hydroxy-5-methylhydantoin.[2, 3, 20]

Addition of OH radical to purines yields C4–OH–, C5–OH– and C8–OH–adduct radicals. The first two radicals dehydrate giving rise to oxidizing purine($-H^\cdot$) radicals, which might be reduced and protonated to reconstitute purines.[15] On the other hand, C8–OH–adduct radicals undergo one-electron oxidation to give 8-hydroxypurines (7,8-dihydro-8-oxopurines), and one-electron reduction to form formamidopyrimidines.[2, 3, 15] These reactions are illustrated in Fig. 2 in the case of guanine. Formation of formamidopyrimidines might occur by opening of the imidazole ring followed by reduction or vice versa. 8-Hydroxypurines and formamidopyrimidines are formed both in the absence and presence of oxygen. Formation of 8-hydroxypurines is preferred in the

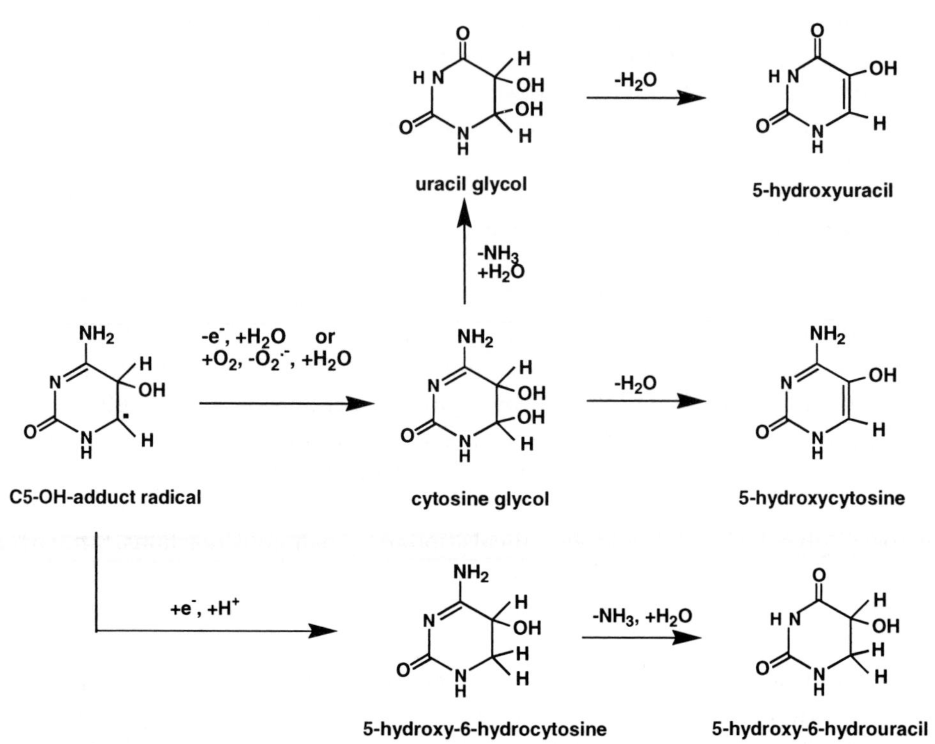

Fig. 1. Mechanism of formation of products from the C5–OH–adduct radical of cytosine.

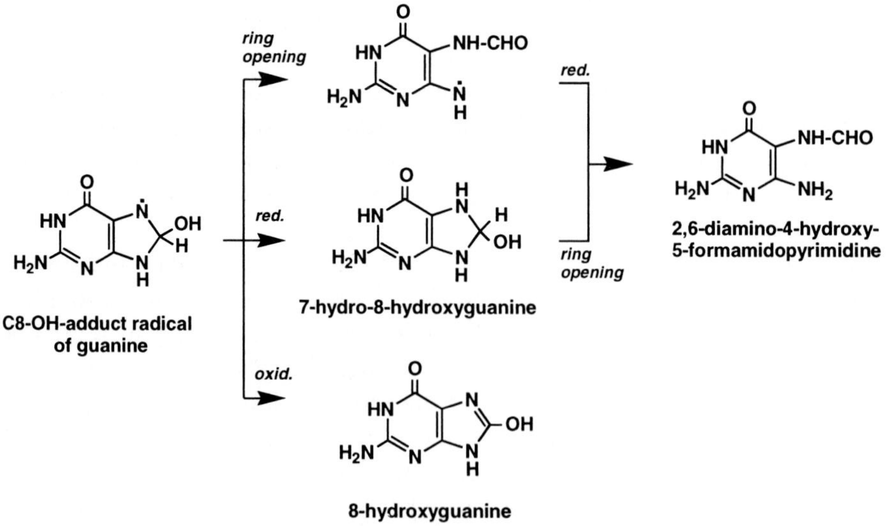

Fig. 2. Mechanism of formation of products from the C8–OH–adduct radical of guanine [adapted from Ref. 15].

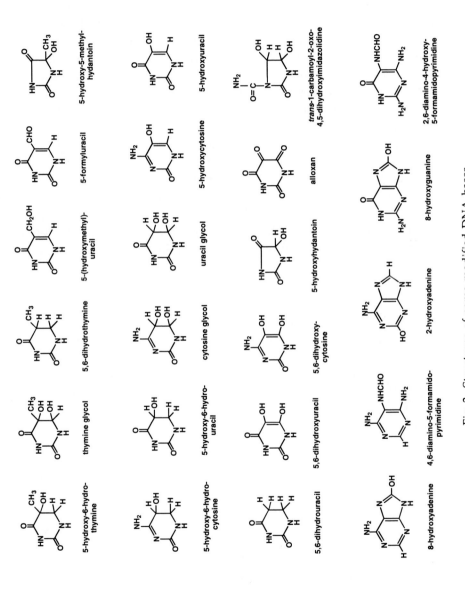

Fig. 3. Structures of some modified DNA bases.

presence of oxygen and that of formamidopyrimidines in the presence of reducing agents.[2, 3]

Reactions of OH radical with pyrimidines and purines yield multiple products in DNA. Structures of some of the major base-derived products in DNA are illustrated in Fig. 3. Many of these lesions have been identified in mammalian cells exposed to free radical-generating systems [reviewed in Ref. 5]. Reactions of carbon-centered sugar radicals lead to DNA strand breaks and a number of sugar products.[14] One unique reaction of the C5'-centered radical of purine nucleosides is addition to the C8-position of the purine ring of the same nucleoside followed by oxidation. This leads to the formation of 8,5'-cyclopurine-2'-deoxynucleosides, which represent a concomitant damage to both the sugar and base moieties of the same nucleoside.[21-24] As an example, Fig. 4 illustrates the formation of 8,5'-cyclopurine-2'-deoxyadenosine from 2'-deoxyadenosine. Both 5'R- and 5'S-diastereomers are formed in DNA.[24] 2'-deoxyguanosine in DNA undergoes analogous reactions to give rise to 5'R- and 5'S-8,5'-cyclopurine-2'-deoxyguanosines.[23] Oxygen inhibits the formation of these compounds because of its diffusion-controlled reaction with the C5'-centered sugar radical.[21-23]

Another type of free radical-induced DNA damage results from DNA-protein crosslinking.[25] Proposed mechanisms involve the addition of a DNA radical to an

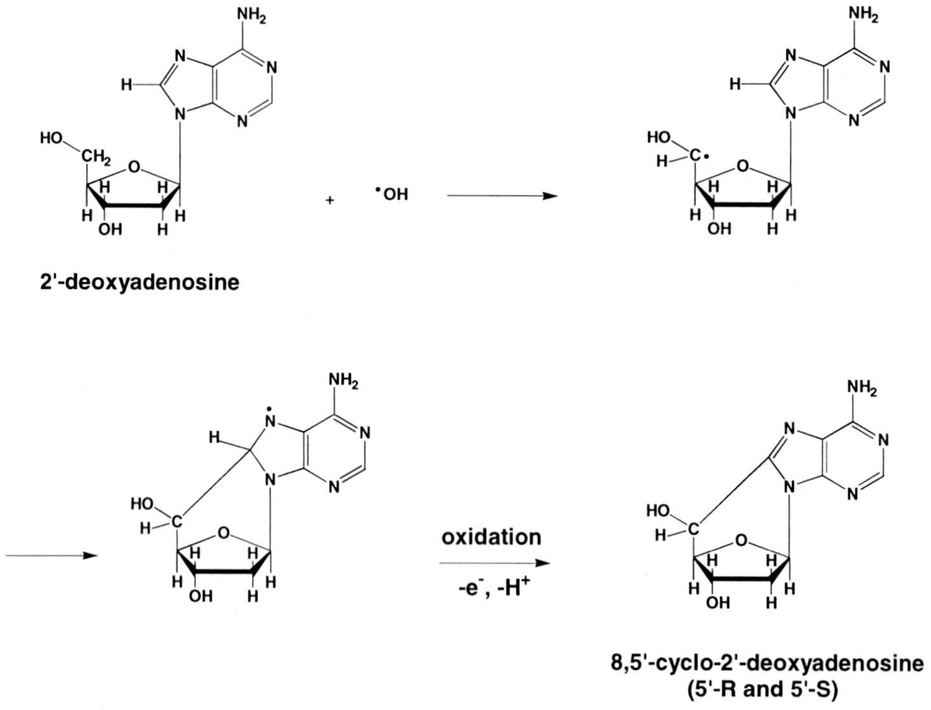

Fig. 4. Mechanism of formation of 8,5'-cyclo-2'-deoxynucleosides.

aromatic amino acid in proteins or vice versa. Crosslinking by combination of DNA and protein radicals might also occur. To this end, a thymine-tyrosine crosslink was identified in mammalian chromatin *in vitro* and in cells exposed to free radical-generating systems.[21, 26, 27] Its formation mechanism has been proposed to involve the addition of the allyl radical of thymine in DNA to the C3-position of the tyrosine ring in a protein followed by oxidation.

3. Measurement of Oxidative Damage to DNA

Oxidative damage to DNA can be measured by a variety of analytical techniques, including immunochemical techniques, postlabeling assays, comet assay, alkaline elution with the use of DNA repair enzymes, high-performance liquid chromatography (HPLC) with electrochemical detection (ECD), gas chromatography/mass spectrometry (GC/MS), liquid chromatography/mass spectrometry (LC/MS) and liquid chromatography/tandem mass spectrometry (LC/MS/MS). It is beyond the scope of this article to review the vast literature in the field of measurement by the techniques mentioned above. Some recent developments in this field will be briefly reviewed. The techniques without MS generally detect one single product of DNA bases at a time without providing any structural evidence. Some techniques such as the comet assay[28] detect global damage to cells only and are of no use for mechanistic studies of DNA damage and for identification and quantification of DNA products.

Of the multiple products generated by free radicals in DNA (Fig. 3), 8-hydroxyguanine (8-OH-Gua, also called 8-oxoGua) has been extensively investigated, because of its mutagenic properties and the availability of a method using HPLC-ECD for the measurement of its nucleoside form 8-OH-dGuo following enzymic hydrolysis of DNA [reviewed in Refs. 4, 5 and 29)]. This product is one of multiple DNA products and is formed by addition of OH radical to the carbon-8 of guanine followed by oxidation of the thus-formed C8–OH–adduct radical (Fig. 2). The latter can also be reduced to yield 2,6-diamino-4-hydroxy-5-formamidopyrimidine (FapyGua).[2, 3, 15] Adenine undergoes analogous reactions to yield 8-hydroxyadenine and 4,6-diamino-5-formamidopyrimidine (FapyAde). Thus the relative yields of 8-hydroxypurines and formamidopyrimidines as well as those of other products might depend on experimental conditions, the redox status of cells and the availability of metal ions.[1, 2] These facts strongly suggest that it might be misleading to measure a single product such as 8-OH-Gua or 8-OH-dGuo for use as a biomarker and/or for determination of the rate of DNA damage. This also means that changes in the yield of a single product might not necessarily reflect the overall rate of DNA damage. A wealth of previous data demonstrated the advantage of measuring multiple DNA products.[5, 30, 31] Moreover, the results from the recent trials by the European Standards Committee for Oxidative damage to DNA (ESCODD) and from other studies showed that the levels of 8-OH-Gua or 8-OH-dGuo measured in various laboratories by a variety of techniques significantly differed from one another, suggesting a laboratory- and technique-dependent variability in measurements of this product.[6, 7, 32]

3.1. Gas Chromatography/Mass Spectrometry

Unlike other techniques, GC/MS is capable of the measurement of base products from all four DNA bases (Fig. 3), 8,5'-cyclopurine-2'-deoxynucleosides (Fig. 4), products of the sugar moiety and DNA-protein crosslinks [reviewed in Ref. 5]. Moreover, this capability of GC/MS permits the measurement of cellular DNA repair, and measurement of substrate specificities and excision kinetics of DNA repair enzymes in terms of a variety of products [reviewed in Ref. 33]. The measurement by GC/MS of multiple DNA products might prevent misleading conclusions drawn from the measurement of a single product as done by other techniques such as HPLC-ECD. For GC/MS analysis, DNA is first hydrolyzed by acid to modified and intact bases or by endo- and exo-nucleases to modified and intact nucleosides. The use of DNA repair enzymes such as *E. coli* Fpg and Nth proteins to release modified bases from DNA prior to GC/MS was also described.[34, 35] Hydrolysates are derivatized by trimethylsilylation and analyzed. Electron-ionization mass spectra of trimethylsilyl derivatives of modified bases and nucleosides yield characteristic mass spectra that can be used for unequivocal identification.[36, 37] Quantitative measurements are best achieved by isotope-dilution mass spectrometry (IDMS) using stable isotope-labeled analogues of modified bases or nucleosides as internal standards.[19, 38]

3.1.1. *Artifacts and Facts*

Recently, an artifactual formation of five modified bases from the corresponding intact DNA bases was alleged to occur during derivatization at high temperature of DNA hydrolysates prior to GC/MS analysis.[39] However, it was not mentioned that hydrolysis and derivatization procedures were significantly different from those previously described and that the artifactual formation was restricted to only a few modified bases.[39] Furthermore, most relevant papers, which reported the levels of modified bases in DNA of various sources, were ignored. Prepurification of acid-hydrolysates using a tedious and time consuming HPLC procedure was proposed to remove intact bases from acid-hydrolysates of DNA, and thus to avoid an artifactual formation during derivatization. The affected products were 8-OH-Gua, 5-hydroxycytosine (5-OH-Cyt), 8-hydroxyadenine (8-OH-Ade), 5-hydroxymethyluracil (5-OHMeUra) and 5-formyluracil. These constitute only a few of those that can be measured by GC/MS (Fig. 3).[5, 33, 40] We discussed the facts about the artifacts and presented additional data on this subject in recently published studies.[5, 32, 41, 42]

We conducted studies to determine whether derivatization conditions generate an artifactual formation of the aforementioned lesions from corresponding intact bases in DNA hydrolysates. For this purpose, calf thymus DNA was freed of possible metal ions to avoid artifacts and trimethylsilylation of acid-hydrolysates of DNA samples was carried out at three different temperatures.[32] As examples, Fig. 5 illustrates the levels of 5-OH-Cyt and 8-OH-Gua found in DNA by GC/IDMS following derivatization at three different temperatures. These data also include those obtained with the use of trifluoroacetic acid (TFA) because of its reported use to dissolve guanine and 8-OH-Gua in derivatization mixtures.[43] On the other hand, TFA was reported to adversely affect the measurement of pyrimidine-derived products.[44] Figure 5 clearly shows that the levels of 5-OH-Cyt were not affected by derivatization temperature ranging from 23°C to 120°C. Neither were those of 8-OH-Ade and 5-OH-Ura (not shown).[32] In the case of 8-OH-Gua, similar levels were observed at 23°C and 60°C; however, the level that was found at 120°C was approximately 2.5-fold greater. TFA increased the level of 5-OH-Cyt [also that of 5-OH-Ura (not shown)],[32] confirming its adverse effect on the pyrimidine-derived products as reported previously.[44] However, it did not affect the levels of 8-OH-Ade and 8-OH-Gua at 23°C and 60°C. No detectable levels of 5-OHMeUra were found, meaning that there was no artifactual formation of this compound under the conditions used. 5-formyluracil had not been measured in our laboratory. Thus it was not discussed.

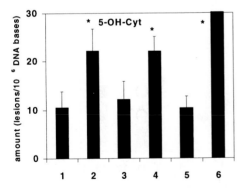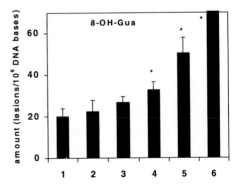

Fig. 5. Levels of modified DNA bases in calf thymus DNA obtained by derivatization of DNA hydrolysates at different temperatures.[32] 1: 23°C, 2: 23°C plus TFA, 3: 60°C, 4: 60°C plus TFA, 5: 120°C, 6: 120°C plus TFA. The stars denote statistical difference from the value to the left. In the case of 8-OH-Gua, the star on column 5 denotes the statistical difference from the values in columns 1 and 3. The values represent the mean ± standard deviation from 10 independent measurements. The values that are out of scale are 58 ± 12 (5-OH-Cyt) and 155 ± 32 (8-OH-Gua).

Levels of 5-OH-Cyt, 8-OH-Ade and 8-OH-Gua found in this work were compared with those reported using calf thymus DNA or DNA from various other sources.[32] As an example, Fig. 6 illustrates the comparison of the levels of 5-OH-Cyt with the data previously published. The levels at 23°C and 120°C obtained in our laboratory were shown, since there was no significant difference between the values obtained by derivatization at 23°C and 60°C. The level of 5-OH-Cyt is 7-fold smaller than that obtained by derivatization at 110°C and even 2.5-fold smaller than that after the so-called prepurification of DNA hydrolysates.[39, 45] This comparison suggests that improper conditions including insufficient removal of oxygen might have been used during derivatization in the previous work. The vast majority of previously published values shown in Fig. 6 are smaller than that obtained following prepurification of DNA hydrolysates.[39, 45] Nevertheless, previous work did not mention this fact and did not reference to corresponding papers, either. A similar trend was observed in the case of 8-OH-Gua.[32] Very high level of 8-OH-Gua observed in previous work[39, 45] was likely due to insufficient removal of oxygen and/or other improper experimental conditions. Such a high level of 8-OH-Gua had not been previously observed. These facts are also true for 8-OH-Ade.[32] In conclusion, these results showed that an increase in derivatization temperature from 23°C to 120°C did not increase the levels of 5-OH-Cyt, 8-OH-Ade and 5-OHMeUra in DNA under our experimental conditions, indicating that these compounds were not formed artifactually during derivatization. This is in contrast to recent claims that these modified bases were artifactually produced during derivatization. On the other hand, the level of

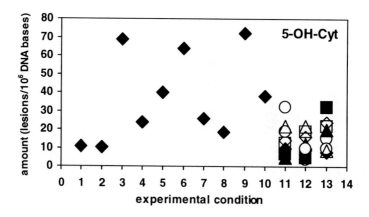

Fig. 6. Levels of 5-OH-Cyt in calf thymus DNA and in DNA of various other sources, obtained by derivatization of hydrolysates under various conditions, 1: at 23°C,[32] 2, at 120°C,[32] 3: at 110°C,[45] 4: at 110°C following prepurification of DNA hydrolysates,[39, 45] 5 and 6: at 23°C and 90°C, respectively,[44] 7 and 8: at 23°C and 23°C plus ethanethiol, respectively,[44] 9 and 10: at 90°C and 90°C plus ethanethiol, respectively,[44] 11 to 13: in DNA of various sources [see references in Ref. 32].

8-OH-Gua increased by an increase in derivatization temperature, indicating an artifact formation of this compound during derivatization. However, its level observed by derivatization at 120°C was much lower than the levels reported previously. These facts indicate that the artifactual formation of 8-OH-Gua can be prevented by derivatization at room temperature for at least 2 hours. The results also showed that a tedious and time-consuming prepurification procedure[39] is not necessary to avoid artifacts during derivatization.

3.1.2. *Use of DNA Repair Enzymes Instead of Acid for Measurements by GC/MS*

DNA repair enzymes such as Fpg and Nth proteins of *E. coli* can be used instead of formic acid to release modified bases from DNA prior to GC/MS analysis. The rationale is that acidic hydrolysis releases modified DNA bases as well as intact bases. The latter may be oxidized during derivatization to yield additional modified bases such as 8-OH-Gua. In contrast, DNA repair enzymes solely release modified bases and no intact bases, excluding a possible artifactual formation of modified bases during derivatization. Recently two papers were published exploring this approach for GC/MS measurements.[34, 35] This approach appears to be more suitable for determination of background levels of modified bases. At higher level of damage, however, the enzymes might not completely excise modified bases.[33] In our laboratory, we studied the measurement of 8-OH-Gua levels in DNA by GC/MS following formic acid hydrolysis or Fpg protein hydrolysis.[35] The aim was to compare two different hydrolysis methods and explore whether derivatization of acid-hydrolysates of DNA generate additional 8-OH-Gua. In the case of Fpg protein hydrolysis, 8-OH-Gua should not be generated because of the absence of free guanine in hydrolysates. Such a comparison should reveal whether the derivatization of acid-hydrolysates of DNA generates 8-OH-Gua from guanine under a given set of conditions. Derivatization was done at room temperature for 2 h, which reportedly prevents artifactual formation of 8-OH-Gua from guanine in acid-hydrolysates of DNA as was outlined above and elsewhere.[43, 44] Figure 7 illustrates levels of 8-OH-Gua in calf thymus DNA measured by GC/IDMS following hydrolysis by either Fpg protein or formic acid under a variety of experimental conditions. Active Nth protein, which does not excise 8-OH-Gua, was also used as a control. The levels of 8-OH-Gua measured using two different hydrolysis procedures were similar with no statistical difference (Fig. 7). These results showed that (1) under the experimental conditions used, 8-OH-Gua was not formed artifactually from free guanine in formic acid-hydrolysates of DNA during derivatization at room temperature, and (2) Fpg protein can remove most if not all 8-OH-Gua residues present in DNA at background levels. This means that it may be used instead of formic acid to determine background levels of 8-OH-Gua in DNA by GC/MS. It should be

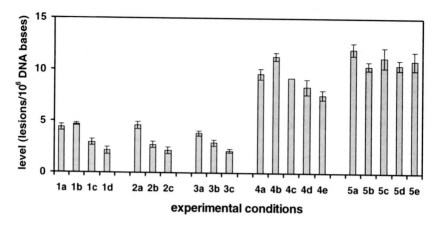

Fig. 7. Levels of 8-OH-Gua in calf thymus DNA as measured by GC/IDMS following hydrolysis by formic acid or Fpg protein.[35] 1: no enzyme, a: 3 h incubation, lot 1 of calf thymus DNA; b: 1 h incubation, lot 1; c: 1 h, lot 2; d: 1 h, lot 2; 2: heat-inactivated Fpg protein; a: 4 µg, 1 h, lot 1 of calf thymus DNA; b: 4 µg, 1 h, lot 2; c: 4 µg, 1 h, lot 2; 3: active Nth protein, a: 4 µg, 1 h, lot 1 of calf thymus DNA; b: 4 µg, 1 h, lot 2; c: 4 µg, 1 h, lot 2; 4: active Fpg protein, a: 5 µg, 1 h plus 5 µg, 2 h (total 10 µg), lot 1 of calf thymus DNA; b: 5 µg, 3 h, lot 1; c: 4 µg, 1 h, lot 1; d: 4 µg, 1 h, lot 2; e: 4 µg, 1 h, lot 2; 5: formic acid hydrolysis, a–c: lot 1 of calf thymus DNA; d: lot 1, 8-OH-Gua-^{18}O as internal standard; e: lot 2. Each bar represents the mean value ± standard deviation from 3 independent measurements. Columns 4a–4e and 5a–5e are statistically different from columns 1a–3c ($p < 0.05$). An aliquot of 100 µg of DNA was used for each experiment.

pointed out that pyridine instead of acetonitrile was used in this study in derivatization mixtures. The results showed that pyridine prevents oxidation of guanine during derivatization of formic acid-hydrolysates of DNA and serves as a proper solvent for 8-OH-Gua and guanine.

Figure 8 illustrates a comparison of the levels of 8-OH-Gua found in our laboratory with the data obtained by recent trials by ESCODD.[6,7] These comparisons suggest that both GC/MS and HPLC-ECD may provide similar levels of 8-OH-Gua, if unbiased experimental conditions are used. Moreover, the data in our study and those by others unequivocally show that a tedious and time-consuming prepurification of DNA hydrolysates[39,45] is not needed to prevent potential artifactual formation of 8-OH-Gua or any other modified bases and that artifacts during derivatization prior to GC/MS may depend on experimental conditions, especially on the complete exclusion of oxygen.

3.2. LC/MS/MS and LC/MS

In recent years, LC/MS/MS with electrospray ionization (ESI) emerged as a useful technique for the measurement of oxidative damage to DNA. First, it was

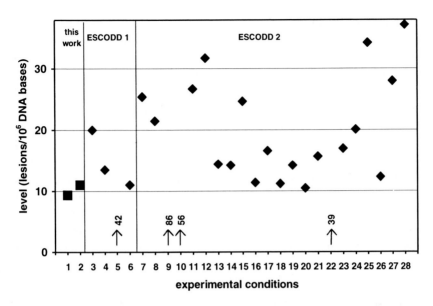

Fig. 8. Levels of 8-OH-Gua in commercial calf thymus DNA measured in this work and by ESCODD trials (numbers with arrows indicate values, which are out of scale in this graph). 1: at room temperature, Fpg protein hydrolysis;[35] 2: at room temperature, formic acid hydrolysis;[35] 3: first ESCODD trial (mean value), GC/MS/HPLC, calf thymus DNA;[6] 4: first ESCODD trial (mean value), HPLC-ECD, enzymatic hydrolysis, calf thymus DNA;[6] 5: first ESCODD trial, GC/MS/HPLC, pig liver DNA;[6] 6: first ESCODD trial, HPLC-ECD, enzymatic hydrolysis, pig liver DNA;[6] 7–11: second ESCODD trial, GC/MS, calf thymus DNA;[7] 12–25: second ESCODD trial, HPLC, calf thymus DNA;[7] 26–28: second ESCODD trial, HPLC/MS/MS, immunological, and ^{32}P-labeling, respectively, calf thymus DNA.[7]

exclusively applied to the measurement of 8-OH-dGuo in DNA following enzymic hydrolysis.[8, 9] Subsequent studies also dealt with several other modified nucleosides in DNA.[10, 11, 46] Furthermore, the content of 8-OH-dGuo in urine was measured by LC/MS/MS.[47, 48] The positive ion ESI mass spectrum of 8-OH-dGuo was first obtained using a quadrupole mass spectrometer with a direct sample introduction by flow injection.[49] It contained a protonated molecular ion (MH$^+$) at m/z 284 and a protonated free base ion (BH$_2^+$) at m/z 168. A sugar ion at m/z 117 was also observed, when enhanced collision-induced dissociation was used. The same paper also described the ESI mass spectrum of 8-hydroxy-2'-deoxyadenosine (8-OH-dAdo) and those of some other modified nucleosides. The former had an MH$^+$ at 268 and a BH$_2^+$ at m/z 152. LC/MS/MS with ESI was applied to the measurement of 8-OH-dGuo in rats.[8] However, the quantification was performed without isotope-dilution technique. Subsequently, isotope dilution-LC/MS/MS using a stable isotope-labeled analog of 8-OH-dGuo as an internal standard was used to measure 8-OH-dGuo in liver DNA and calf thymus DNA as well as in urine.[9] The mass spectrum obtained by LC/MS/MS was similar that described previously by the use of quadrupole mass spectrometer with direct injection of samples.[49] An additional ion at m/z 306 was observed representing a Na-adduct ion (MNa$^+$). A sensitivity level of 20 fmol was obtained, when multiple reaction monitoring (MRM) mode was applied. However, a sensitivity level of only 5 pmol (5000 fmol) was obtained with the use of LC/MS with selected-ion monitoring (SIM) mode of the instrument. Afterward, the same group reported the application of LC/MS/MS to 8-OH-dGuo again and to several other modified nucleosides 8-OH-dAdo, thymidine glycol, 5-hydroxy-2'-deoxycytidine, 5-hydroxy-2'-deoxyuridine, 5-formyl-2'-deoxyuridine and 5-(hydroxymethyl)-2'-deoxyuridine.[10, 46] The measurement of formamidopyrimidines following acidic hydrolysis was also described. This time, a greater sensitivity level of 10 fmol for 8-OH-dGuo was given. The sensitivity levels for other compounds ranged from 5 to 214 fmol. These compounds were measured in DNA isolated from cells and from cells exposed to ionizing radiation at a high dose (90 Gray) and to a chemical generator of singlet oxygen. A simultaneous measurement by LC/MS/MS of 8-OH-dGuo and 8-OH-dAdo was also reported by another group.[11] However, the isotope-dilution technique was not used for quantification. In addition, HPLC-ECD was applied and both techniques yielded similar levels of 8-OH-dGuo in calf thymus DNA. The background level of 8-OH-dGuo in calf thymus DNA was found to be significantly greater than that previously reported using the same technique (85 versus 6 lesions/10^6 DNA bases).[9] However, the latter group subsequently reported a much greater background level of 8-OH-dGuo in calf thymus DNA (circa 100 lesions/10^6 DNA bases),[10] contrasting their previous report.[9]

Just recently, we described the use of LC/MS with atmospheric pressure ionization-electrospray process and isotope-dilution technique for the measurement of 8-OH-dGuo.[12] A quadrupole mass spectrometer was used. The mass spectrum of 8-OH-dGuo was recorded in the total-ion monitoring mode and was similar to those published previously.[8, 9, 49] Enzymic hydrolysis was accomplished using a combination of four enzymes, which were DNase I, phosphodiesterases I and II, and alkaline phosphatase. The level of 8-OH-dGuo was measured in enzymic hydrolysates of DNA using the SIM mode and the isotope-dilution technique with 8-OH-dGuo-^{18}O as an internal standard. The mass spectrum of this compound consisted of a BH$_2^+$ at m/z 170 and an MH$^+$ at m/z 286. Figure 9 illustrates ion-current profiles at m/z 168 and 170 recorded during the LC/MS-SIM analysis of an enzymic hydrolysate of calf thymus DNA. The sensitivity level of the instrument was tested. Figure 10 illustrates the ion-current profile at m/z 168 that represents circa 70 fmol of 8-OH-dGuo on the column. This sensitivity level is 3.5–7-fold lower than that reported for 8-OH-dGuo as measured by LC/MS/MS (10–20 fmol).[9, 10] Previously, a sensitivity level of 5 pmole (5000 fmol) was reported, when LC/MS-SIM mode of the LC/MS/MS instrument was used.[9] This is circa 70-fold lower than the sensitivity level obtained in our study using

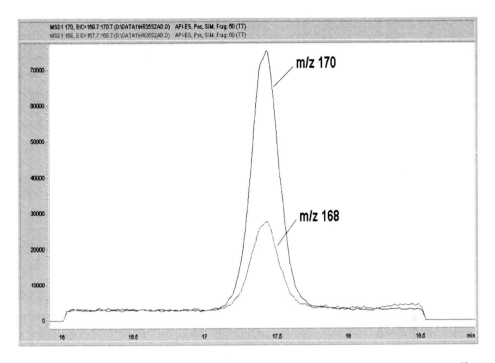

Fig. 9. Ion-current profiles at m/z 168 (8-OH-dGuo) and m/z 170 (8-OH-dGuo-^{18}O) recorded during LC/MS-SIM analysis of calf thymus DNA hydrolyzed by a combination of four enzymes.[12] An aliquot of 10 µg of hydrolyzed DNA was injected on the LC-column.

LC/MS-SIM (70 fmol).[12] The ion-current profile at m/z 168 in Fig. 10 was obtained with 2 µg of DNA and represents 10 lesions per 10^6 DNA bases. This suggests that, if DNA contained 8-OH-dGuo at a level of e.g. 1–2 lesions/10^6 DNA, it would be possible to quantify 8-OH-dGuo by analyzing 10 µg or more of DNA.

For comparison, the sensitivity level of GC/MS-SIM was measured and was found to be circa 3 fmol of the trimethylsilyl derivatives of 8-OH-Gua on the GC column. This sensitivity level is greater than those obtained by LC/MS-SIM and LC/MS/MS.[9, 10] Furthermore, much less DNA amount was used to obtain this sensitivity level[12] than DNA amounts used for LC/MS-SIM[12] or LC/MS/MS.[9, 10] It should be pointed out that the GC/MS equipment used in our laboratory was more than 10 years old. A state-of-the-art equipment might even provide a greater sensitivity level. Recently, a sensitivity level of 300 fmol of GC/MS for the measurement of 8-OH-Gua was reported when used with an HPLC-prepurification of acid-hydrolysates.[50] The same group stated that this procedure was not sensitive enough to detect modified DNA bases.[10] This sensitivity level is two orders of magnitude (100-fold) lower than that of GC/MS obtained in our laboratory as

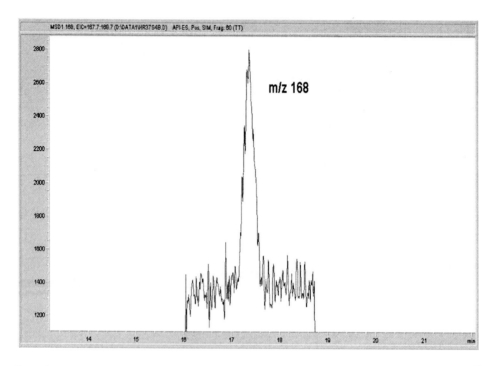

Fig. 10. Ion-current profile at m/z 168, which corresponds to circa 70 fmol of 8-OH-dGuo.[12] The amount of DNA injected onto the LC-column was 2 μg.

described above and elsewhere.[12] This comparison clearly shows that the prepurification of acid-hydrolysates of DNA by HPLC prior to GC/MS analysis as described previously[39, 45] is not only tedious and time-consuming,[12] but also causes a dramatic loss of the sensitivity level of GC/MS.

GC/IDMS-SIM was also used to measure 8-OH-Gua in DNA following its removal from DNA by acidic hydrolysis or by hydrolysis with *E. coli* Fpg protein. The background levels obtained by LC/IDMS-SIM and GC/IDMS-SIM were nearly identical (Fig. 11), indicating that the two techniques can provide similar results. In addition, DNA damaged by ionizing radiation at different radiation doses was analyzed by LC/IDMS-SIM and GC/IDMS-SIM. Again, nearly identical results were obtained by the two techniques (Fig. 12).

Furthermore, we studied the measurement of 8,5'-cdAdo in DNA by LC/ MS.[13] Recent studies suggested that 8,5'-cdAdo may play a role in diseases with defective nucleotide-excision repair, because it is repaired by nucleotide-excision repair rather than by base-excision repair.[51, 52] The LC/MS conditions were similar to those used for the measurement of 8-OH-dGuo.[12] The mass spectrum of 8,5'-cdAdo consisted of an ion at m/z 164 and an MH⁺ at m/z 250. The former ion was assigned to a fragment ion containing the base moiety and

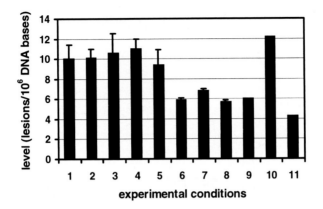

Fig. 11. The level of 8-OH-dGuo measured by LC/IDMS-SIM following enzymatic hydrolysis of DNA by four enzymes and the level of 8-OH-Gua measured by GC/IDMS-SIM following hydrolysis of DNA by formic acid or Fpg protein. Columns 1–3: in calf thymus DNA, measured by HPLC/IDMS-SIM, GC/IDMS-SIM with hydrolysis by formic acid and by GC/IDMS-SIM with hydrolysis by Fpg protein, respectively.[12] Columns 4 and 5: in calf thymus DNA, previously measured by GC/IDMS-SIM with hydrolysis by formic acid and by GC/IDMS-SIM with hydrolysis by Fpg protein, respectively.[35] Columns 6–8: in DNA isolated from cultured HeLa cells, measured by LC/IDMS-SIM, GC/IDMS-SIM with hydrolysis by formic acid and by GC/IDMS-SIM with hydrolysis by Fpg protein, respectively.[12] All values in columns 1–8 represent the average (± standard deviation) of 3–6 independent measurements. Columns 9 and 10: in calf thymus DNA measured by LC/MS/MS.[7,9] Column 11: in rat liver measured by LC/MS/MS.[8] No standard deviation or standard error was given for the values represented by columns 9–11.

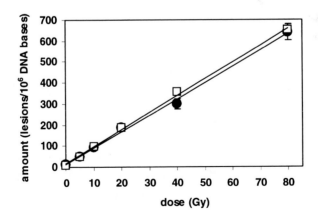

Fig. 12. Radiation dose-yield plots of 8-OH-dGuo measured by LC/IDMS-SIM and 8-OH-Gua measured by GC/IDMS-SIM in DNA exposed to various doses of ionizing radiation.[12] ●: 8-OH-dGuo, LC/IDMS-SIM. □: 8-OH-Gua, GC/IDMS-SIM. All values represent the average (± standard deviation) of 3 independent measurements.

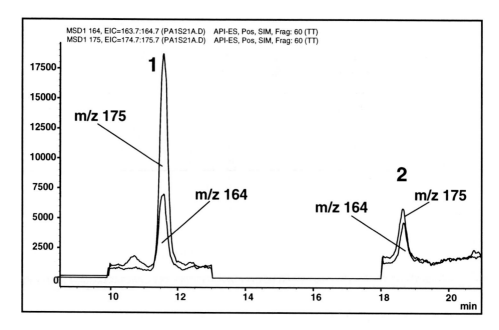

Fig. 13. Ion-current profiles at m/z 164 of 8,5'-cdAdo and at m/z 175 of 8,5'-cdAdo-$^{13}C_{10}$-$^{15}N_5$, which were recorded during the LC/MS-SIM analysis of an enzymatic hydrolysate of 20 µg of DNA irradiated at 20 Gy.[13] Peaks: 1, (5'R)-diastereomers; 2, (5'S)-diastereomers.

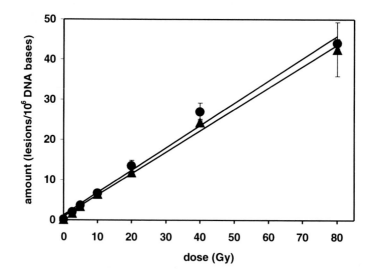

Fig. 14. Radiation dose-yield plots of 8,5'-cdAdo [total of (5'R)- and (5'S)-diastereomers] measured by LC/IDMS-SIM and GC/IDMS-SIM in DNA exposed to various doses.[13] The insert illustrates the dose range from 0 to 10 Gy. The data points represent the mean ± standard deviation from the measurement of 3 independently prepared samples. ●, GC/IDMS-SIM; ▲, LC/IDMS-SIM.

the 5'-CHOH portion of the sugar moiety plus an H atom. Irradiated DNA samples were hydrolyzed by four enzymes as in the case of 8-OH-dGuo. Both (5′R)- and (5′S)-diastereomers of 8,5′-cdAdo were observed. Stable isotope-labeled analogues of these compounds [(5′R)-8,5′-cdAdo-$^{13}C_{10}$-$^{15}N_5$ and (5′S)-8,5′-cdAdo-$^{13}C_{10}$-$^{15}N_5$)] were used for quantification. Figure 13 illustrates ion-current profiles at m/z 164 and 175 obtained using an irradiated DNA sample. Analysis of DNA samples at radiation doses from 2.5–80 Gy yielded linear dose yield-plots (Fig. 14). GC/MS was also used to determine the yield of both (5′R)- and (5′S)-diastereomers of 8,5′-cdAdo. The use of this technique for the measurement of this compound had been reported previously.[24] As Fig. 14 shows, both techniques yielded nearly identical results. The sensitivity level of LC/MS-SIM was determined. Figure 15 illustrates the profile of the m/z 250 ion recorded during the LC/MS-SIM analysis. This profile corresponds to 2 fmol of (5′S)-8,5′-cdAdo on the LC column. This sensitivity level is greater than that for 8-OH-dGuo measured by both LC/MS-SIM (70 fmol)[12] and LC/MS/MS (10–20 fmol).[9, 10] These results show that the sensitivity level of LC/MS-SIM depends on the analyte. The same observation was made when LC/MS/MS was used.[9, 10] The sensitivity level of GC/MS-SIM for the trimethylsilyl derivative of 8,5′-cdAdo was also measured and found to be 1 fmol. This is similar to that recently reported for the trimethylsilyl derivative of 8-OH-Gua.[12] The sensitivity level of LC/MS-SIM for 8,5′-cdAdo suggests that,

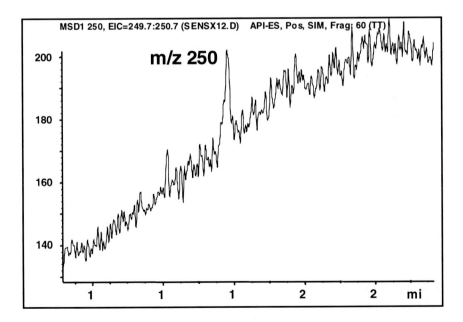

Fig. 15. The ion-current profile at m/z 250 recorded during the LC/MS-SIM analysis of a sample of (5'S)-8,5'-cdAdo.[13] An aliquot of 2 μl of a 1 nM solution of (5'S)-8,5'-cdAdo was injected onto the LC column. The signal corresponds to 2 fmol.

if 8,5′-cdAdo was contained in DNA at a level of one-lesion/10^7 DNA bases, it could be quantified by analyzing less than 10 µg of DNA. Indeed, this was the background level of 8,5′-cdAdo found in calf thymus DNA in our study.[13]

Taken together, these results unequivocally showed that LC/IDMS-SIM is well suited for sensitive and accurate measurement of 8-OH-dGuo and 8,5′-cdAdo in DNA and that both LC/IDMS-SIM and GC/IDMS-SIM can yield similar results. The sensitivity level of LC/IDMS-SIM for 8-OH-dGuo (70 fmol) 3.5–7-fold lower than that of LC/MS/MS (10–20 fmol). However, this difference can be compensated, if more DNA is injected on the LC column. We injected up to 50 µg of DNA on the LC column with no adverse effects on analysis. The use of LC/MS/MS for the measurement of 8,5′-cdAdo has not been reported. Thus, it is not possible to compare LC/MS with LC/MS/MS in terms of the measurement of this compound and the sensitivity level. On the other hand, it should be pointed out that the cost of an LC/MS equipment is much less (circa 2.5-fold) than that of an LC/MS/MS equipment. This could make LC/MS more attractive to many laboratories than LC/MS/MS for measurement of 8-OH-dGuo and 8,5′-cdAdo, or perhaps any other lesions in DNA. Further studies on other DNA lesions will be necessary to extend the application of LC/MS to the measurement of oxidative damage to DNA.

4. Conclusions

Oxidative damage to DNA by free radicals in cells generate multiple products by a variety of mechanisms. Measurement of this type of damage is essential for understanding of its mechanisms, repair and biological consequences. A variety of measurement technique exist for this purpose. They have their own advantages and drawbacks. There is no consensus between laboratories in terms of the measurement of oxidative damage to DNA, especially in the case of a single product, 8-OH-dGuo. Techniques utilizing mass spectrometry (GC/MS, LC/MS, LC/MS/MS) provide positive identification, accurate quantification and versatility. The latter two emerged just recently and were applied to a small number of modifications in DNA, whereas the former has been used for more a decade for the measurement of multiple lesions formed from all four DNA bases. The sensitivities of these techniques are such that levels of modified bases in DNA such as one-lesion/10^6 DNA bases, or even lower levels, can be accurately measured. Measurement of multiple products at the same time and in the same DNA sample might avoid misleading conclusions drawn from the measurement of a single product such as 8-OH-dGuo as was done in the past. The cost of a LC/MS/MS equipment is about 2–2.5-fold greater than that of LC/MS and GC/MS equipments. This prohibitive acquisition cost might limit the use of LC/MS/MS by many research and clinical reference laboratories. Artifacts such as oxidation of intact bases to give additional modified bases are associated with all the techniques used in the past and also with isolation of DNA from cells. However, most artifacts can be avoided, if proper experimental conditions are used. There is a need for a consensus in the measurement of oxidative damage to DNA between laboratories in terms of the techniques and experimental conditions.

Acknowledgment

Certain commercial equipment or materials are identified in this paper in order to specify adequately the experimental procedures. Such identification does not imply recommendation or endorsement by the National Institute of Standards and Technology, nor does it imply that the materials or equipment identified are necessarily the best available for the purpose.

References

1. Halliwell, B. and Gutteridge, J. M. C. (1999). *Free Radicals in Biology and Medicine*, pp. 351–429, Oxford University Press, New York.
2. Dizdaroglu, M. (1992). Oxidative damage to DNA in mammalian chromatin. *Mutat. Res.* **275**: 331–342.
3. Breen, A. P. and Murphy, J. A. (1995). Reactions of oxyl radicals with DNA. *Free Radic. Biol. Med.* **18**: 1033–1077.
4. Collins, A., Cadet, J., Epe, B. and Gedik, C. (1997). Problems in the measurement of 8-oxoguanine in human DNA. Report of a workshop, DNA oxidation, held in Aberdeen, UK, 19–21 January, 1997. *Carcinogenesis* **18**: 1833–1836.
5. Dizdaroglu, M. (1998). Mechanisms of free radical damage to DNA. In "DNA and Free Radicals: Techniques, Mechanisms and Applications" (O. I. Aruoma, and B. Halliwell, B. Eds.), pp. 1–24, OIC International, Saint Lucia.
6. Lunec, J. (1998). ESCODD: European Standards Committee on Oxidative DNA Damage. *Free Radic. Res.* **29**: 601–608.
7. ESCODD (2000). Comparison of different methods of measuring 8-oxoguanine as a marker of oxidative DNA damage. *Free Radic. Res.* **32**: 333–341.
8. Serrano, J., Palmeira, C. M., Wallace, K. B. and Kuehl, D. W. (1996). Determintion of 8-hydroxydeoxyguanosine in biological tissue by liquid chromatography/electrospray ionization-mass spectrometry/mass spectrometry. *Rapid Commun. Mass Spectr.* **10**: 1789–1791.
9. Ravanat, J. L., Duretz, B., Guiller, A., Douki, T. and Cadet, J. (1998). Isotope dilution high-performance liquid chromatography-electrospray tandem mass spectrometry assay for the measurement of 8-oxo-7,8-dihydro- 2′-deoxyguanosine in biological samples. *J. Chromat. B: Biomed. Sci. Appl.* **715**: 349–356.

10. Frelon, S., Douki, T., Ravanat, J. L., Pouget, J. P., Tornabene, C. and Cadet, J. (2000). High-performance liquid chromatography-tandem mass spectrometry measurement of radiation-induced base damage to isolated and cellular DNA. *Chem. Res. Toxicol* **13**: 1002–1010.

11. Podmore, I. D., Cooper, D., Evans, M. D., Wood, M. and Lunec, J. (2000). Simultaneous measurement of 8-oxo-2'-deoxyguanosine by HPLC-MS/MS. *Biochem. Biophys. Res. Commun.* **277**: 764–770.

12. Dizdaroglu, M., Jaruga, P. and Rodriguez, H. (2001). Measurement of 8-hydroxy-2'-deoxyguanosine in DNA by high-performance liquid chromatography-mass spectrometry: comparison with measurement by gas chromatography. *Nucleic Acids Res.* **29**: E12.

13. Dizdaroglu, M., Jaruga, P. and Rodriguez, H. (2001). Identification and quantification of 8,5'-cyclo-2'-deoxyadenosine in DNA by liquid chromatography-mass spectrometry. *Free Radic. Biol. Med.* **30**: 774–784.

14. Von Sonntag, C. (1987). *The Chemical Basis of Radiation Biology*, pp. 116–193, Taylor and Francis, New York.

15. Steenken, S. (1989). Purine bases, nucleosides, and nucleotides: aqueous solution redox chemistry and transformation reactions of their radical cations and e- and OH-adducts. *Chem. Rev.* **89**: 503–520.

16. Dizdaroglu, M., Holwitt, E., Hagan, M. P. and Blakely, W. F. (1986). Formation of cytosine glycol and 5,6-dihydroxycytosine in deoxyribonucleic acid on treatment with osmium tetroxide. *Biochem. J.* **235**: 531–536.

17. Dizdaroglu, M., Laval, J. and Boiteux, S. (1993). Substrate specificity of *Escherichia coli* endonuclease III: excision of thymine- and cytosine-derived lesions in DNA produced by ionizing radiation-generated free radicals. *Biochemistry* **32**: 12 105–12 111.

18. Wagner, J. R. (1994). Analysis of oxidative cytosine products in DNA exposed to ionizing radiation. *J. Chem. Phys.* **91**: 1280–1286.

19. Dizdaroglu, M. (1993). Quantitative determination of oxidative base damage in DNA by stable isotope-dilution mass spectrometry. *FEBS Lett.* **315**: 1–6.

20. Téoule, R. (1987). Radiation-induced DNA damage and its repair. *Int. J. Rad. Biol.* **51**: 573–589.

21. Keck. K. (1968). Bildung von Cyclonucleotiden bei Betrahlung wässriger Lösungen von Purinnucleotiden. *Z. Naturforsch.* **B23**: 1034–1043.

22. Raleigh, J. A., Kremers, W. and Whitehouse, R. (1976). Radiation chemistry of nucleotides: 8,5'-cyclonucleotide formation and phosphate release initiated by hydroxyl radical attack on adenosine monophosphates. *Rad. Res.* **65**: 414–422.

23. Dizdaroglu, M. (1986). Free-radical-induced formation of an 8,5'-cyclo-2'-deoxyguanosine moiety in deoxyribonucleic acid. *Biochem. J.* **238**: 247–254.

24. Dirksen, M. L., Blakely, W. F., Holwitt, E. and Dizdaroglu, M. (1988). Effect of DNA conformation on the hydroxyl radical-induced formation of

8,5'-cyclopurine-2'-deoxyribonucleoside residues in DNA. *Int. J. Rad. Biol.* **54**: 195–204.

25. Oleinick, N. L., Chiu, S., Ramakrishnan, N. and Xue, L. (1987). The formation, identification, and significance of DNA-protein cross-links in mammalian cells. *Brit. J. Cancer* **55(Suppl. VIII)**: 135–140.

26. Margolis, S. A., Coxon, B., Gajewski, E. and Dizdaroglu, M. (1988). Structure of a hydroxyl radical induced cross-link of thymine and tyrosine. *Biochemistry* **27**: 6353–6359.

27. Dizdaroglu, M., Gajewski, E., Reddy, P. and Margolis, S. A. (1989). Structure of a hydroxyl radical induced DNA-protein cross-link involving thymine and tyrosine in nucleohistone. *Biochemistry* **28**: 3625–3628.

28. Collins, A., Dusinská, M., Franklin, M., Somorovská, M., Petrovská, H., Duthie, S., Fillion, L., Panayiotidis, M., Raslová, K. and Vaughan, N. (1997). Comet assay in human biomonitoring studies: reliability, validation, and applications. *Env. Mol. Mutagen.* **30**: 139–146.

29. Grollman, A. P. and Moriya, M. (1993). Mutagenesis by 8-oxoguanine: an enemy within. *Trends Genet.* **9**: 246–249.

30. Rehman, A., Collis, C. S., Yang, M., Kelly, M., Diplock, A. T., Halliwell, B. and Rice-Evans, C. (1998). The effects of iron and vitamin C co-supplementation on oxidative damage to DNA in healthy volunteers. *Biochem. Biophys. Res. Commun.* **246**: 293–298.

31. Podmore, I. D., Griffiths, H. R., Herbert, K. E., Mistry, N., Mistry, P. and Lunec, J. (1998). Vitamin C exhibits pro-oxidant properties. *Nature* **392**: 559.

32. Sentürker, S. and Dizdaroglu, M. (1999). The effect of experimental conditions on the levels of oxidatively modified bases in DNA as measured by gas chromatography-mass spectrometry: how many modified bases are involved? Prepurification or not? *Free Radic. Biol. Med.* **27**: 370–380.

33. Dizdaroglu, M. (2001). Oxidative DNA damage; mechanisms of product formation and repair by base-excision pathway. *In* "Free Radicals in Chemistry, Biology and Medicine" (T. Yoshikawa, S. Toyokuni, Y. Yamamoto, and Y. Naito, Eds.), OICA International, London, pp. 58–76.

34. Jaruga, P., Speina, E., Gackowski, D., Tudek, B. and Olinski, R. (2000). Endogenous oxidative DNA base modifications analyzed with repair enzymes and GC/MS technique. *Nucleic Acids Res.* **28**: E16.

35. Rodriguez, H., Jurado, J., Laval, J. and Dizdaroglu, M. (2000). Comparison of the levels of 8-hydroxyguanine in DNA as measured by gas chromatography mass spectrometry following hydrolysis of DNA by *Escherichia coli* Fpg protein or formic acid. *Nucleic Acids Res.* **28**: E75.

36. Dizdaroglu, M. (1984). The use of capillary gas chromatography-mass spectrometry for identification of radiation-induced DNA base damage and DNA base-amino acid crosslinks. *J. Chromat.* **295**: 103–121.

37. Dizdaroglu, M. (1985). Application of capillary gas chromatography-mass spectrometry to chemical characterization of radiation-induced base damage of DNA: implications for assessing DNA repair processes. *Anal. Biochem.* **144**: 593–603.

38. Watson, J. T. (1985). *Introduction to Mass Spectrometry*, Second Edition, pp. 59–74, Raven Press, New York.

39. Cadet, J., Douki, T. and Ravanat, J. L. (1997). Artifacts associated with the measurement of oxidized DNA bases. *Env. Health Perspect.* **105**: 1034–1039.

40. Dizdaroglu, M. (1991). Chemical determination of free radical-induced damage to DNA. *Free Radic. Biol. Med.* **10**: 225–242.

41. Rodriguez, H., Jurado, J., Laval, J. and Dizdaroglu, M. (2000). Comparison of the levels of 8-hydroxyguanine in DNA as measured by gas chromatography mass spectrometry following hydrolysis of DNA by *Escherichia coli* Fpg protein or formic acid. *Nucleic Acids Res.* **28**: E75.

42. Dizdaroglu, M. (1998). Facts about the artifacts in the measurement of oxidative DNA damage by gas chromatography-mass spectrometry. *Free Radic. Res.* **29**: 551–563.

43. Hamberg, M. and Zhang, L.-Y. (1995). Quantitative determination of 8-hydroxyguanine and guanine by isotope dilution mass spectrometry. *Anal. Biochem.* **229**: 336–344.

44. England, T. G., Jenner, A., Aruoma, O. I. and Halliwell, B. (1998). Determination of oxidative DNA base damage by gas chromatography-mass spectrometry. Effect of derivatization conditions on artifactual formation of certain base oxidation products. *Free Radic. Res.* **29**: 321–330.

45. Douki, T., Delatour, T., Bianchini, F. and Cadet, J. (1996). Observation and prevention of an artefactual formation of oxidized DNA bases and nucleosides in the GC-EIMS method. *Carcinogenesis* **17**: 347–353.

46. Ravanat, J. L., Di Mascio, P., Martinez, G. R., Medeiros, M. H. and Cadet, J. (2000). Singlet oxygen induces oxidation of cellular DNA. *J. Biol. Chem.* **275**: 40 601–40 604.

47. Renner, T., Fechner, T. and Scherer, G. (2000). Fast quantification of the urinary marker of oxidative stress 8-hydroxy-2'-deoxyguanosine using solid-phase extraction and high-performance liquid chromatography with triple-stage quadrupole mass detection. *J. Chromat. B: Biomed. Sci. Appl.* **738**: 311–317.

48. Poulsen, H. E, Loft, S. and Weimann, A. (2000). Urinary measurement of 8-oxodG (8-oxo-2'-deoxyguanosine). *In* "Measuring In Vivo Oxidative Damage: A Practical Approach" (J. Lunec and H. R. Griffiths, Eds.), pp. 69–80, Wiley, New York.

49. Reddy, D. M. and Iden, C. R. (1993). Analysis of modified deoxynucleosides by electrospray ionization mass spectrometry. *Nucleosides and Nucleotides* **12**: 815–826.

50. Cadet, J., Douki, T., Frelon, S., Pouget, J.-P. and Ravanat, J.-L. (2000). Facts and artifacts in the measurement of oxidative base damage to DNA. *Pre-Meeting Workshop "In Vivo Assessment of Oxidative Stress and Antioxidant Status: From Concepts to Validations."* 7th Annual Meeting of the Oxygen Society. San Diego, CA, USA.
51. Brooks, P. J., Wise, D. S., Berry, D. A., Kosmoski, J. V., Smerdon, M. J., Somers, R. L., Mackie, H., Spoonde, A. Y., Ackerman, E. J., Coleman, K., Tarone, R. E. and Robbins, J. H. (2000). The oxidative DNA lesion 8,5'-(S)-cyclo-2'-deoxyadenosine is repaired by the nucleotide excision repair pathway and blocks gene expression in mammalian cells. *J. Biol. Chem.* **275**: 22 355–22 362.
52. Kuraoka, I., Bender, C., Romieu, A., Cadet, J., Wood, R. D. and Lindahl, T. (2000). Removal of oxygen free-radical-induced 5',8-purine cyclodeoxy-nucleosides from DNA by the nucleotide excision-repair pathway in human cells. *Proc. Natl. Acad. Sci. USA* **97**: 3832–3837.

Chapter 10

HPLC-MS/MS Measurement of Oxidative Base Damage to Isolated and Cellular DNA

T. Douki, J.-L. Ravanat, S. Frelon, A.-G. Bourdat,
J.-P. Pouget and J. Cadet*

T. Douki, J.-L. Ravanat, S. Frelon, A.-G. Bourdat, J.-P. Pouget and **J. Cadet** • Laboratoire 'Lésions des Acides Nucléiques'; Service de Chimie Inorganique et Biologique; UMR 5046; Département de Recherche Fondamentale sur la Matière Condensée; CEA/Grenoble; F-38054 Grenoble Cedex 9 (France)

*Corresponding Author.
Tel: (33)-4-38-78-49-87, E-mail: cadet@drfmc.ceng.cea.fr

1. Introduction

The chromatographic measurement of oxidative DNA base damage has mainly involved, until recently, high performance liquid chromatography associated with electrochemical detection (HPLC-EC) and gas chromatography coupled to mass spectrometry (GC-MS). The former technique is specific and sensitive but has been almost exclusively applied to the measurement of 8-oxo-7,8-dihydro-2′-deoxyguanosine (8-oxodGuo),[1, 2] the main oxidation product of guanine bases in DNA. In contrast, GC-MS is more versatile and allows the detection of a wide range of modified bases.[3] However, the latter assay suffers from drawbacks associated with the acidic hydrolysis of DNA and the silylation step aimed at preparing volatile derivatives of the analytes.[4, 5] Recently, the association of liquid chromatography with mass spectrometry has emerged as a promising technique that combines the advantages of both HPLC-EC and GC-MS. The oxidized bases can be analyzed as either nucleosides or nucleotides, which makes possible the use of mild enzymatic hydrolysis procedures. Moreover, the atmospheric pressure sources used are compatible with the direct introduction of the HPLC eluent in the spectrometer. This contrasts with the GC-MS assay that requires a troublesome derivatization step. In addition, like with the latter approach, several lesions can be measured in the same sample. The purpose of this chapter is first to briefly describe the technique. Several examples of application of the HPLC-MS/MS assay to the quantification of oxidative base damage within isolated and cellular DNA are then provided.

2. Principle of the HPLC-MS/MS Method

2.1. HPLC/MS Interface

Several technologies have been developed in the past to associate liquid chromatography with mass spectrometry (thermospray, continuous flow FAB, …). However, the development of atmospheric pressure ion sources (often referred to as "electrospray") has actually allowed the set-up of reliable interfaces between HPLC and MS. In this system, the HPLC eluent is sprayed from a capillary in the presence of a high electric field. This leads to the desorption and the ionization of the analytes which are then extracted towards the high vacuum part of the spectrometer where they are further analyzed. During this process, the overwhelming HPLC mobile phase has to be evaporated. Therefore, flow rates should be keep low, 1 ml·min^{-1} at the most. Typically, 2 mm diameter, or smaller, columns are used at a flow rate of 200 μl·min^{-1}. It can be added that volatile buffers are required to keep the response of the spectrometer constant over long periods of time. Therefore, ammonium acetate or formate are often used in the HPLC mobile phase.

2.2. Detection of the Ions

Most of the reported applications of HPLC-MS to the measurement of modified DNA bases are based on the use of quadrupolar spectrometers. In addition, even though some works have been performed with single quadrupole systems, highly sensitive and specific detection requires triple quadrupolar apparatus. In these tandem mass spectrometry systems, the ions produced in the source are filtered in the first quadrupole. Then, they are fragmented in the second quadrupole used as a collision cell that contains a low pressure of inert gas. The last quadrupole discriminates the different fragments which are then quantified. When the first quadrupole is set on a specific mass and the third is scanning a range of masses, full fragmentation mass spectra are recorded. When both the first and the third quadrupoles are locked on specific masses (multiple reaction monitoring mode or MRM), the spectrometer is used as a specific detector for molecules exhibiting a precise mass and being fragmented into a given daughter ion. In the MRM mode, very high sensitivity may be achieved because of a drastic decrease in the detection background. The detection limit is circa 100 times lower in the MRM mode than with a single quadrupolar detector used in the Selected Ion Monitoring mode.

3. Measurement of Oxidized Bases within DNA

Even though several reports have been published on the HPLC-MS/MS detection of adducts between DNA bases and bulky chemicals (for a review: see Ref. 6), only a few works deal with the measurement of oxidative base damage. 8-OxodGuo has been the first targeted molecule within isolated and cellular DNA.[7, 8] We recently reported the extension of this approach to other nucleosides and bases.[9] In addition, HPLC-MS/MS has been applied to the measurement of several degradation products arising from the photosensitization of 2'-deoxyguanosine.[10]

3.1. Set-Up of the Assay for Oxidized Bases

Oxidative DNA base damage are mostly measured by HPLC-MS/MS as modified nucleosides. Indeed, the latter compounds exhibit good chromatographic properties and are released from DNA by enzymatic digestion under mild conditions. In addition, they are easily fragmented into high intensity ions. This contrasts with most base derivatives which are often very stable and do not yield specific daughter ions. Nucleosides exhibiting an exocyclic amino group (purine and cytosine derivatives) are more sensitively detected as positive protonated ions while thymine and uracil derivatives are detected in the negative mode. Using this approach, six oxidized nucleosides, including thymidine glycols

(ThdGly), 5-hydroxy-2'-deoxyuridine (5-OHdUrd), 5-(hydroxymethyl)-2'-deoxyuridine (5-HMdUrd), 5-formyl-2'-deoxyuridine (5-FordUrd), 8-oxodGuo and 8-oxo-7,8-dihydro-2'-deoxyadenosine (8-oxodAdo), are simultaneously measured. The fragmentation mass spectrum of each nucleoside (Fig. 1) has been recorded in order to determine the specific transitions used for the MRM detection.

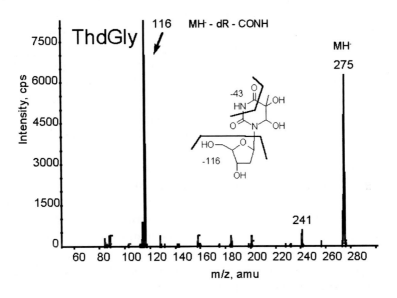

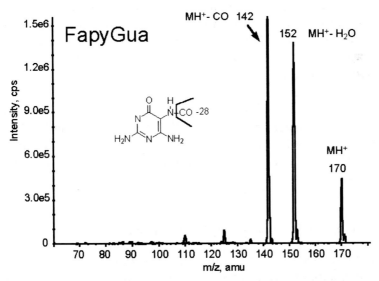

Fig. 1. Fragmentation mass spectra of ThdGly and FapyGua. Analyses were performed in the negative and positive ionization mode, respectively. The proposed fragmentation pathways were confirmed by analysis of the isotopically labeled derivatives.

Prior to the HPLC-MS/MS analysis, DNA is digested by sequential incubation with nuclease P1, a 5'-exonuclease, a 3'-exonuclease and alkaline phosphatase. The use of exonucleases aims at quantitatively releasing ThdGly and 5-FordUrd which are poor substrate for nuclease P1 usually used prior to the HPLC-EC detection of 8-oxodGuo. The nucleosides are separated on a reverse phase silica gel column, using a gradient of acetonitrile in ammonium formate over a period of 30 min. This allows the simultaneous analysis of both highly and less polar compounds such as ThdGly and 8-oxodAdo, respectively. Isotopically labeled derivatives of the six nucleosides have been prepared and are used as internal standards. This is a major advantage for the accuracy of the quantification. The detection is highly sensitive since between 10 and 30 fmol can be quantified. A major exception is 5-OHdUrd which is detected with a 10 times lower sensitivity. HPLC-MS/MS is thus more sensitive than HPLC-EC for the quantification of 8-oxodGuo and allows the detection of other lesions in the same chromatographic run.

The 5-formamidopyrimidine derivatives of guanine and adenine (FapyGua and FapyAde) can also be measured by HPLC-MS/MS (Fig. 1). However, they are analyzed as bases. Indeed, the 2-deoxyribose moiety of the corresponding nucleosides undergoes fast epimerization and rearrangement into a mixture of α and β furanose and pyranose isomers. FapyGua and FapyAde are thus released from DNA by a mild room temperature formic acid hydrolysis.[11] This procedure is used as an alternative to the usual hot formic acid hydrolysis which leads to their conversion into guanine and adenine, respectively.[12] FapyGua and FapyAde are detected by MS/MS in the positive ionization mode following separation on an amino silica gel column using a [9:1] mixture of acetonitrile and ammonium formate as the isocratic eluent. Isotopic dilution is also used for the quantification of these two modified bases. The limit of detection is 5 and 60 fmol for FapyAde and FapyGua, respectively.

The accuracy of the HPLC-MS/MS measurements was further shown by a comparison with other methods. 8-OxodAdo and 8-oxodGuo were quantified within gamma irradiated DNA samples by both HPLC-EC and HPLC-MS/MS. Similarly, the level of FapyGua and FapyAde was determined by HPLC-MS/MS and GC-MS. In all cases, a very good correlation between the two sets of values was observed (Fig. 2).

3.2. Gamma and UV Laser Irradiation of Isolated DNA

The accurate determination of the level of a wide array of modified bases is a powerful tool for mechanistic studies based on the large amount of information available on the radical-induced degradation of DNA bases.[13] Exposure of an aqueous solution of DNA to gamma radiation leads to the overwhelming radiolysis of water (indirect effect). This generates hydroxyl radicals ($^{\bullet}$OH) in high yield.

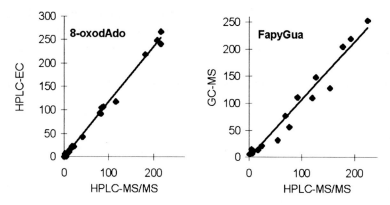

Fig. 2. Comparison of the level of (left panel) 8-oxodGuo determined by either HPLC-EC or HPLC-MS/MS and of (right panel) FapyGua determined by either GC-MS or HPLC-MS/MS. The DNA samples analyzed were exposed to increasing doses of gamma radiation in aerated aqueous solution. Results are expressed in lesions per 10^5 normal bases.

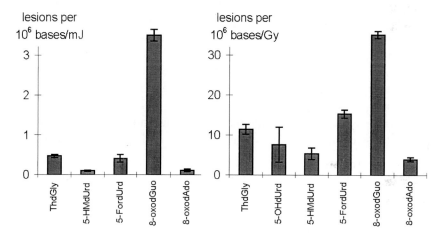

Fig. 3. Yield of oxidized bases within DNA exposed to either gamma radiation or 266 nm laser pulses.

The latter species react unspecifically with the bases and the 2-deoxyribose moieties of DNA. In contrast, exposure of DNA to 266 nm laser pulses leads to the biphotonic ionization of the bases into their radical cations.[14] In a following step, positive charges migrate toward guanine, which exhibits the lowest oxidation potential among DNA bases.[15–17] Consequently, most of the damage is expected to arise from the degradation of guanine bases. The difference in the DNA degradation mechanism is clearly shown by the HPLC-MS/MS analysis of the gamma and UV laser-irradiated samples (Fig. 3). Indeed, oxidation products are observed for the four DNA bases with gamma radiation while 8-oxodGuo is observed as the main product upon exposure to UV laser pulses.

3.3. Degradation of Cellular DNA by Gamma Radiation

DNA damage is thought to be a main cause of the deleterious effects of ionizing radiation to cells. In that respect, the identification of the main lesions involved is a major goal. Therefore, HPLC-MS/MS assay was used to determine the extent of radiation-induced base damage within cultured THP1 human monocytes. A first result is the background level of lesions within control cells which is in the range of 0.2 to 0.8 lesions per 10^6 normal bases for 8-oxodGuo. This value is close to those obtained by HPLC-EC with the same cell line[18] or in other living systems.[19, 20] It is also close to indirect estimation based on the use of comet or alkaline elution assays associated with formamidopyrimidine N-glycosylase repair enzyme (Fpg).[21] HPLC-MS/MS provides thus an additional confirmation that the high background level of 8-oxodGuo previously determined within cellular DNA by the standard GC-MS assay or ^{32}P-post labeling was overestimated.[5, 21] It can be added that HPLC-MS/MS allowed the detection of other oxidized bases within DNA. This was not possible by the accurate but less sensitive HPLC/GC-MS approach[22, 23] previously used in the laboratory to overcome the artifacts associated with the standard GC-MS assay. The low basal level of oxidized bases allowed the determination of the yield of the 8 modified bases upon exposure of cells to gamma radiation. The values obtained were much lower than those obtained by GC-MS analyses[24–26] and agreed with HPLC-EC[18, 27]

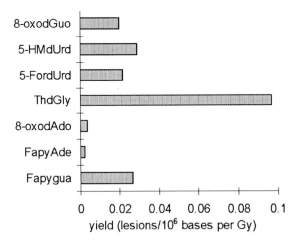

Fig. 4. Yield of oxidized bases determined within the DNA of THP1 cells exposed to gamma radiation. Results are expressed in lesions per 10^6 bases per Gy.

or immunological/capillary electrophoresis measurements.[28] In THP1 cells, 8-oxodGuo is not the major oxidation product, in contrast to isolated DNA. Indeed, ThdGly is the damage measured in the highest amount (Fig. 4). Interestingly, FapyGua is generated at a higher level than 8-oxodGuo, suggesting that cellular DNA is in a more reducing context than isolated DNA in aerated aqueous solution.[13] Indeed, 8-oxodGuo is the major lesion under the latter conditions.[12] This could be accounted for by a lower oxygen concentration and the presence of reducing compounds such as glutathione in cell nucleus. The detection of other modified bases arising from the reduction of the initially produced DNA base radicals would further explain these results. HPLC-MS/MS will be a tool of choice for this task.

4. Detection of Tandem Lesions within DNA

In a pioneering work, Box and coworkers have reported the formation of a lesion carrying both a formamido residue and a 8-oxo-7,8-dihydroguanine moiety (8-oxoGua) upon exposure of the dinucleoside monophosphate 2′-deoxyguanosyl-(3′-5′)-thymidine (dGpT) to X rays in aerated aqueous solution.[29] It appeared to be a primary product, as shown by the linearity of its formation with respect to the dose. This new aspect of the radical chemistry of DNA bases deserves to be further investigated, mostly because these complex lesions are likely to be poorly repaired in cells. Therefore, we developed an HPLC-MS/MS assay for the sensitive detection of the two isomeric formamido/8-oxoGua lesions (arising from either TG or GT sequences) within DNA.[30]

4.1. Detection of Formamido/8-oxoGua within DNA

In addition to the optimization of the HPLC-MS/MS detection, a key step in the set-up of the assay was the release of the tandem lesions from DNA. Indeed, classical hydrolysis procedures would convert the tandem lesion into two individual oxidized bases or nucleosides. However, an extensive study of the enzymatic digestion of oligonucleotides that contained either 8-oxoGua-formamido or formamido-8-oxoGua lesion showed that both calf spleen phosphodiesterase and snake venom phosphodiesterase (5′- and 3′-exonucleases, respectively) were blocked at the site of the damage.[31] Therefore, sequential incubation of DNA with both enzymes allowed the quantitative release of the two isomeric formamido/8-oxodGuo tandem lesions from DNA as dinucleoside monophosphates. The latter compounds were analyzed by HPLC-MS/MS in the negative mode because of the presence of the easily ionized phosphate group. Two transitions were monitored, corresponding to the fragmentation of the parent ion into phosphorylated 8-oxodGuo (8-oxodGuoMP) and deprotonated 8-oxoguanine.

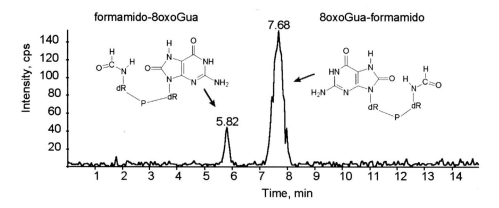

Fig. 5. HPLC-MS/MS chromatogram corresponding to the detection of formamido/8-oxoGua lesions in an hydrolyzed irradiated DNA sample (dose: 10 Gy).

Interestingly, the fragmentation patterns were different between formamido-8-oxoGua and 8-oxoGua-formamido since the signal of 8-oxodGuoMP was very low for the latter lesion. This, in addition to large differences in retention times, allowed the unambiguous individual quantification of the two formamido/8-oxoGua lesions (Fig. 5). The assay was sensitive enough (30 fmol) to allow the detection of the tandem lesions in DNA samples exposed to doses as low as 5 Gy of gamma radiation. This corresponded to low degradation yields and confirmed that the two isomeric formamido/8-oxoGua were primary products.

4.2. Mechanistic Studies

A first series of experiments allowed the determination of the radiolytic yields of formation of the two formamido/8-oxoGua lesions. Interestingly, a drastic sequence effect was observed. Indeed, the yield of formamido-8-oxoGua was one order of magnitude lower than that of 8-oxoGua-formamido. In addition, the formation of both lesions was found to be linear with respect to the dose, as already observed in short oligonucleotides.[29] The ratio between the level of tandem lesion and single 8-oxodGuo measured by HPLC-EC was found to be 0.1 (Fig. 6). This value shows that formamido/8-oxoGua is produced in significant amount within gamma-irradiated DNA. The tandem lesions were then measured within DNA oxidized by photosensitized electron abstraction (type I photosensitization). Under these conditions, the level of formamido/8-oxoGua with respect to single 8-oxodGuo was much lower than in gamma-irradiated samples. This strongly suggests that guanine radicals are not the initial species involved in the formation of the tandem lesion. Indeed, as mentioned above, most radical cations arising from the photosensitization process will be

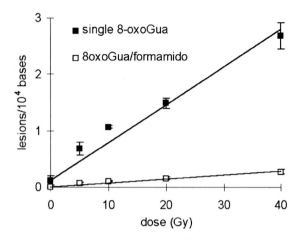

Fig. 6. Formation of single 8-oxodGuo and formamido/8-oxoGua tandem lesions within isolated DNA exposed to gamma radiation in aqueous aerated solution.

located on guanine bases. Altogether, these data allow to propose a mechanism for the formation of formamido/8-oxoGua. The sequential addition of ˙OH and oxygen to the C5 = C6 double bond of thymine leads to the formation of a 5(6)-hydroperoxyl-6(5)-hydroxy-5,6-dihydrothymine radical. The latter species is converted into its hydroperoxide derivative by intramolecular abstraction of an electron to the guanine moiety. This generates the guanine radical cation which yields 8-oxoGua as a final product. The thymine hydroperoxide is reduced into an alkoxyl radical which is the precursor of the formamido residue. Work is in progress to further establish the mechanism of formation of the formamido/8-oxoGua lesion and identify other tandem lesions.

5. Conclusion

The results reported above show the potentials of the HPLC-MS/MS assay for the detection of oxidative DNA damage for both mechanistic and *in vivo* studies. The method requires only limited handling of the sample, in contrast to the GC-MS approach, and allows the detection of a large number of modified bases in the same sample whereas HPLC-EC is mostly limited to 8-oxodGuo. The study of the formation of the formamido/8-oxoGua lesions also illustrates the ability of the HPLC-MS/MS technique to detect more complex types of base damage. In that respect, the measurement of UV-induced dimeric pyrimidine photoproducts is another interesting application.[32] However, even though HPLC-MS/MS appears very promising, some limitations still remain on several aspects of the detection of oxidative DNA damage within DNA. First, standards of the targeted molecules are required in order to optimize and calibrate the detection. Isotopically labeled derivative are also highly recommended, even though the variability of the response of the detection is much lower than with GC-MS. Second, the quantitative aspect of the hydrolysis of DNA remains to be established for every new targeted oxidized base. For example, the procedure used for the detection of 8-oxodGuo was found to be inadequate for ThdGly. However, the most important aspect of the measurement of oxidative damage in cellular DNA remains the possibility of spurious oxidation during the extraction of DNA from the cell.[20, 21] Efforts are still needed to lower the background level of oxidized bases within extracted DNA. Indeed, methods involving the use of repair enzymes without extraction step still provides a slightly lower value for the background level of Fpg and endonuclease III sensitive sites than direct chromatographic measurements of oxidized bases. It can be added HPLC-MS/MS has also been successfully applied to the detection 8-oxodGuo in biological fluid.[8, 33] The approach appears to be specific and straightforward, and may be applied to other nucleosides. Measurement of a panel of oxidized nucleosides within urine and plasma is likely to provide data on the biological significance of this type of analyses which are today mostly focused on 8-oxodGuo.

References

1. Floyd, R. A., Watson, J. J., Wong, P. K., Altmiller, D. H. and Rickard, R. C. (1986). Hydroxyl free radical adduct of deoxyguanosine: sensitive detection and mechanism of formation. *Free Radic. Res. Commun.* **1**: 163–172.
2. Kasai, H. (1997). Analysis of a form of oxidative DNA damage, 8-hydroxy-2′-deoxyguanosine, as a marker of cellular oxidative stress during carcinogenesis. *Mutat. Res.* **387**: 147–163.
3. Dizdaroglu, M. and Gajewski, E. (1990). Selected-ion mass spectrometry: assays of oxidative DNA damage. *Meth. Enzymol.* **186**: 530–544.
4. Cadet, J., Douki, T. and Ravanat, J.-L. (1997). Artifacts associated with the measurement of oxidized DNA bases. *Env. Health Perspect.* **105**: 1034–1039.
5. Cadet, J., D'Ham, C., Douki, T., Pouget, J.-P., Ravanat, J.-L. and Sauvaigo, S. (1998). Facts and artifacts in the measurement of oxidative base damage to DNA. *Free Radic. Res.* **29**: 541–550.
6. Andrews, C. L., Vouros, P. and Harscg, A. (1999). Analysis of DNA adducts using high-performance separation techniques coupled to electrospray ionization mass spectrometry. *J. Chromat.* **A856**: 515–526.
7. Serrano, J., Palmeira, C. M., Wallace, K. B. and Kuehl, D. W. (1996). Determination of 8-hydroxyguanosine in biological tissues by liquid chromatography/electrospray ionization-mass spectrometry/mass spectrometry. *Rapid Commun. Mass Spectr.* **10**: 1789–1791.

8. Ravanat, J. L., Duretz, B., Guiller, A., Douki, T. and Cadet, J. (1998). Isotope dilution high-performance liquid chromatography — electrospray tandem mass spectrometry assay for the measurement of 8-oxo-7,8-dihydro-2'-deoxyguanosine in biological samples. *J. Chromat.* **B715**: 349–356.

9. Frelon, S., Douki, T., Ravanat, J.-L., Pouget, J.-P., Tornabene, C. and Cadet, J. (2000). High performance liquid chromatography — tandem mass spectrometry measurement of radiation-induced base damage to isolated and cellular DNA. *Chem. Res. Toxicol.* **13**: 1002–1010.

10. Ravanat, J.-L., Remaud, G. and Cadet, J. (2000). Measurement of the main photooxidation products of 2'-deoxyguanosine using chromatographic methods coupled to mass spectrometry. *Arch. Biochem. Biophys.* **374**: 118–127.

11. Douki, T., Bretonnière, Y. and Cadet, J. (2000). Protection against radiation-induced degradation of DNA bases by polyamines. *Rad. Res.* **153**: 29–35.

12. Douki, T., Martini, R., Ravanat, J.-L., Turesky, R., J. and Cadet, J. (1997). Measurement of 2,6-diamino-4-hydroxy-5-formamidopyrimidine and 8-oxo-7,8-dihydroguanine in isolated DNA exposed to gamma radiation in aqueous solution. *Carcinogenesis* **18**: 2385–2391.

13. Cadet, J., Berger, M., Douki, T. and Ravanat, J.-L. (1997). Oxidative damage to DNA: formation, measurement and biological significance. *Rev. Physiol. Biochem. Pharmacol.* **131**: 1–87.

14. Angelov, D., Spassky, A., Berger, M. and Cadet, J. (1997). High-intensity UV laser photolysis of DNA and purine 2'-deoxyribonucleosides: formation of 8-oxopurine damage and oligonucleotide strand cleavage as revealed by HPLC and gel electrophoresis studies. *J. Am. Chem. Soc.* **119**: 11 373–11 380.

15. Meggers, E., Michel-Beyerle, M. E. and Giese, B. (1998). Sequence dependent long range hole transport in DNA. *J. Am. Chem. Soc.* **120**: 12 950–12 955.

16. Holmlin, R. E., Dandliker, P. J. and Barton, J. K. (1998). Charge transfer through the DNA base stack. *Angew. Chem. Int. Ed.* **36**: 2715–2730.

17. Schuster, G. B. (2000). Long-range charge transfer in DNA: transient structural distortions control the distance dependence. *Acc. Chem. Res.* **33**: 253–260.

18. Pouget, J.-P., Douki, T., Richard, M.-J. and Cadet, J. (2000). DNA damage induced in cells by gamma and UVA radiations as measured by HPLC/GC-MS, HPLC-EC and comet assay. *Chem. Res. Toxicol.* **13**: 541–549.

19. Nakae, D., Kobayashi, Y., Akai, H., Andoh, N., Satoh, H., Ohashi, K., Tsutsumi, M. and Konishi, Y. (1997). Involvement of 8-hydroxyguanine formation in the initiation of rat liver carcinogenesis by low dose levels of N-nitrosodiethylamine. *Cancer Res.* **57**: 1281–1287.

20. Helbock, H. J., Beckman, K. B., Shigenaga, M. K., Walter, P. B., Woodall, A. A., Yeo, H. C. and Ames, B. N. (1998). DNA oxidation matters: the HPLC-electrochemical detection assay of 8-oxo-deoxyguanosine and 8-oxo-guanine. *Proc. Natl. Acad. Sci. USA* **95**: 288–293.

21. Collins, A., Cadet, J., Epe, B. and Gedik, C. (1997). Problems in the measurement of 8-oxoguanine in human DNA. Report of a workshop, DNA

oxidation, held in Aberdeen, UK, 19–21 January, 1997. *Carcinogenesis* **18**: 1833–1836.

22. Douki, T., Onuki, J., Medeiros, M. H. G., Bechara, E. J. H., Cadet, J. and Di Mascio, P. (1998). Hydroxyl radicals are involved in the oxidation of isolated and cellular DNA bases by 5-aminolevulinic acid. *FEBS Lett.* **428**: 93–96.

23. Bianchini, F., Elmstahl, S., Martinez-Garcia, C., van Kappel, A.-L., Douki, T., Cadet, J., Ohshima, H., Riboli, E. and Kaaks, R. (2000). Oxidative DNA damage in human lymphocytes: correlations with plasma levels of α-tocopherol and carotenoids. *Carcinogenesis* **21**: 321–324.

24. Nackerdien, Z., Olinski, R. and Dizdaroglu, M. (1992). DNA base damage in chromatin of γ-irradiated cultured human cells. *Free Radic. Res. Commun.* **16**: 259–273.

25. Mori, T. and Dizdaroglu, M. (1994). Ionizing radiation causes greater DNA base damage in radiation-sensitive mutant M10 cells than in parent mouse lymphoma L5178Y cells. *Rad. Res.* **140**: 85–90.

26. Zastawny, T. H., Kruszewski, M. and Olinski, R. (1998). Ionizing radiation and hydrogen peroxide induced oxidative DNA base damage in two L5178Y cell lines. *Free Radic. Biol. Med.* **24**: 1250–1255.

27. Pouget, J.-P., Ravanat, J.-L., Douki, T., Richard, M.-J. and Cadet, J. (1999). Measurement of DNA base damage in cells exposed to low doses of γ-radiation: comparison between the HPLC-EC and comet assays. *Int. J. Rad. Biol.* **75**: 51–58.

28. Le, X. C., Xing, J. Z., Lee, J., Leadon, S. A. and Weinfeld, M. (1998). Inducible repair of thymine glycol detected by an ultrasensitive assay for DNA damage. *Science* **280**: 1066–1069.

29. Box, H. C., Dubzinski, E. E., Freund, H. G., Evans, M. S., Patrzyc, H. B., Wallace, J. C. and Maccubin, A. E. (1993). Vicinal lesions in X-irradiated DNA? *Int. J. Rad. Biol.* **64**: 261–263.

30. Bourdat, A.-G., Douki, T., Frelon, S., Gasparutto, D. and Cadet, J. (2000). Tandem base lesions are generated by hydroxyl radicals within isolated DNA in aerated aqueous solution. *J. Am. Chem. Soc.* **122**: 4549–4556.

31. Bourdat, A.-G., Gasparutto, D. and Cadet, J. (1999). Synthesis and enzymatic processing of oligodeoxynucleotides containing tandem base damage. *Nucleic Acids Res.* **27**: 1015–1024.

32. Douki, T., Court, M., Sauvaigo, S., Odin, F. and Cadet, J. (2000). Formation of the main UV-induced thymine dimeric lesions within isolated and cellular DNA as measured by HPLC-MS/MS. *J. Biol. Chem.* **275**: 11 678–11 685.

33. Renner, T., Fechner, T. and Scherer, G. (2000). Fast quantification of the urinary marker of oxidative stress 8-hydroxy-2′-deoxyguanosine using solid phase extraction and high-performance liquid chromatography with triple-stage quadrupole mass detection. *J. Chromat.* **B738**: 311–317.

Chapter 11

The Use of HPLC/EC for Measurements of Oxidative DNA Damage

Mikhail B. Bogdanov, Wayne R. Matson and Ian N. Acworth*

Mikhail B. Bogdanov • Department of Neurology and Neuroscience, Weill Medical College of Cornell University, 525 East 68th St., New York, NY, 10021
Wayne R. Matson and **Ian N. Acworth** • ESA Inc., 22 Alpha Rd., Chelmsford, MA 01824
*Corresponding Author.
Tel: Phone: 978-250-7055, E-mail: inacworth@esainc.com

1. Introduction

Various reactive oxygen species (ROS) are generated endogenously as a result of electron leakage from mitochondria, during transition metal-mediated reduction of oxygen or hydrogen peroxide, and during synthesis of prostaglandins and leukotrienes.[1] Exogenous agents, such as ionizing radiation, UV light, and different carcinogens can also generate ROS. ROS have been implicated in the etiology and pathogenesis of different diseases, including cancer, neuro-degeneration, stroke and diabetes, and also in aging.[1-13] While ROS react with most components of the cell, the key intracellular substrates resulting in cell death following ROS attack are generally unknown. Most studies have been focused on lipid peroxidation, oxidative modification of proteins and oxidative DNA damage.

The hydroxyl radical, produced by homolytic cleavage of water by ionizing radiation, during Fenton reaction involving hydrogen peroxide or hypochlorous acid, or as a hydroxyl radical "tail" of peroxynitrite, is a major free radical contributing to oxidative damage to DNA.[11, 14-16] Following hydroxyl radical attack on DNA different types of DNA lesions are generated, including DNA adducts, abasic sites, lipid-DNA and protein-DNA cross-links, single- and double-strand breaks and intrastrand adducts.[17-22] ROS and reactive nitrogen species (RNS) usually attack the free bases and nucleosides at the C-4 and C-8 positions of purines and at the C-5 and C-6 positions of pyrimidines. More than 100 different oxidative DNA lesions have so far been described.[23, 24] To date only a few DNA adducts are used as markers of oxidative DNA damage, and of these the oxidative C-8 guanine lesion is the most studied as either 8-hydroxy-2'-deoxyguanosine (8OH2'dG) or it's corresponding base, 8-hydroxyguanine (8OHG). 8OH2'dG, first discovered by Kasai in 1984,[7, 25] is the most abundant DNA adduct, and has been used as marker of oxidative damage to DNA in several clinical and epidemiological studies. Increased levels of 8OH2'dG have been seen in aging,[2, 9, 26-28] different forms of cancer,[29-34] diabetes,[8, 35, 36] neurodegenerative disorders,[3, 4, 37-42] and following exposure to different toxic agents and environmental insults.[43-54]

2. Measurement Technologies

During the last decade several techniques have been introduced to measure oxidative DNA damage. Currently, two major approaches are used to estimate levels of oxidative damage to DNA. These can be classified as to whether they measure oxidative base damage in DNA itself (steady-state levels), or whether they measure free levels of DNA adducts in various biological matrices, mostly in urine. Levels of DNA adducts in DNA reflect net effects of DNA damage *in situ*, removal of DNA adducts by repair, availability of adduct nucleotides during DNA replication, and dilution of unrepaired DNA adducts during replication.[55] Free levels of DNA adducts and corresponding deoxynucleotides in biological matrices reflect oxidative damage to DNA, activity of DNA repair pathways, oxidative damage to cytosolic and circulating nucleobase and nucleotide pools, and mitochondrial turnover and repair.[56] Furthermore an additional source of free levels of DNA adducts in biological matrices could be apoptosis. Oxidative damage to DNA has been implicated in apoptotic cell death.[57-62] However, currently there are no direct experimental data on contribution of apoptosis to the free levels of DNA adducts. In the studies of diseases in which role of oxidative DNA damage has been implicated, both free and DNA levels of adducts, as well as the activity of DNA repair pathways, should be addressed. Thus, it has been recently found that in Alzheimer's disease (AD), free levels of 8OH2'dG are decreased in the cerebrospinal fluid from AD patients, while levels of 8OH2'dG in the brain DNA from AD patients are increased as a result of deficiency in DNA repair.[39, 63]

The analysis of DNA adducts within DNA is well established, although there is considerable debate about the protocols used for DNA extraction and digestion, which could contribute to artifactual formation of DNA adducts during sample preparation.[7, 55] Most studies on artifactual production of DNA adducts used 8OH2'dG as a marker of oxidative DNA damage. Much controversy has surrounded the use of phenol for DNA extraction,[7, 64-67] but recently, it was shown that phenol extraction has no profound effect on 8OH2'dG levels.[55, 68, 69] Additional factors which have been implicated in artifactual formation of DNA adducts during DNA extraction include presence of transition metals, oxygen, type of buffer used, the amount and type of tissue, and the amount of DNA.[7, 55, 69, 70] The artifactual production of 8OH2'dG can be minimized by using the chaotropic sodium iodine DNA extraction method, and inclusion of the spin trap 2,2,6,6-tetramethylpiperidine-N-oxyl (TEMPO) into the buffers used during DNA sample preparation.[55, 69] After DNA isolation, bases and nucleotides are liberated using either chemical hydrolysis or enzymatic digestion and the mixture of released DNA components are commonly separated using GC, CE or HPLC, and measured using MS, UV, fluorescence, radioactive or EC detection. Additionally, several methods not requiring DNA isolation have been introduced. These are based on lysis of cells followed by detection of strand breaks using DNA repair enzymes. For example, an alkaline elution assay involving Fpg protein, endonuclease III and exonuclease III, has been used to detect oxidized purine and pyrimidine bases, and abasic sites.[71, 72] Single cell electrophoresis (comet assay) can be used to determine steady-state levels of Fpg protein sensitive DNA lesions within cellular DNA.[73, 74]

3. GCMS

GCMS has been used by many laboratories to measure a variety of DNA adducts, including cytosine glycol, thymine glycol, 8-hydroxyguanine, 2,6-diamino-4-hydroxy-5-formamidopyrimidine, 5-hydroxycytosine, 5-hydroxyuracil, 5-hydroxymethyluracil.[75-78] However the original version of the assay very significantly overestimates levels of 8OHG, 8-hydroxyadenine, 5-hydroxycytosine, 5-formyluracil, and 5-hydroxymethyluracil, as a result of artifactual formation of the adduct during the derivatization of DNA acidic hydrolysate.[75, 76] Use of guanase, ethanethiol, and decreasing temperature during derivatization, as well as use of HPLC to isolate DNA adducts prior to the derivatization (HPLC/GCMS) have been shown to decrease artifactual production of the lesions.[76-78] The HPLC/GCMS approach has been used to simultaneously measure 5-hydroxyuracil, 5-hydroxymethyluracil, 8-hydroxyadenine, 8-OHG and 8-OH2'dG in human urine samples.[79] Recently HPLC/MS-MS methods, devoid of many of the drawbacks of GCMS, have been introduced for the measurement of free levels of DNA adducts.[80, 81]

4. Postlabeling

Another approach to measure DNA adducts uses postlabeling. The ^{32}P-postlabeling assay is based on isolation and digestion of DNA followed by ^{32}P-ATP postlabeling of 3'-monophosphates, which are substrates of polynucleotide kinases.[82] ^{32}P-labeled adducts are separated using TLC or HPLC prior to detection. The assay has been used for measurements of 8OH2'dG and 5-hydroxymethyluracil.[69, 83]

5. HPLC/EC

Probably the most common analytical method is HPLC coupled to electrochemical detection (HPLC/EC). This method was originally introduced by Floyd in 1986,[84] who used single electrode EC detector to measure 8OH2'dG in human granulocytes exposed to the tumor promoter, tetradeconylphorbolacetate. Since then several HPLC/EC techniques have been used for the measurement of different DNA adducts including 7-methylguanine,[85-87] (deoxyguanine-8-yl)-aminofluorene and N-(deoxyguanine-8-yl)-acetylaminofluorene,[88] N7-(2-hydroxyethyl)-guanine,[89] 5-hydroxy-2'-deoxycytidine,[90] 1,N2-propano adducts of 2'-deoxyguanosine with 4-hydroxynonenal,[91] 2,6-diamino-4-hydroxy-5-formamidopyrimidine,[87, 92] 8-hydroxyadenine,[87] N2-methyl-8-oxoguanine,[93] adducts of 4,5-dioxovaleric acid with 2'-deoxyguanosine,[94] and 8-nitroguanine.[95, 96]

Currently, HPLC/EC methods employing coulometric detection offer the most selective and sensitive approach for the measurement of 8OH2'dG within DNA.[80] The coulometric sensor has several advantages over thin-layer amperometric sensor including higher sensitivity, selectivity, better stability, and little maintenance.[97–99] The vast majority of studies using HPLC with dual-channel coulometric detector have focused on the either measurement of 8OH2'dG or 8OHG. Furthermore, using a coulometric electrode array it is not only possible to extend the range bases and adducts measured but also to electrochemically characterize EC-active compounds based on their hydrodynamic voltammetric behavior (analogous to using a photodiode array to spectrally characterize analytes). This offers an opportunity to qualify the analytes of interest as well as to determine possible coelutions. For example, a new method using gradient HPLC separation coupled to a 12-channel coulometric array and a serially placed UV detector was recently developed for the simultaneous measurement of a wide variety of DNA adducts, bases and nucleosides[100] (Fig. 1).

A number of methods have been used to assess free 8OH2'dG levels in different biological matrices with urinary measurements being by far the most common. 8OH2'dG measurement in urine is an inherently challenging analytical problem due to its very low level, polarity, and the complexity of the matrix. The

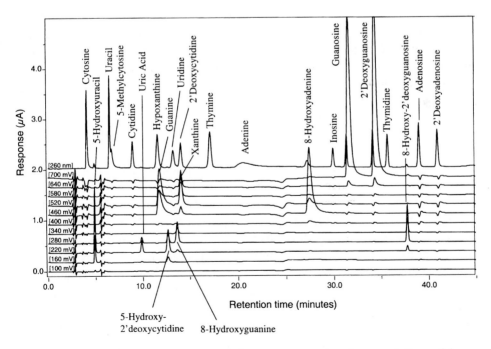

Fig. 1. Gradient HPLC-CoulArray-UV chromatogram showing resolution of bases, nucleosides and DNA adducts. Analytical conditions as in Ref. 100.

assay is additionally confounded by a variability of potential interferences presented in the samples from subjects with different disorders. Furthermore, it is important to note that measurements of 8OH2'dG in urinary samples where the precipitate (formed during freezing-thawing of urine, and containing mostly uric acid) was discarded, should be interpreted with caution. Recently it was shown that the precipitate can carry down significant amounts of 8OH2'dG — up to 60% of its total amount in the sample.[101] Several HPLC/EC methods have been used for urinary 8OH2'dG measurements. All of these methods have employed different approaches to minimize interferences, either prior to, or during the assay. Solid phase extraction (SPE), immunoaffinity columns following SPE, column switching methods, and SPE in conjunction with column switching methods, have been used with HPLC/EC.[102–108] The utility of methods for measurement of 8OH2'dG in urine using SPE protocols followed by HPLC/EC, depends on the type of EC detector used. Although SPE approaches significantly reduce the complexity of chromatograms, the results obtained using a single electrode amperometric sensor could be erroneous, due to the inherent lack of voltammetric resolution. For example, using a gradient coulometric array system, it was shown that the urinary 8OH2'dG peak that appeared to be pure on a single electrode, actually represents several coeluting species.[98] Moreover, the pattern of the interferences showed very high interindividual variability.[98] HPLC/EC methods utilizing an immunoaffinity approach following SPE, are relatively simple and could potentially be useful for analysis of 8OH2'dG in urine.[2, 102–104, 109] However, the efficiency of immunoaffinity columns could be problematic due to decreased binding capacity of the columns and the potential for coelutions developing over time.[55, 109] ELISA methods using monoclonal antibodies have been introduced for measurements of 8OH2'dG[110] and 3-methyladenine.[111] However, the concentrations of urinary 8OH2'dG both in rats and humans obtained using ELISA methods are typically 3–100 times higher than those measured using HPLC.[45, 110, 112] Therefore the applicability of ELISA methods for the measurement of free 8OH2'dG levels requires further investigation.

Column switching methods have been used to generate most of the data on 8OH2'dG levels in urine in normal individuals and in epidemiological studies.[56, 105, 107, 108] The most common column switching methods use a triple-column approach. Here the sample is injected onto the first column and the portion of eluent containing the peak of interest is then transferred from the first column to the second trapping column. Further resolution is obtained when the analyte is finally transferred from the trapping column onto the third (analytical) column. Interestingly, in some studies, a SPE procedure prior to injecting the sample onto the first column was used.[107] Because of the capacity for automation, the column switching method appears to be an attractive approach to the analytical problem. The key factor in using this method for analysis of urine samples is the selectivity of the trapping column for 8OH2'dG. In previous studies, cation exchange[105] or C18[107] columns have been used as trapping columns. When replicating these approaches we found satisfactory performance for the analysis of most urinary samples from healthy control individuals, but not for the analysis of urine obtained from patients with Parkinson's disease, Alzheimer's disease, and amyotrophic lateral sclerosis (ALS). We also encountered significant problems with urine samples from mice and rats, probably because of the inherent contamination of the samples during collection using metabolic cages. A major problem with SPE and column switching methods is that trapping columns used in previous studies (polymer or silica packing materials) have very similar retention characteristics for 8OH2'dG and unknown interferences. These observations suggested an approach based on the use different materials for trapping columns which have properties highly dissimilar to those of polymer or silica materials.

We found that different classes of carbon could be treated by oxidative-reductive cycling and made to be uniquely selective for purines and other aromatic and heterocyclic compounds.[101] For example, in acidic lithium acetate, containing 5% acetonitrile, the capacity factor (k') of the carbon columns for 5-hydroxy-2-deoxycytidine, 8-OHG, guanine, 3-nitrotyrosine, guanosine, 7-methylguanine, and 8OH2'dG are: 10, 68, 99, 121, 164, 203, and 213, respectively. Under the same conditions the k' values for these analytes on C18 columns are all less than 3. The treated carbon materials can be used in two ways: as SPE preparative columns prior to an isocratic or gradient methods coupled to HPLC coulometric array systems, or as trapping columns in column switching methods for analysis of different DNA adducts.

For off-line preparative work we have used the following protocol: 0.25 mL urine sample is placed on a C18 SPE column; the column is then washed with 3 mL of 3% methanol, and the eluent is transferred to a 10 mm × 5 mm carbon column. The carbon column is washed with 3 mL of 25% methanol, and 8OH2'dG is eluted from the carbon column with 0.25 mL of mobile phase containing 5 g/L of adenosine. The resulting samples are then analyzed on an isocratic or gradient coulometric array system. For various disorders, gradient systems are necessary to achieve separation of 8OH2'dG in all samples.

6. Column Switching HPLC/EC Method

We have utilized the selectivity of the carbon columns for purines to develop a column switching HPLC/EC method for analysis of 8OH2'dG in different biological matrices and within DNA. The HPLC/EC setup and column switching arrangement is shown in Fig. 2. The method involves transfer of the band containing 8OH2'dG from the C8 column using basic lithium acetate, containing 4% methanol, to two serially placed carbon columns. After the 8OH2'dG-containing band is trapped on the carbon columns, the C8 column is backflushed to eliminate

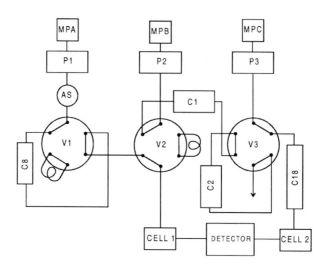

Fig. 2. Valve switching arrangement and HPLC/EC setup of the system for 8OH2'dG. P1, P2 and P3 — pumps delivering mobile phases (MP) A (0.1 M lithium acetate, pH 6.4, methanol 4%), MP B (0.1 M lithium acetate, pH 3.3, acetonitrile 4.5%) and MP C (the same as MP B, but with the 1.5 g/L adenosine), respectively. AS — an autosampler. V1–V3 — high-pressure valves 1-3. C8 — the C8 column, C18 — the C18 column. C1 and C2 — the first and the second carbon columns in series, respectively. Cell 1 and cell 2 — coulometric cells (either dual-channel, or 4- to 12-channel). The detector is either a Coulochem® II with four potentiostats, or a 16-channel Coularray®.

late-eluting peaks and thus minimizing analysis time. Because of the high selectivity of the carbon columns for 8OH2'dG, it is possible to elute potential interferences, while moving 8OH2'dG quantitatively from the first to second carbon column in series, using acidic lithium acetate, containing 4.5% acetonitrile. The 8OH2'dG is then released from the second carbon column to the C18 column using an adenosine-containing mobile phase. A coulometric cell at the output of the C8 column is used for detection of 2'dG in DNA samples, and also to determine the retention time for 8OH2'dG in order to set the times for valve switching. A coulometric cell at the output of the C18 column is used for detection of 8OH2'dG. For analysis of 8OH2'dG in urine we routinely use a 4-channel Coulochem II EC detector which controls coulometric electrodes placed after the C8 column (model 5010 analytical cell), and at the output of the C18 column (serially placed model 5021 conditioning cell, and 5014B analytical cell). Depending on the type of the samples, additional EC resolution may be required to isolate and identify additional analytes, as well as 8OH2'dG, in the final chromatogram. In this case a 16-channel coulometric array detector (Model 5600A) with a 4-channel cell monitoring the output of the C8 column, and 12-channel sensor at the output of the C18 column, is used. Typical chromatograms illustrating the effectiveness of

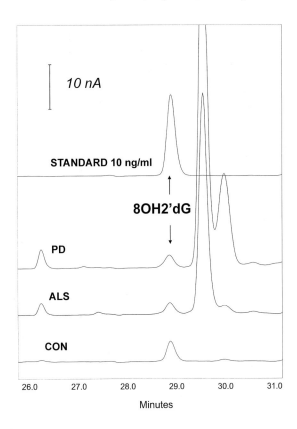

Fig. 3. 8OH2'dG in urine samples from a healthy control subject (CON), and from patients with amyotrophic lateral sclerosis (ALS), and Parkinson's disease (PD). The chromatograms illustrate the selective separation for 8OH2'dG from disorder-related interferences, using carbon column switching method.

carbon columns in achieving selective separation of the 8OH2'dG in urine, are shown in Fig. 3.

Various configurations of the carbon column switching system have been used extensively in this laboratory for over 6000 assays of 8OH2'dG in various matrices. External blind replicate studies on over 300 samples have performed within the published precision of the method. A number of critical issues relate to using 8OH2'dG as a clinical marker have been studied. These are temporal stability, male-female differences and relationships between creatinine corrected spot urine values and whole body output rate per unit body weight. A study of 12 individuals taking every urine void over 72 h showed no net diurnal variation and an average individual variation of 6.51%. Urine samples from 56 control subjects at a 9 month interval showed a net average change of −0.7%, a correlation coefficient of $r = 0.928$ and a standard error of estimating the initial sample from the 9 month sample of

Table 1

	Mobile Phase A*	Mobile Phase B/C†	C1/C2‡	G1/G2§	Plasma (pg/mL)	Urine (μg/g creatinine)	Cervical Extracellular Matrix (pg/g)
8OH2'dG	6% MeOH (B1)	4.5% An (B2)/ADN	2/1	A/A	10.3	3.86	200
O6MG	20% MeOH (B1)	6% An (B2)/ADN	1/1	A/B	0.5	0.32	44
2'dG/7MG	10% MeOH; 3% An (B1)	7% An (B2)/ADN	2/1	A/B	500/3	ND/0.6	2000/15
8OHG	1% MeOH (B1)	1% An (B2)/ADN	1/1	B/B	26	9.83	80
8NG	25% MeOH; 4% An (B2)	18% An (B1)/NBA	1/1	A/B	ND	ND	(?)
8OH2'dA	8% MeOH (B1)	4% An (B2)/DD	2/1	A/A	ND	0.60	30
5OH2'dCy	1% MeOH (B3)	1% An (B2)/ADN	1/1	B/C	ND	ND	30
5OHU	1% MeOH (B3)	1% An (B3)/ADN	1/1	B/C	ND	0.40	56
3NT/3CIT	10% MeOH; 3% An (B1)	7% An (B1)/NBA	2/1	B/B	3.7/1.1	ND	20/10

*Mobile phases A, B/C and buffer (B 1,2,3): MeOH — methanol, An-acetonitrile; B1 — lithium acetate 0.1M, pH 6; B2 — lithium phosphate 0.1M, pH 3; B3 — pentane sulfonic acid 0.1M, pH 4.

†C Mobile phase eluting agent: ADN — adenosine; DD — dodecane; NBA — nitrobenzoic acid.

‡C1/C2 Column types: 1 — TosoHaas C18 ODS 80 TM; 2 — YMC C8 Y02H1.

§G1/G2 Carbon columns: A — 4.6 mm × 4.6 mm B — 4.6 mm × 8 mm; C — 4.6 mm × 13 mm.

ND — not detected.

8OH2'dG — 8-Hydroxy 2'deoxyguanosine, O6MG — O-6-Methylguanosine, 2'dG — 2'deoxyguanosine, 7MG — 7-Methylguanine, 8OHG — 8-Hydroxyguanine, 8NG — 8-Nitroguanine, 8OH2'dA — 8-Hydroxy-2'deoxyadenosine, 5OH2'dCy — 5-Hydroxy-2'deoxycytosine, 5OHU — 5-Hydroxyuracil, 3NT — 3-Nitrotyrosine, 3CIT — 3-Chlorotyrosine.

0.242 ng 8OH2'dG/mg creatinine over a range of 2.018–5.648 ng 8OH2'dG/mg creatinine. In a group of 62 female and 74 male healthy adult volunteers the creatinine corrected values were significantly higher for females than males (4.032 versus 3.866 ng 8OH2'dG/mg creatinine, respectively; $p = 0.039$) and the rate of output was significantly higher for males than females (0.0531 versus 0.492 ng 8OH2'dG/kg/min, respectively; $p = 0.046$). The correlation coefficients were $r = 0.936$ for males and $r = 0.961$ for females between creatinine corrected values and whole body output rate. For the purposes of study design samples can be acquired at any time and spot urines are equivalent to whole body output rate. In the absence of disease progression or intervention 8OH2'dG urinary levels are highly characteristic and stable within an individual.

Because of the selectivity and sensitivity of the carbon column switching method it can be used in the studies of free 8OH2'dG levels in multiple compartments over time. Recently we have found significant elevations of 8OH2'dG in urine, plasma and cerebrospinal fluid from the patients with ALS, and the rate of the increase has correlated with the severity of disease progression.[113] Additionally, the method allows the measurements of free 8OH2'dG levels in the samples collected from the localized areas, such as brain microdialysates, cervical extracellular matrix from PAP smears, pharyngeal swabs, and bronchial alveolar lavage.

The unique selectivity of the carbon columns to other purines and nitro- or chloro-substituted compounds allows its generalization to the measurement of free levels of a variety of DNA damage/repair products and free radical markers. Essentially, the approach involves balancing the variables of selection of mobile phases A and B and the agent used to release a compound of interest from the carbon, selection of the appropriate C18, C8 or ion exchange columns, and selection of the carbon column dimensions. All these variables depend on the specific pattern of the interferences in the sample. Table 1 lists the conditions developed for a number of DNA damage and free radical markers. Methods for 8OH2'dG, 2'dG, 8OHG and O(6)-methylguanine (O6MG) have been run extensively. Other conditions have been established for the purpose of demonstrating feasibility and evaluating levels of DNA damage markers in sample types representing whole body compartments (urine and plasma) and localized area compartments (e.g. extracellular cervical matrix from PAP smears).

7. Conclusions

Coulometric electrode detection with liquid chromatography offers a powerful tool for the study of individual adducts formed in DNA. Coulometric electrode arrays coupled directly with conventional UV detection further offer the ability to simultaneously determine multiple types of adducts and their precursor bases.

The use of the unique selectivity of carbon columns either preparatively or in an automated column switching system allows the rapid sensitive and highly selective separation and assay of free levels of a range of DNA damage/repair products from highly complex matrices in clinically accessible samples. The combination of these techniques holds the promise of helping to establish the nature of the interaction of an individuals innate metabolic processes and environment with their genomic profile.

References

1. Halliwell, B. (1992). Reactive oxygen species and the central nervous system. *J. Neurochem.* **59**(5): 1609–1623.
2. Ames, B. N. (1989). Endogenous oxidative DNA damage, aging, and cancer. *Free Radic Res. Commun.* **7**(3–6): 121–128.
3. Beal, M. F. (1995). Aging, energy, and oxidative stress in neurodegenerative diseases. *Ann. Neurol.* **38**(3): 357–366.
4. Beal, M. F. (2000). Energetics in the pathogenesis of neurodegenerative diseases. *Trends Neurosci.* **23**(7): 298–304.
5. Bowling, A. C. and Beal M. F. (1995). Bioenergetic and oxidative stress in neurodegenerative diseases. *Life Sci.* **56**(14): 1151–1171.
6. Dandona, P. *et al* (1996). Oxidative damage to DNA in diabetes mellitus. *Lancet* **347**(8999): 444–445.
7. Kasai, H. (1997). Analysis of a form of oxidative DNA damage, 8-hydroxy-2'-deoxyguanosine, as a marker of cellular oxidative stress during carcinogenesis. *Mutat. Res.* **387**(3): 147–163.
8. Leinonen, J. *et al.* (1997). New biomarker evidence of oxidative DNA damage in patients with non-insulin-dependent diabetes mellitus. *FEBS Lett.* **417**(1): 150–152.
9. Mecocci, P. *et al.* (1993). Oxidative damage to mitochondrial DNA shows marked age-dependent increases in human brain. *Ann. Neurol.* **34**(4): 609–616.
10. Fiskum, G., Murphy, A. N. and Beal M. F. (1999). Mitochondria in neurodegeneration: acute ischemia and chronic neurodegenerative diseases. *J. Cereb. Blood Flow Metab.* **19**(4): 351–369.
11. Cadet, J. *et al.* (1999). Hydroxyl radicals and DNA base damage. *Mutat. Res.* **424**(1–2): 9–21.
12. Floyd, R. A. (1990). The role of 8-hydroxyguanine in carcinogenesis. *Carcinogenesis* **11**(9): 1447–1450.
13. Hayakawa, M. *et al.* (1991). Age-associated accumulation of 8-hydroxydeoxyguanosine in mitochondrial DNA of human diaphragm. *Biochem. Biophys. Res. Commun.* **179**(2): 1023–1029.

14. Pryor, W. A. (1988). Why is the hydroxyl radical the only radical that commonly adds to DNA? Hypothesis: it has a rare combination of high electrophilicity, high thermochemical reactivity, and a mode of production that can occur near DNA. *Free Radic. Biol. Med.* **4**(4): 219–223.
15. Spencer, J. P. *et al.* (1996). Base modification and strand breakage in isolated calf thymus DNA and in DNA from human skin epidermal keratinocytes exposed to peroxynitrite or 3-morpholinosydnonimine. *Chem. Res. Toxicol.* **9**(7): 1152–1158.
16. Inoue, S. and Kawanishi, S. (1995). Oxidative DNA damage induced by simultaneous generation of nitric oxide and superoxide. *FEBS Lett.* **371**(1): 86–88.
17. Randerath, K. *et al.* (1996). Structural origins of bulky oxidative DNA adducts (type II I-compounds) as deduced by oxidation of oligonucleotides of known sequence. *Chem. Res. Toxicol.* **9**(1): 247–254.
18. Lloyd, D. R., Carmichael, P. L. and Phillips D. H. (1998). Comparison of the formation of 8-hydroxy-2'-deoxyguanosine and single- and double-strand breaks in DNA mediated by fenton reactions. *Chem. Res. Toxicol.* **11**(5): 420–427.
19. Lloyd, D. R., Phillips, D. H. and Carmichael, P. L. (1997). Generation of putative intrastrand cross-links and strand breaks in DNA by transition metal ion-mediated oxygen radical attack. *Chem. Res. Toxicol.* **10**(4): 393–400.
20. Cadet, J. *et al.* (1997). Oxidative damage to DNA: formation, measurement, and biological significance. *Rev. Physiol. Biochem. Pharmacol.* **131**: 1–87.
21. Dizdaroglu, M. *et al.* (1993). Modification of DNA bases in chromatin of intact target human cells by activated human polymorphonuclear leukocytes. *Cancer Res.* **53**(6): 1269–1272.
22. Breen, A. P. and Murphy, J. A. (1995). Reactions of oxyl radicals with DNA. *Free Radic. Biol. Med.* **18**(6): 1033–1077.
23. Cadet, J. *et al.* (1994). Singlet oxygen DNA damage: chromatographic and mass spectrometric analysis of damage products. *Meth. Enzymol.* **234**: 79–88.
24. Dizdaroglu, M. (1994). Chemical determination of oxidative DNA damage by gas chromatography-mass spectrometry. *Meth. Enzymol.* **234**: 3–16.
25. Kasai, H. and Nishimura, S. (1984). Hydroxylation of deoxy guanosine at the C-8 position by polyphenols and aminophenols in the presence of hydrogen peroxide and ferric ion. *Gann.* **75**(7): 565–566.
26. Kaneko, T., Tahara, S. and Matsuo, M. (1996). Non-linear accumulation of 8-hydroxy-2'-deoxyguanosine, a marker of oxidized DNA damage, during aging. *Mutat. Res.* **316**(5–6): 277–285.
27. Greenberg, J. A. *et al.* (2000). Whole-body metabolic rate appears to determine the rate of DNA oxidative damage and glycation involved in aging. *Mech. Ageing Dev.* **115**(1–2): 107–117.

28. Izzotti, A. *et al.* (1999). Age-related increases of 8-hydroxy-2′-deoxyguanosine and DNA-protein crosslinks in mouse organs. *Mutat. Res.* **446**(2): 215–223.
29. Inoue, M. *et al.* (1998). Lung cancer patients have increased 8-hydroxydeoxyguanosine levels in peripheral lung tissue DNA. *Japan J. Cancer Res.* **89**(7): 691–695.
30. Bostwick, D. G. *et al.* (2000). Antioxidant enzyme expression and reactive oxygen species damage in prostatic intraepithelial neoplasia and cancer. *Cancer* **89**(1): 123–134.
31. Romano, G. *et al.* (2000). 8-hydroxy-2′-deoxyguanosine in cervical cells: correlation with grade of dysplasia and human papillomavirus infection. *Carcinogenesis* **21**(6): 1143–1147.
32. Kondo, S. *et al.* (2000). Overexpression of the hOGG1 gene and high 8-hydroxy-2′-deoxyguanosine (8-OHdG) lyase activity in human colorectal carcinoma: regulation mechanism of the 8-OHdG level in DNA. *Clin. Cancer Res.* **6**(4): 1394–1400.
33. Matsui, A. *et al.* (2000). Increased formation of oxidative DNA damage, 8-hydroxy-2′-deoxyguanosine, in human breast cancer tissue and its relationship to GSTP1 and COMT genotypes. *Cancer Lett.* **151**(1): 87–95.
34. Hardie, L. J. *et al.* (2000). The effect of hOGG1 and glutathione peroxidase I genotypes and 3p chromosomal loss on 8-hydroxydeoxyguanosine levels in lung cancer. *Carcinogenesis* **21**(2): 167–172.
35. Suzuki, S. *et al.* (1999). Oxidative damage to mitochondrial DNA and its relationship to diabetic complications. *Diabetes Res. Clin. Pract.* **45**(2–3): 161–168.
36. Ihara, Y. *et al.* (1999). Hyperglycemia causes oxidative stress in pancreatic beta-cells of GK rats, a model of Type-2 diabetes. *Diabetes* **48**(4): 927–932.
37. Lezza, A. M. *et al.* (1999). Mitochondrial DNA 4977 bp deletion and OH8dG levels correlate in the brain of aged subjects but not Alzheimer's disease patients. *FASEB J.* **13**(9): 1083–1088.
38. Browne, S. E., Ferrante, R. J. and Beal, M. F. (1999). Oxidative stress in Huntington's disease. *Brain Pathol.* **9**(1): 147–163.
39. Lovell, M. A., Gabbita, S. P. and Markesbery, W. R. (1999). Increased DNA oxidation and decreased levels of repair products in Alzheimer's disease ventricular CSF. *J. Neurochem.* **72**(2): 771–776.
40. Jovanovic, S. V., Clements, D. and MacLeod, K. (1998). Biomarkers of oxidative stress are significantly elevated in Down syndrome. *Free Radic. Biol. Med.* **25**(9): 1044–1048.
41. Vladimirova, O. *et al.* (1998). Oxidative damage to DNA in plaques of MS brains. *Mult. Scler.* **4**(5): 413–418.
42. Mecocci, P. *et al.* (1998). Oxidative damage to DNA in lymphocytes from AD patients. *Neurology* **51**(4): 1014–1017.

43. Romanenko, A. *et al.* (2000). Increased oxidative stress with gene alteration in urinary bladder urothelium after the Chernobyl accident. *Int. J. Cancer* **86**(6): 790–798.
44. Xu, A. *et al.* (1999). Role of oxyradicals in mutagenicity and DNA damage induced by crocidolite asbestos in mammalian cells. *Cancer Res.* **59**(23): 5922–5926.
45. Toraason, M. *et al.* (1999). Oxidative stress and DNA damage in Fischer rats following acute exposure to trichloroethylene or perchloroethylene. *Toxicology* **138**(1): 43–53.
46. Arimoto, T. *et al.* (1999). Generation of reactive oxygen species and 8-hydroxy-2′-deoxyguanosine formation from diesel exhaust particle components in L1210 cells. *Japan J. Pharmacol.* **80**(1): 49–54.
47. Kamendulis, L. M. *et al.* (1999). Induction of oxidative stress and oxidative damage in rat glial cells by acrylonitrile. *Carcinogenesis* **20**(8): 1555–1560.
48. Matsui, M. *et al.* (1999). The role of oxidative DNA damage in human arsenic carcinogenesis: detection of 8-hydroxy-2′-deoxyguanosine in arsenic-related Bowen's disease. *J. Invest. Dermatol.* **113**(1): 26–31.
49. Takahashi, K. *et al.* (1997). Relationship between asbestos exposures and 8-hydroxydeoxyguanosine levels in leukocytic DNA of workers at a Chinese asbestos-material plant. *Int. J. Occup. Env. Health* **3**(2): 111–119.
50. Hermanns, R. C. *et al.* (1998). Urinary excretion of biomarkers of oxidative kidney damage induced by ferric nitrilotriacetate. *Toxicol. Sci.* **43**(2): 241–249.
51. Canova, S. *et al.* (1998). Tissue dose, DNA adducts, oxidative DNA damage and CYP1A-immunopositive proteins in mussels exposed to waterborne benzo[a]pyrene. *Mutat. Res.* **399**(1): 17–30.
52. Asami, S. *et al.* (1997). Cigarette smoking induces an increase in oxidative DNA damage, 8-hydroxydeoxyguanosine, in a central site of the human lung. *Carcinogenesis* **18**(9): 1763–1766.
53. Takeuchi, T., Nakajima, M. and Morimoto, K. (1999). A human cell system for detecting asbestos cytogenotoxicity in vitro. *Mutat. Res.* **438**(1): 63–70.
54. Stein, T. P. and Leskiw, M. J. (2000). Oxidant damage during and after spaceflight. *Am. J. Physiol. Endocrinol. Metab.* **278**(3): E375–E382.
55. Helbock, H. J. *et al.* (1998). DNA oxidation matters: the HPLC-electrochemical detection assay of 8-oxo-deoxyguanosine and 8-oxo-guanine. *Proc. Natl. Acad. Sci. USA* **95**(1): 288–293.
56. Loft, S. and Poulsen, H. E. (1999). Markers of oxidative damage to DNA: antioxidants and molecular damage. *Meth. Enzymol.* **300**: 166–184.
57. Zhang, D. *et al.* (1997). Vitamin E inhibits apoptosis, DNA modification, and cancer incidence induced by iron-mediated peroxidation in Wistar rat kidney. *Cancer Res.* **57**(12): 2410–2414.
58. Runger, T. M. *et al.* (2000). DNA damage formation, DNA repair, and survival after exposure of DNA repair-proficient and nucleotide excision

repair-deficient human lymphoblasts to UVA1 and UVB. *Int. J. Rad. Biol.* **76**(6): 789–797.

59. Mattson, M. P., Culmsee, C. and Yu, Z. F. (2000). Apoptotic and antiapoptotic mechanisms in stroke. *Cell Tissue Res.* **301**(1): 173–187.

60. Formichi, P. *et al.* (2000). Apoptotic response and cell cycle transition in ataxia telangiectasia cells exposed to oxidative stress. *Life Sci.* **66**(20): 1893–1903.

61. Cui, J. *et al.* Oxidative DNA damage precedes DNA fragmentation after experimental stroke in rat brain. *FASEB J.* **14**(7): 955–967.

62. Mates, J. M. and Sanchez-Jimenez, F. M. (2000). Role of reactive oxygen species in apoptosis: implications for cancer therapy. *Int. J. Biochem. Cell Biol.* **32**(2): 157–170.

63. Lovell, M. A., Xie, C. and Markesbery, W. R. (2000). Decreased base excision repair and increased helicase activity in Alzheimer's disease brain. *Brain Res.* **855**(1): 116–123.

64. Claycamp, H. G. (1992). Phenol sensitization of DNA to subsequent oxidative damage in 8-hydroxyguanine assays. *Carcinogenesis* **13**(7): 1289–1292.

65. Adachi, S., Zeisig, M. and Moller, L. (1995). Improvements in the analytical method for 8-hydroxydeoxyguanosine in nuclear DNA. *Carcinogenesis* **16**(2): 253–258.

66. Nakae, D. *et al.* (1995). Improved genomic/nuclear DNA extraction for 8-hydroxydeoxyguanosine analysis of small amounts of rat liver tissue. *Cancer Lett.* **97**(2): 233–239.

67. Nicotera, T. M. and Fiel, R. J. (1994). Photosensitized formation of 8-hydroxy-2'-deoxyguanosine by a cationic meso-substituted porphyrin. *Adv. Exp. Med. Biol.* **366**: 416–417.

68. Hofer, T. and Moller, L. (1998). Reduction of oxidation during the preparation of DNA and analysis of 8-hydroxy-2'-deoxyguanosine. *Chem. Res. Toxicol.* **11**(8): 882–887.

69. Moller, L., Hofer, T. and Zeisig, M. (1998). Methodological considerations and factors affecting 8-hydroxy-2-deoxyguanosine analysis. *Free Radic. Res.* **29**(6): 511–524.

70. Spear, N. and Aust, S. D. (1998). The effects of different buffers on the oxidation of DNA by thiols and ferric iron. *J. Biochem. Mol. Toxicol.* **12**(2): 125–132.

71. Epe, B. and Hegler, J. (1994). Oxidative DNA damage: endonuclease finger-printing. *Meth. Enzymol.* **234**: 122–131.

72. Pflaum, M., Will, O. and Epe, B. (1997). Determination of steady-state levels of oxidative DNA base modifications in mammalian cells by means of repair endonucleases. *Carcinogenesis* **18**(11): 2225–2231.

73. Collins, A. R. *et al.* (1997). The comet assay: what can it really tell us? *Mutat. Res.* **375**(2): 183–193.

74. Torbergsen, A. C. and Collins, A. R. (2000). Recovery of human lymphocytes from oxidative DNA damage; the apparent enhancement of DNA repair by carotenoids is probably simply an antioxidant effect. *Eur. J. Nutri.* **39**(2): 80–85.

75. Ravanat, J. L. *et al.* (1995). Determination of 8-oxoguanine in DNA by gas chromatography-mass spectrometry and HPLC-electrochemical detection: overestimation of the background level of the oxidized base by the gas chromatography-mass spectrometry assay. *Chem. Res. Toxicol.* **8**(8): 1039–1045.

76. Douki, T. *et al.* (1996). Observation and prevention of an artefactual formation of oxidized DNA bases and nucleosides in the GC-EIMS method. *Carcinogenesis* **17**(2): 347–353.

77. Herbert, K. E. *et al.* (1996). A novel HPLC procedure for the analysis of 8-oxoguanine in DNA. *Free Radic. Biol. Med.* **20**(3): 467–472.

78. Jenner, A. *et al.* (1998). Measurement of oxidative DNA damage by gas chromatography-mass spectrometry: ethanethiol prevents artifactual generation of oxidized DNA bases. *Biochem. J.* **331 (Part 2)**: 365–369.

79. Ravanat, J. L. *et al.* (1999). Simultaneous determination of five oxidative DNA lesions in human urine. *Chem. Res. Toxicol.* **12**(9): 802–808.

80. (2000). Comparison of different methods of measuring 8-oxoguanine as a marker of oxidative DNA damage. ESCODD (European Standards Committee on Oxidative DNA Damage). *Free Radic. Res.* **32**(4): 333–341.

81. Ravanat, J. L., Remaud, G. and Cadet, J. (2000). Measurement of the main photooxidation products of 2'-deoxyguanosine using chromatographic methods coupled to mass spectrometry. *Arch. Biochem. Biophys.* **374**(2): 118–127.

82. Randerath, K., Reddy, M. V. and Gupta, R. C. (1981). 32P-labeling test for DNA damage. *Proc. Natl. Acad. Sci. USA* **78**(10): 6126–6129.

83. Cadet, J. *et al.* (1992). Chemical and biochemical postlabeling methods for singling out specific oxidative DNA lesions. *Mutat Res.* **275**(3–6): 343–354.

84. Floyd, R. A. *et al.* (1986). Hydroxyl free radical adduct of deoxyguanosine: sensitive detection and mechanisms of formation. *Free Radic. Res. Commun.* **1**(3): 163–172.

85. Park, J. W. and Ames, B. N. (1988). 7-methylguanine adducts in DNA are normally present at high levels and increase on aging: analysis by HPLC with electrochemical detection [published erratum appears in *Proc. Natl. Acad. Sci. USA* **85**(24): 9508]. *Proc. Natl. Acad. Sci. USA* **85**(20): 7467–7470.

86. Bianchini, F. *et al.* (1993). Quantification of 7-methyldeoxyguanosine using immunoaffinity purification and HPLC with electrochemical detection. *Carcinogenesis* **14**(8): 1677–1682.

87. Kaur, H. and Halliwell, B. (1996). Measurement of oxidized and methylated DNA bases by HPLC with electrochemical detection. *Biochem. J.* **318 (Part 1)**: 21–23.

88. Bol, S. A. *et al.* (1997). Electrochemical detection and quantification of the acetylated and deacetylated C8-deoxyguanosine DNA adducts induced by 2-acetylaminofluorene. *Anal. Biochem.* **251**(1): 24–31.
89. Van Delft, J. H. *et al.* (1991). Determination of N7-(2-hydroxyethyl)guanine by HPLC with electrochemical detection. *Chem. Biol. Interact* **80**(3): 281–289.
90. Wagner, J. R., Hu, C. C. and Ames, B. N. (1992). Endogenous oxidative damage of deoxycytidine in DNA. *Proc. Natl. Acad. Sci. USA* **89**(8): 3380–3384.
91. Douki, T. and Ames, B. N. (1994). An HPLC-EC assay for 1,N2-propano adducts of 2′-deoxyguanosine with 4-hydroxynonenal and other alpha, beta-unsaturated aldehydes. *Chem. Res. Toxicol.* **7**(4): 511–518.
92. Douki, T. *et al.* (1996). Measurement of oxidative damage at pyrimidine bases in gamma-irradiated DNA. *Chem. Res. Toxicol.* **9**(7): 1145–1151.
93. Helbock, H. J. *et al.* (1996). N2-methyl-8-oxoguanine: a tRNA urinary metabolite-role of xanthine oxidase. *Free Radic. Biol. Med.* **20**(3): 475–481.
94. Douki, T. *et al.* (1998). DNA alkylation by 4,5-dioxovaleric acid, the final oxidation product of 5-aminolevulinic acid. *Chem. Res. Toxicol.* **11**(2): 150–157.
95. Ohshima, H. *et al.* (1998). Antioxidant and pro-oxidant actions of flavonoids: effects on DNA damage induced by nitric oxide, peroxynitrite and nitroxyl anion. *Free Radic. Biol. Med.* **25**(9): 1057–1065.
96. Yermilov, V., Rubio, J. and Ohshima, H. (1995). Formation of 8-nitroguanine in DNA treated with peroxynitrite in vitro and its rapid removal from DNA by depurination. *FEBS Lett.* **376**(3): 207–210.
97. Acworth, I. N., Naoi, M., Parvez, H. and Parvez, S. (1997). Coulometric electrode array detectors for HPLC. *In* "Progress in HPLC-HPCE", Vol. 6. VSP, Netherlands.
98. Acworth, I. N., McCabe, D. R. and Maher, T. J. (1997). The analysis of free radicals, their reaction products and antioxidants. *In* "Oxidants, Antioxidants, and Free Radicals" (S. I. Baskin, and H. Salem, Eds.), pp. 23–77, Taylor and Francis, Washington, DC.
99. Acworth, I. N., Bailey, B. A. and Maher, T. J. (1998). The use of HPLC with electrochemical detection to monitor reactive oxygen and nitrogen species, markers of oxidative damage and antioxidants: application to the neurosciences. *In* "Neurochemical Markers of Degenerative Nervous Diseases and Drug Addiction" (G. A. Qureshi, H. Parvez, P. Caudy and S. Parvez, Eds.), pp. 3–56, Progress in HPLC-HPCE, Vol. 7. VSP, Netherlands.
100. McCabe, D. R., Hensley, K. and Acworth., I. N. (1999). Method for the detection of nucleosides, bases and hydroxylated adducts using gradient HPLC with coulometric array and UV detection. *J. Med. Food.* **2**(3–4): 209–214.
101. Bogdanov, M. B. *et al.* (1999). A carbon column-based liquid chromatography electrochemical approach to routine 8-hydroxy-2′-deoxyguanosine

measurements in urine and other biologic matrices: a one-year evaluation of methods. *Free Radic. Biol. Med.* **27**(5–6): 647–666.

102. Shigenaga, M. K., Gimeno, C. J. and Ames, B. N. (1989). Urinary 8-hydroxy-2'-deoxyguanosine as a biological marker of in vivo oxidative DNA damage. *Proc. Natl. Acad. Sci. USA* **86**(24): 9697–9701.

103. Fraga, C. G. *et al.* (1990). Oxidative damage to DNA during aging: 8-hydroxy-2'-deoxyguanosine in rat organ DNA and urine. *Proc. Natl. Acad. Sci. USA* **87**(12): 4533–4537.

104. Park, E. M. *et al.* (1992). Assay of excised oxidative DNA lesions: isolation of 8-oxoguanine and its nucleoside derivatives from biological fluids with a monoclonal antibody column. *Proc. Natl. Acad. Sci. USA* **89**(8): 3375–3379.

105. Loft, S. *et al.* (1993). 8-Hydroxydeoxyguanosine as a urinary biomarker of oxidative DNA damage. *J. Toxicol. Env. Health* **40**(2–3): 391–404.

106. Lunec, J. *et al.* (1994). 8-hydroxydeoxyguanosine. A marker of oxidative DNA damage in systemic lupus erythematosus. *FEBS Lett.* **348**(2): 131–138.

107. Lagorio, S. *et al.* (1994). Exposure to benzene and urinary concentrations of 8-hydroxydeoxyguanosine, a biological marker of oxidative damage to DNA. *Occup. Env. Med.* **51**(11): 739–743.

108. Tagesson, C. *et al.* Determination of urinary 8-hydroxydeoxyguanosine by automated coupled-column high performance liquid chromatography: a powerful technique for assaying in vivo oxidative DNA damage in cancer patients. *Eur. J. Cancer* **31**A(6): 934–940.

109. Shigenaga, M. K. *et al.* (1994). Assays of oxidative DNA damage biomarkers 8-oxo-2'-deoxyguanosine and 8-oxoguanine in nuclear DNA and biological fluids by high-performance liquid chromatography with electrochemical detection. *Meth. Enzymol.* **234**: 16–33.

110. Toyokuni, S. *et al.* (1997). Quantitative immunohistochemical determination of 8-hydroxy-2'-deoxyguanosine by a monoclonal antibody N45.1: its application to ferric nitrilotriacetate-induced renal carcinogenesis model. *Lab. Invest.* **76**(3): 365–374.

111. Braybrooke, J. P. *et al.* (2000). Evaluation of the alkaline comet assay and urinary 3-methyladenine excretion for monitoring DNA damage in melanoma patients treated with dacarbazine and tamoxifen. *Cancer Chemother. Pharmacol.* **45**(2): 111-119.

112. Erhola, M. *et al.* (1997). Biomarker evidence of DNA oxidation in lung cancer patients: association of urinary 8-hydroxy-2'-deoxyguanosine excretion with radiotherapy, chemotherapy, and response to treatment. *FEBS Lett.* **409**(2): 287–291.

113. Bogdanov *et al.* (2000). Increased oxidative damage to DNA in ALS patients. *Free Radic. Biol. Med.* **29**(7): 652–658.

Chapter 12

Analysis of 8-Hydroxyguanine in DNA and Its Repair

Takeshi Hirano and Hiroshi Kasai*

Takeshi Hirano and **Hiroshi Kasai** • Department of Environmental Oncology, School of Medicine, University of Occupational and Environmental Health, Japan
*Corresponding Author.
Tel: +81-93-691-7468, E-mail: h-kasai@med.uoeh-u.ac.jp

1. Introduction

To study the mechanisms of aging and some diseases, including cancer, evaluations of cellular oxidative DNA damage and its repair capacity are important and useful. Since the discovery in 1984 of 8-hydroxyguanine (8-OH-Gua, 8-oxoguanine) formation in DNA by hydroxyl radicals ($\cdot$OH), generated by Fenton-type reactions[1, 2] or X-irradiation,[3] studies of oxidative DNA damage have made marked advances. 8-OH-Gua is one of the major forms of oxidative DNA damage, and has been well studied because it is relatively easy to detect by using an HPCL-ECD system.[4] It is generally thought that increased 8-OH-Gua may be responsible for carcinogenesis, because it causes GC to TA transversion mutations.[5] Therefore, the levels of 8-OH-Gua in DNA have been assayed to clarify the mechanisms of aging or carcinogenesis and to evaluate the risk of carcinogenesis by chemical materials.[6–10] It was concluded from these studies that 8-OH-Gua is a useful probe for elucidating the role of reactive oxygen species (ROS). However, the measured levels of 8-OH-Gua depend on the balance between formation and repair, and thus the 8-OH-Gua repair systems should also be assayed to evaluate the role of oxidative DNA damage. For example, in an experimental model of aging using cultured human fibroblast cells, we observed an increase of 8-OH-Gua during 20–50 population doubling levels (PDL)

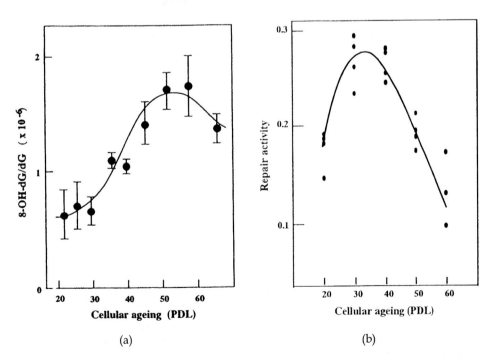

Fig. 1. (a) The 8-OH-Gua level in cells at the ages of 20, 30, 40, 50, and 60 PDL. (b) 8-OH-Gua repair activities in TIG-3 cells at the ages of 20, 30, 40, 50, and 60 PDL.

[Fig. 1(a)],[11] while the 8-OH-Gua repair activity decreased during 30-60 PDL {Fig. 1(b)}.[12] Recent studies have revealed that 8-oxoguanine DNA glycosylase 1 (OGG1), which has glycosylase/lyase activities, plays a role in 8-OH-Gua repair systems.[13–19] To determine whether the OGG1 protein plays a critical role in the prevention of tumorigenesis or carcinogenesis, more data about 8-OH-Gua formation, repair activity, and OGG1 induction are required.

In the present paper, we describe the methods employed in our laboratory, including the measurement of 8-OH-Gua in DNA, the 8-OH-Gua endonuclease nicking assay, and the RT-PCR for OGG1. As examples, we show our recent studies on 8-OH-Gua levels in human leukocyte DNA in relation to physical exercise, and on oxidative stress of the lungs of rats treated with diesel exhaust particles (DEP).

2. Preparation of Samples

For mammalian organs, the samples should be processed immediately. When animals are used, they should be killed by decapitation to avoid the artificial oxidation derived from blood contamination for the measurement of 8-OH-Gua. The samples should be frozen and stored at $-80°C$ until used for the assay. For the nicking assay, although we confirmed that blood contamination does not interfere with the repair activity,[20] decapitation is also recommended. The total crude extract should be prepared, as described in the *Endonuclease Nicking Assay*, immediately after the organs are removed, frozen, and stored at $-80°C$ until used for the assay. Human leukocytes or cultured cells are also practical for both assays. Human leukocytes provide useful information about oxidative stresses in DNA. Twenty ml of whole blood are collected and separated into two portions, of 5 ml and 15 ml. The 5-ml portion of blood is centrifuged for 15 min at 2000 X g, to collect the buffy coat fraction, which is frozen at $-80°C$ until used for the determination of the amount of 8-OH-Gua in the DNA. The 15-ml portion of blood is mixed with a Ficoll-Paque solution for the preparation of the leukocyte fraction. The latter is used for the measurement of 8-OH-Gua repair activity.[21, 22]

3. Measurement of 8-OH-Gua

The 8-OH-Gua level in DNA is determined according to Nakae *et al.*,[23] with some modifications. All tubes or buffer solutions used for DNA isolation and digestion are flushed with nitrogen gas. Fifty to two hundreds mg of removed tissue, isolated leukocytes, or cultured cells are homogenized in lysis buffer with a Potter-type homogenizer, and the nuclear DNA in the homogenate is extracted using the DNA Extractor WB Kit (Wako, Japan). During lysis or contact with organic solvents, exposure to light should be minimized. The extracted DNA is digested with nuclease P1 and acid phosphatase in a 10 mM sodium acetate solution at 37°C for 30 minutes. The mixture is treated with the ion exchange resin, Muromac (Muromachi Kagaku, Tokyo, Japan), to remove the iodide ions, which interfere with the electro-chemical detector (ECD), and is centrifuged at 15 000 rpm for 5 minutes. The supernatant is transferred to an Ultrafree-Probind Filter (Millipore, USA), centrifuged at 10 000 rpm for 1 min, and analyzed by an HPLC-ECD. As standard samples, 20 µl each of deoxyguanosine (0.5 mg/ml) and 8-hydroxydeoxyguanosine (5 ng/ml) solutions are injected. The 8-OH-Gua value is calculated as the number per 10^5 or 10^6 guanine residues. The DNA digest should be analyzed as soon as possible. Six to eight samples are the maximum number that can be analyzed within one day.

4. Endonuclease Nicking Assay

8-OH-Gua repair activity is evaluated as the excision ratio for 8-OH-Gua by an endonuclease nicking assay.[24] Removed tissues, isolated leukocytes, or cultured cells are homogenized in 50 mM Tris-HCl buffer (pH 7.5) containing protease inhibitors (5 µg each of pepstatin, leupeptin, antipain, and chymostatin) with a Potter-type homogenizer. The homogenate is centrifuged at 12 000 X g to obtain the crude extract. The total protein concentration is determined by a protein assay kit (Bio-Rad, Richmond, CA), using bovine serum albumin as the standard, and is adjusted to 5 mg/ml. A 22-mer double-stranded synthetic oligonucleotide containing 8-OH-Gua opposite C (the modified strand: 5'-GGTGGCCTGACG*CATTCCCCAA-3'; G*: 8-OH-Gua), prepared by the method of Bodepudi *et al.*,[25] is used as the substrate for this assay. The crude extract (50 µg protein) is incubated with 0.05 pmol of the ^{32}P or fluorescently end-labeled double-stranded DNA substrate at 25°C for 1 hour. After an ethanol precipitation, the pellet is dried, dissolved in 10 µl of loading buffer (80% formamide, 10 mM NaOH, 1 mM EDTA, 0.1% xylene cyanol, 0.1% bromophenol blue), and denatured by heating at 90°C for 3 minutes. An aliquot of the sample (10 µl) is applied to a 20% denaturing polyacrylamide gel for electrophoresis. If the gel position of the cleaved fragment, generated as a consequence of the excision repair activity, needs to be confirmed, then it can be compared to the hot piperidine treated oligonucleotide as a fragment marker.[26] After electrophoresis, in the case of a ^{32}P labeled oligonucleotide, the autoradiogram is processed and the radioactivity is analyzed using a Bioimage analyzer system (Fujix BAS 2000). For a fluorescently end-labeled double-stranded DNA substrate, the excised fragment is analyzed by a Pharmacia ALF DNA sequencer (Fragment Manager, Ver. 1.1; Amersham Pharmacia Biotech, Uppsala, Sweden).[8]

5. Measurement of OGG1 mRNA Level

Total RNA is prepared from fresh tissue or isolated cells, and the mRNA is isolated on an oligo (dT)-cellulose column (Pharmacia Biotech AB, Uppsala, Sweden). The first strand cDNA is synthesized from mRNA primed with random hexamers using M-MLV reverse transcriptase (Gibco BRL, Grand Island, NY). The primers for OGG1 and GAPDH were designed from the consensus sequences of the human, rat, and mouse OGG1 and GAPDH. The primers for OGG1 are 5′-ATCTGTTCCTCCAACAACAAC-3′ and 5′-GCCAGCATAAGGTCCCCACAG-5′. The primers for GAPDH are 5′-AACGGGAAGCTCACTGGCATG-3′ and 5′-TCCACCACCCTGTTGCTGTAG-3′. GAPDH mRNA is used as an internal standard. Each DNA is amplified using the primers for the OGG1 and GAPDH genes. The amplification for OGG1 consists of 35 cycles at 94°C (60 s), 61°C (60 s), and 72°C (180 s). The PCR products are separated on a 5% polyacrylamide gel and are visualized with ethidium bromide staining.[9, 10]

6. 8-Hydroxyguanine in Human Leukocyte DNA and Physical Exercise

Recently, physical exercise has been found to be one of the most effective means of maintaining or promoting health, as well as preventing diseases such as diabetes mellitus or hypertension. On the other hand, physical exercise has been reported to play an essential role in free radical-mediated damage.[27, 28] It remains to be clarified whether physical exercise confers health benefits. Based on this question, we investigated the effect of physical exercise on the level of 8-OH-Gua and its repair activity in human peripheral leukocytes.[22] Twenty three healthy male volunteers (10 trained athletes and 13 untrained men), aged 19–50 years, performed exercise requiring maximal oxygen uptake on a cycle ergometer, after passing a physical examination. Maximal oxygen uptake was monitored by a symptom limited ramp cycle ergometer method with a gas analyzer and a turbine transducer to detect breath by breath tidal volume, O_2 consumption, and expired CO_2 content. Whole blood samples were collected from these volunteers, both before and after physical exercise, and the levels of 8-OH-Gua and its repair activities in their leukocytes were measured. As a result, the trained athletes showed a lower level of 8-OH-Gua ($2.4 \pm 0.5/10^6$ Gua, $p = 0.0032$) before exercise as compared to that of the untrained men ($6.3 \pm 3.6/10^6$ Gua) [Fig. 2(a)]. The mean 8-OH-Gua levels of the untrained volunteers decreased significantly ($p = 0.0057$), from $6.3 \pm 3.6/10^6$ Gua to $3.3 \pm 1.4/10^6$ Gua, after physical exercise [Fig. 2(a)]. On the other hand, the mean repair activity levels of the untrained volunteers significantly increased after exercise ($p = 0.0093$), from 0.037 ± 0.024 to 0.056 ± 0.036 [Fig. 3(a)]. In the trained athletes,

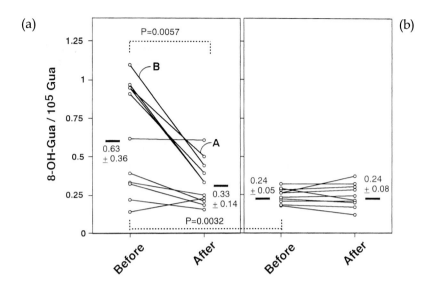

Fig. 2. Effect of physical exercise on the level of 8-OH-Gua in human leukocyte DNA. (a) untrained man; (b) trained athletes. The 8-OH-Gua level was calculated as the number per 10^5 Gua. Mean values ± SD are shown. The significance of the differences between the data of before and after the exercise was evaluated by a paired *t*-test and that between the data of untrained and trained men before the exercise was by an unpaired *t*-test.

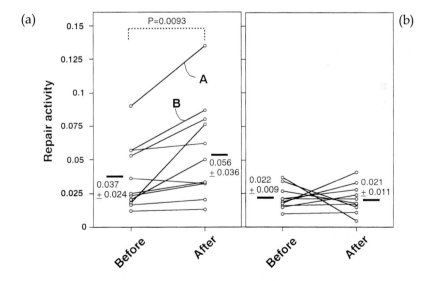

Fig. 3. Effect of physical exercise on the 8-OH-Gua repair activity in human leukocytes. (a) untrained man; (b) trained athletes. Mean values ± SD are shown. The significance of the differences between the data of before and after the exercise was evaluated by a paired *t*-test and that between the data of untrained and trained men before the exercise was by an unpaired *t*-test.

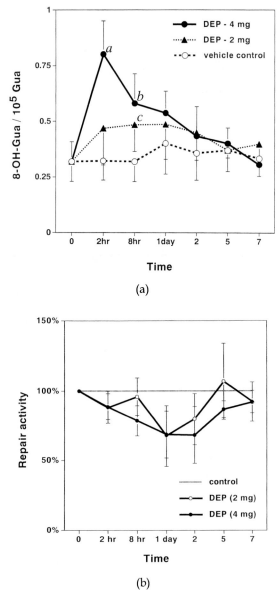

Fig. 4. (a) Time course of 8-OH-Gua levels (mean ± SD) in rat lung DNA after DEP administration. [a]Significantly higher than vehicle control ($p < 0.0001$) and than 2 mg DEP-treated group ($p = 0.0043$). [b]Significantly higher than vehicle control ($p = 0.0025$). [c]Significantly higher than vehicle control ($p = 0.0218$). (b) Time course of repair activities for 8-OH-Gua (mean ± SD) in rat lung after DEP administration. Each value of the repair activity in the DEP-exposed group is represented as a percentage of the vehicle control group at the same time point. [a]Significantly lower than vehicle control ($p = 0.0167$). [b]Significantly lower than vehicle control ($p = 0.0169$). [c]Significantly lower than vehicle control ($p = 0.0216$). (c) OGG1 mRNA levels in rat lung after intratracheal instillation of DEP (4 mg).

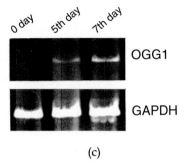

(c)

Fig. 4 (*Continued*)

the 8-OH-Gua level and its repair activity were not changed before and after the exercise [Figs. 2(b) and 3(b)]. Donors A and B showed the highest 8-OH-Gua repair activities. Their 8-OH-Gua levels in DNA were also high. These results suggest that physical exercise causes both rapid and long-range reductions of oxidative DNA damage in human leukocytes, presumably by inducing either repair enzymes or ROS scavengers.

7. 8-Hydroxyguanine, Its Repair Activity, and Rat OGG1 mRNA Induction in the Lungs of Rats Treated with Diesel Exhaust Particles

One of the contemporary concerns about environmental hazards is diesel exhaust particles (DEP)-induced lung cancer. To clarify whether ROS are involved in its carcinogenic mechanism, we examined the levels of 8-OH-Gua, its repair activity, and OGG1 mRNA induction in female Fischer 344 rats (7 weeks old) lungs after DEP were intratracheally instilled.[9] The 8-OH-Gua levels in both DEP-treated groups (2 and 4 mg) were increased during the 2–8 h following exposure to DEP [Fig. 4(a)]. In contrast, the 8-OH-Gua repair activities in the DEP-treated groups decreased during the period from 2 h to 2 days following DEP administration, and then recovered to the level of the control group at 5 days after exposure [Fig. 4(b)]. Moreover, the OGG1 mRNA was induced at least 10-fold in rats treated with 4 mg DEP for 5–7 days after administration [Fig. 4(c)]. Collectively, the 8-OH-Gua level in the rat lung DNA increases during the early period of DEP exposure, as consequences of the generation of ROS and the inhibition of 8-OH-Gua repair activity. In these experiments, the induction of the OGG1 mRNA is also a good marker of cellular oxidative stress during carcinogenesis.

8. Discussion

To clarify the mechanisms of aging or diseases, many different studies have been performed, and it was found that ROS, formed endogenously or by environmental agents, play a critical role. Since we use oxygen to create energy to live, the harmful aspects of ROS remain an important, unavoidable problem. From this point of view, the analysis of oxidative DNA damage, such as 8-OH-Gua, seems to be useful to investigate the etiology of some diseases and the mechanisms of aging. Since atmospheric oxygen may damage the samples during experiments, it is important to minimize the contact of samples with air. The methods described here were very carefully designed to avoid artificial oxidation.

In addition, the repair systems need to be evaluated to reveal the exact role of ROS, because the repair activity is induced in many cases after cellular oxidative stress and the measured levels of 8-OH-Gua are result of the balance between the formation and the repair of the DNA damage. In fact, we found that the 8-OH-Gua repair activity increased in human leukocytes with physical exercise,[22] in the rat kidney after Fe-NTA treatment,[6] in rat and hamster lungs after asbestos treatment,[8] and in the mouse liver after 3'-methyl-4-dimethylaminoazobenzene treatment.[10] However, we also observed that some chemical materials, such as cadmium chloride and DEP, inhibit the repair mechanisms, and thereby cause an accumulation of 8-OH-Gua in DNA.[7, 9] In the case of DEP, 8-OH-Gua was increased in rat lung DNA by the generation of ROS and the inhibition of the 8-OH-Gua repair capacity, while the expression of the OGG1 gene was increased. Taken together, the assays of the repair capacity and the induction of the repair enzyme genes are also important in addition to 8-OH-Gua analysis to assess cellular oxidative stress and to clarify the roles of ROS in aging and carcinogenesis.

References

1. Kasai, H. and Nishimura, S. (1984). Hydroxylation of deoxyguanosine at the C-8 position by ascorbic acid and other reducing agents. *Nucleic Acid Res.* **12**: 2137–2145.
2. Kasai, H. and Nishimura, S. (1984). Hydroxylation of deoxyguanosine at the C-8 position by polyphenols and aminophenols in the presence of hydrogen peroxide and ferric ion. *Gann.* **75**: 565–566.
3. Kasai, H., Tanooka, H. and Nishimura, S. (1984). Formation of 8-hydroxy-guanine residues in DNA by X-irradiation. *Gann.* **75**: 1037–1039.
4. Floyd, R. A., Watson, J. J., Wong, P. K., Altmiller, D. H. and Rickard, R. C. (1986). Hydroxyl free radical adduct of deoxyguanosine: sensitive detection and mechanisms of formation. *Free Radic. Res. Commun.* **1**: 163–172.

5. Cheng, K. C., Cahill, D. S., Kasai, H., Nishimura, S. and Loeb, L. A. (1992). 8-hydroxyguanine, an abundant form of oxidative DNA damage, causes G → T and A → C substitutions. *J. Biol. Chem.* **267**: 166–172.

6. Yamaguchi, R., Hirano, T., Asami, S., Chung, M.H., Sugita, A. and Kasai, H. (1996). Increased 8-hydroxyguanine levels in DNA and its repair activity in rat kidney after administration of renal carcinogen, ferric nitrilotriacetate. *Carcinogenesis* **17**: 2419–2422.

7. Hirano, T., Yamaguchi, Y. and Kasai, H. (1997). Inhibition of 8-hydroxyguanine repair in testes after administration of cadmium chloride to GSH-depleted rats. *Toxicol. Appl. Pharmacol.* **147**: 9–14.

8. Yamaguchi, R., Hirano, T., Ootsuyama, Y., Asami, S., Tsurudome, Y., Fukada, S., Yamato, H., Tsuda, T., Tanaka, I. and Kasai, H. (1999). Increased 8-hydroxyguanine in DNA and its repair activity in hamster and rat lung after intratracheal instillation of crocidolite asbestos. *Japan J. Cancer Res.* **90**: 505–509.

9. Tsurudome, Y., Hirano, T., Yamato, H., Tanaka, I., Sagai, M., Hirano, H., Nagata, N., Itoh, H. and Kasai, H. (1999). Changes in levels of 8-hydroxy-guanine in DNA, its repair and OGG1 mRNA in rat lungs after intratracheal administration of diesel exhaust particles. *Carcinogenesis* **20**: 1573–1576.

10. Hirano, T., Higashi, K., Sakai, A., Tsurudome, Y., Ootsuyama, Y., Kido, R. and Kasai, H. (2000). Analyses of oxidative DNA damage and its repair activity in the livers of 3'-methyl-4-dimethylaminoazobenzene-treated rodents. *Japan J. Cancer Res.* **91**: 681–685.

11. Homma, Y., Tsunoda, M. and Kasai, H. (1994). Evidence for the accumulation of oxidative stress during cellular aging of human diploid fibroblasts. *Biochem. Biophys. Res. Commun.* **203**: 1063–1068.

12. Hirano, T., Yamaguchi, Y., Hirano, H. and Kasai, H. (1995). Age associated change of 8-hydroxyguanine repair activity in cultured human fibroblasts. *Biochem. Biophys. Res. Commun.* **214**: 1157–1162.

13. Lu, R., Nash, H. M. and Verdine, G. L. (1997). A mammalian DNA repair enzyme that excises oxidatively damaged guanines maps to locus frequently lost in lung cancer. *Current Biol.* **7**: 397–407.

14. Rosenquist, T. A., Zharkov, D. O. and Grollman, A. P. (1997). Cloning and characterization of a mammalian 8-oxoguanine DNA glycosylase. *Proc. Natl. Acad. Sci. USA* **94**: 7429–7434.

15. Radicella, J. P., Dherin, C., Desmaze, C., Fox, M. S. and Boiteux, S. (1997). Cloning and characterization of hOGG1, a human homolog of the OGG1 gene of *Saccaromyces cerevisiae*. *Proc. Natl. Acad. Sci. USA* **94**: 8010–8015.

16. Roldán-Arjona, T., Wei, Y. F., Carter, K. C., Klungland, A., Anselmino, C., Wang, R. P., Augustus, M. and Lindahl, T. (1997). Molecular cloning and functional expression of a human cDNA encoding the antimutator enzyme 8-hydroxyguanine DNA glycosylase. *Proc. Natl. Acad. Sci. USA* **94**: 8016–8020.

17. Arai, K., Morishita, K., Shinmura, K., Kohno, T., Kim, S. R., Nohmi, T., Taniwaki, M., Ohwada, S. and Yokota, J. (1997). Cloning of a human homolog of the yeast OGG1 gene that is involved in the repair of oxidative DNA damage. *Oncogene* **14**: 2857–2861.

18. Aburatani, H., Hippo, Y., Ishida, T., Takashima, R., Matsuba, C., Kodama, C., Takao, M., Yasui, A., Yamamoto, K., Asano, M., Fukasawa, K., Yoshinari, T., Inoue, H., Ohtsuka, E. and Nishimura, S. (1997). Cloning and characterization of mammalian 8-hydroxyguanine-specific DNA glycosylase/apurinic, apyrimidinic lyase, a functional mutM homologue. *Cancer Res.* **57**: 2151–2156.

19. Tani, M., Shinmura, K., Kohno, T., Shiroishi, T., Wakana, S., Kim, S.R., Nohmi, T., Kasai, H., Takenoshita, S., Nagashima, Y. and Yokota, J. (1998). Genomic structure and chromosomal localization of the mouse OGG1 gene that is involved in the repair of 8-hydroxyguanine in DNA damage. *Mamm. Genome.* **9**: 32–37.

20. Hirano, T., Yamaguchi, R., Asami, S., Iwamoto, N. and Kasai, H. (1996). 8-hydroxyguanine levels in nuclear DNA and its repair activity in rat organs associated with age. *J. Gerontol.* **51A**: B303–B307.

21. Asami, S., Hirano, T., Yamaguchi, R., Tomioka, Y., Itoh, H. and Kasai, H. (1996). Increase of a type of oxidative DNA damage, 8-hydroxyguanine, and its repair activity in human leukocytes by cigarette smoking. *Cancer Res.* **56**: 2546–2549.

22. Asami, S., Hirano, T., Yamaguchi, R., Itoh, H. and Kasai, H. (1998). Reduction of 8-hydroxyguanine in human leukocyte DNA by physical exercise. *Free. Radic. Res.* **29**: 581–584.

23. Nakae, D., Mizumoto, Y., Kobayashi, E., Noguchi, O. and Konishi, Y. (1995). Improved genomic/nuclear DNA extraction for 8-hydroxydeoxyguanosine analysis of small amount of rat liver tissue. *Cancer Lett.* **97**: 233–239.

24. Hirano, T., Yamaguchi, Y., Hirano, H. and Kasai, H. (1995). Age-associated change of 8-hydroxyguanine repair activity in cultured human fibroblasts. *Biochem. Biophys. Res. Commun.* **214**: 1157–1162.

25. Bodepudi, V., Shibutani, S. and Johnson, F. (1992). Synthesis of 2'-deoxy-7, 8-dihydro-8-oxoguanosine and 2'-deoxy-7, 8-dihydro-8-oxoadenosine and their incorporation into oligomeric DNA. *Chem. Res. Toxicol.* **5**: 608–617.

26. Chung, M. H., Kiyosawa, H., Ohtsuka, E., Nishimura, S. and Kasai, H. (1991). DNA strand cleavage at 8-hydroxyguanine residues by hot piperidine treatment. *Biochem. Biophys. Res. Commun.* **188**: 1–7.

27. Davies, K. J. A., Quintanilha, A. T., Brooks, G. A. and Packer, L. (1982). Free radicals and tissue damage produced by exercise. *Biochem. Biophys. Res. Commun.* **107**: 1198–1205.

28. Novelli, G. P., Bracciotti, G. and Falsini, S. (1989). Spin-trappers and vitamin E prolong endurance to muscle fatigue in mice. *Free Radic. Biol. Med.* **8**: 9–13.

Chapter 13

HPLC-ECD, HPLC-MS/MS (Urinary Biomarkers)

H.E. Poulsen*, S. Loft, B.R. Jensen, M. Sørensen,
A.-M. Hoberg and A. Weimann

Henrik E. Poulsen, B. R. Jensen*, A.-M. Hoberg* and **A. Weimann** • Department of Clinical Pharmacology Q7642, Rigshospitalet, University Hospital Copenhagen, 20 Tagensvej, DK-2200 Copenhagen N, Denmark
Stephen Loft and **M. Sørensen** • Institute of Public Health, University of Copenhagen, 3 Blegdamsvej, DK-2200 Copenhagen N, Denmark
*Corresponding Author.
Tel: +45 3545 7691, E-mail: henrikep@rh.dk

1. Introduction

Mammalian life is based on oxygen and uses oxygen reduction for energy production and synthetic processes. By 4-electron reactions oxygen is reduced to water and the energy released is stored for controlled use. However, one electron reduction occurs in minor amounts giving rise to various reactive oxygen species (ROS).[1, 2] The reactive oxygen species potentially oxidizes important macromolecules and structures in the body. Although posing serious threats of deleterious effects on vital functions, it is now increasingly realized that ROS play important roles as part of defence, signaling and transcription mechanisms.[3–5]

A multiplicity of different oxidative DNA modifications has been described.[6, 7] Still, there is much more scarce data on their occurrence in the *in vivo* situation, and whether they have biological significance or relates to diseases. The exception is the 8-hydroxylation of guanine (8-oxodG), this lesion has been investigated both *in vitro* and *in vivo*, including human studies.

Oxidation processes are prone to occur in the earth's environment, including in test tubes, refrigerators, freezers, laboratories etc. due the ubiquitous oxygen. This poses a major challenge to anybody studying these processes since artefacts can arise from oxidation during sample handling. Particularly, most methods rely on storage or prolonged preparation of samples before the initial analysis. In addition to storage, most procedures are carried out at conditions that clearly makes spontaneous oxidation possible. Often it will be found that immense differences are reported between different laboratories. Consequently published data always should be scrutinized bearing this aspects in mind.

There is no doubt that data considered to reflect the *in vivo* situation with regard to oxidative stress to some extend are contradictory and difficult to interpret, and that this may be because the necessary precautions against oxidation were not taken. This review focus on the methodologies that presently are available, mainly with regard to estimating oxidative DNA products excreted into urine, but also in the context of tissue levels. The methodology is very similar and information from tissue levels and urinary excretion provide information on the oxidative stress to DNA.

2. Principles for HPLC and GC Separation

Two powerful techniques are available for separation of substances in samples in order to enable detection. HPLC, high performance (or pressure) liquid chromatography, is particularly suited for small water-soluble molecules and proteins. Most used for analysis of DNA fragments is the reverse phase HPLC. Gas chromatography (GC) is well suited for volatile compounds, however, with the use of various derivatization procedures it is possible to separate e.g. DNA bases with GC.

In essence chromatography corresponds to a series of organic extractions, e.g. of water and chloroform. However, the theoretical series of extractions corresponds to maybe 5000 or more. Considering this is done within minutes on a HPLC or GC column, it is clear how powerful this separation technique is. The HPLC technique requires few preparative steps of the sample which saves time and work and can also avoid potential problems related to the clean-up procedures. The GC technique requires derivatization and extraction procedures but provides very high and sharp peaks that are particularly suitable for mass spectrometry.

To some degree, theoretical considerations can be a great aid in setting up the analysis. For HPLC computer programs like DryLab® exists and can be of some help. To our experience there are many problems that cannot be foreseen from theoretical considerations, and development/implementation rests to a large degree on trial and error, experience and a little inspiration. Special consideration should be given to factors that reduce noise. Among these, heavy pulse damping of the HPLC system, use of high quality water and solvent and selection of the right column is of particular importance.

Other separation techniques have large potentials, e.g. capillary electrophoresis, but have not been much used for analysis of DNA oxidation.

For HPLC with tandem mass spectrometric detection (HPLC-MS/MS), the HPLC separation is not as demanding as for HPLC-ECD. We have found that a single column is sufficient,[8] however, we must emphasize that unknown substances similar in mass to 8-oxodG needs to be separated from 8-oxodG. For high sensitivity in mass spectrometry the peak height in HPLC is very important. The amount detected is proportional both to the peak height and to the area under the curve.

By derivatization it is possible to use the GC separation procedure coupled with mass spectrometry to measure oxidized DNA products. However, this method has with few exceptions not been used for urinary measurements on DNA, but has been the method used for estimation of 8-oxodG, actually the base after hydrolysis, and other DNA oxidation products in tissue. For urine measurements a semi-preparative HPLC procedure was applied, followed by hydrolysis, derivatization and GC-MS.[9-11]

The immunological methods do not include a prior HPLC separation procedure.

2.1. Electrochemical Detection (ECD)

This method relies on the oxidation of a compound in an electrical field and detection of the change in current by this process. The particular virtue of the ECD detection is the extreme sensitivity making it possible to detect e.g. 8-oxodG in the nM concentration range, i.e. fmol injection. Since the first publication of the HPLC-ECD method,[12] this has been the preferred method and used in a large

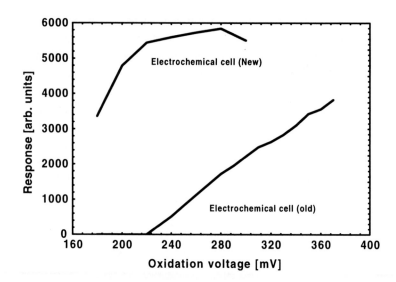

Fig. 1. Dynamic voltamogram of 8-oxodG using a Coulochem Electrochemical detector and a 5011 analytical cell. The graph is constructed from injection of pure 8-oxodG on the system with different voltage setting for oxidation current. The difference between a worn-out and a new cell is depicted.

number of laboratories mostly for detection of 8-oxodG in DNA, but also for detection of 8-oxodG in urine.

The applied electrical potential that can oxidize a compound exhibits several plateaus. The actual plateau for a given compound and brand of apparatus has to be identified for that specific combination. This is done[13] by establishing a dynamic voltamogram by measuring the signal after injection of a known sample and increasing the oxidation voltage. An example is shown in Fig. 1 that depicts a worn out electrochemical cell and a new cell. This figure also illustrates that it is necessary to check the voltamogram with regular intervals and to adjust the oxidation voltage.

There are many different brands of ECD detectors and electrodes/cells. In our laboratory we have found that the ESA Coulochem is working excellent for our purposes and provides excellent sensitivity. Similar experience with other detectors can be found.

We have found that for HPLC-ECD analysis of urine, separation is critical due to electrochemically active peaks eluting close to that of 8-oxodG. Ways to detect a false peak is given in details elsewhere.[14]

The quantification also requires special attention since in HPLC it is not possible to use a true internal standard, i.e. an internal standard that behaves exactly as the substance you want to measure. An internal standard, 2,6-Diamino-8-oxopurine, has been suggested,[15] but is probably only useful in controlling

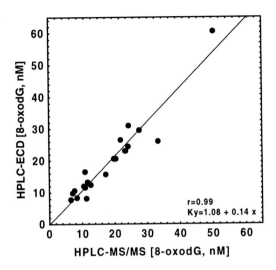

Fig. 2. Correlation between 8-oxodG in 21 different urine samples measured by HPLC-ECD and HPLC-MS/MS. The correlation, $r = 0.99$, the negligible intercept (0.14 nM) and the close to unity slope (1.08) provide evidence for the excellent agreement of the two methods.

variations in the injection volume, and cannot be used for other purposes that poses more severe problems like artificial oxidation, degradation of 8-oxodG on the column etc. Presently there is no experience with the use of this internal standard for urine measurement. We use external standard addition in different concentrations and evaluate the response ratios,[14] and this methodology appears to function satisfactory from comparison with HPLC-MS/MS analysis as seen in Fig. 2.

2.2. Mass Spectrometric Detection (MS)

The detection of a substance from determination of its mass is often stated as a specific methodology, yet many substances have the same mass. With some separation of the substances with GC or HPLC coupled with fragmentation and selection of a suitable fragment, a very high degree of specificity may be achieved. The advancement of the technical side of mass spectrometry and the use of elaborated computer software to control and handle the data output from mass spectrometers have made it possible to use this methodology more widely than previously. For years the dominating methodology has been gas chromatography coupled with mass spectrometry for the detection of oxidized DNA products. However, the development of ionization methods for HPLC, e.g. electrospray has made possible the use of HPLC coupled with mass spectroscopy. For quantification purposes, triple quadrupole mass spectrometers are probably the most useful apparatus as of today. The progress in other techniques like capillary electrophoresis-MS and mass spectrometry based on time of flight may soon develop to a similar or surpassing degree.

Gas-chromatography-mass spectrometry used for quantification of oxidative DNA products has been criticized for errors due to artificial oxidation, however, provided that sufficient precautions are taken, this can be avoided and results similar to those from HPLC-ECD can be provided regarding 8-oxodG in DNA.[16] Presumably this is also valid for other oxidative DNA products, but needs to be validated. In case of 8-oxodA the validity has been questioned[17] regarding 8-oxodA in an experiment with vitamin C and vitamin E intervention[18] and using HPLC-MS/MS it seems likely that the high reported 8-oxodA values relates to artifactual oxidation.[8] Many of the problems regarding artificial oxidation relates to the very high content of non-oxidized dG in DNA hydrolysates, about 1 000 000 times higher. This means that oxidation of only a very minute fraction of dG gives serious artefacts. For urine measurements the levels of oxidized and non-oxidized nucleosides are similar and would *a priori* not present a problem of the same magnitude.

The basis for using mass spectrometry is ionization of the substances. With ion spray this is done from an HPLC outlet, where the mobile phase is evaporated at high temperature and the substances ionized by a high voltage before the ionized molecules enter the vacuum of the mass spectrometer itself. Ionspray is a soft ionization technique that means that the ions do not show extensive fragmentation. The ionization process used in gas chromatography is more energetic resulting in fragmentation of the substances. GC-MS analysis is normally used with a single quadrupole for detection whereas triple quadrupole instruments are used with HPLC separation.

In the triple quadrupole mass spectrometer (Fig. 3), a first quadrupole, Q1, is used to select one or several wanted mass(es). They are passed on to a second quadrupole, Q2, that functions as a collision cell, where the ions collide with a collision gas. The quadrupole arrangement serves to focus the ions. The collisions can be controlled and in case of oxidized nucleosides specific breakage of the pseudomolecularions N-glycosidic bond can be achieved. The product ions are then selected in the third quadrupole, Q3, followed by final detection. This is usually called Multiple Reaction Monitoring (MRM) and is the basis for the particular usefulness of the triple quadrupole when high specificity and sensitivity is needed. Figure 3 depicts this principal build of the triple quadrupole tandem mass spectrometer. Details of how to operate an HPLC-MS/MS system depend on the brand and the software version. Reference is made to the brands manual and for other details to original papers.[8, 14, 19, 20] It should be noted that in HPLC mass spectrometry non-volatile salts should generally be

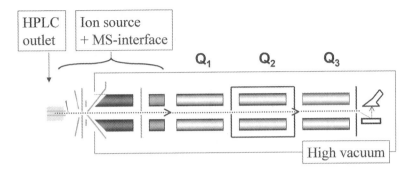

Fig. 3. The principle build of a HPLC-MS/MS instrument. The HPLCC outlet is to a high temperature ion source and MS-interface that provides evaporation and ionization of the substances in the eluent. The ions then enter the vacuum in which the quadrupole select the masses defined, before and after fragmentation.

avoided, mainly because salt suppresses the ionization and also deposits in the apparatus with loss of sensitivity and demand for repeated cleaning procedures.

With careful consideration to these factors, using individual tuning files the HPLC-MS/MS analysis matches the sensitivity of HPLC-ECD, i.e. below 1 nM 8-oxodG in urine with an injection of 20 μL. This opens the possibility for running a parallel column system that with software-controlled valves can give a high sample throughput. The actual practical limit of such a system is still unknown.

One particular virtue of HPLC-MS/MS, as well as GC-MS, is the use of stable isotope labeled internal standards. By labeling e.g. 8-oxodG with stable carbon and/or nitrogen isotopes in non-functional positions, a substance that chemically, chromatographically and in other physical-chemical aspects behaves exactly like the unlabeled compound is provided. In the mass spectrometer the two substances can easily be detected at the same time (for practical purposes) and used for quantification.

Since mass spectrometry does not rely on UV absorption or electrochemical properties, and since mass detection may provide high sensitivity for different compounds the potential for measuring different nucleosides, bases or other DNA products is present. Presently this potential has only been used to a limited degree, but development is ongoing in several laboratories.

As mentioned above GC-MS has been used to measure urinary DNA oxidation products, however, various clean up or up-concentration methods are necessary. The choice for urinary measurement is therefore either HPLC-EC, which is limited mainly to 8-oxodG measurement or to HPLC-MS/MS where multiple products can be measured. Both of these methods can be set up with very little preparation of urine, just a simple centrifugation and dissolving of possible sediments.

The versatility and sensitivity of LC-MS/MS is great. For example, we have developed a method by which it is possible to analyze the promutagenic exocyclic

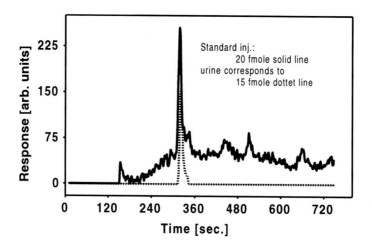

Fig. 4. HPLC-MS/MS analysis of urine from a human volunteer. A Sciex API 3000 mass spectrometer with heated nebulizer (APCI) was used for monitoring m/z 276.2 (MH+) → 160.0 (BH₂+). Urine was separated by column switching between a Zobax Eclipse C_{18} 4.6 * 50 mm, 3.5 μ column and a wacosil C_{18} 4.0 × 50 mm, 3 μ column. Eluents were 5 nM ammonium acetate pH 5.0 and 5 nM ammoniumacetate pH 5.0, 7% acetonitrile. The dotted line is a trace of injection of 20 fmole etheno-deoxyadenosine, the solid line injection of 100 μL urine.

DNA adduct 1,N^6-ethenodeoxyadenosine (εdA), Fig. 4. The analysis was done using atmospheric pressure chemical ionization (APCI) in the positive ion mode with multiple reaction monitoring (MRM) on a Sciex API 3000 triple quadrupole mass spectrometer. Quantification was performed on the characteristic transition m/z 276.3(MH$^+$) → 160.3(BH$_2^+$). Promising results have been obtained with εdA standard with a limit of detection below 1 fmole. Figure 4 shows a chromatogram of urine from a healthy volunteer indicating a concentration of εdA of about 150 fmole/ml.[21]

2.3. Immunological Methods

The use of a specific antibody could be the basis for a fast and effective methodology to measure 8-oxodG. However, it has proven difficult to produce an antibody with sufficient specificity for analysis in urine. Several publications have appeared.[22–25] However, although some characterization of the antibody and epitope is given, it appears not to be tested against the many different DNA and RNA products in urine.[26] Furthermore, testing against the present method of choice HPLC-ECD, GC-MS or HPLC-MSMS has only been stated without data, and at present time the data have not been made available in the literature.[22] One particular problem with the immunologically based assays may relate to the high

Table 1. Studies on Urinary Excretion of Oxidative DNA Products

Disease/Condition/ Intervention	Level	Unit	Number of Individuals	Finding	Method	Lesion	ID
Adriamyacin	$80.8 \pm 8 \rightarrow$ 98.7 ± 6.9	nmol/24 h	20	increased	HPLC+GC/MS	5-OH-me-Uracil	(Ref. 10)
Controls	121 ± 56	pmol/ml	10	concentration only	HPLC-GC/MS	5-OH-Me-Uracil	(Ref. 11)
Adriamycin treatment	$74.4 \pm 9.58 \rightarrow$ 96.3 ± 8.7	nmol/24 h	14	corresponds to normal	HPLC-GC/MS	5-OH-Me-Uracil	(Ref. 46)
Controls	58 ± 23	pmol/ml	10	concentration only	HPLC-GC/MS	5-OH-Uracil	(Ref. 11)
Controls	7 ± 4	pmol/ml	10	concentration only	HPLC-GC/MS	8-oxo-Ade	(Ref. 11)
Controls versus cancer	$1.19 \pm 0.488 \rightarrow$ 2.42 ± 2.28	μmol/mol creatinine	10 + 30	increased in cancer patients	multi-D-HPLC-ECD	8oxodG	(Ref. 47)
Hemochromatosis	1.39 ± 0.40	μmol/mol creatinine	12	corresponds to controls	GC/MS	8-oxodG	(Ref. 48)
Lung cancer	$272 \pm 13/$ 19.4 ± 8.5		14/52	increased in small cell carcinoma	ELISA	8-oxodG	(Ref. 22)
Control	3 ± 1	ng/ml	1	—	solid phase extraction LC-MS	8-oxodG	(Ref. 49)
Smoking	$1.02 \pm 0.35 -$ 3.37 ± 2.84	μmol/mol creatinene	34 controls 10 smokers 24 non-smokers 10 malignant disease	increased in smokers and malignancy	multi-D-HPLC-ECD	8-oxodG	(Ref. 50)
All out' rowing	2.5	nmol/μmol creatinine	10	no change	multi-D-HPLC-ECD	8-oxodG	(Ref. 51)
Repeated exercise	$266 \pm 76 -$ 336 ± 107	pmol /kg/day	10	increased	multi-D-HPLC-ECD	8-oxodG	(Ref. 52)
Single exercise bout	24.7 ± 1.5	nmol/day	28	no change	multi-D-HPLC-ECD	8-oxodG	(Ref. 53)

Table 1 (*Continued*)

Disease/Condition/Intervention	Level	Unit	Number of Individuals	Finding	Method	Lesion	ID
Long distance running	0.12 – 6.45	μmol/mol creatinine	32	no change	multi-D-HPLC-ECD	8-oxodG	(Ref. 54)
Adriamyacin	34.4 ± 5.1 → 35.5 ± 4.6	nmol /24 h	20	no change	HPLC+HPLC-ECD	8-oxodG	(Ref. 10)
Cystic fibrosis	1.51 ± 0.38 versus 2.78 ± 1.21	nmol /mmol creatinine	23	increased in cystic fibrosis	correlation with plasma vit E HPLC-EC Pre-purification	8-oxodG	(Ref. 55)
Workplace exposure	0.5 – 3.0	μmol/mol creatinine	41 + 30	increased	multi-D-HPLC-ECD	8-oxodG	(Ref. 56)
Normal persons	2.7 ± 1.88	μmol/mol creatinine	60	in accordance with literature	multi-D-HPLC-ECD	8-oxodG	(Ref. 57)
Correlation to plasma antioxidants	10 – 105	nmol /24 h	225	no correlation	multi-D-HPLC-ECD	8-oxodG	(Ref. 28)
Vegetable and fruit	49.6 ± 23 → 21.4 ± 19.2		28	not significant	ELISA	8-oxodG	(Ref. 25)
Brussel's sprouts	236 – 1469	pmol/kg/ 24 h	37/52	Reduction	multi-D-HPLC-ECD	8-oxodG	(Ref. 58)
Controls	30 ± 15	pmol /ml	10	concentration only	HPLC-GC/MS	8-oxodG	(Ref. 11)
Comparison of published values ELISA versus HPLC-ECD	2.2 ± 0.9 → 24.3	ng/mg creatinine		3.7–10.1 higher values by ELISA	ELISA and HPLC-ECD	8-oxodG	(Ref. 59)
Exercise ß-carotene intervention	1.5 ± 0.2	nmol/mmol creatinine	14	no change	multi-D-HPLC-ECD	8-oxodG	(Ref. 60)
Children	3.8	μmol/mol creatinine	28 + 14	increase with age, no correlation with urinary malondialdehyde	multi-D-HPLC-ECD	8-oxodG	(Ref. 61)

Table 1 (*Continued*)

Disease/Condition/ Intervention	Level	Unit	Number of Individuals	Finding	Method	Lesion	ID
HIV-infection and AZT treatment ± antioxidants	$110 \pm 79 \rightarrow$ 355 ± 100	pmol/kg/d	8	low in HIV, increased from ATZ, reduced by antioxidants	pre-purification + HPLC-ECD	8-oxodG	(Ref. 62)
Altitude exposure	$23.0 \pm 11.6 \rightarrow$ 30.6 ± 13.9	μmol/L	58	increased with high altitude	ELISA	8-oxodG	(Ref. 63)
Method development	19 – 39	pmol/ml	not given	method development	clean up + HPLC + GC/MS	8-oxodG	(Ref. 64)
Background radiation	$19.5 \pm 1.2 \rightarrow$ 25.3 ± 1.6	nmol/L urine	63	tendency to increase	multi-Dimensional	8-oxodG	(Ref. 65)
Benzyl-8-oxoguanine exposure	1.000 – 9.250	ng/ml urine	1		HPLC-ECD	8-oxoGua	(Ref. 66)
Controls	583 ± 376	pmol /ml	10	concentration only	HPLC-GC/MS	8-oxo-Gua	(Ref. 11)
Allopurinol/children/cancer			2	N2-Me-8oxodGua/ 8-oxodGua ratio increased in infants and cancer patients (22x) only U2-Me-8-oxodGua		N2 methyl-8-oxoguanine	(Ref. 67)

number of DNA/RNA products excreted into urine. In case of RNA products high concentrations of very similar chemical substances are excreted in to urine.[26] A similar myriad of DNA products undoubtedly is also excreted. Together this may make it very difficult to produce a specific antibody. A commercially available kit tested out against the three dimensional HPLC-ECD shoved clear non-specificity.[27] Until clear demonstration of close correlation to the verified HPLC-ECD method the use of immunologically based methods for quantification of 8-oxodG in urine cannot be recommended. Since there is a very close correlation between HPLC-ECD and HPLC-MS/MS (Fig. 2) measurements, presently these methods may be regarded as the golden standard.

3. Comparison of Different Methods

Presently, the only lesion measured in urine in a fair number of studies is 8-oxodG. There is a very close correlation between urinary 8-oxodG measurements with the three dimensional HPLC-ECD method and the HPLC-MS/MS method (Fig. 2). Also, recovery of added 8-oxodG is 100% and storage over many years yields identical results.[13, 27, 28] As stated above there is poor relationship with an immunologically based assay.[27]

With regard to other base modifications the data are so scarce that it is not really possible to establish correlation between methodologies. The various publications can be found in Table 1.

Generally, it can be stated that the huge discrepancy that has been reported for the use of HPLC-ECD and GC-MS to measure tissue levels does not appear to be much of a problem in urinary measurements. This is in agreement with the much lower levels of non-oxidized bases/nucleosides in urine compared with DNA.

4. Interpretation of Tissue Levels and Urinary Excretion

In the following the argumentation is done for 8-oxodG, however, except for specific values, the argumentation is valid for all oxidative DNA products. Specific conditions may exist for the individual products.

The actual unperturbed DNA steady state level (of 8-oxodG) has been the subject of intense debate. While the ESCODD initiative.[29–31] and the extensive work to reduce oxidative artefacts from the GC-MS DNA procedure[16, 32, 33] are steps in the right direction there still is no definite consensus about the "true levels". Very high levels have been reported from GC-MS measurements[33] but also levels close to those measured by HPLC-ECD. Results from the same lab using the same methodology may differ with a factor of 10–100 over the years. The authors do not want to point at single laboratories, since the variation in levels can be pointed out for virtually all laboratories. The plethora of data on 8-oxodG in DNA is not systematically reviewed here, several reviews can be found in the literature list and in other chapters of this book. A large and comprehensive review has been published.[34]

In case of mitochondrial DNA the variation seems to be even greater. A review of the literature showed that the reported level of 8-oxodG in mitochondria spans more than 60 000-fold[35] from 0.035 to 2200 oxidative adducts per 100 000 unaltered bases. Eliminating the most extreme value the level still spans from 0.8 to 2200 oxidative adducts per 100 000 bases, or almost 3.000 fold. The mitochondria genome is 16 KB and the number of mitochondria in cells from various tissues vary from 220 (peritoneal macrophages) to 1720 (lung macrophages) per cell, skeletal muscle cells having less than 400 genomes per cell.[36] An average number of mtDNA copies in mitochondria of the body could well be about 200–400 per cell, giving a total number of base pairs of 16 000 times 200 or 3×10^6 base pairs per cell compared to the 1×10^9 base pairs in the nuclear genome. Although several authors have stated that the mtDNA is much more damaged than the genomic DNA, maybe with a factor of 10 times, these calculation still indicate that only a small portion of the 8-oxodG excreted in urine originates from mitochondria. Furthermore, the mitochondrial genome only contains 16 000 KB corresponding to about 4000 dG. Assuming correctness of data (unpublished) from our own laboratory, which indicate a level of about 1 8-oxodG per 10 000 dG in mtDNA, only an average of one in about 2–3 mitochondria will carry an 8-oxodG. Taking into account that the oxidative DNA damage has been overestimated[37] and that mitochondrial damage persists longer than nuclear damage[38] there is no reason to argue that the urinary 8-oxodG has a major component from mitochondria. Rather it can be concluded that the majority of the excreted 8-oxodG relates to nuclear DNA. Further support for this is our study on 2-nitropropane induced 8-oxodG in various organs where the nuclear DNA increase of 8-oxodG is reflected in urinary excretion.[39] Also it could be argued that mtDNA is less damaged than nuclear DNA, i.e. evaluated per genome. Probably the best evaluation is done as damage for a specific gene. Techniques for this are only emerging.

The urinary excretion of 8-oxodG in pigs following i/v injection follows simple kinetics with a half-life of a about 2.5 h, a clearance of about 4 mL min^{-1} kg BW^{-1} and a volume of distribution close to 1 L kg^{-1} BW,[40] and moreover the urinary excretion rate corresponded to the infusion rate. After liver transplantation we observed an increased urinary excretion of 8-oxodG and in a caval clamp experiment the excretion was temporarily reduced. These experiments indicate that steady state between formation and urinary excretion is obtained rapidly.

The reported values of urinary excretion of 8-oxodG in the literature are in agreement. The reported 8-oxodG urinary excretion rates measured with HPLC-ECD or GC-MS[13] vary from about 100 to 600 pmol kg BW^{-1} 24 h^{-1}, excluding the measurements with immunologically based estimations that vary between 1600–4800 pmol kg BW^{-1} 24 h^{-1} most presumably for the reasons about lack of specificity given above. Classic pharmacokinetic consideration gives a theoretical steady state plasma concentration equal to production (dosing rate) divided by clearance, i.e. between 0.017 and 0.100 nmol/L. The conventional HPLC-ECD and HPLC-MS/MS methods have sensitivity close to that level. Using up-concentrations and a HPLC-ECD system with a non-commercially available carbon column Bogdanov et al.[41] reported plasma values of 0.014–0.070 nmol/L (4–21 pg/ml), i.e. in close agreement with the theoretical values.

Collectively, these data indicate that the 8-oxodG in the urine mainly originates from genomic DNA. However, on a more detailed level the contribution of 8-oxodG from the nucleotide pool cell turnover, cell death, and from inflammatory cells is unknown. Presently, neither direct nor indirect data from the *in vivo* situation are available.

Accepting that the contribution of nuclear DNA reflects the oxidation of nuclear DNA, the urinary excretion is a reflection of the average total oxidative stress to DNA of all body cells. In most experimental situations *in vivo* it is reasonable to argue that a given person is in a steady state, i.e. a constant 8-oxodG level in DNA and a constant repair. Mass conservation will be applicable and consequently the amount of excreted 8-oxodG will equal newly formed 8-oxodG. The urinary measurement is therefore equal to the rate of oxidative stress to DNA. This is depicted in Fig. 5. If an experimental or other form of change happens (say smoking

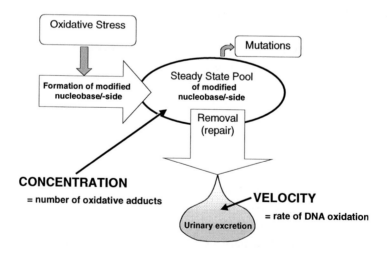

Fig. 5. Model for oxidative stress inducing modification of DNA. A sample taken from tissue will provide information about the <u>concentration</u> of modifications in DNA. Urine measurement is a measure of the amount of oxidative DNA products over a time period and this is a measure of the <u>rate of DNA oxidation</u>, when the system is in steady state. In the latter situation the measurement is independent of repair.

cessation, antioxidant intervention) a new steady state will soon be reached and a change in the rate of oxidation of DNA can be identified. It is important to stress that this measure is independent of DNA repair, a point often not recognized.

The concentration of say 8-oxodG in DNA reflects a balance between newly formed 8-oxodG's and removal. An increased level can consequently reflect either an increased formation (increased oxidative stress) or a decrease in repair or any combination. It is important to note that this cannot be determined from measurement of the level. A similar argumentation can be made for decreased levels.

It can further be argued that comparing two persons with different oxidative stress to DNA, i.e. different urinary excretion rates, the one with the higher stress will statistically have a higher chance for a mutation in DNA. Increased levels cannot necessarily be interpreted in the same way, unless it can be established whether it originates from increased stress or decreased repair.

It should be noted that for urinary excretion studies the preferred design is to collect 24 h urine. In some special designs it can be argued that the use of spot urine samples and correction for urinary creatinine concentration may be a valid measure. A prerequisite for the spot urine — creatinine correction design is a solid argumentation that creatinine excretion is unchanged by the experimental condition or that it is not different between groups. A theoretical example is comparison of lean men versus fat females. Their cell number is comparable but muscle mass very different. Creatinine excretion is mainly dependent on muscle mass, and there can easily be a difference in creatinine excretion of says 3 fold between the two groups. If they have the same oxidative stress to their DNA, females would appear to have 3 times higher values, simply because the male excretion is divided by a three times higher creatinine concentration. The same argumentation can be applied to comparison of catabolic patients versus normal controls, and old versus young adults. Preferentially 24 h urine, overnight urine(s) or at least 8 h urine on a defined period of the day should be collected and the 8-oxodG excretion given as amount per time unit and kg BW, preferentially lean body weight.

5. What is Known from Human Studies of Biomarkers of Oxidative DNA Damage?

The most studied oxidative modification of DNA relates to direct oxidation of DNA, the 8-hydroxylation of guanine being the one most extensively studied, particularly regarding urinary excretion of the repair product 8-oxodG. The excretion of the base, 8-oxoGua, is much less studied, although it is excreted in larger amounts, about 5–10 timer larger than 8-oxodG.[42] There is general agreement that the modifications like 8-oxodG are the result of reactions between DNA and reactive oxygen species. However, other oxidative processes e.g. lipid peroxidation gives rise to reactive intermediates that in turn can modify DNA. Lipid peroxidation leads to formation of malondialdehyde, crotonaldehyde and acrolein that in turn lead to propano- and etheno-DNA adducts, called exocyclic adducts. These adducts are found in lower quantities than e.g. 8-oxodG and require ultra sensitive methods. The urinary excretion of 1,N^6-ethenodeoxyadenosine (εdA) ranges from about 0.1 to 4 fmol/micromol creatinine in human urine.[43] Human studies on the exocyclic adducts and their excretion into urine so far are limited indeed. A comprehensive overview is given in a recent IARC publication.[44]

With regard to 8-oxodG a large number of human studies have been published. They are listed in Table 1. Several pieces of basic information are known (for references see Table 1). The variation among normal people is about 7-fold with a clear sex difference. Smokers have a 50% increased 8-oxodG, which is reduced 4 weeks after smoking cessation, There is a close relationship between individual oxygen uptake (energy expenditure) and 8-oxodG excretion. Extreme longstanding exercise increases the excretion, moderate exercise or extreme bouts do not change the excretion. The majority of antioxidant interventions do not show any effect on the excretion of 8-oxodG, i.e. β-carotene, vitamin E and vitamin C. Yet there is some controversy about the change in lymphocyte levels in antioxidant intervention trials. Occupational exposure to diesel exhaust/air pollution and styrene increase the urinary excretion. Dietary restriction does not influence the excretion to any detectable extent, whereas Mediterranean diet and Brussel's Sprouts seem to reduce excretion.

For a number of diseases there are reports on increased urinary excretion and higher lymphocyte levels, and both radiotherapy and radiomimetic treatment increases excretion. So far there does not seem to be reports on decreased excretion in any disease.

The general picture that emerges is that a variety of environmental and dietary factors can explain some of the variation. However, these influences generally seem to be in the order of about 50% at maximum. The 7-fold variation in the population thus leaves the major determining factors to be found. Probably, if oxidative stress is maintained at a high level throughout life, changes in the order of 10–20% may be highly relevant, particularly on a group basis. Such a difference may reflect an increased risk of cancer or premature ageing.

6. Proper Design of Human Intervention Studies

Several elements are mandatory for a trial, particularly a trial on humans, to adhere to modern scientific demands. The basic elements are an *a priori* defined primary hypothesis and definition of primary and secondary endpoint and in some cases also tertiary endpoints. Once this is defined the design of the trial, the statistical analysis and the control group can be defined.

Only the two major design types will be dealt with here, crossover and parallel groups designs, also called paired and unpaired designs. A large number of studies use comparison of people before and after, say e.g. antioxidant intervention. Such a design is considered based on historical controls and should not be performed. Rather, the persons should be randomized to two different treatments, placebo and active treatment, the randomization gives the random order of the treatments. By such a design effects e.g. due to season is randomly allocated to the groups. The advantage of the paired design is that each person serves as his own control, and the number of subjects in the trial is reduced compared with the unpaired trial. Among the disadvantages are that every time a person drops out the first measurement he/she cannot be included in the analysis. Furthermore, if the variation within individuals is comparable with that between individuals, extra power is not obtained.

The parallel group is a more simple design. A group of people is randomized to two treatments, e.g. active treatment and placebo or two different active treatments, and the primary variable is then compared between the groups.

More complicated designs can be used but is not mentioned here. Most important is to stress the proper use of randomization and controls.

In the planning of a trial it is necessary to calculate the number of persons needed to be able to detect a predefined difference. In many countries, e.g. in Denmark, ethical approval is not given if a proper statistical power analysis is not given.

The power analysis is a calculation of the number of people to enter the trial, provided there is knowledge about the defined Type I error risk (significance

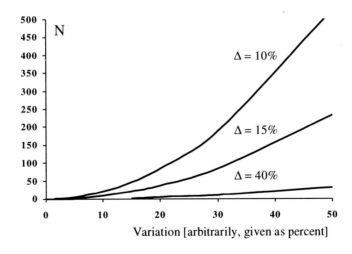

Fig. 6. Relation between the number of persons (N), the difference to detect (Δ) and variation on the measurement used. The graphs are calculated for significance α = 0.05 and power β = 0.90.

level), the Type II error risk (power), the defined difference the trial is supposed to detect (delta Δ) and the variation of the measurement in the trial. A simple mathematical relationship between these factors exists. For details readers should look in statistical textbooks. In Fig. 6, the number of persons to enter a parallel group trial can be read if the variation of the measurement is known. The graphs indicate differences of 10, 15 and 40% between the groups, Types I and II errors are set at 5 and 10%, respectively.

It should be noted that the inter-individual variation e.g. in 24 h 8-oxodG excretion is often 30%. If it is desired to detect a difference of say 10% about 200 persons is required. There are very few studies with that number of persons. This should borne in mind when trials are evaluated. Very often a negative trial can represent a Type II error, not finding a true difference.

7. Future Perspectives

The formation of DNA adducts from endogenous processes and from exogenous factors has emerged as an important factor in the pathogenesis of cancer and ageing. The development of accurate, reliable methods to determine DNA oxidation is essential for understanding the processes. Presently, there has been a fast growing knowledge about the 8-oxodG lesion, and particularly there has been improvement in the knowledge about how to avoid artifacts during the process of quantifying the damage. However, there is only limited knowledge about other lesions than 8-oxodG, particularly *in vivo* in humans. Measurement of single lesions may be misleading and just because one lesion is the most dominating it is not necessarily the most import. Free radicals generate many products at the same time.[45]

Furthermore other free radical induced processes, e.g. lipid peroxidation, produce reactive intermediates that may be important. Examples of such other lesions are for example malondialdehyde induced DNA damage and exocyclic DNA adducts.[44]

Development of methodologies to detects these DNA modifications are in progress. Furthermore, molecular biology methods, e.g. variants of the PCR methods, and newer mass spectrometry methods like time of flight will in the future make it possible to detail the various DNA modifications not only by reliable methods for quantification but also for position in specific genes. Increasingly, we will see animal studies using genetically modified animals, studies that will clarify specific mechanisms, including studies with DNA array techniques to quantify mRNA to give deeper insight into the cellular biology of oxidative stress.

Furthermore, the future will improve the technologies for measurement on smaller samples and for measurement of large number of samples with reasonable use of time and money. This will enable large scale epidemiological and intervention trials with reliable estimates of the precise role of these modification in the pathogenesis of disease and ageing.

8. Conclusions

- Oxidative lesions are implicated in cancer and ageing. Presently the role is based on circumstantial evidence and information in humans *in vivo* is limited.
- Urinary excretion of oxidative DNA products reflects total average DNA oxidation in the body and in the steady state situation it is independent of repair. The urinary excretion reflects mainly the oxidative stress to nuclear DNA.
- DNA levels in tissue of oxidatively modified nucleobases reflect a balance between oxidative stress and repair.
- A timed urine collection, at least 8–16 h preferentially 24 h urine, is recommended and the result should be given as excreted modified nucleobase per time unit and per mass unit body weight (preferentially lean body mass).
- Spot urine samples should only be used if it can be verified that creatinine excretion is unchanged.
- Methods of analysis preferred are HPLC-ECD or HPLC-MS/MS. The latter method has greater potential for measuring multiple oxidation products. Immunologically based methods needs to be verified.
- GC-MS methods can be used provided sufficient measures to counteract artificial oxidation are taken.
- The exact role of DNA oxidation in the pathogenesis of cancer and ageing needs to be established.

References

1. Wiseman, H. and Halliwell, B. (1996). Damage to DNA by reactive oxygen and nitrogen species: role in inflammatory disease and progression to cancer. *Biochem. J.* **313**: 17–29.
2. Chance, B., Sies, H. and Boveris, A. (1979). Hydroperoxide metabolism in mammalian organs. *Physiol. Rev.* **59**: 527–605.
3. Poulsen, H. E., Jensen, B. R., Weimann, A., Jensen, S. A., Sørensen, M. and Loft, S. (2000). Antioxidants, DNA damage and gene expression. *Free Radic. Res.* **33**: S33–S39.
4. Dalton, T. P., Shertzer, H. G. and Puga, A. (1999). Regulation of gene expression by reactive oxygen. *Ann. Rev. Pharmacol. Toxicol.* **39**: 67–101.
5. Karin, M. and Smeal, T. (1992). Control of transcription factors by signal transduction pathways: the beginning of the end. *Trends Biochem. Sci.* **17**: 418–422.

6. Dizdaroglu, M. (1998). Facts about the artifacts in the measurement of oxidative DNA base damage by gas chromatography-mass spectrometry. *Free Radic. Res.* **29**: 551–563.

7. Dizdaroglu, M. (1994). Chemical determination of oxidative DNA damage by gas chromatography-mass spectrometry. *Meth. Enzymol.* **234**: 3–16.

8. Weimann, A., Belling, D. and Poulsen, H. E. (2001). Measurement of 8-oxo-2-deoxyguanosine and 8-oxo-2-deoxyadenosine in DNA and Human urine by high performance liquid chromatography — electrospray tandem mass spectrometry. *Free Radic. Res.* **30**: 757–764.

9. Pourcelot, S., Faure, H., Firoozi, F., Ducros, V., Tripier, M., Hee, J., Cadet, J. and Favier, A. (1999). Urinary 8-oxo-7,8-dihydro-2'-deoxyguanosine and 5-(hydroxymethyl) uracil in smokers. *Free Radic. Res.* **30**: 173–180.

10. Faure, H., Mousseau, M., Cadet, J., Guimier, C., Tripier, M., Hida, H. and Favier, A. (1998). Urine 8-oxo-7,8-dihydro-2-deoxyguanosine versus 5-(hydroxymethyl) uracil as DNA oxidation marker in adriamycin-treated patients. *Free Radic. Res.* **28**: 377–382.

11. Ravanat, J. L., Guicherd, P., Tuce, Z. and Cadet, J. (1999). Simultaneous determination of five oxidative DNA lesions in human urine. *Chem. Res. Toxicol.* **12**: 802–808.

12. Floyd, R. A., Watson, J. J., Wong, P. K., Altmiller, D. H. and Rickard, R. C. (1986). Hydroxyl free radical adduct of deoxyguanosine: sensitive detection and mechanisms of formation. *Free Radic. Res. Commun.* **1**: 163–172.

13. Loft, S. and Poulsen, H. E. (1998). Markers of oxidative damage to DNA: antioxidants and molecular damage. *Meth. Enzymol.* **300**: 166–184.

14. Poulsen, H. E., Loft, S. and Weimann, A. (2000). Urinary measurement of 8-oxodG (8-oxo-2'-deoxyguanosine). *In* "Handbook of Clinical Analysis: In Vivo Damage to Biomolecules" (J. Lunec and H. R. Griffiths, Eds.), pp. 69–80, John Wiley and Sons (Ltd.), London.

15. Ravanat, J. L., Gremaud, E., Markovic, J. and Turesky, R. J. (1998). Detection of 8-oxoguanine in cellular DNA using 2,6-diamino-8-oxopurine as an internal standard for high-performance liquid chromatography with electrochemical detection. *Anal. Biochem.* **260**: 30–37.

16. Rodriguez, H., Jurado, J., Laval, J. and Dizdaroglu, M. (2000). Comparison of the levels of 8-hydroxyguanine in DNA as measured by gas chromatography mass spectrometry following hydrolysis of DNA by *Escherichia coli* Fpg protein or formic acid. *Nucleic Acids Res.* **28**: 4583–4592.

17. Poulsen, H. E., Weimann, A., Salonen, J. T., Nyyssonen, K., Loft, S., Cadet, J., Douki, T. and Ravanat, J. L. (1998). Does vitamin C have a pro-oxidant effect? [letter]. *Nature* **395**: 231–232.

18. Podmore, I. D., Griffiths, H., Herbert, K. and Mistry, N. (1998). Does vitamin C have a pro-oxidant effect? *Nature* **395**: 231–232.

19. Douki, T., Court, M. and Cadet, J. (2000). Electrospray-mass spectrometry characterization and measurement of far- UV-induced thymine photoproducts [In Process Citation]. *J. Photochem. Photobiol.* **B54**: 145–154.
20. Ravanat, J. L., Remaud, G. and Cadet, J. (2000). Measurement of the main photooxidation products of 2'-deoxyguanosine using chromatographic methods coupled to mass spectrometry. *Arch. Biochem. Biophys.* **374**: 118–127.
21. Hoberg, A. M. and Poulsen, H. E. (2000). Analysis of a promutagenic exocyclic DNA adduct in human urine by high performance liquid chromatography API tandem mass spectrometry. *Adv. Mass Spectr. Proc. 15th Int. Mass Spectr. Conf.* **in press**.
22. Erhola, M., Toyokuni, S., Okada, K., Tanaka, T., Hiai, H., Ochi, H., Uchida, K., Osawa, T., Nieminen, M. M., Alho, H. and Kellokumpu-Lehtinen, P. (1997). Biomarker evidence of DNA oxidation in lung cancer patients: association of urinary 8-hydroxy-2'-deoxyguanosine excretion with radiotherapy, chemotherapy, and response to treatment. *FEBS Lett.* **409**: 287–291.
23. Leinonen, J., Lehtimaki, T., Toyokuni, S., Okada, K., Tanaka, T., Hiai, H., Ochi, H., Laippala, P., Rantalaiho, V., Wirta, O., Pasternack, A. and Alho, H. (1997). New biomarker evidence of oxidative DNA damage in patients with non-insulin-dependent diabetes mellitus. *FEBS Lett.* **417**: 150–152.
24. Tsuboi, H., Kouda, K., Takeuchi, H., Takigawa, M., Masamoto, Y., Takeuchi, M. and Ochi, H. (1998). 8-hydroxydeoxyguanosine in urine as an index of oxidative damage to DNA in the evaluation of atopic dermatitis. *Brit. J. Dermatol.* **138**: 1033–1035.
25. Thompson, H. J., Heimendinger, J., Haegele, A., Sedlacek, S. M., Gillette, C., O'Neill, C., Wolfe, P. and Conry, C. (1999). Effect of increased vegetable and fruit consumption on markers of oxidative cellular damage. *Carcinogenesis* **20**: 2261–2266.
26. Schram, K. H. (1998). Urinary nucleosides. *Mass Spectr. Rev.* **17**: 131–251.
27. Prieme, H., Loft, S., Cutler, R. G. and Poulsen, H. E. (1996). Measurement of oxidative DNA injury in humans: evaluation of a commercially available ELISA assay. *In* "Natural Antioxidants and Food Quality in Atherosclerosis and Cancer Prevention" (J. T. Kumpulainen, Ed.), pp. 78–82, The Royal Society of Chemistry.
28. Poulsen, H. E., Loft, S., Prieme, H., Vistisen, K., Lykkesfeldt, J., Nyyssonen, K. and Salonen, J. T. (1998). Oxidative DNA damage in vivo: relationship to age, plasma antioxidants, drug metabolism, glutathione-S-transferase activity and urinary creatinine excretion. *Free Radic. Res.* **29**: 565–571.
29. Collins, A., Cadet, J., Epe, B. and Gedik, C. (1997). Problems in the measurement of 8-oxoguanine in human DNA. Report of a workshop, DNA oxidation, held in Aberdeen, UK, 19–21 January, 1997. *Carcinogenesis* **18**: 1833–1836.

30. Lunec, J. (1998). ESCODD: European Standards Committee on Oxidative DNA Damage. *Free Radic. Res.* **29**: 601–608.
31. Lunec, J. (1999). ESCODD: European Standards Committee on Oxidative DNA Damage. *Free Radic. Res.* **in press**.
32. England, T. G., Jenner, A., Aruoma, O. I. and Halliwell, B. (1998). Determination of oxidative DNA base damage by gas chromatography-mass spectrometry. Effect of derivatization conditions on artifactual formation of certain base oxidation products. *Free Radic. Res.* **29**: 321–330.
33. Halliwell, B. (1998). Can oxidative DNA damage be used as a biomarker of cancer risk in humans? Problems, resolutions and preliminary results from nutritional supplementation studies. *Free Radic. Res.* **29**: 469–486.
34. Kasai, H. (1997). Analysis of a form of oxidative DNA damage, 8-hydroxy-2'-deoxyguanosine, as a marker of cellular oxidative stress during carcinogenesis. *Mutat. Res.* **387**: 147–163.
35. Beckman, K. B. and Ames, B. N. (1999). Endogenous oxidative damage of mtDNA. *Mutat. Res.* **424**: 51–58.
36. Robin, E. D. and Wong, R. (1988). Mitochondrial DNA molecules and virtual number of mitochondria per cell in mammalian cells. *J. Cell Physiol.* **136**: 507–513.
37. Anson, R. M., Hudson, E. and Bohr, V. A. (2000). Mitochondrial endogenous oxidative damage has been overestimated. *FASEB J.* **14**: 355–360.
38. Yakes, F. M. and Van Houten, B. (1997). Mitochondrial DNA damage is more extensive and persists longer than nuclear DNA damage in human cells following oxidative stress. *Proc. Natl. Acad. Sci. USA* **94**: 514–519.
39. Deng, X. S., Tuo, J., Poulsen, H. E. and Loft, S. (1998). Prevention of oxidative DNA damage in rats by Brussels sprouts. *Free Radic. Res.* **28**: 323–333.
40. Loft, S., Larsen, P. N., Rasmussen, A., Fischer-Nielsen, A., Bondesen, S., Kirkegaard, P., Rasmussen, L. S., Ejlersen, E., Tornoe, K., Bergholdt, R. and Poulsen, H. E. (1995). Oxidative DNA damage after transplantation of the liver and small intestine in pigs. *Transplantation* **59**: 16–20.
41. Bogdanov, M. B., Beal, M. F., McCabe, D. R., Griffin, R. M. and Matson, W. R. (1999). A carbon column-based liquid chromatography electrochemical approach to routine 8-hydroxy-2'-deoxyguanosine measurements in urine and other biologic matrices: a one-year evaluation of methods. *Free Radic. Biol. Med.* **27**: 647–666.
42. Shigenaga, M. K., Gimeno, C. J. and Ames, B. N. (1989). Urinary 8-hydroxy-2'-deoxyguanosine as a biological marker of in vivo oxidative DNA damage. *Proc. Natl. Acad. Sci. USA* **86**: 9697–9701.
43. Nair, J. (1999). Lipid peroxidation-induced etheno-DNA adducts in humans. *In* "Exocyclic DNA Adducts in Mutagenesis and Carcinogenesis" (B. Singer and H. Bartsch, Eds.), pp. 55–61, IARC Scientific Publication No. 150.
44. Singer, B. and Bartsch, H. (1999). "Exocyclic DNA Adducts in Mutagenesis and Carcinogenesis", pp. 1–361, IARC Scientific Publications.

45. Dizdaroglu, M. (1992). Oxidative damage to DNA in mammalian chromatin. *Mutat. Res.* **275**: 331–342.
46. Faure, H., Coudray, C., Mousseau, M., Ducros, V., Douki, T., Bianchini, F., Cadet, J. and Favier, A. (1996). 5-hydroxymethyluracil excretion, plasma TBARS and plasma antioxidant vitamins in adriamycin-treated patients. *Free Radic. Biol. Med.* **20**: 979–983.
47. Tagesson, C., Kallberg, M., Klintenberg, C. and Starkhammar, H. (1995). Determination of urinary 8-hydroxydeoxyguanosine by automated coupled-column high performance liquid chromatography: a powerful technique for assaying in vivo oxidative DNA damage in cancer patients. *Eur. J. Cancer* **31A**: 934–940.
48. Holmberg, I., Stal, P. and Hamberg, M. (1999). Quantitative determination of 8-hydroxy-2'-deoxyguanosine in human urine by isotope dilution mass spectrometry: normal levels in hemochromatosis. *Free Radic. Biol. Med.* **26**: 129–135.
49. Moriwaki, H. (2000). Determination of 8-Hydroxy-2'-deoxyguanosine in urine by liquid chromatography — electrospray ionization mass spectrometry. *Anal. Sci.* **16**: 105–106.
50. Tagesson, C., Kallberg, M. and Leanderson, P. (1992). Determination of urinary 8-hydroxydeoxyguanosine by coupled-column high perfomance liquid chromatography with electrochemical detection: a noninvasive assay for *in vivo* oxidative DNA damage in humans. *Toxicol. Meth.* **1**: 242–251.
51. Nielsen, H. B., Hanel, B., Loft, S., Poulsen, H. E., Pedersen, B. K., Diamant, M., Vistisen, K. and Secher, N. H. (1995). Restricted pulmonary diffusion capacity after exercise is not an ARDS-like injury. *J. Sports Sci.* **13**: 109–113.
52. Okamura, K., Doi, T., Hamada, K., Sakurai, M., Yoshioka, Y., Mitsuzono, R., Migita, T., Sumida, S. and Sugawa, K. Y. (1997). Effect of repeated exercise on urinary 8-hydroxy-deoxyguanosine excretion in humans. *Free Radic. Res.* **26**: 507–514.
53. Sumida, S., Okamura, K., Doi, T., Sakurai, M., Yoshioka, Y. and Sugawa-Katayama, Y. (1997). No influence of a single bout of exercise on urinary excretion of 8-hydroxy-deoxyguanosine in humans. *Biochem. Mol. Biol. Int.* **42**: 601–609.
54. Pilger, A., Germadnik, D., Formanek, D., Zwick, H., Winkler, N. and Rudiger, H. W. (1997). Habitual long-distance running does not enhance urinary excretion of 8-hydroxydeoxyguanosine. *Eur. J. Appl. Physiol. Occup. Physiol.* **75**: 469.
55. Brown, R. K., McBurney, A., Lunec, J. and Kelly, F. J. (1995). Oxidative damage to DNA in patients with cystic fibrosis. *Free Radic. Biol. Med.* **18**: 801–806.
56. Toraason, M. (2000). 8-hydroxydeoxyguanosine as a biomarker of workplace exposures. *Biomarkers* **5**: 3–26.
57. Germadnik, D., Pilger, A. and Rudiger, H. W. (1997). Assay for the determination of urinary 8-hydroxy-2'-deoxyguanosine by high performance liquid

chromatography with electrochemical detection. *J. Chromat. B. Biomed. Sci. Appl.* **689**: 399–403.

58. Verhagen, H., de Vries, A., Nijhoff, W. A., Schouten, A., van Poppel, G., Peters, W. H. and van den, B. H. (1997). Effect of Brussels sprouts on oxidative DNA-damage in man. *Cancer Lett.* **114**: 127–130.

59. Cooke, M. S., Evans, M. D., Herbert, K. E. and Lunec, J. (2000). Urinary 8-oxo-2'-deoxyguanosine — source, significance and supplements. *Free Radic. Res.* **32**: 381–397.

60. Sumida, S., Doi, T., Sakurai, M., Yoshioka, Y. and Okamura, K. (1997). Effect of a single bout of exercise and beta-carotene supplementation on the urinary excretion of 8-hydroxy-deoxyguanosine in humans. *Free Radic. Res.* **27**: 607–618.

61. Drury, J. A., Jeffers, G. and Cooke, R. W. (1998). Urinary 8-hydroxydeoxyguanosine in infants and children. *Free Radic. Res.* **28**: 423–428.

62. de la Asuncion, J. G., del Olmo, M. L., Sastre, J., Millan, A., Pellin, A., Pallardo, F. V. and Vina, J. (1998). AZT treatment induces molecular and ultrastructural oxidative damage to muscle mitochondria. Prevention by antioxidant vitamins. *J. Clin. Invest.* **102**: 4–9.

63. Chao, W. H., Askew, E. W., Roberts, D. E., Wood, S. M. and Perkins, J. B. (1999). Oxidative stress in humans during work at moderate altitude. *J. Nutri.* **129**: 2009–2012.

64. Teixeira, A. J., Ferreira, M. R., van Dijk, W. J., van de Werken, G. and de Jong, A. P. (1995). Analysis of 8-hydroxy-2'-deoxyguanosine in rat urine and liver DNA by stable isotope dilution gas chromatography/mass spectrometry. *Anal. Biochem.* **226**: 307–319.

65. Sperati, A., Abeni, D. D., Tagesson, C., Forastiere, F., Miceli, M. and Axelson, O. (1999). Exposure to indoor background radiation and urinary concentrations of 8-hydroxydeoxyguanosine, a marker of oxidative DNA damage. *Env. Health Perspect.* **107**: 213–215.

66. Long, L., McCabe, D. R. and Dolan, M. E. (1999). Determination of 8-oxoguanine in human plasma and urine by high performance liquid chromatography with electrochemical detection. *J. Chromat. B. Biomed. Sci. Appl.* **731**: 241–249.

67. Helbock, H. J., Thompson, J., Yeo, H. and Ames, B. N. (1996). N2-methyl-8-oxoguanine: a tRNA urinary metabolite — role of xanthine oxidase. *Free Radic. Biol. Med.* **20**: 475–481.

Chapter 14

Mapping Oxidative DNA Damage at Nucleotide Resolution in Mammalian Cells

Timothy R. O'Connor*,†, Hsiu-Hua Chen, Shu-Mei Dai,
Steven D. Flanagan, Steven A. Akman, Gerald P. Holmquist,
Henry Rodriguez and Arthur D. Riggs

Timothy R. O'Connor, Hsiu-Hua Chen, Shu-Mei Dai, Steven D. Flanagan, Gerald P. Holmquist and **Arthur D. Riggs** • Beckman Research Institute of the City of Hope, Duarte, California 91010 USA
Steven A. Akman • Department of Cancer Biology, Comprehensive Cancer Center of Wake Forest University, Winston-Salem, North Carolina 27157 USA 20899
Henry Rodriguez • Biotechnology Division, National Institute of Standards and Technology, Gaithersburg, Maryland USA
*On leave from *UMR1772 Physicochimie et Pharmacologie des Macromolecules Biologiques.*
†Corresponding Author.
Tel: 626-301-8220, E-mail: toconnor@coh.org

1. Introduction

Analytical techniques including mass spectrometry have proven invaluable for the detection of DNA adducts, but these methods do not map the sequence context dependent distribution of adducts. The adduct maps are important to understanding the formation of pre-mutagenic lesions that appear in critical growth control genes that are linked to cellular transformation (e.g. P53). In this chapter, we describe some of the methods available for the detection of oxidized base adducts at nucleotide resolution in human cells and some results obtained using this technology.

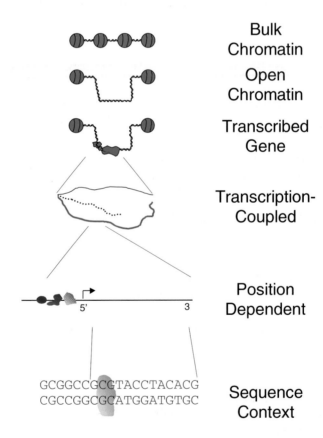

Bulk Chromatin

Open Chromatin

Transcribed Gene

Transcription-Coupled

Position Dependent

GCGGCCGCGTACCTACACG
CGCCGGCGCATGGATGTGC

Sequence Context

Fig. 1. DNA repair heterogeneities found in human cells. The order from bulk chromatin to sequence context dependence is indicated. The schematic is not drawn to scale. From top to bottom, the symbols used are: the spheres represent nucleosomes, the strands DNA, the dotted line refers to an mRNA transcribed from the red strand that is the heavy line, the forms are proteins bound to a promoter, and the arrow refers to the transcription start site, the yellow form represents a bound protein, and the sequence is indicated. References for the different heterogeneities are found in Refs. 1–9.

2. Methods for Genomic DNA Sequencing

In addition to the mutation maps that are found in critical growth control genes of tumor cells, it is also important to discern the origin of those mutations. Databases exist for mutations in critical genes in human tumors or the map of mutations formed following treatment with a specific agent can be generated in the laboratory. One goal in these experiments has been to establish an evidential link between an agent (UV, hydrogen peroxide, alkylating agents, etc.) and the map of mutations. If we can superimpose, for a given agent, the damage or repair map at nucleotide resolution on the mutation map, a connection is established. Such links are best established by the investigation of DNA damage and repair at several levels, from global to nucleotide resolution (Fig. 1).[1-9] One link established between UV radiation and the mutation map in the human P53 gene is illustrated in Fig. 2. In principle, the determination of adduct maps at nucleotide resolution provides the most accurate representation of the bases damaged in the cell. Detection of damaged bases in human DNA requires a genomic sequencing method that is extremely sensitive yet robust and able to quantitatively detect DNA damage reproducibly. Using the techniques described in this chapter, etiological links have been established between at least two agents and two cancers.[5, 10]

Originally, genomic DNA sequencing methods were based on the use of chemical sequencing techniques.[11, 12] These methods were valuable, but were limited in their applications, since the experiments required large amounts of DNA, harsh chemical treatments, transfer of sequencing products to membranes, and sometimes weeks for the results to be visualized using autoradiography. These problems permitted only a few laboratories to employ genomic sequencing as a research tool.

The advent of polymerase chain reaction (PCR) technology led to an improvement in genomic DNA sequencing by permitting amplification. Ligation-mediated PCR genomic sequencing (LMPCR) [Fig. 3(a)] was originally used to obtain an *in vivo* dimethyl sulfate footprint of a muscle specific enhancer region.[13] LMPCR was then adapted for the detection of DNA methylation patterns, and other footprinting methods.[14-16] The successful use of LMPCR in those studies led Pfeifer, Riggs, and Holmquist to use LMPCR for the detection of DNA adducts induced by UVC.[17, 18] Those experiments demonstrated that it was possible to detect adducts in DNA at nucleotide resolution in mammalian cells. As illustrated in Fig. 3(a), DNA polymerase extends the oligodeoxyribonucleotide primer until a break in the phosphodiester backbone of DNA halts synthesis. For LMPCR to detect the signal at a strand break, the original strand must have a 5' phosphoryl group. Subsequently, one strand of a double-stranded oligodeoxyribonucleotide linker is attached to the original strand by DNA ligase. The ligation step is critical for the amplification of the original signal by exponential PCR. Following the ligation step, the original DNA strand with the break is amplified using a second

P53 Gene Mutation Spectrum: Non-Melanoma Skin Cancer

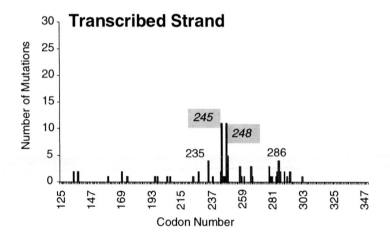

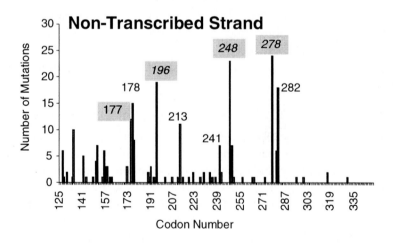

Fig. 2. C→T transition mutation frequencies in non-melanoma skin cancers. The positions of slow repair are indicated in shaded boxes with italic lettering and other frequently observed mutations are indicated next to each position. The positions of slow repair were determined using LMPCR mapping the repair of damage induced in normal human fibroblasts treated with UVC radiation.[5] The majority of the mutations are associated with the non-transcribed strand. The mutation spectrum is shaped by slow DNA adduct repair. In this case, the adducts are cyclobutane pyrimidine dimers at CC, CT, or TT positions.

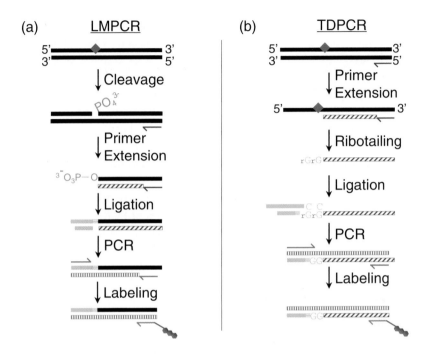

Fig. 3. Schematic describing the steps in ligation-mediated PCR (LMPCR) and terminal transferase-dependent PCR (TDPCR). An adduct position is represented as a red diamond and black lines represent the original strands. Oligodeoxyribonucleotide primers are indicated as arrows. (a) LMPCR detection of adducts. An adduct is removed by chemical or enzymatic reaction leaving a 5'phosphoryl group (red). The product of the Primer Extension reaction is indicated by a green rectangle with slanted lines. Primer Extension is followed by ligation of a linker (orange). The PCR reaction is performed using oligodeoxyribonucleotides that are nested (interior) to the first oligodeoxyribonucleotide primer. The strand formed by the PCR reaction is indicated by a red rectangle with straight lines. (b) TDPCR detection of adducts. The product of the Primer Extension reaction is indicated by a green rectangle with slanted lines. Primer Extension is followed by Ribotailing to add approximately two G ribonucleotides using terminal transferase (orange with ribose indicated by r). A linker is then added to provide a sequence permitting PCR (orange). The PCR reaction is performed using oligodeoxyribonucleotides that are nested (interior) to the first oligodeoxyribonucleotide primer. The strand formed by the PCR reaction is indicated by a red rectangle with straight lines. For both LMPCR and TDPCR, detection of the PCR products is performed using one of at least three methods: (1) Transfer of the products migrated on a DNA sequencing gel to a charged nylon membrane followed by hybridization using a ^{32}P labeled probe. (2) *E. coli* Exonuclease I treatment to remove excess primers from Primer Extension, the PCR and ligation steps. Direct labeling using a ^{32}P-5'end-labeled oligodeoxyribonucleotide prior to an extension reaction and migration on a DNA sequencing gel. (3) *E. coli* Exonuclease I treatment to remove excess primers from Primer Extension, the PCR and ligation steps. Direct labeling using a near infrared dye-5'end-labeled oligodeoxyribonucleotide prior to an extension reaction and migration on a Li-Cor automated DNA sequencing apparatus. Only the third method is indicated in this figure. The infrared dye is represented by the blue hexagons linked to a blue primer.

primer (to the 3′ side of the first primer) and a primer directed against the linker. Several methods can be used to detect the PCR products. These methods will be discussed below, but include either direct labeling or transfer of products to a charged nylon membrane where the products are revealed by radiolabeled probes and autoradiography.

A single-strand ligation-mediated PCR (sslig-PCR) method that requires ligation of single-stranded fragments by RNA ligase has also been reported.[19, 20] The use of sslig-PCR has the advantage that it will detect any lesion that arrests DNA synthesis. However, this method is technically challenging, and that has limited its application.

A more recently developed technique that is widely applicable for genomic sequencing reactions is terminal transferase-dependent PCR (TDPCR) [Fig. 2(b)].[21] This technique also detects adducts that halt or pause DNA synthesis, similar to sslig-PCR. The advantage of this method is that it is more sensitive than that of sslig-PCR. In TDPCR, the first oligodeoxyribonucleotide primer is extended until DNA synthesis terminates by either a strand break or another adduct that terminates DNA synthesis. The second step is to extend from the break site using terminal transferase in the presence of rGTP. Terminal transferase synthesizes phosphodiester bonds for only 2–3 ribonucleotides under these conditions. The terminal 2–3 nucleotides form a scaffold for the ligation of a linker to permit PCR amplification of the different fragments. Therefore, TDPCR bands are not precisely at nucleotide resolution, but in practice, the bands observed are very similar to those found using LMPCR.

3. LMPCR for the Detection of DNA Adducts in Mammalian Cells at Nucleotide Resolution

In practice, prior to performing LMPCR, the extent of global DNA damage must be determined to optimize LMPCR. Global adduct levels can be estimated using denaturing agarose gel electrophoresis.[22] From various LMPCR studies, the adduct levels detected using LMPCR are about 1 adduct in 10^4 bases. As originally introduced, LMPCR relied on chemical cleavage reactions to detect DNA adducts. As demonstrated in subsequent work, chemical cleavage schemes used during preparation of DNA for LMPCR introduce a significant amount of background damage even into control cells (cells not exposed to agents introducing DNA adducts).[23] In particular, the treatment with hot piperidine, which is required to cleave the phosphodiester backbone, can obliterate the detection of modest adduct levels.[23] Reduction of the background is critical to increasing the sensitivity of the technique for studies of DNA adduct repair. To diminish the background Pfeifer *et al.*[18] used a DNA glycosylase to introduce a break in DNA at cyclobutane pyrimidine dimer adducts. DNA glycosylases sever the glycosylic linkage and in some cases the phosphodiester backbone. These enzymes decrease the background generated relative to chemical cleavage schemes. Numerous other DNA glycosylases and the UVRABC Excinuclease have also been applied to LMPCR. The use of these enzymes in cleavage schemes led directly to the detection of adducts induced by oxidizing agents[24-29] and repair at nucleotide resolution.[2, 5, 30-32] Despite advantages presented by LMPCR, this technique is not yet capable of detecting the repair of oxidized bases at non-lethal doses of agent, but the development of LMPCR technology should help attain that goal.

4. Evolution of LMPCR Technology

LMPCR was a distinct improvement over direct genomic sequencing, but was still a tedious procedure that took 5 days to complete. The modification of the protocols, the use of non-radioactive detection, and the use of automation have facilitated the use of LMPCR and permit the collection of large amounts of data. An example of the data obtained using this original protocol for the detection of oxidized DNA bases is shown in Fig. 4.

Several simplifications of the LMPCR reactions have been made to the original LMPCR protocol.[33, 34] One estimation of the number of pipetting steps necessary for 18 LMPCR samples was on the order of 300 coupled with numerous DNA precipitations. The time and complexity of these precipitations was removed by using thermostable Pfu DNA polymerase that functions at higher salt concentrations than Taq DNA polymerase. Pfu polymerase also has the advantage compared to Sequenase and Taq that it does not have terminal transferase activity[35] (Table 1). The lack of terminal transferase activity is important, since overhanging bases block the ligation step of LMPCR. Regardless of the DNA polymerase used, however, the DNA ligase reaction reached completion in less than two hours. These simple modifications permit the completion of the LMPCR reactions in less than 8 h compared to at least two days using an older protocol.[28]

Streamlined detection methods now permit the observation of data on an automated sequencing gel apparatus or by direct-labeling. Although detection of products from the PCR reaction step first used a radiolabeled oligodeoxyribonucleotide primer, this method was abandoned as less sensitive compared to probing a charged nylon membrane. The transfer of the separated PCR products to a charged nylon membrane increased the sensitivity of LMPCR, since each probe could incorporate multiple radiolabeled nucleotides. The transfer of a thin DNA sequencing gel to the membrane, however, was a delicate operation that could result in sample loss. But, another end-labeling method increased the sensitivity of the original LMPCR detection method. The presence of PCR primers at the end of the PCR reaction limit the end-labeling detection sensitivity. However, an Exonuclease I digestion step to remove the excess PCR primers that remain after PCR allows an efficient end-labeling linear PCR step. This obviates the need for transfer of the gel to a charged nylon membrane. Using the conditions described

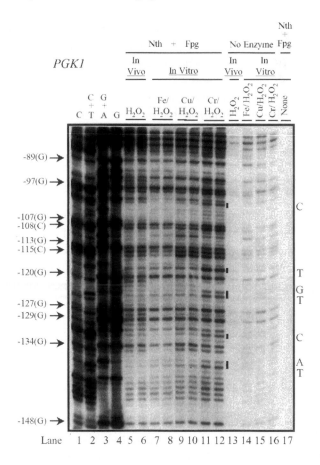

Fig. 4. LMPCR analysis of oxidized DNA base damage induced by Fenton reaction chemistry in the promoter region of the human phosphoglycerate kinase gene using primer set A (transcribed strand). Lanes 1–4, DNA, 0.3 µg of DNA treated with standard Maxam-Gilbert cleavage reactions. G represents a reaction of genomic DNA with dimethyl sulfate, GA a reaction with formic acid, and the TC and C reactions with hydrazine at different salt concentrations. Lanes 5–6, 13, DNA recovered from intact human foreskin fibroblasts that were exposed *in vivo* to 50 mM H_2O_2. Lanes 7–8, 14, genomic DNA treated *in vitro* with 100 µM Fe(III)/100 µM ascorbate/5 mM H_2O_2 in the presence of 0.3 M sucrose to quench hydroxyl radicals that cause strand breaks. Lanes 9–10, 15, DNA treated *in vitro* with 50 µM Cu(II)/100 µM ascorbate/5 mM H_2O_2 in 1 mM potassium phosphate buffer. Lanes 11–12, 16, DNA treated *in vitro* with 50 µM Cr(VI)/100 µM ascorbate/5 mM H_2O_2. Lane 17, DNA digested with Nth and Fpg proteins. The DNA in lanes 13–16 was incubated in digestion buffer alone and shows DNA breaks. The DNA in lanes 5–12 was digested with Nth and Fpg and shows both enzyme-independent breaks and enzyme dependent (base damage dependent) breaks. Positions of high base damage frequency are marked with arrows to the left of lane 1. The sequence of positions heavily damaged in the presence of chromium, but not copper or iron, is denoted by rectangles to the right of lane 12. In general, bound redox cycling transition metals induced *in vitro* the same frequency distribution of oxidized bases as did treating the cells *in vivo* with H_2O_2. [(From Ref. 29). Figure used with permission of the authors and the journal)].

Table 1. Terminal transferase activities found for DNA polymerases used in LMPCR and TDPCR. Addition of an extra base results in a decrease in the DNA ends available for the ligation step of LMPCR. The base added is dependent on the blunt-ended base, but the addition of an extra base is not always observed.

Polymerase	3' End of Blunt Ended DNA	Base Added to 3' End
Sequenase™ II DNA polymerase	T	+T
	G	+G
	C	+C > +A
	A	+A > +AA > +AAA
Taq DNA polymerase	T	+A
	G	+G > +A > +C
	C	+A > +C
	A	+A
Vent™ DNA polymerase	T	+A
	G	+C, +C, +A
	C	+A
	A	+A
Pfu DNA polymerase	T	Blunt
	G	Blunt
	C	Blunt
	A	Blunt

until this point, LMPCR does not require a full day's hybridization step, and the gel can be absorbed onto filter paper and dried prior to autoradiography.

Non-radioactive detection methods are important for reducing exposure to radiochemicals and for developing high-throughput methods for DNA sequencing. We have used infrared dyes (Li-Cor) as a system for the detection of the PCR products.[36] The sequencing system manufactured by Li-Cor has proved versatile for LMPCR purposes. The Li-Cor system uses a near-infrared fluorescence detection system and does not require algorithms to adjust for dye movements at high molecular mass. The near infrared dyes permit the real-time observation of bands that makes this non-radioactive LMPCR method much more rapid than the radioactive labeling methods. Furthermore, using this system, we observe the data as the products are passing through the gel, which permits an investigator to plan a new experiment while collecting data. Moreover, the data are acquired at a unique point on the gel during the migration. Therefore, the bands on the gel are equidistant and not compressed as in non-gradient DNA sequencing gels, making their signals more amenable to quantification using image-processing software. The bands are not high energy such as found with ^{32}P, therefore the data is more easily subject to quantification than data obtained using a phosphorimager.

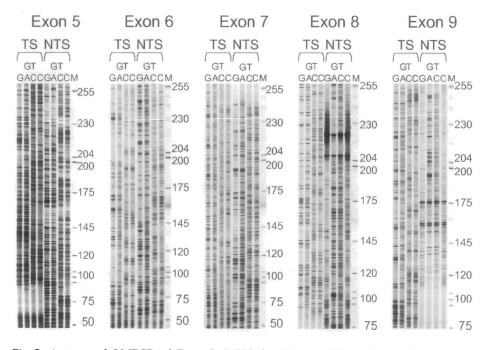

Fig. 5. Automated LMPCR of Exon 5–9 P53 for Maxam-Gilbert chemical sequencing reactions using Li-Cor infrared dyes. The sequencing reactions were performed on both the transcribed (TS) and non-transcribed (NTS) strands using a Beckman BioMek 3000 Workstation. For direct comparison only sequences between 50–255 nucleotides from the start position of the labeled oligodeoxyribonucleotide are indicated. The primers used in this experiment were matched to have identical melting temperatures to permit the LMPCR to be conducted on all the exons simultaneously. The quality of the sequencing data obtained depends on the primer set used. The NTS of Exon 9 yields the poorest quality data, but that may in part be linked to a high AT content. G represents a reaction of genomic DNA with dimethyl sulfate, GA a reaction with formic acid, and the TC and C reactions with hydrazine at different salt concentrations.

The last innovation for LMPCR couples LMPCR and a robotic workstation.[36] In the future, using a robotic workstation will permit investigators to study inter-individual differences and large data sets without requiring an army of workers to perform the LMPCR reactions. Figure 5 shows the latest LMPCR technology using non-radioactive detection for a series of sequencing reactions for exons 5–9 of the human P53 gene.

5. Terminal Transferase-Dependent PCR (TDPCR)

Although LMPCR can detect many types of adducts and there are numerous enzymes available to induce breaks leaving 5′phosphoryl groups, some adducts

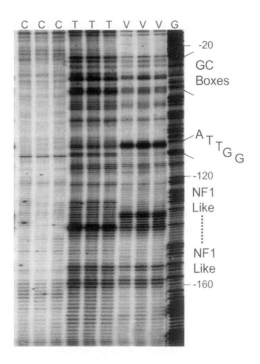

Fig. 6. TDPCR UV photofootprinting reaction in the human PGK1 promoter region. The primer extension step was performed using Vent(exo⁻) DNA polymerase, while the other steps requiring polymerase were performed using AmpliTaq DNA polymerase. All the reactions were performed in triplicate. Lanes marked C are control lanes from cells that were not irradiated. Lanes marked T are from naked DNA that was irradiated. Lanes marked V are from cells that were irradiated with UVC whose DNA was isolated. Treatment conditions of cells for photofootprinting were as previously described (Refs. 53 and 54).

have no known agent capable of incising at the adduct, or a complex cleavage scheme is required. To facilitate the detection of adducts that block DNA synthesis, but are not detectable by other means, Komura and Riggs developed TDPCR.[21] There are other advantages of this technique compared to LMPCR, but the two techniques should both be considered in developing adduct mapping schemes. Unlike LMPCR, TDPCR can perform several primer extension steps, since the terminal transferase will add on the ribonucleotides to the single-stranded DNA. TDPCR is also a robust technique for the detection of DNA damage and repair, but has not been employed as extensively as LMPCR (Fig. 6). The same detection techniques can be used for TDPCR as for LMPCR, therefore non-radiolabeled detection is the method of choice for future use.

Since this book is not focused on methods, the authors would be happy to provide protocols for researchers who are interested in using these techniques.

6. Oxidative Base Damage at Nucleotide Resolution in Normal Human Cells

DNA damage induced by reactive oxygen species (ROS) is an important intermediate in the pathogenesis of human conditions such as cancer and aging.[37] The major species of ROS-induced base modifications are well suited for mapping by LMPCR, because they are substrates for cleavage either by *E. coli* Nth protein (cleaves modified pyrimidines)[38, 39] or *E. coli* Fpg protein (cleaves modified purines).[40, 41] The associated lyase activity of these enzymes cleaves the phosphodiester backbone to produce the 5′phosphoryl groups that are substrates for LMPCR.

One commonly studied model of ROS-induced DNA damage is that caused by H_2O_2 in the presence of transition metal ions. The mutational spectra of H_2O_2[42] and the transition metal ions Fe, Cu,[43] and Cr[44] have been studied in model systems, but at the time the relationship of induced DNA damage to these spectra remained unknown.

The LMPCR technique was applied to map H_2O_2/transition metal-induced DNA damage.[25, 26, 28, 29, 45–47] The *in vivo* and *in vitro* frequencies of DNA base modifications in the human P53 and PGK-1 gene caused by H_2O_2/transition metal treatment of naked DNA or human male fibroblasts were determined using the LMPCR technique.[29] A representative autoradiogram indicating the damage distributions induced in the PGK1 gene[29] is shown in Fig. 4.

It was shown that the frequency distribution of DNA base damage induced *in vitro* in the presence of Cu(II), Fe(III), or Cr(VI) transition metal ions were similar to the *in vivo* base damage induced by H_2O_2-treatment of human male fibroblasts. The *in vitro* similarity suggested a model in which the local binding site occupancy rate and the local geometry of the metal ion-DNA-peroxo coordination complex determine the damage event, with chromatin structure having a limited effect. Additionally, it was shown that the principal determinant of a damaging event occurring at any position was DNA context dependent, suggesting a model in which DNA-metal ion binding domains can accommodate different metal ions. G was the most heavily modified base, with C the next most commonly modified base (G > C ≫ T ≫ A). The triplet d(pCGC) was the principal hotspot sequence.[29]

DNA sequence context was also shown to be the determining factor for another reactive oxygen species, the peroxyl radical (ROO•).[25] Using the LMPCR technique, it was shown that 87% of all G positions in exons 5 and 9 of the P53 gene and the PGK-1 gene were significantly oxidized. The order of reactivity of DNA bases towards oxidation by peroxyl radicals was found to be G ≫ C > T ≫ A. Furthermore, the yield of oxidative base modifications at G and C positions depended upon the exact specification of the 5′ and 3′ flanking bases. For instance, every G in the 5′XGC3′ motif was found to be oxidized, where X is any 5′ neighbor, while 5′ and 3′ purine flanks drastically reduced the extent of peroxyl radical G oxidation. Although G was the most heavily damaged base from H_2O_2 and $ROO^{\cdot-}$ induced DNA damage, it was shown that the pattern of base modification and the influence of nearest neighbors differed substantially from H_2O_2 damage mediated by low valent transition metal ions for the identical DNA sequences.[25]

7. Future Developments in Detection of Damage at Nucleotide Resolution

The major advantage of mapping DNA damage at nucleotide resolution is that the pattern can then be compared to biological endpoints, such as tumor databases to determine if there is a correlation between both data sets. Although these mapping techniques present an enormous advance in the study of the distribution of oxidized bases in DNA, there are some disadvantages of these methods. These methods can detect only bases that either block DNA synthesis or are recognized by enzymes. Fortunately, this covers most adducts, but there is still a possibility that some of the adducts are not detected. Another drawback to these techniques is that due to the broad substrate ranges of the DNA glycosylases and chemical cleavage methods, it is not possible to ascertain the exact adduct detected. Inferences can be made based on the frequency that a damage is observed based on other techniques, but the exact structure of a given base in genomic DNA at a specific site will always have some ambiguity, albeit small, with respect to its precise structure. Despite these limitations, however, this technique remains the only technique available to map adducts at nucleotide resolution.

Automation will provide a degree of freedom for researchers to devise experiments that would have taken months to years previously. A Beckman Biomek Workstation has already been used to perform all the LMPCR reactions (Fig. 5). In the near future, the use of automated LMPCR and TDPCR will provide researchers the ability to design experimental and clinical protocols that were not previously possible using the non-automated procedure.

The detection limit of LMPCR is about 1 base/10 kb in DNA. For detection of oxidized bases, a range 100–1000 fold more sensitive would be more useful for researchers. It should be emphasized that the two techniques described in this chapter do not have sufficient sensitivity to detect endogenous levels of DNA damage (10^7 adducts). Techniques have been developed to enrich the concentration of the DNA target sequence using electrophoresis,[48, 49] but these are technically challenging and yield no more than a 20-fold enrichment. Future work will focus on other methods to augment the concentration of adducted target fragments. These techniques could employ either antibodies against adducts or other proteins that bind to damaged fragments to enhance target sequences for adducted fragments. Use of these methods will permit researchers to attain sensitivities that approach the levels attributed to endogenous damage and to study DNA repair following exposure to sublethal doses of DNA oxidizing agents.

A version of TDPCR has also been employed to map RNA and RNA-protein interactions[50, 51] from *in vivo* samples. Under certain conditions, it is possible to map the RNA in single cells. Some RNA species are long-lived, and it is conceivable that repair mechanisms could function to preserve the integrity of RNA with respect to oxidative damage.[52] TDPCR provides a method to study these processes.

8. Conclusion

Considerable evidence has accrued that implicates ROS-induced DNA damage in adverse clinical outcomes in our increasingly aged population. Concomitantly, the need for and utility of clinical assessment of ROS-induced cellular damage is becoming increasingly apparent. These techniques to map damage at nucleotide resolution will play a role in understanding oxidative damage and the repair of such injury, and will ultimately contribute to our increasing knowledge of aging and disease pathogenesis at the molecular level.

Acknowledgments

Work described in this chapter has been supported by the National Institutes of Health, the Beckman Research Institute, the National Institutes of Standards and Technology, and Wake Forest University. The authors would also like to thank Mr. Steven E. Bates for his competence in cell culture that has provided excellent primary material for these studies.

References

1. Zolan, M. E., Cortopassi, G. A., Smith, C. A. and Hanawalt, P. C. (1982). Deficient repair of chemical adducts in alpha DNA of monkey cells. *Cell* **28**: 613–619.
2. Ye, N., Holmquist, G. P. and O'Connor, T. R. (1998). Heterogeneous repair of N-methylpurines in normal human cells at the nucleotide level. *J. Mol. Biol* **284**: 269–285.
3. Wellinger, R. E. and Thoma, F. (1997). Nucleosome structure and positioning modulate nucleotide excision repair in the non-transcribed strand of an active gene. *EMBO J.* **16**: 5046–5056.
4. Venema, J., Bartosova, Z., Natarajan, A. T., van Zeeland, A. A. and Mullenders, L. H. (1992). Transcription affects the rate but not the extent of repair of

cyclobutane pyrimidine dimers in the human adenosine deaminase gene. *J. Biol. Chem.* **267**: 8852–8856.

5. Tornaletti, S. and Pfeifer, G. P. (1994). Slow repair of pyrimidine dimers at p53 mutation hotspots in skin cancer. *Science* **263**: 1436–1438.

6. Mellon, I., Spivak, G. and Hanawalt, P. C. (1987). Selective removal of transcription-blocking DNA damage from the transcribed strand of the mammalian dhfr gene. *Cell* **51**: 241–249.

7. Mellon, I., Bohr, V. A., Smith, C. A. and Hannawalt, P. C. (1986). Preferential DNA repair of an active gene in human cells. *Proc. Natl. Acad. Sci. USA* **83**: 8878–8822.

8. Li, S. and Smerdon, M. J. (1999). Base excision repair of N-methylpurines in a yeast minichromosome. Effects of transcription, DNA sequence, and nucleosome positioning *J. Biol. Chem.* **274**: 12 201–12 204.

9. Bohr, V. A., Smith, C. A., Okumoto, D. S. and Hanawalt, P. C. (1985). DNA repair in an active gene: removal of pyrimidine dimers from the DHFR gene of CHO cells is much more efficient than in the genome overall. *Cell* **40**: 359–369.

10. Denissenko, M. F., Pao, A., Tang, M. and Pfeifer, G. P. (1996). Preferential formation of benzo[a]pyrene adducts at lung cancer mutational hotspots in P53. *Science* **274**: 430–432.

11. Church, G. M. and Gilbert, W. (1984). Genomic sequencing. *Proc. Natl. Acad. Sci.* **81**: 1991–1995.

12. Saluz, H. P. and Jost, J. P. (1987) *A Laboratory Guide to Genomic Sequencing. The Direct Sequencing of Native Uncloned DNA* (Birkhauser Verlag, Basel).

13. Mueller, P. R. and Wold, B. (1989). In vivo footprinting of a muscle specific enhancer by ligation mediated PCR. *Science* **246**: 780–786.

14. Steigerwald, S. D., Pfeifer, G. P. and Riggs, A. D. (1990). Ligation-mediated PCR improves the sensitivity of methylation analysis by restriction enzymes and detection of specific DNA strand breaks. *Nucleic Acids Res.* **18**: 1435–1439.

15. Pfeifer, G. P., Tanguay, R. L., Steigerwald, S. D. and Riggs, A. D. (1990). In vivo footprint and methylation analysis by PCR-aided genomic sequencing: comparison of active and inactive X chromosomal DNA at the CpG island and promoter of human PGK-1. *Genes Dev.* **4**: 1277–1287.

16. Pfeifer, G. P., Steigerwald, S. D., Mueller, P. R., Wold, B. and Riggs, A. D. (1989). Genomic sequencing and methylation analysis by ligation mediated PCR. *Science* **246**: 810–813.

17. Pfeifer, G. P., Drouin, R., Riggs, A. D. and Holmquist, G. P. (1991). In vivo mapping of a DNA adduct at nucleotide resolution: detection of pyrimidine (6–4) pyrimidone photoproducts by ligation-mediated polymerase chain reaction. *Proc. Natl. Acad. Sci. USA* **88**: 1374–1378.

18. Pfeifer, G. P., Drouin, R., Riggs, A. D. and Holmquist, G. P. (1992). Binding of transcription factors creates hot spots for UV photoproducts in vivo. *Mol. Cell. Biol.* **12**: 1798–1804.

19. Grimaldi, K. A., McAdam, S. R., Souhami, R. L. and Hartley, J. A. (1994). DNA damage by anti-cancer agents resolved at the nucleotide level of a single copy gene: evidence for a novel binding site for cisplatin in cells. *Nucleic Acids Res.* **22**: 2311–2317.

20. Grimaldi, K. A., McAdam, S. R. and Hartley, J. A. (1996). Single-ligation PCR for detection of DNA adducts. *In* "Technologies for the Detection of DNA Damage and Mutations" (G. P. Pfeifer, Ed.), pp. 227–238, Plenum Press, New York.

21. Komura, J. and Riggs, A. D. (1998). A sensitive and versatile method of genomic sequencing: ligation-mediated PCR with ribonucleotide tailing by terminal deoxynucleotidyl transferase. *Nucleic Acids Res.* **26**: 1807–1811.

22. Drouin, R., Gao, S. and Holmquist, G. P. (1996). Agarose gel electrophoresis for DNA damage analysis. *In* "Technologies for the Detection of DNA Damage and Mutations" (G. P. Pfeifer, Ed.), pp. 37–43, Plenum Press, New York.

23. Pfeifer, G. P., Drouin, R. and Holmquist, G. P. (1993). Detection of DNA adducts at the DNA sequence level by ligation-mediated PCR. *Mutat. Res.* **288**: 39–46.

24. Nomoto, M., Yamaguchi, R., Kawamura, M., Kohno, K. and Kasai, H. (1999). Analysis of 8-hydroxyguanine in rat kidney genomic DNA after administration of a renal carcinogen, ferric nitrilotriacetate. *Carcinogenesis* **20**: 837–841.

25. Rodriguez, H., Valentine, M. R., Holmquist, G. P., Akman, S. A. and Termini, J. (1999). Mapping of peroxyl radical induced damage on genomic DNA. *Biochemistry* **38**: 16 578–16 588.

26. Rodriguez, H., Drouin, R., Holmquist, G. P. and Akman, S. A. (1997). A hot spot for hydrogen peroxide-induced damage in the human hypoxia-inducible factor 1 binding site of the PGK-1 gene. *Arch. Biochem. Biophys.* **338**: 207–212.

27. Rodriguez, H. and Akman, S. A. (1999). Measurement of oxidative DNA damage in the human p53 and PGK1 gene at nucleotide resolution. *Ann. New York Acad. Sci.* **893**: 382–385.

28. Rodriguez, H., Drouin, R., Holmquist, G. P., O'Connor, T. R., Boiteux, S., Laval, J., Doroshow, J. H. and Akman, S. A. (1995). Mapping of copper/hydrogen peroxide-induced DNA damage at nucleotide resolution in human genomic DNA by ligation-mediated polymerase chain reaction. *J. Biol. Chem.* **270**: 17 633–17 640.

29. Rodriguez, H., Holmquist, G. P., D'Agostino, R., Jr., Keller, J. and Akman, S. A. (1997). Metal ion-dependent hydrogen peroxide-induced DNA damage is more sequence specific than metal specific. *Cancer Res.* **57**: 2394–2403.

30. Wei, D., Maher, V. M. and McCormick, J. J. (1996). Site-specific excision repair of 1-nitrosopyrene-induced DNA adducts at the nucleotide level in the

HPRT gene of human fibroblasts: effect of adduct conformation on the pattern of site-specific repair. *Mol. Cell Biol.* **16**: 3714–3719.

31. Wei, D., Maher, V. M. and McCormick, J. J. (1995). Site-specific rates of excision repair of benzo[a]pyrene diol epoxide adducts in the hypoxanthine phosphoribosyltransferase gene of human fibroblasts: correlation with mutation spectra. *Proc. Natl. Acad. Sci. USA* **92**: 2204–2208.

32. Gao, S., Drouin, R. and Holmquist, G. P. (1994). DNA repair rates mapped along the human PGK1 gene at nucleotide resolution. *Science* **263**: 1438–1440.

33. Pfeifer, G. P., Chen, H. H., Komura, J. and Riggs, A. D. (1999). Chromatin structure analysis by ligation-mediated and terminal transferase-mediated polymerase chain reaction. *Meth. Enzymol.* **304**: 548–571.

34. Cairns, M. J. and Murray, V. (1996). Influence of chromatin structure on bleomycin-DNA interactions at base pair resolution in the human beta-globin gene cluster. *Biochemistry* **35**: 8753–8760.

35. Hu, G. (1993). DNA polymerase-catalyzed addition of nontemplated extra nucleotides to the 3′ end of a DNA fragment. *DNA Cell Biol.* **12**: 763–770.

36. Dai, S.-M., Chen, H.-H., Chang, C., Riggs, A. D. and Flanagan, S. D. (2000). Ligation-mediated PCR for quantitative in vivo footprinting. *Nature Biotech.* **18**: in press.

37. Cross, C. E., Halliwell, B., Borish, E. T., Pryor, W. A., Ames, B. N., Saul, R. L., McCord, J. M. and Harman, D. (1987). Oxygen radicals and human disease [clinical conference]. *Ann. Int. Med.* **107**: 526–545.

38. Dizdaroglu, M., Laval, J. and Boiteux, S. (1993). Substrate specificity of the *Escherichia coli* Endonuclease III: excision of thymine- and cytosine-derived lesions in DNA produced by radiation-generated free radicals. *Biochemistry* **32**: 12 105–12 111.

39. Hatahet, Z., Kow, Y. W., Purmal, A. A., Cunningham, R. P. and Wallace, S. S. (1994). New substrates for old enzymes. *J. Biol. Chem.* **269**: 18 814–18 820.

40. Tchou, J., Kasai, H., Shibutani, S., Chung, M. H., Laval, J., Grollman, A. P. and Nishimura, S. (1991). 8-oxoguanine (8-hydroxyguanine) DNA glycosylase and its substrate specificity. *Proc. Natl. Acad. Sci. USA* **88**: 4690–4694.

41. Boiteux, S., Gajewski, E., Laval, J. and Dizdaroglu, M. (1992). Substrate specificity of the *Escherichia coli* Fpg protein: excision of purine lesions in DNA produced by ionizing radiation or photosensitization. *Biochemistry* **31**: 106–110.

42. Moraes, E. C., Keyse, S. M., Pidoux, M. and Tyrrell, R. M. (1989). The spectrum of mutations generated by passage of a hydrogen peroxide damaged shuttle vector plasmid through a mammalian host. *Nucleic Acids Res.* **17**: 8301–8312.

43. Loeb, L. A., James, E. A., Waltersdorph, A. M. and Klebanoff, S. J. (1988). Mutagenesis by the autoxidation of iron with isolated DNA. *Proc. Natl. Acad. Sci. USA* **85**: 3918–3922.

44. Tkeshelashvili, L. K., McBride, T., Spence, K. and Loeb, L. A. (1991). Mutation spectrum of copper-induced DNA damage [published erratum appears in *J. Biol. Chem.* 5 July 1992; **267**(19): 13 778]. *J Biol. Chem.* **266**: 6401–6406.

45. Rodriguez, H. and Akman, S. A. (2000). Mapping reactive oxygen-induced DNA damage at nucleotide resolution. *In* "Measuring In Vivo Oxidative Damage: A Practical Approach" (J. Lunec and H. R. Griffiths, Eds.), pp. 125–142, John Wiley and Sons Ltd., London.

46. Drouin, R., Rodriguez, H., Holmquist, G. P. and Akman, S. A. (1996). Ligation-mediated PCR for analysis of oxidative DNA damage. *In* "Technologies for Detection of DNA Damage and Mutations" (G. P. Pfeifer, Ed.), pp. 211–225, Plenum Press, New York.

47. Drouin, R., Rodriguez, H., Gao, S. W., Gebreyes, Z., O'Connor, T. R., Holmquist, G. P. and Akman, S. A. (1996). Cupric ion/ascorbate/hydrogen peroxide-induced DNA damage: DNA-bound copper ion primarily induces base modifications. *Free Radic. Biol. Med.* **21**: 261–273.

48. Rodriguez, H. and Akman, S. A. (1998). Mapping oxidative DNA damage at nucleotide level. *Free Radic. Res.* **29**: 499–510.

49. Rodriguez, H. and Akman, S. A. (1998). Large scale isolation of genes as DNA fragment lengths by continuous elution electrophoresis through an agarose matrix. *Electrophoresis* **19**: 646–652.

50. Buettner, V. L., LeBon, J. M., Gao, C., Riggs, A. D. and Singer-Sam, J. (2000). Use of terminal transferase-dependent antisense RNA amplification to determine the transcription start site of the Snrpn gene in individual neurons. *Nucleic Acids Res.* **28**: E25.

51. Chen, H. H., Castanotto, D., LeBon, J. M., Rossi, J. J. and Riggs, A. D. (2000). In vivo, high-resolution analysis of yeast and mammalian RNA-protein interactions, RNA structure, RNA splicing and ribozyme cleavage by use of terminal transferase-dependent PCR. *Nucleic Acids Res.* **28**: 1656–1664.

52. Rhee, Y., Valentine, M. R. and Termini, J. (1995). Oxidative base damage in RNA detected by reverse transcriptase. *Nucleic Acids Res.* **23**: 3275–3282.

53. Pfeifer, G. P. and Tornaletti, S. (1997). Footprinting with UV irradiation and LMPCR. *Methods* **11**: 189–196.

54. Tornaletti, S. and Pfeifer, G. P. (1995). UV light as a footprinting agent: modulation of UV-induced DNA damage by transcription factors bound at the promoters of three human genes. *J. Mol. Biol.* **249**: 714–728.

Chapter 15

Immunochemical Detection of Oxidative DNA Damage

Marcus S. Cooke* and Joseph Lunec[†]

Marcus S. Cooke and **Joseph Lunec** • Oxidative Stress Group, Division of Chemical Pathology, University of Leicester, Robert Kilpatrick Clinical Sciences Building, Leicester Royal Infirmary NHS Trust, P.O. Box 65, Leicester, LE2 7LX, UK
Tel: +44 (0)116 2525899, Website: www.le.ac.uk/pa/dcp/vac.html

*Corresponding Author. E-mail: jl20@le.ac.uk
[†]Corresponding Author. E-mail: msc5@le.ac.uk

1. Introduction

DNA damage is a seemingly inevitable result in cells exposed to oxidative stress. There is growing support for the hypothesis suggesting that oxidative stress may be involved in a number of diseases, including aging and this means that measurement of DNA damage may indicate the level of stress to which the cells have been exposed and may have significance in terms of risk of disease development.

The studies which ultimately lead to the development of antibodies as probes for DNA damage, began in 1956, at which point the evidence for native DNA being able to elicit an immune response was limited. At Rutgers University, Otto Plescia and colleagues were examining the immunogenicity of DNA, that is, its ability to induce an antibody response upon immunization. Whilst it did indeed appear that unmodified DNA was not immunogenic, the possibility remained

BASE LESION

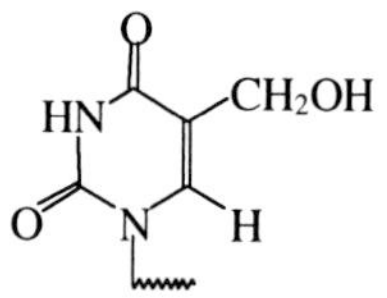

5-Hydroxymethyl-2'-uracil (18.2.1) 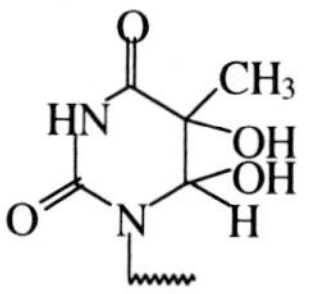   Thymine glycol (18.2.2)

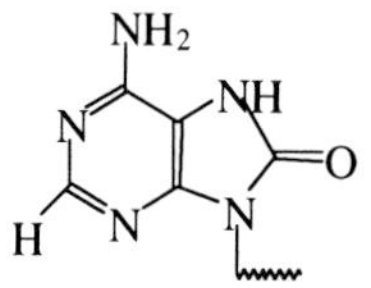

8-Oxoadenine (18.2.3) 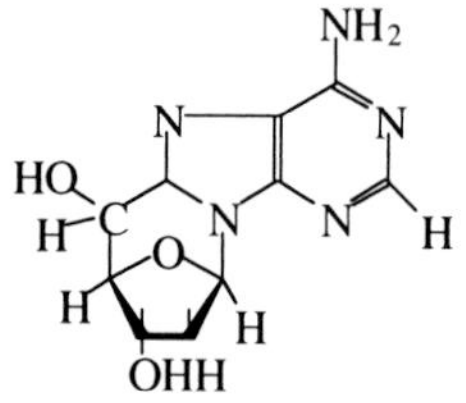   8,5'-cyclo-2'-deoxyadenosine (18.2.4)

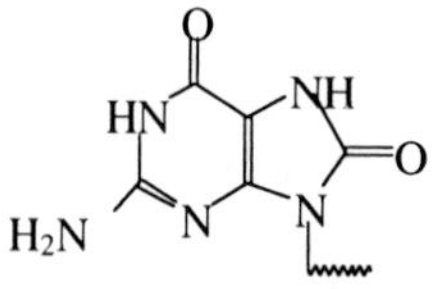

8-Oxoguanine (18.2.5)

Fig. 1. Principal oxidative DNA lesions studied using antibodies.

that DNA-containing complexes, or DNA linked to a carrier molecule, such as a protein, could stimulate the formation of antibodies. This principle was demonstrated for single-stranded DNA, complexed with methylated bovine serum albumin (mBSA), producing an antiserum which showed specific reactivity towards denatured DNA compared to native.[1] The likelihood of producing functional antibodies increased further with the demonstration that synthetic oligodeoxynucleotides could also act as haptens for the production of DNA antibodies.[2]

Related studies investigating modification of DNA utilized the dye, methylene blue, to sensitize the photo-oxidation of guanine residues specifically, generating unknown product(s).[3] As this work pre-dated the expansion of the field of free radical biochemistry by many years, the potential use of these antisera to identify *in vivo* oxidative modification of DNA was not realized. Indeed it was from the work studying active oxygen attack to DNA that the products of methylene blue photo-oxidation were finally identified: 8-oxo-2'deoxyguanosine (Fig. 1) appears to predominate, but with some ring-opened purine (formamidopyrimidine) formation.[4]

2. Antibodies to Oxidation Products of DNA

Early work in the emerging field of free radical biochemistry derived from studies examining the effects of ionizing radiation upon DNA. The techniques available dictated the damage products studied. The alkaline sucrose gradient technique facilitated strand break analysis, although it was acknowledged that DNA base lesions had greater significance in terms of mutation and these began to represent the focus of research interest.

2.1. 5-hydroxymethyldeoxyuridine (5-HMdU)

5-HMdU is a hydroxyl radical (OH$^\bullet$) product of thymidine, derived from the radical generated by hydrogen atom abstraction from the methyl group on thymine, used as a marker of oxidative DNA damage. Lewis *et al.*[5] conjugated 5-hydroxymethyluridine (5-HMUR) to BSA via the periodate linkage of Erlanger and Beiser.[6] This process employs the two –OH groups at positions C2 and C3 of the ribonucleoside to form the link, via a nitrogen-containing group in the protein (Fig. 2), resulting in a conjugate described as bearing more structural resemblance to the deoxyribonucleoside, important for DNA analysis, than the ribonucleoside.[7] Limited characterization of the antiserum was reported, with the lesion preferentially recognized over thymidine, with a limit of detection of one 5-HMdUR in 4×10^3 thymines, applied to a phage neutralization assay. This procedure, of specifically directing antibody production towards the desired lesion,

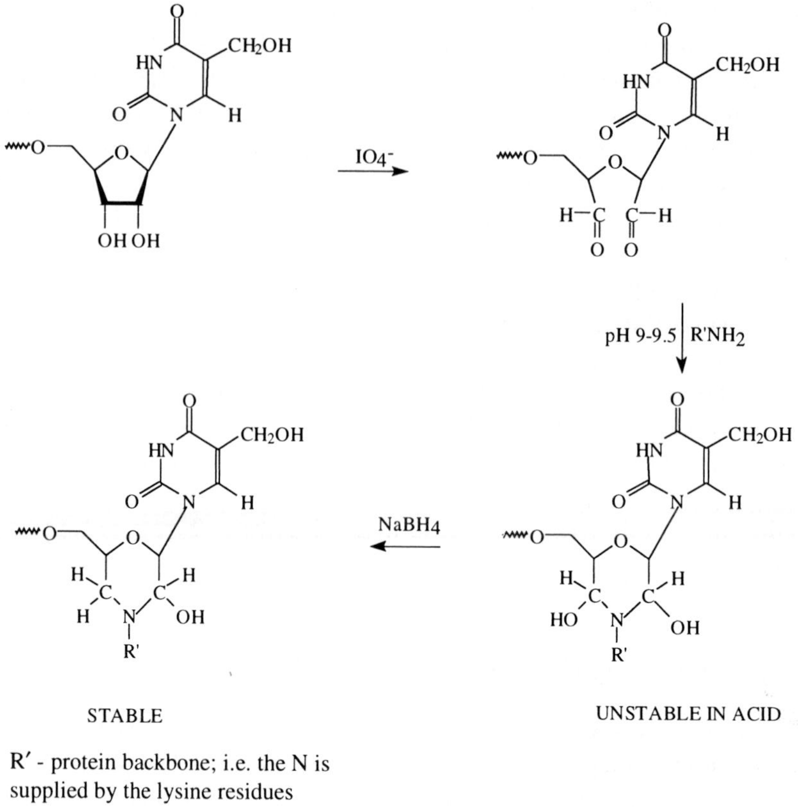

Fig. 2. Periodate linkage of 5-hydroxymethyluridine (5-HMUR) to protein, as described by Erlanger and Beiser.[6]

established an important methodological precedent for the production of future antibodies to various oxidative lesions and indicated the enormous potential of the serological approach to assessing DNA damage.

2.2. 5,6-dihydroxy-5,6-dihydrothymine (Thymine glycol)

Thymine glycol (TG; Fig. 1) is a major product following OH$^{\cdot}$ attack upon thymine. West *et al.*[8] first reported a radioimmunoassay (RIA) for detecting oxidative DNA damage. Immunization with single-stranded, osmium tetroxide treated calf thymus DNA complexed with mBSA produced a polyclonal antiserum highly specific for *cis* thymine glycol (TG) and with a detection limit of 4 femtomoles. However, there did appear to be a requisite for the TG to be clustered. Work with TG was furthered, in 1983, by the development of a monoclonal antibody, using osmium tetroxide treated poly(dT).[9] The use of this antibody to detect damage induction and repair in irradiated cells and a sensitivity of almost one order of magnitude greater than its polyclonal predecessor,[9] emphasized the potential immunochemical assays possessed. Up to this point, immunizations for TG had been with polymeric immunogens, either DNA or poly(dT); Susan Wallace's group at the New York Medical College, in collaboration with Bernard Erlanger, employed a method of carbodiimide linkage[10] to attach thymine glycol monophosphate, via the phosphate, to bovine serum albumin (BSA).[11] The resulting antiserum showed strong specificity for TG, being capable of detecting this lesion in DNA containing multiple modifications.[12] Whilst dose-responsive quantitation of TG was demonstrated, quantitation in terms of numbers of lesions per unit DNA, or unmodified base, was not shown and, to a large extent, remains, as will be discussed later, a problem for immunoassays of oxidative damage and a potential source of criticism when compared to chromatographic techniques. Antibodies to 5-hydroxycytosine, 5-hydroxyuracil and abasic sites have also been reported from Susan Wallace's group.[13, 14]

2.3. 8-oxo-2'-deoxyadenosine (8-oxodA)

In the same year as they reported an RIA for TG, West *et al.*[15] also reported an RIA for 8-oxoadenine (8-oxoA; Fig. 1), at that point a little-studied product of free radical attack to DNA. Using 8-hydroxyadenosine, periodate linked to BSA, West *et al.*[15] generated a polyclonal antiserum which sensitively (limit of detection 4×10^{-14} moles) and specifically recognized 8-oxodA in enzymatically digested DNA, rather than the intact polymer. Ide *et al.*[16] reported the development of a polyclonal antiserum to 8-oxodA. The limit of detection for this antiserum was 4 lesions in 10^4 unmodified nucleotides. Although it was possible to recognize 8-oxodA in DNA, alkaline phosphatase digestion was a prerequisite.

2.4. 8,5'-cycloadenosine-5'-monophosphate (8,5'-cyclo-AMP)

Fuciarelli *et al.*[17] developed polyclonal antisera specific for 8,5'-cyclo-AMP (Fig. 1), an irradiation product of adenosine-5'-monophosphate, in order to assess its formation in DNA. Using the carbodiimide procedure, Fuciarelli *et al.* linked 8,5'-cyclo-AMP to haemocyanin prior to immunization. Comparison of the resulting antiserum with an HPLC method showed a good correlation between the two techniques measuring 8,5'-cyclo-AMP formation in irradiated DNA, although the ELISA method tended to show a two-fold overestimate.[18]

2.5. 8-oxo-2'deoxyguanosine (8-oxodG)

Since its first description in 1984 by Kasai *et al.*,[19] 8-oxoguanine (8-oxoG) and its deoxynucleoside derivative, 8-oxo-2'deoxyguanosine (8-oxodG; Fig. 1), have received much attention from both the free radical community and increasingly, other scientific fields. Having described a novel DNA modification reaction, hydroxylation of guanine at the C-8 position and its resultant product, 8-oxoG, Kasai and co-workers immediately proceeded to prepare monoclonal antibodies, utilizing principles derived from earlier attempts to produce antibodies to oxidized DNA lesions.[20] Just as West *et al.*[15] had periodate-conjugated 8-oxoadenosine to a carrier protein, 8-oxoguanosine (8-oxoGuo) was linked to BSA, prior to immunization. Preliminary results describing a monoclonal antibody with specificity for 8-oxodG were reported, however no further use of this antibody has since been apparent. In retrospect, this might be seen as a portent for difficulties to be encountered by later groups trying to raise antibodies to 8-oxodG for use in biological systems.

In 1990, a report from Bruce Ames' group at Berkeley first mentioned, albeit briefly, the use of antibodies to 8-oxodG in an immunoaffinity column.[21] Detailed description of a polyclonal antiserum to 8-oxodG, again using the periodate linkage to BSA, came the following year with the antiserum having an affinity for 8-oxodG three orders of magnitude greater than the antibody reported by Kasai *et al.*[22] This antibody was incorporated into an immunoaffinity column in order to simplify the prepurification of 8-oxodG from complex biological matrices, such as urine (see Ref. 23), which requires time-consuming solid-phase cleanup or HPLC pre-purification, prior to HPLC-EC or GC-MS analysis. In such urinary assays, it is primarily the deoxynucleoside, 8-oxodG, which is the focus of attention, levels of which have been shown to be independent of dietary influence,[24] contrasting with the base, 8-oxoG, making it a more sensitive marker of oxidative stress.

Development of a radioimmunoassay allowed absolute, rather than relative, quantitation of 8-oxodG, in terms of number of moles of lesion per gram of DNA. Comparison between HPLC-EC and competitive RIA, using this antiserum, revealed an excellent correlation ($r^2 = 0.998$) between the two methods when applied to enzymatically hydrolyzed, native and H_2O_2/ascorbate-modified DNA derived from either rat liver, human lymphocytes or calf thymus.[22] Not only did this support RIA (limit of detection 63 fmol) as an alternative to HPLC-EC (limit of detection ~50 fmol), but also demonstrated that there was no cross-reactivity with unmodified deoxynucleosides, despite each being present in quantities approximately 100 000-fold greater than 8-oxodG. It was noted that the binding affinity for 8-oxodG in intact double- or single-stranded DNA was approximately two orders of magnitude lower than for free 8-oxodG. Clearly, this precludes the usefulness of the antiserum in intact DNA, a desirable facet of any procedure which limits sample pre-processing and hence the potential for artefact formation.

Also raised was the issue of stearic hindrance, that is, the antibody cannot physically reach the lesion due to the structure of surrounding DNA. Furthermore, the role of 8-oxodG conformation in double-stranded DNA versus the free deoxynucleoside was considered and will be discussed in detail later.

A principal drawback of polyclonal antisera, particularly when a most useful one has been developed, is that there is a finite supply. Monoclonal antibodies, derived from an immortalized cell line, should represent a limitless source of antibody and in 1992 Ames' group published a report of a monoclonal antibody to 8-oxodG (designated 15A3) and its application to immunoaffinity clean-up of urine, plasma or cell culture media, prior to HPLC-EC.[25] Using the same immunogen as for the antiserum, 15A3 was characterized as having a comparable affinity for 8-oxodG as the polyclonal, but a greater affinity for 8-oxoG. However, this reactivity also included other 6,8-dioxopurines, resulting in ELISA overestimates for 8-oxodG in biological matrices of ~300-fold, compared to HPLC-EC. Furthermore, cross-reactivity with dG precluded any application to DNA, whether intact or hydrolyzed.[25] The applicability of the polyclonal antiserum was exploited in an immunoslot-blot assay to detect levels of 8-oxodG in unmodified, commercial calf thymus DNA (0.107 +/− 0.024 pmol/g.[26] In contrasts to RIA, this procedure relies upon undigested DNA, immobilized onto a filter, and levels of antiserum binding compared with binding to 8-oxodG-containing DNA standards (evaluated by HPLC-EC).

Comparison of immunoslot-blot data with literature 8-oxodG values for methylene blue-modified DNA illustrated an underestimation by HPLC-EC.[26] The authors suggest this may derive from the need for DNA hydrolysis prior to HPLC analysis, a previous study had shown that damage products, in addition to methylene blue itself, interfere with the nuclease P1 and alkaline phosphatase enzymes, with the hydrolysis issue further discussed by Halliwell and Dizdaroglu.[27]

In 1995, three new monoclonal antibodies to 8-oxodG were reported. Regina Santella's group in New York, again adopted the periodate-linked immunogen to generate their two antibodies (designated 1F7 and 1F11; 28). In contrast, Osawa *et al.*[29] described the characterization of an antibody (N45.1), the immunogen for which, although based upon the principle of an isolated lesion, utilized a novel procedure to link 8-oxodG to the protein carrier. Due to their differing specificities and sensitivities, Yin *et al.*[28] utilized both monoclonal antibodies to study DNA damage with immunoaffinity purification of hydrolysates by 1F7, preceding ELISA by 1F11, abbreviated to IA-ELISA. Direct comparison between HPLC-EC and IA-ELISA using (i) calf thymus DNA spiked with 8-oxodG, (ii) DNA damage *in vitro* with increasing concentrations of H_2O_2 and (iii) human placental DNA, showed, on the whole, good correlation between the two techniques ($r^2 = 0.96$, 0.99 and 0.86, respectively), although actual ELISA values for (ii) and (iii) were 10–15% and up to 600% higher respectively, than HPLC-EC, although with comparable limits of detection. Explanation for this over-estimation might be due to non-specific binding of material by the immuno-columns, either the Sepharose matrix, or the antibody itself or perhaps the samples were too concentrated, promoting non-specific inhibition.[28] Indeed, possible hydrolysis problems, as alluded to earlier, might also apply here. However, as the authors explain, good correlation between IA-ELISA and HPLC-EC would indicate that such inhibitors, for example other modified deoxynucleosides, must be present in proportion to 8-oxodG. It is important to note that in this report, hydrolysates equivalent to 100 g DNA were required for IA-ELISA, whereas 400 g DNA was required for HPLC-EC, highlighting an important benefit of immuno-techniques, particularly where material is limited.

The first report of immunohisto- and cyto-chemically detecting 8-oxodG *in situ* came from Santella's lab, one year later, using the more specific 1F7 antibody.[30] Following a successful workup of the method involving H_2O_2 or aflatoxin B1 (AFB1) treatment of cultured cells and cryostat sections of liver from rats treated with AFB1; the method was applied to the study of damage in oral mucosa cells of smokers. The mean level of relative staining was significantly higher ($p < 0.001$) in the cells of smokers, compared to their matched, non-smoking control, although no relationship was found with number of cigarettes smoked per day.[30] Whilst the need for DNA extraction and hence the potential for artefactual oxidation had been circumvented, it was at the price of absolute quantitation, as described earlier, with assessments only being semiquantitative.

The commercial exploitation of N45.1 began in 1996. A kit was marketed for the measurement of 8-oxodG in urine, serum and cell culture supernatants, the latter as described by Kantha *et al.*[31] This dramatically increased the number of groups performing 8-oxodG analyses in biological matrices, in addition to DNA. Despite this, the challenge to successfully produce an antibody or antiserum which recognized 8-oxodG with appropriate specificity for use in an assay meant that researchers continued to persevere with established methods, whilst also utilizing more sophisticated approaches. Susan Wallace's group in Vermont employed phage display repertoire cloning, generating monoclonal antibodies to 8-oxo-purines.[32] The immunogens consisted of 8-oxoGuo or 8-oxoadenosine, linked via the periodate linkage to BSA and their corresponding deoxyribonucleoside monophosphates, carbodiimide linked to BSA or rabbit serum albumin (RSA). From this approach, monoclonal Phabs were generated which, whilst showing specificity for 8-oxoGuo or 8-oxoguanosine monophosphate over their corresponding, unmodified counterparts, failed to recognize 8-oxoG in DNA.[32] This appeared, in part, to be due to a requirement for BSA to be present in the antigen, seemingly to stabilize the recognition of 8-oxoG. However, amino acid sequence comparisons, chain shuffling studies and homology modeling all revealed information of potential importance to the future generation of antibodies: only a small area of the complimentarity determining regions of the Phabs produced modified base binding specificity, determined primarily by the light chain, the remainder appeared to merely stabilize binding; sequence and structural similarities with other anti-single-stranded DNA antibodies suggested the existence of a DNA binding "canyon", essential for binding to DNA and more dependent on the heavy chain.[32]

The following year, Wallace's group also described highly specific polyclonal antibodies to 8-oxopurines, again immunizing with the modified ribonucleosides periodate-linked to protein.[16] The limit of detection was reported to be 20 lesions in 10^4 unmodified nucleotides for the anti-8-oxoG antiserum, compared to 1 lesion in 10^5 unmodified nucleotides for a previous anti-8-oxodG antibody,[28] recognizing 8-oxodG in X-irradiated DNA following alkaline phosphatase digestion.

The commercial 8-oxodG antibody (N45.1) began to receive increasingly widespread use, principally applied to the measurement of urinary 8-oxodG (Table 1 illustrates some of the reports utilizing this antibody, and their major findings). Takahashi *et al.*[48] successfully demonstrated the applicability of the N45.1 to the immunohistochemical localization of 8-oxodG in paraffin-embedded sections of rat liver, following systemic exposure to carbon tetrachloride, effectively extending the precedent set by Yarborough *et al.*[30] The specificity for 8-oxodG over the native deoxynucleotide is clearly evident in such an application, otherwise all cells would be detected, irrespective of treatment.

In contrast to all the previously described antibody approaches to the detection of 8-oxodG, Struthers *et al.*[49] reported the surprising observation that avidin, a factor routinely used as an amplification step in immunochemistry due to its affinity for biotin, bound to the nuclei of oxidized cells and tissues. Further investigation suggested that it was 8-oxodG, specifically, which was recognized, based upon structural similarities with the imidazolidone group of biotin and, albeit rather limited, competitive ELISA data. However, the associated immunocytochemical studies did appear supportive that cellular changes induced by oxidative stress, were indeed detectable by avidin. Further support for this work was provided by the immunocytochemical studies of Cooke *et al.*,[45] who utilized fluorescently labeled avidin to demonstrate, in addition to UVA-induced free radical damage to human keratinocytes, the amelioration of H_2O_2-induced damage by vitamin E. However, in this study the ELISA-based protocol of Stuthers *et al.*[49] failed to detect modifications in oxidatively modified, extracted DNA.

Degan *et al.*[22] first drew attention to the structural complexities associated with 8-oxodG and its interaction with an antibody. A preliminary report by Cooke *et al.*[50] further discussed this issue, based upon findings with a newly developed monoclonal antibody to 8-oxodG, designated F3/8/1. Subsequent antibody screenings showed a marked preference for 8-oxodG, either in the form of the immunogen, or in single-stranded, modified DNA. The authors suggest that conformational changes in 8-oxodG occur following DNA denaturation which

Table 1. Representative reports utilizing the monoclonal antibody to 8-oxo-2'-deoxyguanosine (N45.1) in biological matrices.

Principal Finding	Application of N45.1	Reference
Elevated systemic level of oxidative DNA damage in non-insulin-dependent diabetes mellitus compared to controls ($68.2 +/- 39.4$ versus $49.6 +/- 37.7$ µg 8-oxodG/24 hours, $p = 0.001$)[1].	Urinary 8-oxodG measurement.	Leinonen et al.[33]
Elevated levels of urinary 8-oxodG in small-cell carcinoma (SCC) patients, compared to non-small cell lung cancer patients (nSCC) and control subjects ($27.2 +/- 17.4$ versus $19.8 +/- 8.6$ and $19.4 +/- 8.5$ ng 8-oxodG/mg creatinine, respectively). SCC patients with a complete or patial response to chemotherapy showed a significant ($p = 0.007$) decrease in urinary 8-oxodG.	Urinary 8-oxodG measurement.	Erhola et al.[34]
Elevated ($p < 0.0001$) levels of DNA damage in atopic dermatitis (AD) which correlated with other disease severity parameters and hence aid clinical evaluation of AD.	Urinary 8-oxodG measurement.	Tsuboi et al.[35]
Lower ($p = 0.008$) levels of urinary 8-oxodG in patients with *Helicobacter pylori* infection, compared to uninfected persons (18.04 versus 14.36 µg 8-oxodG/g creatinine).	Urinary 8-oxodG measurements.	Witherall et al.[36]
Increases in serum and urinary 8-oxodG ($p = 0.001$ and $p = 0.05$, respectively) in healthy subjects undergoing vitamin C (500 mg/day) supplementation. A significant correlation ($p < 0.04$) was also noted, in the same subjects, between serum 8-oxodG and plasma vitamin C.	Serum and urinary 8-oxodG measurements.	Cooke et al.[37]; Lunec et al.[38]
Strong correlation ($r^2 = 0.958$, $p = 0.0002$) between HPLC-EC and monoclonal antibody measurement of 8-oxodG in UVC irradiated DNA.	Measurement of 8-oxodG in UVC-irradiated, enzymatically digested, calf thymus DNA.	Evans et al.[39]
High levels of 8-oxodG in normal human epidermis following a single dose of UVB (λ_{max} 305 nm) radiation (2x MED). Rapid removal over 24–48 h approaching control levels by 72–96 hours.	Immunohistochemical localization of 8-oxodG.	Ahmed et al.[40]

Table 1 (*Continued*)

Principal Finding	Application of N45.1	Reference
Uniform distribution of N45.1 binding in untreated proximal tubule cells, which becomes more intense following treatment with ferric nitrilotriacetate.	Immunocytochemical analysis with fluorescence detection.	Toyokuni *et al.*[41]
No significant difference in urinary 8-oxodG levels, between psoriatic, atopic dermatitis and control groups of subjects (9.36 +/− 7.8, 12.29 +/− 7.72 and 9.78 +/− 3.8pmol 8-oxodG/mmol creatinine, respectively).	Urinary 8-oxodG measurements.	Ahmad *et al.*[42]
Detailed description of methodology to accurately quantify 8-oxodG in DNA using a purely ELISA-based method.	Measurement of 8-oxodG in enzymatically digested DNA.	Cooke *et al.*[43]
Significant ($p < 0.001$), although attenuated, increase in serum 8-oxodG and no increase in urinary 8-oxodG in patients with systemic lupus erythematosus, compared to healthy control subjects.	Serum and urinary 8-oxodG measurements.	Evans *et al.*[44]
Induction of significant levels of 8-oxodG in human keratinocytes treated with biologically relevant doses of UVA or H_2O_2.	ELISA analysis of 8-oxodG levels in DNA extracted from treated cells.	Cooke *et al.*[45]
Vitamin C modulates the repair of 8-oxodG, *in vitro*, via activation of AP-1 and stimulation of nucleotide excision repair.	Measurement of 8-oxodG in tissue culture supernatant.	Holloway *et al.*[46]
Induction and kinetics of repair and excretion of 8-oxodG induced in healthy subjects following a single sub-erythemal dose of UVA.	Urinary 8-oxodG measurements.	Cooke *et al.*[47]

[1]For a full discussion of the units in which urinary 8-oxodG measurements are expressed, see Cooke *et al.*[23]

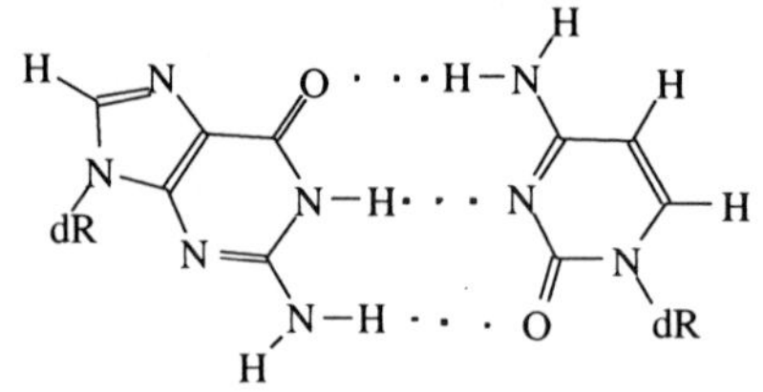

dG (*anti*) pairing deoxycytidine (*anti*)

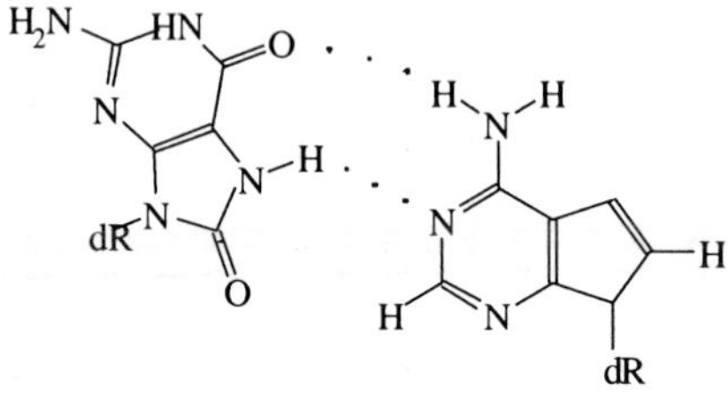

8-oxodG (*syn*) mispairing deoxyadenosine (*anti*)

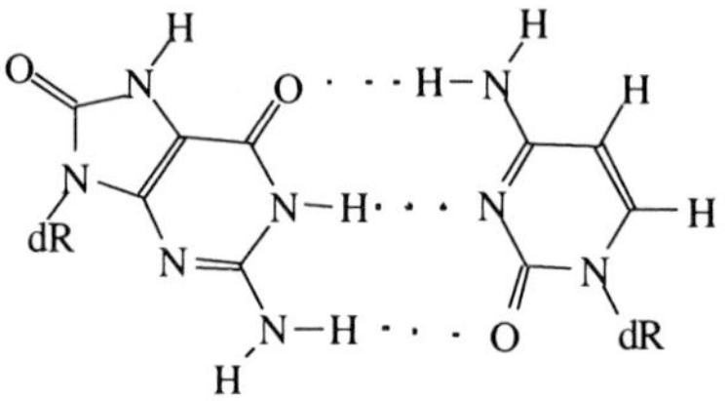

8-oxodG (*anti*) pairing deoxycytidine (*anti*)

Fig. 3. Base pairing of 8-oxo-2′-deoxyguanosine.

allow 8-oxodG to rotate from the *anti* form in which it is held by Watson–Crick base pairing with dC, to the more energetically favourable *syn* glycosidic conformation (Fig. 3). Accepting that denaturation is likely to allow improved access to the lesion, this treatment results in 8-oxodG being in the same conformation as the immunogen, thought to be a contributing factor in its recognition and its specificity over dG, which adopts the *anti* conformation.[50]

   The techniques described previously have all considered cellular DNA as a whole, however higher steady state levels of oxidative DNA damage have been noted in mitochondria, compared to nuclei. This was successfully demonstrated by Soultanakis *et al.*,[51] using Fab166, another recombinant antibody from Susan Wallace's laboratory. Confocal scanning laser microscopy was used to demonstrate the presence of 8-oxodG in both nuclear and mitochondrial DNA, following treatment with H_2O_2 or ionizing radiation. Mitochondrial oxidation was suggested to be greater than nuclear, post-treatment, although accessibility of the Fab to the lesion might account for the differential immunofluorescence.

3. Pitfalls and Problems of Antibody Production

It is clear that oxidatively modified DNA and, more particularly, such lesions in isolation, are not good immunogens. Coupling bases or (deoxy)ribonucleosides to large proteins assists the process and antibodies have been successfully produced. In producing monoclonal antibodies this difficulty is compounded by the hurdles inherent in the cloning process. This is typified in a report of antibodies to TG; Greferath and Nehls[52] used thymidine glycol monophosphate and thymidine glycol to generate monoclonal antibodies, although here it is not the immunogen that is of greatest interest, but the size of the undertaking required to produce a monoclonal antibody. From the 15 mice immunized, 10 000 hybridomas were generated, testing of which revealed only 122 were secreting antibodies with any specificity for TG and of those, only 4 clones produced antibodies with high affinity and specificity. Although rarely reported in the literature, such numbers appear to typify monoclonal antibody production towards oxidative DNA lesions. In light of such a struggle why attempt to produce an antibody to oxidatively modified DNA?

4. Benefits of the Immuno-Approach

From the process of DNA extraction, to the final detection of 8-oxodG, the potential for artefactual oxidation would appear to exist. Whilst the matters of absolute levels and artefact formation will, no doubt, be covered elsewhere in this volume, it has been identified that the potential for oxidation during extraction is great, therefore a method to circumvent extraction, avoids this source of artefact. Antibodies represent the ideal approach to localizing damage *in situ* (illustrated in Fig. 4) and no other technique represents the same level of sensitivity and specificity for this application. Furthermore, the amount of sample workup, in addition to extraction, for example compared to GC-MS, is greatly reduced, hence minimizing further adventitious oxidation. Nevertheless, some limitations do occur, illustrated by the report of Soultanakis *et al.*[51]

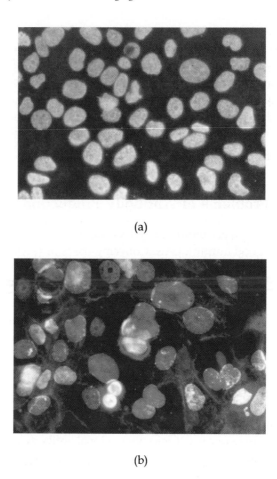

(a)

(b)

Fig. 4. Confocal microscopy of control, untreated human keratinocytes (a) and kera-tinocytes treated with 100 mM H_2O_2 (b), stained with a fluorescently-labeled antiserum raised against ascorbate/H_2O_2-modified DNA, characterized as recognizing glyoxal-mediated DNA damage, a metal-dependent consequence of free radical attack to DNA.[53] No staining is seen in the control cells (a), other than blue fluorescence of the nuclear counterstain (DAPI), however, red (Cy-3) staining is visible on the periphery of treated nuclei and in regions of cytoplasm (b). Discrete areas of intense nuclear staining are also seen, appearing as white due to Cy-3 being superimposed upon the DAPI stain.

5. Conclusions

The place for antibodies in the field of free radical research is clearly defined in terms of well established immunochemical techniques. This position is being extended by a number of emerging techniques which employ antibodies, in conjunction with other techniques, such a capillary electrophoresis, in an effort to minimize artefact, but maintain absolute quantitation of lesions (number of modified per unmodified moieties). Furthermore, rather than assessing global damage, which includes a great deal of "junk" DNA, antibodies are being utilized, in our laboratory, to examine lesion levels within specific gene regions with particular relevance to disease. These approaches are likely to be the basis of assays in future studies of free radical attack to DNA.

Acknowledgments

The authors wish to thank Nalini Mistry (Division of Chemical Pathology, University of Leicester) for the confocal microscopy of keratinocytes shown in Fig. 4. The authors gratefully acknowledge financial support from the UK Food Standards Agency, Scottish Office, Lupus UK, Arthritis and Rheumatism Council and Leicester Dermatology Research Fund. We would also like to thank Dr. Mark Evans for his critical reading of the manuscript.

References

1. Plescia, O. J., Braun, W. and Palczuk, N. C. (1964). Production of antibodies to denatured deoxyribonucleic acid (DNA). *Biochemistry* **52**: 279–285.
2. Plescia, O. J., Palczuk, N. C., Braun, W. and Cora-Figueroa, E. (1965). Antibodies to DNA and a synthetic polydeoxyribonucleotide produced by oligodeoxyribonucleotides. *Science* **148**: 1102–1103.
3. Seaman, E., Levine, L. and Van Vunakis, H. (1966). Antibodies to the methylene blue sensitized photooxidation product in deoxyribonucleic acid. *Biochemistry* **5**: 1216–1223.
4. Floyd, R. A., Wesy, M. S., Eneff, K. L. and Schneider, J. S. (1989). Methylene blue plus light mediates 8-hydroxyguanine formation in DNA. *Arch. Biochem. Biophys.* **273**: 106–111.
5. Lewis, H. L., Muhleman, D. R. and Ward, J. F. (1978). Serologic assay of DNA base damage. *Rad. Res.* **75**: 305–316.
6. Erlanger, B. F. and Beiser, S. M. (1964). Antibodies specific for ribonucleosides and ribonucleotides and their reaction with DNA. *Proc. Natl. Acad. Sci. USA* **52**: 68–74.
7. Mueller, R. and Rajewsky, M. F. (1980). Immunological quantification by high-affinity antibodies of O^6-ethyldeoxyguanosine in DNA exposed to N-ethyl-N-nitrosourea. *Cancer Res.* **40**: 887–896.
8. West, G. J., West I. W.-L. and Ward, J. F. (1982). Radioimmunoassay of a thymine glycol. *Rad. Res.* **90**: 595–608.
9. Leadon, S. A. and Hanawalt, P. C. (1983). Monoclonal antibody to DNA containing thymine glycol. *Mutat. Res.* **112**: 191–200.

10. Halloran, M. J. and Parker, C. W. (1966). The preparation of nucleotide-protein conjugates: carbodiimides as coupling agents. *J. Immunol.* **96**: 373–378.

11. Rajagopalan, R., Melamede, R. J., Laspia, M. F., Erlanger, B. F. and Wallace, S. S. (1984). Properties of antibodies to thymine glycol, a product of the radiolysis of DNA. *Rad. Res.* **97**: 499–510.

12. Hubbard, K., Huang, H., Laspia, M. F., Ide, H., Erlanger, B. F. and Wallace, S. S. (1989). Immunochemical quantitation of thymine glycol in oxidized and X-irradiated DNA. *Rad. Res.* **118**: 257–268.

13. Chen, B. X., Kubo, K., Ide, H., Erlanger, B. F., Wallace, S. S. and Kow, Y. W. (1992). Properties of a monoclonal antibody for the detection of abasic sites, a common DNA lesion. *Mutat. Res.* **273**: 253–261.

14. Hubbard, K., Ide, H., Erlanger, B. F. and Wallace, S. S. (1989). Characterization of antibodies to dihydrothymine, a radiolysis product of DNA. *Biochemistry* **28**: 4382–4387.

15. West, G. J., West, I. W.-L. and Ward, J. F. (1982). Radioimmunoassay of 7,8-dihydro-8-oxoadenine. *Int. J. Rad. Biol.* **42**: 481–490.

16. Ide, H., Kow, Y. W., Chen, B.-X., Erlanger, B. F. and Wallace, S. S. (1997). Antibodies to oxidative DNA damages: characterization of antibodies to 8-oxopurines. *Cell Biol. Toxicol.* **13**: 405–417.

17. Fuciarelli, A. F., Miller, G. G. and Raleigh, J. A. (1985). An immunochemical probe for 8,5′-cycloademosine-5′-monophosphate and its deoxy analog in irradiated nucleic acids. *Rad. Res.* **104**: 272–283.

18. Fuciarelli, A. F., Shum, F. Y. and Raleigh, J. A. (1985). Intramolecular cyclization in irradiated nucleic acids: correlation between high-performance liquid chromatography and an immunochemical assay for 8,5′-cycloademosine-5′-monophosphate. *Rad. Res.* **110**: 35–44.

19. Kasai, H., Tanooka, H. and Nishimura, S. (1984). Formation of 8-hydroxy-guanine residues in DNA by X-irradiation. *Gann.* **75**: 1037–1039.

20. Kasai, H. and Nishimura, S. (1986). Hydroxylation of guanine in nucleosides and DNA at the C-8 position by heated glucose and oxygen radical-forming agents. *Env. Health. Perspects.* **67**: 111–116.

21. Fraga, C. G., Shigenaga, M. K., Park, J. W. and Ames B. N. (1990). Oxidative damage to DNA during ageing: 8-hydroxy-2′-deoxyguanosine in rat organ DNA and urine. *Proc. Natl. Acad. Sci. USA* **87**: 4533–4537.

22. Degan, P., Shigenaga, M. K., Park, E.-M., Alperin, P. E. and Ames, B. N. (1991). Immunoaffinity isolation of urinary 8-hydroxy-2′-deoxyguanosine and 8-hydroxyguanine and quantitation of 8-hydroxy-2′-deoxyguanosine in DNA by polyclonal antibodies. *Carcinogenesis* **12**: 865–871.

23. Cooke, M. S., Evans, M. D., Herbert, K. E. and Lunec, J. (2000). Urinary 8-oxo-2′-deoxyguanosine — source, significance and supplements. *Free Radic. Res.* **32**: 381–397.

24. Shigenaga, M. K., Gimeno, C. J. and Ames, B. N. (1989). Urinary 8-hydroxy-2'-deoxyguanosine as a biological marker of in vivo oxidative DNA damage. *Proc. Natl. Acad. Sci. USA* **86**: 9697–9701.

25. Park, E. M., Shigenaga, M. K., Degan, P., Korn, T. S., Kitzler, J. W., Weher, C. M., Kolachana, P. and Ames, B. N. (1992). Assay of excised oxidative DNA lesions: isolation of 8-oxoguanine and its nucleoside derivatives from biological fluids with a monoclonal antibody column. *Proc. Natl. Acad. Sci. USA* **89**: 3375–3379.

26. Musarrat, J. and Wani, A. A. (1994). Quantitative immunoanalysis of pro-mutagenic 8-hydroxy-2'-deoxyguanosine in oxidized DNA. *Carcinogenesis* **15**: 2037–2043.

27. Halliwell, B. and Dizdaroglu, M. (1992). The measurement of oxidative damage to DNA by HPLC and GC/MS techniques. *Free Radic. Res.* **16**: 75–87.

28. Yin, B., Whyatt, R. M., Perera, F. P., Randall, M. C., Cooper, T. B. and Santella, M. C. (1995). Determination of 8-hydroxyguanosine by an immunoaffinity chromatography-monoclonal antibody-based ELISA. *Free Radic. Biol. Med.* **18**: 1023–1032.

29. Osawa, T., Yoshida, A., Kawakishi, S., Yamashita, K. and Ochi, H. (1995). Protective role of dietary antioxidants in oxidative stress. *In* "Oxidative Stress and Ageing" (R. G. Cutler, L. Packer, J. Bertram, and A. Mori, Eds.), pp. 367–377, Birkhauser Verlag Basel, Switzerland.

30. Yarborough, A., Zhang, Y.-J., Hsu, T.-M. Santella, R. M. (1996). Immuno-peroxidase detection of 8-hydroxydeoxyguanosine in aflatoxin B1-treated rat liver and human oral mucosal cells. *Cancer Res.* **56**: 683–688.

31. Kantha, S. S., Wada, S.-I., Tanaka, H., Takeuchi, M., Watabe, S. and Ochi, H. (1996). Carnosine sustains the retention of morphology in continuous fibroblast culture subjected to nutritional insult. *Biochem. Biophys. Res. Commun.* **223**: 278–282.

32. Bespalov, I. A., Purmal, A. A., Glackin, M. P. Wallace, S. S. and Melamede, R. J. (1996). Recombinant phabs reactive with 7,8-dihydro-8-oxoguanine, a major oxidative DNA lesion. *Biochemistry* **35**: 2067–2078.

33. Leinonen, J., Lehtimäki, T., Toyokuni, S., Okada, K., Tanaka, T., Hiai, H., Ochi, H., Laippala, P., Rantalaiho, V., Wirta, O., Pasternack, A. and Alho, H. (1997). New biomarker evidence of oxidative DNA damage in patients with non-insulin diabetes mellitus. *FEBS* **417**: 150–152.

34. Erhola, M., Toyokuni, S., Okada, K., Tanaka, T., Hiai, H., Ochi, H., Uchida, K., Osawa, T., Neiminen, M. M., Alho, H. and Kellokumpu-Lehtinen, P. (1997). Biomarker evidence of DNA oxidation in lung cancer patients: association of urinary 8-hydroxy-2'-deoxyguanosine excretion with radiotherapy, chemo-therapy and response to treatment. *FEBS* **409**: 287–291.

35. Tsuboi, H., Kouda, K., Takeuchi, H., Takigawa, M., Masamoto, Y., Takeuchi, M. and Ochi, H. 8-hydroxydeoxyguanosine in urine as an index of oxidative

damage to DNA in the evaluation of atopic dermatitis. *Br. J. Dermatol.* **138**: 1033–1035.

36. Witherall, H. L., Hiatt, R. A., Replogle, M. and Parsonnet, J. (1998). Helicobacter pylori infection and urinary excretion of 8-hydroxy-2-deoxyguanosine, an oxidative DNA adduct. *Cancer Epidemiol. Biomark. Prevent.* **7**: 91–96.

37. Cooke, M. S., Evans, M. D., Podmore, I. D., Herbert, K. E., Mistry, N., Mistry, P., Hickenbotham, P. T., Hussieni, A., Griffiths, H. E. and Lunec, J. (1998). Novel repair action of vitamin C upon in vivo oxidative DNA damage. *FEBS* **363**: 363–367.

38. Lunec, J., Cooke, M. S., Podmore, I. D. and Evans, M. D. (2000). Modulation of in vivo DNA repair in humans by vitamin C supplementation. *In* "Human Monitoring after Environmental and Occupational Exposure to Chemical and Physical Agents" (D. Anderson and AE. Karakaya, Eds.), pp. 68–75.

39. Evans, M. D., Cooke, M. S., Podmore, I. D., Zheng, Q., Herbert, K E. and Lunec, J. (1999). Discrepancies in the measurement of UVC-induced 8-oxo-2'deoxyguanosine: implications for the analysis of oxidative DNA damage. *Biochem. Biophys. Res. Commun.* **259**: 374–378.

40. Ahmed, N. U., Ueda, M., Nikaido, O., Osawa, T. and Ichihashi, M. (1999). High levels of 8-hydroxy-2'-deoxyguanosine appear in normal human epidermis after a single dose of ultraviolet radiation. *Br. J. Dermatol.* **140**: 226–231.

41. Toyokuni, S., Iwasa, Y., Kondo, S., Tanaka, T., Ochi, H. and Hiai, H. (1999). Intranuclear distribution of 8-hydroxy-2'-deoxyguanosine: an immunocyto-chemical study. *J. Histochem. Cytochem.* **47**: 833–835.

42. Ahmad, J., Cooke, M. S., Hussieni, A., Evans, M. D., Burd, R. M., Patel, K., Bleiker, T. O., Hutchinson, P. and Lunec, J. (1999). Urinary thymine dimers and 8-oxo-2'deoxyguanosine in psoriasis. *FEBS* **460**: 549–553.

43. Cooke, M. and Herbert, K. (2000). Immunochemical detection of 8-oxodeoxy-guanosine in DNA. *In* "A Handbook of Clinical Analysis" (J. Lunec, Ed.), pp. 63–68, John Wiley and Sons Ltd., Chichester.

44. Evans, M. D., Cooke, M. S., Akil, M., Samanta, A. and Lunec, J. (2000). Aberrant processing of oxidatve DNA damage in systemic lups erythematosus. *Biochem. Biophys. Res. Commun.* **273**: 894–898.

45. Cooke, M. S., Mistry, N., Ladapo, A., Herbert, K. E. and Lunec J. (2000). Immunochemical quantitation of UV-induced dimeric and oxidative DNA damage to human keratinocytes. *Free Radic. Res.* **33**: 369–381.

46. Lunec, J., Holloway, K. A., Cooke, M. S., Faux, S., Griffiths, H. R. and Evans, M. D. 8-oxo-2'deoxyguanosine: redox regulation of repair in vivo. *Free Radic. Biol. Med.* **Submitted**.

47. Cooke, M. S., Evans, M. D., Burd, R. M., Patel, K., Barnard, A., Hutchinson, P. E. and Lunec, J. UVA-induced 8-oxo-2'-deoxyguanosine and thymine dimers: induction and *in vivo* kinetics of repair. *J. Invest. Dermatol.* **116**: 281–285.

48. Takahashi, S., Hirose, M., Tamano, S., Ozaki, M., Orita, S.-I., Ito, T., Takeuchi, M., Ochi, H., Fukada, S., Kasai, H. and Shirai, T. (1998). Immnohisto-chemical detection of 8-hydroxy-2'-deoxyguanosine in paraffin-embedded sections of rat liver after carbon tetrachloride treatment. *Toxicol. Pathol.* **26**: 247–252.

49. Struthers, L., Patel, R., Clark, J. and Thomas, S. (1998). Direct detection of 8-oxodeoxyguanosine and 8-oxoguanine by avidin and its analogues. *Anal. Biochem.* **255**: 20–31.

50. Cooke, M. S., Herbert, K. E., Butler, P. C. and Lunec, J. (1998). Further evidence for a possible role of conformation in the immunogenicity and antigenicity of the oxidative DNA lesion, 8-oxo-2'deoxyguanosine. *Free Radic. Res.* **28**: 459–469.

51. Soultanakis, R. P., Melamede, R. J., Bespalov, I. A., Wallace, S. S., Beckman, K. B., Ames, B. N., Taatjes, D. J. and Janssen-Heininger, Y. M. (2000). Fluorescence detection of 8-oxoguanine in nuclear and mitochondrial DNA of cultured cells using a recombinant Fab and confocal scanning laser microscopy. *Free Radic. Biol. Med.* **28**: 987–998.

52. Greferath, R. and Nehls, P. (1997). Monoclonal antibodies to thymidine glycol generated by different immunization techniques. *Hybridoma* **16**: 189–193.

53. Mistry, N., Evans, M. D., Griffiths, H. R., Kasai, H., Herbert, K. E. and Lunec, J. (1999). Immunochemical detection of glyoxal DNA damage. *Free Radic. Biol. Med.* **26**: 1267–1273.

Chapter 16

Postlabeling Detection of Oxidative DNA Damage

Michael N. Routledge and George D.D. Jones*

Michael N. Routledge • Molecular Epidemiology Unit, Epidemiology and HSR, School of Medicine, Algernon Firth Building, University of Leeds, LS2 9JT, UK
George D.D. Jones • Department of Oncology, University of Leicester, Hodgkin Building, PO Box 138, Lancaster Road, Leicester, LE1 9HN, UK
*Corresponding Author.
Tel: (0)116-252-5527, E-mail: gdj2@leicester.ac.uk

1. Introduction

The ^{32}P-postlabeling assay is a very sensitive, broadly applicable method for the detection of DNA carcinogen adducts, which has undergone many modifications and improvements since first devised by Randerath and co-workers in the early 1980s.[1] The ability to measure very low levels of DNA damage (at least 1 adduct in 10^9 nucleotides), in small quantities of DNA (1–10 μg), has led to different versions of the assay being applied to the analysis of DNA adducts in human tissues of exposed and control populations in numerous molecular epidemiology studies.[2, 3] Bearing in mind the success of the assay in analyzing aromatic carcinogen DNA adducts, it is not surprising that attempts have been made to use the high sensitivity of the assay in measuring oxidative DNA lesions. However, as discussed below, the ^{32}P-postlabeling assay as originally designed, is not always suitable for the measurement of oxidative DNA lesions, particularly when investigating low/endogenous levels of such damage. In this chapter we will review the more recent variations of the postlabeling assay that have been applied to the measurement of oxidative lesions, paying particular attention to the advantages and disadvantages of these methods. Since our last review of this subject,[4] new evidence has been published concerning the problems of adventitious oxidative effects, particularly for the postlabeling detection of 8-oxo-2'-deoxyguanosine (8-oxodG) as a biomarker for oxidative stress. Recent versions of postlabeling methods for detecting 8-oxodG have taken these problems into account, and certain postlabeling methods exist for analyzing other oxidative DNA lesions, some of which may not be easily detected by non-postlabeling methods. In this chapter we review current ^{32}P-postlabeling methods available for analysis of a range of oxidative DNA lesions, including 8-oxodG, thymine glycol, the formamido lesion, oxidative "bulky" adducts and phosphoglycolates (Table 1).

2. Detection of 8-oxo-2'-deoxyguanosine

Because 8-oxodG is a major lesion induced by ROS, and is known to be mutagenic, many assays have concentrated on the measurement of this lesion as a biomarker for oxidative stress. Several postlabeling methods have been reported for the detection of 8-oxodG in DNA with the majority of approaches detecting the lesion as 8-oxo-2'-deoxyguanosine-3'-monophosphate (8-oxodGp) after micrococcal nuclease (MN) and calf spleen phosphodiesterase (CSPD) digestion. Since our previous review of these methods,[4] the problem of the artefactual production of 8-oxodGp as a result of radiolysis occurring during the radiolabeling step has come under scrutiny. In the past not all methods addressed this potential problem, which will be further discussed below. The method of Lu *et al.*[5] separated labeled 8-oxodGp and dGp by anion-exchange thin layer chromatography (TLC), allowing the dGp to serve as an internal standard. However, the sensitivity of this

Table 1. Comparison of recent methods for detecting oxidative DNA lesions by ^{32}P-postlabeling. (CT; calf thymus, HL; human lymphocyte, NR; not reported, P[F]; formamido lesion, Pg; phosphoglycolate, T[g]; thymine glycol).

Reference	Lesion detected	Purification of lesion	Separation of labelled marker	Detected levels in DNA (10^5 dG)	Detection limit
11	8-oxodG	TLC	TLC	0.69–1.76 (rat lung)	NR
16	8-oxodG	HPLC	HPLC	0.5 (CT DNA)	1 fmol
21	P[F] and Tg	Selective digestion (NP1)	TLC + HPLC	5.1 P[F] 2.0 T[g] (CT DNA)	3 fmol (P[F])
39	T[g] and Pg	Selective digestion (SVPD)	PAGE	2.6 (Pg) 1.67 (T[g]) (HL DNA)	1.8 fmol (Pg)

method was very low, at only one lesion in 10^4 nucleotides. The initial method of Devanaboyina and Gupta[6] also involved labeling of the DNA digest without pre-purification of 8-oxodGp, but converted the labeled nucleoside bis-phosphates to 5′-monophosphates prior to TLC separation. This conversion allowed for the separation of the labeled 8-oxodG from the rest of the labeled digest mix by first removing the unmodified labeled nucleoside 5′-monophosphates onto a wick with 1 M formic acid, leaving the labeled 8-oxodG near the origin. This method was more sensitive than that of Lu *et al.*,[5] but as it did not involve pre-purification of the 8-oxodGp prior to labeling it left open the possibility of oxidation of dGp to 8-oxodGp by radiolysis during the labeling procedure (see below). Both Möller and Hofer[7] and Schuler *et al.*[8] showed that 8-oxodGp could be induced during exposure of dGp to [32]P. Möller and Hofer[7] demonstrated that levels of 8-oxodG (as measured by HPLC-ECD) increased in a dose dependent manner, when a solution of dG was mixed with [^{32}P]ATP. From their experiments the authors suggest that the amount of 8-oxodG formed during a postlabeling experiment would be about 25 8-oxodG/10^5 dG molecules. Whilst supporting the suggestion that the higher levels of 8-oxodG reported with the postlabeling method of Devanaboyina and Gupta[6] compared to HPLC-ECD could be due to radiolysis of water molecules by the radioactive ^{32}P, it should also be remembered that the experiments of Möller and Hofer[7] were carried out on pure dG rather than the DNA digest that is actually present during the normal postlabeling procedure. Schuler *et al.*,[8] on the other hand, performed experiments to compare levels of 8-oxodG measured by the postlabeling method of Devanaboyina and Gupta[6] and the HPLC-ECD method, using DNA digests. Schuler *et al.*[8] measured a level of

8-oxodG equivalent to about 26 8-oxodG/10^5 dG by postlabeling, compared to about 1.6 8-oxodG/10^5 dG by HPLC-ECD. This result does appear to support Möller and Hofer's calculated artefactual contribution of radiolysis to levels of measured 8-oxodG in the postlabeling method.[7] However, Schuler *et al.*[8] go on to demonstrate that the discrepancy between methods is not due solely to radiolysis, based on their own experiments in which both the incubation times and amounts of radioactivity in the phosphorylation reaction with dGp were increased, and external irradiation of dG with ^{32}P was investigated. Whatever the precise explanation, the higher levels of 8-oxodG measured by some versions of the postlabeling assay lead to the conclusion that it is necessary to remove dGp from the DNA digest prior to the labeling step. This observation is thought to explain why HPLC-ECD consistently measures levels of 8-oxodG 10–50 times lower than the postlabeling assay of Devanaboyina and Gupta.[6, 9, 10] Recently, the initial method of Devanaboyina and Gupta[6] has been modified to include pre-purification of 8-oxodGp by 1-D PEI cellulose TLC prior to labeling[11] (Fig. 1). Using this

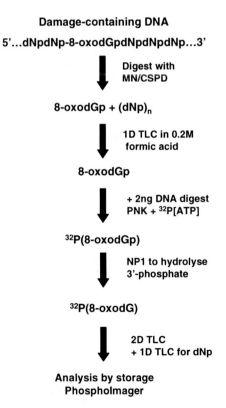

Fig. 1. **Outline of the TLC based ^{32}P-postlabeling method of Arif *et al.*[11] for the detection of 8-oxodG.** After digestion of DNA to release nucleoside 3'-monophosphates, 8-oxodGp is enriched by TLC prior to labeling with ^{32}P. The labeled products are converted to 5'-monophosphates and subsequently separated by TLC.

method, the levels of 8-oxodG in control rat lung DNA were reported to be in the range of 0.69–1.76 lesions per 10^5 dG, which was actually about half the levels of 8-oxodG measured by HPLC-ECD in the same samples. This confirms that the removal of unmodified dGp prior to labeling is crucial if overestimation of 8-oxodG in DNA is to be avoided. Previously, Lutgerink et al.[12] used a chemical decomposition method to remove normal nucleotides prior to labeling of 8-oxodGp, but a more commonly used method to pre-purify 8-oxodGp prior to labeling is HPLC. Indeed, HPLC pre-purification has been used in the past to enhance the sensitivity of detection of 8-oxodGp via postlabeling.[13] This approach of combining HPLC pre-purification with ^{32}P-postlabeling detection (HPLC/PPL) was applied by Wilson et al. to the measurement of 8-oxodG in the peripheral blood leukocytes of patients undergoing radiation therapy.[14] Peripheral blood leukocyte DNA from individuals irradiated with 180–200 cGy were observed to contain 2–4.5 times as much 8-oxodG as that from unexposed individuals. Podmore et al.[15] examined the different stages of the HPLC/PPL procedure, including the release of 8-oxodGp during digestion of the DNA with MN and CSPD, the recovery of the lesion after HPLC, assay sensitivity and removal of contaminating RNA. The final separation of labeled products was by TLC and the limit of detection was 5.4 fmol of 8-oxodGp. The main drawback of the reported procedures of Wilson et al. and Podmore et al. was the requirement for high amounts of DNA (25–100 µg for each analysis). However, Podmore et al. also showed that capillary electrophoresis may be suitable for the pre-purification step.[15] Use of capillary electrophoresis should permit the handling of sub-µg amounts of DNA for the enrichment of 8-oxodGp before postlabeling, which may be of particular value if small biopsy samples are being studied.

More recently, Zeisig et al.[16] have adapted their own postlabeling procedure developed for the analysis of aromatic DNA adducts, which uses HPLC separation of labeled adducts with on-line radioactivity detection,[17] to the measurement of 8-oxodG (Fig. 2). When used to detect aromatic DNA adducts no pre-purification HPLC step was necessary,[17] but for optimal separation of 8-oxodGp and elimination of the problem of artefactual oxidation of dGp via the radiolysis pathway, an extra pre-purification HPLC step was used for detecting 8-oxodG.[16] Following the labeling step and nuclease P1 treatment, the labeled products are again separated by HPLC with on-line radioactivity detection. The use of 8-oxodGp standards showed the labeling efficiency to be $77 \pm 39\%$ and loss of material during the work-up to be 58% (the total recovery being $32 \pm 10\%$). Although the levels of 8-oxodG measured in calf thymus DNA in this study were ~0.5 per 10^5 unmodified dG, the authors report a probable sensitivity for the method of 0.1 8-oxodG in 10^5 dG (at an absolute sensitivity of 1 fmol) when 1 µg of DNA is analyzed, and a standard deviation of repeated analyses of the same sample of $\pm 10\%$.[16] However, in a recent inter-laboratory comparison of different methods of measuring 8-oxodG in identical samples of calf thymus DNA this

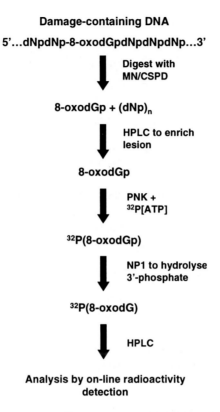

Damage-containing DNA

5'...dNpdNp-8-oxodGpdNpdNpdNp...3'

↓ Digest with
    MN/CSPD

8-oxodGp + (dNp)ₙ

↓ HPLC to enrich
    lesion

8-oxodGp

↓ PNK +
    ³²P[ATP]

³²P(8-oxodGp)

↓ NP1 to hydrolyse
    3'-phosphate

³²P(8-oxodG)

↓ HPLC

**Analysis by on-line radioactivity
detection**

Fig. 2. **Outline of the HPLC based ^{32}P-postlabeling method of Zeisig *et al.*[16] for the detection of 8-oxodG.** After digestion of DNA to release nucleoside 3'-monophosphates, 8-oxodGp is enriched by HPLC prior to labeling with ^{32}P. The labeled products are converted to 5'-monophosphates and separated by HPLC with on-line radioactivity detection.

postlabeling approach failed to detect an experimentally induced dose response and suffered from high coefficients of variation ($\geq 20\%$).[18]

3. Detection of Oxidative Pyrimidine Lesions

Whilst 8-oxodG is a frequently studied marker for oxidative DNA damage, other lesions may be equally as important both biologically and as biomarkers. In addition, there is increasing evidence that different oxidative lesions may respond differently to modulation of oxidative stress,[19, 20] warning against an over reliance on the measurement of one marker of oxidative DNA damage as an index of oxidative stress. It is therefore important that a variety of lesions should be analyzed, and methods that allow the quantification of more than one marker at the same time may be particularly useful in such studies. One such

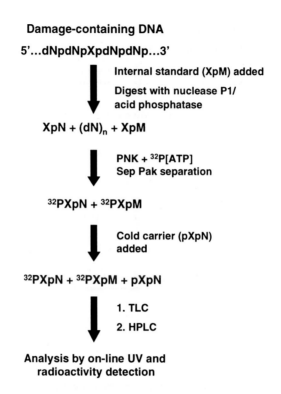

Damage-containing DNA

5'...dNpdNpXpdNpdNp...3'

Internal standard (XpM) added

Digest with nuclease P1/
acid phosphatase

XpN + (dN)ₙ + XpM

PNK + ^{32}P[ATP]
Sep Pak separation

^{32}PXpN + ^{32}PXpM

Cold carrier (pXpN)
added

^{32}PXpN + ^{32}PXpM + pXpN

1. TLC

2. HPLC

**Analysis by on-line UV and
radioactivity detection**

Fig. 3. Outline of the ^{32}P-postlabeling method of Maccubbin *et al.*[21] for the detection of thymine glycol and formamido lesions. The method is based on the selective resistance of certain phosphodiester bonds to nuclease P1 digestion. Lesions are labeled and detected as dinucleotides, using both TLC and HPLC as the means of separation.

postlabeling method is that devised by Maccubbin *et al.*[21] (Fig. 3) to detect thymine glycols and the formamido remnant of pyrimidine bases that is induced when reduction of the peroxyl radical (formed from addition of a hydroxyl radical to the C5–C6 segment of a pyrimidine base followed by addition of oxygen) generates an unstable hydroperoxide. The postlabeling assay that detects both of these lesions simultaneously is based on the fact that the endonuclease, nuclease P1 does not hydrolyze all phosphodiester bonds at an equal rate. Of crucial importance to the assay is the slow hydrolysis of thymine glycol containing dinucleotides (where the thymine glycol is located 5' to the unmodified nucleotide) compared to unmodified nucleotides, and the exceedingly slow rate of hydrolysis of similar dinucleotides that contain the formamido lesion. The differential digestion of modified dinucleotides was first recognized as being useful in postlabeling of bulky aromatic adducts by Randerath and co-workers,[22] and was later developed as a means of postlabeling thymine glycol and formamido lesion containing products by Maccubbin and co-workers.[23] Following digestion of the DNA with nuclease P1 and acid phosphatase, only the dinucleotides containing the lesions are available for labeling by T4 PNK as the remainder of the DNA has been digested to mononucleosides which are not substrates for the phosphorylation process (T4 PNK requires the substrate molecule to ideally contain an intact nucleoside 3'-phosphate moiety). Separation of the labeled dinucleotide species was achieved by anion-exchange TLC, with background levels of the formamido lesion being estimated at ~5 lesions per 10^5 dG in calf thymus DNA. The postlabeling method has been further improved upon to address the key drawbacks of postlabeling methods, namely the positive identification of the final labeled product and the recovery of the lesions through the assay procedure.[21] An additional separation of the labeled products is performed by HPLC after the products have first been separated by TLC. In the modified assay the two lesions are obtained from DNA as the thymine glycol or formamido lesion 5' to adenosine. In order to follow the recovery of the products in the assay, two authenticated standards are added. The first internal standard is the equivalent lesion containing dinucleotide, in which the adenosine has been replaced by N^6-methyl-2'-deoxyadenosine. This allows for the internal standard to be included throughout the assay and separated from the lesion of interest during the final HPLC step. This controls for loss of product during the assay. The second standard is an unlabeled 5'-phosphorylated lesion-containing standard that is added after the labeling reaction, to act as a carrier that will provide authentication of the labeled product in the final HPLC analysis, as this standard can be detected by simultaneous on-line UV detection. Using this version of the assay, it was found that the relative amounts of thymine glycol and formamido lesions in calf thymus DNA irradiated with X-rays varied in the presence or absence of oxygen during the X-irradiation. In the absence of oxygen during irradiation, thymine glycol levels were similar to levels of formamido lesions after 60 Gy irradiation (378 and 385 fmol/μg, respectively) whereas in the presence of oxygen, the levels of formamido lesions were much higher than thymine glycol lesions (3315 versus 586 fmol/μg) at the same dose of radiation. This shows that induction of different oxidative markers may vary depending on the concentrations of oxygen available during the reaction. This assay has also shown that the formamido lesion, which is not detected by other assays, is a major lesion in DNA exposed to ROS.

4. Detection of Oxidative "Bulky" Lesions

The oxidative DNA lesions that have been discussed so far are all similar in molecular weight to unmodified nucleotides, which is why it has been necessary to modify the original Randerath postlabeling method to allow their detection. However, there are certain bulky lesions induced in DNA by oxidative stress that exhibit properties similar to the bulky aromatic hydrocarbon DNA adducts, and which may therefore be detected by the Randerath procedure. The nuclease P1 enhancement version of the assay,[24] uses the observation that bulky adducts are often refractory to 3'-dephosphorylation by nuclease P1. Consequently, when the MN/CSPD digest of adducted DNA is treated with nuclease P1, unmodified nucleotides are dephosphorylated, leaving only the resistant adducted nucleotides as substrates for T4 PNK mediated ^{32}P-labeling. Using this method bulky DNA lesions have been detected in DNA treated with different Fenton-type oxygen radical generating systems[25-28] (Fig. 4). It has been proposed that these lesions may be intrastrand links between adenine and 5' adenine or guanine residues.[26, 27] Indeed, some of these lesions may be the endogenous I-compounds that increase with age in animal tissues.[29, 30] These I-compounds that are related to oxidative DNA damage have been classified as Type II I-compounds to distinguish them

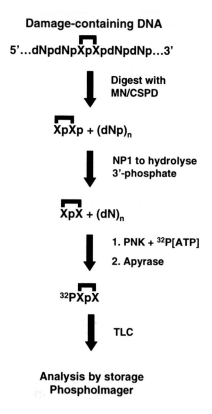

Fig. 4. **Outline of the ^{32}P-postlabeling method of Lloyd *et al.*[27] for the detection of "bulky" oxidative DNA lesions.** The method relies on the enrichment of the DNA digests with respect to bulky-type lesions by treatment of the digested samples with nuclease P1. ▬ represents an intrastrand covalent link between two adjacent nucleotides (X). Lesions are labeled and then detected using TLC.

from other endogenous I-compounds that are not modulated by oxidative stress.[31] Similar bulky adducts have also been detected in the liver DNA of patients with Wilson's disease,[32] a condition in which the accumulation of transition metals in the liver leads to extensive tissue damage, possibly through oxidative mechanisms.

5. The Snake Venom Phosphodiesterase Method

As already stated, some methods are particularly suited to the detection of lesions that other methods may not easily detect. The postlabeling method developed by Weinfeld and co-workers[33-35] is one such method (Fig. 5). Although it cannot be used to detect the commonly measured biomarker 8-oxodG, it is able to detect a range of different lesions at the same time, including thymine glycols, phosphoglycolates and abasic sites. This method, similar to the method of Maccubbin *et al.* described above, relies on the fact that these lesions impede digestion of the phosphodiester bond immediately 5′ to the lesion by the enzyme snake venom phosphodiesterase (SVPD). The DNA is, in fact, digested by a mixture of three enzymes; SVPD, DNase I and an alkaline phosphatase. The products of digestion of DNA by this enzyme mixture are normal nucleosides (which are not substrates for T4 PNK as they lack the 3′-phosphate group)

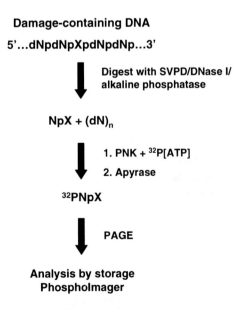

Damage-containing DNA

5'...dNpdNpXpdNpdNp...3'

↓ Digest with SVPD/DNase I/ alkaline phosphatase

NpX + (dN)ₙ

↓ 1. PNK + ³²P[ATP]
2. Apyrase

³²PNpX

↓ PAGE

Analysis by storage Phospholmager

Fig. 5. **Outline of SVPD based ³²P-postlabeling method of Weinfeld *et al.*[35] for the detection of thymine glycol, phosphoglycolate and abasic lesions.** The method is based on the selective resistance of certain phosphodiester bonds to SVPD digestion. Lesions are labeled and detected as damage-containing dimer species and separated by PAGE.

and damage containing "dimer" species consisting of a normal nucleoside-3'-monophosphate moiety (dNp-) 5' to the lesion. The damage containing "dimer" species can be labeled because the 5' normal dNp- moiety is a ready substrate for T4 PNK. Indeed, it is advantageous that it is the normal nucleoside-3'-monophosphate moiety and not the damaged nucleoside-3'-monophosphate moiety that is labeled (as in the method of Maccubbin *et al.*[21]) The efficiency of labeling of thymine glycol as a nucleoside-3'-monophosphate moiety has been shown to be very low (1–2%)[36] and abasic site lesions as the corresponding modified-deoxyribose 3'-monophosphate species are not be substrates for labeling by T4 PNK. Following labeling, the damage containing "dimer" species are separated by high resolution denaturing PAGE on 20% (w/v) polyacrylamide/ 7 M urea gels. The advantages of the SVPD-based protocol are that it allows for the detection of lesions that cannot easily be detected by other methods, and there is no need for pre-purification. Moreover, should there be any adventitious oxidation of the DNA digest occurring during the labeling stage, this does not contribute to (or detract from) the levels of damage that are detected by this protocol.[37] The SVPD method has been used to detect a range of lesions induced by oxidative species, including thymine glycols and phosphoglycolates produced by ionizing radiation and Fenton chemistry,[34, 35] and phosphoglycolates produced in cellular DNA exposed to various types of oxidative stress.[38] The SVPD assay has also been used to estimate DNA damage levels in human lymphocyte DNA[39] and colorectal tissue DNA.[40]

6. Conclusions

There are currently several methods utilizing the sensitive approach of ^{32}P-postlabeling available for the measurement of several different oxidative DNA lesions. Recent advancements have tackled and negated the problems of adventitious oxidative damage formation during the labeling step, and it is hoped that these methods will soon be used to analyze a range of oxidative DNA lesions in experimental systems and human tissues.

Acknowledgments

The authors would like to thank the UK Medical Research Council (MRC), the Ministry of Agriculture Fisheries and Food (MAFF), and the Food Standards Agency (FSA) for supporting their research in this field (MRC project grant: G9527655MA) (MAFF Research Contracts; ANO4/26, ANO4/32 (now with the FSA: N04002) and ANO4/63 (now with the FSA: N04022).

References

1. Randerath, K., Reddy, M. V. and Gupta, R. C. (1981). ^{32}P-labeling test for DNA damage. *Proc. Natl. Acad. Sci. USA* **78**: 6126–6129.
2. Kriek, E., Rojas, M., Alexandrov, K. and Bartsch, H. (1998). Polycyclic aromatic hydrocarbon-DNA adducts in humans: relevance as biomarkers for exposure and cancer risk. *Mutat. Res.* **400**: 215–231.
3. Sram, R. J. and Binkova, B. (2000). Molecular epidemiology studies on occupational and environmental exposure to mutagens and carcinogens, 1997–1999. *Env. Health Perspect.* **108**: 57–70.
4. Routledge, M. N., Weinfeld, M. and Jones, G. D. D. (1998). ^{32}P-postlabeling detection of oxidative DNA damage. *In* "DNA and Free Radicals: Techniques, Mechanisms and Applications" (O. I. Aruoma, and B. Halliwell, Eds.), pp. 301–318, OICA International, Saint Lucia.
5. Lu, L.-J. W., Tasaka, F., Hokanson, J. A. and Khoda, K. (1991). Detection of 8-hydroxy-2′-deoxyguanosine in deoxyribonucleic acid by the ^{32}P-postlabeling method. *Chem. Pharm. Bull.* **39**: 1880–1882
6. Devanaboyina, U. and Gupta, R. C. (1996). Sensitive detection of 8-hydroxy-2′-deoxyguanosine in DNA by ^{32}P-postlabeling assay and the basal levels in rat tissues. *Carcinogenesis* **17**: 917–924.
7. Möller, L. and Hofer, T. (1997). [^{32}P]ATP mediates formation of 8-hydroxy-2′-deoxyguanosine from 2′-deoxyguanosine, a possible problem in the ^{32}P-postlabeling assay. *Carcinogenesis* **18**: 2415–2419.
8. Schuler, D., Otteneder, M., Sagelsdorff, P., Eder, E., Gupta, R. C. and Lutz, W. K. (1997). Comparative analysis of 8-oxo-2′-deoxyguanosine in DNA by ^{32}P and ^{33}P-postlabeling and electrochemical detection. *Carcinogenesis* **18**: 2367–2371.
9. Adachi, S., Zeisig, M. and Möller, L. (1995). Improvements in the analytical method for 8-hydroxydeoxyguanosine in nuclear DNA. *Carcinogenesis* **16**: 253–258.
10. Hirano, T., Homma, Y. and Kasai, H. (1995). Formation of 8-hydroxyguanine in DNA by aging and oxidative stress. *In* "Oxidative Stress and Aging" (A. R. G. Cutler, L. Packer, J. Bertram, and A. Mori, Eds.), pp. 69–76, Brikhauser Verlag, Basel.
11. Arif, J. M., Vadhanam, M. V., de Groot, A. J. L., van Zeeland, A. A., Gairola, C. G. and Gupta, R. C. (2000). Effect of cigarette smoke on 8-oxo-2′-deoxyguanosine (8-oxodG) formation in rat lungs measured by ^{32}P-postlabeling-TLC and HPLC-ECD. *Proc. Am. Assoc. Cancer Res.* **41**: 233.
12. Lutgerink, J. T., de Graaf, E., Hoebee, B., Stavenuitez, H. F. C., Westra, J. G. and Kriek, E. (1992). Detection of 8-hydroxyguanine in small amounts of DNA by ^{32}P-postlabeling. *Anal. Biochem.* **201**: 127–133.

13. Povey, A. C., Wilson, V. L., Taffe, B. G., Wood. M. L., Essigman, J. M. and Harris, C. C. (1989). Detection and quantitation of 8-hydroxyguanine residues by ^{32}P-postlabeling. *Proc. Am. Assoc. Cancer Res.* **30**: 201.

14. Wilson, V. L., Taffe, B. G., Shields, P. G., Povey, A. C. and Harris, C. C. (1993). Detection and quantification of 8-hydroxydeoxyguanosine adducts in peripheral blood of people exposed to ionizing radiation. *Env. Health Perspect.* **99**: 261–263.

15. Podmore, K., Farmer, P. B., Herbert, K. E., Jones, G. D. D. and Martin, E. A. (1997). ^{32}P-postlabeling approaches for the detection of 8-oxo-2'-deoxyguanosine-3'-monophosphate in DNA. *Mutat. Res.* **178**: 139–149.

16. Zeisig, M., Hofer, T., Cadet, J. and Möller, L. (1999). ^{32}P-postlabeling high-performance liquid chromatography (^{32}P-HPLC) adapted for analysis of 8-hydroxy-2'-deoxyguanosine. *Carcinogenesis* **20**: 1241–1245.

17. Möller, L., Zeisig, M. and Vodicka, P. (1993). Optimization of an HPLC method for analysis of ^{32}P-postlabeled DNA adducts. *Carcinogenesis* **14**: 1343–1348.

18. Collins, A. R., Brown, J., Bogdanov, M., Cadet, J., Cooke, M., Douki, T., Dunster, C., Eakins, J., Epe, B., Evans, M., Farmer, P., Gedik, C. M., Halliwell, B., Herbert, K., Hofer, T., Hutchinson, R., Jenner, A., Jones, G. D. D., Kasai, H., Kelly, F., Lloret, A., Loft, S., Lunec, J., McEwan, M., Moller, L., Olinski, R., Podmore, I., Poulsen, H., Ravanat, J. L., Rees, J. F., Reetz, F., Shertzer, H., Spiegelhalder, B., Turesky, R., Tyrrell, R., Vina, J., Vinicombe, D., Weimann, A., de Wergifosse, B. and Wood, S. G. (2000). Comparison of different methods of measuring 8-oxoguanine as a marker of oxidative DNA damage. *Free Radic. Res.* **32**: 333–341.

19. Podmore, I. D., Griffiths, H. R., Herbert, K. E., Mistry, N., Mistry, P. and Lunec, J. (1998). Vitamin C exhibits pro-oxidant properties. *Nature* **392**: 559.

20. Rehman, A. Collis, C. S., Yang, M., Kelly, M., Diplock, A. T., Halliwell, B., Rice-Evans, C. (1998). The effects of iron and vitamin C co-supplementation on oxidative damage to DNA in healthy volunteers. *Biochem. Biophys. Res. Commun.* **246**: 293–298.

21. Maccubbin, A. E., Patrzyc, H. B., Ersing, N., Budzinski, E. E., Dawidzik, J. B., Wallace, J. C., Iijima, H. and Box, H. C. (1999). Assay for reactive oxygen species-induced DNA damage: measurement of the formamido and thymine glycol lesions. *Biochim. Biophys. Acta* **1454**: 80–88.

22. Randerath, K., Randerath, E., Danna, T. F., Vangolen, K. L. and Putman, K. L. (1989). A new sensitive ^{32}P-postlabeling assay based on the specific enzymatic conversion of bulky DNA lesions to radiolabeled dinucleotides and nucleoside 5'- monophosphates. *Carcinogenesis* **10**: 1231–1239.

23. Maccubbin, A. E., Budzinski, E. E., Patrzyc, H., Evans, M. S., Ersing, N. and Box, H. C. (1993). ^{32}P-postlabeling assay for free radical-induced DNA damage: the formamido remnant of thymine. *Free Radic. Res. Commun.* **18**: 17–28.

24. Reddy, M. V. and Randerath, K. (1986). Nuclease P1-mediated enhancement of sensitivity of ^{32}P-postlabeling test for structurally diverse DNA adducts. *Carcinogenesis* **7**: 1543–1551.

25. Randerath, K., Yang, P.-F., Danna, T. F., Reddy, R., Watson, W. P. and Randerath, E. (1991). Base adducts detected by ^{32}P-postlabeling in DNA modified by oxidative damage in vitro. Comparison with rat lung I-compounds. *Mutat. Res.* **250**: 135–144.

26. Carmichael, P. L., She, M. N. and Phillips, D. H. (1992). Detection and characterization by ^{32}P-postlabeling of DNA adducts induced by a Fenton-type oxygen radical-generating system. *Carcinogenesis* **13**: 1127–1135.

27. Lloyd, D. R., Phillips, D. H. and Carmichael, P. L. (1997). Generation of putative intrastrand cross-links and strand breaks in DNA by transition metal ion-mediated oxygen radical attack. *Chem. Res. Toxicol.* **10**: 393–400.

28. Lloyd, D. R. and Phillips, D. H. (1999). Oxidative DNA damage mediated by copper(II), iron(II) and nickel(II) Fenton reactions: evidence for site-specific mechanisms in the formation of double-strand breaks, 8-hydroxydeoxy-guanosine and putative intrastrand cross-links. *Mutat. Res.* **424**: 23–36.

29. Randerath, K., Liehr, J. G., Gladek, A. and Randerath, E. (1989). Age-dependent covalent DNA alterations (I-compounds) in rodent tissues: species, tissue and sex specificities. *Mutat. Res.* **219**: 121–133.

30. Randerath, K., Li, D. and Randerath, E. (1990). Age-related DNA modifications (I-compounds): modulation by physiological and pathological processes. *Mutat. Res.* **238**: 245–253.

31. Randerath, K., Randerath, E., Zhou, G.-D. and Li, D. (1999). Bulky endogenous DNA modifications (I-compounds) — possible structural origins and functional implications. *Mutat. Res.* **424**: 183–194.

32. Carmichael, P. L., Hewer, A., Osborne, M. R., Strain, A. J. and Phillips, D. H. (1995). Detection of bulky DNA lesions in the liver of patients with Wilsons disease and primary hemochromatosis. *Mutat. Res.* **326**: 235–243.

33. Weinfeld, M., Liuzzi, M. and Paterson, M. C. (1990). Response of phage T4 polynucleotide kinase toward dinucleotides containing apurinic sites: design of a ^{32}P-postlabeling assay for apurinic sites in DNA. *Biochemistry* **29**: 1737–1743.

34. Weinfeld, M. and Soderlind, K.-J. M. (1991). ^{32}P-postlabeling detection of radiation-induced DNA damage: identification and estimation of thymine glycols and phosphoglycolate termini. *Biochemistry* **30**: 1091–1097.

35. Weinfeld, M., Liuzzi, M. and Jones, G. D. D. (1996). A postlabeling assay for oxidative damage. *In* "Technologies for Detection of DNA Damage and Mutations" (G. P. Pfeifer, Ed.), pp. 63–71, Plenum Press, New York.

36. Hegi, M. E., Sagelsdorff, P. and Lutz, W. K. (1989). Detection by ^{32}P-postlabeling of thymidine glycol in γ-irradiated DNA. *Carcinogenesis* **10**: 43–47.

37. Jones, G. D. D., Dickinson, L., Lunec, J. and Routledge, M. N. (1999). SVPD-post-labeling detection of oxidative damage negates the problem of adventitious oxidative effects during ^{32}P-labeling. *Carcinogenesis* **20**: 503–507.
38. Bertoncini, C. R. A. and Meneghini, R. (1995). DNA strand breaks produced by oxidative stress in mammalian cells exhibit 3′-phosphoglycolate termini. *Nucleic Acids Res.* **23**: 2995–3002.
39. Routledge, M. N., Lunec, J., Bennett, N., and Jones, G. D. D. (1998). Measurement of basal levels of oxidative damage in human lymphocyte DNA by ^{32}P-postlabeling. *Proc. Am. Assoc. Cancer Res.* **39**: 287.
40. Routledge, M. N., Hinson, F. L., Leveson, S. H. and Jones, G. D. D. (1999). A comparison of aromatic DNA adducts and oxidative DNA lesions in human tissue samples. *Proc. Am. Assoc. Cancer Res.* **40**: 45.

Chapter 17

The Comet Assay: Protective Effects of Dietary Antioxidants Against Oxidative DNA Damage Measured Using Alkaline Single Cell Gel Electrophoresis

Susan J. Duthie

Susan J. Duthie • Rowett Research Institute, Greenburn Road, Bucksburn, Aberdeen AB22 9SB, UK
Tel: 00 44 (1224) 712751, Ext: 2324, E-mail: sd@rri.sari.ac.uk

1. Introduction

1.1. Background

The single cell gel electrophoresis assay or "comet assay" is a sensitive, quick and simple method for measuring DNA damage in single cells (plant or animal) and in small tissue samples.[1-3] The assay we know today is adapted from the microgel electrophoresis technique of Ostling and Johanson[1] who observed that DNA from gamma-irradiated murine lymphoma cells, suspended in an agarose gel on a microscope slide, lysed with neutral detergent and exposed to a weak electric field migrated further towards the anode than DNA from untreated cells. Ostling and Johanson proposed that following irradiation, DNA strand breaks allow relaxation of the DNA supercoiling resulting in a more pronounced migration of DNA in the gel. These "comet" images (following acridine orange staining) represented the extent of DNA damage in the nucleus and were dependent on the dose of irradiation.[1] The assay was modified further by Singh *et al.*,[4] who optimized the alkali DNA denaturing and electrophoresis conditions allowing for evaluation of DNA single strand breaks and alkali-labile sites. It is upon this technique that many recent comet assay modifications are based.

1.2. The Principle of the Comet Assay

In general, the alkaline comet assay consists of several stages beginning with isolation of cells, agarose embedding, lysis with high salt and detergent, alkaline unwinding and electrophoresis, neutralizing, staining and analysis.

When embedded, the cell forms a cavity in the agarose gel. Following lysis, when all the cellular proteins are stripped away and have dissolved through the gel, only the nuclear scaffold proteins remain and the DNA expands to occupy the cavity.[1] This core of supercoiled DNA is termed a "nucleoid". Following alkaline unwinding and electrophoresis, the negatively charged DNA is pulled towards the anode. However only loops of DNA which contain breaks [either overt breaks or breaks resulting from repair enzyme digestion (see below)] are released from the supercoiling and migrate in the gel. These loops extend from the head of the nucleoid to form the comet tail. Comet images can be viewed and analyzed by fluorescence microscopy. The intensity in the comet tail reflects the extent of DNA breakage. However, strand breakage may not be the most useful of lesions to measure, at least not in isolation. The comet assay has been modified to detect specific types of damage, giving more relevant data on the putative mechanisms of action of various genotoxins or chemoprotective agents. By employing appropriate bacterial repair enzymes, oxidized DNA bases, alkylation damage and misincorporated uracil can now be measured.[5-7] We have developed one modification of the comet assay that allows for detection of oxidized pyrimidines and purines in the genome of single cells.[5] Following lysis, the cells suspended in the agarose gel are incubated with a specific repair endonuclease which, at the site of an oxidized base in the DNA molecule makes a break as the first step in repair. This increase in strand breakage owing to the presence of oxidative base damage is measured as increased intensity in the comet tail compared with buffer-only-treated gels. Some of the many applications of this modification will be discussed in a subsequent section.

Since its introduction as a tool for measuring chemotherapy-induced genotoxicity in tumour cells[8, 9] the applications of the assay have increased enormously. The comet assay is widely accepted in biomonitoring studies as a valuable and sensitive tool for measuring DNA damage after occupational exposure to chemicals such as, benzo(a)pyrene, naphthalene and 1,3-butadiene,[10] to ionizing radiation[11] and gases such as radon[12] and to tobacco dust.[13] Environmental contamination can even be monitored using exposed animal species as biomarkers. Frenzille and coworkers observed significant differences in DNA strand breakage in gill cells from *Mytilus galloprovincialis* sampled from several areas of a lagoon with hot spots of damage central to the area.[14] The comet assay can also be used to assess DNA breakage in small tissue samples. Brooks and Winton[3] using EDTA digestion and agitation isolated intact mouse intestinal crypts to examine whether specific regions of the crypt were more susceptible to damage and whether they could undergo differential repair. Intact crypts were treated *in vitro* either with UV radiation, hydrogen peroxide or the drug etoposide. The extent of DNA damage, measured in stem cells, proliferating cells or mature colonocytes was indeed determined by crypt position. This may reflect state of differentiation or metabolic competency of the cells. The comet assay is also being performed on animal germ cells and preimplantation embryos to assess how culture conditions induce oxidative stress and compromise *in vitro* maturation and subsequent pregnancy outcome in women undergoing *in vitro* fertilization.[15, 16]

In this review, I will describe in detail the methodology of the comet assay before summarizing several recent studies investigating the effects of nutrition on oxidative DNA damage. Additionally, some future applications of the assay will be discussed.

2. Method

There are several versions of the comet assay with numerous modifications at every stage in the process. Many of these variations have been described extensively in previous reviews.[17, 18] This section describes the alkaline comet assay as used in our laboratory and serves only to highlight one particular variant of the assay. It is based on the method of Singh *et al.*[4] and is a sensitive method for detecting DNA strand breaks (including those created by alkali-labile sites) in

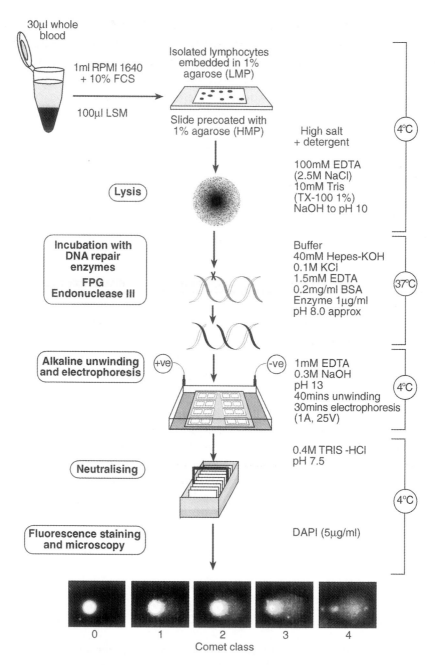

Fig. 1. **Schematic of the comet assay protocol employed in our laboratory.** Cell isolation, agarose embedding, lysis, enzyme digestion, unwinding, electrophoresis, neutralizing and staining are described. Fetal calf serum (FCS), Lymphocyte Separation Medium (LSM), low melting point agarose (LMP), high melting point agarose (HMP), bovine serum albumin (BSA), formamidopyrimidine glycosylase (FPG), 4′,6-diamidine-2-phenylindole dihydrochloride (DAPI).

single cells. It has been modified to detect oxidized pyrimidines or purines, using specific DNA repair endonucleases. A brief description of how to prepare lymphocytes for *ex vivo* analysis is given (Fig. 1).

2.1. Cell Isolation

Human peripheral blood lymphocytes are isolated from a finger prick sample. Whole blood (30 µl) is mixed with 1 ml RPMI 1640 medium supplemented with heat inactivated fetal calf serum [10% HIFCS (v/v)], underlayed with LymphoPrep Lymphocyte Separation medium (100 µl LSM) and centrifuged at 200 xg for 3 min at 4°C. The lymphocytes are seen as a pink layer sitting above the erythrocyte pellet. This pink layer is removed (approximately 100 µl) into 1 ml PBS, pH 7.4 at 4°C and centrifuged at 200 xg for 3 min at 4°C. The PBS is decanted to waste and the lymphocyte pellet resuspended in 85 µl low melting point agarose at 37°C (LMP, 1% w/v in PBS, pH 7.4). The cells (approximately 2×10^4/ml) are immediately pipetted either onto a frosted glass microscope slide pre-gelled with a layer of high melting point agarose (HMP, 1% w/v in PBS, pH 7.4, spread using a 18 × 18 mm glass coverslip) or onto a plain glass slide previously precoated with HMP agarose and dried. The agarose-containing cells are immediately topped with a glass coverslip (18 × 18 mm) to spread the gel. Generally, 2 gels per slide are set. The number of agarose layers (e.g. up to 3) and the composition of the agarose (e.g. 0.5–1% in PBS or water) used differs between labs and can have profound effects on attachment and mobility. Preparing plain glass slides with a layer of agarose before treatment with HMP agarose is cost effective and enables images to be scored some considerable time after the assay (days to weeks) following drying and storage of the slides. Gels cast on frosted slides must be scored quickly to prevent the gels drying and the fluorescent frosting interfering with analysis.

2.2. Lysis

The agarose is allowed to set at 4°C for 10 min and the slides incubated in lysis solution [2.5 M NaCl, 10 mM Tris, 100 mM Na$_2$ EDTA, NaOH to pH 10, with Triton X-100 (1% v/v) added fresh] at 4°C for 1 h to remove cellular proteins. (NB lysis time and lysis solution composition vary between laboratories. We have found that incubating cells in lysis for up to 24 h does not subsequently affect the assay. Certain methods use other detergents such as sarcosyl). This high salt and detergent mix of the lysis solution strips away cellular proteins leaving the DNA as an intact nucleoid.[1] In the basic comet assay (which detects only DNA strand breaks) the cells go straight to the unwinding stage.

2.3. Enzyme Treatment

However, in certain experiments, the cells are incubated with either endonuclease III or formamidopyrimidine glycosylase (FPG) to allow detection of oxidized pyrimidines and purines respectively. Following lysis, the slides are washed for 5 min each with buffer (40 mM HEPES-KOH, 0.1 M KCl, 0.5 mM EDTA, 0.2 mg/ml BSA, pH 8 at 4°C), gently blotted dry with tissue paper and the gel covered with 50 μl of either buffer (to measure strand breaks) or endonuclease III or FPG in buffer (approximately 1 μg/ml) to measure oxidative base damage. The gels are sealed with a glass cover slip (22 × 22 mm) and incubated for 30 min (FPG) or 45 min (endonuclease III) at 37°C in a humidified atmosphere to prevent the gels drying out. An optimum time course for activity against oxidative damage for each DNA repair enzyme must be determined.

2.4. Alkaline Unwinding and Electrophoresis

Following enzyme treatment the slides are aligned in buffer (21 mM Na$_2$ EDTA, 0.3 M NaOH, pH 12.7 at 4°C) in a double row of 8 in a horizontal electrophoresis tank [260 mm wide] for 40 minutes. Spaces between the slides should be avoided. The alkaline allows the DNA to relax and unwind. The slides are electrophoresed for 30 min at 25V, 1 A. [NB this is carried out at an ambient temperature of 4°C, with the temperature of the running buffer approximately 15°C after 30 minutes]. Breaks in the DNA molecule allow DNA loops to be pulled out of the supercoiled structure towards the anode. Variations in time of unwinding and conditions of electrophoresis have significant impact on the images generated.

2.5. Neutralizing and Staining

The slides are washed 3 times at 4°C for 5 min each with neutralizing buffer (0.4 M Tris-HCl, pH 7.5) to remove alkali and detergents and to allow renaturing of the DNA molecule before staining with 20 μl 4′,6-diamidine-2-phenylindole dihydrochloride [DAPI, 5 μg/ml]. (NB many other stains are used to visualize the comet images including ethidium bromide, acridine orange, Hoechst 33258 and propidium iodide.) The slides are stored in the dark at 4°C.

We routinely use visual scoring to classify comet images according to the intensity of fluorescence in the comet tail. Generally, 100 comet images on each gel (scored at random) are classified according to the relative intensity of fluorescence in the tail and given a value of 0, 1, 2, 3 or 4 (from undamaged scoring 0, to maximally damaged scoring 4). In this way, the total score for 100 images can range from 0 (all undamaged) to 400 (all maximally damaged). DNA damage is described in arbitrary units. Figure 1 shows typical examples of all comet classes. DNA damage due to strand breakage is estimated using the score obtained from buffer-treated gels alone. Oxidative base damage is estimated by subtracting the visual score for the buffer-treated gels from the score obtained after incubation with enzyme. This semiquantitative method of visual scoring has been extensively validated by comparison with computerized image analysis [Kinetic Imaging Ltd., Liverpool, UK] and correlates well with more quantitative measures [$R = 0.97$ comparing visual scores with % DNA in the tail and $R = 0.89$ comparing visual scores with tail moment]. There are however, many other ways to analyze comet images. Photomicrographs of single cells can be taken and comet head diameter, tail length and comet area measured. More often, densitometric and geometric comet parameters are obtained by computerized image analysis. Tail moment (the product of the percentage of DNA in the tail and the tail length), the percentage of DNA in the tail or in the head and tail length itself are all routinely used[8]). Skewness and kurtosis, which describe aspects of comet morphology, are also commonly measured.[8]

3. Oxidative DNA Damage and the Impact of Nutrition

Oxidative stress and free radical formation from endogenous respiration and exogenous sources such as pollution, chemicals and radiation have been implicated in the ageing process and diseases associated with age such as cancer, heart disease and cataracts. Oxidative DNA damage accumulates with age, is related to the lifespan of the particular organism and is associated with premature ageing in individuals with metabolic disorders which cause excessive production of reactive oxygen species.[19] Antioxidants protect cellular systems from oxidative damage.[20] Consumption of foods rich in antioxidants such as vitamin C, vitamin E and polyphenols are associated with a decreased risk for cancer and coronary heart disease,[21, 22] while diets high in fat may increase risk.[21] This review will focus on how the comet assay has been used both in molecular epidemiology and experimentally to determine the mechanisms associated with the anticarcinogenic properties of the human diet.

3.1. Molecular Epidemiology

Molecular epidemiology, the study of biomarkers of DNA damage (e.g. strand breaks, oxidized bases) or biochemical markers (e.g. plasma antioxidants) as putative indicators of carcinogenic risk is used increasingly to provide a focused and mechanistic approach to the study of diet, health and disease. Using this approach, Pool-Zobel showed that endogenous DNA strand breakage was reduced in human lymphocytes isolated from subjects given supplemental vegetable juice (tomato juice with lycopene, carrot juice with β-carotene and α-carotene and spinach with lutein in water or milk) for 2 weeks each.[23] Moreover, using the endonuclease III-modified comet assay, endogenous oxidative base damage significantly decreased following ingestion of carrot juice.[23] Conversely, faecal water prepared from volunteers fed a diet low in fruit and vegetables but high in animal fat, meat and sugar was genotoxic to human colon cells in culture. DNA strand breakage was 2-fold higher in HT29 cells incubated with faecal water from subjects fed the deficient diet compared with subjects fed a diet enriched in vegetables.[24] Prolonged supplementation (80 days) with a commercially available fruit and vegetable extract (JuicePlus) decreased DNA strand breakage in elderly volunteers.[25] Moreover, daily supplementation (20 weeks) with an antioxidant cocktail of vitamin C (100 mg), betacarotene (25 mg), and vitamin E (280 mg) significantly reduces endogenous oxidized pyrimidines.[26] Lymphocytes were more resistant to oxidative stress *ex vivo*.[26] Vitamin E also protects against oxidative base damage induced by excess dietary polyunsaturated fatty acids (PUFAs). PUFAs (5% in the diet) were associated with low levels of oxidized pyrimidines and increased resistance to oxidant-induced strand breakage (hydrogen peroxide). However, after 4 weeks on a high PUFA diet (15%), these biomarkers increased significantly. Supplementation with vitamin E (80 mg/day) prevented the genotoxic effects of high PUFA intake.[27] The carotenoid, lycopene, also inhibits oxidant-induced DNA damage. Lymphocytes isolated from human volunteers fed a diet supplemented with lycopene (16.5 mg in tomato puree) for 21 days were more resistant to *ex vivo* hydrogen peroxide treatment compared with untreated controls.[28] In a recent intervention study by Boyle *et al.*,[29] ingestion of a flavonoid-rich onion meal (quercetin) following a low flavonoid diet, caused a significant decrease in endogenous DNA damage (breaks and oxidized pyrimidines) and in hydrogen peroxide-mediated lymphocyte DNA strand breakage after only a few hours.[29] Moreover, flavonoid-mediated cytoprotection was also seen as a decrease in urinary levels of the oxidized purine, 8-hydroxydeoxyguanosine.[29] In a similar study with Type-2 diabetics who exhibit compromised antioxidant defences and elevated oxidative stress, flavonoid supplementation (2 weeks of a low flavonoid diet supplemented with added onions and tea) enhanced lymphocyte resistance to oxidation *ex vivo*.[30]

3.2. *In Vitro* Cell Studies

In vitro studies, either using cultured human cell lines which retain specific differentiated characteristics or isolated primary human cells *ex vivo*, provide important information on the mechanisms of cytoprotection of individual dietary components. Quercetin, (10 μM), substantially inhibits hydrogen peroxide-induced DNA strand breakage (25%) in human lymphocytes *ex vivo*.[31] At a higher concentration (279 μM), approximately 80% of oxidative damage is abolished.[32]

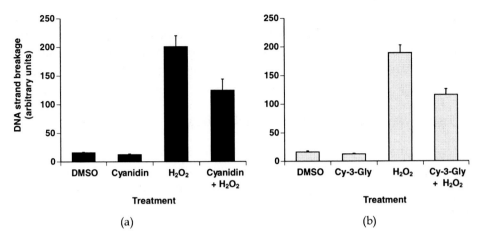

Fig. 2. **Anthocyanin-mediated cytoprotection in human colon epithelial cells (HCEC).** HCEC were exposed to 50 µM cyanidin (a) or 50 µM cyanidin-3-glycoside for 4 h at 37°C before treatment with hydrogen peroxide (100 µM, 5 min on ice). Cells were isolated and DNA strand breaks measured using the comet assay. Results are mean +/– SEM for $n = 8$.

Cytoprotection is dependent upon flavonoid structure and degree of conjugation.[31, 32] Green tea catechins protect human Jurkat T-lymphocytes against Fe^{2+}-induced DNA strand breakage.[33] Epigallocatechin gallate, one of the main flavonoids found in green tea, inhibits hydrogen peroxide- and SIN-1- (a peroxynitrite generator) mediated genotoxicity.[34]

Several studies investigating nutritional cytoprotection in relation to colorectal cancer have used human colon cells as targets. The Caco-2 cell line, derived from a human cancer, retains many specialized normal enterocyte/colonocyte cell functions. The flavonoids, quercetin and myricetin, inhibit oxidative DNA strand breakage in Caco-2 cells. Conversely, quercetin did not alter the level of oxidized pyrimidines following peroxide treatment.[35] Hydrogen peroxide-induced strand breakage is markedly decreased in normal human colonocytes (HCEC) following pretreatment with the anthocyanidin cyanidin and the anthocyanin cyanidin-3-glycoside [Figs. 2(a) and (b)]. Pool-Zobel *et al.* has presented similar findings in HT29 colon cancer cells incubated with several pure anthocyanidins or with anthocyanin-rich fruit extracts.[36] While strand breakage was reduced in this model system, oxidative base damage was unaffected.[36]

While low levels of dietary antioxidants may reduce DNA damage, at higher concentrations they can be genotoxic. Lycopene and betacarotene (10 µM) induce oxidative DNA damage in HT29 colon cells,[37] while quercetin at concentrations above 50 µM is a powerful genotoxin to Caco-2 cells.[38] Curcumin, a common spice, has also been shown to induce DNA damage in lymphocytes and gastric mucosal cells.[39]

These studies highlight the advantages of using *in vitro* mechanistic studies to determine the optimum concentrations of cytoprotective agents. While certain dietary components may be highly effective cytoprotectants at low concentration, this effect may be lost or even reversed at higher, non-physiological doses.

4. Future Applications and Conclusions

One of the most exciting and novel modifications in the comet assay is emerging. That is the use of immunostaining or fluorescent *in situ* hybridization (FISH) combined with single cell gel electrophoresis for the detection of highly specific types of DNA lesions, damage to individual chromosomes and even damage and repair within individual genes. By incorporating CY3-labeled monoclonal antibodies to cyclobutane pyrimidine dimers (CPDs) and pyrimidine[6-4] pyrimidone photoproducts ([6-4]PP) into the comet assay, Sauvaigo *et al.* were able to simultaneously detect DNA strand breakage and 2 of the major types of UV-induced lesions in fibroblasts from normal and xeroderma pigmentosa (XP) individuals.[40] Region-specific genome sensitivity can be investigated using comet-FISH, an amalgamation of a three-layer comet assay and low temperature conventional FISH.[41] Following lysis and electrophoresis the slides are denatured, dehydrated and hybridized with probe. The slides are subsequently renatured, washed and the probe and the counterstained DNA fluorescently detected. Kapp found using this technique with probes to several different human chromosomes, that certain chromosomes were highly resistant to UV irradiation with no or few hybridization signals appearing in the tail of the comet. Others, such as chromosome 8, were found with a high frequency (30%) in the tail indicating that distinct regions of the genome respond differently to damaging radiation. Moreover, using probes for different sequences within 8q, Kapp showed differential sensitivity to UV damage even within the long arm of chromosome 8.[41] This area has advanced even further and it is now possible to measure strand breakage and repair in response to an oxidative stress within a single gene (Collins *et al.*, in preparation). The rate of repair in individual genes may vary e.g. transcribed genes are repaired preferentially compared with non-transcribed genes. Using comet-FISH, Collins and coworkers can measure damage and repair in specific genes (rodent and human alkyltransferase and dihydrofolate reductase). This is outlined in Fig. 3. Different oligonucleotide probes hybridize with the two ends of a target gene in the DNA of a comet. The ends are differentially labeled with either biotin or FITC. Following amplification with appropriate antibodies and fluorescent tags the two ends of the chromosome appear either as red spots (Texas red) or yellow-green spots (fluorescein) against the DAPI background. This is represented in the top panel of Fig. 3 where both alleles of a gene are double-labeled (hatched or black). Both labels in the head of

Gene-specific DNA damage and repair

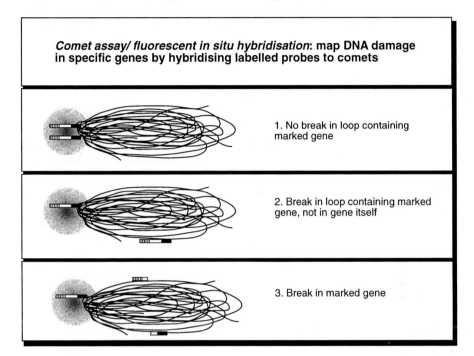

Fig. 3. **The comet assay together with fluorescent *in situ* hybridization (FISH) to detect gene-specific DNA damage and repair.** Modified from Collins *et al.* (unpublished).

the comet indicate little or no damage to or around the gene. If the 2 labels are in the comet tail, but are detected at their normal distance apart (Fig. 3, middle panel) a break near the gene has released the supercoiling of the DNA and the loop of DNA-containing the gene extends into the comet tail. If there is a break within the gene both labels will again appear in the tail but will be separated at a distance unrelated to the length of the gene (Fig. 3, bottom panel). Differential repair can be observed as the rate with which the label (hybridized probe) is restored to the comet head relative to the movement of total DNA from tail to head (DAPI). While still in its infancy, this technique will have a profound impact on the way that DNA repair is detected.

5. Conclusion

Since its conception the comet assay has evolved from a relatively simple procedure for measuring strand breaks in single cells to a versatile and capable technique for human and environmental biomonitoring. In addition, it is proving to be highly effective in determining the complex mechanisms involved in nutritional genotoxicity and cytoprotection both *in vitro* and *in vivo*.

Acknowledgments

The author would like to thank Professor Andrew Collins and Dr. Eva Horvathova for sharing their thoughts on the FISH-COMET assay, Ms Pat Bain for preparing the graphics so expertly, and to acknowledge the financial support of the Scottish Executive Rural Affairs Department and the World Cancer Research Fund.

HCEC were a gift from Dr. E. Offord (Nestle, Lausanne, Switzerland).

References

1. Ostling, O. and Johanson, K. J. (1984). Microelectrophoretic study of radiation-induced DNA damages in individual mammalian cells. *Biochem. Biophys. Res. Commun.* **123**: 291–298.
2. Angelis, J. J., McGuffy, M., Menke, M. and Schubert, I. (1999). Studies of DNA repair in various plants using the comet assay. *Neoplasma* **46(suppl)**: 72–73.
3. Brooks, R. A. and Winton, D. J. (1996). Determination of spatial patterns of DNA damage and repair in intestinal crypts by multicell gel electrophoresis. *J. Cell Sci.* **109**: 2061–2068.
4. Singh, N. P., McCoy, M. T., Tice, R. R. and Schneider, E. L. (1988). A simple technique for quantitation of low levels of DNA damage individual cells. *Exp. Cell Res.* **175**: 184–191.
5. Collins, A. R., Duthie, S. J. and Dobson, V. L. (1993). Direct enzymic detection of endogenous oxidative base damage in human lymphocyte DNA. *Carcinogenesis* **14**: 1733–1735.
6. Collins, A. R., Collins, B. H., Dusinska, M. and Angelis, K. (1999). Three novel applications of the comet assay. *Neoplasma* **46(suppl)**: 82.
7. Duthie S. J. and McMillan. P. (1997). Uracil misincorporation in human DNA detected using single cell gel electrophoresis. *Carcinogenesis* **18**: 1709–1714.
8. Olive, P. L., Banath, J. P. and Durand, R. E. (1990). Heterogeneity in radiation-induced DNA damage and repair in tumour and normal cells measured using the "comet "assay. *Radic. Res.* **122**: 86–94.
9. Tice, R. R., Strauss, G. H. S. and Peters W. P. (1992). High dose combination alkylating agents with autologous bone marrow support in patients with breast cancer: preliminary assessment of DNA damage in individual peripheral blood lymphocytes using the single cell gel electrophoresis assay. *Mutat. Res.* **271**: 101–113.

10. Somorovska. M., Szabova, E., Vodicka, P., Tulinska, J., Barancokova, M., Fabry, R., Liskova, A., Riegerova, Z., Petrovska, H., Kubova, J., Rausova, K., Dusinska, M. and Collins A. (1999). Biomonitoring *f* genotoxic risk in workers in a rubber factory: comparison of the comet assay with cytogenetic methods and immunology. *Mutat. Res.* **45**: 181–192.

11. Wojewodzka, M., Kriszewski, M., Iwanenko, T., Collins, A. R. and Szumile, I. (1998). Application of the comet assay for monitoring DNA damage in workers exposed to chronic low-dose irradiation 1: strand breakage. *Mutat. Res.* **416**: 21–35.

12. Hellman, B., Friis, L., Vaghef, H. and Edling, C. (1999). Alkaline single cell gel electrophoresis and human biomonitoring for genotoxicity: a study on subjects with residential exposure to radon. *Mutat. Res.* **442**: 121–132.

13. Zhu, C. Q., Lam, T. H., Jiang, C. Q., Wei, B. X., Lou, X., Liu, W. W., Lao, X. Q. and Chen, Y. H. (1999). Lymphocyte DNA damage in cigarette factory workers measured by the comet assay. *Mutat. Res.* **444**: 1–6.

14. Frenzelli, G., Scaracelli, V., Taddei, F. and Nigro, M. (1999). Adaptation of SCGE for monitoring marine ecosystems. *Neoplasma* **46(suppl)**: 6–7.

15. Takahashi. M., Takahashi, H., Kanai, Y., Schultz, R. M. and Okano, A. (1999). Assessment of DNA damage in individual hamster embryos by comet assay. *Mol. Reprod. Dev.* **54**: 1–7.

16. Tatemoto, H., Sakurai, N. and Muto, N. (2000). Protection of porcine oocytes against apoptotic cell death caused by oxidative stress during in vitro maturation: role of cumulus cells. *Biol. Reprod.* **63**: 805–810.

17. McKelvey-Martin, V. J., Green, M. H. L., Schmezer, P., Pool-Zobel, B. L., De Meo, M. P. and Collins, A. (1993). The single cell gel electrophoresis assay (comet assay): a European review. *Mutat. Res.* **288**: 47–63.

18. Fairbairn, D. W., Olive, P. L. and O'Neill, K. L. (1995). The comet assay: a comprehensive review. *Mutat. Res.* **339**: 37–50.

19. Beckman, K. B. and Ames, B. N. (1998). The free radical theory of aging matures. *Physiol. Rev.* **78**: 547–581.

20. Krinsky, N. I. (1992). Mechanism of action of biological antioxidants. *Proc. Soc. Exp. Biol. Med.* **299**: 248–254.

21. Byers, T. and Perry, G. (1992). Dietary carotenes, vitamin C and vitamin E as protective antioxidants in human cancers. *Ann. Rev. Nutri.* **12**: 139–159.

22. Hertog, M. G. L., Kromhout, D., Aravanis, C., Blackburn, H., Buzina, R., Fidanza, F., Giampaoli, S., Jansen, A., Menotti, A. and Nedeljkovic, S. (1995). Flavonoid intake and long-term risk of coronary heart disease. *Arch. Int. Med.* **27**: 381–386.

23. Pool-Zobel, B. L., Bub, A., Muller, H., Wollowski. I. and Rechkemmer, G. (1997). Consumption of vegetables reduces genetic damage in humans: first results of a human intervention trial with carotenoid-rich foods. *Carcinogenesis* **18**: 1847–1850.

24. Reiger, M. A., Parlesak, A., Pool-Zobel, B. L., Rechkemmer, G. and Bode, C. (1999). A diet high in fat and meat but low in dietary fibre increases the genotoxic potential of "faecal water". *Carcinogenesis* **20**: 2311–2316.

25. Smith, M. J., Inserra, P. F., Watson, R. R., Wise, J. A. and O'Neill, K. L. (1999). Supplementation with fruit and vegetable extracts may decrease DNA damage in the peripheral lymphocytes of an elderly population. *Nutri. Res.* **19**: 1507–1518.

26. Duthie, S. J., Ma, A.-G., Ross, M. A. and Collins, A. R. (1996). Antioxidant supplementation decreases oxidative DNA damage in human lymphocytes. *Cancer Res.* **56**: 1291–1295.

27. Jenkinson, A. McE., Collins, A. R., Duthie, S. J., Whale, K. W. J. and Duthie, G. G. (1999). The effect of increased intakes of polyunsaturated fatty acids and vitamin E on DNA damage in human lymphocytes. *FASEB J.* **13**: 2138–2142.

28. Riso, P., Pinder, A., Santangelo, A. and Porrini, M. (1999). Does tomato consumption effectively increase the resistance of lymphocyte DNA to oxidative damage? *Am. J. Clin. Nutri.* **69**: 712–718.

29. Boyle, S. P., Dobson, V. L., Duthie, S. J., Kyle, J. A. M. and Collins, A. R. (2000). Absorption and DNA protective effects of flavonoid glycosides from an onion meal. *Eur. J. Nutri.* **39**: 213–223.

30. Lean, M. E. J., Noroozi, M., Kelly, I., Burns, J., Talwar, D. and Sattar, N. (1999). Dietary flavonols protect diabetic human lymphocytes against oxidative damage to DNA. *Diabetes* **48**: 176–181.

31. Duthie, S. J., Collins, A. R., Duthie, G. G. and Dobson, V. L. (1997). Quercetin and myricetin protect against hydrogen peroxide-induced DNA damage (strand breaks and oxidized pyrimidines) in human lymphocytes. *Mutat. Res.* **393**: 223–231.

32. Noroozi, M., Angerson, W. J. and Lean, M. E. J. (1998). Effects of flavonoids and vitamin C on oxidative DNA damage to human lymphocytes. *Am. J. Clin. Nutri.* **67**: 1210–1218.

33. Erba, D., Riso, P., Colombo, A. and Testolin, G. (1999). Supplementation of Jurkat T cells with green tea extract decreases oxidative damage due to iron treatment. *J. Nutri.* **129**: 2130–2134.

34. Johnson, M. K. and Loo, G. (2000). Effects of epigallocatechin gallate and quercetin on oxidative damage to cellular DNA. *Mutat. Res.* **459**: 211–218.

35. Duthie, S. J. and Dobson, V. L. (1999). Dietary flavonoids protect human colonocyte DNA from oxidative attack in vitro. *Eur. J. Nutri.* **38**: 28–34.

36. Pool-Zobel, B. L., Bub, A., Schroder, N. and Rechkemmer, G. (1999). Anthocyanins are potent antioxidants in model systems but do not reduce endogenous oxidative DNA damage in human colon cells. *Eur. J. Nutri.* **38**: 227–234.

37. Lowe, G. M., Booth, L. A., Young, A. J. and Bilton, R. F. (1999). Lycopene and betacarotene protect against oxidative damage in HT29 cells at low concentrations but rapidly lose this capacity at higher doses. *Free Radic. Res.* **30**: 141–151.

38. Duthie, S. J., Johnson, W. and Dobson, V. L. (1997). The effects of dietary flavonoids on DNA damage (strand breaks and oxidized pyrimidines) and growth in human cells. *Mutat. Res.* **390**: 141–151.

39. Blasiak, J., Trzeciak, A., Malecka-Panas, E., Drzewoski, J., Iwanienko, T., Szumiel, I. and Wojewodzka, M. (1999). DNA damage and repair in human lymphocytes and gastric mucosa cells exposed to chromium and curcumin. *Terat. Carcinogen. Mutagen.* **19**: 19–31.

40. Sauvaigo, S., Richard, M.-J. and Cadet, J. (1999). The use of the immunological approach for the detection of DNA damage. *Neoplasma* **46(suppl)**: 97–98.

41. Rapp, A., Bock, C., Dittmar, H. and Greulich, K. O. (1999). Comet-FISH used to detect UV-sensitive regions in the whole genome and on chromosome 8. *Neoplasma* **46(suppl)**: 99–101.

Measurement of Oxidative Stress: Technology, Biomarkers and Applications

(Protein Damage)

Chapter 18

Protein Oxidation Assays

Guohua Cao

Guohua Cao • Jean Mayer USDA Human Nutrition Research Center on Aging at Tufts University, Boston, MA 02111
Tel: 617 563 5820, E-mail: howard.cao@fmr.com

1. Introduction

Considerable progress has been made during the last two decades in the study of oxidative damage to proteins. The physiological and/or pathological importance of the protein oxidation is now well recognized due largely to the demonstration of the chemical effects of reactive oxygen species on proteins[1–5] and to the discovery of the existence of an array of enzymes (proteases) which preferentially degrade oxidatively modified proteins.[6–10]

Recently, the potential role of oxidative damage to proteins in aging has attracted a lot of interest. Many reports have indicated an age-dependent increase in protein carbonyl content in mammalian models.[11–14] There are also reports showing an age-dependent loss of protease function.[11, 15] However, these issues are still controversial.[16, 17] The methodological problems related to the spectrophotometric method, widely used in assessing protein oxidation in those aging studies, have not been fully resolved.[18] Using different methods investigators, in general, failed to achieve the same results.[19, 20] In this chapter, I will review the current methodologies related to the protein carbonyl measurement and discuss their applications in studying oxidative stress in aging and aging-related diseases.

2. Protein Carbonyl Assays Using the Spectrophotometric Method

The spectrophotometric method was described in detail by Levine *et al.*[21] and has to date been the most widely used assay of protein carbonyls in animal tissues. The method is based on the reaction of the carbonyl group with 2,4-dinitrophenylhydrazine (DNPH), which can be determined at 370 (360–390) nm. Briefly, the tissue is homogenized in ice-cold Hepes buffer (pH 7.2–7.4) containing the protease inhibitors leupeptin (0.5 µg/mL), pepstatin (0.7 µg/mL), aprotinin (0.5 µg/mL), and phenylmethylsulfonyl fluoride (40 µg/mL). The soluble protein fraction is separated by centrifugation and then treated with streptomycin sulfate to remove nucleic acids. The protein sample is pipetted into two tubes and dried in a vacuum centrifuge or precipitated with trichloroacetic acid (TCA). DNPH (10 mM in 2 M HCl) is added to one tube and HCl (2 M) to the other tube. The tubes are allowed to stand at room temperature for 1 h, with vortexing every 10–15 minutes. TCA (final concentration: 10%) is added to both tubes and the pellets are recovered by centrifugation. The pellets are washed three times with ethanol-ethyl acetate (1 : 1) to remove free DNPH. The precipitated protein is redissolved in 6 M guanidine solution (pH 2.3, adjusted with trifluoroacetic acid) and the absorbance of both solutions (DNPH and HCl) is measured at 370 nm after removing any insoluble materials by centrifugation. Protein carbonyl content is calculated using a molar absorption coefficient of 22 000 $M^{-1}cm^{-1}$.

Evans *et al.*[22] and Reznick and Packer[23] recommended reaction of the protein with DNPH in solution as opposed to in pellet form whenever possible when the above spectrophotometric method is used. Also, it is recommended that protein levels be determined in the final protein solutions after all washings are finished. Other researchers used these modifications as well in the measurement of protein carbonyls.[18]

Nucleic acids also contain carbonyl groups and react with DNPH. The use of streptomycin to remove nucleic acid could be very important, but this step often seems to have been omitted, particularly in aging-related studies. Another problem often noticed with this spectrophotometric method is how to remove excess or loosely bound DNPH from reaction mixtures or derivatized proteins. Any un-removed free DNPH can cause artificial protein carbonyl signals.

We found the treatment of streptomycin could significantly reduce the nucleic acid content in rat liver protein samples. But the protein carbonyl content measured by the spectrophotometric method in the samples was too low to be reliably quantitated.[18] Dubey and his coworkers then used three to seven replicate determinations when they measured protein carbonyls in gerbil brain using the method. They implied that five measurements of protein carbonyls in each sample are necessary in order to achieve reliable results.[24] However, their data suggest otherwise.

Dubey *et al.* reported an observed within-sample standard deviation (σ_w) of 0.098 nmol/mg and an observed between-sample standard deviation (σ_b) of 0.247 nmol/mg, based on 27 samples each measured five times.[24] Since $\sigma^2_b = \sigma^2 + \sigma^2_w/m$, where σ is the between subject standard deviation and m is the number of measurements per subject, it follows that the between subject standard deviation is 0.243 nmol/mg $\{\sigma = [(0.247)^2 - (0.098)^2/5]^{1/2}\}$. Then, when $m = 1$, $\sigma_b = [(0.243)^2 + (0.098)^2/1]^{1/2} = 0.2621$; when $m = 5$, $\sigma_b = [(0.243)^2 + (0.098)^2/5]^{1/2} = 0.2470$. Therefore, using five replicates reduced between-sample standard deviation only by 5.76% $[(0.2621 - 0.2470)/0.2621]$. Dubey *et al.* correctly pointed out that five measurements can reduce the within subject variability (pure measurement error) by 55%, but the within subject variability contributes so little to overall variability that the effort would be better spent making fewer measurements on more animals.

In similar fashion, Dubey *et al.*[24] mentioned that the within-sample standard deviation is 40% of the between-sample standard deviation. However, for the purpose of determining the effect of repeated measurements on the same sample, it is not the ratio of the standard deviations that matters, but the ratio of the variances.

As we mentioned above, free DNPH and the contamination of nucleic acids could be a problem in the protein carbonyl assay. One way to avoid this problem is to employ HPLC gel filtration to separate proteins from free DNPH and nucleic acids, and measure the carbonyl content and the protein concentration spectrophotometrically on line.[25] The derivatization of proteins can be also carried out in sodium dodecyl sulfate (SDS) instead of guanidine before the separation of excess reagent by HPLC gel filtration.[25] Using this SDS-HPLC system, Cabiscol and Levine did not show a significant age-related change in the carbonyl content of carbonic anhydrase III in the rats between 2 and 18 month old.[19] This is not in agreement with the results obtained from young and old animals using the original spectrophotometric measurement.[11–14, 24]

3. Protein Carbonyl Assays Using the Immunological Method

There are several versions of immunodetection of protein carbonyls. The Western blotting reported by Levine *et al.*[25] uses monoclonal or polyclonal antibodies (e.g. mouse monoclonal IgE) to protein-bound DNPH and a labeled secondary antibody (e.g. biotinylated rat anti-mouse IgE) to the first one. These antibodies are commercially available. The samples are prepared as described for the spectrophotometric method. After derivatization by DNPH and neutralization with 2 M Tris/30% glycerol, samples are directly loaded onto the gel for electrophoresis as usual. The transfer to nitrocellulose and immunodetection (Western blotting) is carried out by standard techniques. Either colorimetric or chemiluminescent detection protocols is used to measure protein carbonyl content.

Cabiscol and Levine modified the above procedure later by transferring the proteins to nitrocellulose before performing the derivatization by DNPH. This modification was thought necessary because of the incompatibility of the strong acid and high SDS concentration with isoelectric focusing. Derivatization on nitrocellulose membranes was performed simply by immersing them in 0.2% DNPH (in 2 M HCl) for 10 min followed by a wash in 2 M HCl for another 10 minutes.[19]

Nakamura and Goto[26] independently developed an immunoblotting method similar to that of Levine *et al.* for measuring protein carbonyls.[26] Antibodies are prepared by injecting subcutaneously the H_2O_2-$FeSO_4$ oxidized and then DNPH derivatized bovine serum albumin (BSA) into rabbits every 2 weeks (three injections altogether). The antibodies are partially purified by ammonium sulfate precipitation at 33% saturation followed by affinity chromatography on a column containing DNPH-coupled Sepharose 4B. The bound antibodies are eluted with 0.1 M glycine-HCl buffer (pH 2.5) containing 0.1% Tween 20. The protein carbonyls in samples are measured using one- or two-dimensional polyacrylamide gel electrophoresis (PAGE). Sample proteins are derivatized using DNPH as described by Levine *et al.*[21] except the last wash is done with acetone. The DNPH derivatized sample protein precipitates are dissolved either in Laemmli's buffer containing 8 M urea and 5% 2-mercaptoethanol for 1-D PAGE or in 8.5 M urea solution containing 2% NP40, 5% 2-mercaptoethanol, and 2% carrier Ampholine-Pharmalite mixture for 2-D PAGE. After the 1- or 2-D PAGE, the proteins are transferred to nitrocellulose or PVDF membranes. The membranes are soaked in PBS containing 3% skim milk, 0.05% Tween 20, and 0.05% sodium azide and treated with anti-DNPH antibodies. After washing with the buffer without the antibodies, the membranes are treated with ^{125}I-Protein A. The radioactive signals are then visualized and quantified. The method can detect concentrations as low as one pmol of carbonyls.[26] Using their immunoblotting method, Nakamura and Goto determined the carbonyl content of proteins in various tissues from young (6–8 months) and old (28–34 months) male rats. No significant difference in nmol of carbonyl per mg protein content was detected between the two age groups in the brain, lung, liver, and heart. The kidney of old animals, however, contained 1.5 times more carbonyls than their young counterparts.[16]

An ELISA version of the immunological measurement of protein carbonyls was described by Winterbourn and Buss recently.[27] The principle of the ELISA method is very similar to the above methods. Sample proteins are reacted with DNPH and nonspecifically adsorbed to an ELISA plate. Unconjugated DNPH and nonprotein constituents are washed away. The adsorbed protein is incubated with a commercial biotinylated anti-DNPH antibody followed by streptavidin-linked horseradish peroxidase. Absorbances are related to a standard curve prepared for BSA containing increased proportions of HOCl-oxidized protein. The standard curves for the ELISA are linear up to 10 nmol carbonyl per mg protein. No data have been published yet regarding the use of this ELISA method in studying the relationship between protein oxidation and aging.

There were no histological markers of protein oxidation until Smith and his coworkers reported a carbonyl assay using *in situ* DNPH labeling linked to an antibody system.[28, 29] Non-aldehyde fixed (methanol : chloroform : acetic acid, 60 : 30 : 10) human hippocampal brain tissues are dehydrated through ascending ethanol and xylene and embedded in paraffin. Sections of 6 μm are prepared and placed on Silane-coated glass slides and sequentially deparaffinized in xylene and rehydrated in descending ethanol. Endogenous peroxidase is inactivated by treating with 30% H_2O_2 in methanol for 20 min to block artificial staining from the endogenous enzyme. A control that omits the H_2O_2 treatment step should be performed with each experimental paradigm, since H_2O_2 might itself, or synergistically with tissue-bound iron, create carbonyl residues. After blocking endogenous peroxidase, sections are hydrated in descending ethanol to 50 mM Tris-HCl buffer containing 0.15 M NaCl (pH 7.6) and then covered with 0.01% DNPH in 2 N HCl for one hour. After rinsing with Tris-HCl buffer followed by incubation with 10% normal goat serum to block nonspecific binding sites, the sections are incubated with a rat monoclonal antibody against DNPH coupled with a peroxidase-linked secondary antibody. The carbonyls are then chromogenically localized with 3,3'-diaminobenzidine/H_2O_2. By using this method, Smith *et al.* found striking differences between Alzheimer's hippocampal sections and age-matched controls. In tissues from Alzheimer's cases, but not in controls, carbonyls were increased in neuronal cytoplasm and nuclei of neurons and glia indicative of a generalized increase in oxidative stress in Alzheimer's.[28]

As with the spectrophotometric method, the cross reaction between protein and nucleic acids in the carbonyl measurement needs to be clearly addressed in all these immunological methods. These immunological methods may also measure the carbonyls contained in lipids or sugars that are attached to proteins.

4. Other Assays for Assessing Protein Oxidation

Levine *et al.*[21, 30] described the protein carbonyl assay using tritiated borohydride when they introduced their original spectrophotometric method using DNPH.[21] Borohydride reduces all carbonyl groups to alcohols. It also reduces Schiff bases to amines so that this method should detect carbonyl groups that have formed a Schiff base with the ε-amino group of lysine or the α-amino group of the amino-terminal residue. When borohydride is tritiated, a stable tritium label will be introduced after these reactions. Compared to the spectrophotometric method using DNPH, the reaction with tritiated borohydride is more sensitive but less specific in measuring protein carbonyls. The tritiated borohydride method never gained popularity in studying protein oxidation and aging.

Oxidative modification of proteins gives rise to a large number of covalent modifications, many of which might serve as qualitative or quantitative markers of oxidative damage to proteins. Histidine frequently functions as a ligand for divalent cations at metal-binding sites in proteins, precisely where metal-catalyzed oxidative modifications will occur. 2-oxohistidine is known to be a product of metal-catalyzed oxidation and thus may serve as a marker of protein oxidation. The measurement of 2-oxohistidine has been described by Alewisch and Levine.[30] Oxidation of phenylalanine residues can produce *o*-tyrosine and *m*-tyrosine. When hydroxyl radical cross-links tyrosine residues, *o*,*o'*-dityrosine is formed. Oxidation of tyrosine by reactive nitrogen species and HOCl or HOCl related reactive species produces 3-nitrotyrosine and 3-chlorotyrosine, respectively. Sensitive and quantitative assays have been reported by Heinecke *et al.* using GC/MS[31] and Kettle using HPLC[32] for measuring these distinct amino acid oxidation products in proteins and tissues. Oxidation of arginine and proline residues can produce γ-glutamyl semialdehyde, which, on reduction and acid hydrolysis, forms 5-hydroxy-2-amino valeric acid (HAVA). A GC/MS technique has been described by Ayala and Cutler for measuring HAVA in proteins and tissues.[33, 34] Using mice of 1 to 30 months of age, Ayala and Cutler failed to detect any significant difference in either HAVA or carbonyl content.[34] The carbonyls were measured by the tritiated borohydride method. However, on using human liver samples a significant decrease from age 16 to 40 years and then an increase to 85 years of age was found for both HAVA and carbonyl group.[34]

5. Summary

The spectrophotometric method using DNPH has to date been the most widely used assay of protein carbonyls in animal tissues, particularly in showing the age-related oxidative damage in proteins. The immunological methods are gaining some popularity due to the problems related to the original spectrophotometric method using DNPH. The assays based on the measurement of amino acid oxidation products have been developed recently for the investigation of oxidative damage of proteins. It appears that more work should be done by using different technologies, other than the spectrophotometric method, in confirming the widely reported increase of protein oxidation with aging.

References

1. Davies, K. J. A. (1987). Protein damage and degradation by oxygen radicals. I. General aspects. *J. Biol Chem.* **262**: 9895–9901.
2. Davies, K. J. A., Delsignore, M. E. and Lin, S. W. (1987). Protein damage and degradation by oxygen radicals. II. Modification of amino acids. *J. Biol. Chem.* **262**: 9902–9907.
3. Davies, K. J. A. and Delsignore, M. E. (1987). Protein damage and degradation by oxygen radicals. III. Modification of secondary and tertiary structure. *J. Biol. Chem.* **262**: 9908–9913.
4. Davies, K. J. A., Lin, S. W. and Pacifici, R. E. (1987). Protein damage and degradation by oxygen radicals. IV. Degradation of denatured protein. *J. Biol. Chem.* **262**: 9914–9920.
5. Stadtman, E. R. (1990). Metal ion-catalyzed oxidation of proteins: biochemical mechanism and biological consequences. *Free Radic. Biol. Med.* **9**: 315–325.
6. Levine, R. L., Oliver, C. N., Fulks, R. M. and Stadtman, E. R. (1981). Turnover of bacterial glutamine synthetase: oxidative inactivation precedes proteolysis. *Proc. Natl. Acad. Sci. USA* **78**: 2120–2124.
7. Rivett, A. J. (1985). Preferential degradation of the oxidatively modified form of glutamine synthetase by intracellular mammalian proteases. *J. Biol. Chem.* **260**: 300–305.
8. Rivett, A. J. (1985). Purification of a liver alkaline protease which degrades oxidatively modified glutamine synthetase. Characterization as a high molecular weight cysteine proteinase. *J. Biol. Chem.* **260**: 12 600–12 606.
9. Pacifici, R. E., Salo, D. C. and Davies, K. J. A. (1989) Macroxyproteinase (M.O.P.): a 670 kDa proteinase complex that degrades oxidatively denatured proteins in red blood cells. *Free Radic. Biol. Med.* **7**: 521–536.
10. Grune, T., Reinheckel, T., Joshi, M. and Davies, K. J. A. (1995). Proteolysis in cultured liver epithelial cells during oxidative stress. Role of the multicatalytic proteinase complex, proteasome. *J. Biol. Chem.* **270**: 2344–2351.

11. Carney, J. M., Starke-Reed, P. E., Oliver, C. N., Landum, R. W., Cheng, M. S., Wu, J. F. and Floyd, R. A. (1991). Reversal of age-related increase in brain protein oxidation, decrease in enzyme activity, and loss in temporal and spatial memory by chronic administration of the spin-trapping compound N-tert-butyl-alpha-phenylnitrone. *Proc. Natl. Acad. Sci. USA* **88**: 3633–3636.
12. Stadtman, E. R. (1992). Protein oxidation and aging. *Science* **257**: 1220–1224.
13. Sohal, R. S. and Weindruch, R. (1996). Oxidative stress, caloric restriction, and aging. *Science* **273**: 59–63.
14. Berlett, B. S. and Stadtman, E. R. (1997). Protein oxidation in aging, disease, and oxidative stress. *J. Biol. Chem.* **272**: 20 313–20 316.
15. Agarwal, S. and Sohal, R. S. (1994). Aging and proteolysis of oxidized proteins. *Arch. Biochem. Biophys.* **309**: 24–28.
16. Goto, S. and Nakamura A. (1997). Age-associated, oxidatively modified proteins: a critical evaluation. *Age* **20**: 81–89.
17. Goto S. (1999). Commentary on AProtein carbonyl accumulation in aging dauer formation-defective (daf) mutants of *Caenorhabditis elegans@. J. Gerontol.* **54A**: B52–B53.
18. Cao, G. and Cutler, R. G. (1995). Protein oxidation and aging. I. Difficulties in measuring reactive protein carbonyls in tissues using 2,4-dinitrophenyl-hydrazine. *Arch. Biochem. Biophys.* **320**: 106–114.
19. Cabiscol, E. and Levine, R. L. (1995). Carbonic anhydrase III. Oxidative modification in vivo and loss of phosphatase activity during aging. *J. Biol. Chem.* **270**: 14 742–14 747.
20. Goto, S., Nakamura, A., Radak, Z., Nakamoto, H., Takahashi, R., Yasuda, K., Sakurai, Y. and Ishii, N. (1999). Carbonylated proteins in aging and exercise: immunoblot approaches. *Mech. Ageing Dev.* **107**: 245–253.
21. Levine, R. L., Garland, D., Oliver, C. N., Amici, A., Climent, I., Lenz, A. G., Ahn. B. W., Shaltiel, S. and Stadtman, E. R. (1990). Determination of carbonyl content in oxidatively modified proteins. *Meth. Enzymol.* **186**: 464–478.
22. Evans, P., Lyras, L. and Halliwell B. (1999). Measurement of protein carbonyls in human brain tissue. *Meth. Enzymol.* **300**: 145–157.
23. Reznick, A. Z. and Packer, L. (1994). Oxidative damage to proteins: spectro-photometric method for carbonyl assay. *Meth. Enzymol.* **233**: 357–363.
24. Dubey, A., Forster, M. J. and Sohal, R. S. (1995). Effect of the spin-trapping compound B tert-butyl-a-phenylnitrone on protein oxidation and life span. *Arch. Biochem. Biophys.* **324**: 249–254.
25. Levine, R. L., Williams, J. A., Stadtman, E. R. and Shacter, E. (1994). Carbonyl assays for determination of oxidatively modified proteins. *Meth. Enzymol.* **233**: 347–357.
26. Nakamura, A. and Goto, S. (1996). Analysis of protein carbonyls with 2,4-dinitrophenyl hydrazine and its antibodies by immunoblot in two-dimensional gel electrophoresis. *J. Biochem. (Tokyo)* **119**: 768–774.

27. Winterbourn, C. C. and Buss, I. H. (1999). Protein carbonyl measurement by enzyme-linked immunosorbent assay. *Meth. Enzymol.* **300**: 106–111.

28. Smith, M. A., Perry, G., Richey, P. L., Sayre, L. M., Anderson, V. E., Beal, M. F., Kowall, N. (1996). Oxidative damage in Alzheimer's. *Nature* **382**: 120–121.

29. Smith, M. A., Sayre, L. M., Anderson, V. E., Harris, P. L., Beal, M. F., Kowall, N. and Perry, G. (1998). Cytochemical demonstration of oxidative damage in Alzheimer disease by immunochemical enhancement of the carbonyl reaction with 2,4-dinitrophenylhydrazine. *J. Histochem. Cytochem.* **46**: 731–735.

30. Lewisch, S. A. and Levine, R. L. (1999). Determination of 2-oxohistidine by amino acid analysis. *Meth. Enzymol.* **300**: 121–124.

31. Heinecke, J. W., Hsu, F. F., Crowley, J. R., Hazen, S. L., Leeuwenburgh, C., Mueller, D. M., Rasmussen, J. E. and Turk, J. (1999). Detecting oxidative modification of biomolecules with isotope dilution mass spectrometry: sensitive and quantitative assays for oxidized amino acids in proteins and tissues. *Meth. Enzymol.* **300**: 124–144.

32. Kettle, A. J. (1999). Detection of 3-chlorotyrosine in proteins exposed to neutrophil oxidants. *Meth. Enzymol.* **300**: 111–120.

33. Ayala, A. and Cutler, R. G. (1996). The utilization of 5-hydroxyl-2-amino valeric acid as a specific marker of oxidized arginine and proline residues in proteins. *Free Radic. Biol. Med.* **21**: 65–80.

34. Ayala, A. and Cutler, R. G. (1996). Comparison of 5-hydroxy-2-amino valeric acid with carbonyl group content as a marker of oxidized protein in human and mouse liver tissues. *Free Radic. Biol. Med.* **21**: 551–558.

Chapter 19

Effect of Oxidative Stress on Protein Synthesis

Antonio Ayala and Alberto Machado*

Antonio Ayala and **Alberto Machado** • Departamento de Bioquimica, Bromatologia, Toxicologia y Medicina Legal, Facultad de Farmacia, Universidad de Sevilla, 41012-Sevilla, Spain

*Corresponding Author. E-mail: machado@us.es

1. Summary

In this report we show the *in vivo* effects of oxidative stress on protein synthesis in rat liver. For this purpose, two experimental groups of animals were used in which the oxidative stress was increased: young rats treated with an oxidant compound (cumene hydroperoxide –CH), and old rats. The approach used was to determine which step of protein synthesis is most affected and the molecular changes in the main protein catalyzing that step. The data provided below shows that elongation is the step most affected in both groups and that the changes in the elongation factor-2 (EF-2) could be responsible for the impairment of translation caused by oxidative stress and aging.

2. Introduction

De novo synthesis of proteins in the cell is a complex process involving multiple steps and numerous components, some of which are an important target for regulating the translation.[1-4] This biochemical process is an integral part of the gene expression pathway and makes important contributions to all cellular functions.

In young organisms, protein synthesis is performed optimally and the process is continuously regulated depending on the metabolic condition. However, the rate of total protein synthesis declines with age.[5-12] The implications and consequences of slower rates of protein synthesis are manifold: including a decrease in the availability of enzymes for the maintenance, repair, and normal metabolic functioning of the cell; an inefficient removal of inactive, abnormal, and damaged macromolecules in the cell; the inefficiency of the intracellular and intercellular signaling pathway.

Different studies have been carried out to try to determine the mechanism responsible for slowing down the rate of translation in old organisms.[13-21] Among them, the study performed in our laboratory[20, 21] has focused on the possible inhibitory role of oxidative stress on protein synthesis. The starting point of this research was based on the fact that (1) many proteins become oxidatively modified in aging,[22-25] and (2) oxidant compounds have been described as inhibiting the overall rate of protein synthesis in cultured cells.[26] This last study was supported by incorporation of labeled amino acids into the protein but no attempt was made to determine the molecular basis of such differences. Therefore, the question was whether or not any of the components of the translation machinery could be modified by oxidative stress.

With the aim of examining the possibility that the mechanism of protein synthesis inhibition caused by an oxidant compound and aging were similar, the following experimental approach was used: firstly, we have studied which of the individual steps in polypeptide synthesis is most affected in the livers of both young rats treated with CH and old rats. This was addressed, by determining

Table 1. **Effect of CH-treatment and aging on MDA level and carbonyl group content of proteins.** MDA was determined by thiobarbituric acid.[31] Carbonyl groups in total protein were determined using radioactive sodium borohydride.[30] The samples (liver 10 000 xg supernatant and microsome suspension) were obtained as described previously.[20, 21]

Age (months)	MDA (nmol/mg prot)		Carbonyl Group (nmol/mg prot)	
	Supernatant	Microsomes	Supernatant	Microsomes
3	22.4 ± 2.3	28.8 ± 10.5	155.2 ± 8.1*	408.5 ± 105.8
3 + CH	53.2 ± 17.9*	97.4 ± 9.6*	249.0 ± 10.7*	831.9 ± 37.8*
24	59.2 ± 7.2*	77.0 ± 6.9*	274.7 ± 10.9*	637.5 ± 47.6

*Significantly different from level in control rats (multifactor ANOVA followed by Tukey's test, $p < 0.05$).

in vivo the ribosome transit time (elongation time) required for the synthesis of an average half-length of a nascent polypeptide chain,[27] and the ribosomal state of aggregation,[28, 29] which is reflected by the polyribosomal profiles. Second, a study of the most relevant protein in the affected step was made in both experimental groups to compare its possible modifications under both conditions.

3. Oxidative Stress Parameter Levels in Young Rats Treated with CH and During Aging

The determination of malondialdehyde (MDA) and carbonyl groups are common assays to study lipid and protein oxidative damage, respectively.[30, 31] The evidence that oxidative stress is increased by CH treatment and during aging is shown by the increased levels of MDA and carbonyl group content in liver extracts (Table 1). CH is an oxidant, which has been used to assess the effects of free radicals and reactive oxygen intermediates on various biological molecules.[32–34] CH treatment produces a significant increase in MDA of 2.7- and 3.4-fold in 10 000 xg supernatant and microsome fraction, respectively (Table 1). A significant increase of carbonyl groups content of liver proteins is also found in both subcellular fractions (Table 1). A similar increase in these parameters is observed in old animals.

4. The Step of Protein Synthesis Most Affected by CH Treatment and Aging

The determination of which step of protein synthesis is most affected by CH-treatment and aging can be assessed by measuring both the polypeptide chain completion time (Tc) and the number of ribosomes actively engaged in protein synthesis.[27-29] Tc is measured by *in vivo* labeling with amino acid precursors followed by separation of released proteins from ribosome-bound (nascent) protein by centrifugation. The method is not sensitive to variations in amino acid pools or specific activities. The number of active ribosomes is determined by sucrose density gradient centrifugation, which separates active ribosomes (polysomes) from nontranslating 80S particles and ribosome subunits.

4.1. Effect of CH-treatment and Aging on Polypeptide Chain Completion Time (Tc)

The time for assembly and release of polypeptide chains (Tc) in the process of protein synthesis in rat liver *in vivo* can be determined by previously described methods.[27] It is assumed that after injection, the radioactive amino acid will meet the ribosomes in the middle of the translation of a mRNA of average size. When a full cycle is completed, the whole peptide on the ribosomes will be labeled, while the chains that have been terminated and released will be only half-labeled. Thus, the ratio between the radioactivity incorporated into nascent peptides in the polyribosomes (Pn) and that incorporated into total peptides (Pt) will be reduced by 50%. After another cycle has elapsed, the pool of released chains increases by one full unit, and the value Pn/Pt should have been reduced to 25%. This seems to be true to the extent that the specific radioactivity of the precursor does not change over the 2-min period.[27, 29] The time required to reduce Pn/Pt from 50 to 25% is taken as the average polypeptide chain completion time, Tc, which can be obtained directly from the slope of a plot of Pn/Pt (Fig. 1). Table 2

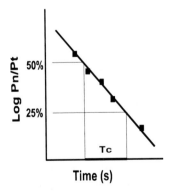

Fig. 1. **Determination of hepatic polypeptide chain completion time (Tc).** The abscissa indicates the time elapsed after the injection of radioactive amino acid. The time required to reduce Pn/Pt from 50 to 25% is taken as the average polypeptide chain completion time, Tc.

Table 2. **Effect of CH-treatment and aging on hepatic polypeptide chain completion time (Tc).** Tc was calculated as described in the text.

Rat Group	Polipeptide Chain Completion Time, Tc (s)
3 months old	60
3 months old + CH	101
12 months old	81
24 months old	169

*Significantly different from level in control rats (multifactor ANOVA followed by Tukey's test, $p < 0.05$).

shows the values of hepatic Tc versus time after injection of the tracer in control, CH-treated rats and old rats. As seen, Tc is increased in CH-treated rats, 12-months and 24-months-old rats (68, 35 and 182%, respectively) with respect to control rats, which would cause an important drop in protein production per unit of time if all other factors were equal. Although these results indicate that CH-treatment and aging affects the elongation and/or termination steps, the possible implication of the initiation step should be addressed by studying the ribosomal pattern of aggregation.

4.2. Effect of CH-treatment and Aging on Hepatic Polyribosomal Profiles

Tc is an expression of the rate of peptide chain elongation and termination. It is independent of the number of ribosomes engaged in the process and it should not be affected by variations in the initiation step.[27, 35] However, in order to determine whether the increase of Tc was accompanied by a significant alteration of the initiation step, the ribosomal state of aggregation should be determined by studying the hepatic polyribosomal profiles.[28, 29]

Under normal conditions, initiation and elongation are synchronized in such a way that any perturbation in only one of them would produce variations in the ribosomal pattern of aggregation.[28, 29, 35] Thus, inhibition of peptide chain initiation would produce an accumulation of monomeric ribosomes and polyribosomal breakdown. When only the elongation process is inhibited (or if it is inhibited to a greater extent than initiation), an increased proportion of polyribosomes and a depletion of the monomeric ribosome pool is obtained.[29] No change of the ribosomal aggregation patterns is found if elongation and initiation are completely inhibited. In both CH-treated and old rat livers, the polyribosomes displayed a higher state of aggregation (more ribosomes per mRNA) with respect to young rats.[20, 21] The ratio polysomes/monomers for the studied groups are shown in Table 3. Concerning the ribosome aggregation in rodent liver it is noteworthy that some studies have shown no change in the polyribosomal profile[36–38] and

Table 3. **Effect of CH-treatment and aging on hepatic ratio polysomes/monomers.** The polyribosomal profiles were obtained as described previously.[27-29]

Rat Group	Polysomes/Monomers	% Control
3 months	17.25 ± 2.9	100
3 months + HC	$29.30 \pm 2.8^*$	169
12 months	$35.80 \pm 1.9^*$	207
24 months	$108.60 \pm 4.0^*$	629

*Significantly different from level in control rats (multifactor ANOVA followed by Tukey´s test, $p < 0.05$).

others found a decrease.[39-42] The results of some of these studies have been revised by Richardson.[5] Because several factors affect the measurement of the polyribosomal profile, we speculate that the cause of this discrepancy could be related to different factors, such as the use of a proper liver cell sap, to inhibit the ribonuclease activity, the step of gradient unloading, the animal strain, time of feeding before the animal was killed, etc.

On the basis of these finding, these results could indicate that changes in peptide chain elongation may be an important mechanism in altering the rate of protein synthesis in both studied groups. The possibility that the chain release mechanism may be rate-limiting (and hence, modulated by CH and aging) cannot be excluded by the model used for determination of Tc.

Although several studies indicate that the initiation step may be the crucial regulatory step during protein synthesis in various biological systems, this does not appear to be the case during aging and CH-treatment. This agrees with previous reports,[9, 15] which also point towards the slowing down of the elongation step becoming rate-limiting for protein synthesis during aging.

These results also suggest that some degree of similarity could exist in the mechanisms underlying the decline of protein synthesis observed in both aging and treatment with oxidants, possibly due to that some component of the elongation step could be affected by some reactive oxygen species.

5. Mechanism Underlying the Alterations in the Elongation Step in CH-treated and Old Rats: Modifications in EF-2

Once it was determined that CH and aging act preferentially on the elongation step, the next questions to investigate were which component of elongation is altered, what kind of alterations occurs and whether the increased oxidative stress could play some role in causing these changes.

EF-2, which catalyzes the translocation of peptidyl-tRNA from the site A to the site P of the ribosome, was chosen for the following reasons: (1) The determination of its activity can be easily performed;[43, 44] (2) It is a relatively abundant protein in the cell,[21] which facilitates its purification; (3) EF-2 contains a histidine at the position 715, which is essential for its activity.[45] This amino acid is one of the most sensitive to oxidations;[22, 46] and (4) Several protein synthesis inhibitor act via a specific interaction with EF-2.[47, 48] In addition, elongation factors such as EF-1α was excluded because even though it had been proposed that in senescent *Drosophila melanogaster* this factor might become limiting and be responsible for the age-related decline in protein synthesis, other reports did not find any evidence of this.[19]

5.1. Comparative Effect of CH Treatment and Aging on the Amount of Absolute and Active EF-2 and on Its Oxidation State in the Rat Liver

The total content of EF-2 can be measured by ELISA using polyclonal antibody against EF-2.[20, 21] As can be seen in Table 4, there is no change in the total amount of EF-2 in both CH-treated and aged rats with respect to control rats.

EF-2 contains an unusual amino acid named diphthamide, whose importance for the function of EF-2 has been reported.[47, 49, 50] The diphthamide residue is specifically ADP-ribosylated by fragment A of diphtheria toxin (DT), which catalyzes the transfer of ADP-ribose from NAD, thus inactivating it.[47] The level of ADP-ribosylatable EF-2, which has been used as an indirect measurement of the amount of active EF-2 in extract,[43] can be determined by using the assay of ribosylation of the diphthamide mediated by DT. As seen in Table 4, a statistically significant reduction of the amount of ADP-ribosylatable EF-2 is observed during the treatment of CH and aging. These data agree with the results found by Riis *et al.*[51] using the same method during cellular aging of human fibroblasts, even though the same authors failed to find this result in aged rat livers.[52] This decrease

Table 4. **Effect of CH-treatment and aging on the amount of total and active EF-2 and on its oxidation state.** The total amount of EF-2 was measured by ELISA.[20, 21] Active amount of EF-2 was measured using radioactive NAD plus DT.[43] Carbonyl groups in EF-2 were determined using radioactive sodium borohydride.[20, 21, 30]

Rat Group	Amount of EF-2 (absorbance at 405 nm)	ADP-ribosylatable EF-2 (pmol/mg protein)	CO Content in EF-2 (nmol CO group/ mg protein)
3 months old	0.173 ± 0.01	13.54 ± 1.48	3.58 ± 0.53
3 months old + CH	0.197 ± 0.016	$8.92 \pm 1.03^*$	$5.60 \pm 0.39^*$
12 months old	0.159 ± 0.01	$8.42 \pm 1.66^*$	$7.94 \pm 1.19^*$
24 months old	0.205 ± 0.01	$4.46 \pm 1.38^*$	$8.60 \pm 1.18^*$

*Significantly different from level in control rats (multifactor ANOVA, Tukey´s test, $p < 0.05$).

in the amount of active EF-2 suggests that changes in elongation and, hence, in protein synthesis could be due to post-translational modifications of EF-2.

Among the post-translational modifications, oxidation has been described to play an important role in inactivating proteins during aging.[23–25, 53] In order to study whether EF-2 is oxidatively damaged with CH-treatment and increasing age, the carbonyl groups of EF-2 immunoprecipitated from hepatic extracts have been measured.[20, 21] The results show that hepatic EF-2 from CH-treated and old rats is oxidized, as indicated by the increase in the carbonyl group content (Table 4). Although there is not a negative linear correlation between EF-2 activity and the EF-2 carbonyl group content, the results show the presence of a phenomenon of oxidative stress in this molecule under both circumstances.

Other possibilities of how reactive oxygen species can affect the amount of active EF-2 are: (a) A chemical modification of diphtamide by oxyradicals, that would abolish the DT-mediated rybosylation; (b) An activation of endogenous mono-ADP-rybosylation of the dipthamide by ribosyltransferases. This mono-ADP-ribosylation seems to regulate the rate of protein synthesis as part of normal cellular metabolism[54, 55] and is activated by oxidative stress;[56] (c) EF-2 amino acid modification by oxidation: EF-2 has a histidine, which is essential for its activity.[45] Because histidine is one of the amino acid residues most susceptible to oxidative damage,[22, 46] it is possible that free radicals produced by CH could affect the activity of EF-2 through a mechanism involving histidine inactivation by oxidation; (d) Conformational changes in EF-2 induced by oxidative modification mediated by free radicals; (e) The synergic effects of all these possibilities.

5.2. Effect of CH-treatment and Aging on State of Fragmentation of EF-2

The molecular weight of hepatic EF-2 from CH-treated and old rats has also been studied.[20, 21] The results show that this protein becomes fragmented being the pattern of fragmentation similar in both groups of rats. A similar EF-2 fragmentation pattern was found in several cells as ox liver, placenta and yeast cells.[57] The data obtained for yeast show that the fragmentation pattern depends on the growing phase.

This fragmentation of EF-2 could also account for the slowing-down of protein synthesis under both circumstances (aging and treatment with oxidants) and could be due to a direct effect of active oxygen species, which has been described to produce directly a fragmentation of polypeptide chain, probably by the peroxyradical-mediated α-amidation pathway.[58] In addition, oxygen radical-mediated oxidation of proteins is a marking step in protein degradation. This is supported by several observations, including the fact that many proteases degrade oxidized proteins more rapidly than unoxidized forms.[59–63] Considering the oxidation state of EF-2 (Table 4), the observed EF-2 fragments could also be due to the fact that oxidized EF-2 becomes more susceptible to proteases. The accumulation of these fragments would occur by the same mechanism described for radical-damaged proteins.[63]

Taken together, these results indicate that the inhibition of protein synthesis during CH treatment and aging could be produced by changes in the elongation step and this drop in the elongation step could be due to the post-translational modifications of EF-2, which result in an increased oxidative damage and a decreased active form of EF-2.

6. Conclusions

De novo protein synthesis is a complex process that is physiologically activated and inhibited by many factors, which affect different components of the translation machinery. The components and pathway of translation have been reviewed by several authors.[1-4] A variety of translational controls have been observed. Hormones, nerve and epidermal growth factors, viruses, plant and bacterial toxins affect protein synthesis by a number of different mechanisms[64-68] so that a variety of translational controls have been observed. Phosphorilation/dephosphorilation of protein components of the translational apparatus appears to play an important role in controlling the overall rate of protein synthesis in mammalian cells.[69-71]

The maintenance of the optimum rate of protein synthesis is very important for the cell as this process is integrated into the overall metabolism. The slowing down of protein synthesis is one of the most commonly observed biochemical changes during aging.[5-12] The possible mechanism underlying this decline has been extensively studied, but only a few investigations have focused on the causative role of reactive oxygen species in the inhibition of protein synthesis observed in old organisms.[21] Even though the effect of oxidative stress on protein metabolism has been reported by various laboratories, most of these studies have focused on protein degradation. However, the effect of reactive oxygen species on protein synthesis has not been so extensively evaluated.

In this chapter, we have shown several similarities in the changes of translation in young animals treated with oxidants and old animals (summarized in Scheme 1). Considering that changes in protein synthesis in aged rats match those found in young rats treated with oxidants, this report suggests that the decline in protein synthesis during aging could be due to an increased endogenous oxidative stress and shows the physiological significance of *in vivo* production of oxidatively damaged proteins such as EF-2. From the point of view of aging control, the question that arises is whether it is possible to prevent or reverse these oxidative changes in EF-2. With regard to this, studies on the effect of different antioxidants are now in progress in our laboratory.

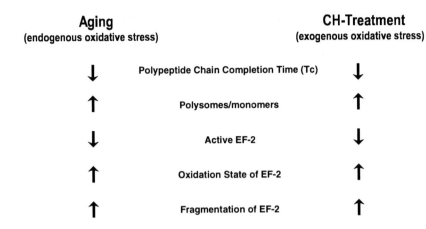

Scheme 1. Similarities in the changes of translation in CH-treated and old rat.

Acknowledgements

Grant FIS 96-1442 from Ministerio de Sanidad y Consumo of Spain supported this work.

References

1. Moldave, K. (1985). Eukaryotic protein synthesis. *Ann. Rev. Biochem.* **54**: 1109–1149.
2. Hershey, J. W. B. (1991). Translational control in mammalian cells. *Ann. Rev. Biochem.* **60**: 717–755.
3. Merrick, W. C. (1990). Overview: mechanism of translation initiation in eukaryotes. *Enzyme.* **44**: 7–16.
4. Rhoads, R. E. (1999). Signal transduction pathways that regulate eukaryotic protein synthesis. *J. Biol. Chem.* **22**: 30 337–30 340.
5. Richardson, A. (1981). The relationship between aging and protein synthesis. *In* "Handbook of Biochemistry in Aging" (J. R. Florini, Ed.), pp. 79–101, CRC Press, FL.
6. Richardson, A. and Birchenall-Sparks, M. C. (1983). Age-related changes in protein synthesis. *In* "Review of Biological Researching in Aging". Vol. 1 (R. Alan, Ed.), pp. 255–273, Liss, Inc., New York.
7. Makrides, S. C. (1983). Protein synthesis and degradation during aging and senescence. *Biol. Rev.* **58**: 343–422.
8. Webster, G. C. (1985). Protein synthesis in aging organisms. *In* "Molecular Biology of Aging: Gene Stability and Gene Expression" (R. S. Sohal, L. S. Birnbaum, and R. G. Cutler, Eds.), pp. 263–289, Raven, New York.

9. Rattan, S. I. S. (1991). Protein synthesis and the components of protein synthetic machinery during cellular ageing. *Mutat. Res.* **256**: 115–125.

10. Rattan, S. I., Derventzi, A. and Clark, B. F. (1992). Protein synthesis, post-translational modifications, and aging. *Ann. NY Acad. Sci.* **663**: 48–62 .

11. Rattan, S. I. S. (1996). Synthesis, modifications and turnover of proteins during aging. *Exp. Gerontol.* **31**: 33–37.

12. Rattan, S. I. and Clark, B. F. (1996). Intracellular protein synthesis, modifications and aging. *Biochem. Soc. Trans.* **24**: 1043–1049.

13. Webster, G. C., Webster, S. L. and Landis, W. A. (1981). The effect of age on the initiation of protein synthesis in *Drosophila melanogaster*. *Mech. Ageing Dev.* **16**: 71–79.

14. Gabius, H. J., Engelhardt, R., Deerberg, F. and Cramer, F. (1983). Age-related changes in different steps of protein synthesis of liver and kidney of rats. *FEBS Lett.* **160**: 115–118.

15. Blazejowski, C. A. and Wester, G. C. (1984). Effect of age on peptide chain initiation and elongation in preparations from brain, liver, kidney and skeletal muscle of the c57b1/6j mouse. *Mech. Ageing Dev.* **25**: 323–333.

16. Khasigov, P. Z. and Nicolaev, A. Y. (1987). Age-related changes in the rates of polypeptide chain elongation. *Biochem. Int.* **15**: 1171–1178.

17. Richardson, A. and Semsei, I. (1987). Effect of aging on translation and transcription. *Rev. Biol. Res. Aging* **3**: 467–483.

18. Shepherd, J. C., Walldorf, U., Hug, P. and Gehring, W. J. (1989). Fruit flies with additional expression of the elongation factor EF-1 alpha live longer. *Proc. Natl. Acad. Sci. USA* **86**: 7520–7521.

19. Shikama, N. and Brack, C. (1996). Changes in the expression of genes involved in protein synthesis during *Drosophila* aging. *Gerontology* **42**: 123–136.

20. Ayala, A., Parrado, J., Bougria, M. and Machado, A. (1996). Effect of oxidative stress, produced by cumene hydroperoxide, on the various steps of protein synthesis. *J. Biol. Chem.* **271**: 23 105–23 110.

21. Parrado, J., Bougria, M., Ayala, A., Castaño, A. and Machado A. (1999). Effects of aging on the various steps of protein synthesis: fragmentation of elongation factor-2. *Free Radic. Biol. Med.* **26**: 362–370.

22. Gordillo, E., Ayala, A., F-Lobato, M., Bautista, J. and Machado, A. (1988). Possible involvement of histidine residues in the loos of enzymatic activity of rat liver malic enzyme during aging. *J. Biol. Chem.* **263**: 8053–8057.

23. Stadtman, E. R. (1992). Protein oxidation and aging. *Science* **257**: 1220–1224.

24. Berlett, B. S. and Stadtman, E. R. (1997). Protein oxidation in aging, disease, and oxidative stress. *J. Biol. Chem.* **272**: 20 313–20 316.

25. Stadtman, E. R. and Levine, R. L. (2000). Protein oxidation. *Ann. NY Acad. Sci.* **899**: 191–208.

26. Poot, M., Verkerk, A., Koster, J. F., Esterbauer, H. and Jongkind, J. F. (1988). Reversible inhibition of DNA and protein synthesis by cumene hydroperoxide and 4-hydroxy-nonenal. *Mech. Ageing Dev.* **43**: 1–9.

27. Scornik, O. A. (1974). In vivo rate of translation by ribosomes of normal and regenerating liver. *J. Biol. Chem.* **249**: 3876–3883.
28. Henshaw, E. H., Hirsch, C. A., Morton, B. E. and Hiatt H. H. (1971). Control of protein synthesis in mammalian tissues through changes in ribosomes activity. *J. Biol. Chem.* **246**: 436–446.
29. Ayuso-Parrilla, M. S., Martin-Requero, A., Perez-Diaz, J. and Parrilla, R. (1976). Role of glucagon on the control of hepatic protein synthesis and degradation in the rat in vivo. *J. Biol. Chem.* **251**: 7785–7790.
30. Lenz, A. G., Costabel, U., Shaltiel, S. and Levine, R. L. (1989). Determination of carbonyl groups in oxidatively modified proteins by reduction with tritiated sodium borohydride. *Anal. Biochem.* **177**: 419–425.
31. Esterbauer, H. and Cheeseman, K. H. (1990). Determination of aldehydic lipid peroxidation products: malonaldehyde and 4-hydroxynonenal. *Meth. Enzymol.* **186**: 407–421.
32. Halliwel, B. and Gutteridge, J. M. C. (1986). *In* "Free Radicals in Biology and Medicine" (B. Halliwel and J. M. C. Gutteridge, Eds.), pp. 139–205, Clarendon Press, Oxford.
33. Poot, M., Verkerk, A., Koster, J. F., Esterbauer, H. and Jongkind, J. F. (1987). Influence of hydroperoxide and 4-hydroxynonenal on the glutathione metabolism during in vitro ageing of human skin fibroblasts. *Eur. J. Biochem.* **162**: 287–291.
34. Halliwell, B. and Gutteridge, J. M. (1990). Role of free radicals and catalytic metal ions in human disease: an overview. *Meth. Enzymol.* **186**: 1–85.
35. Ku, Z. and Thomason, D. B. (1994). Soleus muscle nascent polypeptide chain elongation slows protein synthesis rate during non-weight-bearing activity. *Am. J. Physiol.* **267**: 115–126.
36. Moudgil, P. G., Cook, J. R. and Buetow, D. E. (1979). The proportion of ribosomes active in protein synthesis and the content of polyribosomal poly(A)-containing RNA in adult and senescent rat livers. *Gerontology* **25**: 322–326.
37. Cook, J. R. and Buetow, D. E. (1981). Decreased protein synthesis by polysomes, tRNA and aminoacyl-tRNA synthetases from senescent rat liver. *Mech. Ageing Dev.* **17**: 41–52.
38. Riis, B., Ward, W. F., Clark, B. F. C. and Nygard, O. (1991). Polysome profile and in vitro protein synthesis capacity in liver cell extracts from young and old fischer 344 rats. *Top. Aging Res. Eur.* **16**: 115–124.
39. Layman, D. K., Ricca, G. A. and Richardson, A. (1976). The effect of age on protein synthesis and ribosome aggregation to messenger RNA in rat liver. *Arch. Biochem. Biophys.* **173**: 246–254.
40. Claes-Reckinger, N., Vandenhaute, J., Van Benzooijen, C. F. and Delcour, J. (1982). Functional properties of rat liver protein synthesizing machinery in relation to aging. *Exp. Gerontol.* **17**: 281–286.
41. Vandenhaute, J., Claes-Reckinger, N. and Delcour, J. (1983). Age-related functional alteration of mouse liver ribosomes. *Exp. Gerontol.* **18**: 355–363.

42. Makrides, S. C. and Goldthwaite, J. (1984). The content and size distribution of membrane-bound and free polyribosomes in mouse liver during aging. *Mech. Ageing Dev.* **27**: 111–134.

43. Riis, B., Rattan, S. I. S. and Clark, B. F. C. (1989). Estimating the amounts of ADP-ribosylatable active elongation factor-2 in mammalian cell-free extracts. *J. Biochem. Biophys. Meth.* **19**: 319–326.

44. Riis, B., Rattan, S. I., Cavalius, J. and Clark B. F. (1989). ADP-ribosylatable content of elongation factor-2 changes during cell cycle of normal and cancerous human cells. *Biochem. Biophys. Res. Commun.* **159**: 1141–1146.

45. Omura, F., Kohno, K. and Uchida, T. (1989). The histidine residue of codon 715 is essential for function of elongation factor-2. *Eur. J. Biochem.* **180**: 1–8.

46. Levine, R. L. (1983). Oxidative modification of glutamine synthetase. I. Inactivation is due to loss of one histidine residue. *J. Biol. Chem.* **258**: 11 823–11 827.

47. Perentesis, J. P., Miller, S. P. and Bodley, J. W. (1992). Protein toxin inhibitors of protein synthesis. *Biofactors* **3**: 173–184.

48. Justice, M. C., Hsu, M.-J., Tse, B., Ku, T., Balkovec, J., Schmatz, D. and Nielsen, J. (1998). Elongation factor-2 as a novel target for selective inhibition of fungal protein synthesis. *J. Biol. Chem.* **273**: 3148–3151.

49. Van Ness, B. G., Howard, J. B. and Bodley, J. W. (1980). ADP-ribosylation of elongation factor-2 by diphtheria toxin. Isolation and properties of the novel ribosyl-amino acid and its hydrolysis products. *J. Biol. Chem.* **255**: 10 710–10 716.

50. Fendrick, J. L. and Igleswki, W. J. (1989). Endogenous ADP-ribosylation of elongation factor-2 in polyoma virus-transformed baby hamster kidney cells. *Proc. Natl. Acad. Sci. USA* **86**: 554–557.

51. Riis, B., Rattan, S. I. S., Derventzi, A. and Clark, B. F. C. (1990). Reduced levels of ADP-ribosylatable elongation factor-2 in aged and SV40-transformed human cell cultures. *FEBS Lett.* **266**: 45–47.

52. Rattan, S. I. S., Ward, W. F., Glenting, M., Svendsen, L., Riis, B. and Clark, B. D. F. (1991). Dietary calorie restriction does not affect the levels of protein elongation factors in rat livers during ageing. *Mech. Ageing Dev.* **6**: 379–382.

53. Cabiscol, E. and Levine, R. L. (1995). Carbonic anhydrase III. Oxidative modification in vivo and loss of phosphatase activity during aging. *J. Biol. Chem.* **270**: 14 742–14 747.

54. Iglewski, W. J. and Dewhurst, S. (1991). Cellular mono(ADP-ribosyl) transferase inhibits protein synthesis. *FEBS Lett.* **283**: 235–238.

55. Iglewski, W. J. (1994). Cellular ADP-ribosylation of elongation factor-2. *Mol. Cell. Biochem.* **138**: 131–133.

56. Parrado, J., Bougria, M., Ayala, A. and Machado A. (1999). Induced mono-(ADP)-ribosylation of rat liver cytosolic proteins by lipid peroxidant agents. *Free Radic. Biol. Med.* **26**: 1079–1084.

57. Giovane, A., Servillo, L., Quagliuolo, L. and Balestrieri, C. (1987). Purification of elongation factor-2 from human placenta and evidence of its fragmentation patterns in various eukaryotic sources. *Biochem. J.* **244**: 337–344.
58. Pacifici, R. E. and Davies, K. J. A. (1990). Protein degradation as and index of oxidative stress. *Meth. Enzymol.* **186**: 485–502.
59. Davies, K. J. A. (1987). Protein damage and degradation by oxygen radicals: I. General aspects. *J. Biol. Chem.* **262**: 9895–9901.
60. Dean, R. T., Gieseg, S. and Davies, M. J. (1993). Reactive species and their accumulation on radical-damaged proteins. *Trends Biochem. Sci.* **18**: 437–441.
61. Sahakian, J. A., Szweda, L. I., Friguet, B., Kitani, K. and Levine, R.L. (1995). Aging of the liver: proteolysis of oxidatively modified glutamine synthetase. *Arch. Biochem. Biophys.* **318**: 411–417.
62. Grune, T., Reinheckel, T. and Davies, K. J. A. (1996). Degradation of oxidized proteins in K562 human hematopoietic cells by proteasome. *J. Biol. Chem.* **271**: 15 504–15 509.
63. Dean, R. T., Fu, S., Stocker, R. and Davies, M. J. (1997). Biochemistry and pathology of radical-mediated protein oxidation. *Biochem. J.* **32**: 1–18.
64. Schneider, R. J. and Shenk, T. (1987). Impact of virus infection on host cell protein synthesis. *Ann. Rev. Biochem.* **56**: 317–332.
65. Levenson, R. M., Nairn, A. C. and Blackshear, P. J. (1989). Insulin rapidly induces the biosynthesis of elongation factor-2. *J. Biol. Chem.* **264**: 11 904–11 911.
66. Vary, T. C., Nairn, A. and Lynch, C. J. (1994). Role of elongation factor-2 in regulating peptide-chain elongation in the heart. *Am. J. Physiol.* **266**: E628–E634.
67. Hershey, J. W. (1990). Overview: phosphorylation and translation control. *Enzyme.* **44**: 17–27.
68. Klejin, M., Welsh, G. I., Scheper, G. C., Voorma, H. O., Proud, C. G. and Thomas, A. M. (1998). Nerve and epidermal growth factor induce protein synthesis and eIF2B activation in PC12 cells. *J. Biol. Chem.* **273**: 5536–5541.
69. Hershey, J. W. (1989). Protein phosphorylation controls translation rates. *J. Biol. Chem.* **264**: 20 823–20 826.
70. Venema, R. C., Peters, H. I. and Traugh, J. A. (1991). Phosphorilation of elongation factor-1 (EF-1) and valyl-tRNA synthetase by protein kinase C and stimulation of EF-1 activity. *J. Biol. Chem.* **266**: 12 574–12 580.
71. Calberg, U., Nilson, A. and Nygard, O. (1990). Functional properties of phosphorylated elongation factor-2. *Eur. J. Biochem.* **191**: 639–645.

Chapter 20

Biological Implications of Protein Oxidation

Sataro Goto, Zsolt Radák and Ryoya Takahashi

Sataro Goto and **Ryoya Takahashi** • Department of Biochemistry, School of Pharmaceutical Sciences, Toho University, Funabashi, Japan
Zsolt Radák • Laboratory of Exercise Physiology, School of Physical Education, Semmelweis University, Budapest, Hungary

1. Introduction

Aerobic organisms are dependent on oxygen molecule for the generation of ATP and other essential biological processes. Reactive oxygen species (ROS) derived from molecular oxygen are required in the killing of invading bacteria and in metabolic processes such as prostaglandin synthesis and oxidative removal of noxious substances. Furthermore, in recent years it has become clear that oxygen as a form of ROS is required in signal transduction pathways that mediate signals elicited by cytokines and growth factors. Thus, not only oxygen but also ROS are necessary for the maintenance of life.

On the other hand, ROS as unavoidable by-products of oxygen metabolism or other redox reactions are often damaging to biomolecules and, therefore, can cause aging and a variety of diseases such as cardiovascular disease, arteriosclerosis, diabetes, and neurodegenerative diseases. It has also been stated that ROS are involved in detrimental as well as beneficial effects of exercise.

ROS induce a variety of modifications in nucleic acids, proteins and lipids in the above mentioned situations in cells and tissues or in plasma and cerebrospinal fluid. The oxidative effect on proteins appears to be particularly important because of their involvement in every biological mechanism of catalytic, regulatory and/ or structural functions. In this brief overview, we summarize and discuss the chemistry and biological effects of protein oxidation.

2. Chemistry of Protein Oxidation

Amino acid residues in proteins are oxidatively modified *in vivo* directly by ionizing radiation and metal catalyzed oxidation (MCO)[1] or indirectly by aldehydes derived from lipid peroxidation[2, 3] and glycoxidation products.[4] Ionizing radiation due to X-ray or gamma-ray exposure is not usually problematic except for UV irradiation on the skin exposed to the sun. We, therefore, consider here MCO and oxidation via aldehydes as major oxidative processes *in vivo*.

The transition metals such as iron and copper are involved in MCO because of their chemical property of behaving as electron donors [Fe(II) and Cu(I)] or acceptors [Fe(III) and Cu(II)]. These metals are usually bound *in vivo* to specific metal binding proteins including ferritin, transferrin, metallothionein, and ceruloplasmin or metallo enzymes such as superoxide dismutase (SOD) and proteins like hemoglobin and myoglobin. Some of the iron and copper ions may, however, be released from the primary binding sites and these freed ions can bind nearby proteins and nucleic acids. It has been proposed that transition metals bound at histidine, cysteine or methionine residues can react with hydrogen peroxide to generate highly reactive hydroxyl radicals at the binding sites via Fenton reaction.[1, 5] The hydroxyl radicals can react with nearby amino acid residues giving rise to a variety of oxidation products, e.g. dityrosine from tyrosine, *o-* or *m*-tyrosine from phenylalanine, 2-oxo-histidine from histidine, *N*-formylkynurenine from tryptophan, 4- or 5-hydroxyleucine from leucine, 3- or 4-hydroxyvaline from valine, methionine sulfoxide and methionine sulfone from methionine, and cystine from cysteine.[6-8] Most of these modifications are irreversible except cystine and methionine sulfoxide which are reduced back to their unoxidized forms by enzymatic[9] or non-enzymatic reactions in cells. The proposed scheme of protein oxidation predicts that MCO is site-specific not only because of the specificity of the metal binding amino acid residues but also because of the extremely high reactivity of hydroxyl radical that immediately reacts with molecules that it first encounters. The reaction with hydroxyl radical creates addition products and/or causes hydrogen abstraction that results in formation of a new radical, the latter reaction giving rise to the formation of carbonyl moiety from amino acid residues such as lysine.[6] Such site-specific modifications often result in loss or decrease in biological activity of proteins because metal binding sites or the vicinity are often important for their functions.[10, 11] In addition to non-enzymatic MCO, a variety of enzymatic processes can also give rise to ROS. The electron transport process in mitochondria is believed to be the major source of ROS generation. Others include reactions involved in oxidase in peroxisome, NADPH dehydrogenase in oxygen burst of neutrophils and macrophages, cytochrome P450 reductase in xenobiotic metabolism in endoplasmic reticulum and xanthine oxidase in purine catabolism.

Protein carbonyls are often used to evaluate oxidation status in cells and tissues because of the simplicity of the assay method,[12] although it has been claimed that careful measurements are required to obtain reliable results.[13-17] Protein carbonyls can be formed from lysine, arginine and proline residues by direct MCO[1, 18] or by indirect oxidation due to the reaction with aldehydes derived from lipid peroxidation[2, 19] or glycoxidation[20, 21] products. Lipid hydroperoxides give rise to a variety of aldehydes such as malonaldehyde and 4-hydroxy-2-nonenal which may react with the side chains of lysine, histidine and cysteine residues to generate carbonyl moieties on proteins, contributing indirect oxidation to them.[22]

3. Protein Oxidation as a Physiological Process

It has been implicated that ROS play an essential role in a variety of biological processes such as in proliferation of cancer cells[23, 24] and vascular smooth-muscle cells.[25] These effects are dependent upon reversible activation and inactivation of transcription factors that regulate gene expression responsive to oxidative stress.[26, 27] The transcription factors NFκB, AP-1[28] and c-Jun/c-Fos[29] can serve as biological sensors for oxidative stress, inducing antioxidant enzymes and proteins that protect essential biomolecules from oxidative stress.

Some abundant proteins may function as traps for ROS, being themselves oxidized before other more important but quantitatively minor proteins are damaged by the oxidative stress. Serum albumin that exists in a large amount appears to be one such protein.[30, 31] Ceruloplasmin is the other potential plasma antioxidant protein.[32, 33] Carbonic anhydrase III (CA), an abundant protein with unknown function, exists in the liver (male) and skeletal muscle (male and female) and is highly oxidized in young and old rats.[34] It is noted that CA was found oxidized to a lesser extent after exercise training that appears to reduce oxidative stress in rats[35] while acute aerobic exercise may increase the oxidation.[36] Vitellogenin-6, an egg-yolk protein in a young nematode may also play a role as a sink for ROS since it exists as highly oxidized molecules at high concentration even in old animals.[37] Thus, these proteins appear to have a function against oxidative stress.

4. Protein Oxidation in Diseases

ROS are reported to be involved in the development of a number of age-related diseases. In many of these cases oxidative modifications of proteins are implicated as possible causes. The diseases include diabetes mellitus,[4, 38] arteriosclerosis (Refs. 39 and 40; Kurosaki *et al.* in preparation), ischemia/reperfusion injury,[41, 42] rheumatoid arthritis,[43, 44] renal disease,[45] cataract,[46] cerebral hemorrhage,[47] immobilization stress[48] and neurodegenerative disorders[49] such as Alzheimer's disease (AD)[50–52] and Parkinson's disease (PD).[53]

In both AD[54] and PD[55] iron is supposed to be involved in the oxidative process via Fenton chemistry that causes neuronal cell death in the affected regions. Beta amyloid by itself is implicated in the generation of ROS resulting in oxidative modification of proteins of neural cells.[50] In addition to the increased oxidation of proteins in AD brain, it was recently reported that the plasma of patients showed a significantly elevated level of protein carbonyls.[56] This raises the possibility of the protein oxidation being a biomarker of the disease. The situation in Parkinson's disease is complicated due to the fact that clinical patients are mostly treated with L-dehydroxyl phenylalanine (L-DOPA) that can be a pro-oxidant, generating hydrogen peroxide via reaction by monoamine oxidase.[57]

In arteriosclerosis the deposition of oxidatively modified low density lipoprotein (LDL) in the intima of an artery is generally accepted to play a crucial role in the development of the disease.[58] We have recently shown that carbonylated serum albumin is accumulated in the intima of the artery and the oxidized human serum albumin activates smooth muscle cell migration *in vitro* (Kurosaki *et al.* in preparation). NADPH oxidase of endothelial cells that generate ROS is suggested be involved in this process[59] in addition to the well-known role of the enzyme in neutrophils and macrophages present in the injured arteries.[60]

5. Accumulation of Oxidatively Modified Proteins in Aging

Since the proposal of the pioneering idea of free radical theory of aging by Harman,[61] numerous studies have been conducted which suggest the possible involvement of oxidative modifications of lipids, DNA and proteins in aging, i.e. functional decrease of cellular and tissue functions.[62-64] To evaluate the roles of oxidative damage of these molecules in aging it is important to note whether extent and quality of modification per molecule can cause age-related physiological decline in general. In this regard protein oxidation as measured by carbonyl content may be high enough to explain the decline of cellular and tissue functions with age, e.g. 5% or less in the young versus 10 to 20% in the aged,[65] while the extent of lipid and DNA oxidation is much less than 0.1%. Numerous articles have been published on age-related increase in protein carbonyls in a variety of animals.[65-68] One should be cautious, however, in drawing conclusions since this extent of age-related protein carbonylation may not necessarily be observed in every tissue, at least up to the age of the average life span (Ref. 16, Takahashi *et al.* in preparation). It should also be noted that Dean *et al.* pointed out the possible over-estimation of the reported extent of protein carbonyls in biological materials.[7]

Experimental data supporting the free radical theory of aging have been accumulating.[63, 64, 69, 70] Recent reports suggest that long-lived mutant nematode,[71] genetically manipulated fruit fly,[72] and mouse[73] with a prolonged life span have augmented antioxidant capacities with less oxidative protein damage compared with the control animals with normal life span. These findings appear to provide stronger support for the roles of ROS in aging than positive correlations between accumulation of the oxidative damage and advancing age.[65, 74, 75] Weindruch and his collaborators have demonstrated by DNA micro-array analysis that life-long caloric restriction which is known to retard aging and onset of age-related diseases upregulates the expression of genes for protective functions against accumulation of altered molecules including oxidatively damaged proteins in aged animals: i.e. increased protein turnover and decreased protein damage.[76, 77] These results appear to further support a causative link between oxidative damage of proteins and aging.

Proteins particularly prone to oxidative damage and therefore possibly responsible for age-related physiological decline should be different in individual types of cells and tissues, although the mitochondrial or other proteins that are required for housekeeping functions are obvious common candidates.[78, 79] Two dimensional immunoblot analysis for carbonylated proteins using antibodies against 2,4-dinitrophenylhydrazones is useful to identify such proteins[12, 16, 80] in combination with emerging techniques of proteomics involving mass spectroscopic study.[81]

6. Physical Exercise and Oxidative Stress

Intense exercise is known to induce oxidative stress due to an excessive oxygen uptake, higher consumption of ATP and excretion of catecholamines that can result in elevated ROS formation in mitochondria and other enzymatic systems involving, for example, xanthine oxidase and NADPH dehydrogenase.[82, 83] Increased utilization of oxygen in mitochondria would enhance the generation of superoxide and hence hydrogen peroxide.[84] In fact, it has been reported that a bout of treadmill exercise increased protein oxidation as measured by the carbonyl content in rat skeletal muscles.[41] While a single bout of exercise of sedentary animals is likely to cause increased detrimental oxidative modifications of proteins,[41, 85] moderate daily exercise appears to be beneficial by reducing the damage in rat skeletal muscle.[36] Regular exercise can also improve cognitive functions in rat with concomitant decrease in protein carbonyls of the brain,[86] Interestingly, this latter finding is consistent with the previous observations that age-related decrease in the cognitive function parallels the decrease in the protein carbonyls in the brain of animals treated with a spin trap compound, N-tert-butyl-α-phenylnitrone (PBN).[87, 88] It should be stressed that the general beneficial effects of moderate regular exercise and PBN appear to be at least in part brought about by the increased activity of the proteasome that is believed to be responsible for the degradation of oxidatively or otherwise modified proteins (Refs. 86 and 87; see also the next session).

A role of mitochondrial uncoupling proteins (UCPs) recently investigated extensively deserves a few words here in the regulation of the oxidative stress elicited by mitochondria, a major source of ROS generation, since exercise influences the gene expression of UCPs. Uncoupling of oxidation of respiratory substrates and phosphorylation of ADP lowers the efficiency of ATP formation. At the same time, it may diminish the formation of ROS by decreasing the mitochondrial membrane potential and thereby stimulating the electron transport and oxygen consumption. This is due to shortening of the half-life of semiquinone, an intermediate of the electron transport system, which can transfer the electron to molecular oxygen, generating superoxide radicals and then hydrogen peroxides.[89] A recent report by Vidal-Puig *et al.* demonstrated that oxidative stress is elevated in UCP3 knockout mice as seen by the decrease in the activity of mitochondrial aconitase that is sensitive to ROS.[90] It is noteworthy that an acute bout of exercise up-regulates UCP3 mRNA,[91, 92] mostly expressed in the skeletal muscle, while endurance training down-regulates it[92, 93] or has no effect on it.[91, 94] These results suggest that endurance training adaptation causes increased efficiency in ATP synthesis at the expense of potential increase in oxidative stress that is likely to be compensated by enhanced activities of antioxidant enzymes[95] and proteasome.[36]

7. Degradation of Oxidized Proteins

Proteins of reduced or lost function can be harmful to cells if accumulated. Degradation of such proteins is, therefore, crucial for the maintenance of homeostasis of biological processes in cells.[96-98] Major intracellular proteolytic systems include lysosomal proteases (cathepsins), calcium dependent proteases (calpains) and multicatalytic proteases (20S and 26S proteasomes). It has been demonstrated that oxidatively modified proteins are mostly degraded by 20S proteasome that does not require tagging by ubiquitin of target proteins and ATP for the activity in contrast to 26S proteasome.[98] Activities of three major proteasomal peptidases [chymotrypsin-like, trypsin-like and peptidylglutamyl-peptide hydrolyzing (PGPH)] for artificial fluorogenic substrates decline with age in the liver,[99-102] epidermis,[103] spinal cord,[104] heart, lung, kidney, cerebral cortex,[105] and skeletal muscle (Radak *et al.* in preparation). PGPH activity is decreased most remarkably with age up to nearly 50% of the level of young animals in many of these cases. It should be mentioned that there are some reports in the literature of there being no significant change of the proteasome activity with age (fruit-fly;[106] rat liver[107]). It is argued that natural protein substrates should be used to assess the age-related change in the proteasome activity to better mimic the *in vivo* situation.[108] With respect to substrates of the proteasome it seems important to note that aggregated or too structurally altered proteins would escape proteasomal degradation, probably because the structural constraint does not allow the proteins to reach the catalytic sites located inside the cylinder of the enzyme complex.[108, 109] We have found that oxidatively modified lysozyme introduced into hepatocytes of old mice was degraded at a slower rate than that in younger counterparts and the degradation was inhibited by a proteasome inhibitor (Takahashi *et al.*, in preparation). This finding strengthens the suggestion that the proteasomal degradation of altered proteins, in fact, declines with age.

It appears that the age-related decrease in the proteasome activity is due to post-translational modifications of the enzyme protein since the amount of the enzyme protein remained largely unchanged while the activity was decreased significantly.[99, 101] In accordance with these results, the subunit composition of the proteasome from aged animals is found altered as revealed by two dimensional gel electrophoresis of the purified enzyme.[102, 110] It is conceivable that proteasome by itself is subjected to oxidative stress. In this respect it is interesting to note that *in vitro* studies have suggested that 20S proteasome is much more resistant to oxidative damage than 26S proteasome *in vitro*, making the 20S form of the enzyme capable of functioning under oxidative stress.[111]

8. Conclusion

Generation of ROS and subsequent oxidative modification of protein is inevitable in aerobic life. Since any protein with catalytic, regulatory and structural functions is involved in all aspects of vital biological processes, accumulation of altered forms of proteins could possibly result in loss or decrease of cellular and tissue functions or gain of harmful functions. The modifications can thus cause various detrimental consequences leading to diseases and aging. This does not necessarily mean, however, that any oxidative damage is harmful. Mild oxidative stress such as in the case of moderate regular exercise may even be beneficial by upregulating protective activities of a body to cope with stronger and more harmful stress that may be encountered later. The hormesis effect, a health promoting beneficial effect observed at low doses while higher doses are harmful or even lethal, should be considered as a more general means to retard aging and extend a healthy life span.[112-114] The idea of hormesis deserves future study in the oxidative stress theory of aging. It should be stressed that the basic concept behind this idea is to provoke the intrinsic capability of a body rather than to supply exogenous natural or synthetic antioxidants to try to compensate for age-related decline of physiological activities in the overall maintenance mechanisms of life.

References

1. Stadtman, E. R. (1993). Oxidation of free amino acids and amino acid residues in proteins by radiolysis and by metal-catalyzed reactions. *Ann. Rev. Biochem.* **62**: 797–821.
2. Esterbauer, H., Schaur, R. J. and Zollner, H. (1991). Chemistry and biochemistry of 4-hydroxynonenal, malonaldehyde and related aldehydes. *Free Radic. Biol. Med.* **11**: 81–128.
3. Uchida, K. (1999). Current status of acrolein as a lipid peroxidation product. *Trends Cardiovasc. Med.* **9**: 109–113.
4. Miyata, T., Inagi, R., Asahi, K., Yamada, Y., Horie, K., Sakai, H., Uchida, K. and Kurokawa, K. (1998). Generation of protein carbonyls by glycoxidation and lipoxidation reactions with autoxidation products of ascorbic acid and polyunsaturated fatty acids. *FEBS Lett.* **437**: 24–28.
5. Chevion, M. (1988). A site-specific mechanism for free radical induced biological damage: the essential role of redox-active transition metals. *Free Radic. Biol. Med.* **5**: 27–37.
6. Stadtman, E. R. (1995). Role of oxidized amino acids in protein breakdown and stability. *Meth. Enzymol.* **258**: 373–393.

7. Dean, T. R., Fu, S., Stocker, R. and Davies, M. J. (1997). Biochemistry and pathology of radical-mediated protein oxidation. *Biochem. J.* **324**: 1–18.
8. Davies, M. J., Fu, S., Wang, H. and Dean, R. T. (1999). Stable markers of oxidant damage to proteins and their application in the study of human disease. *Free Radic. Biol. Med.* **27**: 1151–1163.
9. Vogt, W. (1995). Oxidation of methionyl residues in proteins: tools, targets, and reversal. *Free Radic. Biol. Med.* **18**: 93–105.
10. Gordillo, E., Ayala, A., F-Lobato, M., Bautista, J. and Machado, A. (1988). Possible involvement of histidine residues in loss of activity of rat liver malic enzyme during aging. *J. Biol. Chem.* **263**: 8053–8057.
11. Oliver, C. N., Ahn, B., Wittenberger, M. E., Levine, R. L. and Stadtman, E. R. (1985). Age-related alterations of enzymes may involve mixed-function oxidation reactions. *In* "Modification of Proteins During Aging" (R. C. Adelman, and E. E. Dekker, Eds.), pp. 39–52, Alan R. Liss, Inc., New York.
12. Levine, R. L., Williams, J. A., Stadtman, E. R. and Shacter, E. (1994). Carbonyl assays for determination of oxidatively modified proteins. *Meth. Enzymol.* **233**: 365–357.
13. Reznick, A. Z. and Packer, L. (1994). Oxidative damage to proteins: spectrophotometric method for carbonyl assay. *Meth. Enzymol.* **233**: 357–363.
14. Cao, G. and Cutler, R. G. (1995). Protein oxidation and aging. I. Difficulties in measuring reactive protein carbonyls in tissue using 2,4-dinitrophenyl-hydrazine. *Arch. Biochem. Biophys.* **320**: 106–114.
15. Lyras, L., Evans, P. J., Shaw, P. J., Ince, P. G. and Halliwell, B. (1996). Oxidative damage and motor neuron disease. Difficulties in the measurement of protein carbonyls in human brain tissue. *Free Radic. Res.* **24**: 397–406.
16. Goto, S. and Nakamura, A. (1997). Age-associated, oxidatively modified proteins: a critical evaluation. *Age* **20**: 81–89.
17. Evans, P., Lyras, L. and Halliwell, B. (1999). Measurement of protein carbonyls in human brain tissue. *Meth. Enzymol.* **300**: 145–156.
18. Amici, A., Levine, R. L., Tsai, L. and Stadtman, E. R. (1989). Conversion of amino acid residues in proteins and amino acid homopolymers to carbonyl derivatives by metal-catalyzed oxidation. *J. Biol. Chem.* **264**: 3341–3346.
19. Blakeman, D. P., Ryan, T. P., Jolly, R. A. and Petry, T. W. (1998). Protein oxidation: examination of potential lipid-independent mechanisms for protein carbonyl formation. *J. Biochem. Mol. Toxicol.* **12**: 185–190.
20. Kristal, B. S. and Yu, B. P. (1992). An emerging hypothesis: synergistic induction of aging by free radicals and Maillard reactions. *J. Gerontol.* **47**: B107–B114.
21. Wells-Knecht, K. J., Zyzak, D. V., Litchfield, J. E., Thorpe, S. R. and Baynes, J. W. (1995). Mechanism of autoxidative glycosylation: identification of glyoxal and arabinose as intermediates in the autoxidative modification of proteins by glucose. *Biochemistry* **34**: 3702–3709.

22. Uchida, K. and Stadtman, E. R. (1994). Quantitation of 4-hydroxynonenal protein adducts. *Meth. Enzymol.* **233**: 371–380.
23. Burdon, R. (1995). Superoxide and hydrogen peroxide in relation to mammalian cell proliferation. *Free Radic. Biol. Med.* **18**: 775–794.
24. Nose, K. (2000). Role of reactive oxygen species in the regulation of physiological functions. *Biol. Pharm. Bull.* **23**: 897–903.
25. Sundaresan, M., Yu, Z. X., Ferrans, V. J., Irani, K. and Finkel, T. (1995). Requirement for generation of H_2O_2 for platelet-derived growth factor signal transduction. *Science* **270**: 296–299.
26. Gamaley, I. A. and Klyubin, I. V. (1999). Roles of reactive oxygen species: signaling and regulation of cellular functions. *Int. Rev. Cytol.* **188**: 203–255.
27. Allen, R. G. and Tresini, M. (2000). Oxidative stress and gene regulation. *Free Radic. Biol. Med.* **28**: 463–499.
28. Schulze-Osthoff, K., Los, M. and Baeuerle, P. A. (1995). Redox signalling by transcription factors NF-κB and AP-1 in lymphocytes. *Biochem. Pharmacol.* **50**: 735–741.
29. Abate, C., Pastel, L., Rauscher, III, F. J. and Curran, T. (1990). Redox regulation of Fos and Jun DNA-binding activity in vitro. *Science* **249**: 1157–1161.
30. Halliwell, B. (1988). Albumin — an important extracellular antioxidant? *Biochem. Pharmacol.* **37**: 569–571.
31. Loban, A., Kime, R. and Powers, H. (1997). Iron-binding antioxidant potential of plasma albumin. *Clin. Sci. (London)* **93**: 445–451.
32. Krsek-Staples, J. A. and Webster, R. O. (1993). Ceruloplasmin inhibits carbonyl formation in endogenous cell proteins. *Free Radic. Biol. Med.* **14**: 115–125.
33. Tajima, K., Kawanami, T., Nagai, R., Horiuchi, S. and Kato, T. (1999). Hereditary ceruloplasmin deficiency increases advanced glycation end products in the brain. *Neurology* **53**: 619–622.
34. Cabiscol, E. and Levine, R. L. (1995). Carbonic anhydrase III. Oxidative modification in vivo and loss of phosphatase activity during aging. *J. Biol. Chem.* **270**: 14 742–14 747.
35. Radák, Z, Kaneko, T., Tahara, S., Nakamoto, H., Ohono, H., Sasvari, M., Nyakas, C. and Goto, S. (1999). The effect of exercise training on oxidative damage of lipids, proteins and DNA in rat skeletal muscle: evidence for beneficial outcome. *Free Radic. Biol. Med.* **27**: 69–74.
36. Radák, Z., Nakamura, A., Nakamoto, H., Asano, K., Ohno, H. and Goto, S. (1998). A period of exercise increases the accumulation of reactive carbonyl derivatives in the lungs of rats. *Pfluger Arch. Eur. J. Physiol.* **435**: 439–441.
37. Nakamura, A., Yasuda, K., Adachi, H., Sakurai, Y., Ishii, N. and Goto, S. (1999). Vitellogenin-6 is a major carbonylated protein in aged nematode, *C. elegans. Biochem. Biophys. Res. Commun.* **264**: 580–583.

38. Portero-Otin, M., Pamplona, R., Ruiz, M. C., Cabisco, I. E., Prat, J. and Bellmunt, M. J. (1999). Diabetes induces an impairment in the proteolytic activity against oxidized proteins and a heterogeneous effect in nonenzymatic protein modifications in the cytosol of rat liver and kidney. *Diabetes* **48**: 2215–2220.

39. Belkner, J., Wiesner, R., Rathman, J., Barnett, J, Sigal, E. and Kuhn, H. (1993). Oxygenation of lipoproteins by mammalian lipoxygenases. *Eur. J. Biochem.* **213**: 251–261.

40. Uchida, K. (2000). Role of reactive aldehyde in cardiovascular diseases. *Free Radic. Biol. Med.* **28**: 1685–1696.

41. Reznick, A. Z., Kagan, V. E., Ramasey, R., Tsuchiya, M., Khwaja, S., Serbinova, E. A. and Packer, L. (1992). Antiradical effects in L-propionyl carnitine protection of the heart against ischemia-reperfusion injury: the possible role of iron chelation. *Arch. Bichem. Biophys.* **296**: 394–401.

42. He, K., Nukada, H., McMorran, P. D. and Murphy, M. P. (1999). Protein carbonyl formation and tyrosine nitration as markers of oxidative damage during ischaemia-reperfusion injury to rat sciatic nerve. *Neuroscience* **94**: 909–916.

43. Chapman, M. L., Rubin, B. R. and Gracy, R. W. (1989). Increased carbonyl content of proteins in synovial fluid from patients with rheumatoid arthritis. *J. Rheumatol.* **16**: 15–18.

44. Jasin, H. E. (1993). Oxidative modification of inflammatory synovial fluid immunoglobulin G. *Inflammation* **17**: 167–181.

45. Miyata, T., Kurokawa, K. and van Ypersele de Strihou, C. (2000). Relevance of oxidative and carbonyl stress to long-term uremic complications. *Kidney Int.* **76**: S120–S125.

46. Fu, S., Dean, R., Southan, M. and Truscott, R. (1998). The hydroxyl radical in lens nuclear cataractogenesis. *J. Biol. Chem.* **273**: 28 603–28 609.

47. Hall, N. C., Packard, B. J., Hall, C. L., De Courten-Myers, G. and Wagner, K. R. (2000). Protein oxidation and enzyme susceptibility in white and gray matter with in vitro oxidative stress relevance to brain injury from intracerebral hemorrhage. *Cell. Mol. Biol.* **46**: 673–683.

48. Liu, J., Wang, X., Shigenaga, M. K., Yeo, H. C., Mori, A. and Ames, B. N. (1996). Immobilization stress causes oxidative damage to lipd, protein, and DNA in the brain of rats. *FASEB J.* **10**: 1532–1538.

49. Floyd, R. A. (1999). Antioxidants, oxidative stress, and degenerative neurological disorders. *Soc. Exp. Biol. Med.* **222**: 236–245.

50. Smith, C. D., Carney, J. M., Starke-Reed, P. E., Oliver, C. N., Stadtman, E. R., Floyd, R. A. and Markesbery, W. R. (1991). Excess brain protein oxidation and enzyme dysfunction in normal aging and in Alzheimer disease. *Proc. Natl. Acad. Sci. USA* **88**: 10 540–10 543.

51. Butterfield, D. A., Yatin, S. M., Varadarajan, S. and Koppal, T. (1999). Amyloid eta-peptide-associated free radical oxidative stress, neurotoxicity, and Alzheimer's disease. *Meth. Enzymol.* **309**: 746–768.

52. Markesbery, W. R. and Carney, J. M. (1999). Oxidative alterations in Alzheimer's disease. *Brain Pathol.* **9**: 133–146.

53. Jenner, P. and Olanow, C. W. (1996). Oxidative stress and the pathogenesis of Parkinson's disease. *Neurology* **47**: 161–170.

54. Castellani, R. J., Smith, M. A., Nunomura, A., Harris, P. L. and Perry, G. (1999). Is increased redox-active iron in Alzheimer disease a failure of the copper-binding protein ceruloplasmin? *Free Radic. Biol. Med.* **26**: 1508–1512.

55. Riederer, P., Sofic, E., Rausch, W. D., Schmidt, B., Reynolds, G. P., Jellinger, K. and Youdim, M. B. (1998). Transition metals, ferritin, glutathione, and ascorbic acid in parkinsonian brains. *J. Neurochem.* **52**: 515–520.

56. Conrad, C. C., Marshall, P. L., Talent, J. M., Malakowsky, C. A., Choi, J. and Gracy, R. W. (2000). Oxidized proteins in Alzheimer's plasma. *Biochem. Biophys. Res. Commun.* **275**: 678–681.

57. Alam, Z. I., Daniel, S. E., Lees, A. J., Marsden, P. J. and Halliwel, B. (1997). A generalized increase in protein carbonyls in the brain in Parkinson's but not incidental Lewy body disease. *J. Neurochem.* **69**: 1326–1329.

58. Reaven, P. D., Napoli, C., Merat, S. and Witztumc, J. L. (1999). Lipoprotein modification and atherosclerosis in aging. *Exp. Gerontol.* **34**: 527–537.

59. Meyer, J. W. and Schmitt, M. E. (2000). A central role for the endothelial NADPH oxidase in atherosclerosis. *FEBS Lett.* **472**: 1–4.

60. Podrez, E. A., Abu-Soudm, H. M. and Hazen, S. L. (2000). Myeloperoxidase-generated oxidants and atherosclerosis. *Free Radic. Biol. Med.* **28**: 1717–1725.

61. Harman, D. (1956). Aging: a theory based on free radical and radiation chemistry. *J. Gerontol.* **2**: 298–300.

62. Yu, B. P. (1994). Cellular defenses against damage from reactive oxygen species. *Physiol. Rev.* **74**: 139–162.

63. Beckman, K. B. and Ames, B. N. (1998). The free radical theory of aging matures. *Physiol. Rev.* **78**: 547–581.

64. Finkel, T. and Holbrook, N. J. (2000). Oxidants, oxidative stress and the biology of ageing. *Nature* **408**: 239–247.

65. Stadtman, E. R. (1992). Protein oxidation and aging. *Science* **257**: 1220–1224.

66. Sohal, R. S., Agarwal, S., Dubey, A. and Orr, W. C. (1993). Protein oxidative damage is associated with life expectancy of houseflies. *Proc. Natl. Acad. Sci. USA* **90**: 7255–7259.

67. Berlett, B. S. and Stadtman, E. R. (1997). Protein oxidation in aging, disease, and oxidative stress. *J. Biol. Chem.* **272**: 20 313–20 316.

68. Yasuda, K., Adachi, H., Fujiwara, Y. and Ishii, N. (1999). Protein carbonyl accumulation in aging dauer formation-defective (*daf*) mutants of *Caenorhabditis elegans. J. Gerontol.* **54**: B47–B51.

69. Martin, G. M., Austad, S. N. and Johnson, T. E. (1998). Genetic analysis of ageing: role of oxidative damage and environmental stresses. *Nat. Genet.* **13**: 25–34.

70. Sohal, R. S., Mockett, R. J. and Orr, W. C. (2000). Current issues concerning the role of oxidative stress in aging: a perspective. *In* "Results and Problems of Cell Differentiation, Vol. 29. The Molecular Genetics of Aging" (Hekimi, Ed.), pp. 45–66, Springer-Verlag, Berlin Heidelberg.

71. Adachi, H., Fujiwara,Y. and Ishii, N. (1998). Effects of oxygen on protein carbonyl and aging in *Caenorhabditis elegans* mutants with long (age-1) and short (mev-1) life spans. *J. Gerontol.* **53**: B240–B244.

72. Orr W. C. and Sohal, R. S. (1994). Extension of life-span by overexpression of superoxide dismutase and catalase in *Drosophila melanogaster*. *Science* **263**: 1128–1130.

73. Migliaccio, E., Giorgio, M., Mele, S., Pelicci, G., Reboldi, P., Pandolfi, P. P., Lanfrnacone, L. and Pelicci, P. G. (1999). The p66shc adaptor protein controls oxidative stress response and life span in mammals. *Nature* **402**: 309–313.

74. Sohal, R. S. and Weindruch, R. (1996). Oxidative stress, caloric restriction, and aging. *Science* **273**: 59–63.

75. Stadtman, E. R. and Levine, R. L. (2000). Protein oxidation. *Ann. NY Acad. Sci.* **899**: 191–208.

76. Lee, C. K., Klopp, R. G., Weindruch, R. and Prolla, T. A. (1999). Gene expression profile of aging and its retardation by caloric restriction. *Science* **285**: 1390–1393.

77. Lee, C. K., Weindruch, R. and Prolla, T. A. (1999). Gene-expression profile of the aging brain in mice. *Nature Genetics* **25**: 294–297.

78. Agarwal, S. and Sohal, R. S. (1995). Differential oxidative damage to mitochondrial proteins during aging. *Mech. Ageing Dev.* **85**: 55–63.

79. Nagai, M., Takahashi, R. and Goto, S. (2000). Dietary restriction initiated late in life can reduce mitochondrial protein carbonyls in rat livers: Western blot studies. *Biogerontology* **1**: 321–328.

80. Goto, S., Nakamura, A., Radák, Z., Nakamoto, H., Takahashi, R., Yasuda, K., Sakurai, Y. and Ishii, N. (1999). Carbonylated proteins in aging and exercise: immunoblot approaches. *Mech. Ageing Dev.* **107**: 245–253.

81. Toda, T. (2000). Current status and perspectives of proteomics in aging research. *Exp. Gerontol.* **35**: 803–810.

82. Packer L. (1997). Oxidants, antioxidant nutrients and the athlete. *J. Sports. Sci.* **15**: 353–363

83. Ji, L. L. (1999). Antioxidants and oxidative stress in exercise. *Proc. Soc. Exp. Biol. Med.* **222**: 283–292.

84. Boveris, A. and Chance, B. (1972). The mitochondrial generation of hydrogen peroxide: general properties and effect of hyperbaric oxygen. *Biochem. J.* **134**: 707–716.

85. Radák, Z., Asano, K., Nakamura, A., Nakamoto, H. and Goto, S. (1998). Single bout of exercise increases accumulation of reactive carbonyl derivatives in lung of rats. *Pfluger Arch. Eur. J. Physiol.* **435**: 439–441.

86. Radák, Z., Kaneko, T., Tahara, S., Nakamoto, H., Msasvai, M., Nyakas, C. and Goto, S. (2001). Regular exercise improves cognitive function and decreases oxidative damage to proteins in rat brain. *Neurochem. Int.* **38**: 17–23.

87. Carney, J. M., Starke-Reed, P. E., Oliver, C. N., Landum, R. W., Cheng, M. S., Wu J. F. and Floyd, R. A. (1991). Reversal of age-related increase in brain protein oxidation, decrease, in enzyme activity, and loss in temporal and spatial memory by chronic administration of the spin-trapping compound N-tert-butyl-alpha-phenylnitrone. *Proc. Natl. Acad. Sci. USA* **88**: 3633–3636.

88. Forster, M. J., Dubey, A., Dawson, K. M., Stutts, W. A., Lal, H. and Sohal, R. S. (1996). Age-related losses of cognitive function and motor skills in mice are associated with oxidative protein damage in the brain. *Proc. Natl. Acad. Sci. USA* **93**: 4765–4769.

89. Skulachev, V. P. (1998). Uucoupling: new approaches to an old problem of bioenergetics. *Rev. Biochim. Biophys. Acta* **1363**: 100–124.

90. Vidal-Puig, A. J., Grujic, D., Zhang, C. Y., Hagen, T., Boss, O., Ido, Y., Szczepanik, A., Wade, J., Mootha, V., Cortright, R., Muoio, D. M. and Lowell, B. B. (2000). Energy metabolism in uncoupling protein 3 gene knockout mice. *J. Biol. Chem.* **275**: 16 258–16 266.

91. Cortright, R. N., Zheng, D., Jones, J. P., Fluckey, J. D., Di Carlo, S. E., Grujic, D., Lowell, B. B. and Dohm, G. L. (1999). Regulation of skeletal muscle UCP-2 and UCP-3 gene expression by exercise and denervation. *Am. J. Physiol.* **276**: E217–E221.

92. Boss, O., Samec, S., Desplanches, D., Mayet, M. H., Seydoux, J., Muzzin, P. and Giacobino, J. P. (1998). Effect of endurance training on mRNA expression of uncoupling proteins 1, 2, and 3 in the rat. *FASEB J.* **12**: 335–339.

93. Giacobino, J. P. (1998). Effect of endurance training on mRNA expression of uncoupling proteins 1, 2, and 3 in the rat. *FASEB J.* **12**: 335–339.

94. Tonkonogi, M., Krook, A., Walsh, B. and Sahlin, K. (2000). Endurance training increases stimulation of uncoupling of skeletal muscle mitochondria in humans by non-esterified fatty acids: an uncoupling-protein-mediated effect? *Biochem. J.* **351**: 805–810.

95. Hollander, J., Fiebig, R., Gore, M., Bejma, J., Ookawara, T., Ohno, H. and Ji, L. L. (1999) Superoxide dismutase gene expression in skeletal muscle: fiber-specific adaptation to endurance training. *Am. J. Physiol.* **277**: R856–R862.

96. Goto, S., Hasegawa, A., Nakamoto, H., Nakamura, A., Takahashi, R. and Kurochkin, I. V. (1995). Age-associated changes of oxidative modification

and turnover of proteins. *In* "Oxidative Stress and Aging" (R. Cutler, L. Packer, J. Bertram, and A. Mori, Eds.), pp. 151–158, Birkhauser Verlag Basel, Switzerland.

97. Goto, S., Takahashi, R., Kumiyama, A., Radák, Z., Hayashi, T., Takenouchi, M. and Abe, R. (2001). Implications of protein degradation in aging. *Ann. NY Acad. Sci.* 928: 54–64.

98. Grune, T., Reinheckel, T. and Davies, K. J. (1997). Degradation of oxidized proteins in mammalian cells. *FASEB J.* 11: 526–534.

99. Shibatani, T., Nazir, M. and Ward, W. F. (1996). Alteration of rat liver 20S proteasome activities by age and food restriction. *J. Gerontol.* 51: B316–B322.

100. Conconi, M., Szweda, L .I., Levine, R. L., Stadtman, E. R. and Friguet, B. (1996). Age-related decline of rat liver multicatalytic proteinase activity and protection from oxidative inactivation by heat-shock protein 90. *Arch. Biochem. Biophys.* 331: 232–240.

101. Hayashi, T. and Goto, S. (1998). Age-related changes in the 20S and 26S proteasome activities in the liver of male F344 rats. *Mech. Ageing Dev.* 102: 55–66

102. Anselmi, B., Conconi, M., Veyrat Durebex, C., Turlin, E., Biville, F., Alliot, J. and Friguet, B. (1998). Dietary self-selection can compensate an age-related decrease of rat liver 20S proteasome activity observed with standard diet. *J. Gerontol.* 53: B173–B179.

103. Bulteau, A., Petropoulos, I. and Friguet, B. (2000). Age-related alterations of proteasome structure and function in aging epidermis. *Exp. Gerontol.* 35: 767–777.

104. Keller, J. N., Huang, F. F. and Markesbery, W. R. (2000). Decreased levels of proteasome activity and proteasome expression in aging spinal cord. *Neuroscience* 98: 149–156.

105. Keller, J. N., Hanni, K. B. and Markesbery, W. R. (2000). Possible involvement of proteasome inhibition in aging: implications for oxidative stress. *Mech. Ageing Dev.* 113: 61–70.

106. Agarwal, S. and Sohal, R. S. (1994). Aging and proteolysis of oxidized proteins. *Arch. Biochem. Biophys.* 309: 24–28.

107. Sahakian, J. A., Szweda, L. I., Friguet ,B., Kitani, K. and Levine, R. L. (1995). Aging of the liver: proteolysis of oxidatively modified glutamine synthetase. *Arch. Biochem. Biophys.* 318: 411–417.

108. Grune, T. (2000). Oxidative stress, aging and the proteasomal system. *Biogerontology* 1: 31–40.

109. Grant, A. J., Jessup, W. and Dean, R. T. (1993). Inefficient degradation of oxidized regions of protein molecules. *Free Radic. Res. Commun.* 18: 259–267.

110. Bulteau, A., Petropoulos, I. and Friguet, B. (2000). Age-related alterations of proteasome structure and function in aging epidermis. *Exp. Gerontol.* 35: 767–777.

111. Reinheckel, T., Sitte, N., Ullrich, O., Kuckelkorn, U., Davies, K. J. and Grune, T. (1998). Comparative resistance of the 20S and 26S proteasome to oxidative stress. *Biochem. J.* **335**: 637–642.

112. Johnson, T. E. and Bruunsgaard, H. (1998). Implications of hormesis for biomedical aging research. *Hum. Exp. Toxicol.* **17**: 263–265.

113. Masoro, E. J. (2000). Caloric restriction and aging: an update. *Exp. Gerontol.* **35**: 299–305.

114. Minois, N. (2000). Longevity and aging: beneficial effects of exposure to mild stress. *Biogerontology* **1**: 15–29.

Chapter 21

Application of ProteinChip® Array Technology for Detection of Protein Biomarkers of Oxidative Stress

Charlotte H. Clarke, Scot R. Weinberger
and Mark S.F. Clarke*

Charlotte H. Clarke and **Scot R. Weinberger** • Ciphergen Biosystems, Inc., Fremont, CA
Mark S.F. Clarke • Division of Space Life Sciences, Universities Space Research Association,
3600 Bay Area Boulevard, Houston, Texas 77058
*Corresponding Author.
Tel: 281 483 7253, E-mail: mark.s.clarke1@jsc.nasa.gov

1. Introduction

Oxidative stress occurs as a consequence of an imbalance between the production of free radical species and the level of endogenous antioxidants. In the mammalian system, generation of reactive oxygen species (ROS), such as free radicals and peroxides, is a consequence of aerobic respiration. As such, this event is a normal and inescapable metabolic consequence of cellular function. ROS production appears to be part of the normal feedback pathways associated with the control of nutrient utilization and energy production,[1, 2] as well as serving as both mediators and effectors of immune system function.[3] A variety of exogenous stimuli have been associated with ROS production including UV light,[4] ionizing radiation,[5] and tobacco constituents[6] to name but a few. Production of ROS is a continuous event over the period of an individual's lifetime and could therefore have significant cumulative effects. Hence, it is not surprising that ROS-induced cellular damage has been linked to a wide number of diseases associated with the aging process, including Alzheimer's disease,[7, 8] atherosclerosis[9, 10] and cancer.[11] The ubiquitous nature of ROS and their extreme reactivity in the body have led to the development of multiple protective mechanisms designed to prevent ROS damage to cellular components that may come into contact with these compounds unintentionally. These systems are collectively known as antioxidant defenses and include free-radical scavenger enzymes, such as superoxide-dismutases (SODs) and catalases, and anti-oxidants such as alpha-tocopherol (vitamin E) and ascorbate (vitamin C). It is when the balance between the production of ROS and the endogenous antioxidant defenses becomes skewed in favor of ROS production that a state of "oxidative stress" occurs.

Over the past fifty years many analytical techniques have been applied to measuring oxidative stress (for review see Ref. 12). In general, these methodologies focus on either measuring the production of ROS directly or indirectly via intermediates, measuring the effects of ROS production on the anti-oxidant defense system, or measuring the production of oxidized end products. There are three major classes of oxidized "end-products" associated with ROS-induced cellular damage and their nature illustrates the universal effects of ROS activity on cellular components. These are oxidized lipids, DNA and proteins. The purpose of this chapter is not to describe the many published reports detailing the effects of oxidative stress on cellular components, but to describe an emerging technology that specifically shows great promise in the identification and characterization of oxidized proteins. This technology is known as surface enhanced laser desorption/ionization time-of-flight mass spectrometry (SELDI TOF-MS). It not only presents itself as a platform for investigating the effects of ROS on proteins already identified as being sensitive to an increase in oxidative stress, but also in identifying novel proteins that may serve as "biomarkers" of oxidative stress. The ability to identify specific oxidative stress-induced alterations within the complete protein complement of a particular sample makes SELDI TOF-MS an ideal method for identifying novel protein biomarkers of oxidative stress. In addition, the use of this technology presents a very real possibility of being able to identify a "protein expression profile" associated only with one particular type or severity level of oxidative stress.

2. Technology Platform (SELDI TOF-MS)

A commercial version of SELDI TOF-MS is Ciphergen Biosystem's ProteinChip® array technology (Ciphergen Biosystems, Inc., Fremont, CA., USA). This system, which combines the properties of traditional biochemical extraction methods and the powerful detection ability of TOF-MS, consists of ProteinChip® arrays, a TOF-MS chip-reader, and specialized analysis software.

The ProteinChip® arrays are available with a variety of surfaces which can be separated into two major categories: chromatographic surfaces based upon traditional, well understood biochemical properties for the selective capture of proteins via charge, hydrophobicity or metal-chelate interactions and pre-activated surfaces that allow the covalent binding of biomolecular probes (Fig. 1). The pre-activated array surface allows the covalent attachment of a number of different "probe" molecules that exploit highly specific molecular recognition mechanisms to selectively extract target proteins from complex biological fluids. This approach has taken advantage of a number of such "probe-target" interactions, including antibody-antigen, enzyme-substrate, receptor-ligand, and protein-DNA interactions to study complex protein mixtures derived from biological sources. Once selective protein adsorption to the surface of the array has been achieved, contaminating compounds such as buffer salts or detergents can be washed away with reagents that are compatible with the laser desorption process.

After sample purification upon the SELDI surface via selective (i.e. physio-chemical properties) or specific (e.g. antibody recognition) retention, a matrix molecule is deposited onto the surface and the retained protein(s) is/are detected by the use of a TOF-MS equipped with a laser desorption ion source (Fig. 2). When the array is inserted into the instrument, it becomes contiguous with the repellor lens of the ion optic assembly. Desorption energy is provided by a pulsed, UV ($\lambda = 337$ nm) laser. An optical attenuator controls laser fluence, and an optical beam splitter is used to divert some of the laser beam to a high-speed photodetector that is used to trigger TOF measurement. Upon laser activation, the sample becomes irradiated and desorption/ionization proceeds to liberate gaseous ions. These ions are accelerated to a constant final energy by uniform static electric fields and ultimately strike a detector. Signal processing is promoted by a high-speed analog to digital converter (A/D) linked to a personal computer running a system control and data reduction software program. A detected analyte is

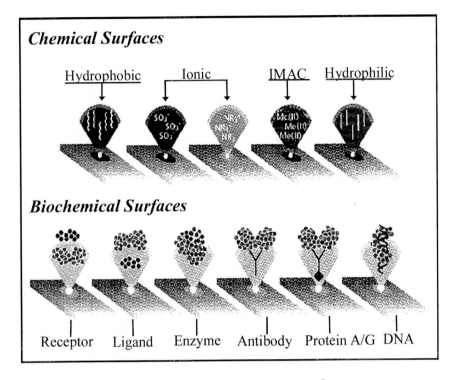

Chemical Surfaces

Hydrophobic   Ionic   IMAC   Hydrophilic

Biochemical Surfaces

Receptor   Ligand   Enzyme   Antibody   Protein A/G   DNA

Fig. 1. **Diagrammatic representation of different ProteinChip® array surfaces based upon either chemically or biochemically modified surfaces.** Individual proteins are selectively adsorbed from a complex protein mixture onto the chemically modified ProteinChip® surfaces by hydrophobic, ionic (cationic or anionic exchange), immobilized metal antigen capture (IMAC) or hydrophilic interactions. Biochemically-modified ProteinChip® surfaces can consist of a range of immobilized biochemical probes that are then used to capture their specific targets from a complex mixture of proteins.

displayed as a peak whose amplitude or area is proportional to abundance; TOF is related to ion mass-to-charge ratio (M/Z) (Fig. 2).

On a technical level, the use of ProteinChip® arrays is a simple and rapid process. Briefly, as little as 1 to 2 μl of sample (e.g. serum, cell lysate, purified protein) can be incubated on an array surface, non-binding molecules washed away, and proteins that specifically interact with the array surface are retained for observation. Protein binding conditions, such as pH, salt concentration and the presence of non-ionic detergents in combination with different surfaces can all be used to control the stringency of protein binding, thereby selecting those proteins that remain attached to the surface of the array. For example, utilizing an experimental scheme in which 1 μl of the same sample is applied in parallel to an anionic exchange array surface, a cationic exchange array surface, a hydrophobic

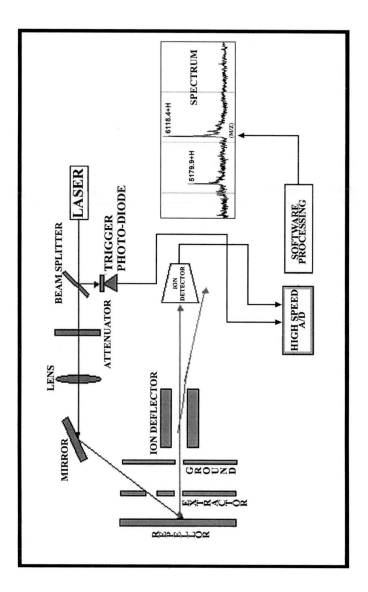

Fig. 2. **Diagrammatic representation of the surface enhanced laser desorption/ionization time-of-flight mass spectrometry (SELDI TOF-MS) utilized as part of the ProteinChip® array system.** The sample becomes contiguous with the repellor lens of the ion optic assembly. When the ProteinChip® array target is irradiated with the laser beam, proteins are liberated as ionized species that are then accelerated to a constant final energy by uniform static electric fields and ultimately strike a detector. Signal processing is promoted by a high-speed analog to digital converter (A/D) linked to a personal computer running a system control and data reduction software program. A detected analyte is displayed as a peak whose amplitude or area is proportional to abundance; TOF is related to ion mass-to-charge ratio (M/Z).

array surface and a hydrophilic array surface, coupled with different buffer formulations that effect binding stringency (i.e. high or low pH, high or low salt concentration, presence of non-ionic detergents), the vast majority of the complete protein complement of a particular sample can be detected. Total protein concentrations from 0.05 to 2.0 mg/ml in the original complex sample are sufficient to reveal femtomole quantities of retained proteins.

A research area to which the ProteinChip® array technology is well suited is that of protein profiling. Protein profiling or protein differential display looks at the protein complement of a biological sample in a given molecular weight range and has become quite popular in the area of disease research.[13-18] Comparison of lysates from normal versus diseased cells can reveal the expression of important marker proteins expressed only during the disease state. Patient serum,[13] crude cell lysates,[15] or purified protein samples,[13, 14] spotted on ProteinChip® arrays with different chromatographic surfaces, allows the creation of a multidimensional binding picture based on different types of interaction. Control samples are read against samples of interest under the same data collection conditions, and subsequent peak subtraction allows the identification of differences. Once a peak of interest has been detected, the molecule can be singled out for further analysis. Frequently, the goal of protein characterization is protein identification. To this end, peptides may be specifically produced from the protein of interest via endoproteolysis by enzymes such as trypsin, lysC and leuC. ProteinChip® array analysis of the resultant peptides provides peptide masses which, when compared to theoretical digests of proteins in databases, can be used to deduce protein identity via peptide mapping and database mining.

3. Preliminary Studies

The purpose of the preliminary studies detailed below is to demonstrate the utility of the ProteinChip® array technology for detecting protein alterations associated with ROS-induced damage. Utilizing both purified protein samples (i.e. purified serum albumin, purified insulin) and complex biological samples (i.e. human serum, human erthyrocyte membranes) we have demonstrated that this technology is capable of detecting protein damage associated with ROS production in a simple and rapid fashion.

When either purified bovine serum albumin (BSA) or purified bovine insulin (INS) are exposed to hydroxyl radicals (generated by an *in vitro* Fenton reaction) for a period of 30 min, subsequent ProteinChip® analysis of the resultant samples discloses the presence of a number of novel protein species not detected in the control samples (Figs. 3 and 4). If complete human serum is exposed to hydroxyl radicals in a similar fashion, a large number of differences in the protein profiles of ROS-exposed serum are also detected compared to control serum profiles (Fig. 5).

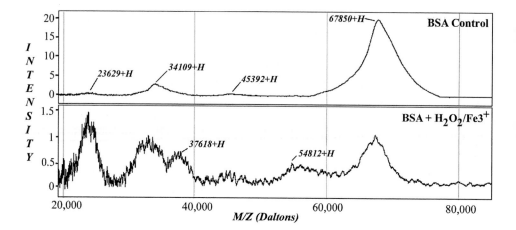

Fig. 3. **The effect of ROS on purified bovine serum albumin (BSA) analyzed using ProteinChip® array technology.** Solubilized BSA was exposed to the hydroxyl radical for a period of 30 min at 37°C by performing a Fenton reaction (i.e. phosphate buffered saline containing 0.2 mg/ml BSA with, or without 5 mM H_2O_2, 100 μM ascorbate and 150 μM Fe^{3+} was incubated for 30 min at 37°C followed by centrifugation at 10 000 xg for 5 min). One-microliter volumes of the resultant supernatants from both control and ROS-exposed samples were spotted onto a normal phase (NP) ProteinChip® array and allowed to dry. Each well of the array was then washed with two, 5 μl volumes of distilled water, allowed to dry and a matrix layer [aqueous buffer containing 50% (v/v) acetonitrile, 0.5% (v/v) trifluoroacetic acid saturated with sinapinic acid — SPA] applied. The array was then placed in the SELDI TOF-MS for analysis. A peak profile was generated using the ProteinChip® Array Software Package. Peaks were identified only in those cases where the signal to noise ratio was greater than a value of 5. ROS treatment induced the appearance of two novel protein species (lower panel, MW = 37 615 and 55 764) that were absent from the control sample (upper panel). Note the overall reduction in the amount of BSA (upper panel, MW = 66 426) binding to the array surface after ROS exposure (reduced intensity value) suggesting either alterations in BSA binding characteristics or loss of parent BSA molecule due to ROS-induced fragmentation.

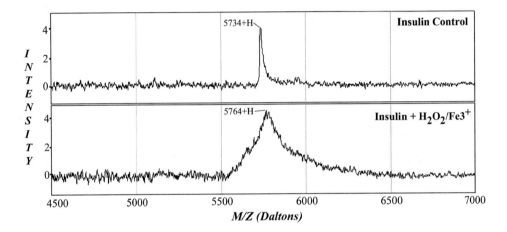

Fig. 4. **The effect of ROS on purified bovine insulin analyzed using ProteinChip®
array technology**. Solubilized insulin was exposed to the hydroxyl radical for a period of
30 min at 37°C by performing a Fenton reaction (i.e. phosphate buffered saline containing
1.15 μg/ml insulin with, or without 5 mM H_2O_2, 100 μM ascorbate and 150 μM Fe^{3+} was
incubated for 30 min at 37°C followed by centrifugation at 10 000 xg for 5 min), applied
to a NP ProteinChip® array and analyzed by SELDI TOF-MS as described in Fig. 3. ROS
treatment induced the appearance of a modified insulin peak (lower panel, MW = 5764)
that was of a higher molecular weight than that detected in the control sample (upper
panel, MW = 5734). Note the overall increase in the amount of insulin binding to the array
surface after ROS exposure (lower panel, MW = 5764, broader peak width) suggesting a
ROS modification of insulin and/or altered binding characteristics of the protein to the
NP ProteinChip® array surface.

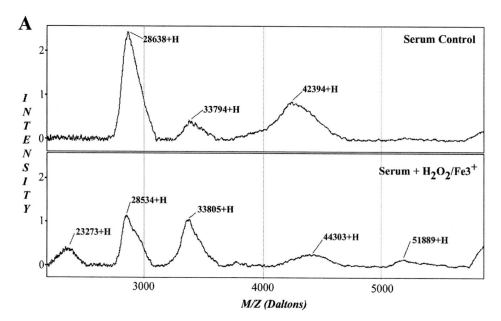

Fig. 5. **The effect of ROS on human serum analyzed using (A) normal phase or (B) anionic exchange ProteinChip® array technology.** Human serum was exposed to the hydroxyl radical for a period of 30 min at 37°C by performing a Fenton reaction (i.e. phosphate buffered saline containing a 1:100 dilution of human serum, with or without 5 mM H_2O_2, 100 μM ascorbate and 150 μM Fe^{3+} was incubated for 30 min at 37°C followed by centrifugation at 10 000 xg for 5 minutes). One-microliter volumes of the resultant supernatants from both control and ROS-exposed samples were spotted onto a normal phase (NP) ProteinChip® array and allowed to dry. Alternatively, each supernatant was diluted 1:9 (v/v) with Tris buffer (pH 10) containing 0.1% (v/v) Triton X-100 and then spotted onto an anionic exchange (SAX) ProteinChip® array and incubated for 30 min in a humidified chamber. Each well of the NP array was washed with distilled water, allowed to dry and a matrix layer (SPA) applied. Each well of the SAX array was washed with two changes of Tris buffer (pH 10) containing 0.1% (v/v) Triton X-100, followed by rapid washing with two changes of distilled water, allowed to dry and a matrix layer (SPA) applied. The arrays were then placed in the SELDI TOF-MS for analysis as described in Fig. 3. In this example, ROS treatment resulted in the loss of a peak at 42 394 as well as the appearance of a number of novel proteins capable of binding to the NP ProteinChip® array (A, lower panel, MW = 23 273, 44 303 and 51 889). ROS treatment also induced the appearance of a number of novel proteins that bound to the SAX ProteinChip® array (B, lower panel, MW = 7680, 6847, 5297, 4353 and 4076), proteins that were not detected in the control sample (B, upper panel).

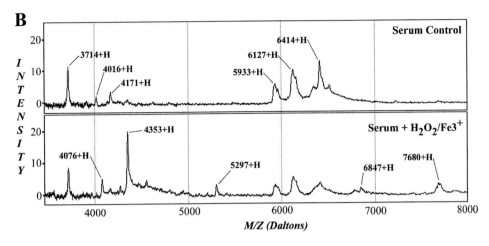

Fig. 5 *(Continued)*

One approach that we are investigating at present is to utilize the protein complement found in the red blood cell (RBC) membrane as a form of biological sensor for oxidative stress. Our rationale for this approach is that these cells are found throughout the body, have a relatively fixed protein complement in their cellular membranes, are exposed to the normal variations in oxygen levels associated with respiration and are easily collected from a subject. In our preliminary studies, we have used whole human blood obtained from a standard venipuncture, which was then exposed to an increasing range of gamma-irradiation as has previously been described.[19] The RBC membranes present in a 50 µl aliquot of whole blood from control and irradiated portions of the same sample were then rapidly isolated by hypotonic lysis and washing by centrifugation. The resulting cell membrane pellet was solubilized in preparation for analysis using the ProteinChip® array. Sample preparation and analysis time (not counting irradiation) was less than 1 hour. When compared, it was found that gamma irradiation exposure, already known to produce a wide range of free radical species,[5] resulted in the appearance of several novel RBC membrane-derived proteins (Fig. 6).

4. Conclusions/Future Directions

As can be seen from the preliminary data presented above, the utility of SELDI TOF-MS coupled with the ProteinChip® array platform indicates that this technology can be utilized for the identification of protein biomarkers of oxidative stress. The ability of this technology to compare control protein complements to that obtained from treated/exposed samples from a wide variety of experimental

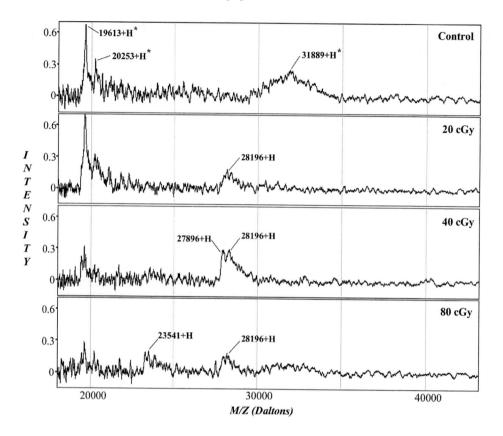

Fig. 6. **The effect of gamma irradiation on human red blood cell membranes (RBC) analyzed using anionic exchange (SAX) ProteinChip® array technology**. Human whole blood was exposed to increasing levels of gamma irradiation as previously described. RBC membranes were isolated from the blood sample by first centrifuging the sample, lysing the RBC pellet with water and washing the resulting membranes by centrifugation with saline to remove contaminating hemoglobin. The membrane pellets were then solubilized by homogenization in an aqueous buffer containing 30% (v/v) acetonitrile, 15% (v/v) isopropanol and 0.1% (v/v) NP-40 followed by centrifugation at 10 000 xg for 5 min to remove any insoluble material. The supernatant was then diluted 1:4 (v/v) with Tris buffer (pH 10) containing 0.1% (v/v) Triton X-100. One-microliter volumes of these solutions were spotted onto an anionic exchange (SAX) ProteinChip® array and incubated for 30 min in a humidified chamber. Each well of the SAX array was washed with two changes of Tris buffer (pH 10) containing 0.1% (v/v) Triton X-100, followed by rapid washing with two changes of distilled water, allowed to dry and a matrix layer (SPA) applied. The arrays were then placed in the SELDI-TOF for analysis as described in Figure 3. Irradiation induced both a reduction of intensity of peaks at 19 613 and 20 253 as well as the appearance of a number of novel RBC membrane proteins that bound to the SAX ProteinChip® Array. In this example, a novel protein (MW = 28 196) was induced at 20, 40 and 80 cGy exposure levels. In addition, novel proteins associated only with a particular dose were detected at both 40 cGy (MW = 27 896) and 80 cGy (MW = 23 541) exposure levels. Furthermore, several RBC membrane proteins became less abundant with increasing exposure levels (Control panel, * = labeled peaks).

models suggests that common biomarkers of oxidative stress could be identified using this approach. The relative ease of use, the short sample analysis time required, the selective/specific protein binding properties of the ProteinChip® array and the unambiguous molecular weight data generated are all advantages of the technology.

The use of the RBC membrane as a biological sensor of oxidative stress is not a new concept.[20-22] However, the application of SELDI-TOF and the ProteinChip® array technology makes this approach a rapid and simple procedure that may find many uses as a monitoring tool in a wide variety of situations including radiation exposure and oxidative stress.

Acknowledgments

This work was made possible by a Space Act Agreement between NASA-Johnson Space Center, Houston, Texas and Ciphergen Biosystems, Inc., Fremont, CA.

References

1. Kagawa, Y., Cha, S. H., Hasegawa, K., Hamamoto, T. and Endo, H. (1999). Regulation of energy metabolism in human cells in aging and diabetes: FoF(1), mtDNA, UCP, and ROS. *Biochem. Biophys. Res. Commun.* **266**(3): 662–676.
2. Fridovich, I. (1979). The biology of oxygen radicals. *Science* **201**(4359): 875–880.
3. Niess, A. M., Dickhuth, H. H., Northoff, H. and Fehrenbach, E. (1999). Free radicals and oxidative stress in exercise — immunological aspects. *Exerc Immunol. Rev.* **5**: 22–56.
4. Zhang, X., Rosenstein, B. S., Wang, Y., Lebwohl, M. and Wei, H. (1997). Identification of possible reactive oxygen species involved in ultraviolet radiation-induced oxidative DNA damage. *Free Radic. Biol. Med.* **23**(7): 980–985.
5. Riley, P. A. (1994). Free radicals in biology: oxidative stress and the effects of ionizing radiation. *Int. J. Rad. Biol.* **65**(1): 27–33.
6. Wells, P. G., Kim, P. M., Laposa, R. R., Nicol, C. J., Parman, T. and Winn, L. M. (1997). Oxidative damage in chemical teratogenesis. *Mutat. Res.* **396**(1–2): 65–78.
7. Gracy, R. W., Talent, J. M., Kong, Y. and Conrad, C. C. (1999). Reactive oxygen species: the unavoidable environmental insult? *Mutat. Res.* **428**(1–2): 17–22.

8. Lovell, M. A., Xie, C., Gabbita, S. P. and Markesbery, W. R. (2000). Decreased thioredoxin and increased thioredoxin reductase levels in Alzheimer's disease brain. *Free Radic. Biol. Med.* **28**(3): 418–427.

9. Dhalla, N. S., Temsah, R. M. and Netticadan, T. (2000). Role of oxidative stress in cardiovascular diseases. *J. Hypertens.* **18**(6): 655–673.

10. Halliwell, B. (1993). The role of oxygen radicals in human disease, with particular reference to the vascular system. *Haemostasis* **23**(S1): 118–126.

11. Abdi, S. and Ali, A. (1999). Role of ROS modified human DNA in the pathogenesis and etiology of cancer. *Cancer Lett.* **142**(1): 1–9.

12. Armstrong, D. (1998). Free radical and antioxidant protocols. *In* "Methods in Molecular Biology" (J. M. Walker, Ed.), Humana Press, Totowa, NJ.

13. Wright, G. L., Jr., Cazares, L. H., Leung, S.-M., Nasim, S., Adam, B.-L., Yip, T.-T., Schellhammer, P. F., Gong, L. and Vlahou, A. (1999). ProteinChip® surface enhanced laser desorption/ionization (SELDI) mass spectrometry: a novel protein biochip technology for detection of prostate cancer biomarkers in complex protein mixtures. *Prostate Cancer and Prostatic Diseases* **2**: 264–276.

14. Xiao, Z., Jiang, X., Beckett, M. L. and Wright, G. L., Jr. (2000). Generation of baculovirus recombinant prostate-specific membrane antigen and its use in the development of a novel protein biochip quantitative immunoassay. *Protein Exp. Pur.* **19**: 12–21.

15. Paweletz, C. P., Gillespie, J. W., Ornstein, D. K., Simone, N. L., Brown, M. R., Cole, K. A., Wang, Q. H., Huang, J., Hu, N., Yip, T.-T., Rich, W. E., Kohn, E. C., Linehan, W. M., Weber, T., Taylor, P., Emmert-Buck, M. R., Liotta, L. A. and Petricoin, E. F. III. (2000). Rapid protein display profiling of cancer progression directly from human tissue using a protein biochip. *Drug Dev. Res.* **49**: 34–42.

16. Davies, H., Lomas, L. and Austen, B. (1999). Profiling of amyloid-beta peptide variants using SELDI ProteinChip® arrays. *Biotechniques* **27**: 1258–1262.

17. Li, Y.-M., Lai, M.-T., Xu, M. Huang, DiMuzio-Mower, J., Sardana, M. K., Shi, X.-P., Yin, K.-C., Shafer, J. A. and Gardell, S. J. (2000). Presenilin 1 is linked with gamma-secretase activity in the detergent solubilized state. *Proc. Natl. Acad. Sci.* **97**: 6139–6143.

18. Austen, B., Davies, H., Stephens, D. J., Frears, E. R. and Walters, C. E. (1999). The role of cholesterol in the biosynthesis of beta-amyloid. *Neuroreport* **10**: 1699–1705.

19. Durante, M., George, K. and Yang T. C. (1997). Biodosimetry of ionizing radiation by selective painting of prematurely condensed chromosomes in human lymphocytes. *Rad. Res.* **148** (**5 Suppl**): S45–S50.

20. Kahane, I. and Rachmilewitz E. A. (1976). Alterations in the red blood cell membrane and the effect of vitamin E on osmotic fragility in beta-thalassemia major. *Isr. J. Med. Sci.* **12**(1): 11–15.

21. Shinar, E., Rachmilewitz, E. A., Shifter, A., Rahamim, E. and Saltman, P. (1989). Oxidative damage to human red cells induced by copper and iron complexes in the presence of ascorbate. *Biochim. Biophys. Acta.* **1014**(1): 66–72.
22. Katz, D., Mazor, D., Dvilansky, A. and Meyerstein, N. (1996). Effect of radiation on red cell membrane and intracellular oxidative defense systems. *Free Radic. Res.* **24**(3): 199–204.

Measurement of Oxidative Stress: Technology, Biomarkers and Applications

(Lipid Damage)

Chapter 22

Quantification of Isoprostanes as an Index of Oxidant Stress Status *In Vivo*

Jason D. Morrow*, Erin E. Reich, L. Jackson Roberts
and Thomas J. Montine

Keywords: Isoprostanes and oxidant stress.

Jason D. Morrow*, Erin E. Reich, L. Jackson Roberts and **Thomas J. Montine** •
Departments of Medicine, Pharmacology and Pathology Vanderbilt University School of
Medicine, Nashville, TN 37232-6602 USA

*Corresponding Author. 526 MRB-1, Vanderbilt University, 23rd and Pierce Aves., Nashville
TN 37232-6602 USA
Tel: 615/343-1124, E-mail: jason.morrow@mcmail.vanderbilt.edu

1. Introduction

The development of specific, reliable and non-invasive methods for measuring oxidative stress in humans is essential for establishing the role of free radicals in human diseases.[1] Lipid peroxidation in a central feature of oxidant stress and can be assessed by a number of methods including the quantification either primary or secondary peroxidation end products. Primary end products include the measurements of conjugated dienes and lipid hydroperoxides. Secondary end products that are quantified include thiobarbituric reactive substances, gaseous alkanes and prostaglandin F_2-like products termed F_2-isoprostanes (F_2-IsoPs).[1, 2] Quantification of these various compounds has proven highly useful for the study of free radical-mediated lipid peroxidation in a number of *in vitro* model systems. On the other hand, the F_2-IsoPs appear to be a far more accurate marker of oxidative stress *in vivo* in humans and animals than other available methods.[3] The purpose of this chapter is to acquaint the reader with the IsoPs from a biochemical perspective and to provide information regarding the utility of quantifying these compounds as indicators of oxidant stress.

2. Mechanism of Formation of the Isoprostanes

IsoPs are prostaglandin (PG)-like compounds formed from the peroxidation of arachidonic acid.[2, 4] Unlike PGs, however, they do not require the cyclooxygenase enzyme for their formation. Figure 1 outlines the mechanism by which IsoPs are generated. Following abstraction of a bis-allylic hydrogen atom and the addition of a molecule of oxygen to arachidonic acid to form a peroxyl radical, endocyclization occurs and an additional molecule of oxygen is added to form PGG_2-like compounds. These unstable bicycloendoperoxide intermediates are then converted to parent IsoPs. Based on this mechanism of formation, four IsoP regioisomers are generated.[2, 4] Compounds are denoted as either 5-, 12-, 8-, or 15-series regioisomers depending on the carbon atom to which the side chain hydroxyl is attached.[5] IsoPs that contain the F-type prostane ring are isomeric to $PGF_{2\alpha}$ and are thus referred to as F_2-IsoPs. An important structural distinction between IsoPs and cyclooxygenase-derived PGs is that the former contain side chains that are predominantly oriented *cis* to the prostane ring while the latter possess exclusively *trans* side chains.[2, 4] A second important difference between IsoPs and PGs is that IsoPs are formed completely *in situ* in phospholipids and are subsequently released by a phospholipase(s).[6] In contrast, PGs are generated only from free arachidonic acid.

Abreviations used: (IsoP), isoprostane; (PG), prostaglandin.

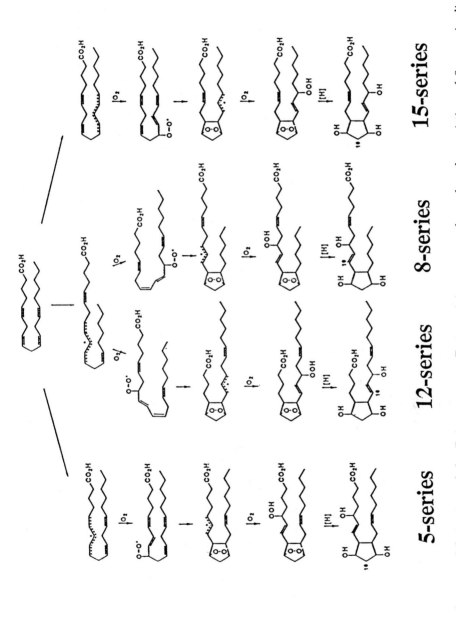

Fig. 1. Mechanism of formation of the F₂-isoprostanes. Four regioisomers are formed each consisting of 8 racemic diastereomers. Stereochemistry is not indicated.

3. Quantification of F$_2$-Isoprostanes

Several methods exist to quantify the F$_2$-IsoPs. They can be measured using gas chromatographic/negative ion chemical ionization mass spectrometric approaches employing stable isotope dilution.[3] 15-F$_{2t}$-IsoP (8-iso-[^{2}H$_4$] PGF$_{2\alpha}$) and [^{2}H$_4$]-PGF$_{2\alpha}$ are available from commercial sources for use as internal standards for these assays. Mass spectrometry is a highly sensitive method to measure IsoPs and yields quantitative results in the low picogram range. Its drawbacks are that it is labor intensive and requires substantial capital expenditure.

Alternative methods have recently been developed to quantify IsoPs using immunological approaches.[7] Antibodies have been generated against 15-F$_{2t}$-IsoP and immunoassay kits are commercially available. A potential drawback of these methods is that limited information is currently available regarding their precision and accuracy. In addition, little data exists comparing IsoP levels determined by immunoassay to mass spectrometry. On the other hand, it is likely that the quantification of IsoPs using immunoassays will expand research in this area since these techniques are affordable and relatively easy to perform.

4. Isoprostane Formation *In Vivo*

Normal levels of F$_2$-IsoPs in human biological fluids such as plasma (35 ± 6 pg/ml) have been defined.[3] Interestingly, quantities of these compounds exceed those of cyclooxygenase derived PGs and thromboxane by an order of magnitude, suggesting that the formation of IsoPs is a major pathway of arachidonic acid disposition. Further, it is important to consider the relevance of the finding that levels of F$_2$-IsoPs are sufficient to be detected in every normal biological fluid that has been assayed including plasma, urine, bronchoalveolar lavage fluid, cerebrospinal fluid, and bile.[7] Previously, using other methods to assess lipid peroxidation, there was little definitive evidence indicating lipid peroxidation occurs *in vivo* except under situations of oxidant stress. However, the finding of detectable levels of F$_2$-IsoPs in normal animal and human biological fluids and tissues indicates there is ongoing lipid peroxidation that is incompletely suppressed by antioxidant defenses, even in normal humans. This finding may lend some support to the hypothesis that the normal aging process is due to enhanced oxidant damage of important biological molecules over time. In this regard, a previous study has suggested that IsoP levels in normal humans increase with age, although a more recent report refutes this.[8, 9] Additional studies will be necessary to definitively determine whether normal aging is associated with increased IsoP formation *in vivo*.

Over the past several years, a number of studies involving the quantification of F$_2$-IsoPs as an index of lipid peroxidation or oxidant stress both *in vitro* and *in vivo* have been performed. This work is particularly important with regards to *in vivo* oxidant stress since it has been previously recognized that one of the greatest needs in the field of free radical research is a reliable non-invasive method to assess lipid peroxidation *in vivo* in humans.[1] In this respect, most methods available to assess oxidant stress *in vivo* previously have suffered from a lack of sensitivity and/or specificity or are unreliable. However, a substantial body of evidence indicates that measurement of IsoPs in body fluids such as plasma provides a reliable approach to assess lipid peroxidation *in vivo* and represents a major advance in our ability to assess oxidative stress status in animals and humans.[7, 10] Specific examples where IsoPs have been utilized as measures of oxidant stress in association with human disease or animal models of human disease are discussed below.

5. Measurement of the Major Urinary Metabolite of 15-F_{2t}-Isop as an Index of Systemic Oxidant Status

Unmetabolized IsoPs can be quantified either as free compounds in biological fluids or esterified in tissue lipids.[7, 10] While applicable to animal studies, the measure of these compounds in human trials is potentially limited by invasive procedures necessary to obtain tissues or biological fluids. In addition, levels of IsoPs in a particular organ or body cavity likely do not represent an index of systemic or "whole body" oxidation. Whereas unmetabolized urinary IsoPs can be quantified, the interpretation of measuring these compounds as an index of total systemic production of IsoPs is confounded by the potential contribution of local IsoP production in the kidney.[7, 10] It is well established in the PG field that measurement of the urinary excretion of metabolites of prostaglandins represents the most reliable approach to assess total endogenous PG production.[11] Accordingly, quantification of the urinary excretion of an F_2-IsoP metabolite should also afford an accurate measure of endogenous production of IsoPs and allow for urine collected over a number of hours to provided an integrated index of IsoP production.

Thus, we recently undertook a study to identify a urinary metabolite of an F_2-IsoP that could be measured. One of the F_2-IsoPs which we have shown is formed *in vivo* is 15-F_{2t}-IsoP (8-iso-$PGF_{2\alpha}$).[12] Metabolism of PGs has usually has been found to produce a number of metabolites. However, we found that a single metabolite predominated in the profile of compounds produced from metabolism of 15-F_{2t}-IsoP. This metabolite was identified by mass spectrometry as 2,3-dinor-5,6-dihydro-15-F_{2t}-IsoP. We have recently synthesized this compound and converted it to an [$^{18}O_4$] derivative for use as an internal standard and developed a stable isotope dilution negative ion chemical ionization mass spectrometric assay. The urinary excretion of this metabolite in normal humans was found to be 0.39 ± 0.18 ng/mg creatinine (mean $\pm$ 2 SD). Excretion of the metabolite increases markedly in animal models of oxidant stress.[13] Further, urinary levels of the metabolite increased by a mean of 2.5 above levels in humans with polygenic hypercholesterolemia, a condition associated with enhanced IsoP formation.[13] These increases were suppressed by a mean of 54% following eight weeks of treatment with a combination of vitamin E, vitamin C, and beta-carotene. These data support the contention that measurement of the urinary excretion of 2,3-dinor-5,6-dihydro-15-F_{2t}-IsoP will contribute importantly to the assessment of free radical-mediated lipid peroxidation *in vivo*. This approach may also provide an important tool to assess oxidative stress status in large clinical trials where the logistics of obtaining blood or tissues are limited. Development of an immunoassay method for the measurement of this metabolite is currently being undertaken. If the accuracy of the immunoassay for this compound can be validated by mass spectrometry, this may result in the wide availability of quantification of this metabolite as a facile, non-invasive means to assess oxidant stress.

6. Formation of Isoprostanes in Animal Models of Oxidant Stress

It has been previously shown that the formation of IsoPs increases dramatically in animal models of oxidant stress. When rats were administered CCl_4, they undergo free-radical-induced injury with the major site of toxicity being the liver. Esterified levels of IsoPs in liver tissue increased by 200-fold within 1 h of treatment and subsequently decline over 24 hours.[14] Plasma free and lipid esterified IsoP concentrations increased after liver levels up to 50-fold greater than levels in untreated animals. Formation of IsoPs was proportional to CCl_4 dose given. Administration of the antioxidant lazaroid U78517 to CCl_4-treated animals significantly blunted the enhanced formation of IsoPs in this model.[10]

Diquat is a dipyridyl herbicide that undergoes redox cycling *in vivo* generating large amounts of the superoxide anion. This compound causes hepatic and renal injury in rats and this effect is markedly augmented in animals deficient in Se, a trace element that is required for the enzymatic activities of glutathione peroxidase and other antioxidant proteins. Previous studies have suggested that lipid peroxidation might be involved in the tissue damage associated with this agent. To study whether F_2-IsoPs were generated in increased amounts in association with diquat administration to Se-deficient rats, levels of compounds were quantified in plasma and tissues from Se-deficient rats following diquat administration. Se-deficient rats administered diquat showed 10- to 200-fold increases in plasma F_2-IsoPs and the sources of the IsoPs were determined to be the kidney and liver.[15] Further studies disclosed that the extent of tissue injury and IsoP formation directly correlated with the degree of Se depletion.[16] Taken together, these studies suggest that quantification of F_2-IsoPs in animal models of oxidant injury represents an accurate method to assess lipid peroxidation *in vivo*.

More recently, we have examined IsoP formation in mice with a targeted deletion of the apolipoprotein E gene *(apoE)*.[17] These animals, when aged, serve as a limited experimental model for human Alzheimer's disease based on numerous studies that have associated inheritance of the aplipoprotein E gene (APOE4) with an increased risk and earlier onset of the disorder in humans.[17] As discussed below, we have found that IsoP formation is significantly increased in the central nervous system of patients with Alzheimer's disease, supporting the hypothesis that increased oxidant stress plays a role in this disorder. In this study, we examined male and female *apoE* −/− and control mice at 10–12 months of age and quantified F_2-IsoPs in cerebral tissue. Compared to control mice, levels of IsoPs were significantly increased in *apoE* −/− animals ($p < 0.05$), supporting a role for increased oxidant stress in this animal model of neurodegeneration.

7. Quantification of F_2-Isops in Association with Human Disease-Studies in Patients with Alzheimer's Disease

A number of studies have been reported examining the utility of quantifying IsoPs as an index of oxidant stress in association with human disease. Although as noted above, it is not entirely clear whether IsoP formation is increased in normal human aging, a number of diseases whose incidence increases with age have been associated with increased tissue and body fluid levels of these compounds.[7, 10, 18] One disorder that we have studied extensively is Alzheimer's disease. Studies on patients with this disease are summarized below so as to provide an example of the importance of the measurement of F_2-IsoPs to delineate the role of lipid peroxidation in human pathophysiology. The reader is referred to additional cited references for other examples regarding the quantification of IsoPs in human diseases.[7, 10, 18]

Regional increases in oxidative damage and lipid peroxidation are a feature of brain tissue obtained post mortem from patients with Alzheimer's disease.[19] However, an objective index of oxidative damage associated with Alzheimer's disease that can be assessed during life is lacking. Such a biomarker could have an important impact on the ability to test hypotheses concerning oxidative damage in Alzheimer's patients by permitting repeated evaluation to follow disease progression or responses to therapeutic interventions. Toward such a goal, we have undertaken several studies to determine whether F_2-IsoP levels are increased in cerebrospinal fluid from patients with Alzheimer's disease compared to individuals without evidence of neurodegenerative disorders.

In the first study, we obtained ventricular fluid from 11 patients with a pathological diagnosis of Alzheimers and 11 control patients post mortem.[20] All subjects participated in a rapid autopsy protocol such that fluid was collected within 3 h of death. Average ventricular fluid F_2-IsoP levels were significantly increased $(72 \pm 7 \text{ pg/ml, mean} \pm \text{S.E.M.})$ compared to control individuals $(46 \pm 4 \text{ pg/ml})$, $p = 0.01$. A significant correlation existed between increases in IsoP levels and higher Braak stage and decreased brain weight, two indices of Alzheimer's disease severity. Subsequently, in a larger study, we have shown that ventricular fluid levels of F_2-IsoPs correlates with the extent of pathological neurodegeneration but not with density of neuritic plaques or neurofibrillary tangles. In addition, increased in IsoP levels did not correlate with ApoE genotype.[21]

Subsequently, we undertook a study in living patients with probable Alzheimer's disease to determine whether cerebrospinal fluid F_2-IsoP levels were altered.[22] Cerebrospinal fluid was obtained from the lumbar cistern in 27 patients with Alzheimer's disease and 25 controls without neurodegenerative disorders. Subjects were matched for age and sex. Interestingly, lumbar cerebrospinal fluid levels of F_2-IsoPs were significantly increased $(31.0 \pm 2.6 \text{ pg/ml})$ in Alzheimer's patients compared to control subjects $(22.9 \pm 1.0 \text{ pg/ml})$, $p < 0.05$. Levels of IsoPs did not correlate with age. Taken together, these studies suggest that quantification of IsoPs in cerebrospinal fluid of patients with Alzheimer's disease may be of use as an *intra vitum* index of disease progression or of response to therapeutic interventions.

8. Conclusions

The discovery of IsoPs as products of non-enzymatic lipid peroxidation has opened up new areas of investigation regarding the role of free radicals in human physiology and pathophysiology. The quantification of these compounds as markers of oxidative stress status appears to be an important advance in our ability to explore the role of free radicals in the pathogenesis of human disease. Although considerable information has been obtained since the initial report of the discovery of IsoPs, much remains to be understood about the role of these molecules as not only markers, but as mediators, of oxidant stress *in vivo*. It is anticipated that additional research in this area will provide these important insights.

Acknowledgments

Supported by NIH Grants DK48831, CA77839, GM42056, GM15431 and DK26657 from the United States Public Health Service. Dr. Morrow is the recipient of a Burroughs-Wellcome Fund Clinical Scientist Award in Translational Research.

References

1. Halliwell, B. and Grootveld M. (1987). The measurement of free radical reactions in humans. *FEBS Lett.* **213**: 9–14.
2. Morrow, J. D., Hill, K. E., Burk, R. F., Nammour, T. M., Badr, K. F. and Roberts, L. J. (1990). A series of prostaglandin F_2-like compounds are produced *in vivo* in humans by a non-cyclooxygenase, free radical-catalyzed mechanism. *Proc. Natl. Acad. Sci. USA* **87**: 9383–9397.
3. Morrow, J. D. and Roberts, L. J. (1999). Mass spectrometric quantification of F_2-isoprostanes in biological fluids and tissues as measure of oxidant stress. *Meth. Enzymol.* **300**: 3–12.
4. Morrow, J. D., Harris, T. M. and Roberts, L. J. (1990). Noncyclooxygenase oxidative formation of a series of novel prostaglandins: analytical ramifications for measurement of eicosanoids. *Anal. Biochem.* **184**: 1–10.
5. Taber, D. F., Morrow, J. D. and Roberts, L. J. (1997). A nomenclature system for the isoprostanes. *Prostaglandins* **53**: 63–67.
6. Morrow, J. D., Awad, J. A., Boss, H. J., Blair, I. A. and Roberts, L. J. (1992). Non-cyclooxygenase-derived prostanoids (F_2-isoprostanes) are formed in situ on phospholipids. *Proc. Natl. Acad. Sci. USA* **89**:10 721–10 725.
7. Morrow, J. D., Chen, Y., Brame, C. J., Yang, J., Sanchez, S. C., Xu, J., Zackert, W. E., Awad, J. A. and Roberts, L. J. (1999). The isoprostanes: unique prostaglandin-like products of free radical-initiated lipid peroxidation. *Drug Metab. Rev.* **31**: 117–139.
8. Pratico, D., Reilly, M., Lawson, J., Delanty, N. and FitzGerald, G. A. (1995). Formation of 8-iso-prostaglandin $F_{2\alpha}$ by human platelets. *Agents Actions* **45 (Suppl.)**: 27–31.
9. Feillet-Coudray, C., Tourtauchaux, R., Niculescu, N., Rock, E., Tauveron, I., Alexandre-Gouabau, M., Rayssiguier, R., Jalenques, Y. I. and Mazur, A. (1999). Plasma levels of 8-epi-PGF2α, an in vivo marker of oxidative stress, are not affected by aging or Alzheimer's disease. *Free Radic. Biol. Med.* **27**: 463–469.
10. Morrow, J. D. and Roberts, L. J. (1997). The isoprostanes: unique bioactive products of lipid peroxidation. *Prog. Lipid Res.* **36**: 1–21.
11. Patrono, C. (1989). Measurement of thromboxane biosynthesis in man. *Eicosanoids* **2**: 249–251.
12. Morrow, J. D., Minton, T. A., Badr, K. F. and Roberts, L. J. (1994). Evidence that the F_2-isoprostane, 8-epi-prostaglandin $F_{2\alpha}$, is formed in vivo. *Biochim. Biophys. Acta.* **1210**: 244–248.
13. Roberts, L. J., Morrow, K. P., Zackert, W. E., Oates, J. A. and Morrow, J. D. (1996). Identification of the major urinary metabolite of the F_2-isoprostane 8-iso-prostaglandin $F_{2\alpha}$ in humans. *J. Biol. Chem.* **271**: 20 617–20 620.
14. Morrow, J. D., Awad, J. A., Kato, T., Takahashi, K., Badr, K. F., Roberts, L. J. and Burk, R. F. (1992). Formation of novel non-cyclooxygenase-derived

prostanoids (F$_2$-isoprostanes) in carbon tetrachloride hepatotoxicity. *J. Clin. Invest.* **90**: 2502–2507.

15. Awad, J. A., Morrow, J. D., Hill, K. E., Roberts, L. J. and Burk, R. F. (1994). Detection and localization of lipid peroxidation in selenium- and vitamin-E deficient rats using F$_2$-isoprostanes. *J. Nutri.* **124**: 810–816.

16. Burk, R. F., Hill, K. E., Awad, J. A., Morrow, J. D., Kato, T., Cockell, K. A. and Lyons, P. R. (1995). Pathogenesis of diquat-induced liver necrosis in selenium-deficient rats: assessment of the roles of lipid peroxidation and selenoprotein P. *Hepatology* **21**: 561–569.

17. Montine, T. J., Montine, K. S., Olson, S. J., Graham, D. G., Roberts, L. J., Morrow, J. D., Linton, M. F., Fazio, S. and Swift, L. L. (1999). Increased cerebral cortical lipid peroxidation and abnormal phospholipids in aged homozygous apoE-deficient C57BL/6J mice. *Exp. Neurol.* **158**: 234–241.

18. Awad, J. A., Roberts, L. J., Burk, R. F. and Morrow, J. D. (1996). Isoprostanes-prostaglandin-like compounds formed in vivo independently of cyclooxygenase. *Gastroenterol. Clin. NA* **25**: 409–427.

19. Markesbery, W. R. (1997). Oxidative stress hypothesis in Alzheimer's disease. *Free Radic. Biol. Med.* **23**: 137–147.

20. Montine, T. J., Markesbery, W. R., Morrow, J. D. and Roberts, L. J. (1998). Cerebrospinal fluid F$_2$-isoprostane levels are increased in Alzheimer's disease. *Ann. Neurol.* **44**: 410–413.

21. Montine, T. J., Markesbery, W. R., Zackert, W. E., Sanchez, S. C., Roberts, L. J. and Morrow, J. D. (1999). The magnitude of brain lipid peroxidation correlates with the extent of degeneration but not with density of neuritic plaques or neurofibrillary tangles or with ApoE genotype in Alzheimer's disease patients. *Am. J. Path.* **155**: 863–868.

22. Montine, T. J., Beal, M. F., Cudkowicz, M. E., O'Donnell, H., Margolin, R. A., McFarland, L., Bachrach, A. F., Zackert, W. E., Roberts, L. J. and Morrow, J. D. (1999). Increased CSF F$_2$-isoprostanes in probable AD. *Ann. Neurol.* **52**: 562–565.

Chapter 23

Analysis of Phospholipid Oxidation by Electrospray Mass Spectrometry

Corinne M. Spickett

Corinne M. Spickett • Department of Immunology, Strathclyde Institute for Biomedical Science, University of Strathclyde, 27 Taylor Street, Glasgow, G4 0NR, UK
Tel: +44 141 5483827, E-mail: c.m.spickett@strath.ac.uk

1. Introduction

Oxidative damage to biomolecules is becoming an established facet of the process of aging and age-related diseases, so the development of improved and more informative methods for monitoring oxidative damage is an important area of research. One subject which has long attracted much interest is that of lipid oxidation, as most biological systems contain significant amounts of mono- or poly-unsaturated lipids which are susceptible to oxidative damage. Lipids are essential components of cell membranes and lipoproteins, and their oxidation can lead to changes in lipid packing, dysfunction or loss of cell integrity. The study and detection of the specific processes involved is therefore extremely important in understanding age-related disease.

Methods for the analysis of oxidative damage to lipids can be categorized into those that measure early oxidation products, such as conjugated dienes or lipid hydroperoxides, and those that measure small breakdown products, such as aldehydes. Gas chromatography mass spectrometry has emerged as a sensitive and specific technique for detecting the formation of both types of oxidation product.[1-3] Its main disadvantage is the requirement for volatile analytes, and therefore for sample derivatization; this complicates sample preparation and may result in loss of information about the native lipid. On the other hand, the more recent application to the field of lipid oxidation of the soft ionization technique electrospray mass spectrometry (ESMS) has advantages for the analysis of oxidized derivatives of phospholipids, as these can be detected directly without hydrolysis or derivatization. ESMS has been used in a wide range of applications, including the analysis of lipids, isoprostanes, and fatty acids from biological samples, studies of lipid remodelling in cell membranes, and more recently for the detection of oxidized lipids in LDL, atherosclerotic plaques and oxidatively stressed cells. This review will describe the application of ESMS to studies of biological lipids and will review the progress in detection of various types of oxidized phospholipids in biological and clinical samples by this technique. While the main emphasis is on intact phospholipid analysis, some work on fatty acid derivatives and cholesterol will also be included where relevant.

2. Principles of Electrospray Mass Spectrometry (ESMS)

Electrospray ionization is a soft ionization technique for mass spectrometry, which, unlike electron impact or chemical ionization, does not necessarily induce fragmentation of the molecular ion, and is thus particularly suitable for the analysis of biomolecules. It was initially used in studies of small proteins and peptides because of this property.[4, 5] Subsequently it has also become popular for analysis of lipids and fatty acids, and is ideal for the study of complex biological mixtures of lipids, as the individual components can be determined from their mass without the complication of fragmentation.

The majority of biological molecules contain readily ionizable groups, and are therefore suitable for detection with ESMS. It can be carried out in either positive or negative ion mode, depending on the properties of the desired analyte: negative ion is suitable for the detection of fatty acids, isoprostanes, phosphatidylglycerols, and phosphatidylserines, while positive ion electrospray is particularly good for the detection of phosphatidylcholines and sphingomyelins, and can also detect phosphatidylethanolamines, phosphatidylserines and cholesterol.

The functioning of electrospray mass spectrometers has been well reviewed elsewhere,[4, 6] and will not be described at length here, but some basic details will be given to highlight important aspects of the technique. The sample, in a volatile solvent, is introduced into the source through a charged capillary. The source is usually heated to 70–200°C, depending on the machine in question, to aid desolvation. Introduction of a sheath gas (usually nitrogen) at the tip of the capillary assists in the formation of a spray containing solvated molecular ions, which is then dried by a further gas stream and heating to yield individual molecular ions. These are focused into a beam while still within the source, and the beam is refined further before entering the detector. The most common detector in electrospray mass spectrometers is the quadrupole detector, although time of flight (TOF) and quadrupole ion trap detectors are also becoming more common. Fragmentation of particular molecular ions, and thus greater power of identification, can be achieved with tandem instruments or MS^n ion traps. In both cases this requires selection of the molecular ion of interest, followed by introduction of gas (usually argon or helium) into the collision cell to cause collision-induced-decomposition.

3. Analysis of Phospholipids in Biological Samples

By the mid-1990s, electrospray ionization had begun to take over from other ionization techniques such as fast atom bombardment (FAB), chemical ionization and thermospray for the analysis of intact phospholipids in biological systems, owing to the soft ionization mechanism and the minimal sample preparation required. Since then ESMS has been used to investigate the lipid composition of a variety of different cells and tissues. Some of these studies will be described briefly in order to introduce the nature of the information available and facilitate understanding of the detection of oxidized species.

Phosphatidylcholines, phosphatidylethanolamines and phosphatidylserines all appear at even mass to charge ratios (m/z), whether in positive ion mode or negative ion mode. Sphingomyelins, phosphatidylinositols, phospatidylglycerols and diphosphatidyl-glycerols all appear at odd mass to charge ratios, owing to the odd number of phosphorus and nitrogen atoms they contain (Table 1 and Fig. 1). In tandem MS the phospholipid class can readily be determined from characteristic fragment ions, such as 184 m/z in positive ion for PCs and sphingomyelins, or 241 m/z for phosphatidylinositols and 87 m/z for phosphatidylserines in negative ion.[7] Information about the fatty acyl chains can also be obtained from fragmentation.[8]

Early work on human red blood cell lipids showed the appearance of a large number of different lipid species.[9] Gycerophosphocholines were observed in their sodiated forms in positive ion mode, while negative ion mode was used to observe glycerophosphoethanolamines, serines and inositols. Subsequently ESMS was used to investigate the thrombin-induced release of arachidonic acid from platelet phospholipids, and plasmenylethanolamines were identified as the major source.[8] ESMS has also been used by Hsu *et al.*[10] to show that differentiation of the monocytic cell line U937 results in an increase in the arachidonic acid content of membrane phosphatidylcholines, which facilitates phospholipase A2 activation. In pancreatic islets it has been shown using positive ion ESMS that neither IL-1 or TNFα cause sphingomyelin hydrolysis,[11] unlike in other cell types where it has been indicated as a signalling mechanism. Bacterial cell membranes tend to contain predominantly phosphatidylethanolamines and phosphatidylglycerols which can be observed better in negative ion mode, as shown in *Escherichia coli* and three species of *Bacillus*.[12]

The advantage of coupling HPLC to ESMS (LC-MS) was also demonstrated early on when studying the remodelling of mammalian cell membranes of cells

Table 1. Molecular Ion Mass to Charge Ratios (m/z) for Some Typical Phospholipids Observed by ESMS.

Total Fatty Acid Carbon Number and Double Bonds (Possible Distribution)	Phophatidy-cholines (M+H)⁺	Phosphatidyl-ethanolamine (M+H)⁺	Phosphatidyl-inositols (M–H)⁻	N-acyl Chain Carbon Number + Unsaturation	Sphingo-myelins (M+H)⁺
C 32: 1 (e.g. 16:0, 16:1)	732	690	807	C 16: 0	703
C 34: 2 (e.g. 16:0, 18:2)	758	716	833	C 18: 0	731
C 34: 1 (e.g. 16:0, 18:1)	760	718	835	C 20: 0	759
C 36: 4 (e.g. 16:0, 20:4)	782	740		C 22: 0	787
C 36: 2 (e.g. 18:0, 18:2)	786	744	861	C 24: 1	813
C 38: 6 (e.g. 16:0, 22:6)	806	764		C 24: 0	815
C 38: 4 (e.g. 18:0, 20:4)	810	768	885	C 26: 0	843

- Sodiated phospholipids (M+Na)⁺ give signals at +22 m/z compared to the (M+H)⁺ form.
- Plasmenyl phospholipids give signals at –16 m/z compared to the diacyl forms.
- For phospholipids observable in both positive and negative mode, (M+H)⁺ appears at +2 m/z compared to (M–H)⁻.

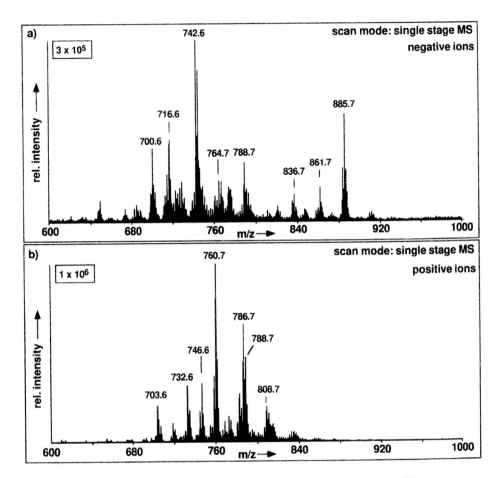

Fig. 1. ESMS of the total lipid extract from Chinese hamster ovary cells. (a) Negative ion mass spectrum, showing $(M-H)^-$ ions of phosphatidylinositols, plasmenyl and acyl phosphatidylethanolamines, and phosphatidylserines. (b) Positive ion mass spectrum, showing $(M+H)^+$ ions of phosphatidylcholines and sphingomyelins, with minor contributions from phosphatidylethanolamines. The boxed numbers indicate the full scale intensity in arbitary units. Assignments of the major signals are given in Table 1. [From Brügger *et al.* (1999). *Proc. Natl. Acad. Sci. USA* **94**: 2339–2344. Copyright © 1997 National Academy of Sciences USA.]

grown in the presence of docosahexenoic acid.[13] Reverse phase HPLC allows the separation of phospholipids mainly based on fatty acyl chain length, although some separation according to headgroup can also be achieved. Normal phase HPLC can also be used, and allows the separation of different phospholipid classes.

A current trend in ESMS is towards decreasing sample sizes by the use of a nanospray source. This has allowed the identification of membrane phospholipids[7] and cholesterol[14] from as few as 1000 Chinese hamster ovary cells. This advance introduces the possibility of monitoring lipid changes in membranes of specific organelles, for example mitochondria, which could be of particular interest in studies of cellular changes during ageing.

4. Detection of Phospholipid Peroxidation in Model Lipid Systems

Oxidation of phospholipids results in a change in their mass, for example due to incorporation of an oxygen molecule or cleavage of an oxidized fatty acyl chain, and this allows the modification to be observed conveniently by ESMS. Experiments by Maffei-Facino *et al.*[15] had already shown that heavier, oxidized lipid species could be observed using FAB mass spectrometry. They reported the formation of multiple peroxide and hydroxy species of the linoleoyl and linolenoyl chains of phosphatidylcholine, although interpretation of their spectra is complicated by the possibility of sodiated lipid forms. However, FAB is not an ideal method for phospholipids, as the low ionization energies required to prevent sample fragmentation lead to low sensitivity, and high backgrounds can result from the desorption of the matrix.

An early study which established the potential of ESMS for detecting lipid peroxidation focused on fatty acid oxidation, using negative ion electrospray for detection of the carboxylate ions.[16] Hydroperoxy derivatives of arachidonic and linoleic acids were generated *in vitro* by lipoxygenases and observed by negative ion ESMS. It was noted that loss of water was common for these oxidation products under the conditions used. Tandem ESMS was used to obtain information about the typical fragmentation characteristics of these molecules, as fragment ions indicative of the position of the hydroperoxide were observed. Reverse phase HPLC was also coupled to ESMS and used to separate 15-hydroperoxyeicosatetraenoic acid (HPETE), 12- HPETE and 5-HPETE. Subsequently this approach has been developed to investigate the oxidation of glycerophosphocholine and glycerophosphoethanolamine plasmalogens, involving initial detection of the intact phospholipids, followed by hydrolysis of the alkyl or plasmenyl substituents and identification of the oxidation products by negative ion electrospray tandem MS.[17, 18]

The peroxidation of stearoylarachidonoyl-phosphatidylcholine in vesicles prepared from egg yolk lecithin by *tert*-butylhydroperoxide and Fe^{2+} has also been investigated.[19] Products at +32, +64 , and +96 mass units were observed by positive ion ESMS, corresponding to the addition of 1, 2, or 3 molecules of dioxygen to the phospholipid (Fig. 2). Under the conditions used in this study the hydroperoxides were the major primary oxidation products; no clear evidence for the formation of hydroxides was observed.

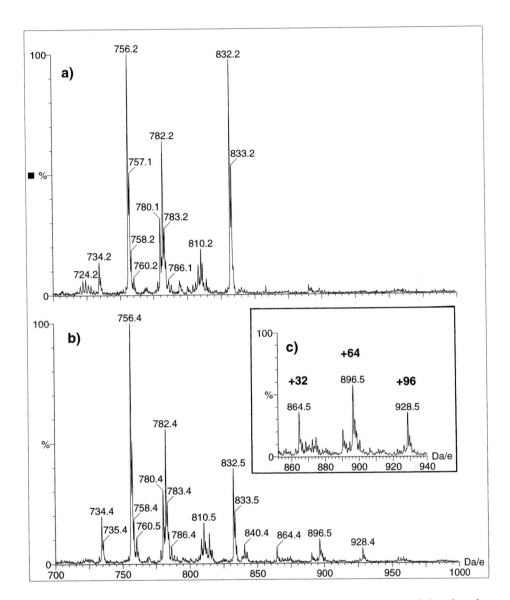

Fig. 2. Positive ion ESMS spectra of phospholipid vesicles. The majority of the phosphatidylcholines are in the sodiated form, giving signals at +22 m/z compared to the masses
in Table 1. (a) shows the spectrum of control vesicles with a large signal corresponding
to arachidonylstearoyl phosphatidylcholine at 832 m/z. (b) shows the vesicles after
incubation with 100 mM *t*-BHP and 0.5 mM Fe^{2+}, with depletion of the signal at 832
m/z. [From Spickett *et al.* (1998). *Free Radic. Biol. Med.* **25**: 613–620. Copyright © 1998
Elsevier.]

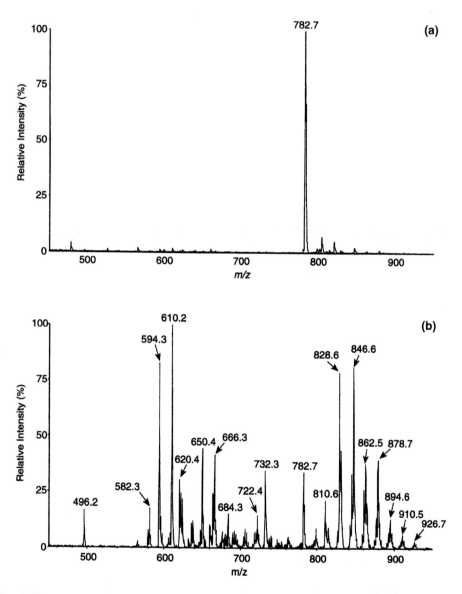

Fig. 3. Positive ion ESMS spectrum of (a) unoxidized palmitoylarachidonoyl phosphatidylcholine (PAPC) at 782 m/z, and (b) PAPC that had been allowed to autooxidize for 24–48 h, showing the appearance of multiple new species. [From Watson *et al.* (1997). *J. Biol. Chem.* **272**: 13 597–13 607. Copyright © 1997 by the American Society for Biochemistry and Molecular Biology, Inc.]

Chain-shortened oxidized phospholipids have also been detected by positive ion ESMS (Fig. 3), following autooxidation of palmitoylarachidonoyl phosphatidylcholine, which has been used as a model of the oxidized LDL phospholipids that induce monocyte binding to endothelial cells. The three main oxidation products were identified as 1-palmitoyl-2-oxovaleryl-GPC, 1-palmitoyl-2-glutaryl-GPC, and 1-palmitoyl-2-(5,6-epoxyisoprostane E_2)-GPC by tandem ESMS analysis together with chemical modifications or ^{1}H-NMR.[20, 21]

5. Halogenated Lipid Oxidation Products Detected by ESMS

In addition to lipid hydroperoxides and their products, other types of lipid oxidation products can also occur. The phagocytic enzyme myeloperoxidase uses hydrogen peroxide and halide ions to generate hypohalous acids (most commonly HOCl), which are strong oxidants. HOCl is known to react readily with unsaturated fatty acids to form chlorohydrins, in addition to many other reactions with biologically relevant groups.[22] Such products could be important effectors of phagocyte microbicidal activity, and are likely to contribute to tissue damage in inflammatory conditions which often occur in age-related diseases.

Carr *et al.*[23] have studied the formation of bromohydrins from oleic, linoleic and arachidonic acids by the myeloperoxidase system (using bromide), and identified a variety of brominated and nonbrominated products by negative ion ESMS. This method was found to be an improvement over GC-MS for products of linoleic and arachidonic acids, which were too unstable to be detected by the latter technique. The major species were monobromohydrins, but bisbromohydrins of the polyunsaturated fatty acids were also observed (Fig. 4), while dibromo-derivatives, hydroperoxides and dihydroxides occurred as minor products.

There has also been interest in halogenated products of cholesterol resulting from the activity of the myeloperoxidase system. Hazen *et al.*[24] reported that a family of chlorinated sterols results from treatment of human low density lipoprotein (LDL) with myeloperoxidase-hydrogen peroxide-chloride, including three different chlorohydrins and a dichlorinated cholesterol derivative. In this study the cholesterol oxidation products were initially separated by thin layer chromatography before analysis by positive ion ESMS. Similar results were obtained using cholesterol-containing liposomes and red blood cell membranes incubated with HOCl or the myeloperoxidase system, and dehydration products of both native cholesterol and cholesterol chlorohydrins were also observed.[25]

Another study has looked at the effect of HOCl and the myeloperoxidase system on human LDL phospholipids;[26] following treatment the LDL was diluted with methanol and analyzed directly using ESMS interfaced to reverse phase HPLC. Separation of the oxidized phospholipids prior to detection greatly increased the sensitivity of the method as this reduced ion suppression by salts and the native lipids. Monochlorohydrins of palmitoyloleoyl PC, palmitoyllinoleoyl

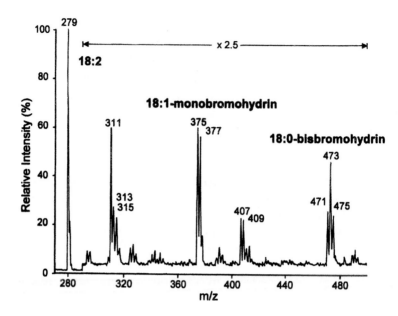

Fig. 4. Negative ion ESMS spectrum of linoleic acid (C18:2) treated with the MPO/H$_2$O$_2$/ Br$^-$ system at pH 5.5. The peak at m/z 279 represents unreacted fatty acid, the peaks at m/z 375/377 (1:1) represent monobromohydrins, and the peaks at m/z 471/473/475 (1:2:1) represent bisbromohydrins. [From Carr *et al.* (1996). *Arch. Biochem. Biophys.* **327**: 227–233. Copyright © 1996 by Academic Press, Inc.]

PC, stearoyllinoleoyl PC and stearoylarachidonoyl PC were observed, but there was no evidence for phospholipid hydroperoxy, hydroxy, or dichlorinated derivatives.

6. Lipid Peroxidation Products Observed in Biological and Clinical Samples

A major aim in the development of ESMS for the analysis of lipid oxidation products is to establish a sensitive method which can be used to detect specific oxidation products in disease conditions and may provide information on the source of the oxidant and/or the native lipid. Research on oxidative damage or oxidative lipid metabolism in isolated cells or tissues is also an important area. Some studies have used simple systems such as red blood cells, but more reports on detection of oxidized species in clinical samples or animal models are now emerging.

One approach which has been much exploited involves the base hydrolysis of the extracted phospholipids and subsequent analysis of the oxidized fatty acyl chains by negative ion ESMS or LC-MS. For example, epoxyeicosatetraneoic acids

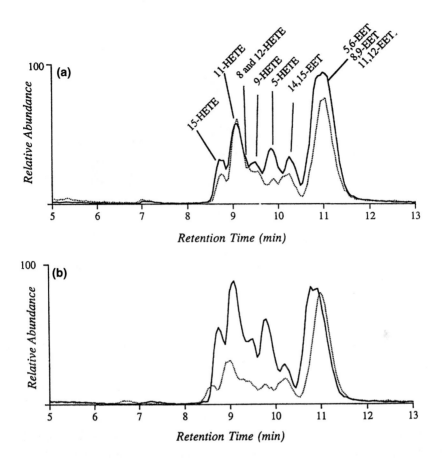

Fig. 5. LC/MS/MS analysis of HETEs and EETs from lung glycerophospholipids isolated from mice that (a) were untreated, and (b) had *t*-BHP administered intratracheally. Lipids extracted from lung homogenates were separated into free eicosanoid and phospholipid fractions, and the phospholipid fraction was then saponified with NaOH, before analysis by negative ion ESMS. The solid lines correspond to ester-derived HETEs and EETs, and the dotted lines correspond to free HETEs and EETs. [From Nakamura *et al.* (1998). *Anal. Biochem.* **262**: 23–32. Copyright © 1998 by Academic Press, Inc.]

(EETs) and monohydroxytetraenoic acids (HETEs) esterified in phospholipids have been investigated in human erythrocyte membranes, and the levels were found to increase on incubation with tert-butylhydroperoxide.[27] A similar range of oxidized products have been observed in normal murine lung tissue, and it was found that infusion of the lungs with *t*-BHP resulted in an increase only in esterified HETEs (Fig. 5), while levels of esterified EETs and all unesterified oxidized fatty acids were unaltered.[28] The relationship between the presence of HETEs, EETs, oxo-ETEs or F_2-isoprostanes and the stability of atherosclerotic plaques has also been investigated using the same approach.[29] Elevated levels of

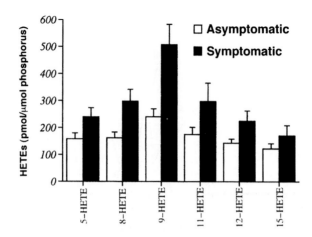

Fig. 6. Distribution of HETE compounds in plaques retreived from asymptomatic (18 stable plaques) and symptomatic (12 unstable plaques) patients. HETE concentrations are expressed on a per-mole-of-lipid-phosphorus basis. Each HETE was significantly more abundant in unstable, compared to stable, plaques ($p < 0.01$). [From Mallat *et al.* (1999). *J. Clin. Invest.* **103**: 421–427. Copyright © 1999 by the American Society for Cinical Investigation.]

all these compounds were found in plaques, with a small but significant increase of HETEs in plaques from symptomatic patients, compared to plaques from asymptomatic patients (Fig. 6). The most abundant oxidized derivative was 9-HETE, for which no enzymatic synthetic pathway is known, suggesting that the development of plaques involves a non-enzymatic free radical oxidation process.

While this approach has provided much useful information, it has the disadvantage of increased sample manipulation and that information about the intact phospholipid is lost unless individual phospholipid classes are separated prior to hydrolysis. Alternatively, the intact oxidized phospholipids can be detected and information about the fatty acyl chains obtained by fragmentation in tandem mode. Oxidized GPCs and GPE from red blood cell membranes treated with *t*-BHP were detected by normal phase HPLC coupled to ESMS, and the esterified HETEs and HPETEs were identified from the fragment ions.[30]

A number of oxidized phospholipid derivatives have been found to exhibit biological activity which could be relevant to the progression of atherosclerosis, and their structure has been elucidated by experiments involving ESMS. Platelet activating factor (PAF) and several C4-PAF analogues have been detected in oxidized LDL by tandem MS; high PAF-like activity was found for analogues containing an sn-1 ether bond.[31] Several chain-shortened derivatives of PAPC which are known to induce monocyte-endothelial binding[20, 21, 32] have also been found in aortic lesions of rabbits fed on an atherogenic diet by positive ion ESMS (Fig. 7).[20]

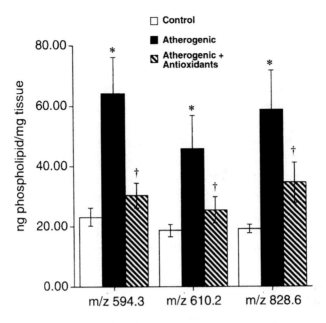

Fig. 7. Presence of biologically active oxidized phospholipids in atherosclerotic lesions. Rabbits were fed on a normal chow diet (Control) or an atherogenic diet with or without antioxidant supplementation (vitamin E or probucol). Following extraction and separation, the phospholipid fraction was analyzed by positive ion ESMS and the molecular ion abundance was quantitated by comparison with dimyristoyl phosphatidylcholine added as an internal standard. * = $p < 0.01$ compared to Control; † = $p < 0.05$ compared to Atherogenic. [From Watson *et al.* (1997). *J. Biol. Chem.* **272**: 13 597–13 607. Copyright © 1997 by the American Society for Biochemistry and Molecular Biology, Inc.]

7. Quantitation in ESMS

In ESMS the intensity of the signal observed is dependent on several factors, including the ionizability of the analyte, ion suppression by other components present in the sample, especially salts, and the tuning of the instrument. These conspire to make quantitation non-trivial. Information can be obtained about the relative levels of the components of a mixture only if these have comparable ionization characteristics: for example within phosphatidylcholines the fatty acyl chain length makes a small but significant contribution,[7] oxidative modification has little effect, but levels of GPCs and GPEs cannot be compared. Absolute quantitation requires the rigorous use of carefully chosen standards, and often stable isotope labelled analogues of the analyte of interest are used, for example d_8-5-HETE[28] and $[^{18}O]$-12-HETE[29] have been used as internal standards for the quantitation of HETEs and EETs.

8. Conclusions

ESMS is rapidly becoming an important and informative technique in biomedical research. It has already been used to detect the formation of a variety of phospholipid oxidation products both in cells and in clinical samples, thus providing information on the nature and mechanisms of oxidative damage. It has been of great value in the characterization of oxidized phospholipids with activity as inflammatory mediators, and has been used to demonstrate the presence of oxidized lipids in atherosclerotic lesions. Much of the research on phospholipids using ESMS has focused on atherosclerosis, as this is a disease where oxidative damage to lipids clearly plays an important role in the disease progression, but there is no reason why ESMS should not be applied to other age-related diseases with equal success.

Structural identification of intact phospholipids requires the use of tandem ESMS and fragmentation; however, this introduces a further layer of complexity into the analysis and the instruments are correspondingly considerably more expensive. For many purposes where detection of previously characterized oxidized species is required, single analyzer ESMS or LC-MS also offer valuable information with greater simplicity and accessibility.

Abbreviations

EET, epoxyeicosatetraneoic acid; ESMS, electrospray mass spectrometry; (G)PC, (glycero)phosphocholine; (G)PE, (glycero)phosphoethanolamine; HETE, monohydroxy-tetraenoic acid; HPETE, hydroperoxytetraenoic acid; oxo-ETE, ketoeicosatetraenoic acid; LDL, low density lipoprotein; m/z, mass to charge ratio; PAF, platelet-activating factor; PAPC, palmitoylarachidonoyl phosphatidylcholine; t-BHP, tert-butylhydroperoxide.

Acknowledgments

CMS would like to thank the University of Strathclyde Glaxo-Jack Endowment for financial support.

References

1. Hughes, H., Smith, C.V., Tsokos-Kuhn, J. O. and Mitchell, J. R. (1986). Quantitation of lipid peroxidation products by gas chromatography-mass spectrometry. *Anal. Biochem.* **152**: 107–112.

2. Van Kuijk, F. J. G. M., Thomas, D. W., Stephens, R. J. and Dratz, E. A. (1990). Gas chromatography-mass spectrometry assays for lipid peroxides. *Meth. Enzymol.* **186**: 388–398.

3. Van Kuijk, F. J. G. M., Thomas, D. W., Stephens, R. J. and Dratz, E. A. (1990). Gas chromatography-mass spectrometry of 4-hydroxynonenal in tissues. *Meth. Enzymol.* **186**: 399–406.

4. Mann, M. (1990). Electrospray: its potential and limitations as an ionization method for biomolecules. *Org. Mass Spectr.* **25**: 575–587.

5. Fenn, J. B., Mann, M., Meng, C. K., Wong, S. F. and Whitehouse C. M. (1990). Electrospray ionization: principles and practice. *Mass Spectr. Rev.* **9**: 37–70.

6. Pitt, A. R. (1998). Application of electrospray mass spectrometry in biology. *Nat. Prod. Rep.* **15**: 59–72.

7. Brügger, B., Erben, G., Sandhoff, R., Wieland, F. T. and Lehmann, W. D. (1997). Quantitative analysis of biological membrane lipids at the low picomole level by nano-electrospray ionization tandem mass spectrometry. *Proc. Natl. Acad. Sci. USA* **94**: 2339–2344.

8. Han, X., Gubitosi-Klug, R. A., Collins, B. J. and Gross, R. W. (1996). Alterations in individual molecular species of human platelet phospholipids during thrombin stimulation: electrospray ionization mass spectrometry-facilitated identification of the boundary conditions for the magnitude and selectivity of thrombin-induced platelet phospholipid hydrolysis. *Biochemistry* **35**: 5822–5832.

9. Han, X. and Gross, R. W. (1994). Electrospray ionization mass spectrometric analysis of human erythrocyte plasma membrane phospholipids. *Proc. Natl. Acad. Sci. USA* **91**: 10 635–10 639.

10. Hsu, F. F., Ma, Z. M., Wohltmann, M., Bohrer, A., Nowatzke, W., Ramanadham, S. and Turk, J. (2000). Electrospray ionization/mass spectrometric analyses of human promonocytic U937 cell glycerophospholipids and evidence that differentiation is associated with membrane lipid composition changes that facilitate phospholipase A(2) activation. *J. Biol. Chem.* **275**: 16 579–16 589.

11. Kwom, G., Bohrer, A., Han, X., Corbett, J. A., Ma, Z., Gross, R. W., McDaniel, M. L. and Turk, J. (1996). Characterization of the sphingomyelin content of isolated pancreatic islets. Evaluation of the role of sphingomyelin hydrolysis in the action of interleukin-1 to induce islet overproduction of nitric oxide. *Biochim. Biophys. Acta* **1300**: 63–72.

12. Smith, P. B. W., Snyder, A. P. and Harden, C. S. (1995). Characterization of bacterial phospholipids by electrospray ionization tandem mass spectrometry. *Anal. Chem.* **67**: 1824–1830.

13. Kim, H. Y., Wang, T. C. L. and Ma, Y. C. (1994). Liquid chromatography/ mass spectrometry of phospholipids using electrospray ionization. *Anal. Chem.* **66**: 3977–3982.

14. Sandhoff, R., Brügger, B., Jeckel, D., Lehmann, W. D. and Wieland F. T. (1999). Determination of cholesterol at the low picomole level by nano-electrospray ionization tandem mass spectrometry. *J. Lipid Res.* **40**: 126–132.

15. Maffei-Facino, R., Carini, M., Aldini, G., and Colombo, L. (1996). Characterization of the intermediate products of lipid peroxidation in phosphatidylcholine liposomes by fast-atom bombardment spectrometry and tandem mass spectrometry techniques. *Rapid Commun. Mass Spectr.* **10**: 1148–1152.

16. MacMillan, D. K. and Murphy, R. C. (1995). Analysis of lipid hydroperoxides and long-chain conjugated keto acids by negative ion electrospray mass spectrometry. *J. Am. Soc. Mass Spectr.* **6**: 1190–1201.

17. Khaselev, N. and Murphy, R. C. (1999). Susceptibility of plasmenyl glycero-phosphoethanolamine lipids containing arachidonate to oxidative stress. *Free Radic. Biol. Med.* **26**: 275–284.

18. Khaselev, N. and Murphy, R. C. (2000). Structural characterization of oxidized phospholipid products derived from arachidonate-containing plasmenyl glycerphosphocholine. *J. Lipid Res.* **41**: 564–572.

19. Spickett, C. M., Pitt, A. R. and Brown, A. J. (1998). Direct observation of lipid hydroperoxides in phospholipid vesicles by electrospray mass spectrometry. *Free Radic. Biol. Med.* **25**: 613–620.

20. Watson, A. D., Leitinger, N., Navab, M., Faull, K. F., Hörkkö, S., Witztum, J. L., Palinski, W., Schwenke, D., Salomon, R. G., Sha, W., Subbanagounder, G., Fogelman, A. M. and Berliner, J. A. (1997). Structural identification by mass spectrometry of oxidized phospholipids in minimally oxidized low density lipoprotein that induce monocyte/endothelial interactions and evidence for their presence in vivo. *J. Biol. Chem.* **272**: 13 597–13 607.

21. Watson, A. D., Subbanagounder, G., Welsbie, D. S., Faull, K. F., Navab, M., Jung, M. E., Fogelman, A. M. and Berliner, J. A. (1999). Structural identification of a novel pro-inflammatory epoxyisoprostane phospholipid in mildly oxidized low density lipoprotein. *J. Biol. Chem.* **274**: 24 787–24 798.

22. Schaur, R. J., Jerlich, A. and Stelmaszynska, T. (1998). Hypochlorous acid as reactive oxygen species. *Curr. Top. Biophys.* **22B**: 176–185.

23. Carr, A. C., Winterbourn, C. C. and van den Berg, J. J. M. (1996). Peroxidase-mediated bromination of unsaturated fatty acids to form bromohydrins. *Arch. Biochem. Biophys.* **327**: 227–233.

24. Hazen, S. L., Hsu, F. F., Duffin, K. and Heinecke, J. W. (1996). Molecular chlorine generated by the myeloperoxidase-hydrogen peroxide-chloride system of phagocytes converts low density lipoprotein cholesterol into a family of chlorinated sterols. *J. Biol. Chem.* **271**: 23 080–23 088.

25. Carr, A. C., van den Berg, J. J. M. and Winterbourn, C. C. (1996). Chlorination of cholesterol in cell membranes by hypochlorous acid. *Arch. Biochem. Biophys.* **332**: 63–69.

26. Jerlich, A., Pitt, A. R., Schaur, R. J. and Spickett, C. M. (2000). Pathways of phospholipid oxidation by HOCl in human LDL detected by LC-MS. *Free Radic. Biol. Med.* **28**: 673–682.

27. Nakamura, T., Bratton, D. L. and Murphy, R. C. (1997). Analysis of epoxyeicosatetraenoic and monohydroxyeicosatetraenoic acids esterified to phospholipids in human red blood cells by electrospray tandem mass spectrometry. *J. Mass Spectr.* **32**: 888–896.

28. Nakamura, T., Henson, P. M. and Murphy, R. C. (1998). Occurrence of oxidized metabolites of arachidonic acid esterified to phospholipids in murine lung tissue. *Anal. Biochem.* **262**: 23–32.

29. Mallat, Z., Nakamura, T., Ohan, J., Leseche, G., Tedgui, A., Maclouf, J. and Murphy, R. C. (1999). The relationship of hydroxyeicosatetraenoic acids and F_2-isoprostanes to plaque instability in human carotid atherosclerosis. *J. Clin. Invest.* **103**: 421–427.

30. Hall, L. M. and Murphy, R. C. (1998). Analysis of stable oxidized molecular species of glycerophospholipids following treatment of red blood cell ghosts with t-butylhydroperoxide. *Anal. Biochem.* **258**: 184–194.

31. Marathe, G. K., Davies, S. S., Harrison, K. A., Silva, A. R., Murphy, R. C., Castro-Faria-Neto, H., Prescott, S. M., Zimmerman, G. A. and McIntyre, T. M. (1999). Inflammatory platelet-activating factor-like phospholipids in oxidized low density lipoproteins are fragmented alkyl phosphatidylcholines. *J. Biol. Chem.* **274**: 28 395–28 404.

32. Watson, A. D., Berliner, J. A., Hama. S. Y., La Du, B. N., Faull, K. F., Fogelman, A. M. and Navab, M. (1995). Protective effect of high density lipoprotein associated paraoxonase. Inhibition of the biological activity of minimally oxidized low density lipoprotein. *J. Clin. Invest.* **96**: 2882–2891.

Measurement of Oxidative Stress: Technology, Biomarkers and Applications

(Carbohydrate Damage)

Chapter 24

Glycoxidative and Carbonyl Stress in Aging and Age-Related Diseases

Vincent M. Monnier*, David R. Sell, Amit Saxena, Poonam Saxena,
Ram Subramaniam, Frederic Tessier and Miriam F. Weiss

**Vincent M Monnier, David R. Sell, Amit Saxena, Poonam Saxena, Ram Subramaniam,
Frederic Tessier** and **Miriam F. Weiss** • Institute of Pathology and Department of
Medicine, Case Western Reserve University, Cleveland, OH 44106

*Corresponding Author.
Tel: 216-368-6613, E-mail: vmm3@po.cwru.edu

1. Introduction

As suggested by the content of this book, the free radical/oxidative stress theory of aging[1] has become as one of the most prominent stochastic theories of aging. The theory postulates that an array of chronic debilitative insults inflicted by free radicals are a major cause of life span limiting processes. Over the years, however, a number of discoveries have led to the necessity to revise and expand this theory in several ways.

Besides the deleterious effects of reactive oxygen species (ROS) on cells, it is increasingly recognized that oxidatively modified molecules have cell signaling properties necessary for physiological cell function. These pathways affect the processing of oxidatively damaged molecules. In addition, considerable molecular damage in aging also originates non-oxidatively from carbonyl compounds that are generated both by the Maillard reaction and metabolic pathways. These observations led to the proposition of a Maillard reaction theory of aging[2] and that "carbonyl stress" should be considered an equally important source of molecular damage in aging and age-related diseases.[3] Undoubtedly both oxidative and carbonyl stress mechanisms are highly intertwined. We propose that the concept of "stress", although convenient and easily understood, is misleading when applied to the biology of aging. It arose primarily from experiments in which large quantities of actinic, oxidant or chemical stimuli were applied acutely to a cellular or molecular system. In most of these experiments, addition of e.g. catalase, superoxide dismutase or any free radical scavenger readily prevented the damage. In contrast, the aging process evolves very slowly without impairment of homeostasis at rest or the necessity to invoke biological or chemical "stresses". However, a major difference between the old and the young organism is that response time and amplitude to a particular stress (e.g. hypothermic,[4] hyperglycemic[5] or immunologic[6]) are impaired in the old organism. The logical conclusion from these observations is that death will occur when even the ability to maintain homeostasis, i.e. the ability to respond to a *physiological* stimulus or minor stress in a critical organ is impaired. In that regard, Kohn found no overt cause of death in at least 30% of subjects dying after the age of 85 years, suggesting aging itself is a disease process which predisposes to death induced by a stress which otherwise would not kill a 20 year old.[7]

It is becoming increasingly important to consider oxidative and carbonyl "stressors" as entities inseparable from metabolic factors, and to refer more generally to "metabolic stress" as suggested by Baynes and Thorpe.[3] In the case of diabetes or atherosclerosis, for example, development of certain lesions is strongly dependent on a total increase in "fuel" molecules and their metabolites or catabolites (e.g. glucose, glycolytic intermediates and lipids) without necessarily invoking an increase in ROS. Accordingly, there is no evidence for a "net" increase in the level of oxidized proteins in the connective tissue proteins in diabetic

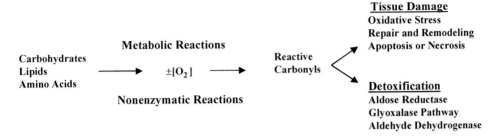

Fig. 1. Interface between metabolic, oxidative and carbonyl stress as a basis for molecular and tissue damage and cell signaling in aging and age-related diseases from Ref. 3, with permission.

human.[8] However, this latter observation may not be applicable to intracellular events, as recent evidence suggests increased intracellular ROS formation occurs in cells exposed to high glucose environment.[9]

The source of, and interplay between, the various metabolic, oxidative and carbonyl stressors has been elegantly summarized by Baynes and Thorpe[3] (Fig. 1). Baynes' concept restores metabolism to the forefront of the biology of aging and age-related diseases by linking metabolic "stressors" with the disease process rather than restricting the latter to the molecular damage as the major factor in the disease. Not surprisingly, a recent study of gene expression profiling in old and calorically restricted animals revealed more pronounced differences in metabolic enzymes than those involved in antioxidant defense.[10] Importantly, expression of aldehyde dehydrogenase II was 2.4-fold increased in old mice, suggesting a need for increased detoxification of aldehydes in aging. Several other enzymes related to glucose homeostasis were decreased.

This introduction thus emphasizes the need to take a global view of the nature of stochastic damage in the biology of aging instead of considering oxidative stress as the only culprit in the aging process. In that sense, this chapter will examine the question of the interface between glycation, glycoxidation and oxidant stress, i.e. how oxidation leads to formation of reactive carbonyls from glucose and its metabolites, and how such carbonyls are linked to oxidant stress, either directly or indirectly via formation of advanced glycation endproducts (AGEs). The chapter is not meant to be a comprehensive review of the Maillard reaction field, which has grown exponentially in recent years.

2. Chemical Pathways of the Maillard Reaction *In Vivo*

Considerable progress has been achieved in the past 10 years in the characterization of glucose degradation pathways, the structure of AGEs and

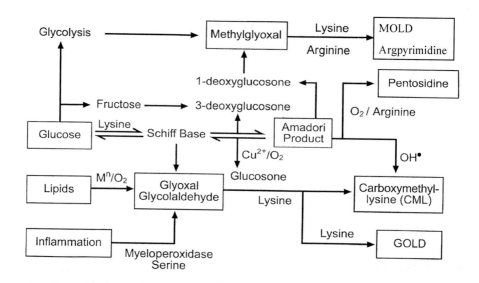

Fig. 2. Chemical pathways of the Maillard reaction *in vitro*.

their precursors. The discovery that some AGEs can have multiple precursors, including lipoxidation and ascorbylation products, has greatly increased the complexity of the pathways potentially involved in their formation (Fig. 2).

Glucose can undergo three types of transformations (Fig. 2). It can be biologically transformed through glycolysis, autoxidatively modified by redox active metals by the Wolff pathway of the Maillard reaction (see below) or transformed, oxidized and degraded via the Maillard reaction, i.e. after engagement with a primary amine group. A fourth pathway, not depicted in Fig. 2, is its enzymatic reduction to sorbitol, a pathway hypothesized to play a role in the pathogenesis of diabetic complications.[11] Due to its glycosidic nature and slow anomerization rate, glucose is relatively inert, both as a glycating agent and toward metal catalyzed oxidation. However, some of its intermediate compounds, such as the Amadori product, have ascorbate-like reactivity toward metals (see below).

Glycation is initiated by condensation of glucose with an amine. The Schiff base intermediate becomes highly susceptible to oxidant attack and forms the reactive oxoaldehydes glyoxal and methylglyoxal (Namiki pathway). These readily react with aminoguanidine to form triazines[12] which can be detected in diabetic rat plasma.[13] Both glyoxal and methylglyoxal react with lysyl residues to form AGE products such as carboxymethyl-lysine (CML)[14] and carboxyethyl-lysine (CEL),[15] and the imidazole crosslinks glyoxal lysine dimer (GOLD) and methylglyoxal lysine dimer (MOLD), respectively.[16, 17] Both oxoaldehydes are also reactive with arginine residues. Structures like carboxymethyl-arginine (CMA),[18] hydroimidazolones[19] and the fluorescent argpyrimidine modification have been described. The identification of 5'-methylimidazolone as a fluorescent molecule was subsequently disproved, and was likely due to contamination by argpyrimidine.[20] All of the above compounds have been detected in biological tissues.

Another important oxoaldehyde arising during glycation is 3-deoxyglucosone (3-DG) (Fig. 2). Importantly, no oxidation is required for its formation! Its reactivity is somewhat kept in check due to the hydration of its carbonyl group in aqueous solution.[12] Nevertheless, it can form pyrroles (pyrraline) and pyrrole crosslinks with lysine residues,[21, 22] as well as the crosslink 3-deoxyglucosone lysine dimer DOLD.[23] *In vivo*, possible contributors of 3-DG formation are glucose, Amadori products, fructose and fructose 3-phosphate. The latter can form through the aldose reductase pathway.[24] A novel kinase, fructosamine-3-kinase, can phosphorylate both fructose as well as Amadori products in proteins.[25] Recently, fructosamine-3-kinase was cloned.[26] The chemistry and biology of 3-deoxyglucosone were the subject of a recent review.[27]

Glycolysis itself is a major source of carbonyl "stressors". The foremost is methylglyoxal which forms spontaneously from glyceraldehyde 3-phosphate and dihydroxyacetone-phosphate through "leakage" at the triose isomerase step.[28] The physiological blood concentration of methylglyoxal ranges from 25 to 900 pmol/g blood and is increased two-fold in diabetes.[28] The plasma concentration is estimated at 1 µM. Most of the methylglyoxal is reversibly bound to protein via hemithioacetal bonds.[19] The highest tissue levels are found in lens and nerve.

Finally, ascorbic acid has emerged as a powerful glycating ("ascorbylating") agent, in particular in tissues rich in vitamin C, such as the lens, where most postsynthetic modifications of aging crystallins can be reproduced through incubation of the protein with ascorbic acid.[29] To this date no single AGE product has been found that is entirely specific for ascorbic acid. However, a few have been identified with ascorbic acid as a major precursor, such as oxalate monoamide[30, 31] and vesperlysine A.[32] Interestingly, both ascorbic acid and methylglyoxal are precursors of argpyrimidine, a fluorescent amino acid of the aging human lens.[33] Thus, some of the methylglyoxal found in the lens may originate from ascorbic acid degradation.

3. Metal- and Non-Metal Catalyzed Oxidation During Protein Glycation

3.1. Glucose Autoxidation and Glycoxidation

Research on protein glycation has traditionally involved a standard incubation protocol consisting of 0.1 to 1 M glucose concentration with 10 to 50 mg/ml protein dissolved in 100 to 200 mM sodium phosphate buffer, i.e. with sufficient buffering capacity to prevent a drop in pH due to proton release during the glycation reaction. Short incubation time (2 to 7 days) generally favors a protein rich in Amadori compound whereas longer incubation time (14 to 60 days) and highly reactive sugars (e.g. ribose, threose, glyceraldehyde) or ascorbic acid lead to extensive browning, crosslinking and oxidation.

Experiments by Wolff and colleagues demonstrated that glucose which was preincubated in phosphate buffer prior to addition of protein caused greatly accelerated protein fluorescence and fragmentation which could be suppressed with a chelating agent.[34] In the process, extensive oxidative modifications of the sugar adduct and/or amino acid residues such as sulfhydryl,[35] histidine,[36] tyrosine and tryptophan[37, 38] were observed. The mechanism underlying these changes involves autoxidation of the enol form of the reducing sugar through binding of redox active metals (Cu^{2+} or $Fe3^{+}$) to form ROS (O_2^- and $OH^{\cdot}$) and the sugar breakdown products glyoxal and arabinose[39, 40] (Fig. 2). The latter are precursors of the AGEs carboxymethyl-lysine and pentosidine, respectively. The glucose autoxidation pathway of the Maillard reaction is now also called the Wolff's pathway.

Some controversy exists concerning glucosone, the formation of which was reported in glycation mixtures with Cu^{2+}.[41] However, Wells-Knecht *et al.* failed to detect it or any of its degradation products when primary amines were omitted from the incubation mixture.[16] Most likely, glucosone is generated during oxidation of the Amadori product or an immediate precursor thereof. Indeed, incubation of Amadori products with Cu^{2+}, nitroblue tetrazolium or the deglycating FAD-linked amadoriase enzymes generates glucosone quantitatively.[42, 43]

Baynes introduced the general concept of "glycoxidation" in order to denote any form of sugar oxidation in the course of the Maillard reaction.[44] The overall extent of glycoxidation is strongly dependent on the concentration of the phosphate in the buffer, not only because of its metal content, but also because the phosphate ion itself catalyzes Amadori product formation. When the phosphate concentration increases from 5 to 200 mM, accelerated CML formation and protein oxidation occurs.[37] Substitution of phosphate by an organic ion or addition of chelating agent strongly inhibits glycoxidation and protein oxidation.

3.2. Metal-Binding and Free Radical Generation by Glycated Proteins

The formation of superoxide anion and free radicals in incubation mixtures containing glycated proteins has been observed by several authors.[45, 46] In-depth studies have been carried out on the roles of Cu^{2+} and Fe^{3+} on free radical generation in incubation systems consisting of glycated (Amadori-product rich) protein and added metal. On one end of the spectrum, incubation of glycated serum protein (albumin, superoxide dismutase) in very low concentration (e.g. 1 mg/ml) in phosphate buffered saline (PBS) with added Cu^{2+} leads to rapid H_2O_2 mediated depolymerization of the proteins.[47, 48] On the other end, addition of serum albumin (or lens crystallins) to a Cu^{2+}/ascorbic acid system totally suppresses oxidation.[49, 50] Moreover, when serum albumin that was glycated with 0 to 50 mM glucose was incubated with lipopotein in PBS, the antioxidant properties of albumin on LDL oxidation were progressively suppressed as glycation of albumin increased, i.e. albumin was transformed from an antioxidant into a pro-oxidant protein.[51] Potential explanations include the impairment of the physiological binding site of Cu^{2+} on albumin, the formation of Amadori-Cu^{2+} complexes,[52] conformational changes of the protein and exposure of novel redox active sites,[51] the formation of AGE products that can bind redox active metals, such as CML,[53] as well free radical formation by a novel mechanism in absence of metals.[54]

In addition to copper, Amadori products may also bind iron and form redox active complexes with Fe^{3+}. Qian *et al.*[55] incubated glycated gelatin, albumin or elastin with either Cu^{2+} or Fe^{3+} and found concentration dependent metal binding and formation of redox active complexes in presence of ascorbic acid.

3.3. Carboxymethyl-Lysine: A Molecular "Sink" of the Aging Process

Carboxymethyl-lysine (CML) has emerged as one of the most important AGE products for a number of reasons. Firstly, it can be formed at any stage during the Maillard reaction. Glomb and Monnier showed 40–50% originate from Schiff base or a pre-Amadori step and the rest from Amadori product autoxidation, and less than 10% from glucose autoxidation.[55] The higher the phosphate concentration in the buffer, the higher CML formation.[37] Fenton chemistry is required for CML formation from Amadori product.[56] Secondly, CML can be formed not only from glucose but also from any reducing sugar and the oxoaldehyde glyoxal. In contrast, methylglyoxal itself is a poor precursor of CML (REF). The most active sugars are glyceraldehyde and glycoladehyde. Thirdly, it can be formed from ascorbic acid,[57] and through lipid peroxidation (lipoxidation) via glycolaldehyde, a precursor common to both glycoxidation and lipoxidation.[58] Finally, CML can be formed through myeloperoxidase catalyzed serine oxidation at sites of inflammation.[59] Thus, CML emerges as a molecular "sink" of protein aging[53] in that a large variety of metabolites will generate a CML precursor upon degradation. Therefore it may not be too surprising that proteins rich in CML have a number of unwanted biological properties, and that there are cell surface receptors which recognize CML modified proteins with high affinity.

The observation that CML has a glycine/EDTA-like configuration lead the authors to investigate the potential ability of CML rich proteins to bind divalent metals.[53] The extent of Cu^{2+}, Zn^{2+} and Ca^{2+} binding increases with the amount of CML present in the protein. Cu–CML protein complexes are redox active and oxidize ascorbic acid. Protein depolymerization in the presence of H_2O_2 is observed. A similar process may explain the extensive glycoxidative damage observed during

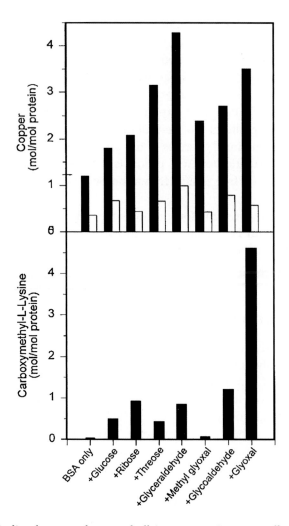

Fig. 3. Copper binding by an carboxymethyllsine content in serum albumin modified by various sugars and oxoaldehydes. The Cu^{2+} ligand in methylglyoxal modified albumin is expected to be carboxyethyl-instead of carboxymethyl-lysine (author's unpublished data).

long-term incubation of albumin and other proteins with glucose. In this setting, Cu^{2+} from the phosphate buffer is likely activated by binding to CML residues, leading thus to accelerated superoxide formation, glucose autoxidation and glycoxidation of Amadori products. Redox active proteins that bind increased amounts of Cu^{2+} could also be obtained by incubating protein with other sugars and oxoaldehydes, all of which had increased amounts of CML (Fig. 3). However, there is no strict stoichiometric relationship between CML content and the amount of metal bound, suggesting only few CML sites on the protein are able to bind

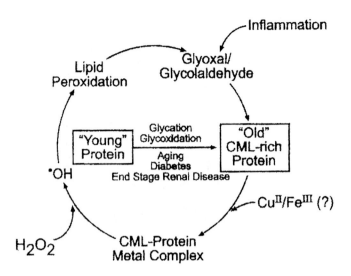

Fig. 4. Hypothesized existence of a vicious cycle consisting of binding of redox active transition metal by carboxymethyl-lysine (CML) rich protein and generation of free radicals which oxidize lipids into glycoladehye, itself a potent precursor of CML.

Cu^{2+}. Interestingly, a strong discrepancy between CML content and copper binding is apparent in the case of methylglyoxal-modified protein. One likely explanation is that Cu^{2+} is bound by carboxy*ethyl*-lysine (CEL) which is one of the major modification found in methylglyoxal-treated proteins[60] and is isosteric with CML and CEL. The significance of these AGE-metal chelates is that their formation is likely to engender a vicious cycle, whereby the redox active complex can lead to lipoxidation and generation of glycoladehyde, which itself is a CML precursor (Fig. 4).

The potential *in vivo* relevance of redox active CML-metal-protein complexes was investigated.[53] When CML rich tendons were inserted into the peritoneal cavity of diabetic rats and retrieved after 28 days, increased copper content and redox activity was noted, compared to unmodified tendons or normal tendons implanted into control rats. Similarly, when immunoprecipitated CML-rich proteins from plasma of patients with end stage renal disease were exposed to H_2O_2, a DMPO radical could be observed by ESR (electron spin resonance) spectroscopy. Finally, in a study involving lenses from cataractous patients, a correlation was observed between the amount of protein bound Cu^{2+}, the overall extent of CML and AGE product formation and the presence of proxidant activity in extracted proteins.[61] All changes could be duplicated by glycating the lens proteins with ascorbic acid. Interestingly, although a large part of the redox activity of the CML-rich protein that was obtained by immunoprecipitation with a CML antibody could be suppressed with the chelating agent DTPA (also called

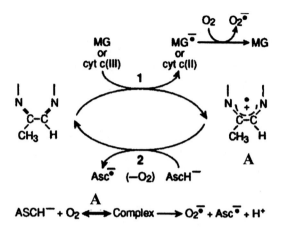

Fig. 5. Proposed structure for methylglyoxal-derived free radical which can mimick metal catalyzed reactions according to Lee *et al.*[54] with permission.

DETAPAC), the more pigmented the cataractous or ascorbate-treated lens protein was, the less suppressible was its redox activity through chelation.

3.4. Non-Metal Catalyzed Generation of Free Radicals

The inability of chelating agents to completely suppress the pro-oxidant effects of heavily glycated proteins suggests some of the structures formed by the Maillard reaction have non-metal dependent pro-oxidant activities. Lee *et al.*[54] found that methylglyoxal modified protein is redox active and can oxidize cytochrome C or ascorbic acid in absence of metal. These studies have likely been inspired by the earlier work of Albert Szent-Gyorgyi[62] who first described free radical formation from methylglyoxal and protein. A mechanism for free radical formation by methylglyoxal modified protein was proposed by Lee *et al.* (Fig. 5). Although the *in vivo* existence of these structures has not yet been demonstrated, they are likely to be of significance in states of accelerated methylglyoxal formation such as cataract,[63] diabetes,[64] end stage renal disease,[65] and perhaps cancer as well.[62]

4. Induction of Cellular Oxidative Stress by Advanced Glycation Endproducts

4.1. RAGE Activation, Oxidative Stress and NF-κB Activation

AGE-albumin added to cultured cells triggers a number of biological responses that are thought to be receptor mediated. Typically, endothelial cells exposed to

AGE-albumin respond with increased monolayer permeability,[66] tissue procoagulant factor upregulation,[67] increased expression or release of VCAM-I, ICAM-I,[68] VEGF[69] and PDGF.[70] Monocyte/macrophages show chemotactic activity and release cytokines such as TNFα, IL-1β.[70,71] The field has literally exploded in recent years and a large number of groups have documented various deleterious effects of AGEs on cells implicated in diabetes, atherosclerosis and other age-related diseases. A number of reviews are available.[72, 73]

Several receptors have been identified which may serve as signal transducers for AGEs. These include the receptor for advanced glycation products (RAGE),[73] the macrophage scavenger receptor A (MSR-A)[74] and the AGE-1, 2 and 3 receptor complex which involves OST48, p60/90 and galectin-3.[75]

Among the various AGE receptors that have been proposed, RAGE is currently the best characterized receptor.[73] It is a 35 kD protein from the immunoglobulin superfamily which has a single transmembrane spanning domain and a phosphorylation site in the C-terminal cytoplasmic domain. What makes it particularly interesting is its ability to be activated not only by AGE products but also by amyloid β peptide and the s100/calgranulin family of proteins. Various effects have been attributed to RAGE activation. One pathway of RAGE activation includes activation of $p21^{ras}$ followed by MAP kinase (ERK1/2) and NF-κB translocation, resulting in transcription of target genes. Oxidant stress appears to be involved in this pathway since glutathione depletion enhanced the $p21^{ras}$ activation.

Evidence that RAGE activation is critically involved in e.g. diabetic nephropathy, accelerated atherosclerosis, impaired wound healing and diabetes-related periodontitis comes from experiments in which progression of these diseases were blocked in diabetic mice treated with sRAGE, i.e. a soluble truncated form of RAGE which can bind AGE products. The major AGE-ligand identified so far is CML.[76]

Currently, most data indicate RAGE engagement is deleterious to the organism. The authors propose a two-hit model in which activation of RAGE with its ligands "resets the baseline, resulting in a state of cellular activation on which other stimuli are superimposed".[73] Thus, in light of the biology of aging, RAGE should be considered a late acting (deleterious) pleiotropic gene, the activation of which will not only accelerate age-related phenomena, but also cancer and metastasis.[77]

The evolutionary significance of RAGE remains to be established. Why would evolution have retained a signal transduction system that so far appears to have only adverse properties? Other questions that need to be addressed include the mechanism by which RAGE activation interfaces with cellular oxidant stress and the relative role of RAGE versus that of other AGE receptors in the physiology and pathology of carbonyl stress and age-related diseases.

4.2. Methylglyoxal Modification of Albumin and Cellular Oxidative Stress

Whereas a large number of studies have addressed the biological role of AGE proteins, relatively few studies have been carried to characterize the structure-function relationship of AGE-proteins. Protein-bound CML is a recognized RAGE ligand which can also activate protein kinase C βII in certain cultured cells.[78] Such a mechanism could be important in end stage renal disease in which circulating CML levels are high. However, plasma protein-bound CML is not increased in diabetes.[79]

Ongoing studies in the authors' laboratories show that incubation of RAW 264.7 macrophage-like cells, mesangial or tubular cells with AGE-albumins induces intracellular oxidative stress as measured by the dichlorofluorescein assay (DCF). The oxidative response follows the general reactivity order methylglyoxal > glyceraldehyde ~ glyoxal ≫ ribose > glucose for albumin incubated for 7 days with 20 mM or reactant. The oxidative response could be duplicated by modifying the protein with phenylglyoxal, a reagent specific for arginine residues. Most interestingly, it was found that native albumin downregulates the oxidative stress, but the proteolytically digested albumin induced a vigorous proxidant response. This suggests that postsynthetic modification of albumin abrogates its recognition by the yet-to-be cloned albumin receptor and that the modified albumin is recognized by a novel non-specific receptor that has affinity for "foreign" or altered proteins, since similar effects were induced by unmodified ovalbumin, ribonuclease A and lens crystallins.

5. Advanced Glycation Products as Markers of Longevity

From the above review it is apparent that Maillard reaction products can be markers of metabolic and oxidative processes. At equivalent formation rate, and assuming similar tissue concentrations of the AGE product precursors (e.g. glucose, methylglyoxal, ascorbic acid), Maillard reaction products can also be markers of tissue turnover rate, as recently demonstrated by Verzil *et al.*[80] These authors estimated the half life of human skin and cartilage collagen at 15 years and 117 years respectively, based on pentosidine and CML determination in relationship to aspartic acid racemization. Thus, the determination of AGE products in long-lived proteins offers the unique ability to reflect cumulative trends in metabolism and tissue turnover, and to investigate the question of whether these parameters can predict longevity. In a cross-sectional study involving 8 mammalian species, the authors found an overall inverse relationship between skin collagen pentosidine formation rate and longevity.[81] Of interest was the fact that there appeared to be an inverse relationship with the metabolic rate of the animal, similar to that formerly described by Sacher.[82] This relationship was confirmed in a follow-up longitudinal study with C57/6NNia mice. The glycation (furosine) and glycoxidation products of collagen (pentosidine and CML) were determined at 20 mos and at death, and a correlation between rate of formation with maximum life span was investigated.[83] Furosine was a strong inverse predictor of life span in both ad libitum (P < 0.0001) and caloric restricted mice (P < 0.028), higher levels being associated with early death. Similarly, CML and pentosidine formation rates were inversely related with longevity in ad libitum (P = 0.067, NS, and P < 0.0001) and calorically restricted mice (P = 0.0015 and P = 0.09, NS), respectively. Glycation and glycoxidation products were also found lower in skin of dietary restricted animals by Cefalu *et al.*[84] These data suggest that glucose tolerance is impaired in aging mice, leading to early death. Alternatively, a process responsible for decreased collagen turnover rate may be responsible for decreased removal rate of damaged cellular organelles from critical life span controlling tissues. Thus, it is likely carbonyl stress, with or without oxidative stress, plays an important role in controlling tissue turnover rate.

6. Conclusions

Considerable progress has been achieved in recent years in our understanding of the chemistry and pathobiology of advanced Maillard/glycation products. The studies reviewed above focused primarily on protein modifications and their biological roles. However, the recent demonstration that specific DNA modifications involving carboxymethylguanine and carboxymethylguanine[89, 90] are associated with single-strand breaks[85] will likely catalyze the research on DNA glycation that was pioneered by Bucala *et al.*[91] The existence of a tight link between oxidative and carbonyl stress has been shown in the recent demonstration that high glucose levels mediate mitochondrial superoxide and methylglyoxal formation.[9] Despite these elegant mechanistic results, the sempiternal question remains: are these lesions (protein/DNA adducts, crosslinks, molecular fragments etc.) important to the disease of aging or to life span limiting events? Answering that question might be very difficult, because one would need to find pharmacological agents that selectively block AGE formation, without independent effects on cellular function. Aminoguanidine, for example, is not only an AGE inhibitor, but also a nitric oxide synthase inhibitor.[86] Pyridoxamine, a novel AGE and advanced lipoxidation inhibitor[87] blocks AGE formation *in vivo* and normalizes diabetic ketoacidosis.[88] Furthermore, caloric restriction which may prolong lifespan by decreasing tissue AGE levels also has broad activity in delaying age-related processes. The authors' working hypothesis is that any intervention that would successfully prevent both carbonyl and oxidant stress may be expected to prolong life span.

Acknowledgments

The authors thank the National Eye Institute, the National Institute on Aging, the NIDDK, the Juvenile Diabetes Research Foundation and the American Diabetes Foundation for support.

References

1. Harman, D. (1988). Free radicals in aging. *Mol. Cell Biochem.* **84**: 155–161.
2. Monnier, V. M. (1989). Toward a Maillard reaction theory of aging. *Prog. Clin. Biol. Res.* **304**: 1–22.
3. Baynes, J. W. and Thorpe, S. R. (1999). Role of oxidative stress in diabetic complications: a new perspective on an old paradigm. *Diabetes* **48**: 1–9.
4. Hensley, J. C., McWilliams, P. C. and G. E. O. (1964). Effect of cold stress on physiological capacitance in the aging mouse. *J. Gerontol.* **19**: 317–321.
5. Crockford, P. M., Harbeck, R. J. and Williams, R. H. (1966). Influence of age on intravenous glucose tolerance and serum immunoreactive insulin. *Lancet* **1**: 465–467.
6. Zheng, B., Han, S., Takahashi, Y. and Kelsoe, G. (1997). Immunosenescence and germinal center reaction. *Immunol. Rev.* **160**: 63–77.
7. Kohn, R. R. (1982). Cause of death in very old people. *JAMA* **247**: 2793–2797.
8. Wells-Knecht, M. C., Lyons, T. J., McCance, D. R., Thorpe, S. R. and Baynes, J. W. (1997). Age-dependent increase in ortho-tyrosine and methionine sulfoxide in human skin collagen is not accelerated in diabetes. Evidence against a generalized oxidative stress in diabetes. *J. Clin. Invest.* **100**: 839–846.
9. Nishikawa, T., Edelstein, D., Du, X. L., Yamagishi, S., Matsumura, T., Kaneda, Y., Yorek, M. A., Beebe, D., Oates, P. J., Hammes, H. P., Giardino, I. and Brownlee, M. (2000). Normalizing mitochondrial superoxide production blocks three pathways of hyperglycaemic damage. *Nature* **404**: 787–790.
10. Lee, C. K., Klopp, R. G., Weindruch, R. and Prolla, T. A. (1999). Gene expression profile of aging and its retardation by caloric restriction [see comments]. *Science* **285**: 1390–1393.
11. King, G. L. and Brownlee, M. (1996). The cellular and molecular mechanisms of diabetic complications. *Endocrinol. Metab. Clin. North Am.* **25**: 255–270.
12. Thornalley, P. J., Yurek-George, A. and Argirov, O. K. (2000). Kinetics and mechanism of the reaction of aminoguanidine with the alpha-oxoaldehydes glyoxal, methylglyoxal, and 3-deoxyglucosone under physiological conditions. *Biochem. Pharmacol.* **60**: 55–65.
13. Araki, A., Glomb, M., Takahashi, M. and Monnier, V. M. (1998). Amino-guanidine traps α-dicarbonyl compounds during Maillard reaction in vitro and in experimental diabetes. *Diabetes* **47s1**: A121.

14. Glomb, M. A. and Monnier, V. M. (1995). Mechanism of protein modification by glyoxal and glycolaldehyde reactive intermediates of the Maillard reaction. *J. Biol. Chem.* **270**: 10 017–10 026.

15. Ahmed, M. U., Brinkmann Frye, E., Degenhardt, T. P., Thorpe, S. R. and Baynes, J. W. (1997). N-epsilon-(carboxyethyl)lysine, a product of the chemical modification of proteins by methylglyoxal, increases with age in human lens proteins. *Biochem J.* **324**: 565–570.

16. Wells-Knecht, K. J., Zyzak, D. V., Litchfield, J. E., Thorpe, S. R. and Baynes, J. W. (1995). Mechanism of autoxidative glycosylation: identification of glyoxal and arabinose as intermediates in the autoxidative modification of proteins by glucose. *Biochemistry* **34**: 3702–3709.

17. Nagaraj, R. H., Shipanova, I. N. and Faust, F. M. (1996). Imidazolysine crosslinks of the Maillard reaction between methylglyoxal and ε-aminolysine residues. *J. Biol. Chem.* **271**: 19 338–19 345.

18. Iijima, K., Murata, M., Takahara, H., Irie, S. and Fujimoto, D. (2000). Identification of N(omega)-carboxymethylarginine as a novel acid-labileadvanced glycation end product in collagen. *Biochem. J.* **347**: 23–27.

19. Lo, T. W. C., Westwood, M. E., McLellan, A. C., Selwood, T. and Thornalley, P. J. (1994). Binding and modification of proteins by methylglyoxal under physiological conditions — a kinetic and mechanistic study with N-alpha-acetylarginine, N-alpha-acetylcysteine, and N-alpha-acetyllysine, and bovine serum albumin. *J. Biol. Chem.* **269**: 32 299–32 305.

20. Oya, T., Hattori, N., Mizuno, Y., Miyata, S., Maeda, S., Osawa, T. and Uchida, K. (1999). Methylglyoxal modification of protein. Chemical and immunochemical characterization of methylglyoxal-arginine adducts. *J. Biol. Chem.* **274**: 18 492–18 502.

21. Hayase, F., Nagaraj, R. H., Miyata, S., Njoroge, F. G. and Monnier, V. M. (1989). Aging of proteins: immunological detection of a glucose-derived pyrrole formed during Maillard reaction in vivo. *J. Biol. Chem.* **264**: 3758–3764.

22. Nagaraj, R. H., Portero-Otin, M. and Monnier, V. M. (1996). Pyrraline ether crosslinks as a basis for protein crosslinking by the advanced Maillard reaction in aging and diabetes. *Arch. Biochem. Biophys.* **325**: 152–158.

23. Skovsted, I. C., Christensen, M., Breinholt, J. and Mortensen, S. B. (1998). Characterization of a novel AGE-compound derived from lysine and 3-deoxyglucosone. *Cell Mol. Biol. (Noisy-le-grand)* **44**: 1159–1163.

24. Lal, S., Randall, W. C., Taylor, A. H., Kappler, F., Walker, M., Brown, T. R. and Szwergold, B. S. (1997). Fructose-3-phosphate production and polyol pathway metabolism in diabetic rat heart. *Metabolism* **46**: 1333–1338.

25. Szwergold, B. S., Kappler, F. and Brown, T. R. (1990). Identification of fructose 3-phosphate in the lens of diabetic rats. *Science* **247**: 451–454.

26. Delpierre, G., Rider, M. H., Collard, F., Stroobant, V., Vanstapel, F., Santos, H. and Van Schaftingen, E. (2000). Identification, cloning, and heterologous

expression of a mammalian fructosamine-3-kinase [In Process Citation]. *Diabetes* **49**: 1627–1634.

27. Niwa, T. (1999). 3-deoxyglucosone: metabolism, analysis, biological activity, and clinical implication. *J. Chromat. B: Biomed. Sci. Appl.* **731**: 23–36.

28. Thornalley, P. J. (1993). The glyoxalase system in health and disease. *Mol. Aspects. Med.* **14**: 289–371.

29. Nagaraj, R. H., Sell, D. R., Prabhakaram, M., Ortwerth, B. J. and Monnier, V. M. (1991). High correlation between pentosidine protein crosslinks and pigmentation implicates ascorbate oxidation in human lens senescence and cataractogenesis. *Proc. Natl. Acad. Sci. USA* **88**: 10 257–10 261.

30. Pischetsrieder, M., Larisch, B. and Seidel, W. (1997). Immunochemical detection of oxalic acid monoamides that are formed during the oxidative reaction of L-ascorbic acid and proteins. *J. Agric. Food Chem.* **45**: 2070–2075.

31. Nagaraj, R. H., Shamsi, F. A., Huber, B. and Pischetsrieder, M. (1999). Immunochemical detection of oxalate monoalkylamide, an ascorbate-derived Maillard reaction product in the human lens. *FEBS Lett.* **453**: 327–330.

32. Tessier, F., Obrenovich, M. and Monnier, V. M. (1999). Structure and mechanism of formation of human lens fluorophore LM-1. Relationship to vesperlysine A and the advanced Maillard reaction in aging, diabetes, and cataractogenesis. *J. Biol. Chem.* **274**: 20 796–20 804.

33. Shipanova, I. N., Glomb, M. A. and Nagaraj, R. H. (1997). Protein modification by methylglyoxal: chemical nature and synthetic mechanism of major fluorescent adduct. *Arch. Biochem. Biophys.* **344**: 29–36.

34. Hunt, J. V. and Wolff, S. P. (1990). Is glucose the sole source of tissue browning in diabetes mellitus? *FEBS* **269**: 258–260.

35. Monnier, V. M., Stevens, V. J. and Cerami, A. (1979). Nonenzymatic glycosylation, sulfhydryl oxidation and aggregation of lens proteins in experimental sugar cataracts. *J. Exp. Med.* **150**: 1098–1107.

36. Cheng, R. Z. and Kawakishi, S. (1994). Site-specific oxidation of histidine residue in glycated insulin mediated by copper ion. *Eur. J. Biochem.* **223**: 759–764.

37. Wells-Knecht, M. C., Thorpe, S. R. and Baynes, J. W. (1995). Pathways of formation of glycoxidation products during glycation of collagen. *Biochemistry* **34**: 15 134–15 141.

38. Munch, G., Schicktanz, D., Behme, A., Gerlach, M., Riederer, P., Palm, D. and Schinzel, R. (1999). Amino acid specificity of glycation and protein-AGE crosslinking reactivities determined with a dipeptide SPOT library. *Nat. Biotech.* **17**: 1006–1010.

39. Cheng, R.-Z., Tsunehiro, J., Uchida, K. and Kawakishi, S. (1991). Oxidative damage of glycated protein in the presence of transition metal ion. *Agric. Biol. Chem.* **55**: 1993–1998.

40. Wells-Knecht, K. J., Zyzack, D. V., Litchfield, J. E., Thorpe, S. R. and Baynes, J. W. (1995). Mechanism of autoxidative glycosylation: identification of glyoxal and arabinose as intermediates in the autoxidative modification of proteins by glucose. *Biochemistry* **34**: 3702–3709.

41. Cheng, R. Z., Uchida, K. and Kawakishi, S. (1992). Selective oxidation of histidine residues in proteins or peptides through the copper(II)-catalyzed autoxidation of glucosone. *Biochem. J.* **285**: 667–671.

42. Baker, J. R., Zyzak, D. V., Thorpe, S. R. and Baynes, J. W. (1994). Chemistry of the fructosamine assay: d-glucosone is the product of oxidation of Amadori compounds. *Clin. Chem.* **40**: 1950–1955.

43. Takahashi, M., Pischetsrieder, M. and Monnier, V. M. (1997). Isolation, purification and characterization of amadoriase isoenzymes (fructosyl amine-oxygen oxidoreductase EC 1.5.3) from Aspergillus. *J. Biol. Chem.* **272**: 3437–3443.

44. Baynes, J. W. (1991). Role of oxidative stress in development of complications in diabetes. *Diabetes* **40**: 405–412.

45. Mullarkey, C. J., Edelstein, D. and Brownlee, M. (1990). Free radical generation by early glycation products: a mechanism for accelerated atherogenesis in diabetes. *Biochem. Biophys. Res. Commun.* **173**: 932–939.

46. Sakurai, T., Sugioka, K. and Nakano, M. (1990). O^{2-} generation and lipid peroxidation during the oxidation of a glycated polypeptide, glycated polylysine, in the presence of iron-ADP. *Biochim. Biophys. Acta* **1043**: 27–32.

47. Coussons, P. J., Jacoby, J., McKay, A., Kelly, S. M., Price, N. C. and Hunt, J. V. (1997). Glucose modification of human serum albumin: a structural study. *Free Radic. Biol. Med.* **22**: 1217–1227.

48. Ookawara, T., Kawamura, N., Kitagawa, Y. and Taniguchi, N. (1992). Site-specific and random fragmentation of Cu, Zn-superoxide dismutase by glycation reaction. *J. Biol. Chem.* **267**: 18 505–18 510.

49. Ortwerth, B. J. and James, H. L. (1999). Lens proteins block the copper-mediated formation of reactive oxygen species during glycation reactions in vitro. *Biochem. Biophys. Res. Commun.* **259**: 706–710.

50. Rowley, D. A. and Halliwell, B. (1983). Superoxide-dependent and ascorbate-dependent formation of hydroxyl radicals in the presence of copper salts: a physiologically significant reaction? *Arch. Biochem. Biophys.* **225**: 279–284.

51. Bourdon, E., Loreau, N. and Blache, D. (1999). Glucose and free radicals impair the antioxidant properties of serum albumin. *FASEB J.* **13**: 233–244.

52. Mossine, V. V., Linetsky, M., Glinsky, G. V., Ortwerth, B. J. and Feather, M. S. (1999). Superoxide free radical generation by Amadori compounds: the role of acyclic forms and metal ions. *Chem. Res. Toxicol.* **12**: 230–236.

53. Saxena, A. K., Saxena, P., Wu, X., Obrenovich, M., Weiss, M. F. and Monnier, V. M. (1999). Protein aging by carboxymethylation of lysines generates redox active and divalent metal binding sites: relevance to diseases of glycoxidative stress. *Biochem. Biophys. Res. Commun.* **260**: 332–338.

54. Lee, C., Yim, M. B., Chock, P. B., Yim, H. S. and Kang, S. O. (1998). Oxidation-reduction properties of methylglyoxal-modified protein in relation to free radical generation. *J. Biol. Chem.* **273**: 25 272–25 278.

55. Qian, M., Liu, M. and Eaton, J. W. (1998). Transition metals bind to glycated proteins forming redox active "glycochelates": implications for the pathogenesis of certain diabetic complications. *Biochem. Biophys. Res. Commun.* **250**: 385–389.

56. Nagai, R., Ikeda, K., Higashi, Y., Sano, H., Jinnouchi, Y., Araki, T. and Horiuchi, S. (1997). Hydroxyl radical mediates Ne-(carboxymethyl)lysine formation from Amadori product. *Biochem. Biophys. Res. Commun.* **234**: 167–172.

57. Dunn, J. A., Ahmed, M. U., Murtiashaw, M. H., Richardson, J. M., Walla, M. D., Thorpe, S. R. and Baynes, J. W. (1990). Reaction of ascorbate with lysine and protein under autoxidizing conditions: formation of N-epsilon-(carboxymethyl)lysine by reaction between lysine and products of autoxidation of ascorbate. *Biochemistry* **29**: 10 964–10 970.

58. Fu, M. X., Requena, J. R., Jenkins, A. J., Lyons, T. J., Baynes, J. W. and Thorpe, S. R. (1996). The advanced glycation end product, Ne-(carboxymethyl)lysine, is a product of both lipid peroxidation and glycoxidation reactions. *J. Biol. Chem.* **271**: 9982–9986.

59. Anderson, M. M., Requena, J. R., Crowley, J. R., Thorpe, S. R. and Heinecke, J. W. (1999). The myeloperoxidase system of human phagocytes generates N-epsilon-(carboxymethyl)lysine on proteins: a mechanism for producing advanced glycation end products at sites of inflammation. *J. Clin. Invest.* **104**: 103–113.

60. Degenhardt, T. P., Thorpe, S. R. and Baynes, J. W. (1998). Chemical modification of proteins by methylglyoxal. *Cell. Mol. Biol. (Noisy-le-grand)* **44**: 1139–1145.

61. Saxena, P., Saxena, A. K., Cui, X. L., Obrenovich, M., Gudipaty, K. and Monnier, V. M. (2000). Transition metal-catalyzed oxidation of ascorbate in human cataract extracts: possible role of advanced glycation end products. *Invest. Ophthalmol. Vis. Sci.* **41**: 1473–1481.

62. Szent-Gyorgyi, A. (1980). The living state and cancer. *Physiol. Chem. Phys.* **12**: 99–110.

63. Shamsi, F. A., Lin, K., Sady, C. and Nagaraj, R. H. (1998). Methylglyoxal-derived modifications in lens aging and cataract formation. *Invest. Ophthalmol. Vis. Sci.* **39**: 2355–2364.

64. Beisswenger, P. J., Howell, S. K., Touchette, A. D., Lal, S. and Szwergold, B. S. (1999). Metformin reduces systemic methylglyoxal levels in Type-2 diabetes. *Diabetes* **48**: 198–202.

65. Miyata, T., van Ypersele de Strihou, C., Kurokawa, K. and Baynes, J. W. (1999). Alterations in nonenzymatic biochemistry in uremia: origin and

significance of "carbonyl stress" in long-term uremic complications [In Process Citation]. *Kidney Int.* **55**: 389–399.

66. Esposito, C., Gerlach, H., Brett, J., Stern, D. and Vlassara, H. (1989). Endothelial receptor-mediated binding of glucose-modified albumin is associated with increased monolayer permeability and modulation of cell surface coagulant properties. *J. Exp. Med.* **170**: 1387–1407.

67. Ichikawa, K., Yoshinari, M., Iwase, M., Wakisaka, M., Doi, Y., Iino, K., Yamamoto, M. and Fujishima, M. (1998). Advanced glycosylation end products induced tissue factor expression in human monocyte-like U937 cells and increased tissue factor expression in monocytes from diabetic patients. *Atherosclerosis* **136**: 281–287.

68. Vlassara, H., Fuh, H., Donnelly, T. and Cybulsky, M. (1995). Advanced glycation endproducts promote adhesion molecule (VCAM-1, ICAM-1) expression and atheroma formation in normal rabbits. *Mol. Med.* **1**: 447–456.

69. Lu, M., Kuroki, M., Amano, S., Tolentino, M., Keough, K., Kim, I., Bucala, R. and Adamis, A. P. (1998). Advanced glycation end products increase retinal vascular endothelial growth factor expression. *J. Clin. Invest.* **101**: 1219–1224.

70. Kirstein, M., Brett, J., Radoff, S., Ogawa, S., Stern, D. and Vlassara, H. (1990). Advanced protein glycosylation induced transendothelial human monocyte chemotaxis and secretion of platelet-derived growth factor: role in vascular disease of diabetes and aging. *Proc. Natl. Acad. Sci.* **87**: 9010–9014.

71. Vlassara, H., Brownlee, M., Manogue, K. R., Dinarello, C. A. and Pasagian, A. (1988). Cachectin/TNF and IL-1 induced by glucose-modified proteins: role in normal tissue remodeling. *Science* **240**: 1546–1548.

72. Thornalley, P. J. (1998). Cell activation by glycated proteins. AGE receptors, receptor recognition factors and functional classification of AGEs. *Cell. Mol. Biol. (Noisy-le-grand)* **44**: 1013–1023.

73. Schmidt, A. M., Yan, S. D., Wautier, J. L. and Stern, D. (1999). Activation of receptor for advanced glycation end products: a mechanism for chronic vascular dysfunction in diabetic vasculopathy and atherosclerosis. *Circ. Res.* **84**: 489–497.

74. Araki, N., Higashi, T., Mori, T., Shibayama, R., Kawabe, Y., Kodama, T., Takahashi, K., Shichiri, M. and Horiuchi, S. (1995). Macrophage scavenger receptor mediates the endocytic uptake and degradation of advanced glycation end products of the Maillard reaction. *Eur. J. Biochem.* **230**: 408–415.

75. Stitt, A. W., Burke, G. A., Chen, F., McMullen, C. B. and Vlassara, H. (2000). Advanced glycation end product receptor interactions on microvascular cells occur within caveolin-rich membrane domains. *FASEB J.* **14**: 2390–2392.

76. Kislinger, T., Fu, C., Huber, B., Qu, W., Taguchi, A., Du Yan, S., Hofmann, M., Yan, S. F., Pischetsrieder, M., Stern, D. and Schmidt, A. M. (1999). N-epsilon-(carboxymethyl)lysine adducts of proteins are ligands for receptor for advanced glycation end products that activate cell signaling pathways and modulate gene expression. *J. Biol. Chem.* **274**: 31 740–31 749.

77. Taguchi, A., Blood, D. C., del Toro, G., Canet, A., Lee, D. C., Qu, W., Tanji, N., Lu, Y., Lalla, E., Fu, C., Hofmann, M. A., Kislinger, T., Ingram, M., Lu, A., Tanaka, H., Hori, O., Ogawa, S., Stern, D. M. and Schmidt, A. M. (2000). Blockade of RAGE-amphoterin signalling suppresses tumour growth and metastases [see comments]. *Nature* **405**: 354–360.
78. Scivittaro, V., Ganz, M. B. and Weiss, M. F. (2000). AGEs induce oxidative stress and activate protein kinase C-beta(II) in neonatal mesangial cells. *Am. J. Physiol. Renal. Physiol.* **278**: F676–F683.
79. Weiss, M. F., Erhard, P., Kader-Attia, F. A., Wu, Y. C., Deoreo, P. B., Araki, A., Glomb, M. A. and Monnier, V. M. (2000). Mechanisms for the formation of glycoxidation products in end-stage renal disease. *Kidney Int.* **57**: 2571–2585.
80. Verzijl, N., DeGroot, J., Thorpe, S. R., Bank, R. A., Shaw, J. N., Lyons, T. J., Bijlsma, J. W., Lafeber, F. P., Baynes, J. W. and TeKoppele, J. M. (2000). Effect of collagen turnover on the accumulation of advanced glycation endproducts. *J. Biol. Chem.* **275**: 39 027– 39 031.
81. Sell, D. R., Lane, M. A., Johnson, W. A., Masoro, E. E., Mock, O. B., Reiser, K. M., Fogarty, J. F., Cutler, R. C., Ingram, D. K., Roth, G. S. and Monnier, V. M. (1996). Longevity and the genetic determination of collagen glycoxidation kinetics in mammalian senescence. *Proc. Natl. Acad. Sci. USA* **93**: 485–490.
82. Sacher, G. A. and Duffy, P. H. (1979). Genetic relation of life span to metabolic rate for inbred mouse strains and their hybrids. *Fed. Proc.* **38**: 184–188.
83. Sell, D. R., Kleinman, N. R. and Monnier, V. M. (2000). Longitudinal determination of skin collagen glycation and glycoxidation rates predicts early death in C57BL/6NNIA mice. *FASEB J.* **14**: 145–156.
84. Cefalu, W. T., Bell-Farrow, A. D., Wang, Z. Q, Sonntag, W.E , Fu, M. X., Baynes, J. W. and Thorpe, S. R. (1995). Caloric restriction decreases age-dependent accumulation of the glycoxidation products Ne-(carboxymethyl))lysine and pentosidine. *J. Gerontol.* **50A**: B337–B341.
85. Pischetsrieder, M., Seidel, W., Munch, G. and Schinzel, R. (1999). N(2)-(1-carboxyethyl)deoxyguanosine, a nonenzymatic glycation adduct of DNA, induces single-strand breaks and increases mutation frequencies. *Biochem. Biophys. Res. Commun.* **264**: 544–549.
86. Corbett, J. A., Tilton, R. G., Chang, K., Hasan, K. S., Ido, Y., Wang, J. L., Sweetland, M. A., Lancaster, J. R., Jr., Williamson, J. R. and McDaniel, M. L. (1992). Aminoguanidine, a novel inhibitor of nitric oxide formation, prevents diabetic vascular dysfunction. *Diabetes* **41**: 552–556.
87. Onorato, J. M., Jenkins, A. J., Thorpe, S. R. and Baynes, J. W. (2000). Pyridoxamine, an inhibitor of advanced glycation reactions, also inhibits advanced lipoxidation reactions. Mechanism of action of pyridoxamine. *J. Biol. Chem.* **275**: 21 177–21 184.
88. Degenhardt, T. P., Alderson, N. L., Thorpe, S. R. and Baynes, J. W. (1988). Pyridorin preserves renal function and corrects dyslipidemia and redox

imbalances in STZ-diabetic rats. Conference on Diabetic Complications, Joslin Diabetes Center (Symposium Proceedings), p. 46.

89. Papoulis, A., Al-Abed, Y. and Bucala, R. (1995). Identification of N2-(1-carboethyl)guanine (CEG) as a guanine advanced glycosylation end product. *Biochemistry* **34**: 648–655

90. Al-Abed Y., Schleicher E., Voelter W., Liebich H., Papoulis A. and Bucala R. (1998). Identification of N2-(1-carboxymethyl)guanine (CMG) as a guanine advanced glycation end product. *Bioorg. Med. Chem. Lett.* **18**: 2109–2110.

91. Bucala R., Model P. and Cerami A. (1984). Modification of DNA by reducing sugars: a possible mechanism for nucleic acid aging and age-related dysfunction in gene expression. *Proc. Natl. Acad. Sci. USA* **81**: 105–109.

Measurement of Oxidative Stress: Technology, Biomarkers and Applications

(Mitochondrial Damage)

Chapter 25

Mitochondrial Mutations in Vertebrate Aging

Aubrey D.N.J. de Grey

Aubrey D. N. J. de Grey • Department of Genetics, University of Cambridge, Downing Street, Cambridge CB2 3EH, UK
Tel: +44 1223 333963, E-mail: ag24@gen.cam.ac.uk

1. Scope of This Review

The mitochondrial DNA theory of aging is among the longest-standing mechanistic proposals for why we age at the rate we do. It has also been among the most popular such ideas throughout its existence, giving rise to thousands of publications. Thus it is impossible, in a short review, to do justice to all aspects of the theory and its implications. Here I will discuss the development of the theory, the key experiments that have eliminated certain elaborations of it, and the possibilities that remain for how mitochondrial DNA mutations may drive vertebrate aging. For a fuller treatment, the reader is recommended to my recent book;[1] a discussion of the intervention which I see as having most promise to reverse mitochondrial DNA-driven aspects of aging has also recently appeared.[2]

2. Origins of the Mitochondrial DNA Theory of Aging

The idea that mitochondrial DNA (mtDNA) mutations might be a primary force in determining the rate of human aging was proposed by Harman in 1972.[3] He wrote:

> "Free radicals 'escaping' from the respiratory chain ... would be expected to produce deleterious effects mainly in the mitochondria ... Are these effects mediated in part through alteration of mitochondrial DNA functions?"

This was the first significant elaboration of Harman's original free radical theory of aging, published in 1956,[4] which was itself the first biochemically-based explanation of the "rate of living theory" that Rubner, Pearl and others put forward in the early decades of the 20th century.[5, 6] There is a remarkable and unfortunate tendency nowadays to attribute this insight to an article published fully 17 years later[7] which failed to cite Harman's work; it is to be hoped that this amnesia will cease.

This idea was rapidly embraced by the few researchers engaged in free radical gerontology at that time. Notably, however, it was roundly dismissed in a seminal review[8] written in 1974 by Alex Comfort, subsequently better known in a more recreational capacity but at that time the prominent and hugely respected editor of *Experimental Gerontology*. He drew attention to the fact, known since 1961,[9] that mitochondria are not static entities but are perpetually turned over in non-dividing cells, some being replicated while others are engulfed by the vacuolar apparatus and degraded. Depending on the tissue, the half-life of a rat mitochondrion is roughly a week to a month;[10, 11] recent work indicates that the human mitochondrial half-life is two or three months,[12] though more detailed measurements are urgently needed. This led Comfort to reason that mtDNA mutations could never accumulate during aging, since whenever a mutant mitochondrion arose it would be destroyed and replaced by replication of a non-mutant one. And if they did not accumulate, perforce they could not be a driving force in aging.

We now know that Comfort's reasoning was wrong. However, this was not in fact demonstrated for another 19 years, during which the only mechanistic elaborations of the mtDNA theory of aging were ones which absolutely ignored Comfort's objection.

3. Elaborations Founded on Implausible Population Dynamics

In 1980, Miquel proposed a specific way in which mitochondrial free radical damage could cause aging.[13] He suggested that the mtDNA was progressively altered to a state where it could not be replicated, and that this led to a diminution of mitochondrial numbers per cell (or per unit volume), especially in postmitotic cell types, with the result that ATP supply was compromised.

The fundamental weakness of this proposal is, of course, that if a cell contains a population of mitochondria with a range of abilities to be replicated, the ones which *can* be replicated will be and the ones which cannot will not, with the result that the ones which cannot be replicated will be eliminated from the cell as fast as they arise. This objection is simple, and it is highly curious that the proposal was published without any accompanying attempt to rebut the objection — particularly since Miquel cited Comfort[8] in his opening paragraph[13] and even published the article in Comfort's own journal, *Experimental Gerontology*. There are only two formal possibilities that might allow Miquel to be right. One is that all mitochondria in a given cell simultaneously transform to the non-replicable state; the other is that the non-replicable state is also a non-degradable state. Neither of these proposals has much intuitive merit — though a version of the latter has recently been revived in much more plausible form.[14]

This model was succeeded, a decade later,[15] by one which is very little better. It was shown in various laboratories in the early 1970s[16, 17] that the quantity of free radicals escaping from the respiratory chain was greatly increased by certain antibiotics that blocked the flow of electrons along the chain, such as antimycin A. This led to the idea that mitochondrial mutations might have the same effect, since 11 of the 13 proteins encoded in the mtDNA are subunits of respiratory chain enzymes. (The other two are subunits of the ATP synthase.) On this basis it was proposed that mitochondrial free radical production and mtDNA mutations participate in a positive feedback loop, or vicious cycle, whereby mutations in some mitochondria cause a rise in free radical flux throughout the cell, resulting in an accelerated production of more mtDNA mutations in previously intact mitochondria. This hypothesis accordingly became known as the "vicious cycle" theory (Fig. 1).

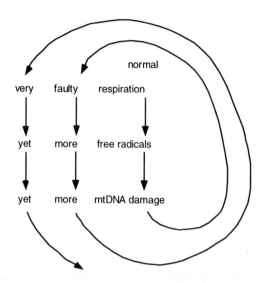

Fig. 1. The vicious cycle theory for accumulation of mutant mtDNA. If mutant mitochondria produce more free radicals than wild-type ones, and if these free radicals are mutagenic to other mitochondria in the cell. there is a positive feedback loop which rapidly mutates all the mtDNA in the cell. Neither prerequisite seems in fact to be true.

Inherent in the vicious cycle theory is the assumption that replication and destruction of mitochondria are broadly unbiased as between mutant and intact mitochondria. Again, therefore, this model ignores Comfort's objection of 1974. Additionally, it relies on the idea that mtDNA mutations would increase free radical production, when in fact (as was, to the authors' credit, pointed out in the original article[15] but nonetheless ignored by essentially everyone) the mutations mainly detected *in vivo*, large deletions,[18] would be predicted to *reduce* free radical production. Finally it has the problem that the feedback loop is very simple, and thus the acceleration of the process would be predicted to be rapid, with the result that cells would be entirely depleted of intact mtDNA shortly after a mutation arose, whereas aging is in fact an extremely slow process. Despite these profound weaknesses, however, the vicious cycle held sway over the mitochondrial gerontology community for the next decade and is only now beginning to loosen its grip.

4. Clonal Expansion of Mutant mtDNA

In the same year that the vicious cycle theory was published, another suggestion was also made.[19] It sought to explain a finding of the previous year[20] which can, in fact, be reconciled quite economically with both the vicious cycle theory and the "Comfort problem";[21] no such link was made at the time, however. The finding was that mutant mitochondria are distributed very focally within muscle, with a few fibre segments becoming totally devoid of cytochrome c oxidase (COX) activity while surrounding fibres appear normal. The frequency of COX-negative fibres increased with age, but was always under 1%. The proposal to explain this[19] was that some mutations of the mtDNA confer a selective advantage during mitochondrial turnover, such that copies of a single mutational event take over the whole cell. This was therefore a head-on challenge to the Comfort problem. It is certainly reasonable to presume that mutant mitochondria would be preferentially eliminated during turnover, but, reasonable or not, it might be wrong. In particular, it was noted that deletions (which are indeed found in aging — see above) might have a selective advantage simply because a smaller genome can be replicated more rapidly. The clonal expansion theory is very cleanly distinguishable experimentally from the vicious cycle theory, because in the latter each COX-negative cell is populated by a spectrum of different mtDNA mutations, whereas in the former each cell contains copies of only one mutation.

This proposal did not achieve popular acclaim. More extraordinarily, it still failed to gain acceptance even after the publication in 1993[22] of a cast-iron refutation of the vicious cycle theory, by the experiment just described — *in situ* hybridization to portions of the mtDNA, which revealed conclusively that COX-negative cells contained copies of a single mutation. The mechanism of clonal expansion is almost certainly not the one originally proposed,[19] faster replication of smaller molecules; we still do not know what it is, though a proposal[23] based on slower degradation, rather than faster replication, of mutant mitochondria appears to fit the evidence well.

It should be noted here that the phenomenon of clonal expansion is the main reason why this chapter is restricted in scope to vertebrate, as opposed to metazoan, aging. *Drosophila* has been reported to show no significant rise with age in the level of mutant mtDNA during adulthood;[24] this is no surprise if amplification during turnover is a requirement for such a rise, because flies do not live long enough (a dozen mitochondrial generations at absolute minimum) to allow a mutation to take over a cell. The role of free radicals in aging of *Drosophila* has been compellingly demonstrated,[25, 26] but it evidently does not work via mtDNA damage. The difference between flies and mammals is further underscored by the unaltered longevity of mice lacking cytosolic superoxide dismutase,[27] a condition which reduces *Drosophila* lifespan by a factor of five.[28]

5. Clonal Expansion Challenges the "Tip of the Iceberg" Theory

It is now — at last — broadly accepted, however, that mutant mtDNA accumulates by clonal expansion of rare somatic mutations. What has been less widely recognized is that this poses a huge problem for the idea that mtDNA mutations could possibly play a significant role in aging.

When efforts were first made to quantify the proportion of mtDNA that becomes mutant by old age,[29] it rapidly became apparent that their abundance was an issue of utmost importance to their role (or lack of one) in aging. Levels of a particular deletion known to occur in various myopathies were measured as about 0.1% of total mtDNA.[29] It was never claimed that such levels could be significant on their own, but it was rapidly noted[30, 31] that the number of possible deletions (even ignoring point mutations) was enormous, such that if they were all present at about 0.1% then the total "mutation load" would be a high proportion of total mtDNA, thus very plausibly contributing to age-related bioenergetic insufficiency. Similar arguments have been made[32] regarding point mutations: their accumulation during aging remains controversial,[33, 34] though the observation of no rise with age[34] was in a region encoding an ATPase subunit, which has been argued[23] to be atypical.

This reasoning became known as the "tip of the iceberg" hypothesis. It was very hard to test at the time, because the only available method for quantifying the levels of mutant mtDNA was quantitative PCR, which could only amplify DNA segments a few hundred base pairs long and thus could only detect a small minority of possible mtDNA deletions (namely, those that removed nearly but not quite all of the interval between the primers).

Two approaches have since been used to test the "tip of the iceberg" theory, but they have both been robustly challenged on methodological grounds. One[35] is to use a very large number of PCR primers in the same reaction mixture, allowing the amplification of essentially any mtDNA deletion (because some pair of primers will always be in the right region). The other[36] is to use newer variants of PCR which can amplify sequences as long as the entire mtDNA, 16.5 kilobases, and so can detect nearly all possible deletions using just one pair of primers. The main challenge to both these methods has been that PCR is notorious for its sensitivity to two artefacts that complicate — some would say abolish — its utility for quantifying rare sequence variants: it generates novel sequences spontaneously, and it amplifies some sequences much more efficiently than others.

The proponents of these methods are continuing to refine them, while their critics continue to dismiss them;[37] the debate seems set to continue for some time. One recent study may be the beginning of a resolution, however. It was shown[38] that in cardiomyocytes the majority of molecules identified by PCR as mtDNA deleteions are in fact duplications, which when amplified using divergent primers in the duplicated region are indistinguishable from heteroplasmic deletions (Fig. 2). The presence of duplications was shown because the sensitivity of the PCR reaction was sufficient to allow use of only part of a single cell's mtDNA; thus, having identified the breakpoints of the mutant molecule, a second sample from the same cell could be analyzed using primers which lay within the putatively deleted region, and which thus would amplify

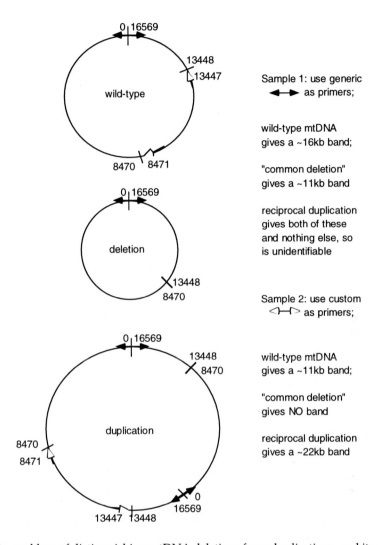

Fig. 2. The problem of distinguishing mtDNA deletions from duplications, and its solution. The tandem duplication, when amplified with primers in the duplicated region, gives PCR products identical to those which would derive from a 1:1 mixture of a deletion and a wild-type molecule. When a second sample from the same cell (so containing the same mutant molecule) can be analyzed, primers in the unduplicated region can be used which amplify a unique sequence from the duplication but nothing from the deficiency.

nothing at all if the molecule were a deletion but would amplify a characteristic signature fragment if it were a duplication (Fig. 2).

Two items of evidence having nothing to do with methodology weigh very heavily, however, against the "tip of the iceberg" hypothesis. The first is that a high proportion of the deletions detected by the new methods are lacking in one or both of the mitochondrial origins of replication, which are thought to be absolutely essential for replication of the mtDNA [though one of them may be dispensable in some unusual circumstances[39]]. The second is perhaps even stronger. There is no doubt that nearly all cells retain cytochrome c oxidase activity even in very old age, as revealed by numerous histochemical analyses.[20, 40, 41] Thus, if there is indeed a lot of deleted mtDNA in tissues of elderly individuals, most of it must be in cells that also contain a lot of intact mtDNA to encode three core subunits of the functioning cytochrome c oxidase. But then, since it is established that deleted mtDNA has a selective advantage, such cells should become completely devoid of wild-type mtDNA — and so COX-negative — very soon after reaching this highly heteroplasmic state. This is not easily reconcileable with the observed very slow increase with age in the number of COX-negative cells. (If most of the mutant molecules are actually duplications,[38] on the other hand, amplification may occur but be phenotypically silent: this could reconcile the rarity of COX-negative cells with the high level of PCR-detectable mutations, but it also implies that the duplicated mtDNA does not contribute to aging.)

6. Once You Have Eliminated the Impossible ...

The foregoing logic led the present author to the view that, if the accumulation of mtDNA mutations with age really has a causative role in determining the rate of mammalian aging, it cannot be simply by virtue of cellular bioenergetic insufficiency, because too few cells suffer that insufficiency. It is, however, far from easy to see how else mtDNA mutations could matter. Accordingly, one must first revisit the evidence that mtDNA mutations really do matter in aging.

The evidence that mitochondrial free radical damage is a key determinant of the rate of vertebrate aging is very strong indeed. The extensive interspecies studies of Cutler, Sohal and Barja have clearly determined that mitochondrial superoxide production varies inversely with maximum lifespan across a wide range of taxa. This, when combined with the non-effect on mouse maximum lifespan of eliminating the cytosolic isoform of superoxide dismutase, strongly implies that — just as Harman had suggested in 1972[3] — the damage resulting from mitochondrial superoxide production is restricted to the mitochondria themselves. Elimination of mitochondrial superoxide dismutase in mice causes perinatal death,[42] and heterozygotes show a range of hallmarks of oxidative damage within the mitochondria but none in the cytosol or nucleus.[43]

So, if mitochondrial damage matters, does mitochondrial DNA damage matter? To address this we can profitably return to Comfort's 1974 argument.[8] Everything except the mtDNA is subject to a regular dilution of any damage which may have accumulated, as a result of the replication of mitochondria and the accompanying incorporation of pristine, undamaged proteins and lipids. Thus, the mtDNA is the only component of the mitochondria which can accumulate damage. As so often in biology, however, what sounds like a proof is not one. The flip-side of mitochondrial replication in non-dividing cells is mitochondrial destruction by the vacuolar apparatus, and that process is not 100% effective: a tiny proportion of the autophagocytosed material is resistant to lysosomal degradation and accumulates over time as age pigment, or lipofuscin. In 1992, Brunk suggested[44] that lipofuscin might, despite its apparent chemical inertness and its sequestration within lysosomes, be a key player in the gradual decline of cellular efficiency, by progressively inhibiting the efficacy of the vacuolar apparatus in which it resides. His group have since demonstrated very clearly that heavily lipofuscin-loaded lysosomes are less effective at many aspects of the degradative process, and also that they are more prone to rupture, which puts the cell under considerable stress.[45–47] However, all this work has been done in cell culture; it remains un-certain whether the *in vivo* levels of lipofuscin can be considered to have this effect, especially in view of reports that vitamin E deficiency causes a substantial acceleration in the accumulation of pigment but does not detectably reduce lifespan.[48] That objection has been challenged on the grounds that the extra pigment may not be truly lipofuscin,[49] but the model proposed by Brunk seems clearly to predict that accumulation of anything lipofuscin-like would have the same effect, and indeed "artificial lipofuscin"[50] has been shown to impair lysosomal function *in vitro* in his experiments.[46]

One is thus compelled to seek a mechanism whereby rare mutant mtDNA can be systemically harmful. The only such mechanism so far proposed[1, 51, 52] is now termed the "reductive hotspot hypothesis" (Fig. 3). Since the mutant mtDNA *in vivo* is localized so non-uniformly, taking over a few cells while being essentially absent in most, it is conceivable that the affected cells might exert systemic toxicity by alterations to circulating material that mitochondrially healthy cells subsequently import. A specific mechanism for such toxicity, once sought, is in fact rather easy to identify in the context of what has been known for over a decade about how mitochondrially mutant cells can be maintained *in vitro*:[53] they absolutely require an external sink for electrons, in order to maintain a steady-state $NAD^+/NADH$ ratio. It was shown in 1993[54] that the electron acceptor need not enter the cell, implying (since NADH cannot leave the cell) that there is a trans-plasma membrane electron export system involved. This system, commonly termed the plasma membrane oxidoreductase (PMOR), remains only poorly characterized, though work on its structure and mechanism is gathering pace.[55] It was observed in 1994[56] that this might be the way in which mitochondrially mutant cells survive (and thence accumulate) *in vivo*. The insight underlying the reductive hotspot hypothesis was to identify oxygen as the likely electron acceptor *in vivo* and to suggest that it might often undergo one-electron reduction forming extracellular superoxide. Skeletal muscle, which by virtue of its contribution to total body mass is the major consideration in any hypothesis involving systemic toxicity, has very little extracellular superoxide dismutase;[57]

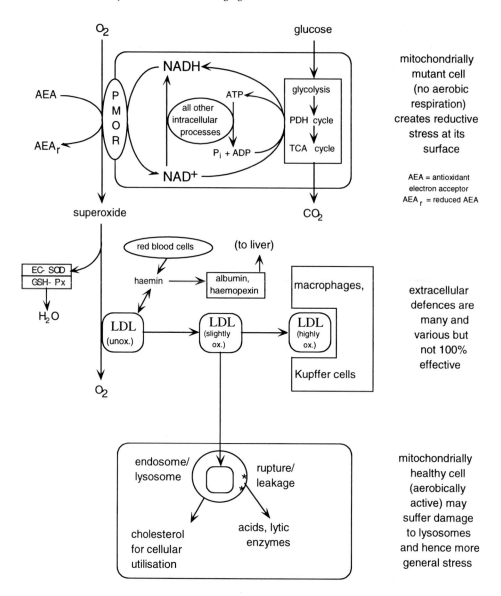

Fig. 3. The reductive hotspot hypothesis. Reproduced from Ref. 52, with permission.

thus, there may be ample opportunity for this extracellular superoxide to react with traces of redox-active transition metals, particularly hemin,[58, 59] so initiating Fenton chemistry that may oxidize circulating material, such as low-density lipoprotein (LDL). LDL is of particular interest because it is the vehicle whereby most cells acquire cholesterol, an indispensable component of membranes. Thus, the reductive hotspot hypothesis is that the gradually increasing number of mitochondrially mutant cells during aging leads to a rise in the contamination of LDL with peroxidation products such as lipid hydroperoxides, which are imported into mitochondrially healthy cells but then have the capacity to cause further peroxidation chain reactions within the recipient cell, so increasing oxidative stress there.

The reductive hotspot hypothesis is widely regarded — not least by its originator — as unattractively elaborate and implausible. However, we must face the facts: aging happens. It is not an experimental artefact. Something, therefore, drives it. It remains my view that, elaborate though it may be, the mechanism proposed by the reductive hotspot hypothesis is the most economical interpretation of existing data that has yet been proposed. It is thus urgent to engage in tests of the numerous specific predictions inherent in that model: without such work, the mitochondrial free radical theory of aging is in danger of remaining in the slow lane of biology research that it has occupied for so long.

References

1. de Grey, A. D. N. J. (1999). *The Mitochondrial Free Radical Theory of Aging.* R.G. Landes Company, Austin.
2. de Grey, A. D. N. J. (2000). Mitochondrial gene therapy: an arena for the biomedical use of inteins. *Trends Biotech.* **18**: 394–399.
3. Harman, D. (1972). The biologic clock: the mitochondria? *J. Am. Geriatr. Soc.* **20**: 145–147.
4. Harman, D. (1956). Aging: a theory based on free radical and radiation chemistry. *J. Gerontol.* **11**: 298–300.
5. Rubner, M. (1908). *Das Problem der Lebensdauer und seine Beziehungen zu Wachstum und Ernährung.* Oldenburg, Munchen, Germany.
6. Pearl, R. (1928). *The Rate of Living.* Knopf, New York.
7. Linnane, A. W., Marzuki, S., Ozawa, T. and Tanaka, M. (1989). Mitochondrial DNA mutations as an important contributor to ageing and degenerative diseases. *Lancet* **1**: 642–645.
8. Comfort, A. (1974). The position of aging studies. *Mech. Ageing Dev.* **3**: 1–31.
9. Fletcher, M. J. and Sanadi, D. R. (1961). Turnover of rat-liver mitochondria. *Biochim. Biophys. Acta* **51**: 356–360.
10. Gross, N. J., Getz, G. S. and Rabinowitz, M. (1969). Apparent turnover of mitochondrial deoxyribonucleic acid and mitochondrial phospholipids in the tissues of the rat. *J. Biol. Chem.* **244**: 1552–1562.
11. Menzies, R. A. and Gold, P. H. (1971). The turnover of mitochondria in a variety of tissues of young adult and aged rats. *J. Biol. Chem.* **246**: 2425–2429.
12. Rooyackers, O. E., Adey, D. B., Ades, P. A. and Nair, K. S. (1996). Effect of age on *in vivo* rates of mitochondrial protein synthesis in human skeletal muscle. *Proc. Natl. Acad. Sci. USA* **93**: 15 364–15 369.

13. Miquel, J., Economos, A. C., Fleming. J. and Johnson, J. E. (1980). Mitochondrial role in cell aging. *Exp. Gerontol.* **15**: 575–591.

14. Kowald, A. and Kirkwood, T. B. L. (2000). Accumulation of defective mitochondria through delayed degradation of damaged organelles and its possible role in the ageing of post-mitotic and dividing cells. *J. Theor. Biol.* **202**: 145–160.

15. Bandy, B. and Davison, A. J. (1990). Mitochondrial mutations may increase oxidative stress: implications for carcinogenesis and aging? *Free Radic. Biol. Med.* **8**: 523–539.

16. Loschen, G., Flohé, L. and Chance, B. (1971). Respiratory chain-linnked H_2O_2 production in pigeon heart mitochondria. *FEBS Lett.* **18**: 261–226.

17. Boveris, A., Oshino, N. and Chance, B. (1972). The cellular production of hydrogen peroxide. *Biochem. J.* **128**: 617–630.

18. Lee, C. M., Lopez, M. E., Weindruch, R. and Aiken, J. M. (1998). Association of age-related mitochondrial abnormalities with skeletal muscle fiber atrophy. *Free Radic. Biol. Med.* **25**: 964–972.

19. Kadenbach, B. and Müller-Höcker, J. (1990). Mutations of mitochondrial DNA and human death. *Naturwissenschaften* **77**: 221–225.

20. Müller-Höcker, J. (1989). Cytochrome c oxidase deficient cardiomyocytes in the human heart — an age-related phenomenon. A histochemical ultracyto-chemical study. *Am. J. Pathol.* **134**: 1167–1173.

21. de Grey, A. D. N. J. (2002). Mechanisms underlying the age-related accumula-tion of mutant mitochondrial DNA: a critical review. *In* "Genetics of Mitochondrial Diseases" (I. J. Holt, Ed.), **in press**.

22. Müller-Höcker, J., Seibel, P., Schneiderbanger, K. and Kadenbach, B. (1993). Different in situ hybridization patterns of mitochondrial DNA in cytochrome c oxidase-deficient extraocular muscle fibres in the elderly. *Virchows Arch.* **A422**: 7–15.

23. de Grey, A. D. N. J. (1997). A proposed refinement of the mitochondrial free radical theory of aging. *BioEssays* **19**: 161–166.

24. Calleja, M., Pena, P., Ugalde, C., Ferreiro, C., Marco, R. and Garesse, R. (1993). Mitochondrial DNA remains intact during *Drosophila* aging, but the levels of mitochondrial transcripts are significantly reduced. *J. Biol. Chem.* **268**: 18 891–18 897.

25. Parkes, T. L., Elia, A. J., Dickinson, D., Hilliker, A. J., Phillips, J. P. and Boulianne, G. L. (1998). Extension of *Drosophila* lifespan by overexpression of human SOD1 in motorneurons. *Nature Genetics* **19**: 171–174.

26. Sun, J. and Tower, J. (1999). FLP recombinase-mediated induction of Cu/Zn-superoxide dismutase transgene expression can extend the life span of adult *Drosophila melanogaster* flies. *Mol. Cell. Biol.* **19**: 216–228.

27. Reaume, A. G., Elliott, J. L., Hoffman, E. K., Kowall, N. W., Ferrante, R. J., Siwek, D. F., Wilcox, H. M., Flood, D. G., Beal, M. F., Brown, R. H., Scott,

R. W. and Snider, W. D. (1996). Motor neurons in Cu/Zn superoxide dismutase-deficient mice develop normally but exhibit enhanced cell death after axonal injury. *Nature Genetics* **13**: 43–47.

28. Phillips, J. P., Campbell, S. D., Michaud, D., Charbonneau, M. and Hilliker, A. J. (1989). Null mutation of copper/zinc superoxide dismutase in *Drosophila* confers hypersensitivity to paraquat and reduced longevity. *Proc. Natl. Acad. Sci. USA* **86**: 2761–2765.

29. Cortopassi, G. A. and Arnheim, N. (1990). Detection of a specific mitochondrial DNA deletion in tissues of older humans. *Nucleic Acids Res.* **18**: 6927–6933.

30. Arnheim, N. and Cortopassi, G. A. (1992). Deleterious mitochondrial DNA mutations accumulate in aging human tissues. *Mutat. Res.* **275**: 157–167.

31. Wallace, D. C. (1992). Mitochondrial genetics: a paradigm for aging and degenerative diseases? *Science* **256**: 628–632.

32. Nagley, P. and Wei, Y. H. (1998). Ageing and mammalian mitochondrial genetics. *Trends Genetics* **14**: 513–517.

33. Münscher, C., Müller-Höcker, J. and Kadenbach, B. (1993). Human aging is associated with various point mutations in tRNA genes of mitochondrial DNA. *Biol. Chem. Hoppe Seyler* **374**: 1099–1104.

34. Pallotti, F., Chen, X., Bonilla, E. and Schon, E. A. (1996). Evidence that specific mtDNA point mutations may not accumulate in skeletal muscle during normal human aging. *Am. J. Human Genetics* **59**: 591–602.

35. Hayakawa, M., Katsumata, K., Yoneda, M., Tanaka, M., Sugiyama, S. and Ozawa, T. (1996). Age-related extensive fragmentation of mitochondrial DNA into minicircles. *Biochem. Biophys. Res. Commun.* **226**: 369–377.

36. Kovalenko, S. A., Kopsidas, G., Kelso, J. M. and Linnane, A. W. (1997). Deltoid human muscle mtDNA is extensively rearranged in old age subjects. *Biochem. Biophys. Res. Commun.* **232**: 147–152.

37. Lightowlers, R. N., Jacobs, H. T. and Kajander, O. A. (1999). Mitochondrial DNA — all things bad? *Trends Genetics* **15**: 91–93.

38. Bodyak, N. D., Nekhaeva, E., Wei, J. Y. and Khrapko, K. (2001). Quantification and sequencing of somatic deleted mtDNA in single cells: evidence for partially duplicated mtDNA in aged human tissues. *Hum. Mol. Genet.* **10**: 17–24.

39. Holt, I. J., Lorimer, H. E. and Jacobs, H. T. (2000). Coupled leading- and lagging-strand synthesis of mammalian mitochondrial DNA. *Cell* **100**: 515–524.

40. Müller-Höcker, J. (1990). Cytochrome c oxidase deficient fibres in the limb muscle and diaphragm of man without muscular disease: an age-related alteration. *J. Neurol. Sci.* **100**: 14–21.

41. Shoubridge, E. A. (1994). Mitochondrial DNA diseases: histological and cellular studies. *J. Bioenerg. Biomembr.* **26**: 301–310.

42. Li, Y., Huang, T. T., Carlson, E. J., Melov, S., Ursell, P. C., Olson, J. L., Noble, L. J., Yoshimura, M. P., Berger, C., Chan, P. H., Wallace, D. C. and Epstein,

C. J. (1995). Dilated cardiomyopathy and neonatal lethality in mutant mice lacking manganese superoxide dismutase. *Nature Genetics* **11**: 376–381.

43. Williams, M. D., van Remmen, H., Conrad, C. C., Huang, T. T., Epstein, C. J. and Richardson, A. (1998). Increased oxidative damage is correlated to altered mitochondrial function in heterozygous manganese superoxide dismutase knockout mice. *J. Biol. Chem.* **273**: 28 510–28 515.

44. Brunk, U. T., Jones, C. B. and Sohal, R. S. (1992). A novel hypothesis of lipofuscinogenesis and cellular aging based on interactions between oxidative stress and autophagocytosis. *Mutat. Res.* **275**: 395–403.

45. Brunk, U. T and Terman, A. (1997). The mitochondrial-lysosomal axis theory of cellular aging. *In* "Understanding the Basis of Aging: Mitochondria, Oxidants and Aging" (E. Cadenas, and L. Packer, Eds.), pp. 229–250, Marcel Dekker, New York.

46. Sundelin, S., Wihlmark, U., Nilsson, S. E. and Brunk, U. T. (1998). Lipofuscin accumulation in cultured retinal pigment epithelial cells reduces their phagocytic capacity. *Curr. Eye Res.* **17**: 851–857.

47. Terman, A., Abrahamsson, N. and Brunk, U. T. (1999). Ceroid/lipofuscin-loaded human fibroblasts show increased susceptibility to oxidative stress. *Exp. Gerontol.* **34**: 755–770.

48. Blackett, A. D. and Hall, D. A. (1981). Tissue vitamin E levels and lipofuscin accumulation with age in the mouse. *J. Gerontol.* **36**: 529–533.

49. Porta, E. A., Sablan, H. M., Joun, N. S. and Chee, G. (1982). Effects of the type of dietary fat at two levels of vitamin E in Wistar male rats during development and aging. IV. Biochemical and morphometric parameters of the heart. *Mech. Ageing Dev.* **18**: 159–199.

50. Nilsson, E. and Yin, D. (1997). Preparation of artificial ceroid/lipofuscin by UV-oxidation of subcellular organelles. *Mech. Ageing Dev.* **99**: 61–78.

51. de Grey, A. D. N. J. (1998). A mechanism proposed to explain the rise in oxidative stress during aging. *J. Anti-Aging Med.* **1**: 53–66.

52. de Grey, A. D. N. J. (2000). The reductive hotspot hypothesis: an update. *Arch. Biochem. Biophys.* **373**: 295–301.

53. King, M. P. and Attardi, G. (1989). Human cells lacking mtDNA: repopulation with exogenous mitochondria by complementation. *Science* **246**: 500–503.

54. Martinus, R. D., Linnane, A. W. and Nagley. P. (1993). Growth of rho0 human Namalwa cells lacking oxidative phosphorylation can be sustained by redox compounds potassium ferricyanide or coenzyme Q10 putatively acting through the plasma membrane oxidase. *Biochem. Mol. Biol. Int.* **31**: 997–1005.

55. de Grey, A. D. N. J. (2000). Forum editorial: biomedical aspects of plasma membrane redox. *Antiox. Redox Signal.* **2**: 155–156.

56. Larm, J. A., Vaillant, F., Linnane, A. W. and Lawen, A. (1994). Up-regulation of the plasma membrane oxidoreductase as a prerequisite for the viability of human Namalwa rho-0 cells. *J. Biol. Chem.* **269**: 30 097–30 100.

57. Marklund, S. L. (1984). Extracellular superoxide dismutase and other superoxide dismutase isoenzymes in tissues from nine mammalian species. *Biochem. J.* **222**: 649–655.

58. Miller, Y. I., Felikman, Y. and Shaklai, N. (1995). The involvement of low-density lipoprotein in hemin transport potentiates peroxidative damage. *Biochim. Biophys. Acta* **1272**: 119–127.

59. Miller, Y. I. and Shaklai, N. (1999). Kinetics of hemin distribution in plasma reveals its role in lipoprotein oxidation. *Biochim. Biophys. Acta* **1454**: 153–164.

Chapter 26

Mitochondria, Oxidative Stress and Mitochondrial Diseases

Keshav K. Singh*, Britt M. Luccy and Steven J. Zullo

Keshav K. Singh and **Britt M. Luccy** • Johns Hopkins Oncology Center, Bunting-Blaustein Cancer Research Building, 1650 Orleans Street, Room 143, Baltimore, MD 21231-1000, and Division of Toxicological Sciences, Department of Environmental Health Sciences, The Johns Hopkins School of Public Health, 615 N. Wolfe Street, Baltimore, MD 21205-2179

Steven J. Zullo • Advanced Technology Program, National Institute of Standards and Technology, 100 Bureau Drive, A239, Gaithersburg MD 20899-4730

*Corresponding Author.
Tel: 410-614-5128, E-mail: singhke@jhmi.edu

1. Introduction

The mitochondrion is a major source of endogenous oxidative damage because it produces about 10^7 reactive oxygen species (ROS)/mitochondrion/day[1-2] during normal oxidative phosphorylation. Mitochondrial DNA (mtDNA) contains no introns, has no protective histones and is constantly exposed to ROS produced within the mitochondria.[1,3] Thus the mitochondrial genome is extremely susceptible to genetic changes occurring spontaneously and occurring at increased frequency by exogenous oxidants.

Mitochondrial dysfunction is involved in pathogenesis of both adult and childhood diseases. Adulthood diseases are linked to diseases of aging. These include diabetes, heart disease, stroke, osteoporosis, and neurodegenerative diseases such as Parkinson, Alzheimer's, Huntington's diseases, and cancer.[1-5] More than 50 million people in the United States suffer from diseases in which mitochondrial dysfunction is involved (Fig. 1, for more details visit http://biochemgen.ucsd.edu/). There are more than 50 inherited metabolic childhood diseases in which mitochondrial dysfunction is reported. Taken, together, it is estimated that of four million children born each year in the United States, up to 4000 are born with mitochondrial disease (Figs. 2 and 3). Thus mitochondria play a central role in numerous diseases that can affect any organ, at any age.[5] The mitochondrial genetic changes have been observed during aging,

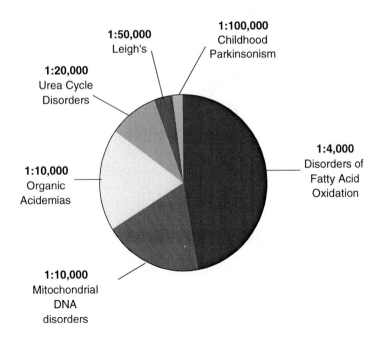

Fig. 1. Incidence of childhood mitochondrial disease in the United States.

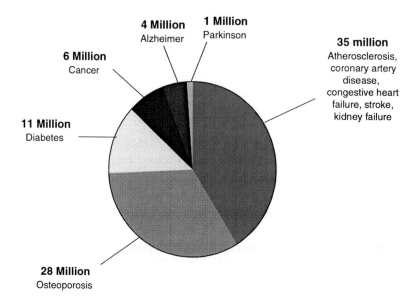

Fig. 2. Number of adult diseases with mitochondrial dysfunction in the United States.

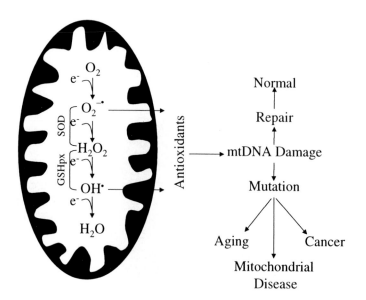

Fig. 3. Mitochondria as the major site of ROS production. ROS are continuously generated in mitochondria due to one electron transfer during oxidative phosphorylation. SOD, superoxide dismutase; GSHpx, glutathione peroxidase.

in mitochondrial diseases and in cancers (see Sec. 6.7–6.9 Fig. 3). The genetic changes in these diseases range from point mutations to large deletions or insertions in the mtDNA.[1, 3, 6] These genetic changes occur due to relentless damage to mtDNA by ROS produced within the mitochondria (Fig. 3).

2. Mitochondria Perform Multiple Cellular Functions

Mitochondria produce more than 90% of the energy (ATP) needs of a cell. Therefore, they are regarded as the "power house". However, mitochondria perform multiple functions that are essential for the cell. Mitochondria maintain intracellular homeostasis of inorganic ions such Ca^{2+}, mitochondrial membrane potential, adequate antioxidants and mineral support for SOD and GSX, membrane fluidity, entry of fatty acids for beta oxidation, cardiolipin, and structural support of enzymes of the electron transport chain. Mitochondria also maintain adequate intake of carbohydrate into mitochondria and adequate performance of Complex 1-NADH Dehydrogenase, Complex 2-Succinate Dehydrogenase, Complex 3-Cytochrome C Reductase, Complex 4-Cytochrome C oxidase and Complex 5-ATP Synthase. Mitochondria carry out cholesterol, estrogen and testosterone synthesis and ROS production and detoxification. Mitochondria also initiate programmed cell death.[3, 5, 7, 8] There are about 250 different cell types in human body. Mitochondrial function is tailored to meet the demands of each cell types. Thus, there are about 250 different mitochondria with specialized metabolic function.[10]

3. Mitochondrial DNA is Maternally Inherited

All human cells with the exception of erythrocytes contain mitochondria. Each mitochondrion contains several copies of mtDNA. Human mtDNA is 16 569 base pairs long double-stranded circular molecule.[9] About 3000 genes participate in the biogenesis of a single mitochondrion. Of these 3000 genes, mtDNA encodes only 37 genes. It encodes thirteen proteins, all of which are subunits of the electron transport chain. In addition, mtDNA encodes a minimal set of 22 transfer RNAs and two ribosomal RNAs, necessary for translation in mitochondria.[5, 10, 11] The other genes needed for mitochondrial biogenesis are encoded by the nuclear genome and the proteins are transported to the mitochondria.[11]

MtDNA is maternally inherited. A heteroplasmic condition exists when a heterogeneous mtDNA population, mutant and wild-type mtDNA, coexist in the same cell. Homoplasmic condition exists when all wild-type or all mutant mtDNA exist in a cell. The heteroplasmic condition is frequently associated with mitochondrial disease. Heteroplasmy can occur within and between cells and between different organs in a patient with mitochondrial disease. Mitochondrial pathology is manifested when the mutant mtDNA exceed a threshold. The threshold levels for mutant mtDNA may vary with the energy demand of the tissue. The type of mutation may also influence whether symptoms of the mitochondrial disease are present.[3, 11, 12] Unlike quantal transmission of nuclear genes, the distribution of mtDNA to daughter cells is graded from 0–100% and dependent on factors such as germ line mosaicism, intracellular and intracellular heteroplasmy, postnatal selection of certain mitochondrial genotype and oocyte heteroplasmy.[4] There is evidence that in cancer cells certain pathogenic mtDNA mutations may have a selective advantage over the wild-type.[13]

4. Mitochondrial DNA is Susceptible to Oxidative Damage

Mitochondrial oxidative phosphorylation is the predominant source of energy for various tissues. Unfortunately, mitochondria are also the most important source of endogenous ROS in the cell, because they carry the electron transport chain that during oxidative phosphorylation, reduces oxygen to water by addition of electrons.[1, 6, 11] ROS production can be considerably enhanced under certain pathological conditions and by environmental oxidants.[14]

ROS-induced oxidative damage is more extensive and more persistent in mtDNA than in nuclear DNA.[15] The endogenous rate of oxidation is estimated to be ~150 000 events/cell/ day in human cells.[16] To date, an estimate for the oxidation rate of bases in mtDNA is lacking. Oxidation of DNA can result in strand interruptions and inter- and intra-strand cross links that block DNA replication. Oxidation of DNA bases can also lead to production of apurinic/apyrimidinic (AP) sites in DNA. AP sites in DNA can be either mutagenic or cytotoxic. Since mtDNA does not contain protective histones and the majority of ROS are produced in mitochondria, it is likely that a greater proportion of bases are modified in mtDNA than in nuclear DNA.[1] Furthermore, it is important to note that modified bases in the regulatory region of mtDNA, if not repaired, will have profound effects on cellular function because mtDNA gene expression and mtDNA replication is controlled by one common regulatory element (the D-loop region). Any mutation in this regulatory region will effect expression of all mtDNA encoded genes as well as replication of mtDNA.[17]

Other factors that contribute to the vulnerability of mtDNA include:

(1) Proximity of mtDNA to the site of continuous ROS in mitochondria.[14]
(2) The lipid-rich nature of mitochondria, resulting in selective accumulation of lipophilic environmental oxidants in mitochondria.[14]

(3) The mitochondrial respiratory chain system that metabolizes several chemicals resulting in redox-cycling. For example quinonoid compounds and aromatic amines are redox cycled to semiquinone and reactive free radicals in mitochondria.[14]

(4) The action of enzymes such as steroid hydroxylases, cytochrome P450, and monoamine oxidase, whose metabolic pathways lead to generation of ROS.[14]

Taken together, mtDNA can be extremely susceptible to mutations by ROS. Consistent with this, mtDNA is reported to accumulate mutations at least 10–16 times greater than the nuclear DNA.[2, 3]

5. Oxidative Damage to Mitochondrial DNA and Its Repair

Reactive oxygen species produce more than 100 types of base modifications in DNA.[18] MtDNA is rich in guanine. Oxidative damage to guanine leads to 8-oxoguanine (8-oxoG), a DNA lesion that is mutagenic since during DNA replication it can basepair with cytosine as well as adenine with almost equal efficiency resulting in GC to TA transversion mutations. 8-oxo-G is repaired by the base excision repair pathway.[19] The first step in this process involves the removal of damaged base by 8-oxo-G DNA glycosylase (encoded by *hOGG1*). The *hOGG1* gene product cleaves the N-glycosylic bond linking the damaged base to the deoxyribose sugar-phosphate backbone of DNA. Thus, the action of hOGG1 protein leads to an abasic site in the DNA. In the next step apyrimidinic and apurinic endonuclease (*APE*) cleaves the phosphodiester bond forming a nucleotide gap. This gap is then filled and sealed by DNA polymerase and DNA ligase, respectively. Like the nucleus, mitochondria also contain the base excision DNA repair pathway.[19] More importantly, mammalian mitochondria contain hOGG1 protein that is involved in the repair of 8-oxo-G lesion in mtDNA.[19, 20] More detail about the proteins involved in DNA repair in mitochondria can be found in Chap. 43.

6. Mitochondrial Damage Checkpoint: Monitoring the Functional State of Mitochondria

Previous studies conducted in yeast *Saccharomyces cerevisiae* and other organisms have noted the differential transcription of nuclear genes in response to mito-chondrial dysfunction.[21] This type of response has been described as retrograte regulation.[22] Using cDNA microarrays, we have analyzed the gene expression profile of HeLa cells and a breast cancer cell line completely devoid of mtDNA (Rho°). The Rho° cells lack 13 proteins involved in the electron transport chain

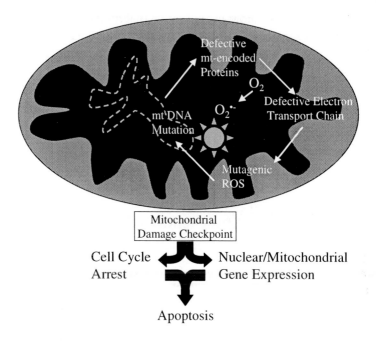

Fig. 4. The mitochondrial damage checkpoint and theory of vicious cycle. Mitochondrial damage checkpoint monitors the functional state of mitochondria. For details see text.

and respiratory function. Thus these cells contain dysfunctional mitochondria, experience oxidative stress and require pyruvate as an antioxidant for growth.[7] Analysis of approximately 2000 known genes that are involved in cancer revealed differential regulation of a group of genes in Rho° cells. This group of genes are involved in the cell cycle (cyclin E, cyclin F, replication factor C), mitotic checkpoint (*BUB3*) and DNA repair *ERCC5* (Delsite, Kacchap and Singh, unpublished results). Using cDNA microarrays, we also analyzed the gene expression profile in yeast *S. cerevisiae* Rho° cells. This approach yielded similar results (Rasmussen, Rasmussen and Singh, in preparation). These analyses indicate that cells may contain a *"mitochondrial damage checkpoint"* which monitors the functional state of mitochondria and accordingly adjusts the cell cycle response and nuclear and mitochondrial gene expression to protect the cells from mitochondria mediated apoptosis (Fig. 4). This hypothesis is consistent with a previous study reporting G1 cell cycle arrest of human cells in response to respiratory inhibitors.[23] It is likely that *"mitochondrial damage checkpoint"* is analogous to DNA damage checkpoint.[24] Furthermore, a link between nuclear DNA and mtDNA replication has been demonstrated recently in *Drosophila*.[25, 26] DREF, a transcription factor coordinates the transcription of genes involved in nuclear DNA replication, cell cycle, and the expression of genes encoding mitochondrial single stranded DNA binding protein and accessory subunit of polymerase γ involved mtDNA replication. If the mitochondria are continuously damaged due to oxidative stress resulting in a "vicious cycle" (for details see Chap. 25 by Aubrey de Gray), the damage can be extensive and the *"mitochondrial damage checkpoint"* genes may not be able to protect the cells. Such an event would lead to the death (apoptosis) of a cell (Fig. 4). *"Mitochondrial damage checkpoint"* must coordinate and maintain proper balance between apoptotic and anti-apoptotic signals and is likely to be involved in aging as well as other diseases.

7. mtDNA Mutations and Aging

Gradual loss of mitochondrial function due to oxidative stress is a common feature of aging. Consistently, accumulation of mutations in mtDNA during aging has been reported.[27, 28] It has been shown that the number of deletions formed in mtDNA parallels the content of 8-oxo-G with age suggesting that oxidative damage is one of the major factors related to induction of deletions in mtDNA.[29] Various types of deletions are formed and increase with age.[30, 31] Point mutations associated with mitochondrial diseases (3243A-to-G, 8344G-to-A and 10006A-to-G) are also frequently found in old individuals.[32] Accumulation of somatic mutations of mtDNA in heart, brain, liver and other postmitotic tissues appears to be a constant feature of normal aging in all vertebrates thus far examined. This accumulation appears to contribute to age-related disorders such as Parkinson's, Alzheimer's and Huntington's disease.[3, 33]

Activities of enzymes involved in electron transport chain decrease with age in lymphocyte, skeletal muscle and cardiomyocytes. This decrease in enzyme activities correlates with an increased ROS production in a variety of tissues.[34] There are several lines of evidence that suggest oxidative stress leading to mitochondrial dysfunction is a causal event in aging. These include:

(1) Young rats, when injected with mitochondria isolated from old rats undergo senescence.[27, 28]
(2) Mitochondrial hydroperoxide and the maximum life span of species is inversely related.[35, 36]
(3) Changes in morphology of mitochondria and decreased mitochondrial membrane potential due to oxidative stress have been reported in aged tissues.[37]
(4) Antioxidant compounds such as N-acetyl cysteine[38] GSH[39] and vitamin C[40] reduce mitochondrial oxidative stress and increase life span.
(5) Some mtDNA germ line mutations have been found at higher frequency in Centenarians.[41] These mutations are suggested to reduce the rate of aging associated accumulation of mtDNA mutations in somatic cells.[41]

8. mtDNA Mutations and Mitochondrial Diseases

Diseases caused by disorders of mitochondrial functions are termed mitochondrial diseases. Human mitochondrial diseases are often multisystem disorders. Many different mutations of mitochondrial DNA have been found to be pathogenic in humans. Individuals with mitochondrial disease either inherit their mutation from the mother or experience a mutation during oogenesis or early in embryogenesis. Several general classes of mutations in mtDNA have been reported. These include point mutations, deletions, duplication and insertions and rearrangements. To date over 100 pathogenic point mutation and 200 deletions, insertions and rearrangements have been reported.[4] Deletions removing tRNAs and protein coding genes give rise to the Kearns-Sayre/Chronic Progressive External Ophthalmoplegia (KSS/CPEO) phenotype.[42] Generally point mutations in protein coding genes are found in homoplasmic form and lead to Leber's Hereditary Optic Associated with Wolff-Parkinson-White conduction anomalies.[42] New mutations in mtDNA are described every year and catalogued at www.emory.gen.mitomap.html. Since mitochondria perform a variety of different functions in different tissues, there are literally hundreds of different mitochondrial diseases. Each disorder produces a wide spectrum of dysfunction that has proven to be extremely perplexing to the scientists and physicians. Mitochondrial diseases are severely debilitating, often fatal and characteristically complex in nature. They normally are inherited through the mother, but can also be inherited from either parent. They can also be sporadic or induced by the environment. Many mitochondrial diseases are so new that they have not yet been mentioned in the medical textbooks or in the medical literature.[10]

9. mtDNA Mutations and Cancer

Mitochondria, being the major intracellular source of ROS, are likely to play a significant role in carcinogenesis.[43] Mutations in mtDNA have been reported in a variety of cancers. These cancers include leukemia,[44–47] ovarian,[48] thyroid,[49, 50] salivary,[51, 52] kidney,[53–55] liver,[56, 57] lung,[58] colon,[14, 59] gastric,[60] brain cancers,[61, 62] bladder, head and neck cancers[63] as well as breast cancer.[64] MtDNA mutations could contribute to neoplastic transformation by changing cellular energy capacity, increasing mitochondrial oxidative stress, and activating Ca^{2+} activated proteases, endonucleases, and cellular cascade.[43] MtDNA mutations may also activate mitogenesis, oncogene and inactivate tumor suppressor genes and/or modulate apoptosis (Fig. 5). Increase in mtDNA mutations with aging may also accelerate production of ROS, and may in turn further damage nuclear DNA and mtDNA. Indeed, our study indicates that mitochondrial dysfunction, is mutagenic which may cause inactivation of nuclear genes (Rasmussen *et al.*, in preparation).

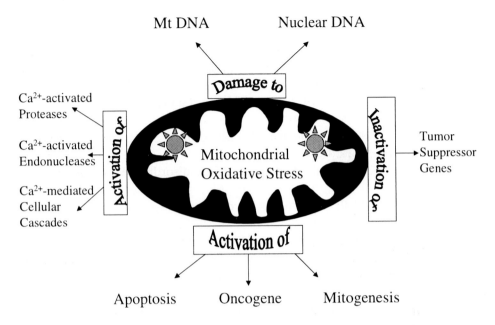

Fig. 5. Mitochondrial damage: the activation of mitogenesis and oncogenes and the inactivation of tumor suppressor genes and/or modulate apoptosis.

10. Perspective

The process of aging and the pathogenesis of mitochondrial diseases and cancer are among the most complex diseases known. Mitochondrial production of ROS is an established fact. In all of these processes oxidative stress and mitochondrial dysfunction is deeply involved. Protective mechanisms from antioxidant to DNA repair enzymes exist in mitochondria. However, damage to mitochondria and mtDNA leading to mitochondrial dysfunction is significant in aging, mitochondrial diseases, and cancer. Identifying genes that play a central role in "*mitochondrial damage check point*" may be critical in understanding the pathogenesis of mitochondrial diseases and future understanding of aging and cancer.

Acknowledgment

We thank Robert Naviaux for providing the Figs. 1 and 2 used in this article. Research in our laboratory is supported by the grants from National Institutes of Health RO1-097714, P50 CA88843, P20 CA86346 and American Heart Association Scientist Development Award 9939223N.

References

1. Rassmussen, L. and Singh, K. K. Genetic integrity of mitochondrial genome. *In* "Mitochondrial DNA Mutations in Aging, Disease, and Cancer" (K. K. Singh, Ed.), Springer, New York.
2. Richter, C. (1992). Reactive oxygen and DNA damage in mitochondria. *Mutat. Res.* **275**: 249–255.
3. Singh, K. K. (1998). *Mitochondrial DNA Mutations in Aging, Disease, and Cancer*, Springer, New York, USA.
4. Naviaux, R. K. (2000). Mitochondrial DNA disorders. *Eur. J. Pediatr.* **159**: S219–S226.
5. Singh, K. K. (2000). Mitochondrial me and the mitochondrion journal. *Mitochondrion* **1**: 1–2.
6. Kang, D., Takeshig, K., Sekiguchi, M. and Singh, K. K. Mitochondrial DNA mutations in aging, disease, and cancer: an introduction. *In* "Mitochondrial DNA Mutations in Aging, Disease, and Cancer" (K. K. Singh, Ed.), Springer, New York, USA.
7. Singh, K. K., Russell, J., Sigala, B., Zhang, Y., Williams, J. and Keshav, K. (1999). Mitochondrial DNA determines the cellular response to cancer therapeutic agents. *Oncogene* **18**: 6641–6646.
8. Kroemer, G. and Reed, J. C. (2000). Mitochondrial control of cell death. *Nat. Med.* **6**: 513–519.
9. Anderson, S., Bankier, A. T., Barrell, B. G., de Bruijn, M. H. L., Couloson, A. R., Drouin, J., Eperon, Ic. *et al.* (1981). Sequence and organization of the human mitochondrial genome. *Nature* **290**: 457–465.
10. Naviaux, R. K. (1997). Spectrum of mitochondrial diseases. Exceptional parent. http://biochemgen.ucsd.edu/mmdc/ep-toc.htm
11. Scheffler, I. E. (2000). A century of mitochondrial research: achievements and perspectives. *Mitochondrion* **1**: 3–33.
12. Wallace, D. C. (1999). Mitochondrial diseases in man and mouse. *Science* **283**: 1482–1488.
13. Polyak, K., Li, Y., Lengauer, C., Wilson, J. K. V., Markowitz, S. D., Trush, M. A., Kinzler, K. W. and Vogelstein, B. (1998). Somatic mutation of the mitochondrial genome in human colorectal cancer cells and tumors. *Nature Genetics* **20**: 291–293.
14. Bandy, B. and Davison, A. J. (1990). Mitochondrial mutations may increase oxidative stress: implication for carcinogenesis and aging? *Free Radic. Biol. Med.* **8**: 523–539.
15. Yakes, F. M. and van Houten, B. (1997). Mitochondrial DNA damage is more extensive and persists longer than nuclear DNA damage in human cells following oxidative stress. *Proc. Natl. Acad. Sci. USA* **94**: 514–519.
16. Beckman, K. B. and Ames, B. N. (1997). Oxidative decay of DNA. *J. Biol. Chem.* **272**: 19 633–19 636.

17. Shadel, G. S. and Clayton, D. A. (1997). Mitochondrial DNA maintenance in vertebrates. *Ann. Rev. Biochem.* **66**: 409–435.
18. Wallace, S. (1998). Enzymatic processing of radiation-induced free radical damage. *Rad. Res.* **150(Suppl.)**: S60–S79.
19. Singh, K. K., Sigala, B., Sikder, H. A. and Schwimmer, C. (2001). Inactivation of *Saccharomyces cerevisiae* OGG1 DNA repair gene leads to a frequency of mitochondrial mutants. *Nucleic Acids Res.* **29**: 1381–1388.
20. Takao, M., Aburatani, H., Kobayashi, K. and Yasui, A. (1998). Mitochondrial targeting of human DNA glycosylases for repair of oxidative DNA damage. *Nucleic Acid Res.* **26**: 2917–2922.
21. Poyton, R. O. and McEwen, J. E. (1996). Crosstalk between nuclear and mitochondrial genomes. *Ann. Rev. Biochem.* **65**: 563–607.
22. Sekito, T., Thornton, J. and Butow, R. A. (2000). Mitochondria-to-nuclear signaling is regulated by the subcellular localization of the transcription factors Rtg1p and Rtg3p. *Mol. Biol. Cell* **11**: 2103–2115.
23. Sweet, S. and Singh, G. (1995). Accumulation of human promyelocytic leukemic (HL-60) cells at two energetic cell cycle checkpoints. *Cancer Res.* **55**: 5164–5167.
24. Weinert, T. (1998). DNA damage and checkpoint pathways: molecular anatomy and interactions with repair. *Cell* **94**: 555–558.
25. Lefai, E., Fernandez-Moreno, M. A., Alahari, A., Kaguni, L. S. and Garesse, R. (2000). Differential regulation of the catalytic and accessory subunit genes of *Drosophila* mitochondrial DNA polymerase. *J. Biol. Chem.* **275**: 33 123–33 133.
26. Ruiz de Mena, I., Lefai, E., Garesse, R. and Kaguni, L. S. (2000). Regulation of mitochondrial single-stranded DNA-binding protein gene expression links nuclear and mitochondrial DNA replication in *Drosophila*. *J. Biol. Chem.* **275**: 13 628–13 636.
27. Michikawa, Y., Mazzucchelli, F., Bresolin, N., Scarlato, G. and Attardi, G. (1999). Aging-dependent large accumulation of point mutations in the human mtDNA control region for replication. *Science* **286**: 774–779.
28. Cortopassi, G. A., Shibata, D., Soong, N. W. and Arnheim, N. (1992). A pattern of accumulation of a somatic deletion of mitochondrial DNA in aging human tissues. *Proc. Natl. Acad. Sci. USA* **89**: 7370–7374.
29. Hayakawa, M., Hattori, K., Sugiyama, S. *et al.* (1992). Age-associated oxygen damage and mutations in mitochondrial DNA in human hearts. *Biochem. Biophys. Res. Commun.* **189**: 979–985.
30. Katsumata, K., Hayakawa, M., Tanaka, M. *et al.* (1994). Fragmentation of human heart mitochondrial DNA associated with premature aging. *Biochem. Biophys. Res. Commun.* **202**: 102–110.
31. Wallace, D. C. (1992). Mitochondrial genetics: a paradigm for aging and degenerative diseases? *Science* **256**: 628–632.

32. Kadenabach, B., Münscher, C., Frank, V. *et al.* (1995). Human aging is associated with stochastic somatic mutations of mitochondrial DNA. *Mutat. Res.* **338**: 161–172.

33. Shigenaga, M. K., Hagen, T. M. and Ames, B. N. (1994). Oxidative damage and mitochondrial decay in aging. *Proc. Natl. Acad. Sci. USA* **91**: 10 771–10 778.

34. Wei, Y. H., Lu, C. Y., Lee, H. C., Pang, C. Y. and Ma, Y. S. (1998). Oxidative damage and mutation to mitochondrial DNA and age-dependent decline of mitochondrial respiratory function. *Ann. NY Acad. Sci.* **854**: 155–170.

35. Sohal, R. S. and Weindruch, R. (1996). Oxidative stress, caloric restriction, and aging. *Science* **273**: 59–63.

36. Barja, G. (1999). Mitochondrial oxygen radical generation and leak: sites of production in states 4 and 3, organ specificity, and relation to aging and longevity. *J. Bioenerg. Biomemb.* **31**: 347–366.

37. Beckman, K. B. and Ames, B. N. (1998). The free radical theory of aging matures. *Physiol. Rev.* **78**: 547–581.

38. Martinez B. M. (2000). N-acetylcysteine elicited increase in complex I activity in synaptic mitochondria from aged mice: implications for treatment of Parkinson's disease. *Brain Res.* **859**: 173–175.

39. Vina, J., Sastre, J., Anton, V., Bruseghini, L., Esteras, A. and Asensi, M. (1992). *In* "Free Radicals and Aging" (I. Emerit, and B. Chance, Eds.), pp. 136–144, Birkhauser Verlag, Basel.

40. Ghosh, M. K., Chattopadhyay, D. J. and Chatterjee, I. B. (1996). Vitamin C prevents oxidative damage. *Free Radic. Res.* **25**: 173–179.

41. Tanaka, M., Gong, J., Zhang, J., Yamada, Y., Borgeld, H. J. and Yagi, K. (2000). Mitochondrial genotype associated with longevity and its inhibitory effect on mutagenesis. *Mech. Ageing Dev.* **116**: 65–76.

42. Shoffner, J. M. and Wallace, D. C. (1994). Oxidative phosphorylation diseases and mitochondrial DNA mutations: diagnosis and treatment. *Ann. Rev. Nutri.* **14**: 535–568

43. Bandy, B. and Davison, A. J. (1998). Perspective on mitochondrial carcinogenesis. *In* "Mitochondrial DNA Mutations in Aging, Disease, and Cancer" (K. K. Singh, Ed.), pp. 319–325, Springer, New York, USA.

44. Clayton, D. A. and Vinograd, J. (1967). Circular dimer and catenate forms of mitochondrial DNA in human leukemic leucocytes. *Nature* **216**: 652–657.

45. Clayton, D. A. and Vinograd, J. (1969). Complex mitochondrial DNA in leukemic and normal human myeloid cells. *Proc. Natl. Acad. Sci. USA* **62**: 1077–1084.

46. Boultwood, J., Fidler, C., Mills, K. I., Frodsham, P. M., Kusec, R., Gaiger, A., Gale, R. E., Linch, D. C., Littlewood, T. J. and Moss, P. A. (1996). Amplification of mitochondrial DNA in acute myeloid leukemia. *Br. J. Haematol.* **95**: 426–431.

47. Inanova, R., Lepage, V., Loste, M. N., Schachter, F., Wijnen, E., Busson, M., Cayuela, J. M., Sigaux, F. and Charron, D. (1998). Mitochondrial DNA sequence variation in human leukemic cells. *Int. J. Cancer* **76**: 495–498.

48. Hudson, B. and Vinograd (1967). Catenated circular DNA molecules in HeLa cells. *Nature* **216**: 647–652.

49. Welter, C., Kovacs, G., Seitz, G. and Blin, N. (1989). Alteration of mitochondrial DNA in human oncocytomas. *Genes Chromosome Cancer* **1**: 79–82.

50. Tallini, G. (1997). Oncocytic tumors. *Virchow Arch.* **433**: 5–12.

51. Tallini, G., Ladanyi, M., Rosai, J. and Jhanwar, S. C. (1994). Analysis of nuclear and mitochondrial DNA alterations in thyroid and renal oncocytic tumors. *Cytogenet. Cell Genet.* **66**(4): 253–259.

52. Heddi, A., Faure-Vigny, H., Wallace, D. C. and Stepien, G. (1996). Coordinate expression of nuclear and mitochondrial genes involved in energy production in carcinoma and oncocytoma. *Biochem. Biophys. Acta* **1316**(3): 203–209.

53. Kovacs, A., Storkel, S., Thoenes, W. and Kovacs, G. (1992). Mitochondrial and chromosomal DNA alterations in human chromophobe renal cell carcinomas. *J. Pathol.* **167**(3): 273–277.

54. Horton, T. M. *et al.* (1996). Novel mitochondrial DNA deletion found in renal cell carcinoma. *Genes Chromosomes Cancer* **15**: 95.

55. Selvanayagam, P. and Rajaraman, S. (1996). Detection of mitochondrial genome depletion by a novel cDNA in renal cell carcinoma. *Lab. Invest.* **74**(3): 592–599.

56. Yamamoto, H., Tanaka, M. *et al.* (1992). Siginficance of existence of deleted mitochondrial DNA in cirrhotic liver surrounding hepatic tumor. *Biochim. Biophys. Res. Commun.* **182**: 913–920.

57. Fukushima, S., Honda, K., Awane, M., Yamamoto, E., Takeda, R., Kaneko, I., Tanaka, A., Morimoto, T., Tanaka, K. and Yamaoka, Y. (1995). The frequency of 4977 base pair deletion of mitochondrial DNA in various types of liver disease and in normal liver. *Hepatology* **21**(6): 1547–1551.

58. El Meziane, A., Lehtinen, S. K., Holt, I. J. and Jacobs, H. T. (1998). Mitochondrial tRNALeu isoforms in lung carcinoma cybrid cells containing the np 3243 mtDNA mutation. *Human Mol. Genet.* **7**: 2141–2147.

59. Savre-Train, I., Piatyszek, M. A. and Shay, J. W. (1992). Transcription of deleted mitochondrial DNA in human colon adenocarcinoma cells. *Human Mol. Genet.* **1**: 203–204.

60. Burgart, L. J., Zheng, J., Shu, Q., Strickler, J. G. and Shibata, D. (1995). Somatic mitochondrial DNA mutations in gastric cancer. *Am. J. Pathol.* **147**: 1105–1111.

61. Liang, B. C. (1996). Evidence for association of mitochondrial DNA sequence amplification and nuclear localization in human low-grade gliomas. *Mutat. Res.* **354**: 27–33.

62. Liang, B. C. and Hays, L. (1996). Mitochondrial DNA copy number changes in human gliomas. *Cancer Lett.* **105**: 167–173.

63. Fliss, M. S., Usadel, H., Caballero, Wu, L., Buta, M., Eleff, S. M., Jen, J. and Sidransky, D. (2000). Facile detection of mitochondrial DNA mutations in tumors and bodily fluids. *Science* **287**: 2017–2019.
64. Richard, S. M., Bailliet, G., Paez, G. L., Bianchi, S., Peltomaki, P. and Bianchi, N. O. (2000). Nuclear and mitochondrial genome instability in human breast cancer. *Cancer Res.* **60**: 4231–4237.

Measurement of Oxidative Stress: Technology, Biomarkers and Applications

(Standards for Oxidative Stress Measurements)

Chapter 27

Oxidative DNA Damage Biomarkers:
A Need for Quality Control

Andrew R. Collins*

Andrew R. Collins • DNA Instability Group, Rowett Research Institute, Greenburn Road, Bucksburn, Aberdeen AB21 9SB, Scotland, UK
Tel: +44 (0) 1224 716634, E-mail: a.collins@rri.sari.ac,uk

1. Introduction

Oxidative damage to DNA is implicated in the etiology of cancer, as DNA damage is clearly the precursor of mutation, and mutations are ultimately responsible for the changes that occur as cells are transformed and start the process of tumorigenesis. Oxidation of DNA, and other biomolecules, is also thought to play an important part in the aging process. Essentially, nucleic acids, protein and lipids are supposed to accumulate damage during life until the amount of damage significantly affects biological function.[1] There are several steps in the argument. First, DNA is known to be relatively susceptible to oxidation *in vitro*; numerous oxidation products have been identified in DNA treated with ionizing radiation, or other oxidizing agents.[2] A mechanism exists to account for the oxidative damage, since a small proportion of the oxygen passing through the mitochondria during normal respiration is released as the free radical superoxide.[3] Superoxide is converted by superoxide dismutase to H_2O_2, which then can give rise to hydroxyl radicals via the Fenton reaction, catalyzed by transition metal ions such as Fe^{2+} or Cu^+

$$Fe^{2+} + H_2O_2 \rightarrow Fe^{3+} + {}^{\bullet}OH + OH^-.$$

Hydroxyl (${}^{\bullet}OH$) radicals are reactive but short-lived, and so can cause oxidative damage only if they are produced close to the target molecule. They are potentially dangerous because of the proximity to the DNA of ions capable of catalyzing the Fenton reaction. Copper is a component of the nuclear matrix — a kind of scaffold supporting the DNA.[4] The occurrence of free radical damage is confirmed by the demonstration of significant amounts of 8-oxoguanine in DNA from normal human cells.[5] In animal experiments, the amount of oxidative damage to DNA was shown to increase with age.[6] Finally, there is the circumstantial evidence from epidemiology; high consumption of fruit and vegetables is associated with lower incidence of cancer and heart disease,[7,8] and antioxidants are common components of these foods. Their health-protective properties are perhaps attributable to a decrease in oxidative damage.

In the past 3 or 4 years, attention has been directed to the enormous discrepancies in values of 8-oxoguanine in normal human DNA reported in the literature, using various techniques. HPLC and GC-MS appear to be prone to a serious artefact of oxidation occurring during preparation of DNA for analysis, and special precautions must be taken to avoid spuriously high levels of 8-oxoguanine being detected. At the same time, methods based on a different approach — measuring oxidized bases using specific endonucleases to convert them to DNA breaks — have given values even lower than the lowest of the estimates from HPLC. The European Standards Committee on Oxidative DNA Damage (ESCODD) was set up in 1997 to identify the causes of error in measurement of DNA oxidation, and — ultimately — to reach a consensus on the background level of damage in humans. Only when this is achieved will we be in a position to assess the significance of reactive oxygen species — and antioxidants — for human health.

2. Measuring Oxidative DNA Damage

2.1. Different Analytical Approaches

Gas Chromatography-Mass Spectrometry (GC-MS) was used with great success to identify the various products of ionizing radiation damage to DNA. Radiolysis of water produces reactive oxygen species, and so radiation is a good model for oxidative damage *per se*. In these experiments, which involved very high doses of radiation causing large amounts of damage, numerous products were formed — over 20 products of thymine alone.[9] GC-MS requires a derivatization step, to convert the bases (following acid hydrolysis) to a form that is volatile and amenable to separation by gas chromatography. The advantage of GC-MS is that it allows identification of oxidation products, according to the mass spectrum produced. (See Chap. 9 by Dizdaroglu.) *High Performance Liquid Chromatography* (HPLC), on the other hand, identifies oxidized bases (as nucleosides following hydrolyisis of DNA) by reference to standards, using the method of electrochemical detection (superior in sensitivity to ultraviolet detection), and not all bases can be detected. Identification of an analyte is confirmed by plotting a voltammogram, and comparing this with a standard, but it is not unequivocal, and cannot be applied to the small amounts of material normally obtained from biological samples. The nucleosides are released from the DNA by digestion with mixtures of nucleases, phosphodiesterases and phosphatases. (See Chap. 11 by Bogdanov.) *Liquid Chromatography-Mass Spectrometry* combines the ease of separation of HPLC with the ability of GC-MS to identify altered bases. (See Chap. 10 by Cadet.) [32]*P-postlabeling* can be carried out on enzymically hydrolyzed DNA; in one variant, the hydrolysate is enriched for 8-oxo-dG-3'-phosphate by HPLC, and then post-labeled with ^{32}P-ATP and T4 polynucleotide kinase. 5'-^{32}P-labeled modified nucleotides are then separated by HPLC and quantified. (See Chap. 16 by Jones.) *Antibodies* raised to 8-oxo-dG-containing DNA have been used, for instance in a competitive ELISA, to detect 8-oxoguanine; the problem with this approach seems to be that antibodies are not specific to the one oxidized base. (See Chap. 15 by Cooke.)

A completely different approach has been followed by several laboratories. *Lesion-specific endonucleases* are employed to detect oxidized bases and convert them to strand breaks, which can then be measured by a variety of techniques. Endonuclease III detects oxidized pyrimidines;[10] formamidopyrimidine DNA glycosylase (FPG), as its name suggests, was first characterized on the basis of activity on ring-opened purines (formamidopyrimidines, fapyGua and fapyAde).

However, it also detects 8-oxoguanine, and it is likely that the latter is a major substrate *in vivo*.[2] The methods used to measure the breaks include:

(1) Alkaline elution; cells are lysed in alkali above a filter, and the released, single-stranded DNA is washed through the filter. The rate at which it elutes depends on the frequency of DNA breaks.[11] The analogy usually given is that of spaghetti in a sieve; short pieces escape more readily than long.

(2) Alkaline unwinding; at high pH, DNA strands separate, but the extent of denaturation depends on the frequency of unwinding points (strand breaks) as well as time and pH.[12] The partially denatured DNA is neutralized and broken into short pieces which are separated into single- and double-stranded DNA fractions by hydroxyapatite chromatography.

(3) Single cell gel electrophoresis (the comet assay); cells embedded in agarose are lysed with detergent and high salt to remove membranes, cytoplasm, and most nuclear proteins. Nucleoids are left, consisting of supercoiled loops of DNA attached to the nuclear matrix. Alkaline unwinding and electrophoresis follow; DNA is attracted to the anode, but little movement occurs unless breaks are present; they relax supercoiling, and loops of DNA then extend to form a "comet tail". The intensity of tail DNA reflects the frequency of breaks.[13] (See Chap. 17 by Duthie.)

All three methods depend on calibration against an agent (normally radiation) that breaks DNA at known frequency. They are indirect assays compared with HPLC or GC-MS. They also have the limitation that their specificity is dependent on the specificity of the endonuclease (as discussed above). However, a major advantage of these assays is that very little processing of the DNA occurs and so artefactual oxidation of DNA is minimized.

Brief mention should also be made of the measurement of 8-oxo-dG in urine (see Chap. 13 by Poulsen). Measured in different human subjects as a biomarker of oxidative stress, excreted 8-oxo-dG (measured by an ELISA method) correlates with 8-oxo-dG measured in lymphocyte DNA by HPLC or with the comet assay (our unpublished results), but it may simply represent the oxidation of DNA from dead cells as it passes through the kidneys, as suggested by Lindahl,[14] rather than the product of cellular repair, as is often claimed. The major repair pathway acting on oxidative damage, after all, removes the base, *not the nucleotide*, so 8-oxo-dG is not the expected product.

2.2. The Range of Values of 8-oxoguanine in Human Cells

Table 1 illustrates the range of background levels of 8-oxoguanine determined with various techniques in human cells. GC-MS typically gives values between 100 and 1000 per million unaltered guanines in DNA. Determinations by HPLC

Table 1. Representative Estimates of 8-oxoguanine (8-oxo-dG) in Human Cells

Method	8-oxoguanine/ 10^6 Guanines	Cell Type	Reference
GC-MS	100	lymphoblastoid	36
GC-MS	1650	breast cancer	37
GC-MS	330	lymphocytes (cancer patients)	38
GC-MS	300	lymphocytes	39
HPLC-EC	40	leukocytes	40
HPLC-EC	4.3	leukocytes	41
HPLC-EC	37	leukocytes	30
HPLC-EC	2.4	leukocytes	21
Alkaline elution	0.6	lymphocytes	15
Alkaline unwinding	0.6	lymphocytes	15
Comet assay	0.4	lymphocytes	41

are generally lower, ranging from 2 to 40 per million. Thus, comparing laboratories using essentially the same method, there is at least a 10-fold range in values, and between the highest GC-MS determination of normal human white blood cells and the lowest estimate by HPLC the difference is more than 100-fold. When the enzymic methods are included in the comparison, there is a further 10-fold jump, to even lower values. Included in the table are mean values, from my own laboratory, applying HPLC and the comet assay to cells isolated from the same blood samples donated by over 30 volunteers. The lack of agreement demands some explanation.

The problems in measuring 8-oxoguanine (or 8-oxo-dG, depending on technique) were highlighted at a meeting held in Aberdeen, Scotland, in January 1997,[15] and as a result a consortium of laboratories came together to form ESCODD — the European Standards Committee on Oxidative DNA Damage. Our aim was to resolve methodological problems, establish standard protocols, and reach agreement on the basal level of oxidative DNA damage in human cells.

2.3. Reasons for the Discrepancies in Measurement of 8-oxoguanine

Obviously, some variation between laboratories could result from trivial technical errors in preparing standards, measuring absorbance, or making calculations. But there seem to be systematic differences indicating that different methods may have intrinsic problems that lead to over- or under-estimation of 8-oxoguanine. For example, HPLC requires enzymic digestion of DNA to nucleosides; the extent of digestion is assessed on the basis of the recovery of released nucleosides relative to the original DNA (using absorbance measurements). But the enzymes might be less effective at breaking bonds involving the abnormal base 8-oxoguanine — in which case incomplete digestion (not detectable in the measurements on total DNA) would lead to under-estimation. A problem that seems to apply to both GC-MS and HPLC is the *in vitro* oxidation of guanine in DNA. The high temperature derivatization step required for GC-MS is particularly susceptible. Ravanat *et al.*[16] pre-purified the DNA hydrolysate by HPLC and then derivatized the 8-oxoguanine; the yield of 8-oxoguanine was much lower than when the whole hydrolysate was derivatized (and was similar to the HPLC determination on the same material). Digestion of DNA with FPG to release 8-oxoguanine achieves the same objective as prepurification, since no guanine is released to be oxidized during derivatization. Rodriguez *et al.*[17] found very similar yields of 8-oxoguanine using this approach compared with acid hydrolysis. Prepurification is therefore probably unnecessary — but it should be noted that in this comparative study derivatization was at room temperature (with pyridine as solvent to inhibit oxidation). Performing derivatization at lower temperature and for shorter times eliminates spurious oxidation[18] — as does performing the derivatization reaction in the presence of ethanethiol.[19]

HPLC also involves lengthy preparation procedures, during which spurious oxidation might occur. Inclusion of antioxidants is now standard, and attempts are made to replace oxygen with nitrogen or inert gas. Performing the entire DNA isolation and hydrolysis procedure under anaerobic conditions[20] certainly resulted in the lowest estimates to date of background levels of 8-oxo-dG in human cells.[21]

Enzymic methods are not without potential problems. The clustering of damage is one. Two breaks in close proximity are no more effective than a single break at allowing either denaturation of the double helix (in the alkaline elution and alkaline unwinding assays) or relaxation of supercoiled loops (in the comet assay). Any multiple breaks will therefore lead to underestimation of damage. Clustering of breaks has been reported to occur after ionizing radiation,[22] but whether it is significant in DNA damaged by endogenous oxidative processes is not known. A second possibility is that all lesions are not equally susceptible to attack by the endonuclease. Steric effects within the chromatin, or local sequence effects, might lead to underestimation of damage. Arguing against this are the observations that supercoiled DNA (as occurs in the nucleoid) is an especially good substrate for endonuclease III;[10] and that the analogous enzyme endonuclease V, active against UV-induced pyrimidine dimers, detects the expected amount of damage after a particular dose of UV.[23] On the other hand, FPG does not detect only 8-oxoguanine, but also ring-opened purines, so we might even be *over*-estimating damage with FPG.

In summary, the discrepancies between measures of basal levels of DNA damage using different approaches may represent either spurious oxidation during sample preparation, or systematic over/underestimation. Figure 1 illustrates

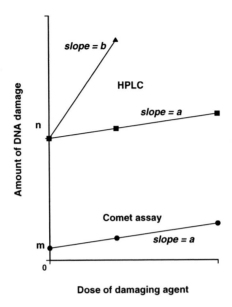

Fig. 1. A test for over/underestimation of DNA base oxidation. The comet assay gives a baseline level of damage, m, and a dose response for induced damage, slope a. HPLC gives a higher baseline level of damage, n. If m is the true baseline level (and $n - m$ is spurious oxidation), then the expected dose response will have the same slope, a. But if the comet assay systematically underestimates damage, and n is correct, then the expected dose response with HPLC will have slope b, such that $b/a = n/m$.

a way of deciding between these possibilities, by performing a dose-response experiment. DNA oxidation is induced — above a certain basal level — by treatment of isolated DNA or cells with a suitable DNA-damaging agent. If spurious oxidation is the problem, but lesions are detected equally effectively by two techniques, then the y-axis intercepts will differ, but dose-response slopes should be identical. Alternatively, if systematic over/underestimation is responsible for the differences, then both slopes and basal levels will appear to differ — by the same factor.

2.4. The ESCODD Approach

In the first phase of ESCODD, we were probably over-optimistic that the differences over how to measure 8-oxoguanine would be easily resolved. Participating laboratories were sent standard samples of 8-oxo-dG, 8-oxo-dG containing oligonucleotides, calf thymus DNA and liver tissue to analyze using their normal procedures. Great variation was found even in the determination of the 8-oxo-dG standard, and so it was impossible to draw conclusions from assays

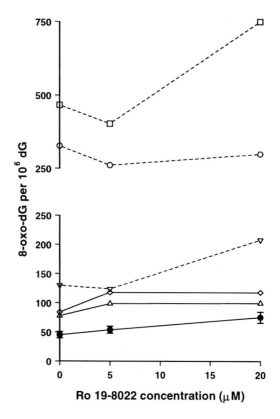

Fig. 2. GC-MS determinations of 8-oxoguanine in calf thymus DNA samples; baseline damage and damage induced by photosensitizer Ro 19-8022 and light. The line with solid circles represents the dose response for these samples established in the source laboratory (by HPLC). Open symbols represent five different laboratories. Procedures showing a positive dose response are indicated by solid lines. (From Ref. 25, with permission.)

on the more biologically relevant samples.[24] The next stage was simpler; aliquots of standard 8-oxo-dG solution were distributed, together with calf thymus DNA — untreated, or given one of two doses of an agent that induces 8-oxoguanine (the photosensitiser Ro 19-8022 plus visible light). 16 HPLC analyses were completed in 15 laboratories (one laboratory employing two different procedures); most gave values of 8-oxo-dG standard within 40% of the target value (from absorbance measurements on the original 8-oxo-dG solution). LC-MS-MS (used in two laboratories) also gave values close to the target. Two of the four GC-MS analyses were also within 40% of target. Determination of the calf thymus DNA samples was an instructive exercise. Figure 2 shows results from the laboratories using GC-MS, and Fig. 3 results of some of the HPLC procedures. Solid lines indicate laboratories/procedures successfully detecting the dose response in the DNA samples, i.e. reporting more 8-oxoguanine in the sample given the low

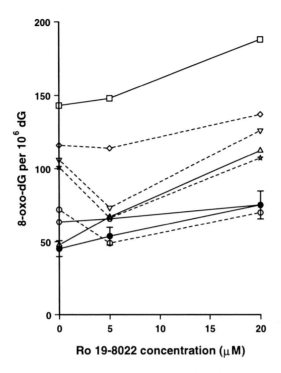

Fig. 3. HPLC determinations of 8-oxo-dG in calf thymus DNA samples; baseline damage and damage induced by photosensitizer Ro 19-8022 and light. The line with solid circles represents the dose response for these samples established in the source laboratory. Open symbols represent seven different laboratories. Procedures showing a positive dose response are indicated by solid lines. (From Ref. 25, with permission.)

concentration of Ro 19-8022 than in the untreated DNA, and a further increase in the sample given the high concentration. In all, two out of five GC-MS procedures and seven of 14 HPLC procedures met these not very rigorous criteria. Coefficients of variation were calculated from 6 determinations of identical samples; two GC-MS laboratories and six HPLC procedures showed a CV of 10% or less. We concluded[25] that:

(1) Certain GC-MS procedures consistently report concentrations of 8-oxoguanine (in both the standard and calf thymus DNA) substantially higher than target values, probably indicating a calibration problem.
(2) Coulometric detection is more sensitive than the amperometric method, and the former is probably essential to measure low background levels of oxidation.
(3) The non-standard methods (^{32}P-postlabeling, LC-MS-MS, ELISA) are not yet sufficiently reliable to recommend their use. Two of the three failed to detect the dose response, and all suffered from high CVs.

Since the beginning of 2000, ESCODD has been supported by European Commission finances as a Concerted Action. It now has 27 laboratories as members, and the first year has been spent in repeating the tests of ability to analyze standard 8-oxo-dG solutions and calf thymus DNA, as well as oligonucleotides containing 8-oxoguanine. Results are not yet published; but it seems that an increasing proportion of laboratories agree with target 8-oxo-dG values and can detect the dose response in calf thymus DNA. Future activities will include the analysis of biological samples — liver tissue, cultured cells, and finally human lymphocytes — and will therefore be more demanding in terms of ability to detect low levels of damage. They will also allow comparison with the enzymic methods (which require intact cellular DNA and cannot measure damage in calf thymus DNA).

2.5. Optimizing 8-oxoguanine Estimation; What do We Know So Far?

As already mentioned, the derivatization step in preparation of DNA for GC-MS analysis seems to be particularly prone to oxidation of guanine. It is now customary to perform derivatization at lower temperature and to include ethanethiol as an antioxidant. Antioxidants are also employed in preparation for HPLC. TEMPO (a nitroxide that maintains transition metals in an unreactive oxidized form) was claimed by Hofer and Möller[26] to decrease spurious oxidation, though we found that neither this, nor deferoxamine mesylate (DF — a chelator of free iron), with or without histidine (an antioxidant), consistently decreased the yield of 8-oxo-dG when present during homogenization of liver or hydrolysis of the isolated nuclear DNA.[27] A major cause of problems was dialysis, which increased 8-oxo-dG levels up to 10-fold; however, dialysis through EDTA-pretreated tubing maintained a low level of oxidation. Freeze-drying or vacuum-drying also led to an increase in 8-oxo-dG.[27] When isolating DNA from tissue such as liver, it is advisable to prepare nuclei first. In contrast to HeLa cells (which gave the same amount of 8-oxo-dG regardless of whether nuclei were isolated first), DNA isolated from whole hepatocytes had 10 times as much 8-oxodG as DNA from hepatocyte nuclei — presumably because the high content of iron in hepatocyte cytoplasm promotes free radical production.[27] Replacement of EDTA in the homogenization buffer with 5 mM DF reduced the amount of 8-oxoguanine detected by a factor of 2.8 (C. M. Gedik, unpublished).

Enzymic hydrolysis is a potentially critical step, if the presence of 8-oxo-dG leads to incomplete digestion. We do not know for certain whether such an effect operates; but two hydrolysis regimes give very similar results — one making use of P1 nuclease and alkaline phosphatase, the other DNase I, phosphodiesterases I and II and alkaline phosphatase.[28] Oligonucleotides with defined 8-oxoguanine content are being used in the present phase of ESCODD to check the completeness of digestion. Results — at least from our laboratory — are promising, as the expected yield of 8-oxoguanine is obtained.

2.6. How Do the Different Approaches Compare?

In a direct comparison of 8-oxo-dG determined by HPLC, and FPG-sensitive sites measured with the comet assay, in samples from 37 individuals, we found no significant correlation.[27] Although FPG may detect some sites other than 8-oxoguanine, there is generally a good correlation between FPG-sensitive sites and endonuclease III-sensitive sites, both measured with the comet assay (Ref. 27 and unpublished observations) — implying that oxidative damage is a major substrate for FPG. As oxidation artefacts are eliminated in the HPLC procedure (see Sec. 2.5), determinations by FPG and by HPLC have become closer; they are now separated by a factor of approximately 5.

While this discrepancy is worrying, and efforts are continuing to eliminate it, there is plenty of evidence that biologically meaningful differences can be detected using either HPLC or the comet assay. The use of the latter as a biomarker for oxidative DNA damage in human population studies is described in detail in the book chapter by Duthie. The level of 8-oxoguanine in white blood cells — measured by HPLC — has been shown to increase with exercise,[29] to be higher in diabetics

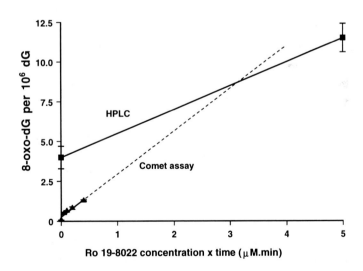

Fig. 4. Comparison of dose response curves for 8-oxo-dG determined by HPLC or the comet assay (with FPG). HeLa cells were treated with Ro 19-8022 and light; different doses were needed for the two assays, which have very different detection ranges. Samples for HPLC showed a linear response at all doses (up to 25 μM. min); only the lowest dose is accommodated on this scale.

than in normal controls,[30] and to vary with geographical distribution within Europe as well as with sex.[31] If there is simply a systematic over- or under-estimation of oxidized bases, then we would expect both approaches to be equally sensitive to biological differences — only the absolute amount of damage would be in dispute. However, if the problem is an artificially high basal level of damage resulting from oxidation during preparation, then HPLC should be much less able to detect differences in levels of truly endogenous damage. The dose response experiment described in Fig. 1 is therefore crucial. Pflaum *et al.*[32] found that, for cultured cells, the amount of 8-oxoguanine per unit dose of damaging agent was similar for HPLC compared with FPG/alkaline elution. Figure 4 shows the result of a recent experiment (C. M. Gedik and A. R. Collins, unpublished) comparing HPLC with the comet assay. Slopes differ, but only by a factor of 2, and that for the comet assay is *greater* than that for HPLC — the opposite of what might have been expected from the differences in determinations of background damage. It appears that, if anything, HPLC *under*estimates 8-oxoguanine in comparison with the enzymic approach, or FPG *over*estimates it. The basic conundrum is not yet solved.

3. Conclusions

The ESCODD exercise has highlighted the need for extreme care in the analysis of 8-oxoguanine — and has also shown up quite basic problems such as poor calibration of the assay against standards, poor reproducibility, and insensitivity. So far, our tests on DNA have employed calf thymus DNA, known to contain far more 8-oxoguanine than cellular DNA. Inter-laboratory variability seems to be decreasing, but we have a long way still to go. When we investigate DNA from cultured cells, liver and lymphocytes, requirements for sensitivity and reproducibility will be a great deal more demanding.

If, as seems likely to many of us, the background level of oxidative damage in human cells turns out to be in the range of the lowest of the HPLC determinations and the various enzymic assays, then we shall have to reconsider whether it can pose such a great threat to human health as has been supposed. The endogenous exposure to free radicals is an inevitable consequence of aerobic respiration, and efficient defenses against excessive damage — antioxidant enzymes and DNA repair — have evolved. The moderating effect of dietary antioxidants on an already low level of oxidative damage may not be so important. Recent evidence on the cancer-protective effects of antioxidants or even fruit and vegetables is, to say the least, equivocal; β-carotene supplementation in smokers was associated with *increased* risk of lung cancer,[33, 34] and in a large prospective study no link was seen between consumption of fruits and vegetables and lower risk of colorectal cancer.[35]

Acknowledgments

The author thanks the Scottish Executive Environment and Rural Affairs Department for support, and Mrs. Catherine Gedik for helpful comments (and agreement to present unpublished data).

References

1. Ames, B. N., Shigenaga, M. K. and Hagen, T. M. (1993). Oxidants, antioxidants, and the degenerative diseases of aging. *Proc. Natl. Acad. Sci.* **90**: 7915–7922.
2. Boiteux, S., Gajewski, E., Laval, J. and Dizdaroglu, M. (1992). Substrate specificity of the *Escherichia coli* Fpg protein (formamidopyrimidine-DNA glycosylase): excision of purine lesions in DNA produced by ionizing radiation or photosensitization. *Biochemistry* **31**: 106–110.
3. Chance, B., Sies, H. and Boveris, A. (1979). Hydroperoxide metabolism in mammalian organs. *Physiol. Rev.* **59**: 527–605.
4. Lewis, C. D. and Laemmli, U. K. (1982). Higher order metaphase chromosome structure: evidence for metalloprotein interactions. *Cell* **29**: 171–181.
5. Kiyosawa, H., Suko, M., Okudawa, H., Murata, K., Miyamoto, T., Chung, M. H., Kasai, H. and Nishimura, S. (1990). *Free Radic. Res. Commun.* **11**: 23–27.
6. Helbock, H. J., Beckman, K. B., Shigenaga, M. K., Walter, P. B., Woodall, A. A., Yeo, H. C. and Ames, B. N. (1998). DNA oxidation matters: the HPLC-electrochemical detection assay of 8-oxo-deoxyguanosine and 8-oxo-guanine. *Proc. Natl. Acad. Sci.* **95**: 288–293.
7. Gramenzi, A., Gentile, A., Fasoli, M., Negri, E., Parazzini, F. and La Vecchia, C. (1990). Association between certain foods and risk of acute myocardial infarction in women. *Br. Med. J.* **300**: 771–773.
8. Anonymous. (1997). *Food, Nutrition and the Prevention of Cancer: A Global Perspective*, WCRF/AICR, Washington, DC.
9. Téoule, R. and Cadet, J. (1971). Radiolysis of thymine in aerated aqueous solution. *J. Chem. Soc.: Chem. Commun.* 1269–1270.
10. Doetsch, P. W., Henner, W. D., Cunningham, R. P., Toney, J. H. and Helland, D. E. (1987). A highly conserved endonuclease activity present in *Escherichia coli*, bovine, and human cells recognizes oxidative DNA damage at sites of pyrimidines. *Mol. Cell. Biol.* **7**: 26–32.
11. Kohn, K. W., Ewig, R. A. G., Erickson, L. C. and Zwelling, L. A. (1981). Measurement of strand breaks and cross-links by alkaline elution. *In* "DNA Repair. A Laboratory Manual of Research Procedures" (E. C. Friedberg, and P. C. Hanawalt, Eds.), pp. 379–401, Marcel Dekker, New York.
12. Ahnstrom, G. and Erixon, K. (1981). Measurement of strand breaks by alkaline denaturation and hydroxyapatite chromatography. *In* "DNA Repair. A

Laboratory Manual of Research Procedures" (E. C. Friedberg, and P. C. Hanawalt, Eds.), pp. 403–418, Marcel Dekker, New York.

13. Collins, A. R., Dušinská, M., Gedik, C. and Štetina, R. (1996). Oxidative damage to DNA: do we have a reliable biomarker? *Env. Health Perspect.* **104 (Suppl. 3)**: 465–469.

14. Lindahl, T. (1993). Instability and decay of the primary structure of DNA. *Nature* **362**: 709–714.

15. Collins, A., Cadet, J., Epe, B. and Gedik, C. (1997). Problems in the measurement of 8-oxoguanine in human DNA. Report of a workshop, DNA Oxidation, held in Aberdeen, UK, 19–21 January, 1997. *Carcinogenesis* **18**: 1833–1836.

16. Ravanat, J.-L., Turesky, R. J., Gremaud, E., Trudel, L. J. and Stadler, R. H. (1995). Determination of 8-oxoguanine in DNA by gas chromatography-mass spectrometry and HPLC-electrochemical detection: overestimation of the background level of the oxidized base by the gas chromatography-mass spectrometry assay. *Chem. Res. Toxicol.* **8**: 1039–1045.

17. Rodriguez, H., Jurado, J., Laval, J. and Dizdaroglu, M. (2000). Comparison of the levels of 8-hydroxyguanine in DNA as measured by gas chromatography mass spectrometry following hydrolysis of DNA by *Escherichia coli* Fpg protein or formic acid. *Nucleic Acids Res.* **28**: E75.

18. Hamberg, M. and Zhang, L.-Y. (1995). *Anal. Biochem.* **229**: 336–344.

19. Jenner, A., England, T. G., Aruoma, O. I. and Halliwell, B. (1998). *Biochem. J.* **331**: 365–369.

20. Nakajima, M., Takeuchi, T. and Morimoto, K. (1996). Determination of 8-hydroxydeoxyguanosine in human cells under oxygen-free conditions. *Carcinogenesis* **17**: 787–791.

21. Nakajima, M., Takeuchi, T., Takeshita, T. and Morimoto, K. (1996). 8-hydroxydeoxyguanosine in human leukocyte DNA and daily health practice factors: effects of individual alcohol sensitivity. *Env. Health Perspect.* **104**: 1336–1338.

22. Ward, J. F., Webb, C. F., Limoli, C. L. and Milligan, J. R. (1990). DNA lesions produced by ionizing radiation: locally multiply damaged sites. *In* "Ionizing Radiation Damage to DNA: Molecular Aspects" (S. S. Wallace, and R. B. Painter, Eds.), pp. 43–50, Wiley-Liss, New York.

23. Collins, A. R., Mitchell, D. L., Zunino, A., de Wit, J. and Busch, D. (1997). UV-sensitive rodent mutant cell lines of complementation groups 6 and 8 differ phenotypically from their human counterparts. *Env. Mol. Mutagen.* **29**: 152–160.

24. Lunec, J. (1998). ESCODD: European Standards Committee on Oxidative DNA Damage. *Free Radic. Res.* **29**: 601–608.

25. ESCODD. (2000). Comparison of different methods of measuring 8-oxoguanine as a marker of oxidative DNA damage. *Free Radic. Res.* **32**: 333–341.

26. Hofer, T. and Möller, L. (1998). Reduction of oxidation during the preparation of DNA and analysis of 8-hydroxy-2'-deoxyguanosine. *Chem. Res. Toxicol.* **11**: 882–887.

27. Gedik, C. M., Wood, S. G. and Collins, A. R. (1998). Measuring oxidative damage to DNA; HPLC and the comet assay compared. *Free Radic. Res.* **29**: 609–615.

28. Wood, S. G., Gedik, C. M. and Collins, A. R. (2000). Controlled oxidation of calf thymus DNA to produce standard samples for 8-oxo-deoxyguanosine analysis; effects of freeze-drying, storage and hydrolysis conditions. *Free Radic. Res.* **32**: 327–332.

29. Inoue, T., Mu, Z., Sumikawa, K., Adachi, K. and Okochi, T. (1993). Effect of physical exercise on the content of 8-hydroxydeoxyguanosine in nuclear DNA prepared from human lymphocytes. *Japanese J. Cancer Res.* **84**: 720–725.

30. Dandona, P., Thusu, K., Cook, S., Snyder, B., Makowski, J., Armstrong, D. and Nicotera, T. (1996). Oxidative damage to DNA in diabetes mellitus. *Lancet* **347**: 444–445.

31. Collins, A. R., Gedik, C. M., Olmedilla, B., Southon, S. and Bellizi, M. (1998). Oxidative DNA damage measured in human lymphocytes; large differences between sexes and between countries, and correlations with heart disease mortality rates. *FASEB J.* **12**: 1397–1400.

32. Pflaum, M., Will, O. and Epe, B. (1997). Determination of steady-state levels of oxidative DNA base modifications in mammalian cells by means of repair endonucleases. *Carcinogenesis* **18**: 2225–2231.

33. The Alpha-Tocopherol, Beta Carotene Cancer Prevention Study Group (1994). The effect of vitamin E and beta carotene on the incidence of lung cancer and other cancers in male smokers. *New England J. Med.* **330**: 1029–1035.

34. Omenn, G. S., Goodman, G. E., Thornquist, M. D., Balmes, J., Cullen, M. R., Glass, A., Keogh, J. P., Meyskens, F. L., Valanis, B., Williams, J. H., Barnhart, S. and Hammar, S. (1996). Effects of a combination of beta carotene and vitamin A on lung cancer and cardiovascular disease. *New England J. Med.* **334**: 1150–1155.

35. Michels, K. B., Giovannucci, E., Joshipura, K. J., Rosner, B. A., Stampfer, M. J., Fuchs, C. S., Colditz, G. A., Speizer, F. E. and Willett, W. C. (2000). Prospective study of fruit and vegetable consumption and incidence of colon and rectal cancers. *J. Natl. Cancer Inst.* **92**: 1740–1752.

36. Jaruga, P. and Dizdaroglu, M. (1996). Repair of products of oxidative DNA base damage in human cells. *Nucleic Acids Res.* **24**: 1389–1394.

37. Malins, D. C. and Haimanot, R. (1991). Major alterations in the nucleotide structure of DNA in cancer of the female breast. *Cancer Res.* **51**: 5430–5432.

38. Olinski, R., Zastawny, T. H., Foksinski, M., Windorbska, W., Jaruga, P. and Dizdaroglu, M. (1996). DNA base damage in lymphocytes of cancer patients undergoing radiation therapy. *Cancer Lett.* **106**: 207–215.

39. Podmore, I. D., Griffiths, H. R., Herbert, K. E., Mistry, N., Mistry, P. and Lunec, J. (1998). Vitamin C exhibits pro-oxidant properties. *Nature* **392**: 559–559.

40. Degan, P., Bonassi, S., de Caterina, M., Korkina, L. G., Pinto, L., Scopacasa, F., Zatterale, A., Calzone, R. and Pagano, G. (1995). In vivo accumulation of 8-hydroxy-2′-deoxyguanosine in DNA correlates with release of reactive oxygen species in Fanconi's anaemia families. *Carcinogenesis* **16**: 735–742.

41. Collins, A. R., Duthie, S. J., Fillion, L., Gedik, C. M., Vaughan, N. and Wood, S. G. (1997). Oxidative DNA damage in human cells: the influence of antioxidants and DNA repair. *Biochem. Soc. Trans.* **25**: 326–331.

SECTION 5

Dietary Antioxidants

Chapter 28

Dietary Antioxidants — Human Studies Overview

Paul Milbury and Jeffrey B. Blumberg*

Paul Milbury and **Jeffrey B. Blumberg** • Antioxidants Research Laboratory, Jean Mayer USDA Human Nutrition Research Center on Aging, Tufts University

*Corresponding Author.
Tel: 617-556-3334, E-mail: blumberg@hnrc.tufts.edu

1. Introduction

Dietary antioxidants have been defined as substances in foods that significantly decrease the adverse effects of reactive species, such as reactive oxygen and nitrogen species, on normal physiological function in humans.[1] This is a reasonably stringent definition based on specific mechanisms of action which may or may not be related to the variety of beneficial health outcomes associated with their generous intakes from foods and/or dietary supplements. Defining the role and clinical significance of the non-antioxidant mechanisms of these substances to the promotion of health and the prevention of chronic disease is an important challenge. However, there are available now a number of reliable technologies for the assessment of antioxidant defenses and intervention strategies utilizing biomarkers of oxidative stress which can elucidate the contributions of these substances to prolonging the healthspan. Unfortunately, many human studies examining the efficacy and safety of antioxidants do not employ these measures. Thus, opportunities are lost for determining whether the success or failure of a study resulted from an antioxidant impact on oxidative stress status, making it difficult to better design follow-up studies. Translating research results to clinical application and public policy are hampered by "insufficient evidence available on which to base a firm conclusion that antioxidants are capable of reducing risk of disease".[2] Understanding the apparent inconsistency between and within epidemiologic studies and randomized controlled trials (RCT) is almost always frustrated by the absence of established technologies for assessment of antioxidant status and action. Indeed, the failure to apply these approaches in human studies largely accounts for the conclusion of the National Academy of Sciences that oxidative damage to biomolecules "have not yet been adequately validated as markers of the onset, progression, or regression of any chronic diseases".[1]

Meta-analyses of observational studies and RCT of the efficacy and safety of drug therapies indicate a close correlation between these two approaches to examining human populations.[3, 4] This data is in marked contrast to studies of antioxidants like β-carotene and vitamins C and E where results fitfully suggest significant or modest benefit, no effect or harm. While differences in the dose of the antioxidant(s), duration of the intervention, health status or disease severity of the subjects, and other factors confound comparisons between these studies, the lack of any consistent assessment relevant to an antioxidant mechanism in most all these studies prevents a clear elucidation of the basis for efficacious versus nil outcomes. Moreover, the failure to appreciate the fundamentally different nature of primary versus secondary versus tertiary interventions often appears to confuse the conclusions drawn from these studies. This chapter reviews the principal approaches to assessing antioxidant defenses and oxidative stress status available for human studies and RCT and the context of their results within the prevailing standards of evidence.

2. Assessment of Antioxidant Intake and Status

Epidemiologic studies of antioxidant nutrients typically estimate intake of these substances from foods and supplements (but sometimes ignore supplement intake altogether) using questionnaire or food record techniques and/or rely upon a biochemical measures of exposure to the nutrient of interest. For some antioxidants, dietary intake estimation is problematic due to the nature of the assessment tool or the quality and completeness of the nutrient database; in these cases, the use of biological markers of exposure, i.e. blood or tissue concentrations, are preferred. In contrast, for other antioxidants, biological markers may be more problematic than estimates of intake. Generally, quantifying nutrient intake and status utilizing both approaches produces complementary and more valid information. While not specific to particular antioxidants, determination of the antioxidant capacity of foods or biological samples can serve to further complement these assessments, e.g. by use of the oxygen radical absorbance capacity (ORAC), total radical trapping antioxidant potential (TRAP), and/or trolox equivalent activity capacity (TEAC) assays.[5]

2.1. Vitamin E

Most nutrient databases do not distinguish between the different forms of vitamin E in foods, but instead present the data as α-tocopherol equivalents which consider differences in antioxidant capacity of the different isomers. Supplements are formulated with natural and/or synthetic sources of α-tocopherol, although questionnaires which include their use generally do not make this distinction. Although the supra-dietary intakes of vitamin E associated with supplement use are relatively easy to determine, the limitations associated with assessing dietary vitamin E intake suggest a greater value by examining tocopherol status via HPLC. Indeed, plasma α-tocopherol correlates poorly with vitamin E intake data.[6] Importantly, assessment of plasma or serum vitamin E should include adjustment for cholesterol or total lipids. Adipose tissue, obtained via needle biopsy, can be used to estimate long-term exposure to vitamin E.[7]

2.2. Carotenoids

Many food composition databases have recently been updated to include α- and β carotene, β-cryptoxanthin, lutein + zeaxanthin, and lycopene.[8] Carotenoids increase in plasma and serum in response to dietary interventions and these changes can be readily measured by HPLC.[9] Plasma carotenoid concentrations generally correlate modestly with estimates dietary intake as they are markedly influenced by host factors, including adiposity, gender, smoking status, alcohol consumption, and plasma cholesterol, as well as the effect of cooking on bioavailability.[10] Correlations between plasma and adipose tissue are generally reasonable and exceed those between diet and plasma.[11] Adipose represents a logical tissue to reflect long-term exposure to lipid-soluble antioxidants like carotenoids and tocopherols; however, fat depots may not reflect physiologically active pools for combating oxidative challenges in other tissues. The impact of the invasive nature and cost of adipose biopsies is self-evident with regard to subject recruitment and research funding.

2.3. Vitamin C

While vitamin C intakes are readily determined by dietary assessment tools, there is a significant impact on actual ingestion by food handling and cooking. HPLC methods for vitamin C analysis are well established but the procedures necessary for blood collection and processing to prevent degradation have discouraged assessment of ascorbate status.[12] Further, plasma ascorbate fluctuates in response to intake, making fasting blood samples essential for assessment of status. Pharmacokinetic studies indicate that plasma ascorbate is directly related to intakes between 50–90 mg/d, but at much greater intakes renal clearance increases and plasma is saturated at doses about 200 mg/d. Thus, plasma ascorbate reflects and can predict low intakes but not those levels associated with high dietary or supplement consumption.

2.4. Phenolic Antioxidants

Many of the more than 5000 food phenolics, including flavonoids, cinnamic acid derivatives and coumarins, are multifunctional antioxidants acting as free radical terminators, metal chelators, and singlet oxygen quenchers. Total daily intake of flavonoids in the United States has been calculated to be as high as 1 g, although this is probably an overestimate.[13] While food phenolics are largely absent from nutrient databases, Hertog *et al.*[14] calculated the intake of five principal flavonols and flavones at 23 mg/d. Although a variety of electrochemical and chromatographic methods are available for the determination of phenolic antioxidants, population based references on blood levels of flavonoids are unavailable. Epidemiologic studies have indicated an inverse association between the intake of selected flavonoids and cancer, coronary heart disease, and stroke among older adults, but nutrient databases remain inadequate to provide a broader reference for intakes of these dietary antioxidants.[15] RCT employing flavonoid interventions are limited but do confirm they are effective in reducing the concentration of biomarkers of oxidative stress.[16, 17]

3. Biomarkers of Oxidative Stress Status

Bray[18] has provided a scheme of the strategies and approaches employed in different studies to assess oxidative stress status as a: (i) chemical approach relying on measurement of oxidative damage to cellular macromolecules; (ii) "balanced" approach where reactive oxygen species generation and antioxidant defenses are determined to provide a ratio between overall pro- and anti-oxidant reactions; (iii) "potential" approach through characterizing the susceptibility of a biological sample to resist oxidation *in vitro*; (iv) "molecular" approach to evaluate early transcription factor activation or signal transduction events which may precede oxidant-mediated damage; and (v) clinical approach employing non-invasive imaging technology such as proton MRI to monitor immediate/early signs of tissue damage. In practice, single elements or combined parameters from these approaches are employed, although not infrequently only one analyte is measured and incorrect conclusions drawn that it satisfactorily reflects oxidative stress status. Regrettably, many human studies appear to employ a "black box" approach, assessing antioxidant intake in a cohort or administering an antioxidant supplement in a RCT and then measuring disease outcome(s) without determining any of the biological actions which were postulated to underlie the efficacy of the antioxidant. Regardless of the approaches used by investigators, it is important to recognize the need to account for other confounding variables including the intake of other antioxidant nutrients, medications, and lifestyle factors like physical activity and use of alcohol and tobacco. While the capabilities for adequately assessing relevant genomic factors is limited, this facet to research approaches will become increasingly important to determining which individuals are most likely to benefit from antioxidant interventions.

3.1. Biomarkers of Lipid Oxidation

Lipid peroxidation initiates chain redox reactions forming conjugated dienes, lipid hydroperoxides, and other degradation products such as alkanes, aldehydes, and isoprostanes. The thiobarbituric acid reactive substances (TBARS) spectro-photometric assay has been commonly employed to assess lipid peroxidation based on the chromophore formed between malondialdehyde (MDA) with thiobarbituric acid (TBA).[19] Although useful in some *in vitro* systems and simple to run, the TBARS assay is not specific to lipid peroxidation and not appropriate to the assessment blood or other tissues from human studies. HPLC separation of the MDA-TBA from other interfering TBA adducts improves the sensitivity, specificity and reproducibility of the TBARS assay but still fails to address MDA reactivity with other biomolecules *in vivo*. The ability of MDA to react with the ε-amino group of lysine (which is not detected by TBA) serves as the basis for the assay of a stable MDA-lysine adduct.[20] Lipid hydroperoxides can be determined in plasma via chemiluminescence HPLC and the ferrous xylenol orange (FOX) assay but have not been widely employed in human studies.

F_2-isoprostanes are formed in phospholipids as radical-induced peroxidation products of arachidonic acid and, especially isoprostaglandin $F_{2\alpha}$ Type III and VI isoforms, appear to be one of the better biomarkers of lipid peroxidation in human studies.[21-23] F_2-isoprostanes are usually analyzed by GC/MS, but immunoassays are also available though their antibodies may cross-react with other prostanoids.[24] α-lipoic acid and vitamin E in healthy adults[25] and vitamin C in heavy smokers[26] have been found to reduce urinary F_2-isoprostanes although the isoflavones genistein and daidzein had no impact on this measure of lipid peroxidation in hypertensive patients.[27] Urinary F_2-isoprostanes are correlated with increasing age and elevated concentrations have been associated with Alzheimer's disease, cystic fibrosis, ischemic heart disease, diabetes, and hepatic cirrhosis.[28-30] The accuracy of F_2-isoprostane analysis in plasma requires flash freezing upon collection and limited periods of storage while urine concentrations are more stable but best obtained from 24 h collections and adjusted for creatinine content.

Volatile alkane products of polyunsaturated fatty acid (PUFA) metabolism can be detected in expired breath, e.g. pentane from n-6 PUFA and ethane from n-3 PUFA.[31] Elevated levels of expired pentane have been reported in smokers and patients with HIV and inflammatory bowel disease and are responsive to antioxidant intervention with vitamins C and E.[32] Gottlieb *et al.*[33] reported a dose-dependent reduction by β-carotene of pentane but not ethane in smokers.

Oxidatively modified low density lipoprotein (LDL) appears to play a critical role in artherosclerosis and, thus, its detection in the circulation or vascular tissue may serve as an intermediary biomarker of heart disease risk. Pro-oxidant challenges to LDL *ex vivo* with copper ion, 2,2'-azo-bis (2-amidinopropane) HCl (AAPH), activated macrophages or other radical generators reveals the susceptibility of LDL to oxidation and, thus, may serve as a proxy measure of oxidative stress and cardiovascular disease processes.[34] As the choice of radical generator may affect the results, it is recommended that more than one agent be tested before conclusions are drawn. The concentration of antioxidant nutrients within LDL may also present a measure of potential resistance to oxidation. RCT of vitamin E,[35] flavonoids,[36] and antioxidant cocktails[37] but not carotenoids[38] have demonstrated an increased resistance of LDL to oxidation. It is worth noting that hydrophilic antioxidants such as the flavonoids may be oxidized or lost during the preparation of LDL samples. It is also important to appreciate the influence of nutrient–nutrient interactions in this model, e.g. dietary fiber intake reducing the impact of antioxidants[39] and deficiencies *in vitro* of ascorbate or coenzyme Q converting tocopherol to a pro-oxidant.[40]

3.2. Biomarkers of Protein Oxidation

Protein oxidation, often catalyzed by copper or iron cations, is usually targeted to amino acid side chains with arginine, histidine, lysine, and/or proline and converts them to carbonyls, a modification presumably more susceptible to the action of proteases. However, carbonyls may be stabilized or further modified in the pathogenesis or progression of some chronic diseases, particularly those associated with protein deposition or alteration of matrix components, including amyotrophic lateral sclerosis, Alzheimer's disease, cataract, Parkinson's disease, progeria, and rheumatoid arthritis.[41-43] Protein carbonyls can be assayed by a dinitrophenylhydrazine reaction monitored colorimetrically or by enzyme-linked immunosorbent assay (ELISA) and each yield correlating but not comparable results.[44-46] Supplementation with α-lipoic acid but not vitamin E in healthy subjects was shown to decrease AAPH-induced protein carbonylation *ex vivo* although both antioxidants affected LDL oxidation and urinary F_2-isoprostanes.[47]

Proteins are also modified by nitration and can be detected by HPLC.[48] Nitrotyrosine is a stable end product formed when reactive nitrogen species, such as peroxynitrite, react with either free or bound tyrosine. Assessment of 3-nitrotyrosine or any other single modified amino acid, while a useful measure of protein oxidation, can underestimate the total impact of reactive nitrogen species on protein oxidation.[49] Antibodies raised against peroxynitrite-treated proteins are available for use in immunohistochemical analyses.

3.3. Biomarkers of DNA Oxidation

DNA can be oxidatively damaged through modification of nucleotide bases or sugars and via formation of crosslinks and may reflect a risk for cancer.[50] The most prominent oxidative DNA adducts studied to date are 8-hydroxy-2′-deoxyguanosine (8-OHdG) and 5-hydroxymethyl-2′-deoxyuridine (5-HMdU). Their determination, most commonly from white blood cells or urine, has been accomplished via HPLC with electrochemical detection, GC/MS, and ELISA.[51-53] HPLC and ELISA results are generally well correlated with values from the latter usually being higher, probably due to the cross-reactivity of the antibody. A higher throughput HPLC/MS method with an initial solid-phase extraction to concentrate the analyte has been reported.[54] Marked within- and between-laboratory variations in DNA oxidation products may result from artifacts generated during sample collection[55, 56] and processing as well as from subject differences in metabolic rate or DNA excision repair mechanisms arising from non-oxidative processes.[57]

Single cell microgel electrophoresis is employed to determine the degree of strand breakage in a DNA sample. Breaks in DNA result in an alteration of supercoiled DNA loops that alter its electrophoretic motility. The resulting gel exhibits a staining pattern reminiscent of a comet, so the method is often called the Comet assay. The Comet assay has been refined with an endonuclease III pretreatment so that additional DNA breaks are made at sites with oxidized pyrimidines.[58] Increased antioxidant consumption through vegetable consumption or dietary supplementation has been reported to decrease oxidative base damage in lymphocyte DNA.[59]

4. Standards of Evidence

Several criteria are essential to establish a causative relationship between a dietary factor and a chronic disease. While the strength of the association is usually indicated by the magnitude of the relative risk, particularly in linking cause and effect, the attributable risk (the size of the difference in risk) may often be a more important measure of the impact of the relationship on public health. The association should also be consistently observed in different studies with different populations, even though variations in the magnitude of the relative risk are to be expected, to provide confidence in the observation. The temporal relationship of the association must also be logical with exposure of the suspected causal factor preceding the onset of the disease and usually accompanying the entire pathogenic process. While specificity of the association is a usual criterion for determining causality, the known risk factors for chronic diseases seldom have a single effect so lack of specificity should not rule out a contributing causal effect by a nutrient. Finally, there must be a biological plausibility in support of the association.

Epidemiologic approaches establish antioxidants as affecting the risk of a disease when a difference in intake or status between groups is significantly associated with a change in the incidence of the condition. The complexity that arises in identifying which antioxidants or even which classes of antioxidants influence disease etiology becomes clear when the issues that must be addressed are considered. Some risk factors are readily amenable to comparison or intervention, e.g. cigarette smoking, where there are populations with varying degrees of exposure and others with none and at least some of the former can be engaged to stop the behavior. The no exposure situation obviously does not exist with dietary antioxidants and assessing their precise intake is seriously limited by the accuracy of self-reports, the extent of the nutrient databases, and the long latent period associated with chronic diseases. The close association in foods between many dietary antioxidants as well as other nutrients further confounds these analyses. While there is an increasing appreciation of the impact genetic differences within and between populations can have on antioxidant parameters, few convenient tools are available to manage this situation in large scale human studies. It is interesting to note the postulation by Brown[60] that differences in the frequency of a polymorphism in the gene for endothelial nitric oxide synthase between British and Italian populations might, in part, underlie the discrepant results between the CHAOS and GISSI trials.

Clinical signs and symptoms of chronic disease are not generally apparent until middle age although they are a consequence of pathogenic events occurring in early adulthood or even in adolescence. Thus, age at the initiation of the disease cannot be determined and this situation makes it difficult to ascertain whether exposure to the suspected cause actually preceded the disease. This situation also complicates planning an intervention intended to prevent the initiating or early promotion and progression events of the disease. In fact, chronic diseases have several interacting environmental and heritable risk factors that influence their pathogenesis so research efforts must be directed at identifying the characteristics of individuals who are most susceptible.

In most cases, decreases in disease risk associated with the generous intake of antioxidants are modest, oftentimes being only half or less that of the group with the lowest intake. Nonetheless, a weak association between antioxidants and diseases having a major impact on the health of a population can have a significant impact on health promotion and disease prevention. One reason for the apparent modest impact of some antioxidants on chronic disease is that levels of intake, particularly within a national population or culture, are not large enough to allow for strong associations. Studies of the intake of vitamins C and E offer an advantage in this regard as the use of supplements can create substantial differences in intake between users and non-users, e.g. median intakes of vitamin E from 6 to 420 IU/d across quintiles in the Health Professionals Follow-up Study revealed significant trends for reduced coronary heart disease risk but only the suggestion of an inverse relationship within the range of dietary intakes.[52]

5. Hierarchy of Research Designs

RCT were introduced over fifty years ago when streptomycin was evaluated in the treatment of tuberculosis and have become the gold standard for assessing the effectiveness of therapeutic agents.[61] Advocates of evidence-based medicine classify studies according to grades of evidence on the basis of the research design, using internal validity (i.e. the correctness of the results) as the criterion for hierarchical rankings. The highest grade is reserved for RCT with lower rankings provided to cohort and case-control observational studies. Yet epidemiologic studies have provided compelling, if potentially confounded, results suggesting the benefits of generous intakes of dietary antioxidants. Nonetheless, RCT are now held as the standard for evidence of the impact of dietary antioxidants and other food ingredients in the primary and secondary prevention of chronic disease. However, the actual number of RCT which test single and, especially, combinations of antioxidants is very limited due, in part, to their cost for government agencies and the absence of a proprietary advantage to be derived from a successful outcome for companies. Further, for both practicality and scientific rigor, eligibility criteria for enrollment in RCT are narrowly defined; thus, the ability to extrapolate the results from RCT to the general population is also limited.

RCT were developed to test the effect of a selected dose(s) of a drug(s) on a limited set of outcomes for treatment of a particular disease but antioxidant substances act within a dynamic interrelationship to affect oxidative stress and slow the pathogenesis and progression of many chronic conditions. While potential interactions between antioxidants can be readily examined in epidemiologic studies, factorial designs within RCT are required to fully account for the efficacy of antioxidant cocktails, but few such studies have been conducted due to their size and cost. Novel drugs can readily be studied against placebo controls while it is impossible to have a strict control group with no exposure to a dietary antioxidant. Although epidemiologic studies suggest the most promising benefit of antioxidants lies in their ability to reduce the risk and/or delay the onset of several chronic diseases in the general population, most RCT actually test their efficacy in patient populations, in part due to the requirement of secondary prevention studies for much smaller sample sizes. For example, while the Nurses Health,[62] Health Professionals Follow-up,[63] and EPESE[64] studies suggested vitamin E reduces the risk of coronary heart disease, they stimulated RCT like CHAOS,[65] GISSI,[66] and HOPE[67] which administered vitamin E as a treatment to high risk cardiac patients. While it is not unreasonable to hypothesize that an intervention that is successful in primary prevention may also be successful in secondary prevention (or vice versa), there is no reason to presume this must be the case. Moreover, even when this is the case, extrapolation from one to the other is risky as the effective dose and duration of treatment as well as mechanism of action may be different. For example, over decades, moderate intakes of vitamin E may retard the formation of atherosclerotic plaque by inhibiting the oxidative modification of LDL, but much higher doses may be necessary over a few years to prevent recurrence of myocardial, infarct by promoting endothelial function and preventing plaque rupture. Further, unlike the situation in observational studies of primary prevention, the typical polypharmacy regimen in patients with chronic disease can substantially confound the determination of efficacy of an antioxidant intervention. For example, RCT of vitamin E in patients with coronary artery disease are complicated by the continuing and concomitant use of antiplatelet agents, beta-blockers, calcium channel blockers, diuretics, and/or hypocholesterolemic drugs. While the need for long-term trials in low risk populations is self-evident, only the ongoing Physicians Health Study II,[68] Womens Health Study,[69] and SU.VI.MAX[70] RCT meet this criterion, and even so, most of the subjects are older adults.

6. Conclusion

The principal hypothetical basis for employing antioxidants in primary prevention and disease treatment is their capacity to significantly decrease the adverse effects of reactive oxygen and nitrogen species thought to be intimately involved in the initiation and/or progression of the condition. Without measuring parameters relevant to the status of antioxidant defenses and oxidative stress in both epidemiologic studies and RCT, it is not be possible to determine whether the dose and duration of the antioxidant exposure actually achieves its intended biochemical endpoint. Conclusions that antioxidant substances are or are not efficacious cannot be reached if an antioxidant action is not assessed as part of human studies. Finally, it is worth recognizing that inherent in every research approach are substantial limitations and, thus, no single study can truly be definitive. Thus, conclusions about the value of dietary antioxidants in the promotion of health and the treatment of patients with chronic disease will always require a considered scientific judgment of the totality of the available evidence without undue reliance on only one standard of evidence.

References

1. Institute of Medicine. (2000). *Dietary Reference Intakes for Vitamin C, Vitamin E, Selenium, and Carotenoids*, National Academy Press. Washington, D.C.
2. Diplock, A. T., Charleux, J. L., Crozier-Willi, G. Kok, F. J., Rice-Evans, C., Roberfroid, M., Stahl, W. and Vina-Ribes, J. (1998). Functional food science and defense against reactive oxidative species. *Br. J. Nutri.* **80**: S77–S112.
3. Benson, K. and Hartz, A. J. (2000). A comparison of observatonal studies and randomized, controlled trials. *New England J. Med.* **342**: 1878–1886.
4. Concato, J., Shah, N. and Horwitz, R. I. (2000). Randomized, controlled trials, observational studies, and the hierarchy of research designs. *New England J. Med.* **342**: 1887–1892.
5. Prior, R. L. and Cao, G. (1999). In vivo total antioxidant capacity: comparison of different analytical methods. *Free Radic. Biol. Med.* **27**: 1173–1181.
6. Ford, E. S. and Sowell, A. (1999). Serum alpha-tocopherol status in the United States population: findings from the Third National Health and Nutrition Examination Survey. *Am. J. Epidemiol.* **150**: 290–300.
7. Su, L. C., Bui, M., Kardinaal, A., Gomez-Aracena, J., Martin-Moreno, J., Martin, B., Thamm, M., Simonsen, N., van Veer, P., Kok, F., Strain, S. and Kohlmeier, L. (1998). Differences between plasma and adipose tissue biomarkers of carotenoids and tocopherols. *Cancer Epidemiol. Biomark. Prev.* **7**: 1043–1048.

8. US Department of Agriculture, Agricultural Research Service. (1999). USDA Nutrient Database for Standard Reference, Release 13. Nutrient Data Laboratory Home Page, http://www.nal.usda.gov/fnic/foodcomp

9. McEligot, A. J., Rock, C. L., Flatt, S. W., Newman, V., Faerber, S. and Pierce, J. P. (1999). Plasma carotenoids are biomarkers of long-term high-vegetable intake in women with breast cancer. *J. Nutri.* **129**: 2258–2263.

10. Brady, W. E., Mares-Perlman, J. A., Bowen, P. and Stacewicz-Sapuntzakies, M. (1996). Human serum carotenoid concentrations are related to physiologic and lifestyle factors. *J. Nutri.* **126**: 129–137.

11. Peng, Y.-M., Peng, Y.-S., Lin, Y., Moon, T., Roe, D. J. and Ritenbaugh, C. (1995). Concentrations and plasma-tissue-diet relationships of carotenoids, retinoids, and tocopherols, in humans. *Nutri. Cancer* **23**: 233–246.

12. Margolis, S. A. and Duewer, D. L. (1996). Measurement of ascorbic acid in human plasma and serum: stability, intra-laboratory repeatability, and inter-laboratory reproductibility. *Clin. Chem.* **42**: 1257–1262.

13. Hertog, M. G. L. and Katan, M. B., (1997). Quercetin in foods, cardiovascular disease, and cancer. *In* "Flavonoids in Health and Disease" (C. A. Rice-Evans, and L. Packer, Eds.), p. 447, Marcel Dekker, New York,

14. Hertog, M. G. L., Hollman, P. C. H., Katan, M. B. and Kromhout, D. (1993a). Intake of potentially anticarcinogenic flavonoids and their determinants in adults in The Netherlands, *Nutri. Cancer* **20**: 9.

15. Keli, S. O., Hertog, M. G. L., Feskens, E. J. M. and Kromhout, D. (1996). Flavonoids, antioxidant vitamins and risk of stroke, The Zutphen Study. *Arch. Int. Med.* **154**: 637.

16. Manuel, Y., Keenoy, B., Vertommen, J. and de Leeuw, I. (1999). The effect of flavonoid treatment on the glycation and antioxidant status in Type-1 diabetic patients. *Diabetes, Nutr. Metab. — Clin. Exp.* **12**: 256–263.

17. Stein, J. H., Keevil, J. G., Wiebe, D. A., Aeschlimann, S. and Folts, J. D. (1999). Purple grape juice improves endothelial function and reduces the susceptibility of LDL cholesterol to oxidation in patients with coronary artery disease. *Circulation* **100**: 1050–1055.

18. Bray, T. M. (2000). Dietary antioxidants and assessment of oxidative stress. *Nutrition* **16**: 578–581

19. Gutteridge, J. M. and Haliwell, B. (1990). The measurement and mechanism of lipid perioxidation in biological systems. *Trends Biochem. Sci.* **15**: 129–134.

20. Wong, S. H. Y., Knight, J. A., Hopfer, S. M., Zaharia, O., Leach, C. N., Jr. and Sunderman, F. W., Jr. (1987). Lipoperoxides in plasma as measured by liquid-chromatographic separation of malondialdehyde-thiobarbituric acid adduct. *Clin. Chem.* **33**: 214–220.

21. Morrow, J. D., Hill, K. E., Burk, R. F., Nammour, T. M., Bradr, K. F. and Roberts, L. J. II. (1990). A series of prostaglandin F_2-like compounds are produced in vivo in humans by a non-cyclooxygenase, free radical-catalyzed mechanism. *Proc. Natl. Acad. Sci. USA* **87**: 9383–9387.

22. Roberts, L. J. II. and Morrow, J. D. (1997). The generation and actions of isoprostanes. *Biochim. Biophys. Acta.* **1345**: 121–135.
23. Pratico, G., Iuliano, L., Mauriello, A., Spagnoli, L., Lawson, J. A., Rokach, J., Maclouf, J., Violi, F. and FitzGerald, G. A. (1997). Localization of distinct F_2-isoprostanes in human atherosclerotic lesions. *J. Clin. Invest.* **100**: 2028–2034
24. Walter, M., Blumberg, J. B., Dolnikowski, G. and Handelman, G. (2000). Streamlined F_2-isoprostane analysis in plasma and urine with HPLC and GC/MS. *Anal. Biochem.* **280**: 73–79.
25. Marangon, K., Devaraj, S., Tirosh, O., Packer, L. and Jialal, I. (1999). Comparison of the effect of alpha-lipoic acid and alpha-tocopherol supplementation on measures of oxidative stress. *Free Radic. Biol. Med.* **27**: 1114–1121.
26. Reilly, M., Delanty, N., Lawson, J. A. and FitzGerald, G. A. (1996). Modulation of oxidant stress in vivo in chronic cigarette smokers. *Circulation* **94**: 19–25.
27. Hodgson, J. M., Puddey, I. B., Croft, K. D., Mori, T. A., Rivera, J., Beilin, L. J. (1999). Isoflavonoids do not inhibit in vivo lipid peroxidation in subjects with high-normal blood pressure. *Atherosclerosis* **145**: 167–172.
28. Cracowski, J. L., Stanke-Labesque, F., Souvignet, C. and Bessard, G. (2000). Isoprostanes: new markers of oxidative stress in human diseases. *Presse Medicale.* **29**: 604–610.
29. Davi, G., Alessandrini, P., Mezzetti, A., Minotti, G., Bucciarelli, T., Costantini, F., Cipollone, F., Bon, G. B., Ciabattoni, G. and Patrono, C. (1997). In vivo formation of 8-epiprostaglandin- F_2[alpha] is increased in hypercholesterolaemia. *Arterioscler. Thromb. Vasc. Biol.* **17**: 3230–3235.
30. Collins, C. E., Quaggiotto, P., Wood, L., O'Loughlin. E.V., Henry, R. L. and Garg, M. L. (1999). Elevated plasma levels of F_2 alpha isoprostane in cystic fibrosis. *Lipids* **34**: 551–556.
31. Knutson, M. D., Handelman, G. J. and Viteri, F. E. (2000). Methods for measuring ethane and pentane in expired air from rats and humans. *Free Radic. Biol. Med.* **28**: 514–519.
32. Aghdassi, E. and Allard, J. P. (2000). Breath alkanes as markers of oxidative stress in different clinical conditions. *Free Radic. Biol. Med.* **28**: 880–886.
33. Gottlieb, K., Zarling, E. J., Mobarhan, S., Bowen, P. and Sugerman, S. (1993). β-carotene decreases markers of lipid peroxidation in healthy volunteers. *Nutri. Cancer* **19**: 207–212.
34. Esterbauer, H., Puhl, M., Dieber-Rotheneder, M., Waeg, G. and Rabl, H. (1991). Effect of antioxidants on oxidative modification of LDL. *Ann. Med.* **23**: 573–581.
35. Jialal, I. and Fuller, C. J. Effect of vitamin E, vitamin C and beta-carotene on LDL oxidation and atherosclerosis. *Can. J. Cardiol.* **11(Suppl. G)**: 97G–103G.
36. Chopra, M., Fitzsimons, P. E., Strain, J. J., Thurnham, D. I. and Howard, A. N. (2000). Nonalcoholic red wine extract and quercetin inhibit LDL oxidation

without affecting plasma antioxidant vitamin and carotenoid concentrations. *Clin. Chem.* **46**: 1162–1170.

37. Woodside, J. V., Young, I. S., Yarnell, J. W., Roxborough, H. E., McMaster, D., McCrum E. E., Gey, K. F. and Evans, A. (1999). Antioxidants, but not B-group vitamins increase the resistance of low-density lipoprotein to oxidation: a randomized, factorial design, placebo-controlled trial. *Atherosclerosis* **144**: 419–427.

38. Carroll, Y. L., Corridan, B. M. and Morrissey, P. A. (2000). Lipoprotein carotenoid profiles and the susceptibility of low density lipoprotein to oxidative modification in healthy elderly volunteers. *Eur. J. Clin. Nutri.* **54**: 500–507.

39. Hoffmann, J., Linseisen. J., Riedl, J. and Wolfram, G. (1999). Dietary fiber reduces the antioxidative effect of a carotenoid and alpha-tocopherol mixture on LDL oxidation ex vivo in humans. *Eur. J. Clin. Nutri.* **38**: 278–285.

40. Upston, J. M., Terentis, A. C. and Stocker, R. (1999). Tocopherol-mediated peroxidation of lipoproteins: implications for vitamin E as a potential antiatherogenic supplement. *FASEB J.* **13**: 9077–9094.

41. Fucci, L., Oliver, C. N., Coon, M. J. and Stadtman, E. R., (1983). Inactivation of key metabolic enzymes by mixed-function oxidation reactions: possible implications in protein turnover and aging. *Proc. Natl. Acad. Sci. USA* **80**: 1521.

42. Stadtman, E. R. and Berlett, B. S. (1998). Reactive oxygen-mediated protein oxidation in aging and disease. *Drug Metab. Rev.* **30**: 225–243.

43. Witt, E .H., Reznick, A. Z., Viguie, C. A., Starke-Reed, P. and Packer, L. (1992). Exercise, oxidative damage and effects of antioxidant manipulation. *J. Nutri.* **122**: 766–773.

44. Starke-Reed, P. (1998). Protein oxidation. *In* "Methods in Aging Research" (B. P. Yu, Ed.), pp. 637–655, CRC Press. Boca Raton.

45. Levine, R. L., Garland, D., Oliver, C. N., Amici, A., Climent, I., Lenz, A. G., Ahn, B. W., Shaltiel, S. and Stadtman, E. R. (1990). Determination of carbonyl content in oxidatively modified proteins. *Meth. Enzymol.* **186**: 464–478.

46. Winterbourn, C. C. and Buss, I. H. (1999). Protein carbonyl measurement by enzyme linked immunosorbent assay. *Meth. Enzymol.* **300**: 106–111.

47. Marangon, K., Devaraj, S., Tirosh, O., Packer, L. and Jialal, I. (1999). Comparison of the effect of alpha-lipoic acid and alpha-tocopherol supplementation on measures of oxidative stress. *Free Radic. Biol. Med.* **27**: 1114–1121.

48. Shigenaga, M. K., Lee, H. H., Blount, B.C., Christen, S., Shigeno, E. T., Yip, H. and Ames, B. N. (1997). Inflammation and NO(X)-induced nitration: assay for 3-nitrotyrosine by HPLC with electrochemical detection. *Proc. Natl. Acad. Sci. USA* **94**: 3211–3216.

49. Whiteman, M. and Halliwell, B. (1999). Loss of 3-nitrotyrosine on exposure to hypochlorous acid: implications for the use of 3-nitrotyrosine as a bio-marker in vivo. *Biochem. Biophys. Res. Commun.* **258**: 168–172.

50. Halliwell, B. (1998). Can oxidative DNA damage be used as a biomarker of cancer risk in humans? Problems, resolutions and preliminary results from nutritional supplementation studies. *Free Radic. Res.* **29**: 469–486.

51. Kasai, H. (1997). Analysis of a form of oxidative DNA damage, 8-hydroxy-2'-deoxyguanosine, as a marker of cellular oxidative stress during carcinogenesis. *Mutat. Res.* **387**: 147–163.

52. Santella, R. M. (1999). Immunological methods for detection of carcinogen-DNA damage in humans. *Cancer Epidemiol. Biomark. Prev.* **8**: 733–739.

53. Frenkel, K., Karkoszka, J., Glassman, T., Dubin, N., Toniolo, P., Taioli, E., Mooney, L. A. and Kato, I. (1998). Serum autoantibodies that recognize 5-hydroxymethyl-2'-deoxyuridine, an oxidized DNA base, as biomarkers of cancer risk in women. *Cancer Epidemiol. Biomark. Prev.* **7**: 49–57.

54. Renner, T., Fechner, T. and Scherer G. (2000). Fast quantification of the urinary marker of oxidative stress 8-hydroxy-2'-deoxyguanosine using solid-phase extraction and high performance liquid chromatography with triple-stage quadrupole mass detection. *J. Chromat.: B. Biomed. Sci. Appl.* **738**: 311–317.

55. Bogdanov, M. B., Beal, M. F., McCabe, D. R., Griffin, R. M. and Matson, W. R. (1999). A carbon column-based liquid chromatography electrochemical approach to routine 8 hydroxy-2'-deoxyguanosine measurements in urine and other biologic matrices: a one-year evaluation of methods. *Free Radic. Biol. Med.* **27**: 647–666

56. Helbock, H. J. Beckman, K. B. and Ames, B. N. (1999). 8-hydroxydeoxyguanosine and 8-hydroxyguanine as biomarkers of oxidative DNA damage. *Meth. Enzymol.* **300**: 156–166.

57. Halliwell, B. (1999). Establishing the significance and optimal intake of dietary antioxidants: the biomarker concept. *Nutri. Rev.* **57**: 104–113.

58. Collins, A. R., Duthie, S. J. and Dobson, V. L. (1993). Direct enzymic detection of endogenous oxidative base damage in human lymphocyte DNA. *Carcinogenesis* **14**: 1733–1735.

59. Pool-Zobel, B. L., Bub, A., Muller, H., Wollowski, I. and Rechkemmer, G. (1997). Consumption of vegetables reduces genetic damage in humans: first results of a human intervention trial with carotenoid-rich foods. *Carcinogenesis* **18**: 1847–1850.

60. Brown, M. (1999). Do vitamin E and fish oil protect against ischaemic heart disease? *Lancet* **354**: 441–442.

61. Abel, U. and Koch, A. (1999). The role of randomization in clinical studies: myths and beliefs. *J. Clin. Epidemiol.* **52**: 487–497.

62. Stampfer, M. J., Hennekens, C. H., Manson, J. E., Colditz, G. A., Rosner, B. and Willett, W. C. (1993). Vitamin E consumption and the risk of coronary disease in women. *New England J. Med.* **328**: 1444–1449.

63. Rimm, E. B., Stampfer, M., Ascherio, A., Giovannucci, E., Colditz, G. A. and Willett, W. C. (1993). Vitamin E consumption and the risk of coronary heart disease in men. *New England J. Med.* **328**: 1450–1456.

64. Losonczy, K. G., Harris, T. B. and Havlik, R. J. (1996). Vitamins E and C supplement use and risk of all-cause and coronary heart disease mortality in older persons: the establish populations for epidemiological studies of the elderly. *Am. J. Clin. Nutri.* **64**: 190–196.
65. Stephens, N. G., Parsons, A., Schofield, P. M., Kelly, F., Cheeseman, K. and Mitchinson, M. J. (1996). Randomised controlled trial of vitamin E in patients with coronary disease: Cambridge Heart Antioxidant Study (CHAOS). *Lancet* **347**: 781–786.
66. GISSI-Prevenzione Investigators. (1999). Dietary supplementation with *n*-2 polynsaturated fatty acids and vitamin E after myocardial infarction: results of the GISSI-Prevenzione Trial. *Lancet* **354**: 447–455.
67. HOPE (Hear Outcomes Prevention Evaluation) Study Investigators. (2000). Vitamin E supplementation and cardiovascular events in high-risk patients. *New England J. Med.* **342**: 154–160.
68. Christen, W. G., Gaziano, J. M. and Hennekens, C. H. (2000). Design of Physicians' Health Study II — a randomized trial of beta-carotene, vitamins E and C, and multivitamins, in prevention of cancer, cardiovascular disease, and eye disease, and review of results of completed trials. *Ann. Epidemiol.* **10**: 125–134.
69. Rexrode, K. M., Lee, I. M., Cook, N. R., Hennekens, C. H. and Buring, J. E. (2000). Baseline Characteristics of Participants in the Women's Health Study. *J. Womens Health Gender-Based Med.* **9**: 19–27.
70. Hercberg, S., Galan, P., Preziosi, P., Roussel, A. M., Arnaud, J., Richard, M. J., Malvy, D., Paul-Dauphin, A., Briancon, S. and Favier, A. (1998). Background and rational behind the SU. VI.MAX Study, a prevention trial using nutritional doses of a combination of antioxidant vitamins and minerals to reduce cardiovascular diseases and cancers. Supplementation en vitamines et Mineraux AntioXydants Study. *Int. J. Vitamin. Nutri. Res.* **68**: 3–20.

Chapter 29

Current Status of the Potential Role of Flavonoids in Neuroprotection

Hagen Schroeter, Jeremy P.E. Spencer and Catherine Rice-Evans*

Hagen Schroeter, Jeremy PE Spencer and **Catherine Rice-Evans** • Wolfson Centre for Age-Related Diseases, Guy's King's and St Thomas' School of Biomedical Sciences, King's College, London SE1 9RT

*Corresponding Author. Wolfson Centre for Age-Related Diseases, Antioxidant Research Group, Guy's King's and St Thomas's School of Biomedical Sciences, Hodgkin Building — 3rd Floor, King's College London
E-mail: catherine.rice-evans@kcl.ac.uk

1. Introduction

Evidence supporting the involvement of oxidative stress mediated by reactive oxygen and nitrogen species (ROS/ RNS) as a contributory factor to the pathology of neurodegeneration and ageing is accumulating. Several findings substantiate the susceptibility of the CNS to oxidative stress: the high density of mitochondria, a major cellular source of ROS, in brain tissue, the high rate of oxygen consumption, a pool of autoxidizable compounds such as dopamine, and noradrenalin, the presence of locally high levels of potentially excitotoxic glutamate, high Ca^{2+} traffic across membranes and the lipid-composition of the neuronal membrane.

Oxidative stress has been shown to contribute to the neuropathology of a number of neurodegenerative disorders (reviewed in Ref. 1), including Alzheimer's disease,[2] Parkinson's disease[3] and Huntington's diseases,[4] as well as being implicated in neuronal loss associated with age-related cognitive decline, cerebral ischemia and seizures[5] and neuroinflammation. In particular, increased iron levels in the substantia nigra,[6] elevation of lipid peroxidation[7,8] and a decline in glutathione concentrations[9] have been associated with neuronal dysfunction.

Consequently, there is a growing interest in the establishment of therapeutic strategies to combat oxidative stress-induced damage to the CNS and attention is turning towards the potential neuroprotective effects of dietary antioxidants, especially flavonoids. Flavonoids, a large group of plant derived polyphenolic compounds, are widely distributed throughout the plant kingdom and therefore common in a great variety of fruit, vegetables and beverages. Many studies have described the efficacy of flavonoids to act as antioxidants *in vitro* due to their hydrogen donating and metal chelating properties.

In the recent past flavonoids have been shown to possess neuroprotective properties, thus these compounds may play a useful role in preventing age-related cognitive, motoric and mood decline[10–14] and protect against oxidative stress[15,16] as well as cerebral ischemia/reperfusion injuries.[17] Studies in animal models have shown that dietary supplementation with flavonoid-rich extracts can retard the onset of age-related loss of neuronal function.[10,11,18] Furthermore, the oral administration of catechin protected against ischemia/reperfusion-induced neuronal death in gerbils[17] and grape polyphenols blocked neurodegenerative changes caused by chronic ethanol administration.[19]

2. Oxidative Stress and the Brain

The molecular mechanisms underlying oxidative stress-induced neuronal damage are complex and involve a variety factors. Oxidative damage to neuronal cells has often been implicated in the pathogenesis of many neurodegenerative diseases.[20–22] This evidence is supported by the detection of a number of bio-chemical markers of oxidative stress in post-mortem brain tissue from patients with neurodegenerative diseases such as Parkinson's disease.[23,24] The potential role of reactive oxygen- and reactive-nitrogen species in the pathophysiology of neurodegenerative disorders is relevant as nitric oxide (NO$^{\cdot}$) and superoxide (O$_2^{\cdot-}$) are generated in the brain and their formation is often colocalized in specific neurons. Recently, inflammation has been discussed as an important contributor to neuronal damage in neurodegenerative disorders such as Alzheimer's disease, Parkinson's disease, multiple sclerosis and amyotrophic lateral sclerosis.[25, 26] It is believed that an initial toxic insult promotes neuronal damage and the resulting deposit of cellular debris activates microglial cells leading to the release of cytotoxic agents, such as peroxynitrite (ONOO$^-$), and the initiation of the classical complement cascade.[27] As a result of this inflammatory mechanism there is further neuronal injury which in turn increases further inflammatory responses. This reaction is a self-sustaining autodestructive force in which vast numbers of local neurons may be damaged, producing large lesions in the surrounding tissue.

It is possible that large amounts of ONOO$^-$ could be formed during glutamate excitotoxicity and mitochondrial dysfunction similar to that reported in amyptrophic lateral sclerosis (ALS), Parkinson's disease and other neurodegenerative diseases.[28] Hyperactivity of glutamate neurotransmission leads to large increases

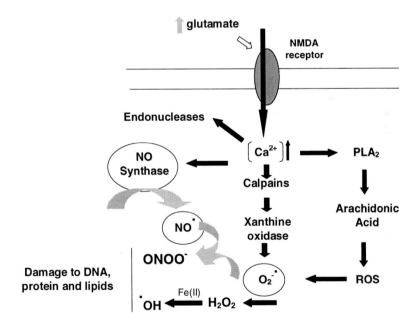

Fig. 1. Sequence of events initiated by the hyperactivity of glutamate neurotransmission. Resulting increases in intracellular Ca(II) activate calcium-dependent pathways, capable of generating both ROS and NOS, ultimately lead to the modification of lipids, proteins and DNA.

in intracellular Ca^{2+} which activates calcium-dependent pathways capable of generating both ROS and $NO^{\bullet}$, and consequently $ONOO^-$ [29] (Fig. 1). For example, $O_2^{\bullet-}$ may be generated via the activation of calpains and xanthine oxidase intracellularly or via phospholipase A_2 and arachidonic acid, whereas the activation of Type-I NOS will generate large amounts of $NO^{\bullet}$. Furthermore microglia, the resident macrophages in the brain, which account for ~ 20% of the total glial population in the CNS, become activated in response to inflammatory triggers such as pro-inflammatory cytokines and the β-amyloid peptide (β-A).[30, 31] During inflammation, they may produce large quantities of $O_2^{\bullet-}$ which could react with locally formed $NO^{\bullet}$ to generate peroxynitrite. However, *in vivo*, $ONOO^-$ may combine with CO_2[32] to generate very reactive intermediates capable of greater nitrating potential and in the CNS this reactivity could be a factor which exacerbates nitration of protein tyrosine residues, particularly in pathologies which promote acidosis. The nitration of tyrosine may cause neuronal injury by a series of mechanisms including initiation of autoimmune processes due to the antigenic nature of nitrated phenolic molecules, alteration of tyrosine phosphorylation-dependent signaling as well as the potential for modification of protein conformation and enhancement of proteolysis (reviewed in Ref. 33). Recent studies have detected an increased ratio of 5-nitro-γ-tocopherol to γ-tocopherol in the brains of individuals with Alzheimer's disease and of ALS patients, implicating peroxynitrite as a contributing factor to the pathogenesis of such neurological diseases.[34]

Furthermore, recently emerging findings appear to involve an oxidative stress induced apoptotic mode of death in which extracellular signal-regulated kinases-1 and -2 (ERK-1/2)[35–37] and *c-Jun* N-terminal kinase (JNK)[38] have been strongly implicated. In addition, it is becoming clear that products of lipid peroxidation such as 4-hydroxy-2,3-nonenal (4-HNE), lipid-hydroperoxides (LOOH) and oxysterols, found in oxLDL as well as products of plasma membrane damage, are important mediators of oxidative stress-induced apoptosis in the CNS,[7, 39–41] possibly through activation of the transcription factor activator protein-1 (AP-1) complex.[38]

3. Flavonoids as Free Radical Scavengers

The precise mechanisms by which flavonoids might exert their neuroprotective actions *in vivo* are presently unknown and it is currently unclear whether or not these compounds function as electron-donating antioxidants or exert their neuroprotective actions independently of such properties. Before discussing this issue the structural families of dietary flavonoids and the relationship with antioxidant activities will be described.

It is possible to divide the flavonoid family into groups reflecting the differences in the saturation of the basic flavan ring system and the hydroxylation pattern of

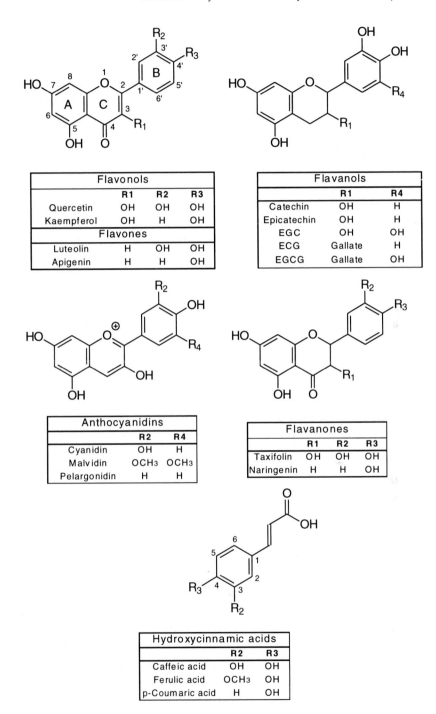

Fig. 2. Structures of the 5 main classes of dietary flavonoids. The major differences between individual flavonoids are the hydroxylation patterns of the ring-structure, the degree of saturation of the C-ring and the substitution in the 3-position.

the molecule. The most common groups of flavonoids (Fig. 2) include flavanols, flavanones, anthocyanidins, flavones and flavonols. The electron-donating properties of flavonoids have been intensively investigated and are well-defined to explain their antioxidant properties.[42-46] In addition, their ability to act as antioxidants *in vitro* is based on metal-chelating capacity[47,48] and as quenchers of singlet oxygen.[49,50] Structurally important features defining the reduction potential of flavonoids are believed to be the hydroxylation pattern, a 3',4'-dihydroxy catechol structure in the B-ring, the planarity of the molecule and the presence of 2,3 unsaturation in conjugation with a 4-oxo-function in the C-ring.

Many studies have described the antioxidant efficacy of flavonoids in inhibiting the oxidation of LDL[45,51,52] and other biomolecules such as proteins and DNA[53-55] *in vitro*. Scavenging of peroxynitrite by dietary phenolics, such as flavonoids and hydroxycinnamates, has also been observed *in vitro* with results suggesting that these compounds can inhibit the nitration of tyrosine by peroxynitrite either by directing the nitration to their own structures, in the case of monohydroxylated compounds such as kaempferol and *p*-coumaric acid, or by deactivating ONOO$^-$ by electron donation as with the catechol-rich structures.[56,57] Furthermore, flavonoids are capable of inhibiting ONOO$^-$-mediated oxidation of dopamine *in vitro* similarly by a structure-dependent mechanism.[58] The inhibition of dopamine oxidation may be of particular relevance to the pathology of Parkinson's disease as dopamine oxidation in the presence of thiols can lead to the formation of specific neurotoxins which act by inhibiting complex I of the mitochondrial electron transport chain.[59,60] However, although flavonoids react rapidly which peroxynitrite in chemical systems *in vitro*, their reaction *in vivo* will be dependent on the form that is bioavailable to the brain.

Experiments undertaken in cell culture systems demonstrate both antioxidant and pro-oxidant effects, often depending on the flavonoid, its concentration and the design of the experiments. Numerous studies describe the protective effects of flavonoids against various oxidative insults in several different cell types, including cells originating from the CNS. In contrast, other studies demonstrate the cytotoxic effects of flavonoids and link this to pro-oxidant effects.[61]

Whether or not flavonoids play a significant role as hydrogen-donating antioxidants *in vivo* is still a controversial issue. Epidemiological data associate flavonoid intake with reduction in risk of coronary heart disease, certain cancers and inflammatory diseases in all of which oxidative stress and the formation of ROS/RNS is implicated. Indeed *in vivo* studies in humans and animals have demonstrated that flavonoids protect against oxidative stress-related damage in a variety of pathologies including cardiac and cerebral ischemia/reperfusion injuries, coronary heart disease,[62-64] cancers[65-68] and arthritis.[69] In addition, supplementation studies in humans and animals demonstrate positive effects on vascular function, increase of antioxidant capacity of blood plasma,[70,71] activity against platelet aggregation,[72,73] increased resistance against oxidative insults of erythrocytes[74] and LDL.[52] The findings *in vivo* are equivocal, other investigators finding no change in plasma antioxidant status[53, 75] or the oxidizability of LDL.

Although it is known that flavonoids are potent antioxidants, whether or not polyphenols influence the pathological events as classical hydrogen donors or by other means needs further investigation.

4. Biological Properties Independent of Free Radical Scavenging

Many studies have investigated the antioxidant effects of flavonoids and phenolic compounds in the context of a variety of cell functions. These include modulation of cell signalling,[76, 77] suppression of TNFα expression,[78] the down-regulation of the expression of adhesion molecules[79–81] and the improvement of vascular dysfunction induced by oxidative stress.[74] Thus, flavonoids also participate in mechanisms of action independently of their conventional hydrogen-donating free radical scavenging properties by modulating enzyme activity, interfering with numerous pathways of intermediary metabolism and acting at various sites within signal transduction cascades by mimicking substrates for various binding sites.[82] The potential role flavonoids play in such intracellular signalling events is becoming increasingly clear especially with regard to the influence they have on protein kinase A, protein kinase C, phosphatidyl-inositol-3 kinase and nuclear factor kappa-B.[76, 77, 83–85] Examples include quercetin, kaempferol, the catechins and morin, reported to interact with proteins such as the mitochondrial ATPase,[86] calcium plasma membrane ATPase,[87] protein kinase A,[88] protein kinase C[83, 89] and topoisomerase.[90] In addition, resveratrol and and several flavonoids, including the citrus flavanones hesperetin and naringenin, inhibit protein kinases by binding to the ATP binding site.[67, 82, 91] It has been demonstrated that certain flavonoids exhibit significant steroid hormone activity,[92] alter the expression and activity of proteins and modulate leukotriene/prostacyclin synthesis.[93, 94] For example, resveratrol, the phenolic stilbene from grape seed, as well as flavonoids and caffeic acid analogues, have been reported to exert inhibitory effects on cyclooxygenase I and lipoxygenases.

Accumulating evidence also suggests that flavonoids interact selectively within mitogen-activated protein kinase (MAPK) signalling cascades.[95, 96] This could have important implications with regard to their possible sites of action in neurons since members of the MAPK family are believed to be involved in signalling to neuronal survival, regeneration and death.[97, 98]

A variety of other mechanisms of action have been demonstrated. Green tea polyphenols are implicated in the induction of phase II enzymes, such as glutathione transferases, quinone reductase, UDP-glucuronyl transferase, and the antioxidant enzymes glutathione peroxidase, glutathione reductase, superoxide dismutase and catalase.[99, 100]

However, it has been suggested that resveratrol is glucuronidated as it crosses the small intestine[101] and epicatechin, one of the green tea polyphenols is methylated and glucuronidated. The question as to whether the conjugates might exert the above-described properties or whether they might be hydrolyzed or de-methylated prior to exerting such actions has recently been addressed. Spencer *et al.*[102] have provided the first evidence that 3'-O-methyl epicatechin inhibits cell death induced by peroxides and that the mechanism involves suppression of caspase-3 activity as a marker for apoptosis. Furthermore, the protection elicited by methylated epicatechin was not significantly different from that of epicatechin, suggesting that the H-donating antioxidant activity is not the primary mechanism. The mode of protection by both native compound and the O-methylated metabolite remains unclear. 3'-O-methyl epicatechin would be expected to enter cells more freely than epicatechin itself due to its higher lipophilicity and higher partition coefficient. However, the possibility cannot be excluded that demethylation might occur intracellularly by the action of cytochrome P450 or etherases cleaving the O-methylated metabolite to epicatechin. In contrast, it is conceivable that epicatechin may be methylated intracellularly by catechol-O-methyl transferases. Consequently, the redox potential *per se* may not be the fundamental feature when determining the ability of specific phenolics to protect against either oxidative or other insults to cells. The findings support the notion that studies investigating the action of flavanols, and other flavonoids, should be undertaken with the *in vivo* conjugated and/or metabolized forms of the compounds, as well as native forms, in order to assess potential mechanisms of action in cell protection.

These findings exemplify the possibility that the ability of flavonoids, such as epicatechin, to provide health benefits may not necessarily be dependent on the ability of the native compound to act as a scavenger of free radicals or reactive oxygen- and reactive nitrogen-species, but by the ability of their metabolites to interact with cell signalling cascades, to influence the cell at a transcriptional level and to down-regulate pathways leading to cell death.

5. Biotransformation of Flavonoids

Although the pool of data demonstrating the *in vitro* effects of flavonoids as antioxidants or modulators of protein functions is large, only little is known about the antioxidant potential and bioactivity of *in vivo* flavonoid metabolites. This is surprising since early investigations in the 1950s/1960s in mammals already indicated that most of the flavonoids are conjugated, metabolized or degraded mainly in the liver or by the colonic microflora (reviewed in Ref. 103).

More recently, investigations of the sites and mechanisms of metabolism of flavonoids (Fig. 3) have demonstrated that flavonoids can be metabolized in the small intestine (reviewed in Ref. 104). Deglycosylation of monoglucosides via glucosidases has been described,[105, 106] although cyanidin-3-glucoside has been

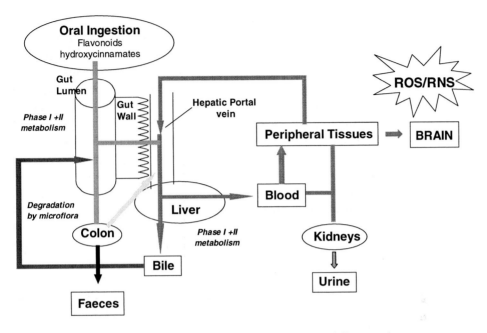

Fig. 3. Sites of metabolism of orally ingested flavonoids.

shown to be directly absorbed and distributed to the blood and excreted in the urine as the glucoside.[107–110] Studies using the isolated rat small intestine have also demonstrated the glucuronidation of flavonoids and phenolics by small intestinal UDP-glucuronyl transferases.[105, 111] In addition, it has been established that intestinal catechol-O-methyltransferases are capable of methylating flavonoids with an orthodihydroxy catechol structure and that the major metabolites of epicatechin include methylated and glucuronidated forms.[101] As well as small intestinal transformations, metabolism in the colon can result in extensive modifications including hydrolysis, oxidation and ring cleavage, forming secondary phenolic metabolites.[103] Thus, the direct effects of flavonoids on the gut microflora and the potential bioactivity of the small phenolic degradation products might have to be taken into account when considering the bioactivity of flavonoids *in vivo*.

Current interest in the metabolism of flavonoids and hydroxycinnamates is leading to many publications reporting the occurrence of methylated, glucuronidated and sulphated products in blood plasma, urine and faeces. Thus, the potential bioactivity of flavonoids *in vivo* may derive from their conjugates and metabolites and not necessarily from the ingested dietary forms.

In view of their powerful antioxidant properties *in vitro*,[45, 46, 57] many studies have examined the absorption and bioavailability of the (epi)catechin flavanols, monomeric members of the flavanol family of polyphenols, components of green tea and red grapes. Their oligomeric forms, the procyanidins, are key constituents of *theobroma cacao* (chocolate), red wine and apples. Recent work by Okushio *et al.*[112] reported that both the O-methylated form and glucuronidated conjugates could be detected in rat urine after oral administration of epicatechin. Furthermore, epicatechin-5-O-β-glucuronide and catechin-5-O-β-glucuronide were measured in plasma, bile and urine of rats after consumption of epicatechin and catechin respectively.[113] In humans, increases in 3'-O-methyl-catechin, sulphate and glucuronide metabolites in plasma were observed after consumption of either red wine or de-alcoholized red wine.[114] Procyanidin oligomers of epicatechin/catechin would not be expected to be absorbed via the small intestine by virtue of their size, rather cleavage in the large intestine is predicted to be more likely. It has also been proposed on the basis of model studies that procyanidin oligomers might be cleaved in the acidic environment of the gastric lumen to epicatechin and its dimers, the more absorbable forms.[115] Ingestion of chocolate, rich in the oligomers of epicatechin, led to the appearance of epicatechin glucuronide and sulphate in plasma and urine, with small amounts of the methylated compound.[70] Thus the question arises as to whether these metabolites cross the blood-brain barrier, whether glucuronides have access to the brain due to their polarity or are hydrolyzed prior to uptake, or whether the methylated compounds are more likely to enter the brain due to their higher lipophilicity. Studies applying the radio-labelled tea polyphenol, epigallocatechin gallate, have demonstrated wide distribution in mouse tissues, including the brain.[116]

6. Interactions of Flavonoids with Cells Derived from the CNS

Antioxidant therapy has been shown to be beneficial in neurological disorders. While flavonoids *per se* have been implicated in protection against central nervous system disorders, knowledge of their abilities to cross the blood/brain barrier and their accessibility to different regions of the brain is scant. Recently, oral administration of (–)-epicatechin and resultant examination of the consequences for protection against neuronal death following transient ischaemia and reperfusion in gerbil brains has been undertaken. The results suggest that (–)-catechin, or a conjugate or metabolite derived therefrom crossed the blood-brain barrier and delayed neuronal death by antioxidant mechanisms.[17] Information is, however, accumulating on the potential mechanisms of action of dietary phenolic compounds in neuroprotection in *ex vivo* cell models. Experiments on oxLDL-induced neuronal injury in primary (striatal) neuronal cultures demonstrated the neuroprotective effects of epicatechin, kaempferol and cyanidin (Fig. 4). These flavonoid aglycones were able to attenuate neuronal damage including DNA fragmentation, loss of mitochondrial function and membrane integrity. In the same study ascorbic acid and hydroxycinnamates were shown to be only weakly neuroprotective even when applied at ten times higher concentrations than the flavanoids. (+)-Catechin,

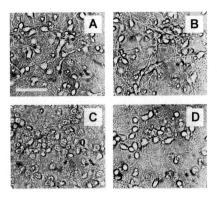

Fig. 4

quercetin and the stilbene phenolic *trans* resveratrol are all capable of attenuating hippocampal cell death and the accumulation of intracellular reactive oxygen species, via a mechanism that does not involve the inhibition of cyclooxygenases or lipoxygenases.[117]

Quercetin, the aglycone of the polyphenol found in many kinds of fruit and vegetables, has been extensively studied for its effects in protecting cells against neurotoxic insults induced by oxidative stress. Wang and Joseph[118] have investigated structure-activity relationships of flavonoids in protecting cells against oxidative stress. Specifically, the demonstration that, of the compounds studied, quercetin was the most effective in reducing oxidative stress in PC12 cells induced by hydrogen peroxide and in protecting against calcium dysregulation has led to the interpretation that the catechol structure in the B ring, the feature most responsible for the H-donating activity, along with the 2,3-double bond in the C ring were crucial for protection. Other studies suggesting a role for the antioxidant function of this polyphenol have demonstrated the ability of quercetin to increase the amount of available nitric oxide by the elimination of superoxide radical during reperfusion, applying a model of global forebrain ischaemia and reperfusion model.[119] Another approach to cellular oxidative stress has been the induction of glutathione depletion by buthionine sulphoximine. Exposure of sensory ganglion neurons to quercetin prevented cell death provoked by glutathione depletion, reducing oxidative injury while the GSH levels remained depressed.[120]

The phenolic constituents of Gingko biloba extract (Egb 761) have received enormous attention in recent years as neuroprotective agents. Its therapeutic value in reducing symptoms of decline in mental function has been a subject of many investigations. For example, it has been suggested that Gingko biloba extract protects against neuronal decline by facilitating the uptake of neurotransmitters,[121] through its role as a platelet activating factor antagonist,[121, 122] by reducing episodes of ischaemia-reperfusion and in preventing dysfunction of the mitochondrial respiratory chain.[123] Its preventive effects on apoptosis in neuronal cells induced by oxidative stress have also been reported.[123, 124] Numerous studies have shown that Gingko biloba extract is a nitric oxide scavenger with neuroprotective properties, but the underlying mechanisms are unclear. Studies involving hippocampal cells exposed to nitric oxide-induced toxicity have revealed that the protective properties of Egb 761 are not only due to the antioxidant properties of its flavonoid constituents but also to their ability to inhibit nitric oxide-induced protein kinase C activity.[16] Furthermore, pre-treatment of hippocampal neurons with Gingko biloba extract prior to beta-amyloid exposure, protected against cell death, reactive oxygen species accumulation and apoptosis, events induced by beta-amyloid. This again implicates, at least partially, the antioxidant properties of the extract (and its flavonoid components) in neuroprotection and highlights its potential against neurodegeneration in diseases such as Alzheimer's disease.[117]

Recent studies have suggested kaempferol as a potential neuroprotectant derived from Gingko biloba leaves through its activity as a monoamine oxidase inhibitor *in vitro*.[125] However it is not clear the extent to which this flavonol is absorbed from ingested Gingko, the nature of its conjugates and metabolites, whether or not these might cross the blood-brain barrier, in what forms and whether these forms exert similar neuroprotective properties to the native aglycone.

7. Mechanisms of Lipid Peroxide-Induced Neuronal Dysfunction Involving Mitogen Activated Kinases

Investigations involving oxidized low-density lipoprotein (oxLDL), which contains typical products of lipid peroxidation, such as LOOH, 4-HNE or oxysterols, have established their neurotoxic potential.[39, 40, 126, 127] Our recent studies support this notion by demonstrating that oxLDL enters primary cultured striatal neurons and induces neuronal cell death.[128] Pre-treatment of neurons with low micromolar concentrations of flavonoids, in particular epicatechin, kaempferol and cyanidin, protected against oxLDL-induced neuronal death.[129] Furthermore, 3'-O-methyl epicatechin, one of the metabolites of epicatechin *in vivo*, also elicited protective effects against oxidative stress-induced apoptotic cell death confirming that the mechanism of neuronal protection is not necessarily due to the conventional hydrogen-donating properties.[130]

More recent research has begun to define possible sites of action for flavonoids and focused on the potential roles of MAP kinase signalling cascades. Early data demonstrate that oxLDL mediates neuronal cell death involving a calcium-dependent activation of ERK-1/2 and JNK and that these effects are attenuated by pre-treatment with low micromolar concentrations of epicatechin.[128] Thus, the question as to whether the mechanisms of the neuroprotective effects of flavonoids are due to actions involving cell signalling components that have been linked to neurotoxicity is now being addressed.

Acknowledgments

This research was supported by a European Union Fifth Framework RTD Programme Grant (grant no. QLK4-1999-01590) and the Biological and Biotechnological Sciences Research Council.

References

1. Halliwell, B. (1992). Reactive oxygen species and the central nervous system. *J. Neurochem.* **59**: 1609–1623.
2. Behl, C. (1998). Alzheimer's disease and oxidative stress: implications for novel therapeutic approaches. *Prog. Neurobiol.* **57**: 301–323.
3. Zhang, Y., Dawson, V. L. and Dawson, T. M. (2000). Oxidative stress and genetics in the pathogenesis of Parkinson's disease. *Neurobiol. Dis.* **7**: 240–250. Yuan, J. and Yanker, B. A. (2000). Apoptosis in the nervous system. *Nature* **407**: 802–809.
4. Alexi. T., Borlongan, C. V., Faull, R. L. F., Williams, C. E., Clark, R. G., Gluckman, P. D. and Hughes, P. E. (2000). Neuroprotective strategies for basal ganglia degeneration: Parkinson's and Huntington's diseases. *Prog. Neurobiol.* **60**: 409–470.
5. Coyle, J. T. and Puttfarcken, P. (1993). Oxidative stress, glutamate and neuro-degenerative disorders. *Science* **262**: 689–695.
6. Mohanakumar, K. P., de Bartolomeis, A., Wu, R. M., Yeh, K. J., Sternberger, L. M., Peng, S. Y., Murphy, D. L. and Chiueh, C. C. (1994). Ferrous-citrate complex and nigral degeneration: evidence for free-radical formation and lipid peroxidation. *Ann. NY Acad. Sci.* **738**: 392–399.
7. Keller, J. N. and Mattson, M. P. (1998). Roles of lipid peroxidation in modulation of cellular signalling pathways, cell dysfunction, and death in nervous system. *Rev. Neurosci.* **9**: 105–116.
8. Salsman, S., Gabita, S. P., Nguyen, X., Mou, S., Szweda, L., Humphries, K., Markesbery, W. R., Floyd, R. A. and Hensley, K. (2000). Characterization of 4-hydroxynonenal-reactive proteins in the Alzheimer's disease brain. *Free Radic. Biol. Med.* **29**: S128.
9. Ben-Yoseph, O., Boxer, P. A. and Ross, B. D. (1996). Assessment of the role of the glutathione and pentose phosphate pathways in the protection of primary cerebrocortical cultures from oxidative stress. *J. Neurochem.* **66**: 2329–2337.

10. Joseph, J. A., Shukitt-Hale, B., Denisova, N. A., Prior, R. L., Cao, G. H., Martin, A., Taglialatela, G. and Bickford, P. C. (1998a). Long-term dietary strawberry, spinach, or vitamin E supplementation retards the onset of age-related neuronal signal-transduction and cognitive behavioural deficits. *J. Neurosci.* **18**: 8047–8055.

11. Joseph, J. A., Denisova, N. A., Fisher, D., Shukitt-Hale, B., Bickford, P. C., Prior, R. and Cao, G. H. (1998b). Age-related neurodegeneration and oxidative stress — putative nutritional intervention. *Neurol. Clin.* **16**: 747–756.

12. Joseph, J. A., Shukitt-Hale, B., Denisova, N. A., Bielinski, D., Martin, A., McEwen, J. J. and Bickford, P. C. (1999). Reversals of age-related declines in neuronal signal transduction; cognitive and behavioural deficits with blueberry, spinach or strawberry dietary supplementation. *J. Neurosci.* **19**: 8114–8121.

13. Cantutui-Castelvetri, I., Shukitt-Hale, B. and Joseph, J. A. (2000). Neuro-behavioral aspects of antioxidants in aging. *Int. J. Develop. Neurosci.* **18**: 367–381.

14. Cockle, S. M., Kimbe, S. and Hindmarch, I. (2000). The effects of Gingko biloba extract (LI 1370) supplementation on activities of daily living in free living older volunteers: a questionnaire survey. *Human Psychoparm. Clin.* **15**: 227–235.

15. Bastianetto, S., Zheng, W. H. and Quirion, R. (2000). Neuroprotective abilities of resveratrol and other red wine constituents against nitric oxide-related toxicity in cultured hippocampal neurons. *Br. J. Pharmacol.* **131**: 711–720.

16. Bastianetto, S., Zheng, W. H. and Quirion, R. (2000). The Gingko biloba extract (Egb 761) protects and rescues hippocampal cells against nitric oxide-induced toxicity: involvement of its flavonoids constituents and protein kinase C. *J. Neurochem.* **74**: 2268–2277.

17. Inanami, O., Watanabe, Y., Syuto, B., Nakano, M., Tsuji, M. and Kuwabara, M. (1998). Oral administration of (–)catechin protects against ischemia re-perfusion-induced neuronal death in the gerbil. *Free Radic. Res.* **29**: 359–365.

18. Bickford, P. C., Gould, T., Briederick, L., Chadman, K., Pollock, A., Young, D., Shukitt-Hale, B. and Joseph, J. (2000). Antioxidant-rich diets improve cerebellar physiology and motor learning in aged rats. *Brain Res.* **866**: 211–217.

19. Sun, G. Y., Xia, J. M., Draczynska-Lusiak, B., Simonyi, A. and Sun, A. Y. (1999). Grape polyphenols protect neurodegerative changes induced by chronic ethanol administration. *Neuroreport* **10**: 93–96.

20. Gotz, M. E., Kunig, G., Riederer, P. and Youdim, M. B. H. (1994). Oxidative stress: free radical production in neuronal degeneration. *Pharmacol. Ther.* **63**: 37–122.

21. Jenner, P. and Olanow, W. C. (1996). Oxidative stress and the pathogenesis of Parkinson's disease. *Neurology* **47**: S161–S170.

22. Jenner, P. and Olanow, W. C. (1998). Understanding cell death in Parkinson's disease. *Ann. Neurol.* **44**: S72–S84.
23. Dexter, D., Carter, C., Agid, F., Agid, Y., Lees, A. J., Jenner, P. and Marsden, C. D. (1986). Lipid-peroxidation as a cause of nigral cell death in Parkinson's disease. *Lancet* **2**: 639–640.
24. Jenner, P., Dexter, D. T., Sian, J., Schapira, A. H. V. and Marsden, C. D. (1992). Oxidative stress as a cause of nigral cell-death in Parkinson's disease and incidental Lewy body disease. *Ann. Neurol.* **32**: S82–S87.
25. McGeer, E. G. and McGeer, P. L. (1998). The importance of the inflammatory mechanisms in Alzheimer's disease. *Exp. Gerontol.* **33**: 371–378.
26. McGeer, E. G. and McGeer, P. L. (1995). The inflammatory response system of brain: implications for therapy of Alzheimer and other neurodegenerative diseases. *Brain Res. Rev.* **21**: 195–218.
27. Stahel, P. F., Morganti-Kossmann, M. C. and Kossmann, T. (1998). The role of the compliment system in traumatic brain injury. *Brain Res. Rev.* **27**: 243–256.
28. Nakanishi, S., Nakajima, Y., Masu, M., Ueda, Y., Nakahara, K., Watanabe, D., Yamaguchi, S., Kawabata, S. and Okada, M. (1998). Glutamate receptors: brain function and signal transduction. *Brain Res. Rev.* **26**: 230–235.
29. Bonfoco, E., Krainc, D., Ankarcrona, M., Nicotera, P. and Lipton, S. A. (1995). Apoptosis and necrosis: two distinct events induced, respectively, by mild and intense insults with N-methyl-D-aspartate or nitric oxide/superoxide in cortical cell cultures. *Proc. Natl. Acad. Sci. USA* **92**: 7162–7166.
30. Sturchler-Pierrat, C., Abramowski, D., Duke, M., Wiederhold, K. H., Mistle, C., Rothacher, S., Ledermann, B., Burki, K., Frey, P., Paganetti, P. A., Waridel, C., Calhoun, M. E., Jucker, M., Probst, A., Staufenbiel, M., Sugawa, M., Ikeda, S., Kushima, Y., Takashima, Y. and Cynshi, O. (1997). Oxidized low density lipoprotein caused neuron cell death. *Brain Res.* **761**: 165–172.
31. Weldon, D. T., Rogers, S. D., Ghilardi, J. R., Finke, M. P., Cleary, J. P., O'Hare, E., Elser, W. P., Maggio, J. E. and Mantyh, P. W. (1998). Fibrillar beta-amyloid induces microglial phagocytosis, expression of inducible nitric oxide synthase, and loss a select population of neurons in the rat CNS in vivo. *J. Neurosci.* **18**: 2161–2173.
32. Squadrito, G. L. and Pryor, W. A. (1998). Oxidative chemistry of nitric oxide: the roles of superoxide, peroxynitrite and carbon dioxide. *Free Radic. Biol. Med.* **25**: 392–403.
33. Halliwell, B., Zhao, K. and Whiteman, M. (1999). Nitric oxide and peroxynitrite. The ugly, the uglier and the not so good: a personal view of recent controversies. *Free Radic. Res.* **31**: 651–669.
34. Williamson, K. S., Markesbery, W., Floyd, R. A., Cudkowicz, M. and Hensley, K. (2000). 5-nitro-γ-tocopherol is a biomarker for nitartive stress in human neurological disease. *Free Radic. Biol. Med.* **29**: S129.

35. Samanta, S., Morgan, M., Perkinton, M. S. and Williams, R. J. (1998). Hydrogen peroxide enhances signal-responsive arachidonic acid release from neurons: role of mitogen-activated protein kinase. *J. Neurochem.* **70**: 2082–2090.

36. Satoh, T., Nakatsuka, D., Watanabe, Y., Nagata, I., Kikuchi, H. and Namura, S. (2000). Neuroprotection by MAPK/ERK kinase inhibition with U0126 against oxidative stress in a mouse neuronal cell line and rat primary cortical neurons. *Neurosci Lett.* **288**: 163–166.

37. Stanciu, M., Wang, Y., Kentor, R., Burke, N., Watkins, S., Kress, G., Reynolds, I., Klann, E., Angiolieri, M. R., Johnson, J. W. and de Franco, D. B. (2000). Persistent activation of ERK contributes to glutamate-induced oxidative toxicity in a neuronal cell line and primary cortical neuron cultures. *J. Biol. Chem.* **275**: 12 200–12 206.

38. Camandola, S., Poli, G. and Mattson, M. P. (2000). The lipid peroxidation product 4-hydroxy-2,3-nonenal increases AP-1 binding activity through caspase activation in neurons. *J. Neurochem.* **74**: 159–168.

39. Kruman, I., Bruce-Keller, A. J., Bredesen, D., Waeg, G. and Mattson, M. P. (1997). Evidence that 4-hydroxynonenal mediates oxidative stress-induced neuronal apoptosis. *J. Neurosci.* **17**: 5089–5100.

40. Mark, R. J., Lovell, M. A., Markesbery, W. R., Uchida, K. and Mattson, M. P. (1997). A role of 4-hydroxynonenal in disruption of ion homeostasis and neuronal death induced by amyloid β-peptide. *J. Neurochem.* **68**: 255–264.

41. Soh, Y. J., Jeong, K. S., Lee, I. J., Bae, M. A., Kim, Y. C. and Song, B. J. (2000). Selective activation of the *c-jun* n-terminal protein kinase pathway during 4-hydroxynonenal-induced apoptosis of PC12 cells. *Mol. Pharmacol.* **58**: 535–541.

42. Sichel, G., Corsaro, C., Scalia, M., di Billio, A. and Bonomo, R. P. (1991). In vitro scavenging activity of some flavonoids and melanin against superoxide. *Free Radic. Biol. Med.* **11**: 1–8.

43. Bors, W., Michel, C. and Schikora, S. (1995). Interaction of flavonoids with ascorbate and determination of their univalent reduction potentials: a pulse radiolysis study. *Free Radic. Biol. Med.* **19**: 45–52.

44. Jovanovic, S. V., Steenken, S., Simic, M. G. and Hara, Y. (1998). Antioxidant properties of flavonoids: reduction potentials and electron transfer reactions of flavonoid radicals. *In* "Flavonoids in Health and Disease" (C. Rice-Evans, and L. Packer, Eds.), Marcel Dekker, New York.

45. Salah, N., Miller, N. J., Paganga, G., Tijburg, L., Bolwell, G. P. and Rice-Evans, C. (1995). *Arch. Biochem. Biophys.* **332**: 339–346.

46. Rice-Evans, C. A., Miller, N. J. and Paganga, G. (1996). Structure-antioxidant activity relationship of flavonoids and phenolic acids. *Free Radic. Biol. Med.* **20**: 933–956.

47. Brown, J. E., Khodr, H., Hider, R. C. and Rice-Evans, C. (1998). Structural dependence of flavonoid interactions with copper ions: implications for their antioxidant properties. *Biochem. J.* **330**: 1173–1178.
48. Morel, I., Cillard, P. and Cillard, J. (1998). Flavonoid-metal interactions in biological systems. *In* "Flavonoids in Health and Disease" (C. Rice-Evans, and L. Packer, Eds.), Marcel Dekker, New York.
49. Tournaire, C., Croux, S., Maurette, M.-T., Beck, I., Hocquaux, M., Braum, A. M. and Oliveros, E. (1993). Antioxidant activity of flavonoids: efficiency of singlet oxygen quenching. *J. Photochem. Photobiol.* **B19**: 205–208.
50. Jovanovic, S. V., Hara, Y., Steenken, S. and Simic, M. G. (1995). Antioxidant potential of gallocatechins. A pulse radiolysis and laser photolysis study. *J. Am. Chem. Soc.* **117**: 9881–9889.
51. Castelluccio, C., Paganga, G., Melikian, N., Bolwell, G. P., Pridham, J., Sampson, J. and Rice-Evans, C. (1995). Antioxidant potential of intermediates in phenyl propanoid metabolism in higher plants. *FEBS Lett.* **368**: 188–192.
52. Hirano, R., Osakabe, N., Iwamoto, T., Matsumoto, A., Natsume, M., Takizawa, T., Igarashi, O., Itakura, H. and Kondo, K. (2000). Antioxidant effects of polyphenols in chocolate on low desity lipoprotein both in vitro and in vivo. *J. Nutri. Sci. Vitaminol.* **46**: 199–204.
53. Boyle, S. P., Dobson, V. L., Duthies, S. J., Kyle, J. and Collins, A. R. (2000). Absorption and DNA protective effects of flavonoid glycosides from an onion meal. *Eur. J. Nutri.* **39**: 213–223.
54. Anderson, R., Amarasinghe, C., Fisher, L. J., Mak, W. B. and Packer, J. E. (2000). Reduction in free radical-induced strand breaks and base damage through fast chemical repair by flavonoids. *Free Radic. Res.* **33**: 91–103.
55. Russo, A., Acquaviva, R., Campisi, A., Sorrenti, V., di Giacomo, C., Virgata, G., Barcellona, M. L. and Vanella, A. (2000). Bioflavonoids as antiradicals, antioxidants and DNA cleavage protectors. *Cell Biol. Toxicol.* **16**: 91–98.
56. Pannala, A., Rice-Evans, C., Halliwell, B. and Singh, S. (1997). Inhibition of peroxynitrite-mediated tyrosine nitration by catechin polyphenols. *Biochem. Biophys. Res. Commun.* **232**: 164–168.
57. Pannala, A., Razzaq, R., Halliwell, B. and Singh, S. and Rice-Evans, C. (1998). Inhibition of peroxynitrite dependent tyrosine nitration by hydroxycinnamic acids: nitration or electron donation. *Free Radic. Biol. Med.* **24**: 594–606.
58. Kerry, N. and Rice-Evans, C. (1999). Inhibition of peroxynitrite-mediated oxidation of dopamine by flavonoid and phenolic antioxidants and their structural relationships. *J. Neurochem.* **73**: 247–253.
59. Li, H., Shen, X. M. and Dryhurst, G. (1998). Brain mitochondria catalyze the oxidation of 7-(2-aminoethyl)-3,4-dihydro-5-hydroxy-2H-1,4-benzothiazine-3-carboxylic acid (DHBT-1) to intermediates that irreversibly inhibit complex I and scavenge glutathione, potential relevance to the pathogenesis of Parkinson's disease. *J. Neurochem.* **71**: 2049–2062.

60. Spencer, J. P. E., Jenner, P., Daniel, S. E., Lees, A. J., Marsden, C. D and Halliwell, B. (1998). Conjugates of catecholamines with cysteine and GSH in Parkinson's disease. Possible mechanisms of formation involving reactive oxygen species. *J. Neurochem.* **71**: 2112–2122.

61. Dangles, O., Dufour, C. and Fargeix, G. (2000). Inhibition of lipid peroxidation by quercetin and quercetin derivatives: antioxidant and prooxidant effects. *J. Chem. Soc. Perkin Trans. 2.* **6**: 1215–1222.

62. Giugliano, D. (2000). Dietary antioxidants for cardiovascular prevention. *Nutri. Metab. Cardiovasc. Dis.* **10**: 38–44.

63. Liebgott, T., Miollan, M., Berchadsky, Y., Drieu, K., Culcasi, M. and Pietri, S. (2000). Complementary cardioprotective effects of flavonoid metabolites and terpenoid constituents of Gingko biloba extract during ischemia and reperfusion. *Basic Res. Cardiol.* **95**: 368–377.

64. Sasazuki, S., Kodama, H., Yoshimasu, K., Liu, Y., Washio, M., Tanaka, K. *et al.* (2000). Relation between green tea consumption and severity of coronary atherosclerosis among Japanese men and women. *Ann. Epidemiol.* **10**: 401–408.

65. Caltarigone, S., Rossi, C., Poggi, A., Ranelletti, F. O., Natali, P. G., Brunetti, M., Aiello, F. B. and Piantelli, M. (2000). Flavonoids apigenin and quercetin inhibit melanoma growth and metastatic potential. *Int. J. Cancer* **87**: 595–600.

66. Zi, X. l., Zhang, J. C., Agarwal, R. and Pollak, M. (2000). Silibinin up-regulates insulin-like growth factor-binding protein-3 expression and inhibits proliferation of androgen-independent prostate cancer cells. *Cancer Res.* **60**: 5617–5620.

67. Fischer, P. M. and Lane, D. P. (2000). Inhibitors of cyclin-dependent kinases as anti-cancer therapeutics. *Curr. Med. Chem.* **7**: 1213–1245.

68. Elattar, T. and Virji, A. S. (2000). Effect of tea polyphenols on growth of oral squamous carcinoma cells in vitro. *Anticancer Res.* **20**: 3459–3465.

69. Hanninen, O., Kaartinen, K., Rauma, A. L., Nenonen, M., Torronen, R., Hakkinen, S., Aldercreutz, H. and Laasko, J. (2000). Antioxidants in vegan diet and rheumatic disorders. *Toxicology* **155**: 45–53.

70. Baba, S., Osakabe, N., Yasuda, A., Matsume, M., Takizawa, T., Nakamura, T. and Terao, J. (2000). Bioavailability of (−)-epicatechin upon intake of chocolate and cocoa in human volunteers. *Free Radic. Res.* **33**: 635–641.

71. Langley-Evans, S. C. (2000). Consumption of black tea elicits an increase in plasma antioxidant potential in humans. *Int. J. Food Sci. Nutri.* **51**: 309–315.

72. Rein, D., Paglieroni, T. G., Pearson, D. A., Wun, T., Scmitz, H., Gosselin, R. and Keen, C. (2000). Cocoa and wine polyphenols modulate platelet activation and function. *J. Nutri.* **130**: 2120S–2126S.

73. Pignatelli, P., Pulcinelli, F. M., Celestini, A., Lenti, L., Ghiselli, A., Gazzianiga, P. P. and Violo, F. (2000). The flavonoids quercetin and catechin synergistically inhibit platelet function by antagonizing the intracellular production of hydrogen peroxide. *Am. J. Clin. Nutri.* **72**: 1150–1155.

74. Youdim, K. A., Martin, A. and Joseph, J. A. (2000). Incorporation of the elderberry anthocyanins by endothelial cells increases protection against oxidative stress. *Free Radic. Biol. Med.* **29**: 51–60.

75. De Maat, M. P. M., Pijl, H., Kluft, C. and Princen, H. M. G. (2000). Consumption of black and green tea has no effect on inflammation, haemostasis and endothelial markers in smoking healthy adults. *Eur. J. Clin. Nutri.* **54**: 757–763.

76. Musconda, C. A. and Chipman, J. K. (1998). Quercetin inhibits hydrogen peroxide-induced NF-κB DNA binding activity and DNA damage in HepG2 cells. *Carcinogenesis* **19**: 1583–1589.

77. Tsai, S. H., Lin-Shiau, S. and Lin, J. K. (1999). Suppression of nitric oxide synthase and the down-regulation of the activation of NF-κB in macrophages by resveratrol. *Br. J. Pharmacol.* **126**: 673–680.

78. Manthey, J. A., Grohmann, K., Montanari, A., Ash, K. and Manthey, C. L. (1999). Polymethoxylated flavones derived from citrus suppress tumour necrosis factor-alpha expression in human monocytes. *J. Nat. Prod.* **62**: 441–444.

79. Gerritsen, M. E., Carley, W. W., Ranges, G. E., Shen, C. P., Phan, S. A., Ligon, G. F. and Perry. C. A. (1995). Flavonoids inhibit cytokine-induced endothelial cell adhesion protein gene expression. *Am. J. Pathol.* **147**: 278–292.

80. Panes, J., Gerritsen, M. E., Anderson, D. C., Miyasaka, M. and Granger, D. M. (1996). Apigenin inhibits tumour necrosis factor-induced intercellular adhesion molecule-1 upregulation in vivo. *Microcirculation* **3**: 279–286.

81. Soriani, M., Rice-Evans, C. A. and Tyrrell R. M. (1998). Modulation of the UVA activation of haem oxygenase, collagenase and cyclooxygenase gene expression by epigallocatecgin in human skin cells. *FEBS Lett.* **439**: 253–257.

82. Huang, Y. T., Huang, J. J., Lee, P. P., Ke, F. C., Huang, J. H., Kandaswani, C., Middleton, E. and Lee, M. T. (1999). Effects of luteolin and quercetin, inhibitors of tyrosine kinase, on cell growth and metastasis-associated properties in A431 cells over-expressing epidermal growth factor receptor. *Br. J. Pharmacol.* **128**: 999–1010.

83. Gamet-Payrastre, L., Maneati, S., Gratacap, M. P., Jacues, T., Chap, H. and Payrastre, B. (1999). Flavonoids and inhibition of PKC and PI3 kinase. *Genet. Pharmacol.* **32**: 279–286.

84. Agullo, G., Remesy, C. and Payrastre, B. (1997). Relationship between flavonoid structure and inhibition of phosphatidylinositol-3 kinase: a comparison with tyrosine kinase and protein kinase C inhibition. *Biochem. Pharmacol.* **53**: 1649–1657.

85. Agarwal, R., Katiyar, S. K., Lindgren, D. W. and Mukhtar, H. (1994). Inhibitory effect of silymarin, an antihepatotoxic flavonoid, on phorbol ester induced epidermal ornithine decarboxylase activity amd mRNA in SENCAR mice. *Carcinogenesis* **15**: 1099–1103.

86. Di Pietro, A., Godinot, C., Bouillant, M. L. and Gautheron, D. C. (1975). Pig heart mitochondrial ATPase: properties of purified and membrane-bound enzyme. Effects of flavonoids. *Biochimie* **57**: 959–967.

87. Barzilai, A. and Rahamimoff, H. (1983). Inhibition of Ca^{2+}-transport from synaptosomal vesicles by flavonoids. *Biochim. Biophys. Acta* **730**: 245–254.

88. Revuelta, M. P., Cantabrana, B. and Hidalgo A. (1997). Depolarization dependent effect of flavonoids in rat uterine smooth muscle contraction elicted by $CaCl_2$. *Genet. Pharmacol.* **29**: 847–857.

89. Lee, S. F. and Lin, J. K. (1997). Inhibitory effect of phytopolyphenols on TPA induced transformation, PKC activation, and *c-jun* expression in mouse fibroblast cells. *Nutri. Cancer* **28**: 177–183.

90. Boege, F., Straubt, T., Kehr, A., Boesenberg, C., Christiansen, K., Andersen, A., Jakob, F. and Koehrle, J. (1996). Selected novel flavones inhibit the DNA binding or the DNA religation step eukaryotic topoisomerase I. *J. Biol. Chem.* **271**: 2262–2270.

91. So, F. V., Guthrie, N., Chambers, A. F., Carroll, K. K. (1996). Inhibition of human breast cancer cell proliferation and delay of mammary tumorigenesis by flavonoids and citrus juices. *Nutri. Cancer* **26**: 167–181.

92. Zand, R., Jenkins, D. and Diamandis, E. (2000). Steroid hormone activity of flavonoids and related compounds. *Breast Cancer Res. Treatment* **62**: 35–49.

93. MacCarrone, M., Lorenzon, T., Guerrieri, P. and Agro, A. (1999). Resveratrol prevents apoptosis in K562 cells by inhibiting lipoxygenase and cyclooxygenase. *Eur. J. Biochem.* **265**: 27–34.

94. Jang, M., Cai, L., Udeani, G. O., Slowing, K., Thomas, C., Beecher, C., Fonmg, H. S., Farnsworth, N. R., Kinghorn, A. D., Mehta, R., Moon, R. C. and Pezzuto, J. M. (1997). Cancer chemoprotective activity of resveratrol, a natural product derived from grapes. *Science* **275**: 218–220.

95. Kobuchi, H., Roy, S., Sen, C. K., Nguyen, H. G. and Packer, L. (1999). Quercetin inhibits inducible ICAM-1 expression in human endothelial cells through the JNK pathway. *Am. J. Physiol. (Cell)* **277**: 403–411.

96. Kong, A.-N. T., Yu, R., Chen, C., Mandlekar, S. and Primiano, T. (2000). Signal transduction events elicited by natural products: role of MAPK and caspase pathways in homeostatic response and induction of apoptosis. *Arch. Pharm. Res.* **23**: 1–16.

97. Herdegen, T., Skene, P. and Bahr, M. (1997). The *c-jun* transcription factor — bipotential mediator of neuronal death, survival and regeneration. *Trends Neurosci.* **20**: 227–231.

98. Mielke, K. and Herdegen, T. (2000). JNK and p38 stress kinases — degenerative effectors of signal transduction-cascades in the nervous system. *Prog. Neurobiol.* **61**: 45–60.

99. Bu-Abbas, A., Clifford, M. N., Walker, R. and Ioannides, S. (1998). Contribution of caffeine and flavanols in the induction of phase II activites by green tea. *Food Chem. Toxicol.* **36**: 617–621.

100. Lee, S. F., Liang, Y. C. and Lin, J. K. (1995). Induction of phase II enzymes by green tea polyphenols. *Chem. Biol. Interact.* **98**: 283–301.
101. Kuhnle, G., Spencer, J. P. E., Schroeter, H., Shenoy, B., Debnam, E. S., Srai, S. K. S., Rice-Evans, C. A. and Hahn, U. (2000). Epicatechin and catechin are O-methylated and glucoronidated in the small instestine. *Biophys. Biochem. Res. Commun.* **277**: 507–512.
102. Spencer, J. P. E., Chaudry, F., Pannala, A. S., Srai, S. K., Debnam, E. and Rice-Evans, C. (2000). Decomposition of cocoa procyanidins in the gastric milieu. *Biochem. Biophys. Res. Commun.* **272**: 236–241.
103. Scheline, R. R. (1991). *CRC Handbook of Mammalian Metabolism of Plant Compounds*, pp. 267–290, CRC Press Inc., Baco Raton, FL.
104. Rice-Evans, C., Spencer, J. P. E., Schroeter, H. and Andreas, Rechner. (2000). Bioavailability of flavonoids and potential bioactive forms in vivo. *Drug Metab. Drug Interact.* **17**: 291–310.
105. Spencer, J. P. E., Chowrimootoo, G., Choudhury, R., Debnam, E. S., Srai, S. K. and Rice-Evans, C. (1999). The small intestine can both absorb and glucuronidate luminal flavonoids. *FEBS Lett.* **458**: 224–230.
106. Shimoi, K., Okada, H., Furugori, M., Goda, T., Takase, S., Suzuki, M., Hara, Y., Yamamoto, H. and Kinae, N. (1998). Intestinal absorption of luteolin and luteolin-7-O-β-glucoside in rats and humans. *FEBS Lett.* **438**: 220–224.
107. Matsumoto, H., Inaba, H., Kishi, M., Tominaga, S., Hirayama, M. and Tsuda, T. (2001). Orally administered delphinidin 3-rutinoside and cyaniding 3-rutinoside are directly absorbed in rats and humans and appear in the blood as the intact forms. *J. Agric. Food Chem.* **49**: 1546–1551.
108. Miyazawa, T., Nakagawa, K., Kudo, M., Muraishi, K. and Someya, K. (1999), Direct intestinal absorption of red fruit anthocyanins, cyaniding-3-glucoside and cyaniding 3,5-diglucoside in rats and humans. *J. Agric. Food Chem.* **47**: 1083–1091.
109. Tsuda, T., Horio, F. and Osawa. T. (1999). Absorption and metabolism of cyanidin-3-O-β-D-glucoside in rats. *FEBS Lett.* **449**: 179–182.
110. Lapidot, T., Harel, S., Granit, R. and Kanner, J. (1998). Bioavailability of red wine anthocyanins as detected in human urine. *J. Agric. Food Chem.* **46**: 4297–4302.
111. Kuhnle, G., Spencer, J. P. E. S., Chowrimootoo, G., Schroeter, H., Debnam. E. S., Srai, S. K., Rice-Evans, C. and Hahn, U. (2000). Resveratrol is absorbed in the small intestine as resveratrol glucuronide. *Biochem. Biophys. Res. Commun.* **272**: 212–217.
112. Okushio, K., Suzuki, M., Matsumoto, N., Nanjo, F. and Hara, Y. (1999). Identification of (–)-epicatechin metabolites and their metabolic fate in the rat. *Drug Metab. Dispos.* **27**: 309–316.
113. Harada, M., Kan, Y., Naoki, H., Fukui, Y., Kageyama, N., Nakai, M., Miki, W. and Kiso, Y. (1999). Identification of the major antioxidative metabolites in

biological fluids of the rat with ingested (+)-catechin and (−)-epicatechin. *Biosci. Biotechnol. Biochem.* **63**: 973–977.

114. Donovan, J. L., Bell., J. R., Kasim-Karakas, S., German, J. B., Walzem, R. L., Hansen, R. J. and Waterhouse, A. L. (1999). Catechin is present as metabolites in human plasma after consumption of red wine *J. Nutri.* **129**: 1662–1668.

115. Spencer, J. P. E., Schroeter, H., Kuhnle, G., Srai, S. K. S., Tyrrell, R. M. T., Hahn, U. and Rice-Evans, C. (2001). Epicatechin and its in vivo metabolite, 3′-O-methylepicatechin, protects human fibroblasts from oxidative stress-induced cell death involving caspase-3 activation. *Biochem. J.* **354**: 493–500.

116. Suganuma, M., Oktabe, S., Oniyama, M., Tada, Y., Ito, H. and Fujiki, H. (1998). Wide distribution of [3H](−)-epigallocatechin gallate, a cancer preventive tea polyphenol in mouse tissue. *Carcinogen* **19**: 1771–1776.

117. Bastianetto, S., Ramassamy, C., Dore, S., Christen, Y., Poirier, J. and Quirion, R. (2000). The Gingko biloba extract (Egb 761) protects hippocampal neurons against cell death induced by beta-amyloid. *Eur. J. Neurosci.* **12**: 1882–1890.

118. Wang, H. and Joseph, J. A. (1999). Structure-activity relationships of quercetin in antagonizing hydrogen peroxide-induced calcium dysregulation in PC12 cells. *Free Radic. Biol. Med.* **27**: 683–694.

119. Shutenko, Z., Henry, Y., Pinard, E., Seylaz, J., Potier, P., Berthet, F., Girard, P. and Sercombe, R. (1999). Influence of the antioxidant quercetin in vivo on the level of nitric oxide determined by electron paramagnetic resonance in rat brain during global ischemia and reperfusion. *Biochem. Pharmacol.* **57**: 199–208.

120. Skaper, S. D., Fabris, M., Ferrari, V., Carbonare, M. D. and Leon, A. (1997). Quercetin protects cutaneous tissue-associated cell types including sensory neurons from oxidative stress induced by glutathione depletion: cooperative effects of ascorbic acid. *Free Radic. Biol. Med.* **22**: 669–678.

121. Logani, S., Chen, M. C., Tran, T., Le, T. and Raffa, R. B. (2000). Actions of Gingko biloba related to potential utility for treatment of conditions involving cerebral hypoxia. *Life Sci.* **67**: 1389–1396.

122. Smith, M. A., Richey Harris, P. L., Sayre, L. M., Beckman, J. S. and Perry, G. (1997). Widespread peroxynitrite-mediated damage in Alzheimer's disease. *J. Neurosci.* **17**: 2653–2657.

123. Droy-Lefaix, M. T. (1997). Effect of the antioxidant action of Gingko biloba extract (Egb 761) on ageing and oxidative stress. *Age* **20**: 141–149.

124. Ni, Y. C., Zhao, B. L., Hou, J. W. and Xin, W. J. (1996). Preventive effect of Gingko biloba extract on apoptosis in rat cerebellar neuronal induced cells induced by hydroxyl radicals. *Neurosci. Lett.* **214**: 115–118.

125. Sloley, B. D., Urichuk, L., Morley, P., Durkin, J., Shan, J. J., Pang, P. K. T. and Coutts, R. T. (2000). Identification of kaempferol as a monoamine oxidase inhibitor and potential neuroprotectant in extracts of Gingko biloba leaves. *J. Pharm. Pharmacol.* **52**: 451–459.

126. Sugawa, M., Ikeda, S., Kushima, Y., Takashima, Y. and Cynshi, O. (1997). Oxidized low density lipoprotein caused neuron cell death. *Brain Res.* **761**: 165–172.
127. Keller, J. N., Hanni, K. B. and Markesbery, W. R. (1999). Oxidized low-density lipoprotein induces neuronal death: implications for calcium, reactive oxygen species and caspases. *J. Neurochem.* **72**: 2601–2609.
128. Schroeter, H., Williams, R. J., Matin, R., Iversen, L. and Rice-Evans, C. A. (2000a). Phenolic antioxidants attenuate neuronal cell death following uptake of oxidized low-density lipoprotein. *Free Radic. Biol. Med.* **29**: 1222–1233.
129. Schroeter, H., Williams, R. J. and Rice-Evans, C. (2000b). Activation of MAP kinase signalling cascades by oxidative stress and their modulation by flavonoids. *Free Radic. Biol. Med.* **29**: S128.
130. Schroeter, H., Spencer, J. P. E., Rice-Evans, C. and Williams, R. J. (2001). Flavonoids protect neurons from oxidized low-density lipoprotein — induced apoptosis involving *c-jun* N-terminal kinase (JNK), *c-jun* and caspase-3. *Biochem. J.* **358**: 547–557.

Chapter 30

Neurological Aging and Nutritional Intervention

Barbara Shukitt-Hale, Kuresh A. Youdim and James A. Joseph*

Barbara Shukitt-Hale, Kuresh A. Youdim and **James A. Joseph** • USDA-ARS, Human Nutrition Research Center on Aging at Tufts University, 711 Washington Street, Boston, MA 02111

*Corresponding Author.
Tel: (617) 556-3178, E-mail: jjoseph@hnrc.tufts.edu

1. Introduction

There has been a great deal of research which has documented the neuronal and behavioral changes that take place as a function of aging, even in the absence of neurodegenerative diseases such as Alzheimer's disease (AD) and Parkinson's disease (PD). The neuronal changes may include decrements in calcium homeostasis (e.g. see Ref. 1) and in the sensitivity of several neurotransmitter receptor systems, particularly: (a) adrenergic;[2] (b) dopaminergic;[3, 4] (c) muscarinic;[5–7] and (d) opioid.[8, 9] These decrements can be expressed, ultimately, as alterations in both motor[10–12] and cognitive behaviors that require the use of spatial learning and memory.[12–17] Age-related deficits in motor performance are thought to be the result of alterations in the striatal dopamine (DA) system[18] or in the cerebellum,[19, 20] while age-related memory decrements can occur from alterations in either the hippocampus (which mediates allocentric spatial navigation or place learning) or the striatum (which mediates egocentric spatial orientation or response/cue learning).[21–23] The alterations in motor function may include decreases in balance, muscle strength and coordination,[10, 11] while memory deficits appear to occur primarily in secondary memory systems and are reflected in the storage and retrieval of newly acquired information.[18, 24]

The mechanisms involved in these age-related neuronal and behavioral changes have yet to be determined. However, growing evidence appears to implicate increased susceptibility to the long-term effects of oxidative stress (OS) and inflammatory insults as major contributing factors. Age-related deficits in brain functions due to OS may be due, in part, to a decline in the normal antioxidant defense mechanisms in the brain with age[25–29] and to the vulnerability of the brain to the deleterious effects of oxidative damage.[30–32] With respect to inflammation, increases in inflammatory mediators (e.g. cytokines) known to be involved in the activation of glia cells and perivascular/parenchymal macrophages, as well as increased mobilization and infiltration of peripheral inflammatory cells into the brain, have been shown to produce deficits in behavior similar to those observed during aging.[33]

As the aged population continues to increase, the number exhibiting cognitive impairments is also increasing. With the rising health care costs required to care for this aged population, scientists have been searching to find ways to alleviate vulnerability to age-related oxidative and inflammatory insults, and thereby the age-related decrements in neuronal and behavioral functions. Thus, it is extremely important to explore methods to retard or reverse age-related neuronal deficits as well as their subsequent behavioral manifestations. With this in mind, previous findings in our laboratory and others found that protection against these aforementioned insults in the brain might be accomplished by increasing the dietary intake of fruits and vegetables, since they contain numerous compounds, i.e. phytochemicals, that exhibit a multitude of antioxidant and anti-inflammatory properties. This chapter will focus on the consequences of OS and inflammation in the aging brain and the effectiveness of nutritional interventions in reducing the effects of neuronal aging, specifically the role played by the phytochemical components found in foods. Recent reviews have highlighted the neuroprotective functions of vitamins E and C,[34, 35] alpha-lipoic acid,[34] Chinese remedies[34, 36] and dietary fatty acids,[37] and as such they will not be covered in this section.

2. Phytochemicals in Fruits and Vegetables

It is known that a poor diet can adversely impact cognitive performance in elderly persons.[38] In one study, correlations were observed in seniors between low cognitive scores and a low intake of fruits, vegetables, and overall energy, and high intakes of cholesterol, monosaturated and saturated fatty acids.[39] Although initially it was assumed that the vitamin component of fruits and vegetables was the primary source of dietary antioxidants, it is now well established that the phytochemical components also contribute substantially to the overall dietary antioxidant intake.[38] Fruits and vegetables are known to contain numerous phytochemicals, and until recently, their beneficial health effects on aging brain function had not been scientifically studied. One class of phytochemicals are polyphenolics, of which over 4000 different structures have been identified,[40] and these occur ubiquitously in foods of plant origin (e.g. fruits, vegetables, nuts, seeds, and grains), and have been shown to possess antioxidant, antiallergic, anti-inflammatory, antiviral, antiproliferative, and anticarcinogenic activities.[41–47] There are numerous experiments in the literature which have shown that the combinations of polyphenolic compounds found in fruits and vegetables may reduce the incidence of cardiovascular and cerebrovascular diseases.[48–51] and cancer.[52, 53] There is even evidence that extracts of single foods, such as lycopene from tomatoes,[54] can have some antitumor properties. Although there are not many studies which have investigated the role of dietary polyphenolics on brain function, the ones that are in the literature have found positive benefits following consumption.

Polyphenolics found in dietary supplements have been shown to provide beneficial effects following ingestion, although the literature in this area is also sparse. As examples, indirect evidence has been provided that shows that the consumption of the flavonoid glycosides of Ginkgo biloba increase antioxidant protection in aged animals and humans, thereby improving cognitive performance,[55] difficulties in concentration,[56, 57] and calcium-induced increases in the oxidative metabolism of brain neurons.[58, 59] More recent evidence suggests that administration of Ginkgo biloba (Egb 761) slowed the progression of AD.[60] There is also evidence that Panax ginseng and Ding lang root extract have positive benefits on behavior in aging (for review see Refs. 34 and 36). However, it is the effectiveness of foods in reducing the behavioral consequences of neuronal aging that is the focus of this chapter.

3. Fruits and Vegetables: Antioxidants and Anti-Inflammatories

Research at our institute has shown that fruit and vegetable extracts that have high levels of polyphenolics also displayed high total antioxidant activity, as assessed via the Oxygen Radical Absorbance Capacity Assay (ORAC).[61] Results found that foods with the highest ORAC activity included spinach and strawberries[62, 63] and blueberries.[64] Given that these fruits and vegetables have high antioxidant properties, our laboratory, together with colleagues at the University of Colorado, studied the effects of supplementation with these foods for their ability to forestall or reverse age-related changes in performance.[65–69]

3.1. Spinach and Strawberries

In our first study, using age-valid tests,[3, 12, 20] we examined whether long-term feeding from adulthood (6 months) to middle age (15 months) of Fischer 344 rats with a control diet (AIN-93) or supplemented (strawberry extract or spinach extract or vitamin E) diet would prevent age-related decrements in motor and cognitive behavior, as well as brain function.[67] These various extracts were supplemented into the diet so that each would supply the same antioxidant activity based upon mmol Trolox equivalents. A number of different parameters known to be sensitive to oxidative stress were prevented by the antioxidant diets including: (1) receptor sensitivity, i.e. oxotremorine-enhanced DA release in isolated striatal slices and cerebellar Purkinje cell activity; (2) calcium buffering capacity, i.e. the ability of striatal synaptosomes to extrude calcium following depolarization, deficits of which ultimately result in reduced cellular signaling and eventually cell death, and (3) changes in signal transduction assessed by carbachol-stimulated GTPase coupling/uncoupling in striatal membranes.[67] Efficient coupling/uncoupling is necessary to evoke a controlled on/off stimulatory signal or inhibitory signal, dependent upon G protein-receptor interactions. Cognition, which also has been shown to decline under models of oxidative stress,[70] was examined using performance in the Morris water maze (MWM), which measures spatial learning and memory by requiring the rat to use spatial cues to find a hidden platform that is located just below the surface of the water. Results indicated that the supplemented diets could prevent the onset of age-related deficits in brain function, as well as cognitive behavior, with spinach having the greatest effects.[67] There was no effect of the diets on motor behavior. These results suggested that phytochemicals present in antioxidant rich foods might be effective in forestalling functional age-related deficits due to oxidative stress.[67]

3.2. Blueberries

However, while prevention of these decrements is important, it was also critical to determine if these interventions would be effective in reversing these deficits once they occur, since the mean human lifespan is continuing to increase. Therefore, determinations of whether dietary supplementations might be effective in aged organisms is of paramount importance. Thus, in subsequent experiments[66, 68] we examined whether dietary supplementations (for 8 weeks) with spinach, strawberry or blueberry extracts in a control diet (AIN-93) would be effective in reversing age-related deficits in brain and behavioral function in aged (19 months) Fischer 344 rats. Blueberries were added to this study because they were found to have the highest antioxidant capacity (ORAC) of all fruits and vegetables tested.[64]

Overall, the results showed that these extracts, particularly blueberry, were effective in reversing age-related deficits in the various neuronal and behavioral parameters which improved following the long-term feeding mentioned earlier.[66, 68] The most striking finding, however, was that the blueberry diet also improved motor behavioral performance on the rod walking test, which assesses psychomotor coordination and the integrity of the vestibular system, and the accelerod task, which assesses balance and coordination. In fact, the changes seen with respect to motor behavior are particularly interesting since age-related decrements in motor behavior have been very resistant to reversal.[71] Blueberry extract appeared the most effective at reversing age-related deficits, in particular oxotremorine-enhanced DA release in isolated striatal slices, which was over 5-fold higher in the blueberry group than that released in stimulated striatal slices obtained from control rats.[68] This observation is particularly interesting since studies performed in our laboratory[6, 10, 72] have shown that maintaining the functional integrity of the striatum, which is high in dopaminergic neurons, may have a major impact on a number of behavioral parameters, especially motor behavior.

An important point in these initial studies was that these diets were supplemented based on equal antioxidant activity, as determined by the ORAC assay.[62] However, since the diets were not equally effective in preventing/reversing age-related changes, it seems that antioxidant activity alone was not predictive in assessing the potency of certain compounds at being beneficial against certain disorders affected by aging. In this regard, when two indices of oxidative stress, i.e. reactive oxygen species production as measured by DCF fluorescence and glutathione (GSH) levels in the brain, were measured,[68] there were no differences in GSH levels as a function of diet and DCF fluorescence was only modestly reduced by the diets. These results suggest that the supplementations only had moderate effects on oxidative stress parameters, which could not account for the efficacy of these treatments on the functional measures. Hence, there may be other properties of the polyphenolics aside from antioxidative, and differences in the polyphenolic composition of these extracts could account for the positive effects observed.

Based upon the idea that different components may afford different protective roles, we decided to examine the potency of extracts from two blueberry cultivars (tif-blue which is a cultivated Rabbiteye blueberry, and wild blueberry) which have different phytochemical compositions.[69] Furthermore, since earlier studies[65–68, 73] utilized dietary fruits and vegetable extracts supplemented in a purified AIN-93 diet that essentially lacked many natural ingredients, we reasoned that it was necessary to show that the same beneficial effects would be present if the blueberry was superimposed on an already well-balanced rodent diet ("chow") which was more representative of a balanced human diet. We found that addition of the blueberry extract to this rat diet composed of natural ingredients was also effective in reversing the effects of aging,[69] and the tif-blue diet had an additional beneficial effect on reference memory (i.e. long-term memory) that was not seen when the purified diet alone was used,[68] further showing the robustness of blueberry supplementation.[69] Therefore, the beneficial effects of fruits and vegetables on neuronal and behavioral function were seen even when superimposed on an already well-fortified, healthy diet.

The above findings suggest that, in addition to their known beneficial effects on cancer and heart disease, polyphenolics present in antioxidant rich foods may be beneficial in reversing the course of brain and behavioral aging. The mechanisms involved, as well as the polyphenolic "families" that produce the most beneficial effects on neuronal aging and behavior, are being investigated further in our laboratory, but it is still too early to identify the particular properties of the polyphenolics contained in these fruits and vegetables that may be involved in these effects. There is even some indication from recent findings (unpublished) that blueberry supplementation may have direct beneficial effects on skeletal muscles because certain calf muscles, i.e. gastrocnemius and soleus, taken from the aged animals receiving blueberry extract, displayed morphology similar to that seen in young animals, while those from aged controls displayed significant leukocyte invasion and fat accumulation.

The findings also suggest that, in addition to antioxidant properties, other actions may be elicited by the polyphenolics to ameliorate cognitive impairments, which may include anti-inflammatory actions. The potential anti-inflammatory actions of polyphenolics are supported by a number of studies that have shown them be able to antagonize arachidonic acid transport[74] and suppress the 5-lipoxygenase pathway[75] and thus reduce inflammatory responses,[76, 77] as well as regulate signal transduction processes/transcription factors involved in the regulation of inflammatory genes.[45] Furthermore, we have shown that the polyphenolics contained in blueberry extract are effective in reducing tumor necrosis factor induced increases in release of inflammatory mediators, such as adhesion molecules and chemoattractants, involved in regulation of leukocyte invasion to sites of inflammatory/oxidative insult, as well as enhancing protection against cell death in vascular endothelial cells (unpublished). Moreover, in preliminary experiments in which behavioral assessments were made in rats given dietary supplementation with blueberry extract and given chronic central administration of lipopolysaccharide to induce localized brain inflammation, the supplemented rats appeared to show less behavioral deficits on some motor tasks than non-supplemented rats (unpublished). There are also data which suggest that polyphenolics can increase membrane fluidity,[78-80] and a previous experiment in our laboratory[81] has shown that experimental decreases in membrane rigidity (via s-adenosyl-l-methionine) can ameliorate deficits in striatal signal transduction in old animals, a process thought to be important in the performance of spatial memory tasks. Taken together, these various other actions/properties elicited by polyphenolics may also be contributing factors in protection against impairments in brain functions.

3.3. Aged Garlic Extract

Aside from these studies discussed above, very few reports have investigated nutritional supplementation from fruits and vegetables with regard to their protection against oxidative stress and ultimately in their prevention against age-related memory and cognitive impairments. However, aged garlic extract (*Allium sativum*, which contains S-allycisteine, S-allymercaptocysteine, allicin and diallosulfides) administered at 2% (w/w) of the diet has been reported to exhibit beneficial effects towards cognitive impairments in a novel strain of senescence accelerated mouse (SAM).[82-85] Even though motor activity was not affected by garlic extract treatment, the SAM animals did show increased survival, less atrophy in the frontal brain areas, and better performance in both passive and active avoidance, which the authors attribute to the antioxidant properties of the aged garlic extract. Although only tested *in vitro*, garlic extract has also been shown to protect against oxidative stress-induced increases in thiobarbituric-acid-reactive substances (TBARS),[86, 87] and to promote the survival of neurons derived from various regions of the neonatal brain (i.e. increase neurogenesis).[88] Until these effects are tested in the whole animal, and until the aged garlic extract components are actually isolated and identified within the brain, the mechanisms involved in its protective effects are difficult to determine.

3.4. Red Bell Pepper

Red bell pepper (*Capsicum annuum* L.) has also been reported to evoke a beneficial effect on learning performance in SAM.[89] Following supplementation of an experimental diet which contained 20% (w/w) lyophilized powder of red bell pepper, SAMP8 mice showed greater acquisition in passive avoidance tasks compared to a control group given a non-supplemented diet. However, the authors did not examine any other parameters, which makes it difficult to begin to delineate possible mechanisms by which red bell pepper ameliorated the learning impairment in these senescence accelerated mice.

3.5. Tomatoes

One study[90] which investigated the relationship of plasma antioxidants to reduced functional capacity in the elderly found a strong positive association between plasma lycopene and the ability of the elderly nuns to perform self-care (e.g. dressing, feeding). Lycopene is a red pigment found in a small number of vegetables such as tomatoes, watermelons, and pink grapefruit, with the majority of lycopene being consumed in tomatoes and tomato-based products such as pizza.

4. Beverages: Antioxidants and Anti-Inflammatories

4.1. Wine

People who are moderate wine drinkers (3–4 glasses/day) have been shown to have a significantly lower incidence of dementia and AD compared to non-drinkers.[91–93] Orgogozo and colleagues[93] propose several possible protective mechanisms of action regarding moderate wine consumption, including antioxidant and/or anti-inflammatory properties, as well as elevation of plasma apolipoprotein E levels. Although a direct demonstration of the protective effects of wine, or a delineation of its mechanisms, has not been shown, moderate wine drinking can be seen as a possible preventive measure against senile dementia or AD.

4.2. Tea

Polyphenolics from tea, known to be good cancer preventative agents,[94, 95] have also been suggested to elicit potentially beneficial effects toward improving brain function. Tea catechins were shown to protect mice against the deleterious effects on memory induced by injection of glucose oxidase or cerebral ischemia,[96] possibly due to the potent antioxidant[97] and anti-inflammatory properties[98–100] of the tea polyphenolics. A recent pilot study showed that kombucha, a lightly fermented tea beverage, significantly inhibited weight gain, increased environmental awareness and responsiveness, and prolonged life in mice.[101] However, there is a health safety issue with drinking kombucha since it has been linked to health problems and fatalities in humans, and chronic drinking by mice in the current study contributed to longer spleens and enlarged livers.[101]

Furthermore, although not examined in the elderly but rather in young adults (mean age 29.2 years), black tea ingestion was associated with rapid increases in alertness and information processing capacity, and tea drinking throughout the day largely prevented diurnal patterns of performance decrements.[102] The authors concluded that the effects of tea consumption could not be entirely explained as the acute effects of caffeine ingestion, and felt that other factors, particularly the polyphenolics in the tea, might play a significant role in mediating these responses.

5. Conclusions

There are numerous studies showing that there is a significant relationship between fruit and vegetable intake, cardiovascular disease[48–51] and cancer.[52, 53] The findings reviewed here suggest that, in addition to these known beneficial effects, polyphenolics present in antioxidant rich foods may be beneficial in forestalling or reversing the course of neuronal and behavioral aging.

One other factor that may be important here is that many of the changes that we have reviewed above that occur with aging suggest that there is increased vulnerability to inflammatory and oxidative stress insults as a function of age. Interestingly, these changes in sensitivity in aging parallel, and are exaggerated in, neurodegenerative diseases such as Alzheimer's disease (for review see Ref. 103). As alluded to above, the polyphenolic compounds present in Ginkgo biloba and wine may be effective in forestalling the onset or reducing the severity of AD. Recent evidence also suggests that individuals with the ApoE phenotype 3/4 are more sensitive to the dietary intake of fats (R. P. Freidland, personal communication). Taken together, these findings suggest that these dietary considerations may be especially important with regard to aging and ApoE polymorphisms. They further suggest that the mechanisms and polyphenolic families involved in producing the positive effects in aging and behavior should be investigated further, with a view towards increasing the protection in both neurodegenerative disease and aging. Considering that, at present, the world population comprised of people over 65 years of age represents over 50% of all those who have ever lived who have attained this age, the determinations of whether these dietary supplementations might be effective in aged organisms are of paramount importance.

Acknowledgments

The authors would like to thank Aleksandra Szprengiel for her helpful contributions in the preparation of this manuscript.

References

1. Landfield, P. W. and Eldridge, J. C. (1994). The glucocorticoid hypothesis of age-related hippocampal neurodegeneration: role of dysregulated intraneuronal Ca2$^+$. *Ann. NY Acad. Sci.* **746**: 308–321.
2. Gould, T. J. and Bickford, P. C. (1997). Age-related deficits in the cerebellar beta adrenergic signal transduction cascade in Fischer 344 rats. *J. Pharmacol. Exp. Ther.* **281**: 965–971.
3. Joseph, J. A., Berger, R. E., Engel, B. T. and Roth, G. S. (1978). Age-related changes in the nigrostriatum: a behavioral and biochemical analysis. *J. Gerontol.* **33**: 643–649.
4. Levine, M. S. and Cepeda, C. (1998). Dopamine modulation of responses mediated by excitatory amino acids in the neostriatum. *Adv. Pharmacol.* **42**: 724–729.
5. Egashira, T., Takayama, F. and Yamanaka, Y. (1996). Effects of bifemelane on muscarinic receptors and choline acetyltransferase in the brains of aged rats following chronic cerebral hypoperfusion induced by permanent occlusion of bilateral carotid arteries. *Japan J. Pharmacol.* **72**: 57–65.
6. Joseph, J. A., Kowatch, M. A., Maki, T. and Roth, G. S. (1990). Selective cross-activation/inhibition of second messenger systems and the reduction of age-related deficits in the muscarinic control of dopamine release from perifused rat striata. *Brain Res.* **537**: 40–48.
7. Yufu, F., Egashira, T. and Yamanaka, Y. (1994). Age-related changes of cholinergic markers in the rat brain. *Japan J. Pharmacol.* **66**: 247–255.
8. Kornhuber, J., Schoppmeyer, K., Bendig, C. and Riederer, P. (1996). Characterization of [3H] pentazocine binding sites in post-mortem human frontal cortex. *J. Neural. Trans.* **103**: 45–53.
9. Nagahara, A. H., Gill, T. M., Nicolle, M. and Gallagher, M. (1996). Alterations in opiate receptor binding in the hippocampus of aged Long-Evans rats. *Brain Res.* **707**: 22–30.
10. Joseph, J. A., Bartus, R. T., Clody, D., Morgan, D., Finch, C., Beer, B. and Sesack, S. (1983). Psychomotor performance in the senescent rodent: reduction of deficits via striatal dopamine receptor up-regulation. *Neurobiol. Aging* **4**: 313–319.
11. Kluger, A., Gianutsos, J. G., Golomb, J., Ferris, S. H., George, A. E., Frannssen, E. and Reisberg, B. (1997). Patterns of motor impairment in normal aging, mild cognitive decline, and early Alzheimer's disease. *J. Gerontol.* **52**: 28–39.
12. Shukitt-Hale, B., Mouzakis, G. and Joseph, J. A. (1998). Psychomotor and spatial memory performance in aging male Fischer 344 rats. *Exp. Gerontol.* **33**: 615–624.
13. Barnes, C. A. (1988). Aging and physiology of spatial memory. *Neurobiol. Aging* **9**: 563–568.

14. Bartus, R. T. (1990). Drugs to treat age-related neurodegenerative problems. The final frontier of medical science? *J. Am. Geriatr. Soc.* **38**: 680–695.

15. Brandeis, R., Brandys, Y. and Yehuda, S. (1989). The use of the Morris water maze in the study of memory and learning. *Int. J. Neurosci.* **48**: 29–69.

16. Gallagher, M. and Pelleymounter, M. A. (1988). Spatial learning deficits in old rats: a model for memory decline in the aged. *Neurobiol. Aging* **9**: 549–556.

17. Ingram, D. K., Jucker, M. and Spangler, E. (1994). Behavioral manifestations of aging. *In* "Pathobiology of the Aging Rat", Vol. 2 (U. Mohr, D. L. Cungworth, and C. C. Capen, Eds.), pp. 149–170, ILSI, Washington.

18. Joseph, J. A. (1992). The putative role of free radicals in the loss of neuronal functioning in senescence. *Integ. Physiol. Behav. Sci.* **27**: 216–227.

19. Bickford, P., Heron, C., Young, D. A., Gerhardt, G. A. and de La Garza, R. (1992). Impaired acquisition of novel locomotor tasks in aged and norepinephrine-depleted F344 rats. *Neurobiol. Aging* **13**: 475–481.

20. Bickford, P. (1993). Motor learning deficits in aged rats are correlated with loss of cerebellar noradrenergic function. *Brain Res.* **620**: 133–138.

21. Devan, B. D., Goad, E. H. and Petri, H. L. (1996). Dissociation of hippocampal and striatal contributions to spatial navigation in the water maze. *Neurobiol. Learn. Mem.* **66**: 305–323.

22. McDonald, R. J. and White, N. M. (1994). Parallel information processing in the water maze: evidence for independent memory systems involving dorsal striatum and hippocampus. *Behav. Neural Biol.* **61**: 260–270.

23. Oliveira, M. G. M., Bueno, O. F. A., Pomarico, A. C. and Gugliano, E. B. (1997). Strategies used by hippocampal- and caudate-putamen-lesioned rats in a learning task. *Neurobiol. Learn. Mem.* **68**: 32–41.

24. Bartus, R. T., Dean, R. L., Beer, B. and Lippa, A. S. (1982). The cholinergic hypothesis of geriatric memory dysfunction. *Science* **217**: 408–417.

25. Halliwell, B. (1994). Free radicals and antioxidants: a personal view. *Nutri. Rev.* **52**: 253–265.

26. Harman, D. (1981). The aging process. *Proc. Natl. Acad. Sci. USA* **78**: 7124–7128.

27. Youdim, K. A. and Deans, S. G. (2000). Effect of thyme oil and thymol dietary supplementation on the antioxidant status and fatty acid composition of the ageing rat brain. *Br. J. Nutri.* **83**: 87–93.

28. Yu, B. P. (1994). Cellular defenses against damage from reactive oxygen species. *Physiol. Rev.* **74**: 139–162.

29. Zhang, J. R., Andrus, P. K. and Hall, E. D. (1993). Age-related regional changes in hydroxyl radical stress and antioxidants in gerbil brain. *J. Neurochem.* **61**: 1640–1647.

30. Carney, J. M., Starke-Reed, P. E., Oliver, C. N., Landum, R. W., Cheng, M. S., Wu, J. F. and Floyd, R. A. (1991). Reversal of age-related increase in brain

protein oxidation, decrease in enzyme activity, and loss in temporal and spatial memory by chronic administration of the spin-trapping compound N-tert-butyl-alpha-phenylnitrone. *Proc. Natl. Acad. Sci.* **88**: 3633–3636.

31. Olanow, C. W. (1992). A scientific rationale for protective therapy in Parkinson's disease. *J. Neural Transm.* **91**: 161–180.

32. Olanow, C. W. (1993). A radical hypothesis for neurodegeneration. *Trends Neurosci.* **16**: 439–444.

33. Hauss-Wegrzyniak, B., Vannucchi, M. G. and Wenk, G. L. (2000). Behavioral and ultrastructural changes induced by chronic neuroinflammation in young rats. *Brain Res.* **859**: 157–166.

34. Cantuti-Castelvetri, I., Shukitt-Hale, B. and Joseph, J. A. (2000). Neurobehavioral aspects of antioxidants in aging. *Int. J. Dev. Neurosci.* **18**: 367–381.

35. Martin, A., Youdim, K., Szprengiel, A., Shukitt-Hale, B. and Joseph, J. A. (2001). Roles of vitamins E and C on neurodegenerative diseases and cognitive performance. *Nutri. Clin. Care,* **in press**.

36. Youdim, K. A. and Joseph, J. A. (2001). A possible emerging role of phytochemicals in improving age-related neurological dysfunctions: a multiplicity of effects. *Free Radic. Biol. Med.,* **30**: 583–594.

37. Youdim, K. A., Martin, A. and Joseph, J. A. (2000). Essential fatty acids and the brain: possible health implications. *Int. J. Dev. Neurosci.* **18**: 383–399.

38. Greenwood, C. E. and Winocur, G. (1999). Decline in cognitive function with aging: impact of diet. *Mature Med.* **2**: 205–209.

39. Ortega, R. M., Requejo, A. M., Andres, P., Lopez-Sobaler, A. M., Quintas, M. E., Redondo, M. R., Navia, B. and Rivas, T. (1997). Dietary intake and cognitive function in a group of elderly people. *Am. J. Clin. Nutri.* **66**: 803–809.

40. Macheix, J.-J., Fleuriet, A. and Billot, J. (1990). *Fruit Phenolics.* CRC Press, Inc., Boca Raton.

41. Middleton, E. (1998). Effect of plant flavonoids on immune and inflammatory cell function. *Adv. Exp. Med. Biol.* **439**: 175–182.

42. Middleton, E. and Kandaswami, C. (1992). Effects of flavonoids on immune and inflammatory cell functions. *Biochem. Pharmacol.* **43**: 1167–1179.

43. Eastwood, M. A. (1999). Interaction of dietary antioxidants in vivo: how fruit and vegetables prevent disease? *Q. J. Med.* **92**: 527–530.

44. Edenharder, R., von Peterdorff, I. and Rauscher, R. (1993). Antimutagenic effects of flavonoids, chalcones and structurally related compounds on the activity of 2-amino-3-methylimidazo[4,5-f]quinoline (IQ) and other heterocyclic amine mutagens from cooked foods. *Mutat. Res.* **287**: 261–274.

45. Gerritsen, M. E. (1998). Flavonoids inhibitors of cytokine induced gene expression. *In* "Flavonoids in the Living System" (J. A. Manthey, and B. S. Buslig, Eds.), pp. 182–190, Plenum Press, New York.

46. Hollman, P. C. and Katan, M. B. (1999). Health effects and bioavailability of dietary flavonols. *Free Radic. Res.* **31**: S75–S80.

47. Fotsis, T., Pepper, M. S., Aktas, E., Breit, S., Rasku, S., Adlercreutz, H., Wahala, K., Montesano, R. and Schweigerer, L. (1997). Flavonoids, dietary-derived inhibitors of cell proliferation and in vitro angiogenesis. *Cancer Res.* **57**: 2916–2921.

48. Armstrong, B. K., Mann, J. I., Adlestein, A. M. and Eskin, F. (1975). Commodity consumption and ischemic heart disease mortality, with special reference to dietary practices. *J. Chron. Dis.* **28**: 455–469.

49. Hughes, K. (1995). Diet and coronary heart disease — a review. *Ann. Acad. Med. Singapore* **24**: 224–229.

50. Mayne, S. T. (1996). Beta-carotene, carotenoids, and disease prevention in humans. *FASEB J.* **10**: 690–701.

51. Verlangieri, A. J., Kapeghian, J. C., El-Dean, S. and Bush, M. (1985). Fruit and vegetable consumption and cardiovascular mortality. *Med. Hypotheses* **16**: 7–15.

52. Doll, R. (1990). An overview of the epidemiological evidence linking diet and cancer. *Proc. Nutri. Soc.* **49**: 119–131.

53. Willett, W. C. (1994). Diet and health: what should we eat? *Science* **264**: 532–537.

54. Sharoni, Y., Giron, E., Rise, M. and Levy, J. (1997). Effects of lycopene-enriched tomato oleoresin on 7,12-dimethyl-benz[a]anthracene-induced rat mammary tumors. *Cancer Detect. Prev.* **21**: 118–123.

55. Rai, G. S., Shovlin, C. and Wesnes, K. A. (1991). A double-blind, placebo controlled study of Ginkgo biloba extract ("tanakan") in elderly outpatients with mild to moderate memory impairment. *Curr. Med. Res. Opin.* **12**: 350–355.

56. Kleijnen, J. and Knipschild, P. (1992). Ginkgo biloba. *Lancet* **340**: 1136–1139.

57. Kleijnen, J. and Knipschild, P. (1992). Ginkgo biloba for cerebral insufficiency. *Br. J. Clin. Pharmacol.* **34**: 352–358.

58. Oyama, Y., Hayashi, A. and Ueha, T. (1993). Ca(2+)-induced increase in oxidative metabolism of dissociated mammalian brain neurons: effect of extract of Ginkgo biloba leaves. *Japan J. Pharmacol.* **61**: 367–370.

59. Oyama, Y., Fuchs, P. A., Katayama, N. and Noda, K. (1994). Myricetin and quercetin, the flavonoid constituents of Ginkgo biloba extract, greatly reduce oxidative metabolism in both resting and Ca(2+)-loaded brain neurons. *Brain Res.* **635**: 125–129.

60. Kanowski, S., Herrmann, W. M., Stephan, K., Wierich, W. and Horr, R. (1996). Proof of efficacy of the Ginkgo biloba special extract EGb 761 in outpatients suffering from mild to moderate primary degenerative dementia of the Alzheimer type or multi-infarct dementia. *Pharmacopsychiatry* **29**: 47–56.

61. Cao, G., Verdon, C. P., Wu, A. H. B., Wang, H. and Prior, R. L. (1995). Automated assay of oxygen radical absorbance capacity with the COBAS FARA II. *Clin. Chem.* **41**: 1738–1744.

62. Cao, G., Sofic, E. and Prior, R. L. (1996). Antioxidant capacity of tea and common vegetables. *J. Agric. Food Chem.* **44**: 3426–3431.

63. Wang, H., Cao, G. and Prior, R. L. (1996). Total antioxidant capacity of fruits. *J. Agric. Food Chem.* **44**: 701–705.

64. Prior, R. L., Cao, G., Martin, A., Sofic, E., McEwen, J., O'Brien, C., Lischner, N., Ehlenfeldt, M., Kalt, W., Krewer, G. and Mainland, M. (1998). Antioxidant capacity as influenced by total phenolic and anthocyanin content, maturity and variety of vaccinium species. *J. Agric. Food Chem.* **46**: 2586–2593.

65. Bickford, P. C., Shukitt-Hale, B. and Joseph, J. (1999). Effects of aging on cerebellar noradrenergic function and motor learning: nutritional interventions. *Mech. Ageing Dev.* **111**: 141–154.

66. Bickford, P. C., Gould, T., Briederick, L., Chadman, K., Pollock, A., Young, D., Shukitt-Hale, B. and Joseph, J. (2000). Antioxidant-rich diets improve cerebellar physiology and motor learning in aged rats. *Brain Res.* **866**: 211–217.

67. Joseph, J. A., Shukitt-Hale, B., Denisova, N. A., Prior, R. L., Cao, G., Martin, A., Taglialatela, G. and Bickford, P. C. (1998). Long-term dietary strawberry, spinach, or vitamin E supplementation retards the onset of age-related neuronal signal-transduction and cognitive behavioral deficits. *J. Neurosci.* **18**: 8047–8055.

68. Joseph, J. A., Shukitt-Hale, B., Denisova, N. A., Bielinski, D., Martin, A., McEwen, J. J. and Bickford, P. C. (1999). Reversals of age-related declines in neuronal signal transduction, cognitive, and motor behavioral deficits with blueberry, spinach, or strawberry dietary supplementation. *J. Neurosci.* **19**: 8114–8121.

69. Youdim, K. A., Shukitt-Hale B., Martin A., Wang H., Denisova N., Bickford P. C. and Joesph, J. A. (2000). Short-term dietary supplementation of blueberry polyphenolics: beneficial effects on aging brain performance and peripheral tissue function. *Nutri. Neurosci.* **3**: 383–397.

70. Shukitt-Hale, B. (1999). The effects of aging and oxidative stress on psychomotor and cognitive behavior. *Age* **22**: 9–17.

71. Shukitt-Hale, B., Smith, D. E., Meydani, M. and Joseph, J. A. (1999). The effects of dietary antioxidants on psychomotor performance in aged mice. *Exp. Gerontol.* **34**: 797–808.

72. Joseph, J. A., Shukitt-Hale, B., McEwen, J. and Rabin, B. (1999). Magnesium activation of GTP hydrolysis or incubation in S-adenosyl-l-methionine reverses iron-56-particle-induced decrements in oxotremorine enhancement of K^+-evoked striatal release of dopamine. *Radic. Res.* **152**: 637–641.

73. Bickford, P. C., Shukitt-Hale, B., Gould, T. J., Breidrick, L., Denisova, N., Bielinkski, D. and Joseph, J. A. (1998). Reversal of age-related declines in CNS neurotransmission with diets supplemented with fruit or vegetable extracts. *Soc. Neurosci. Abs.* **24**: 2157.

74. Krischer, S. M., Eisenmann, M., Bock, A. and Mueller, M. J. (1997). Protein-facilitated export of arachidonic acid from pig neutrophils. *J. Biol. Chem.* **272**: 10 601–10 607.

75. Mirzoeva, O. K. and Calder, P. C. (1996). The effect of propolis and its components on eicosanoid production during the inflammatory response. *Prostaglandins Leukot. Essent. Fatty Acids* **55**: 441–449.

76. Bors, W., Heller, W., Michel, C. and Stettmaier, K. (1996). Flavonoids and polyphenols: chemistry and biology. In "Handbook of Antioxidants" (E. Cadenas, and L. Packer, Eds.), pp. 409–466, Marcel Dekker Inc., New York.

77. Formica, V. J. and Regelson, W. (1995). Review of the biology of quercetin and related bioflavonoids. *Food Chem. Toxicol.* **33**: 1061–1080.

78. Halder, J. and Bhaduri, A. N. (1998). Protective role of black tea against oxidative damage of human red blood cells. *Biochem. Biophys. Res. Commun.* **244**: 903–907.

79. Ramassamy, C., Girbe, F., Christen, Y. and Costentin, J. (1993). Ginkgo biloba extract (EGb 761) or trolox C prevent the ascorbic acid/Fe2$^+$ induced decrease in synaptosomal membrane fluidity. *Free Radic. Res. Commun.* **19**: 341–350.

80. Stoll, S., Scheuer, K., Pohl, O. and Muller, W. E. (1996). Ginkgo biloba extract (EGb 761) independently improves changes in passive avoidance learning and brain membrane fluidity in the aging mouse. *Pharmacopsychiatry* **29**: 144–149.

81. Joseph, J. A., Villalobos-Molina, R., Yamagami, K., Roth, G. S. and Kelly, J. (1995). Age-specific alterations in muscarinic stimulation of K$^+$-evoked dopamine release from striatal slices by cholesterol and S-adenosyl-L-methionine. *Brain Res.* **673**: 185–193.

82. Zhang, Y., Moriguchi, T., Saito, H. and Nishiyama, N. (1998). Functional relationship between age-related immunodeficiency and learning deterioration. *Eur. J. Neurosci.* **10**: 3869–3875.

83. Nishiyama, N., Moriguchi, T. and Saito, H. (1997). Beneficial effects of aged garlic extract on learning and memory impairment in the senescence-accelerated mouse. *Exp. Gerontol.* **32**: 149–160.

84. Moriguchi, T., Saito, H. and Nishiyama, N. (1997). Anti-ageing effect of aged garlic extract in the inbred brain atrophy mouse model. *Clin. Exp. Pharmacol. Physiol.* **24**: 235–242.

85. Moriguchi, T., Takashina, K., Chu, P. J., Saito, H. and Nishiyama, N. (1994). Prolongation of life span and improved learning in the senescence accelerated mouse produced by aged garlic extract. *Biol. Pharm. Bull.* **17**: 1589–1594.

86. Horie, T., Awazu, S., Itakura, Y. and Fuwa, T. (1992). Identified diallyl polysulfides from an aged garlic extract which protects the membranes from lipid peroxidation. *Planta Med.* **58**: 468–469.

87. Horie, T., Murayama, T., Mishima, T., Itoh, F., Minamide, Y., Fuwa, T. and Awazu, S. (1989). Protection of liver microsomal membranes from lipid peroxidation by garlic extract. *Planta Med.* **55**: 506–508.
88. Moriguchi, T., Matsuura, H., Itakura, Y., Katsuki, H., Saito, H. and Nishiyama, N. (1997). Allixin, a phytoalexin produced by garlic, and its analogues as novel exogenous substances with neurotrophic activity. *Life Sci.* **61**: 1413–1420.
89. Suganuma, H., Hirano, T. and Inakuma, T. (1999). Amelioratory effect of dietary ingestion with red bell pepper on learning impairment in senescence-accelerated mice (SAMP8). *J. Nutri. Sci. Vitaminol.* **45**: 143–149.
90. Snowdon, D. A., Gross, M. D. and Butler, S. M. (1996). Antioxidants and reduced functional capacity in the elderly: findings from the Nun Study. *J. Gerontol. A: Biol. Sci. Med. Sci.* **51**: M10–M16.
91. Letenneur, L., Dartigues, J. F. and Orgogozo, J. M. (1993). Wine consumption in the elderly. *Ann. Int. Med.* **118**: 317–318.
92. Launer, L. J., Feskens, E. J., Kalmijn, S. and Kromhout, D. (1996). Smoking, drinking, and thinking. The Zutphen Elderly Study. *Am. J. Epidemiol.* **143**: 219–227.
93. Orgogozo, J. M., Dartigues, J. F., Lafont, S., Letenneur, L., Commenges, D., Salamon, R., Renaud, S. and Breteler, M. B. (1997). Wine consumption and dementia in the elderly: a prospective community study in the Bordeaux area. *Rev. Neurol. (Paris)* **153**: 185–192.
94. Suganuma, M., Okabe, S., Sueoka, N., Sueoka, E., Matsuyama, S., Imai, K., Nakachi, K. and Fujiki, H. (1999). Green tea and cancer chemoprevention. *Mutat. Res.* **428**: 339–344.
95. Suganuma, M., Okabe, S., Oniyama, M., Tada, Y., Ito, H. and Fujiki, H. (1998). Wide distribution of [3H](−)-epigallocatechin gallate, a cancer preventive tea polyphenol, in mouse tissue. *Carcinogenesis* **19**: 1771–1776.
96. Matsuoka, Y., Hasegawa, H., Okuda, S., Muraki, T., Uruno, T. and Kubota, K. (1995). Ameliorative effects of tea catechins on active oxygen-related nerve cell injuries. *J. Pharmacol. Exp. Ther.* **274**: 602–608.
97. Nanjo, F., Mori, M., Goto, K. and Hara, Y. (1999). Radical scavenging activity of tea catechins and their related compounds. *Biosci. Biotech. Biochem.* **63**: 1621–1623.
98. Hofbauer, R., Frass, M., Gmeiner, B., Handler, S., Speiser, W. and Kapiotis, S. (1999). The green tea extract epigallocatechin gallate is able to reduce neutrophil transmigration through monolayers of endothelial cells. *Wien. Klin. Wochenschr.* **111**: 278–282.
99. Lin, Y. L. and Lin, J. K. (1997). (−)-epigallocatechin-3-gallate blocks the induction of nitric oxide synthase by down-regulating lipopolysaccharide-induced activity of transcription factor nuclear factor kappa-B. *Mol. Pharmacol.* **52**: 465–472.

100. Lin, Y. L., Tsai, S. H., Lin-Shiau, S. Y., Ho, C. T. and Lin, J. K. (1999). Theaflavin-3,3'-digallate from black tea blocks the nitric oxide synthase by down-regulating the activation of NF-κB in macrophages. *Eur. J. Pharmacol.* **367**: 379–388.

101. Hartmann, A. M., Burleson, L. E., Holmes, A. K. and Geist, C. R. (2000). Effects of chronic kombucha ingestion on open-field behaviors, longevity, appetitive behaviors, and organs in c57-bl/6 mice: a pilot study. *Nutri.* **16**: 755–761.

102. Hindmarch, I., Quinlan, P. T., Moore, K. L. and Parkin, C. (1998). The effects of black tea and other beverages on aspects of cognition and psychomotor performance. *Psychopharmacol.* **139**: 230–238.

103. Joseph, J. A., Shukitt-Hale, B., Denisova, N. A., Martin, A., Perry, G. and Smith, M. A. (2001). Copernicus revisted: Amyloid beta in Alzheimer's Disease. *Neurobiol. Aging* **22**: 131–146.

Chapter 31

Comparative Study on Antioxidant Activity Against Lipid Peroxidation

Etsuo Niki* and Noriko Noguchi

Etsuo Niki • Human Stress Signal Research Center, AIST 1-8-31 Midorigaoka, Ikeda Osaka 563-8577, Japan
Noriko Noguchi • University of Tokyo, Research Center for Advanced Science and Technology, 4-6-1 Komaba, Tokyo 153-8904, Japan
*Corresponding Author.
Tel: +81-727-51-9991, E-mail: etsuo-niki@aist.go.jp

1. Introduction

The radical-scavenging antioxidants play an important role in the total defense system *in vivo* against oxidative stress. Numerous natural and synthetic antioxidants have been studied and their capacities have been evaluated both *in vitro* and *in vivo*. It has been shown that the antioxidant capacity is determined by many factors even in the model systems *in vitro*. Obviously, the antioxidant capacity *in vivo* must be determined by complex mechanisms and dynamics and depend on conditions and circumstances. In this presentation, the antioxidant action and capacity will be considered, especially how they vary with the conditions and circumstances.

2. Inhibition of Lipid Peroxidation by Antioxidant

The free radical-mediated lipid peroxidation can be inhibited by suppressing the chain initiation or by accelerating the chain termination. The chain initiation, that is, the formation of active free radicals, can be inhibited by sequestering metal ions such as iron and copper, reducing hydroperoxides and hydrogen peroxide, or by quenching active species such as radicals, peroxynitrite, superoxide and singlet oxygen. The stabilization of light and inactivation of lipoxygenases and cyclooxygenases may be also effective.

Many free radicals with different reactivities may attack lipids to give lipid radicals which initiate the lipid peroxidation chain reaction. The resulting chain propagation is carried by the lipid peroxyl radicals independent of the initial attacking species. Various types of chain initiation reaction may take place *in vivo*, therefore, many antioxidants with different functions are required to inhibit the chain initiation. In order to enhance the chain termination, the radical-scavenging antioxidants should scavenge the lipid peroxyl radicals before the radicals attack lipids. The capacity of the antioxidants is determined by the competition between chain inhibition reaction 1 and chain propagation reaction 2,

$$LO_2^{\bullet} + IH \xrightarrow{\ k_{inh}\ } LOOH + I^{\bullet} \tag{1}$$

$$LO_2^{\bullet} + LH \xrightarrow{\ k_p\ } LOOH + L^{\bullet} \tag{2}$$

where $LO_2^{\bullet}$, IH, LH and LOOH are lipid peroxyl radical, antioxidant, lipid, and lipid hydroperoxide and k_{inh} and k_p are the rate constants for the reactions 1 and 2 respectively. The ratio of the rate of the reaction 1 to that of 2, R_1/R_2, determines the antioxidant capacity.

$$\frac{R_1}{R_2} = \frac{k_{inh}[LO_2^{\bullet}][IH]}{k_p[LO_2^{\bullet}][LH]} = \frac{k_{inh}[IH]}{k_p[LH]}. \tag{3}$$

Thus, the antioxidant capacity is proportional to k_{inh}, that is, the reactivity of the antioxidant toward peroxyl radicals and the concentration of antioxidant.

3. Reactivity of the Antioxidant Toward Radicals

It is not easy to measure the reactivities of antioxidants toward peroxyl radicals, especially in the membranes and lipoproteins. Thus, the stable radicals are often used to measure the reactivities of antioxidants toward radicals. Among them, galvinoxyl and DPPH are the typical examples. Galvinoxyl may be better than DPPH since the former is an oxygen radical, while the latter is a nitrogen-centered radical. Another kind of hindered phenoxyl radical has been also used.[1] If the conditions are suitably chosen, both the reactivities and stoichiometry can be measured from the reaction of the antioxidants and galvinoxyl. The example is shown in Fig. 1.[2] The reactivity can be measured from the initial rate while the stoichiometry from the amount of galvinoxyl reacted with the antioxidant. The rate constants for the reactions of α-tocopherol (TOH), α-tocopherylhydroquinone (TQH$_2$), ubiquinol-10 (UQH$_2$) and quercetin (QH) with galvinoxyl are summarized in Table 1, from which the relative reactivities of TOH, TQH$_2$, UQH$_2$ and QH toward galvinoxyl are obteined as 1.0, 4.2, 2.5, and 2.5. As for the stoichiometry, TOH has one active hydrogen while TQH$_2$ and UQH$_2$ have two and quercetin has four active hydrogens. It must be noteworthy that reactivity and stoichiometry determine the rate of oxidation and duration of inhibition respectively, and they should be separately evaluated.

The absolute rate constant for the reaction of antioxidant with the peroxyl radical is not easy to measure, but the relative reactivities of the antioxidants toward peroxyl radicals can be obtained by a competition method. The compounds such as N,N-diphenyl-p-phenylenediamine,[3] β-carotene,[4] ABTS[5] and cis-parinaric acid[6] have been used as a reference compound. It has been found that the reactivities toward peroxyl radicals decrease in the order of TQH$_2$ > UQH$_2$ > TOH,[3] which is the same order toward galvinoxyl.

The above data suggest that TQH$_2$ and UQH$_2$ are more reactive and should be stronger antioxidant than TOH since TQH$_2$ and UQH$_2$ are capable of scavenging redicals faster and twice as many radicals than TOH. However, as shown in Table 2,[3] the capacity of TOH, TQH$_2$ and UQH$_2$ as antioxidant against the peroxidation of methyl linoleate in acetonitrile solution decreases in the order of TOH > UQH$_2$ > TQH$_2$. Thus, the reactivities of antioxidants toward radicals and the extent of inhibition of oxidation by antioxidants are in the reverse order. Furthermore, TOH which has smaller stoichiometric number toward galvinoxyl than TQH$_2$ and UQH$_2$ produces longer inhibition period or lag phase. TOH has just one active hydrogen but it can scavenge two molecules of peroxyl radicals. Furthermore, both TQH$_2$ and UQH$_2$ undergo autoxidation and the apparent

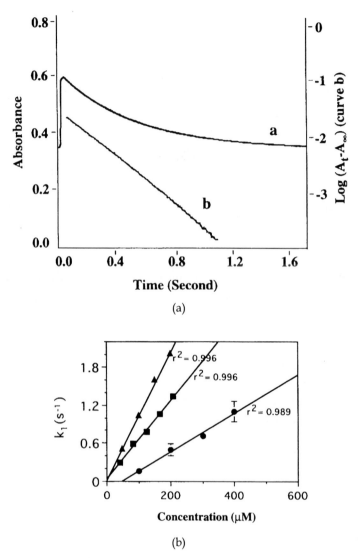

(a)

(b)

Fig. 1. Kinetic study of the reaction of antioxidants with galvinoxyl. (a) Representative experiment of a stopped-flow study: 400 μM TQH_2 was mixed with 20 μM galvinoxyl and the absorption at 428 nm was followed (curve a). Curve b is a typical semilog plot versus time, which gives a first-order rate constant ($2.0\,s^{-1}$), from the experimental curve a. (b) Plots of the first-order rate constants (k_1) as a function of concentration of the antioxidants, TQH_2 (triangle), UQH_2 (square), and TOH (circle). The data points represent means ±SD of five individual experiments. When not visible, error bars lie within data symbols. (c) Plots of absorbance falls of galvinoxyl at 428 nm against concentrations of hydrogen-donating antioxidants. Upper abscissa represents the concentration of Gingko biloba extract (GBE) whereas lower abscissa represents concentrations of pure compounds tested. Line a: quercetin; b: propyl gallate; c: GBE; d: kaempferol; and e: α-tocopherol. Data points represent means ±SD of five individual experiments.

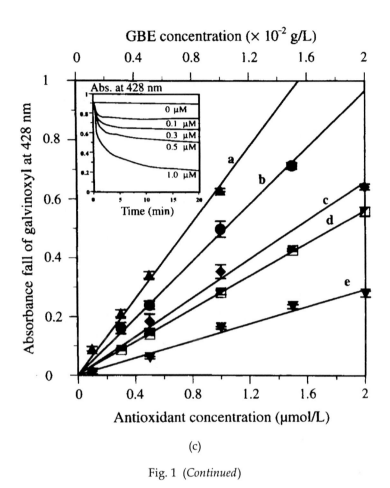

(c)

Fig. 1 (*Continued*)

Table 1. Activities of α-tocopherol (TOH), α-tocopherylhydroquinone (TQH$_2$), Ubiquinol-10 (UQH$_2$) and Quercetin (QH) as Antioxidant

	TOH	TQH$_2$	UQH$_2$	QH
a				
k_2 (M^{-1}s^{-1})	2.4×10^3	1.4×10^4	6.0×10^3	5.9×10^3
b				
n	1.0	1.9	2.0	4.0

a: k_2 is the second-order rate constant for the reaction of antioxidant with galvinoxyl at 25°C in ethanol.
b: stoichiometric number.

Table 2. Inhibitory Effect of TQH$_2$, UQH$_2$ and TOH on Oxidation of Methyl Linoleate

Conc. (μM)	TQH$_2$			UQH$_2$			TOH		
	t_{inh} (min)	R_{inh} (nM/s)	n	t_{inh} (min)	R_{inh} (nM/s)	n	t_{inh} (min)	R_{inh} (nM/s)	n
1	—	—	—	17 ± 1	13.2 ± 0.6 (14%)	2.0	19 ± 3	3.4 ± 1.2 (78%)	2.2
2	—	—	—	29 ± 1	11.4 ± 0.9 (26%)	1.7	36 ± 3	2.0 ± 0.6 (87%)	2.1
3	—	—	—	41 ± 2	8.0 ± 0.9 (48%)	1.6	46 ± 5	1.9 ± 0.0 (88%)	1.8
4	—	—	—	54 ± 3	5.8 ± 0.9 (62%)	1.6	65 ± 4	1.5 ± 0.4 (90%)	1.9
5	13 ± 3	7.6 ± 1.0 (51%)	0.30	65 ± 2	4.3 ± 1.4 (72%)	1.5	78 ± 1	1.3 ± 0.0 (92%)	1.8
10	21 ± 4	6.7 ± 1.1 (56%)	0.24	—	—	—	—	—	—
15	29 ± 4	5.4 ± 1.0 (65%)	0.23	—	—	—	—	—	—
20	34 ± 4	3.9 ± 0.0 (75%)	0.20	—	—	—	—	—	—
25	48 ± 1	3.4 ± 0.4 (78%)	0.22	—	—	—	—	—	—

Methyl linoleate (12.5 mM) was mixed with antioxidants at various concentration in acetonitrile. AMVN (0.5 mM) was finally added to start the reaction. The measurements were carried out under air at 37°C. The data are means ± standard deviation of 3 or 4 individual experiments. The oxidation rate in the absence of an antioxidant was 15.4 nM/s. The numbers in the parentheses are the percentage of inhibition. t_{inh}, induction period, R_{inh}, oxidation rate during induction period. n, stoichiometric number.

stoichiometric number decreases with increasing antioxidant concentration as observed for ascorbate.[7]

These results show that, although the reactivities of the antioxidants toward radicals are indeed the major factor which determines the antioxidant activity in solution, the over-all antioxidant capacity is dependent on other factors as well.

4. Fate of Antioxidant-Derived Radical

When the antioxidant scavenges radical, the antioxidant is converted to a radical and the fate of this antioxidant-derived radical is important in determining its antioxidant capacity. The reverse orders of the reactivities toward radical and antioxidant capacity, observed for TOH, TQH_2 and UQH_2 shown above are due primarily to the autoxidation of TQH_2 and UQH_2. The semiquinone radicals derived from TQH_2 and UQH_2 react readily with oxygen to give quinone, α-tocopherylquinone and ubiquinone respectively, and hydroperoxyl radical, which attack hydroquinones to induce autoxidation. This accelerates the consumption of hydroquinones and reduces inhibition period. The hydroperoxyl radicals may attack lipids to induce chain oxidation. Thus, the autoxidation of antioxidants results in decreased antioxidant capacity, both in rate and duration of inhibited oxidation. Stocker proposes that the hydroperoxyl radicals escape from the lipophilic domain into the aqueous phase,[8] which enhances the antioxidant efficacy.

Another example is β-carotene. β-carotene scavenges peroxyl radicals not by donating hydrogen like α-tocopherol but by addition reaction.[9] The resulting β-carotene carbon-centered radical is stabilized by polyene-resonance and not reactive. If the β-carotene radical scavenges another radical, β-carotene acts as an efficient antioxidant. However, the β-carotene radical reacts rapidly with oxygen to give peroxyl radical, which is not stable but capable of attacking lipids like lipid peroxyl radicals. Under such case, β-carotene simply transfer chain and does not suppress lipid peroxidation. The reaction of β-carotene radical and oxygen is reversible and the antioxidant capacity of β-carotene increases with increasing oxygen concentration.[4, 9] Such an effect of oxygen concentration has been also observed for bilirubin.[10]

The compounds called spin trap react with active radicals to yield stable radicals and hence have a potential to act as an antioxidant.[11] For example, phenyl-N-tert-butylnitrone (PBN) is known to act as an antioxidant and its application to *in vivo* system has been studied. The reactivity of PBN toward peroxyl radicals is moderate.[11, 12] A more important issue may be the stability of the spin adduct. PBN reacts with peroxyl radical to give an adduct, which is not stable but undergoes cleavage reaction to give alkoxyl radical (reaction 4).

$$LO_2^{\bullet} + PhCH{=}\overset{\overset{\displaystyle O}{|}}{N}{-}C(CH_3)_3 \rightarrow PhCH{-}\underset{\underset{\displaystyle OOL}{|}}{\overset{\overset{\displaystyle O^{\bullet}}{|}}{N}}{-}C(CH_3)_3$$

$$\rightarrow PhCH{-}\overset{\overset{\displaystyle O^{\bullet}}{|}}{\underset{\underset{\displaystyle O}{||}}{N}}{-}C(CH_3)_3 + LO^{\bullet} \rightarrow \rightarrow . \tag{4}$$

This alkoxyl radical is more reactive than peroxyl radical and may attack lipids to induce lipid peroxidation chain. This also results in the decrease in antioxidant capacity of PBN.

The three examples shown above clearly show that the fate of antioxidant-derived radical is important in determining antioxidant capacity. Thiols such as glutathione and cysteine may scavenge oxygen radicals. The complicated reactions of the resulting thiyl radicals determine the antioxidant capacity.

5. Location of Antioxidant

The radical-scavenging antioxidants are localized unevenly *in vivo*. The lipid-soluble antioxidants are present in the lipophilic domain of the membranes and lipoproteins, while the water-soluble antioxidants are localized in cytosols or extracellular fluids. The apparent antioxidant potency depends very much on the type of oxidants and the place of their formation. For example, the aqueous peroxyl radicals formed in extracellular fluids attack plasma membranes first, while nitric oxide formed in extracellular fluids is permeable into cytosol and induces oxidation inside the cells. Therefore, α-tocopherol in the plasma membranes acts as an important antioxidant against aqueous peroxyl radical-induced oxidation, whereas it is less important against NO-induced oxidation. These points should be considered carefully when assessing the antioxidant activities in the oxidative stress upon cultured cell system. The action of vitamin E and vitamin C most important lipophilic and hydrophilic antioxidants, respectively, *in vivo* is interesting.

They have roughly the similar reactivities toward radicals. Vitamin E, α-tocopherol, is present in the membranes or lipoproteins, while vitamin C, ascorbic acid, is present in the aqueous phase. Ascorbate acts as a primary radical-scavenging antioxidant in the plasma. However, it cannot scavenge radicals within the membranes and lipoproteins efficiently and it has been demonstrated that the efficacy of radical scavenging decreases as the radicals go deeper into the interior of the membranes[13] and LDL.[14] Thus, if one compares the relative activities of ascorbate and α-tocopherol, the results depend on where the radicals are formed initially. Ascorbate is superior to α-tocopherol toward aqueous radicals, while α-tocopherol is superior to ascorbate toward lipophilic radicals. The consumption of antioxidant is often followed to assess the antioxidant activity. When, for example, the oxidation of the membranes is induced by the lipophilic radicals generated within the membranes, ascorbate is not consumed. However, if α-tocopherol is present in the membranes, ascorbate is consumed rapidly while α-tocopherol is spared. Such a sparing effect by ascorbate has been observed for the oxidation of both membranes and LDL. Under these conditions, α-tocopherol acts as a primary antioxidant although ascorbate is consumed faster than α-tocopherol.

α-tocopherol is capable of scavenging both aqueous radicals attacking from outside of the membranes and lipophilic radicals within the membranes. The active site of α-tocopherol, the phenolic hydrogen, is localized at the surface of the membranes and the efficacy for α-tocopherol to scavenge lipophilic radicals has been shown to decrease as the radicals go deeper into the interior. This decrease in the efficacy of radical scavenging is due to a rigid membrane structure and a long phytyl side chain of α-tocopherol. Cholesterol which makes membranes rigid decreases α-tocopherol's activity. The shorter the side chain of chromanol, the higher the mobility within and between the membranes. Thus, although α-tocopherol and 2,2,5,7,8-pentamethyl-6-chromanol (PMC) exert the same antioxidant activities against lipid peroxidation in solution, PMC is often more potent than α-tocopherol against the oxidation of membranes and LDL.[15] α-tocopherol appears to be a poor antioxidant when the multilamellar liposomal membranes are oxidized by an aqueous radicals.

Thus, the activities of antioxidants depend on the conditions and there is no simple answer to the question which is a more potent antioxidant.

6. Interactions between Antioxidants

In order to assess and understand the activity and action of the antioxidant, the *in vitro* model system is used. This is necessary indeed but it has to be well appreciated that many antioxidants interact each other *in vivo*, which affects the capacity of antioxidant. As shown above, the apparent antioxidant activity may vary remarkably in the presence of other antioxidants.

7. Pro-Oxidant Action of Antioxidant

It is sometimes reported that under certain conditions, the antioxidant act as a pro-oxidant. This is not surprising since it is easy to set up an *in vitro* system where the antioxidant acts as a pro-oxidant.

A potent antioxidant often acts as a potent reducing agent. For example, ascorbate reduces Fe(III) and Cu(II) to Fe(II) and Cu(I), which decompose hydroperoxides and hydrogen peroxide much more rapidly than Fe(III) and Cu(II) respectively. Therefore, ascorbate may easily act as a pro-oxidant in the presence of metal ions and in fact ascorbate-iron has been used as an oxidant to the *in vitro* experiments. Similar effects have been also observed for α-tocopherol. For example, α-tocopherol in LDL reduces Cu(II) to Cu(I), which induces oxidation of LDL. Under such cases, α-tocopherol may act as a pro-oxidant. Other reducing agents such as thiols may also act as a pro-oxidant. However, such a pro-oxidant action depends on the presence of transition metal ions which can be reduced by the antioxidants. The metal ions *in vivo* are sequestered by proteins and not readily oxidized. Therefore, such a pro-oxidant action of antioxidants by reduction of metal ions is probably not important *in vivo*.

It has been accepted that α-tocopherol is a poorer antioxidant than α-tocopherol and that under certain circumstances α-tocopherol accelerates the oxidation of foods. It has been also known that such an effect of α-tocopherol is diminished by ascorbate. More recently, Stocker and his colleagues[16] proposed that α-tocopherol acts as a pro-oxidant against LDL oxidation by phase transfer and chain transfer mechanism, that is, α-tocopherol reacts with aqueous radicals and the resulting α-tocopheroxyl radical acts as an oxidant for LDL lipids. Such a pro-oxidant action of α-tocopherol may be important under certain conditions in the oxidation of isolated LDL *in vitro*. However, LDL is not isolated *in vivo* and this effect may not be important *in vivo* in the presence of physiological concentration of vitamin C.

As shown above, hydroquinones are potent antioxidants. They are redily oxidized *in vitro* to give semiquinone radicals, which reacts with oxygen rapidly to give hydroperoxyl/superoxide radical. It has been also reported that ubisemiquinone radical decomposes hydrogen peroxide and organic hydroperoxides to give hydroxyl and alkoxyl radicals. Thus, hydroquinones may also act as a pro-oxidant under certain circumstances.

The above discussion suggests that the antioxidants may exert a pro-oxidant effect in the *in vitro* test system which is not necessarily relevant to the *in vivo* situation. A great care should be taken in determining the experimental conditions and interpretation of the results for the assessment of antioxidant activities *in vivo*.

8. Metabolites

The antioxidants may be metabolized to give different compounds. A well-known example is metabolites of polyphenols such as glucuronide. Some metabolites may be a stronger antioxidant while others may be a weaker antioxidant than the parent compound. The oxidation product from the antioxidant may also act as an antioxidant. For example, α-tocopherol is oxidized to α-tocopheryl quinone, which is reduced to α-tocopheryl hydroquinone. The metabolites may have different hydrophilicity or lipophilicity, which affect the localization.

9. Concluding Remarks

Many methods have been used to assess the *in vitro* antioxidant activities but as described above, the antioxidant capacity *in vivo* is determined by many factors and there is no simple method by which the antioxidant activity *in vivo* is accurately measured. Most of the methods reported so far simply measure the activities of the antioxidants toward specific radical or total amount of antioxidants under specific condition. The typical examples are to measure the lag phase observed in the oxidation of plasma or isolated LDL induced by an oxidant. This measures the total amount of antioxidants, not the reactivity toward oxidants, that is the time required for exhaust of antioxidants is measured. However, it is doubtful if this is physiologically important, since the antioxidants would never be exhausted *in vivo*. What is probably more important is the extent of reduction of oxidation rate which is determined by the amount of antioxidants, the reactivity of each antioxidant involved, and other factors discussed above. It is important to make clear which factor is being measured by the method employed. It should become possible, however, to evaluate the antioxidant capacity *in vivo* if the experimental system and conditions are suitably set up and the results are carefully interpreted.

References

1. Mukai, K., Yokoyama, S., Fukuda, K. and Uemoto, Y. (1987). Kinetic studies of antioxidant activity of new tocopherol model compounds in solution. *Bull. Chem. Soc. Japan* **60**: 2163–2167.
2. Shi, H. and Niki, E. (1998). Stoichiometric and kinetic studies on Ginkgo biloba extract and related antioxidants. *Lipids* **33**: 365–370.
3. Shi, H., Noguchi, N. and Niki, E. (1999). Comparative study on dynamics of antioxidative action of α-tocopheryl hydroquinone, ubiquinol, and α-tocopherol against lipid peroxidation. *Free Radic. Biol. Med.* **27**: 334–346.
4. Tsuchihashi, H., Kigoshi, M., Iwatsuki, M. and Niki, E. (1995). Action of β-carotene as an antioxidant against lipid peroxidation. *Arch. Biochem. Biophys.* **323**: 137–147.
5. Rice-Evans, C. and Miller, N. J. (1994). Total antioxidant status in plasma and body fluids. *Meth. Enzymol.* **234**: 279–293.
6. Tsuchiya, M., Kagan, V. E., Freisleben, H.-J., Manabe, M. and Packer, L. (1994). Antioxidant activity of α-tocopherol, β-carotene, and ubiquinol in membranes: *cis*-parinaric acid-incorporated liposomes. *Meth. Enzymol.* **234**: 371–383.
7. Wayner, D. D. M., Burton, G. W., Ingold, K. U., Barclay, L. R. L. and Locke, S. J. (1987). The relative contributions of vitamin E, urate, ascorbate and proteins to the total peroxyl radical-trapping antioxidant activity of human blood plasma. *Biochim. Biophys. Acta* **924**: 408–419.

8. Neuzil, J., Thomas, S. R. and Stocker, R. (1997). Requirement for promotion, or inhibition by α-tocopherol of radical-induced initiation of plasma lipoprotein lipid peroxidation. *Free Radic. Biol. Med.* **22**: 57–71.

9. Burton, G. W. and Ingold, K. U. (1984). β-carotene: an unusual type of lipid antioxidant. *Science* **224**: 569–573.

10. Stocker, R., Yamamoto, Y., McDonagh, A. F., Glazer, A. N. and Ames, B. N. (1987). Bilirubin is an antioxidant of possible physiological importance. *Science* **235**: 1043–1046.

11. N. Ohto, N., E. Niki, E. and Kamiya, Y, (1977). Study of autoxidation by spin trapping. Spin trapping of peroxyl radicals by phenyl N-t-butyl nitrone. *J. Chem. Soc. Perkin. II* **13**: 1770–1774.

12. Barclay, L. R. C. and Vinqvust, M. R. (2000). Do spin trap also act as classical chain-breaking antioxidant? A quantitative kinetics study of phenyl tert-butylnitrone (PBN) in solution and in liposomes. *Free Radic. Biol. Med.* **28**: 1079–1090.

13. Takahashi, M., Tsuchiya, J., Niki, E. and Urano, S. (1988). Action of vitamin E as antioxidant in phospholipid liposomal membranes as studied by spin label technique. *J. Nutri. Sci. Vitaminol.* **34**: 25–34.

14. Gotoh, N., Noguchi, N., Tsuchiya, J., Morita, K., Sakai, H., Shimasaki, H. and Niki, E. (1996). Inhibition of oxidation of low density lipoprotein by vitamin E and related compounds. *Free Radic. Res.* **24**: 123–134.

15. Noguchi, N. and Niki, E. (1998). Dynamics of vitamin E action against LDL oxidation. *Free Rad. Res.* **28**: 561–572.

16. Bowry, V. W. and Stocker, R. (1993). Tocopherol-mediated peroxidation. The pro-oxidant effect of vitamin E on the radical-initiated oxidation of human low-density lipoprotein. *J. Am. Chem. Soc.* **115**: 6029–6044.

Chapter 32

α-Tocopherol: Beyond the Antioxidant Dogma

Theresa Visarius and Angelo Azzi*

Theresa Visarius • Department of Molecular Toxicology, Faculty of Biology, University of Konstanz, P.O. Box 911, 78457 Konstanz, Germany
Angelo Azzi • Institute of Biochemistry and Molecular Biology, University of Bern, Bühlstrasse 28, 3012 Bern, Switzerland

*Corresponding Author.
Tel: +41316314131, E-mail: angelo.azzi@mci.unibe.ch

1. Introduction

The discovery of vitamin E in 1922 framed the family of compounds we now know to be the tocopherols and tocotrienols as a nutritional factor essential for normal mammalian reproduction.[1] Since then, efforts to chemically characterize vitamin E resulted in the isolation of four tocopherols (α-, β-, γ-, δ-) and the corresponding tocotrienols,[2] all lipophilic chain-breaking antioxidants. The ensuing burst of research activity involving these molecules, which commenced about 30 years after their discovery, shifted the focus of investigation to the biological significance of the newly attributed antioxidant properties. Radical-mediated processes and a multitude of dysfunctions linked to oxidative stress were hampered or prevented by α-tocopherol more significantly than the other tocopherols or tocotrienols, thus vitamin E, a mixture of tocopherols and tocotrienols comprising α-tocopherol, became known as the most potent, biologically relevant antioxidant (see for a review, Ref. 3). It took another 30 years before the first evidence was provided profiling α-tocopherol as a signal transduction molecule.[4, 5] Protein kinase C, protein phosphatase 2A, cell proliferation, and gene expression were found to be regulated by α-tocopherol,[5–12] thus, the α-tocopherol-antioxidant concept required modification. Today α-tocopherol is recognized as having two distinct functions and in this sense has been identified as a Janus molecule[a]. The non-antioxidant functions of α-tocopherol, presently considered to be the primary intracellular role for this molecule, are herein discussed and differentiated from its antioxidant property.[6]

2. Uptake and Transport

To function effectively, vitamin E must be absorbed, transported, and delivered to cells or cellular compartments. In humans, vitamin E is taken up in the jejunum; the proximal part of the small intestine, where this first phase of uptake is dependent upon the amount of other lipids, bile, and pancreatic esterases present. Unspecific absorption occurs at the intestinal brush border membrane by passive diffusion where, together with triglycerides, phospholipids, cholesterol, and apolipoproteins, the tocopherols are re-assembled into chylomicrons by the Golgi of the mucosa cells. The chylomicrons are then stored as secretory granula and eventually excreted by exocytosis into the lymphatic system from where they in turn reach the blood stream. Intravascular degradation of the chylomicrons proceeds via endothelial lipoprotein lipase, a prerequisite for the hepatic uptake

---

[a]Janus, the Roman god of beginnings and the custodian of the Universe. His chief function was to guard gates and doors and was usually represented with two heads placed back to back which allowed a simultaneous view of opposite (α-tocopherol has two faces in one molecular body).

of tocopherols.[7] α-Tocopherol transfer protein (α-TTP) governs the hepatic uptake of vitamin E, a tightly regulated process.[8] The bioavailability and biopotency of the various tocopherol derivatives and stereoisomers have accordingly been extensively studied where a preference for the most abundant natural derivative of vitamin E, namely RRR-α-tocopherol, was observed.[9-11] α-TTP in the liver specifically sorts out RRR-α-tocopherol from all incoming tocopherols for incorporation into plasma lipoproteins.[12] Relative affinities (RRR-α-tocopherol = 100%) calculated from the degree of competition are as follows: β-tocopherol 38%, γ-tocopherol 9%, δ-tocopherol 2%, α-tocopherol acetate 2%, α-tocopherol quinone 2%, SRR-α-tocopherol 11%, α-tocotrienol 12%, Trolox® 9%.[13] Recently, The Food and Nutrition Board of the National Academies has suggested that α-tocopherol alone should be used for estimating vitamin E requirements and for recommending daily vitamin E intake since the other naturally occurring forms of vitamin E are not converted to α-tocopherol in humans and are poorly recognized by α-TTP.[14]

In addition to α-tocopherol selection, and to the establishment of the quantity absorbed, α-TTP transfers α-tocopherol to plasma lipoproteins for stable transport of tocopherol in the circulation. Following its systemic delivery in plasma, tissue-specific distribution and specific regulation of α-tocopherol have been described.[15, 16] Since α-TTP is expressed almost exclusively in the liver, it is unlikely that this protein mediates the α-tocopherol tissue and intracellular distribution. Recently however, a novel 46 kDa tocopherol associated protein (TAP) has been identified in cytosol containing structural motifs, which suggesting it belongs to the class of hydrophobic ligand-binding proteins CRAL-TRIO.[17, 18] Thus far, TAP has been found in all cells studied, but is predominantly expressed in adult liver, prostate, and brain tissues. Ubiquitous localization of this newly described protein underscores the importance of regulated α-tocopherol transport and further implies specific key cellular functions rather than non-specific use as a cellular antioxidant. The emerging concept, that TAP is involved in the local (cellular and tissue) homeostasis of α-tocopherol while α-TTP is responsible for the global (systemic) maintenance of α-tocopherol concentration is now considered an attractive hypothesis.

3. Vitamin E Deficiency

Lipid malabsorption, deterioration of lipoprotein metabolism, or genetic defects in α-TTP result in vitamin E deficiency (VED).[19-25] In consideration of its possible roles in α-tocopherol cellular distribution and/or in α-tocopherol associated signaling, it is anticipated that TAP genetic defects may be a source of pathological events. Commonly, VED is associated with ataxia, peripheral neuropathy, Friedreich ataxia-like symptoms, and often coincides with mutations on the α-TTP gene (Table 1). In cases of α-TTP mutation, α-tocopherol absorption

Table 1. Pathophysiological States Linked to Vitamin E Deficiency

Condition*	References
Neurodegenerative diseases	
Friedreich ataxia	(1–3)
Alzheimer disease	(4, 5)
Multiple sclerosis	(6–8)
Amyotrophic lateral sclerosis	(9–11)
Familial isolated vitamin E deficiency	(12–14)
Peripheral neuropathy	(15, 16)
Cardiovascular disease	(17, 18)
Coronary artery spasm/angina	(19)
Diabetes	(20–24)
Compromised immune response	(25–27)

*Supplementation with vitamin E lessens the severity or abolishes impairment.

1. Ben Hamida, C., Doerflinger, N., Belal, S., Linder, C., Reutenauer, L., Dib, C., Gyapay, G., Vignal, A., Le Paslier, D. and Cohen, D. *et al.* (1993). Localization of Friedreich ataxia phenotype with selective vitamin E deficiency to chromosome 8q by homozygosity mapping [see comments]. *Nature Genetics* 5: 195–200.
2. Ben Hamida, M., Belal, S., Sirugo, G., Ben Hamida, C., Panayides, K., Ionannou, P., Beckmann, J., Mandel, J. L., Hentati, F. and Koenig, M. *et al.* (1993). Friedreich's ataxia phenotype not linked to chromosome 9 and associated with selective autosomal recessive vitamin E deficiency in two inbred Tunisian families [see comments]. *Neurology* 43: 2179–2183.
3. Yokota, T., Shiojiri, T., Gotoda, T., Arita, M., Arai, H., Ohga, T., Kanda, T., Suzuki, J., Imai, T. and Matsumoto, H. *et al.* (1997). Friedreich-like ataxia with retinitis pigmentosa caused by the His101Gln mutation of the alpha-tocopherol transfer protein gene [see comments]. *Ann. Neurol.* 41: 826–832.
4. Jimenez-Jimenez, F. J., de Bustos, F., Molina, J. A., de Andres, C., Gasalla, T., Orti-Pareja, M., Zurdo, M., Porta, J., Castellano-Millan, F. and Arenas, J. *et al.* (1998). Cerebrospinal fluid levels of alpha-tocopherol in patients with multiple sclerosis. *Neurosci. Lett.* 249: 65–67.
5. Koppal, T., Subramaniam, R., Drake, J., Prasad, M.R., Dhillon, H. and Butterfield, D. A. (1998). Vitamin E protects against Alzheimer's amyloid peptide (25–35)-induced changes in neocortical synaptosomal membrane lipid structure and composition. *Brain Res.* 786: 270–273.
6. Karg, E., Klivenyi, P., Nemeth, I., Bencsik, K., Pinter, S. and Vecsei, L. (1999). Nonenzymatic antioxidants of blood in multiple sclerosis. *J. Neurol.* 246: 533–539.
7. Langemann, H., Kabiersch, A. and Newcombe, J. (1992). Measurement of low-molecular-weight antioxidants, uric acid, tyrosine and tryptophan in plaques and white matter from patients with multiple sclerosis. *Eur. Neurol.* 32: 248–252.
8. Elovaara, I., Ukkonen, M., Leppakynnas, M., Lehtimaki, T., Luomala, M., Peltola, J. and Dastidar, P. (2000). Adhesion molecules in multiple sclerosis: relation to subtypes of disease and methylprednisolone therapy. *Arch. Neurol.* 57: 546–551.
9. Tohgi, H., Abe, T., Saheki, M., Yamazaki, K. and Takahashi, S. (1996). Alpha-tocopherol quinone level is remarkably low in the cerebrospinal fluid of patients with sporadic amyotrophic lateral sclerosis. *Neurosci. Lett.* 207: 5–8.
10. Gurney, M. E. (1997). Transgenic animal models of familial amyotrophic lateral sclerosis. *J. Neurol.* 244(Suppl. 2): S15–S20.

Table 1 (*Continued*)

11. Terro, F., Lesort, M., Viader, F., Ludolph, A. and Hugon, J. (1996). Antioxidant drugs block in vitro the neurotoxicity of CSF from patients with amyotrophic lateral sclerosis. *Neuroreport* 7: 1970–1972.
12. Ouahchi, K., Arita, M., Kayden, H., Hentati, F., Ben Hamida, M., Sokol, R., Arai, H., Inoue, K., Mandel, J. L. and Koenig, M. (1995). Ataxia with isolated vitamin E deficiency is caused by mutations in the alpha-tocopherol transfer protein. *Nature Genetics* 9: 141–145.
13. Cavalier, L., Ouahchi, K., Kayden, H. J., di Donato, S., Reutenauer, L., Mandel, J. L. and Koenig, M. (1998). Ataxia with isolated vitamin E deficiency: heterogeneity of mutations and phenotypic variability in a large number of families. *Am. J. Human Genet.* 62: 301–310.
14. Hentati, A., Deng, H. X., Hung, W. Y., Nayer, M., Ahmed, M. S., He, X., Tim, R., Stumpf, D. A., Siddique, T. and Ahmed, M. S. (1996). Human alpha-tocopherol transfer protein: gene structure and mutations in familial vitamin E deficiency. *Ann. Neurol.* 39: 295–300.
15. Traber, M. G., Sokol, R. J., Ringel, S. P., Neville, H. E., Thellman, C. A. and Kayden, H. J. (1987). Lack of tocopherol in peripheral nerves of vitamin E-deficient patients with peripheral neuropathy. *New England J. Med.* 317: 262–265.
16. Martinello, F., Fardin, P., Ottina, M., Ricchieri, G. L., Koenig, M., Cavalier, L. and Trevisan, C. P. (1998). Supplemental therapy in isolated vitamin E deficiency improves the peripheral neuropathy and prevents the progression of ataxia. *J. Neurol. Sci.* 156: 177–179.
17. Devaraj, S., Li, D. and Jialal, I. (1996). The effects of alpha tocopherol supplementation on monocyte function. Decreased lipid oxidation, interleukin-1 beta secretion, and monocyte adhesion to endothelium. *J. Clin. Invest.* 98: 756–763.
18. Islam, K. N., Devaraj, S. and Jialal, I. (1998). Alpha-tocopherol enrichment of monocytes decreases agonist-induced adhesion to human endothelial cells. *Circulation* 98: 2255–2261.
19. Miwa, K., Igawa, A., Nakagawa, K., Hirai, T. and Inoue, H. (1999). Consumption of vitamin E in coronary circulation in patients with variant angina. *Cardiovasc. Res.* 41: 291–298.
20. Kunisaki, M., Bursell, S. E., Umeda, F., Nawata, H. and King, G. L. (1998). Prevention of diabetes-induced abnormal retinal blood flow by treatment with d-alpha-tocopherol. *Biofactors* 7: 55–67.
21. Bursell, S. E., Clermont, A. C., Aiello, L. P., Aiello, L. M., Schlossman, D. K., Feener, E. P., Laffel, L. and King, G. L. (1999). High-dose vitamin E supplementation normalizes retinal blood flow and creatinine clearance in patients with Type-1 diabetes [see comments]. *Diabetes Care* 22: 1245–1251.
22. Koya, D., Haneda, M., Kikkawa, R. and King, G. L. (1998). d-alpha-tocopherol treatment prevents glomerular dysfunctions in diabetic rats through inhibition of protein kinase C-diacylglycerol pathway. *Biofactors* 7: 69–76.
23. Gazis, A., White, D. J., Page, S. R. and Cockcroft, J. R. (1999). Effect of oral vitamin E (alpha-tocopherol) supplementation on vascular endothelial function in Type-2 diabetes mellitus. *Diabet. Med.* 16: 304–311.
24. Ferber, P., Moll, K., Koschinsky, T., Rosen, P., Susanto, F., Schwippert, B. and Tschope, D. (1999). High dose supplementation of RRR-alpha-tocopherol decreases cellular hemostasis but accelerates plasmatic coagulation in Type-2 diabetes mellitus [In Process Citation]. *Horm. Metab. Res.* 31: 665–671.
25. Meydani, S. N., Meydani, M., Blumberg, J. B., Leka, L. S., Siber, G., Loszewski, R., Thompson, C., Pedrosa, M. C., Diamond, R. D. and Stollar, B. D. (1997). Vitamin E supplementation and in vivo immune response in healthy elderly subjects. A randomized controlled trial [see comments]. *JAMA* 277: 1380–1386.
26. Fulop, T., Jr., Wagner, J. R., Khalil, A., Weber, J., Trottier, L. and Payette, H. (1999). Relationship between the response to influenza vaccination and the nutritional status in institutionalized elderly subjects. *J. Gerontol. A: Biol. Sci. Med. Sci.* 54: M59–M64.
27. Fulop, T., Jr., Gagne, D., Goulet, A. C., Desgeorges, S., Lacombe, G., Arcand, M. and Dupuis, G. (1999). Age-related impairment of p56lck and ZAP-70 activities in human T lymphocytes activated through the TcR/CD3 complex. *Exp. Gerontol.* 34: 197–216.

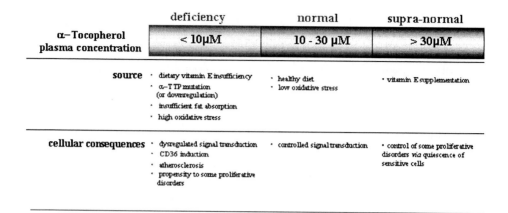

Fig. 1. The roles of α-tocopherol in the human organism relative to its plasma concentration.

is normal, but clearance results in the removal of vitamin E more rapid than its supply. This phenomenon is explained by the chylomicron form in which vitamin E finds itself, which is more susceptible to degradation and elimination rather than the stable, lipoprotein-associated form. Authentic α-TTP mutations result in both cellular and sub-cellular α-tocopherol depletion.

A cellular defense network[26] consisting of the antioxidants ascorbate, glutathione, protein thiols and ubiquinol, work together toward maintaining a high intracellular level of α-tocopherol. Thus, a cell utilizes α-tocopherol as an environmental sensor, signaling the existence of harshly oxidizing conditions that have led to α-tocopherol consumption, despite all defenses.[27] In the absence of sufficient α-tocopherol, cell signaling becomes altered and a plethora of deleterious phenomena may emerge (cf. Refs. 28 and 29). Recent studies indicate that the rank order of intracellular antioxidants depleted is ascorbate, glutathione, other thiols, ubiquinols, and lastly α-tocopherol.[26] Protection of α-tocopherol in this manner by both the water and lipid soluble intracellular antioxidants would allow α-tocopherol to maintain its primary function in the orchestration of cell signaling (Fig. 1).

4. Non-Antioxidant Molecular Mechanisms

That α-tocopherol possesses functions independent of its antioxidant/radical scavenging ability does not represent an unusual property. A number of bio-molecules, including but not limited to estrogens, retinol, melatonin, polyphenols, flavonoids, carotenoids, vitamin A, vitamin C, and α-tocopheryl quinone are provided with antioxidant as well as additional properties, the latter being sometimes of greater importance than the former.[28] The specific, known non-antioxidant functions of α-tocopherol are discussed below.

Table 2. Beneficial Effects of α-tocopherol Attributed to Non-Antioxidant Action

Condition	References
Cancer	(1–4)
Cardiovascular disease	(5–7)
Diabetes	(8–12)
Immune response	(13–16)

1. El Attar, T. M. and Lin, H. S. (1992). Effect of vitamins C and E on prostaglandin synthesis by fibroblasts and squamous carcinoma cells. *Prostaglandins Leukot. Essent. Fatty Acids* **47**: 253–257.
2. Wu, C. G., Hoek, F. J., Groenink, M., Reitsma, P. H., van Deventer, S. J. and Chamuleau, R. A. (1997). Correlation of repressed transcription of alpha-tocopherol transfer protein with serum alpha-tocopherol during hepatocarcinogenesis. *Int. J. Cancer* **71**: 686–690.
3. Pastori, M., Pfander, H., Boscoboinik, D. and Azzi, A. (1998). Lycopene in association with alpha-tocopherol inhibits at physiological concentrations proliferation of prostate carcinoma cells. *Biochem. Biophys. Res. Commun.* **250**: 582–585.
4. Sigounas, G., Anagnostou, A. and Steiner, M. (1997). d-alpha-tocopherol induces apoptosis in erythroleukemia, prostate, and breast cancer cells. *Nutri. Cancer* **28**: 30–35.
5. Devaraj, S., Li, D. and Jialal, I. (1996). The effects of alpha tocopherol supplementation on monocyte function. Decreased lipid oxidation, interleukin-1 beta secretion, and monocyte adhesion to endothelium. *J. Clin. Invest.* **98**: 756–763.
6. Islam, K. N., Devaraj, S. and Jialal, I. (1998). alpha-tocopherol enrichment of monocytes decreases agonist-induced adhesion to human endothelial cells. *Circulation* **98**: 2255–2261.
7. Devaraj, S. and Jialal, I. (1998). The effects of alpha-tocopherol on critical cells in atherogenesis. *Curr. Opin. Lipidol.* **9**: 11–15.
8. Kunisaki, M., Bursell, S. E., Umeda, F., Nawata, H. and King, G. L. (1998). Prevention of diabetes-induced abnormal retinal blood flow by treatment with d-alpha-tocopherol. *Biofactors* **7**: 55–67.
9. Bursell, S. E., Clermont, A. C., Aiello, L. P., Aiello, L. M., Schlossman, D. K., Feener, E. P., Laffel, L. and King, G. L. (1999). High-dose vitamin E supplementation normalizes retinal blood flow and creatinine clearance in patients with Type-1 diabetes [see comments]. *Diabetes Care* **22**: 1245–1251.
10. Koya, D., Haneda, M., Kikkawa, R. and King, G. L. (1998). d-alpha-tocopherol treatment prevents glomerular dysfunctions in diabetic rats through inhibition of protein kinase C-diacylglycerol pathway. *Biofactors* **7**: 69–76.
11. Gazis, A., White, D. J., Page, S. R. and Cockcroft, J. R. (1999). Effect of oral vitamin E (alpha-tocopherol) supplementation on vascular endothelial function in Type-2 diabetes mellitus. *Diabet. Med.* **16**: 304–311.
12. Ferber, P., Moll, K., Koschinsky, T., Rosen, P., Susanto, F., Schwippert, B. and Tschope, D. (1999). High dose supplementation of RRR-alpha-tocopherol decreases cellular hemostasis but accelerates plasmatic coagulation in Type-2 diabetes mellitus [In Process Citation]. *Horm. Metab. Res.* **31**: 665–671.
13. Meydani, S. N., Meydani, M., Blumberg, J. B., Leka, L. S., Siber, G., Loszewski, R., Thompson, C., Pedrosa, M. C., Diamond, R. D. and Stollar, B. D. (1997). Vitamin E supplementation and in vivo immune response in healthy elderly subjects. A randomized controlled trial [see comments]. *JAMA* **277**: 1380–1386.

Table 2 (*Continued*)

14. Fulop, T., Jr., Wagner, J. R., Khalil, A., Weber, J., Trottier, L. and Payette, H. (1999). Relationship between the response to influenza vaccination and the nutritional status in institutionalized elderly subjects. *J. Gerontol. A: Biol. Sci. Med. Sci.* **54**: M59–M64.

15. Fulop, T., Jr., Gagne, D., Goûlet, A. C., Desgeorges, S., Lacombe, G., Arcand, M. and Dupuis, G. (1999). Age-related impairment of p56lck and ZAP-70 activities in human T lymphocytes activated through the TcR/CD3 complex. *Exp Gerontol.* **34**: 197–216.

16. Lim, T. S., Putt, N., Safranski, D., Chung, C. and Watson, R. R. (1981). Effect of vitamin E on cell-mediated immune responses and serum corticosterone in young and maturing mice. *Immunology* **44**: 289–295.

5. Protein Kinase C Regulation

The first non-antioxidant role for α-tocopherol described was protein kinase C regulation in brain and vascular smooth muscle cells.[5, 30] The inhibition of protein kinase C activity and cell proliferation by α-tocopherol are parallel events in vascular smooth muscle cells,[31] where inhibition has been observed to occur at concentrations of α-tocopherol close to those measured in healthy adults.[32] While α-tocopherol inhibits protein kinase C activity, β-tocopherol is ineffective, however, when both tocopherols are present, β-tocopherol prevents the inhibitory effect of α-tocopherol.[33] The inhibition of cell proliferation and protein kinase C activity by α-tocopherol and the lack of inhibition by β-tocopherol allows the conclusion that the mechanism involved is not related to the radical scavenging properties of these two molecules, which are essentially equal.[34] These findings have been subsequently confirmed in monocytes, macrophages, neutrophils, fibroblasts, mesangial cells,[35–40] and *in vivo*.[41]

In monocytes, the activity of PKC is inhibited by α-tocopherol in a specific manner.[35] This event leads to inhibition of phosphorylation and translocation of the cytosolic factor p47(phox) and impaired assembly of NADPH-oxidase and of superoxide production. Additionally, α-tocopherol produces a specific and significant decrease in monocyte superoxide anion release, lipid oxidation, interleukin-1 β (IL-1 β) release and adhesion to endothelium. This reaction, which appears to be mediated by the inhibition of the 5-lipoxygenase pathway,[36] may account for the known antiinflammatory properties of the compound.

5.1. Inhibition by α-Tocopherol is not Caused by a Direct Interaction with Protein Kinase C

Addition of α-tocopherol to recombinant protein kinase C in a chemically isolated system does not result in inhibition of protein kinase C.[42] If α-tocopherol is added in the G_0-phase of the cycle, no inhibition of PKC is observed. When, in the presence of α-tocopherol, cells are stimulated to enter in the G_1-phase, inhibition of protein kinase is observed but, if α-tocopherol is added to cells in the G_1-phase, no inhibition of PKC is observed.[33]

5.2. α-Tocopherol Inhibits Protein Kinase C Phosphorylation-State and Activity, not Expression

When smooth muscle cells are supplemented in the G_0 phase with fetal calf serum, a time-dependent α-tocopherol sensitive increase in protein kinase C activity is observed. The protein levels expressed during the transition are essentially the same, thus, α-tocopherol does not affect the mRNA of protein kinase C. Instead, the protein kinase C phosphorylation required for activity is inhibited by α-tocopherol. Inhibition of PKCα activity can be observed *in vitro* by incubating the enzyme with PP_2A but not with PP_1. Further, PP_2A *in vitro* is activated by α-tocopherol, an effect not produced by β-tocopherol.[42]

5.3. Effect on Gene Expression

Differences in gene expression profiles of cells treated with α-tocopherol have been recently characterized using cDNA microarray chips. Several candidate cDNAs of differentially expressed mRNAs have been detected. Among them, genes under α-tocopherol control are the TMBr-2 isoform of tropomyosin, the scavenger receptor CD36 and collagenase. In smooth muscle cells and macrophages, the scavenger receptor CD36 mRNA and protein expression are down regulated by α- but not β-tocopherol. Therefore, the use of α-tocopherol for the prevention of atherosclerosis[43–45] can be explained, at least in part, by its effect of lowering the uptake of oxidized lipoproteins, with consequent reduction of foam cell formation.[46, 47]

In vitro experiments with skin fibroblasts show that α-tocopherol decreases the level of collagenase expression. Through this event, α-tocopherol may protect against skin damage, a damage known to be induced by various environmental insults and has been linked to the process of aging.[48]

6. Effects of Tocopherols on Cell Proliferation

α-Tocopherol [50 μM] inhibits rat A7r5 smooth muscle cell proliferation, while β-tocopherol at the same concentration is ineffective. When α-tocopherol and β-tocopherol are added together, no inhibition of cell growth is seen *in vitro* despite the fact that both compounds are transported equally in cells and do not compete with each other for uptake.[28, 49, 50] The prevention by β-tocopherol of the proliferation inhibition by α-tocopherol suggests a site-directed event rather than a general radical scavenging reaction. The oxidized product of α-tocopherol, α-tocopherylquinone, is not inhibitory, indicating that the effects of α-tocopherol are not related to its antioxidant properties. α-Tocopherol is not only responsible for the proliferation control of smooth muscle cells but also exhibits similar functions in a number of different cell lines (cf. Refs. 50–52). Further, since α-tocopherol has been shown to slow tumor growth in a number of cell lines derived from human prostate,[53] it will become interesting to investigate the therapeutic benefit of vitamin E in preventing prostate cancer.[54]

7. Conclusions

The development of aging associated disorders, such as atherosclerosis, neurodegenerative diseases, skin damage, and, cancer appears to be prevented or retarded by α-tocopherol. Such an observation has prompted research on the mechanism of action of this vitamin. The understanding of vitamin E function has moved from a mechanism of α-tocopherol action based on the elimination of toxic radicals towards the role of α-tocopherol as a cell regulator. Since the α-tocopherol amount in a cell is modulated by the oxidant environment, the molecule has the property of sensing its cellular ambient and of signaling, through its concentration change, important information to the cell. The signal transduction and gene expression responses discussed above may be at the basis of the observed clinically relevant outcomes.

Acknowledgments

Some of the studies reviewed in this article were rendered possible by the support of the Swiss National Science Foundation; by a grant of Cognis Corporation and by the Stiftung für die Ernährungsforschung in der Schweiz. T. V. is recipient of a Swiss National Science Foundation fellowship.

References

1. Evans, H. M. and Bishop, K. S. (1922). Fetal resorption. *Science* **55**: 650.
2. Evans, H. M., Emerson, O. H. and Emerson, G. A. (1936). Isolation of tocopherols. *J. Biol. Chem.* **113**: 319.
3. Packer, L. (1991). Protective role of vitamin E in biological systems. *Am. J. Clin. Nutri.* **53**: 1050S–1055S.

4. Boscoboinik, D., Szewczyk, A. and Azzi, A. (1991). Alpha-tocopherol (vitamin E) regulates vascular smooth muscle cell proliferation and protein kinase C activity. *Arch. Biochem. Biophys.* **286**: 264–269.
5. Boscoboinik, D., Szewczyk, A., Hensey, C. and Azzi, A. (1991). Inhibition of cell proliferation by alpha-tocopherol. Role of protein kinase C. *J. Biol. Chem.* **266**: 6188–6194.
6. Azzi, A. and Stocker, A. (2000). Vitamin E: non-antioxidant roles. *Prog. Lipid Res.* **39**: 231–255.
7. Brigelius-Flohe, R. and Traber, M. G. (1999). Vitamin E: function and metabolism. *FASEB J.* **13**: 1145–1155.
8. Arita, M., Sato, Y., Miyata, A., Tanabe, T., Takahashi, E., Kayden, H. J., Arai, H. and Inoue, K. (1995). Human alpha-tocopherol transfer protein: cDNA cloning, expression and chromosomal localization. *Biochem. J.* **306(Part 2)**: 437–443.
9. Burton, G. W., Cheeseman, K. H., Doba, T., Ingold, K. U. and Slater, T. F. (1983). Vitamin E as an antioxidant in vitro and in vivo. *Ciba Found. Symp.* **101**: 4–18.
10. Weiser, H., Vecchi, M. and Schlachter, M. (1986). Stereoisomers of alpha-tocopheryl acetate. IV. USP units and alpha-tocopherol equivalents of all-rac-, 2-ambo- and RRR-alpha-tocopherol evaluated by simultaneous determination of resorption-gestation, myopathy and liver storage capacity in rats. *Int. J. Vitam. Nutri. Res.* **56**: 45–56.
11. Kamal-Eldin, A. and Appelqvist, L. (1996). The chemistry and antioxidant properties of tocopherols and tocotrienols. *Lipids* **31**: 671–701.
12. Traber, M. G. and Kayden, H. J. (1989). Preferential incorporation of alpha-tocopherol versus gamma-tocopherol in human lipoproteins. *Am. J. Clin. Nutri.* **49**: 517–526.
13. Hosomi, A., Arita, M., Sato, Y., Kiyose, C., Ueda, T., Igarashi, O., Arai, H. and Inoue, K. (1997). Affinity for alpha-tocopherol transfer protein as a determinant of the biological activities of vitamin E analogs. *FEBS Lett.* **409**: 105–108.
14. Krinsky, N. I., Beecher, G. R., Burk, R. F., Chan, A. C., Erdman, J. W., Jacob, R., Jialal, I., Kolonel, L. N., Marshall, J. R., Taylor Mayne, S., Messing R. B., Prentice, R. L., Schwarz, K., Steinberg, D. and Traber, M. G. (2000). Dietary Antioxidants and related compounds. http://www4.nationalacademies.org/ IOM/IOMHome.nsf/Pages/FNB+Antioxidants
15. Desrumaux, C., Deckert, V., Athias, A., Masson, D., Lizard, G., Palleau, V., Gambert, P. and Lagrost, L. (1999). Plasma phospholipid transfer protein prevents vascular endothelium dysfunction by delivering alpha-tocopherol to endothelial cells. *FASEB J.* **13**: 883–892.
16. Lagrost, L., Desrumaux, C., Masson, D., Deckert, V. and Gambert, P. (1998). Structure and function of the plasma phospholipid transfer protein. *Curr. Opin. Lipidol.* **9**: 203–209.

17. Zimmer, S., Stocker, A., Sarbolouki, M., Spycher, S., Sassoon, J. and Azzi, A. (2000). A novel human tocopherol-associated protein: cloning, in vitro expression and characterization. *J. Biol. Chem.*, **275**: 2567–2580.

18. Stocker, A., Zimmer, S., Spycher, S. E. and Azzi, A. (1999). Identification of a novel cytosolic tocopherol-binding protein: structure, specificity, and tissue distribution. *IUBMB Life* **48**: 49–55.

19. Morisaki, N., Yokote, K. and Saito, Y. (1992). Atherosclerosis from a viewpoint of arterial wall cell function: relation to vitamin E. *J. Nutri. Sci. Vitaminol. (Tokyo)* **Spec**: 196–169.

20. Ouahchi, K., Arita, M., Kayden, H., Hentati, F., Ben Hamida, M., Sokol, R., Arai, H., Inoue, K., Mandel, J. L. and Koenig, M. (1995). Ataxia with isolated vitamin E deficiency is caused by mutations in the alpha-tocopherol transfer protein. *Nature Genet.* **9**: 141–145.

21. Hentati, A., Deng, H. X., Hung, W. Y., Nayer, M., Ahmed, M. S., He, X., Tim, R., Stumpf, D. A., Siddique, T. and Ahmed. (1996). Human alpha-tocopherol transfer protein: gene structure and mutations in familial vitamin E deficiency. *Ann. Neurol.* **39**: 295–300.

22. Beharka, A., Redican, S., Leka, L. and Meydani, S. N. (1997). Vitamin E status and immune function. *Meth. Enzymol* **282**: 247–263.

23. Adachi, N., Migita, M., Ohta, T., Higashi, A. and Matsuda, I. (1997). Depressed natural killer cell activity due to decreased natural killer cell population in a vitamin E-deficient patient with Shwachman syndrome: reversible natural killer cell abnormality by alpha-tocopherol supplementation [see comments]. *Eur. J. Pediatr.* **156**: 444–448.

24. Cavalier, L., Ouahchi, K., Kayden, H. J., Di Donato, S., Reutenauer, L., Mandel, J. L. and Koenig, M. (1998). Ataxia with isolated vitamin E deficiency: heterogeneity of mutations and phenotypic variability in a large number of families. *Am. J. Human Genet.* **62**: 301–310.

25. Hammans, S. R. and Kennedy, C. R. (1998). Ataxia with isolated vitamin E deficiency presenting as mutation negative Friedreich's ataxia. *J. Neurol. Neurosurg. Psychiatry.* **64**: 368–370.

26. Haramaki, N., Stewart, D. B., Aggarwal, S., Ikeda, H., Reznick, A. Z. and Packer, L. (1998). Networking antioxidants in the isolated rat heart are selectively depleted by ischemia-reperfusion. *Free Radic. Biol. Med.* **25**: 329–339.

27. Azzi, A., Boscoboinik, D., Marilley, D., Özer, N.K., Stäuble, B. and Tasinato, A. (1995). Vitamin E: a sensor and an information transducer of the cell oxidation state. *Am. J. Clin. Nutri.* **62(Suppl.)**: 1337S–1346S.

28. Azzi, A., Breyer, I., Feher, M., Pastori, M., Ricciarelli, R., Spycher, S., Staffieri, M., Stocker, A., Zimmer, S. and Zingg, J. M. (2000). Specific cellular responses to alpha-tocopherol. *J. Nutri.* **130**: 1649–1652.

29. Traber, M. G., Sokol, R. J., Ringel, S. P., Neville, H. E., Thellman, C. A. and Kayden, H. J. (1987). Lack of tocopherol in peripheral nerves of vitamin E-deficient patients with peripheral neuropathy. *New England J. Med.* **317**: 262–265.

30. Mahoney, C. W. and Azzi, A. (1988). Vitamin E inhibits protein kinase C activity. *Biochem. Biophys. Res. Commun.* **154**: 694–697.

31. Chatelain, E., Boscoboinik, D. O., Bartoli, G. M., Kagan, V. E., Gey, F. K., Packer, L. and Azzi, A. (1993). Inhibition of smooth muscle cell proliferation and protein kinase C activity by tocopherols and tocotrienols. *Biochim. Biophys. Acta* **1176**: 83–89.

32. Meydani, S. N., Meydani, M., Blumberg, J. B., Leka, L. S., Pedrosa, M., Diamond, R. and Schaefer, E. J. (1998). Assessment of the safety of supplementation with different amounts of vitamin E in healthy older adults. *Am. J. Clin. Nutri.* **68**: 311–318.

33. Tasinato, A., Boscoboinik, D., Bartoli, G. M., Maroni, P. and Azzi, A. (1995). d-α-tocopherol inhibition of vascular smooth muscle cell proliferation occurs at physiological concentrations, correlates with protein kinase C inhibition, and is independent of its antioxidant properties. *Proc. Natl. Acad. Sci. USA* **92**: 12 190–12 194.

34. Pryor, W. A. (2000). Vitamin E and heart disease: basic science to clinical intervention trials. *Free Radic. Biol. Med.* **28**: 141–164.

35. Cachia, O., Benna, J. E., Pedruzzi, E., Descomps, B., Gougerot-Pocidalo, M. A. and Leger, C. L. (1998). Alpha-tocopherol inhibits the respiratory burst in human monocytes. Attenuation of p47(phox) membrane translocation and phosphorylation. *J. Biol. Chem.* **273**: 32 801–32 805.

36. Devaraj, S. and Jialal, I. (1999). Alpha-tocopherol decreases interleukin-1 beta release from activated human monocytes by inhibition of 5-lipoxygenase. *Arterioscler. Thromb. Vasc. Biol.* **19**: 1125–1133.

37. Freedman, J. E., Farhat, J. H., Loscalzo, J. and Keaney, J. F., Jr. (1996). Alpha-tocopherol inhibits aggregation of human platelets by a protein kinase C-dependent mechanism. *Circulation* **94**: 2434–2440.

38. Koya, D. and King, G. L. (1998). Protein kinase C activation and the development of diabetic complications. *Diabetes* **47**: 859–866.

39. Martin-Nizard, F., Boullier, A., Fruchart, J. C. and Duriez, P. (1998). Alpha-tocopherol but not beta-tocopherol inhibits thrombin-induced PKC activation and endothelin secretion in endothelial cells. *J. Cardiovasc. Risk* **5**: 339–345.

40. Studer, R. K., Craven, P. A. and DeRubertis, F. R. (1997). Antioxidant inhibition of protein kinase C-signaled increases in transforming growth factor-beta in mesangial cells. *Metabolism* **46**: 918–925.

41. Sirikci, Ö., Özer, N. K. and Azzi, A. (1996). Dietary cholesterol-induced changes of protein kinase C and the effect of vitamin E in rabbit aortic smooth muscle cells. *Atherosclerosis* **126**: 253–263.

42. Ricciarelli, R., Tasinato, A., Clement, S., Ozer, N. K., Boscoboinik, D. and Azzi, A. (1998). Alpha-tocopherol specifically inactivates cellular protein kinase C alpha by changing its phosphorylation state. *Biochem. J.* **334**: 243–249.

43. Stephens, N. (1997). Anti-oxidant therapy for ischaemic heart disease: where do we stand? [see comments]. *Lancet* **349**: 1710–1711.

44. Stephens, N. G., Parsons, A., Schofield, P. M., Kelly, F., Cheeseman, K. and Mitchinson, M. J. (1996). Randomized controlled trial of vitamin E in patients with coronary disease. Cambridge Heart Antioxidant Study (CHAOS) [see comments]. *Lancet* **347**: 781–786.

45. Davey, P. J., Schulz, M., Gliksman, M., Dobson, M., Aristides, M. and Stephens N. G. (1998). Cost-effectiveness of vitamin E therapy in the treatment of patients with angiographically proven coronary narrowing (CHAOS trial). Cambridge Heart Antioxidant Study. *Am. J. Cardiol.* **82**: 414–417.

46. Zingg, J. M., Ricciarelli, R. and Azzi, A. (2000). Scavenger receptors and modified lipoproteins: fatal attractions? *Iubmb Life* **49**: 397–403.

47. Ricciarelli, R., Zingg, J. M. and Azzi, A. (2000). Vitamin E reduces the uptake of oxidized LDL by inhibiting CD36 scavenger receptor expression in cultured aortic smooth muscle cells. *Circulation* **102**: 82–87.

48. Ricciarelli, R., Maroni, P., Ozer, N., Zingg, J. and Azzi, A. (1999). Age-dependent increase of collagenase expression can be reduced by alpha-tocopherol via protein kinase C inhibition. *Free Radic. Biol. Med.* **27**: 729–737.

49. Azzi, A., Boscoboinik, D., Clement, S., Ozer, N., Ricciarelli, R. and Stocker, A. (1999). Vitamin E mediated response of smooth muscle cell to oxidant stress. *Diabetes Res. Clin. Prac.* **45**: 191–198.

50. Azzi, A., Boscoboinik, D., Clement, S., Ozer, N. K., Ricciarelli, R., Stocker, A., Tasinato, A. and Sirikci, O. (1997). Signalling functions of alpha-tocopherol in smooth muscle cells. *Int. J. Vitam. Nutri. Res.* **67**: 343–349.

51. Azzi, A., Boscoboinik, D., Clement, S., Marilley, D., Ozer, N. K., Ricciarelli, R. and Tasinato, A. (1997). Alpha-tocopherol as a modulator of smooth muscle cell proliferation. *Prostaglandins Leukot. Essent. Fatty Acids* **57**: 507–514.

52. Azzi, A., Boscoboinik, D., Fazzio, A., Marilley, D., Maroni, P., Ozer, N. K., Spycher, S. and Tasinato, A. (1998). RRR-alpha-tocopherol regulation of gene transcription in response to the cell oxidant status. *Z. Ernahrungswiss* **37**: 21–28.

53. Pastori, M., Pfander, H., Boscoboinik, D. and Azzi, A. (1998). Lycopene in association with alpha-tocopherol inhibits at physiological concentrations proliferation of prostate carcinoma cells. *Biochem. Biophys. Res. Commun.* **250**: 582–585.

54. Fleshner, N., Fair, W. R., Huryk, R. and Heston, W. D. (1999). Vitamin E inhibits the high-fat diet promoted growth of established human prostate LNCaP tumors in nude mice. *J. Urol.* **161**: 1651–1654.

Chapter 33

γ-Tocopherol, the Neglected Form of Vitamin E: New Functions, New Interest

Qing Jiang[†], Stephan Christen[¶],
Mark K. Shigenaga[§] and Bruce N. Ames[*,‡]

Qing Jiang, and **Bruce N. Ames** • University of California, Department of Molecular and Cell Biology, Berkeley, CA; Children's Hospital Oakland Research Institute (CHORI), 5700 Martin Luther King, Jr. Way, Oakland, CA 94706-1673
Mark K. Shigenaga • Children's Hospital Oakland Research Institute (CHORI), 5700 Martin Luther King, Jr. Way, Oakland, CA 94706-1673
Stephan Christen • Institute for Infectious Diseases, University of Berne, CH-3010, Berne, Switzerland

*Corresponding Author.
[†]E-mail: Qjiang@uclink4.berkeley.edu, or qjiang@chori.org
[§]E-mail: mshigenaga@chori.org
[‡]E-mail: bnames@uclink4.berkeley.edu, or bames@chori.org
[¶]E-mail: stephan.christen@ifik.unibe.ch

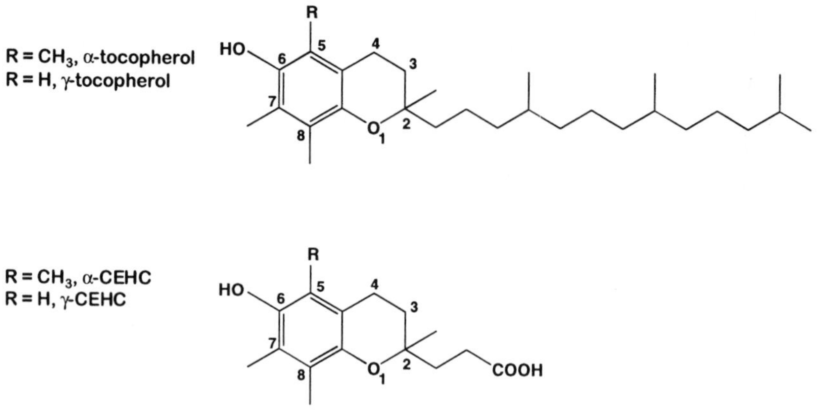

R = CH₃, α-tocopherol
R = H, γ-tocopherol

R = CH₃, α-CEHC
R = H, γ-CEHC

Scheme I. **Structure of α- and γ-tocopherol and its degradation products found in plasma and urine.**

1. Introduction

γ-Tocopherol (γT) is one of the eight naturally occurring forms of the lipophilic antioxidant vitamin E. Despite the fact that γT is rich in US diets, it has drawn relatively little attention since the discovery of vitamin E in 1922. This is mainly because the bioavailability and biological potency of γT assessed in animal studies is much lower than that of α-tocopherol (αT), quantitatively the major form of vitamin E present in humans and animals. In contrast to αT, γT is unsubstituted at the C-5-position of the chromanol ring (Scheme I), which distinguishes γT from αT in many regards, such as metabolism, chemical reactivity and possibly also biological activity. 50% of γT is estimated to be metabolized to the water-soluble γ-CEHC (2,7,8-trimethyl-2-[β-carboxyethyl]-6-hydroxychroman), presumably by the action of a cytochrome P-450. γ-CEHC was originally identified in human urine as a natriuretic factor, whose activity is not shared by the corresponding metabolite derived from αT. Recent evidence indicates that, in contrast to αT, γT and γ-CEHC possess anti-inflammatory activity. A few epidemiological studies also suggest that not only αT, but also γT may be important for human health and disease prevention. In this chapter, we would like to review the current knowledge of γT's chemistry, bio-availability, metabolism, and role in human health with an emphasis on aspects that distinguish it from αT.

2. Bio-Availability and Bio-Activity

Humans and animals primarily acquire their vitamin E from plants, which are the only species capable of synthesizing tocopherols and tocotrienols. Since γT is often the most prevalent form of vitamin E in plant seeds and products derived from them,[1] it comprises 70–80% of vitamin E in the typical US diet.[1] On the other hand, αT is the predominant form in most tissues, including blood plasma. In rats, for example, concentrations of αT are generally 10 to 100 times higher than those of γT.[2,3] In humans, plasma concentrations of γT are about 4–10 times lower than those of αT.[4] Studies reporting human tissue levels of γT are sparse. Burton *et al.*,[5] however, recently reported that in humans γT can be as high as 30–50% of total vitamin E in certain tissues, such as skin, muscle, and adipose. Importantly, γT levels in these tissues are approximately 20–40 fold greater than those in plasma.[5] Furthermore, concentrations of γT in human tissues, e.g. > 100 nmol per gram of muscle or skin, are 20–30 fold higher than those measured in the corresponding rodent tissues, i.e. 3–4 nmol/g of rat muscle,[6] and 2–3 nmol/ g of mouse skin.[7] These data suggest that humans and rodents may differ in their retention of γT by (an) as yet unknown mechanism(s).

The biological activity of vitamin E has traditionally been determined in the rat fetal resorption assay, where bio-activity is defined as the ability of supplemented tocopherols to prevent embryo death in mothers depleted of vitamin E.[8] In this assay, αT exhibits the highest biological vitamin E activity, whereas γT shows only about 10–30% of the activity of αT.[8] This difference in activity in rodents, however, is in all likelihood primarily due to a lower retention of γT in plasma and tissues compared to αT, as a consequence of differences in the manner by which these two tocopherols are metabolized.

3. Absorption and Metabolism

The topic of tocopherol absorption and transport has been excellently reviewed by Kayden and Traber.[9] We herein summarize the current knowledge about absorption and metabolism of αT and γT in Scheme II. Both αT and γT are non-discriminatorily taken up by the intestine along with dietary fat and secreted in chylomicron particles together with triacylglycerol and cholesterol.[8,10] Some of the chylomicron-bound vitamin E appears to be transported and transfered to peripheral tissues including muscle and adipose (and possibly brain) during lipoprotein lipase-mediated catabolism of chylomicron particles.[11] The resulting chylomicron remnants are subsequently taken up by the liver, where αT is preferentially reincorporated into nascent very-low-density lipoprotein (VLDL) by a tocopherol-transfer protein (TTP),[10] which enables further distribution of αT throughout the body. γT, on the other hand, appears to be degraded in large part by a cytochrome P450-dependent process[12] to the hydrophilic γ-CEHC,[13] which is primarily excreted into urine.[14] Although α-CEHC, the corresponding metabolite of αT, can also be detected in urine,[15] catabolism of αT appears to be quantitatively much less important compared to that of γT, since the relative amount of α-CEHC was lower than that of γ-CEHC even after supplementation with αT.[16,17]

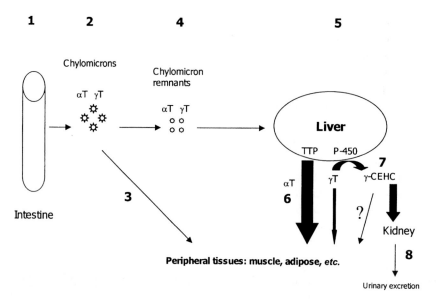

Scheme II. **Summary of absorption, transport and metabolism of αT and γT.** Both αT and γT are similarly absorbed by the intestine along with dietary fat (1), and secreted into chylomicron particles (2). Portions of the chylomicron-bound vitamin E are transported, via the lymph system, to peripheral tissues with the aid of lipoprotein lipase (3). The resulting chylomicron remnants re-enter the circulation (4) and are subsequently taken up by the liver (5). In the liver, most of the remaining αT is reincorporated by TTP into nascent VLDL (6), which then, together with LDL, delivers vitamin E to the tissues through the blood stream. Substantial amounts of γT, on the other hand, appear to be degraded to γ-CEHC (7) while only a small proportion is packed into VLDL along with αT. γ-CEHC is mainly excreted to the urine by way of the kidney (8).

γ-CEHC was originally discovered by Wechter and his coworkers[13] in the search of an endogenous natriuretic factor in human urine. They found that while γ-CEHC possesses potent natriuretic activity, α-CEHC does not exhibit any appreciable activity.[13, 18] They also unambiguously established in a radio-isotope tracing study that γ-CEHC is derived from naturally occurring RRR-γT.[18] Although 5'-carboxychroman was subsequently identified as another metabolite in the supernatant of cultured cells and human urine, γ-CEHC appears to be by far the most predominant metabolite.[19] Swanson *et al.* estimated that as much as 50% of γT may be converted to γ-CEHC and excreted into urine.[14] Plasma concentrations of γ-CEHC were reported to be ~ 50–100 nM in humans[17] and > 300 nM in rats.[20] In human urine, γ-CEHC exists predominantly as a glucuronide conjugate and is present at concentrations ranging from 3 to 36 μM,[14] which increase to > 100 μM after supplementation with γT.[19]

Recent work by Parker *et al.*[12] strongly suggests that the degradation of tocopherols is a cytochrome P450 3A-dependent process, since specific inhibition of this enzyme markedly reduced accumulation of tocopherol metabolites in the supernatant of cultured hepatocytes supplemented with tocopherols. The co-detection of both 3'-(γ-CEHC) and 5'-carboxylate metabolites, products thought to be derived from ω-oxidation followed by β-oxidation of the phytyl side chain, is also consistent with a P450-mediated mechanism.[15, 19] Parker *et al.* also showed that sesamin, the sesame lignan, inhibited γ-CEHC formation in this system, most likely also due to inhibition of P450 activity.[12] This observation provides an explanation for the previous finding by Yamashita *et al.*[21, 22] that rats fed with a diet containing γT together with sesame seeds or sesame lignans, have γT plasma and liver concentrations comparable to those of αT. In the sesame seed or sesame lignan-treated rats, γT exerted a vitamin E activity very similar to that afforded by αT with regard to inhibition of lipid peroxidation, erythrocyte hemolysis and liver necrosis.[22]

In summary, the biological disposition and retention of αT and γT are regulated by differential transport and metabolic mechanism. Chylomicron-associated tissue uptake, which occurs prior to liver metabolism, is possibly important for the accumulation of γT in adipose and muscle tissue. This could explain the strong correlation in humans between dietary γT uptake and γT concentration in these tissues.[5] However, hepatic tocopherol catabolism and TTP-mediated transfer appears to play a key role in the preferential enrichment of αT in plasma and other tissues. It is possible that TTP maintains αT levels not only by facilitating its re-incorporation into nascent VLDL, but also by preventing it from being catabolized, in contrast to γT which appears to be rapidly degraded by cytochrome P450 once it enters the liver.[10, 23, 24] Based solely on the preferential retention of αT, this form of vitamin E is currently the only one considered in dietary intake recommendations.[25] Finally, "normal" levels of γT appear to be eliminated primarily through urinary excretion (as γ-CEHC) and not biliary excretion, as has been proposed earlier.[26] This is also supported by the fact that γT level in the bile was several fold lower than that of αT.[21, 26]

4. Biochemistry of γT

γT is less potent in donating electrons compared to αT, due to the lack of one of the electron-donating methyl groups on the chromanol ring, and is thus a less powerful antioxidant.[27] In this respect, αT is generally considered to be the most potent chain-breaking antioxidant inhibiting lipid peroxidation.[27] However, the unsubstituted 5-position appears to make γT a better nucleophile for trapping electrophiles such as reactive nitrogen oxide species (NO_x). Excess generation of NO_x is associated with chronic inflammation-related diseases including cancer, atherosclerosis and neurodegenerative disorders.[28, 29] NO_x formed during inflammation include peroxynitrite,[30] NO_2 and NO_2-like species generated from myeloperoxidase (or SOD)-H_2O_2–NO_2^-.[31–33] In a pioneering study, Cooney *et al.* found that γT was superior in detoxifying NO_2 compared with αT.[34, 35] They then demonstrated that, analogous to the nitration of tyrosine, the reaction of γT with NO$_2$ was accompanied by the formation of 5-nitro-γT (5-NγT).[35] Later, we[36] and others[37] showed that the reaction of peroxynitrite with γT also resulted in the formation of 5-NγT. Since the chromanol ring of αT is fully substituted, a stable nitro adduct cannot be formed unlike with γT.[34, 36]

We observed that the yield of 5-NγT generated during liposomal lipid peroxidation initiated by peroxynitrite or SIN-1 was independent of the presence of αT, suggesting that γT complements αT in scavenging membrane-soluble NO$_x$.[36] However, this conclusion was later questioned by Goss *et al.*,[38] who found that NγT could only be detected after αT had been almost completely consumed. Although the reasons for these apparently discrepant findings are not entirely clear, it is likely that they are a reflection of the differences in experimental conditions employed, such as the use of saturated[38] versus unsaturated liposomes.[36] It is clear however, 5-NγT is formed with higher yield than that of 3-nitrotyrosine when LDL was treated with peroxynitrite[36] or when nitration was initiated by SOD/H$_2$O$_2$/nitrite.[31] This is most likely the consequence of a much higher reactivity of γT toward electrophilic NO$_x$[36] and the increased solubility of NO$_x$ in membranes. 5-NγT was therefore proposed as another marker in addition to 3-nitrotyrosine for detecting the formation of NO$_x$.[39, 40] Whether nitration of γT is a physiologically relevant process and occurs even in the presence of αT, can only be answered by *in vivo* experiments.

Hensley *et al.* recently reported an HPLC method for measuring 5-NγT using coulometric array detection.[39] By using this method, they reported an increase in 5-NγT (both unadjusted and adjusted for γT) in rat astrocytes stimulated with bacterial lipopolysaccharide.[39] We have recently also developed a highly sensitive HPLC assay with electrochemical detection, in which tissue 5-NγT can be measured simultaneously with αT, γT, and unesterified cholesterol (Christen *et al.*, unpublished results). Using this method, we were able to detect a significant two-fold increase in γT-adjusted plasma NγT in a rat model of zymosan-induced peritonitis.[41] Surprisingly, the level of nitration of γT (even under basal conditions) was in the low percentage range, as compared to parts per million (!) in the case of protein-bound tyrosine, suggesting that γT may indeed act as an NO$_x$ scavenger *in vivo*. Clearly, further studies are needed to address this question. Especially, the metabolism of nitrated γT (e.g. urinary excretion) could be an important aspect.

5. Non-Antioxidant Activity

It has been well established that at 50–100 μM, αT inhibits smooth muscle cell (SMC) proliferation by inhibiting protein kinase C activity.[42] While both γT and δ-tocopherol exhibit a similar anti-proliferative effect, β-tocopherol does not share this activity,[43] indicating that this effect is independent of antioxidant activity. Because SMC proliferation plays an important role in the development of artherosclerosis, the potential benefit of tocopherols in preventing heart disease may be partially rooted in their ability to inhibit SMC proliferation.

Recently, we found that both γT and γ-CEHC exhibited anti-inflammatory activity,[44] as γT and γ-CEHC inhibit prostaglandin E_2 (PGE_2) synthesis in LPS-stimulated macrophages and IL-1β-activated epithelial cells with an IC_{50} of 4–10 μM and ~30 μM respectively, while αT is without effect at these physiologically relevant concentrations. We further found that γT and γ-CEHC directly inhibit cyclooxygenase-2 (COX-2) activity in intact cells, but do not affect COX-2 protein expression. Similar to the anti-proliferative effect of αT, this anti-inflammatory property of γT is yet another effect of vitamin E that is independent of antioxidant activity. Since chronic inflammation contributes to the development of degenerative diseases, the anti-inflammatory activity of γT and its major water-soluble metabolite may be important in human disease prevention. Human colon cancer, for example, is associated with increased expression of COX-2 and formation of PGE_2[45] and, frequent use of non-steroidal anti-inflammatory drugs reduces the incidence of colon cancers.[46–48] Interestingly, Cooney *et al.* found that γT is superior to αT in inhibiting neoplastic transformation of $C_3H_{10}T_{1/2}$ cells.[34] The anti-inflammatory activity of γT could partially explain this effect.

6. γT and Coronary Heart Disease

The role of vitamin E in coronary heart disease (CHD) has been intensely studied in numerous epidemiological and intervention studies.[49, 50] These studies have been recently summarized in an excellent book chapter in the latest edition of the Dietary Reference Intakes for vitamin C, vitamin E, Selenium and Carotenoids.[25] While these studies, for the most part, have focused mainly on αT, they have not yet produced conclusive results as to whether αT protects from CHD.[49, 50] Although much less is known compared to αT, several lines of evidence suggest that γT may also be important in CHD. First, several independent studies have reported that plasma γT concentrations are inversely associated with morbidity and mortality of CHD.[51–53] Secondly, in a seven-year follow-up study in 34 486 postmenopausal women, Kushi *et al.*[54] concluded that the intake of vitamin E from diets (which consist mostly of γT), but not from supplements (which contain predominantly αT) was significantly inversely associated with the risk of death due to CHD. Stampfer *et al.*,[55] however, reported that a high intake of αT from supplements, but not from diets is associated with a significantly reduced risk of CHD. While the reason for this discrepancy is not clear, it is worth mentioning that in both studies the overall level of vitamin E intake from the diet (which presumably consisted mainly of γT) was much lower than the total intake among supplement users. Finally, it has been reported that regular consumption of nuts, which are an excellent source of γT, is associated with a lower risk of developing and dying from CHD.[56]

In addition to the above-cited human studies, several animal studies also provide some evidence that γT might be beneficial. Thus, Saldeen and his coworkers found that γT supplementation in Sprague Dawley rats, compared with αT, led to a more potent decrease in platelet aggregation and delay of arterial thrombogenesis.[57] γT supplementation also resulted in stronger *ex vivo* inhibition of superoxide generation, lipid peroxidation and LDL oxidation. In a follow-up study, they reported that γT was significantly more potent than αT in enhancing SOD activity in plasma and arterial tissue, as well as increasing arterial protein expression of both MnSOD and Cu/Zn SOD.[58] Furthermore, although both tocopherols increased NO generation and endothelial nitric oxide synthase (eNOS) activity, only γT supplementation resulted in increased protein expression of this enzyme.[58] Since endothelium-derived NO is a key regulator of vascular homeostasis, up-regulation of eNOS and NO formation by γT could result in enhancement of arterial endothelial function.[59] Together, the animal studies and the aforementioned human studies warrant further investigations into potential benefits of γT in human health.

7. Summary

Although mostly ignored in the past, γT is unique in many aspects. Compared to αT, γT is a less potent antioxidant with regard to electron-donating propensity, but is superior in detoxifying electrophiles such as NO_x, partially because of its ability to form a stable nitro adduct, 5-NγT. γT is well absorbed, accumulates to a significant degree in some human tissues, but is also rapidly metabolized to the water-soluble metabolite γ-CEHC. γ-CEHC, but not α-CEHC, exhibits natriuretic activity, which may be physiologically important. Both γT and γ-CEHC, but not αT, also possess anti-inflammatory activity. Results from recent epidemiological studies suggest a potential protective effect of γT in CHD. The unique properties of γT and its major metabolite raise significant questions as to the current definition of vitamin E activity, which has been almost exclusively based on the results obtained from the rat fetal resorption assay and has been used as the primary argument that αT is the only important form of vitamin E. We propose that although αT is certainly a very important, if not the most important, component of vitamin E, γT may be necessary to unleash the full power of vitamin E. Since large doses of αT are known to deplete plasma and tissue γT, it is our opinion that this possibility should be acknowledged and carefully evaluated.

In the future, controlled intervention studies in human will be required to clearly establish the benefits from supplementation of this tocopherol isomer. Cellular research combined with animal supplementation study should be valuable to help in the understanding of the mechanism behind the biological effects of γT. In addition, potential synergistic effects between γT and other antioxidants should also be explored. These efforts should clarify the role of γT in human disease prevention.

Acknowledgments

This chapter is adapted from a recently published review.[60] This work was supported by a Postdoctoral Fellowship from the American Heart Association-Western Affiliates Grant 98-24 (to QJ), the Wheeler Fund for the Biological Sciences at the University of California Berkeley, the Department of Energy Grant DE-FG03-00ER62943 and the National Institute of Environmental Health Sciences Center Grant P30-ES01896 (to BNA).

References

1. McLaughlin, P. J. and Weihrauch, J. L. (1979). Vitamin E content of foods. *J. Am. Diet. Assoc.* **75:** 647–665.
2. Clement, M. and Bourre, J. M. (1997). Graded dietary levels of RRR-gamma-tocopherol induce a marked increase in the concentrations of alpha- and gamma-tocopherol in nervous tissues, heart, liver and muscle of vitamin E-deficient rats. *Biochim. Biophys. Acta* **1334:** 173–181.
3. Behrens, W. A. and Madere, R. (1987). Mechanisms of absorption, transport and tissue uptake of RRR-alpha-tocopherol and d-gamma-tocopherol in the white rat. *J. Nutri.* **117:** 1562–1569.
4. Behrens, W. A. and Madere, R. (1986). Alpha- and gamma-tocopherol concentrations in human serum. *J. Am. Coll. Nutri.* **5:** 91–96.
5. Burton, G. W., Traber, M. G., Acuff, R. V., Walters, D. N., Kayden, H., Hughes, L. and Ingold, K. U. (1998). Human plasma and tissue alpha-tocopherol concentrations in response to supplementation with deuterated natural and synthetic vitamin E. *Am. J. Clin. Nutri.* **67:** 669–684.
6. Bieri, J. G. and Evarts, R. P. (1974). Gamma tocopherol: metabolism, biological activity and significance in human vitamin E nutrition. *Am. J. Clin. Nutri.* **27:** 980–986.
7. Weber, C., Podda, M., Rallis, M., Thiele, J. J., Traber, M. G. and Packer, L. (1997). Efficacy of topically applied tocopherols and tocotrienols in protection of murine skin from oxidative damage induced by UV- irradiation. *Free Radic. Biol. Med.* **22:** 761–769.
8. Bieri, J. G. and Evarts, R. P. (1974). Vitamin E activity of gamma-tocopherol in the rat, chick and hamster. *J. Nutri.* **104:** 850–857.
9. Kayden, H. J. and Traber, M. G. (1993). Absorption, lipoprotein transport, and regulation of plasma concentrations of vitamin E in humans. *J. Lipid Res.* **34:** 343–358.

10. Traber, M. G., Burton, G. W., Hughes, L., Ingold, K. U., Hidaka, H., Malloy, M., Kane, J., Hyams, J. and Kayden, H. J. (1992). Discrimination between forms of vitamin E by humans with and without genetic abnormalities of lipoprotein metabolism. *J. Lipid Res.* **33**: 1171–1182.

11. Traber, M. G., Olivecrona, T. and Kayden, H. J. (1985). Bovine milk lipoprotein lipase transfers tocopherol to human fibroblasts during triglyceride hydrolysis in vitro. *J. Clin. Invest.* **75**: 1729–1734.

12. Parker, R. S., Sontag, T. J. and Swanson, J. E. (2000). Cytochrome P4503A-dependent metabolism of tocopherols and inhibition by sesamin. *Biochem. Biophys. Res. Commun.* **277**: 531–534.

13. Wechter, W. J., Kantoci, D., Murray, E. D., Jr., D'Amico, D. C., Jung, M. E. and Wang, W. H. (1996). A new endogenous natriuretic factor: LLU-α. *Proc. Natl. Acad. Sci. USA* **93**: 6002–6007.

14. Swanson, J. E., Ben, R. N., Burton, G. W. and Parker, R. S. (1999). Urinary excretion of 2,7, 8-trimethyl-2-(beta-carboxyethyl)-6-hydroxychroman is a major route of elimination of gamma-tocopherol in humans. *J. Lipid Res.* **40**: 665–671.

15. Schultz, M., Leist, M., Elsner, A. and Brigelius-Flohe, R. (1997). alpha-Carboxyethyl-6-hydroxychroman as urinary metabolite of vitamin E. *Meth. Enzymol.* **282**: 297–310.

16. Traber, M. G., Elsner, A. and Brigelius-Flohe, R. (1998). Synthetic as compared with natural vitamin E is preferentially excreted as alpha-CEHC in human urine: studies using deuterated alpha-tocopheryl acetates. *FEBS Lett.* **437**: 145–148.

17. Stahl, W., Graf, P., Brigelius-Flohe, R., Wechter, W. and Sies, H. (1999). Quantification of the alpha- and gamma-tocopherol metabolites 2,5,7, 8-tetramethyl-2-(2'carboxyethyl)-6-hydroxychroman and 2,7, 8-trimethyl-2-(2'-carboxyethyl)-6-hydroxychroman in human serum. *Anal. Biochem.* **275**: 254–259.

18. Murray, E. D., Jr., Wechter, W. J., Kantoci, D., Wang, W. H., Pham, T., Quiggle, D. D., Gibson, K. M., Leipold, D. and Anner, B. M. (1997). Endogenous natriuretic factors-7: biospecificity of a natriuretic gamma-tocopherol metabolite LLU-alpha. *J. Pharmacol. Exp. Ther.* **282**: 657–662.

19. Parker, R. S. and Swanson, J. E. (2000). A novel 5'-carboxychroman metabolite of gamma-tocopherol secreted by HepG2 cells and excreted in human urine. *Biochem. Biophys. Res. Commun.* **269**: 580–583.

20. Hattori, A., Fukushima, T. and Imai, K. (2000). Occurrence and determination of a natriuretic hormone, 2,7,8-trimethyl-2-(beta-carboxyethyl)-6-hydroxy chroman, in rat plasma, urine, and bile. *Anal. Biochem.* **281**: 209–215.

21. Yamashita, K., Takeda, N. and Ikeda, S. (2000). Effects of various tocopherol-containing diets on tocopherol secretion into bile. *Lipids* **35**: 163–170.

22. Yamashita, K., Nohara, Y., Katayama, K. and Namiki, M. (1992). Sesame seed lignans and gamma-tocopherol act synergistically to produce vitamin E activity in rats. *J. Nutri.* **122:** 2440–2446.

23. Traber, M. G., Rudel, L. L., Burton, G. W., Hughes, L., Ingold, K. U. and Kayden, H. J. (1990). Nascent VLDL from liver perfusions of cynomolgus monkeys are preferentially enriched in RRR-compared with SRR-alpha-tocopherol: studies using deuterated tocopherols. *J. Lipid Res.* **31:** 687–694.

24. Traber, M. G., Burton, G. W., Ingold, K. U. and Kayden, H. J. (1990). RRR- and SRR-alpha-tocopherols are secreted without discrimination in human chylomicrons, but RRR-alpha-tocopherol is preferentially secreted in very low density lipoproteins. *J. Lipid Res.* **31:** 675–685.

25. Panel on Dietary Antioxidants and Related Compounds, Food and Nutrition Board (2000). *In* "Dietary Reference Intakes for Vitamin C, Vitamin E, Selenium and Carotenoids" (I. O. Medicine, Eds.), pp. 186–283, National Academy Press, Washington, D.C.

26. Traber, M. G. and Kayden, H. J. (1989). Preferential incorporation of alpha-tocopherol versus gamma-tocopherol in human lipoproteins. *Am. J. Clin. Nutri.* **49:** 517–526.

27. Kamal-Eldin, A. and Appelqvist, L. A. (1996). The chemistry and antioxidant properties of tocopherols and tocotrienols. *Lipids* **31:** 671–701.

28. Ames, B. N., Shigenaga, M. K. and Hagen, T. M. (1993). Oxidants, antioxidants, and the degenerative diseases of aging. *Proc. Natl. Acad. Sci. USA* **90:** 7915–7922.

29. Christen, S., Hagen, T. M., Shigenaga, M. K. and Ames, B. N. (1999). *In* "Microbes and Malignancy: Infection as a Cause of Human Cancers" (J. Parsonnet, Eds.), pp. 35–88, Oxford University Press, New York, Oxford.

30. Beckman, J. S., Beckman, T. W., Chen, J., Marshall, P. A. and Freeman, B. A. (1990). Apparent hydroxyl radical production by peroxynitrite: implications for endothelial injury from nitric oxide and superoxide. *Proc. Natl. Acad. Sci. USA* **87:** 1620–1624.

31. Singh, R. J., Goss, S. P., Joseph, J. and Kalyanaraman, B. (1998). Nitration of gamma-tocopherol and oxidation of alpha-tocopherol by copper-zinc superoxide dismutase/H2O2/NO2-: role of nitrogen dioxide free radical. *Proc. Natl. Acad. Sci. USA* **95:** 12 912–12 917.

32. Jiang, Q. and Hurst, J. K. (1997). Relative chlorinating, nitrating, and oxidizing capabilities of neutrophils determined with phagocytosable probes. *J. Biol. Chem.* **272:** 32 767–32 772.

33. Eiserich, J. P., Hristova, M., Cross, C. E., Jones, A. D., Freeman, B. A., Halliwell, B. and van der Vliet, A. (1998). Formation of nitric oxide-derived inflammatory oxidants by myeloperoxidase in neutrophils. *Nature* **391:** 393–397.

34. Cooney, R. V., Franke, A. A., Harwood, P. J., Hatch-Pigott, V., Custer, L. J. and Mordan, L. J. (1993). γ-Tocopherol detoxification of nitrogen dioxide: superiority to α-tocopherol. *Proc. Natl. Acad. Sci. USA* **90:** 1771–1775.
35. Cooney, R. V., Harwood, P. J., Franke, A. A., Narala, K., Sundström, A.-K., Berggren, P.-O. and Mordan, L. J. (1995). Products of γ-tocopherol reaction with NO$_2$ and their formation in rat insulinoma (RINm5F) cells. *Free Radic. Biol. Med.* **19:** 259–269.
36. Christen, S., Woodall, A. A., Shigenaga, M. K., Southwell-Keely, P. T., Duncan, M. W. and Ames, B. N. (1997). γ-Tocopherol traps mutagenic electrophiles such as NO$_X$ and complements alpha-tocopherol: physiological implications. *Proc. Natl. Acad. Sci. USA* **94:** 3217–3222.
37. Hoglen, N. C., Waller, S. C., Sipes, I. G. and Liebler, D. C. (1997). Reactions of peroxynitrite with gamma-tocopherol. *Chem. Res. Toxicol.* **10:** 401–407.
38. Goss, S. P., Hogg, N. and Kalyanaraman, B. (1999). The effect of alpha-tocopherol on the nitration of gamma-tocopherol by peroxynitrite. *Arch. Biochem. Biophys.* **363:** 333–340.
39. Hensley, K., Williamson, K. S. and Floyd, R. A. (2000). Measurement of 3-nitrotyrosine and 5-nitro-gamma-tocopherol by high-performance liquid chromatography with electrochemical detection. *Free Radic. Biol. Med.* **28:** 520–528.
40. Ischiropoulos, H., Zhu, L., Chen, J., Tsai, M., Martin, J. C., Smith, C. D. and Beckman, J. S. (1992). Peroxynitrite-mediated tyrosine nitration catalyzed by superoxide dismutase. *Arch. Biochem. Biophys.* **298:** 431–437.
41. Shigenaga, M. K., Christen, S., Lykkesfeldt, J., Jiang, Q., Shigeno, E. T., Chang, H. and Ames, B. N. (1998). Time course of tyrosine and gamma-tocopherol nitration and antioxidant status in zymosan-induced peritonitis. *Free Radic. Biol. Med.* **25:** S67.
42. Tasinato, A., Boscoboinik, D., Bartoli, G. M., Maroni, P. and Azzi, A. (1995). d-alpha-tocopherol inhibition of vascular smooth muscle cell proliferation occurs at physiological concentrations, correlates with protein kinase C inhibition, and is independent of its antioxidant properties. *Proc. Natl. Acad. Sci. USA* **92:** 12 190–12 194.
43. Chatelain, E., Boscoboinik, D. O., Bartoli, G. M., Kagan, V. E., Gey, F. K., Packer, L. and Azzi, A. (1993). Inhibition of smooth muscle cell proliferation and protein kinase C activity by tocopherols and tocotrienols. *Biochim. Biophys. Acta* **1176:** 83–89.
44. Jiang, Q., Elson-Schwab, I., Courtemanche, C. and Ames, B. N. (2000). γ-Tocopherol and its major metabolite, in contrast to alpha-tocopherol, inhibit cyclooxygenase activity in macrophages and epithelial cells [In Process Citation]. *Proc. Natl. Acad. Sci. USA* **97:** 11 494–11 499.
45. Levy, G. N. (1997). Prostaglandin H synthases, nonsteroidal anti-inflammatory drugs, and colon cancer. *FASEB J.* **11:** 234–247.

46. Giovannucci, E., Egan, K. M., Hunter, D. J., Stampfer, M. J., Colditz, G. A., Willett, W. C. and Speizer, F. E. (1995). Aspirin and the risk of colorectal cancer in women. *New England J. Med.* **333**: 609–614.
47. Smalley, W. E. and DuBois, R. N. (1997). Colorectal cancer and nonsteroidal anti-inflammatory drugs. *Adv. Pharmacol.* **39**: 1–20.
48. Thun, M. J., Namboodiri, M. M., Calle, E. E., Flanders, W. D. and Heath, C. W., Jr. (1993). Aspirin use and risk of fatal cancer. *Cancer Res.* **53**: 1322–1327.
49. Jha, P., Flather, M., Lonn, E., Farkouh, M. and Yusuf, S. (1995). The antioxidant vitamins and cardiovascular disease. A critical review of epidemiologic and clinical trial data. *Ann. Int. Med.* **123**: 860–872.
50. Marchioli, R. (1999). Antioxidant vitamins and prevention of cardiovascular disease: laboratory, epidemiological and clinical trial data. *Pharmacol. Res.* **40**: 227–238.
51. Kontush, A., Spranger, T., Reich, A., Baum, K. and Beisiegel, U. (1999). Lipophilic antioxidants in blood plasma as markers of atherosclerosis: the role of alpha-carotene and gamma-tocopherol. *Atherosclerosis* **144**: 117–122.
52. Kristenson, M., Zieden, B., Kucinskiene, Z., Elinder, L. S., Bergdahl, B., Elwing, B., Abaravicius, A., Razinkoviene, L., Calkauskas, H. and Olsson, A. G. (1997). Antioxidant state and mortality from coronary heart disease in Lithuanian and Swedish men: concomitant cross sectional study of men aged 50. *BMJ* **314**: 629–633.
53. Ohrvall, M., Sundlof, G. and Vessby, B. (1996). Gamma, but not alpha, tocopherol levels in serum are reduced in coronary heart disease patients. *J. Int. Med.* **239**: 111–117.
54. Kushi, L. H., Folsom, A. R., Prineas, R. J., Mink, P. J., Wu, Y. and Bostick, R. M. (1996). Dietary antioxidant vitamins and death from coronary heart disease in postmenopausal women. *New England J. Med.* **334**: 1156–1162.
55. Stampfer, M. J., Hennekens, C. H., Manson, J. E., Colditz, G. A., Rosner, B. and Willett, W. C. (1993). Vitamin E consumption and the risk of coronary disease in women. *New England J. Med.* **328**: 1444–1449.
56. Sabate, J. (1999). Nut consumption, vegetarian diets, ischemic heart disease risk, and all-cause mortality: evidence from epidemiologic studies. *Am. J. Clin. Nutri.* **70**: 500S–503S.
57. Saldeen, T., Li, D. and Mehta, J. L. (1999). Differential effects of alpha- and gamma-tocopherol on low-density lipoprotein oxidation, superoxide activity, platelet aggregation and arterial thrombogenesis [published erratum appears in *J. Am. Coll. Cardiol.* 2000 January; **35**(1): 263]. *J. Am. Coll. Cardiol.* **34**: 1208–1215.
58. Li, D., Saldeen, T., Romeo, F. and Mehta, J. L. (1999). Relative effects of alpha- and gamma-tocopherol on low-density lipoprotein oxidation and superoxide dismutase and nitric oxide synthase activity and protein expression in rats. *J. Cardiovasc. Pharmacol. Ther.* **4**: 219–226.

59. Carr, A. and Frei, B. (2000). The role of natural antioxidants in preserving the biological activity of endothelium-derived nitric oxide. *Free Radic. Biol. Med.* **28:** 1806–1814.
60. Tiang, Q., Christen, S., Shigenaga, U.K. and Ames, B.N. (2001). γ-Tocopherol, the major form of vitamin E in the US diet, deserves more attention. *Am. J. Clin. Nutr.* **74:** 714–722.

Chapter 34

Ascorbic Acid DNA Damage and Repair

Joseph Lunec

Joseph Lunec • Oxidative Stress Group; Division of Chemical Pathology, University of Leicester, Robert Kilpatrick Clinical Sciences Building, Leicester Royal Infirmary NHS Trust, P.O. Box 65, Leicester LE2 7LX, UK
Tel: 0044 (0) 116 252 5890, Website: www.le.ac.uk/pa/dcp/vac.html, E-mail: jl20@le.ac.uk

1. Introduction

Ascorbic acid is present in almost all foods of plant origin. Particularly rich foods include citrus and soft fruits. In the human diet potatoes are also a particularly rich source. Ascorbic acid is readily lost during the cooking process, therefore maximal intakes are obtained from raw plant food. Ascorbic acid rapidly decomposes in water, due to rapid oxidation and this in fact highlights one of its fundamental properties as an antioxidant, i.e. in a mixture of biological origin ascorbic acid will be preferentially oxidized to *L*-dehydro ascorbic acid.[1] Additional to its ubiquitous natural occurrence in plants, ascorbic acid is used as a chemical preservative in foods because of its considerable antioxidant properties. In most European countries, Canada and the US, ascorbic acid is consumed as a supplement in doses ranging from 25 → 1500 mg, and in some instances much higher.

Ascorbic acid consumption is regulated under the medicines order 1984 (S1 1984/769). There is no restriction as to the maximum daily dose, however, when used as an active ingredient in multi-nutrient products and used in combination with analgesics such as paracetamol and decongestants to relieve symptoms of colds and flu the maximum daily dose specified in the licences is up to 3000 mg. The optimal ascorbic acid requirements for humans are unknown.[2] A daily intake of 40–60 mg per day is suggested as the minimal dose range to combat dietary deficiency. It is well recognized that a low ascorbic acid status is associated with ageing, smoking and with chronic diseases such as rheumatoid arthritis and cancer. The Food and Nutritions Board of the National Academy of Sciences recently reviewed the recommendations for ascorbic acid intake. They proposed a Recommended Dietary Allowance (RDA) of 120 mg/day compared to the 1989 figure of 60 mg on the basis of recent scientific literature (post 1989).

2. Recommendations on Maximum Intake

In 1991, COMA[3] noted that the adverse effects of ascorbic acid occurred at g/day quantities, but did not make any specific recommendations. The EU Scientific Committee on Food (1993) suggested a maximum intake of 1000–10 000 mg/person per day. In 1991, the Department of Health (UK) and MAFF noted that 6 g per day was undesirable as a chronic dose and suggested maximum daily intake should not exceed 600 mg per day.[4] The Council for Responsible Nutrition, recommended 2 g/day (CRN 1999). This was stated as precautionary as no maximum level had been determined. This upper level of 2 g/day for adults was, independently, also recommended by the National Academy of Sciences in April 2000.

3. Metabolism of Ascorbic Acid

Ascorbic acid is absorbed from the intestine by a sodium — dependent active transport process, and also in the form of dehydro ascorbic acid (DHA) facilitated through the ubiquitous glucose transporter proteins (GLUTS).[5] Metabolism of ascorbic acid in humans occurs via the irreversible hydrolysis of DHA to diketo gulonic acid, and threonic acids.

However, once inside the cell as DHA it is rapidly reduced back to ascorbic acid via either reduced glutathione or NADPH. Since ascorbate is not actually transported by the glucose transporter, it is trapped inside the cell. Excretion of ascorbic acid occurs mainly through the urine as ascorbic acid and dehydro ascorbic acid. 60 to 90% of supplementation doses of 200 mg per day and above is excreted in the form of ascorbic acid itself. Oxidation of ascorbic acid can occur *in vivo* and results ultimately in the excretion of carbon dioxide in expired air.[6]

4. Properties

Ascorbic acid has a profound antioxidant capacity because of its reducing properties. Its established roles are in the synthesis of collagen, neurotransmitters and carnitine; and in the intestinal absorption of iron in its non-haem form. It is also essential for the detoxification of many foreign compounds and is believed to be part of the armoury of defences against the pathological effects of free radicals, however, it has yet to be proven that oxidative damage *in vivo* can be ameliorated by supplementation with large doses of ascorbic acid.[7] Many epidemiological studies have been carried out in order to establish any beneficial effects in recommending supplements to the healthy population at intakes and tissue levels greater than those needed to prevent or treat scurvy (COMA, 1991). Although much *in vitro* data exists to support the proposed *in vivo*, antioxidant effect, very little *in vivo* data has been forthcoming to support this hypothesis. Benefits have been claimed for osteoarthritis, reduction in cataracts by 50%, diabetes, Parkinsons's disease and oesophageal cancer. Manic depressives and depressives also appear to improve significantly on a high dose of 3 g/day.[8]

5. Adverse Effects

The medicines control agency yellow card scheme forms the basis of the known toxic effects of ascorbic acid. Fifteen suspected reactions to ascorbic acid have been reported to date. Toxic effects reported include metabolic acidosis; oxaluria; renal stones; renal tubular disease, gastrointestinal disturbances, sensitivity reactions, pro-thrombin and cholesterol disturbances, vitamin B12 destruction, fatigue and sterility. These effects are associated with daily doses of up to 1 g taken over a period of several weeks. However, higher doses taken over longer periods of time may give rise to serious toxic effects in susceptible individuals. For example patients with a pre-existing hyperxaluria may have an increased risk of nephrolithiasis at ascorbic acid doses of 1 g or more. In patients who are not able to regulate iron absorption (those with idiopathic haemachromatosis, thalassemia major and sideroblastic anaemia), ascorbic acid may substantially increase the already excessive absorption of iron.[9]

An increased risk of haemolysis may occur with excessive ascorbic acid intake, particularly in those with a genetic predisposition e.g. with glucose-6-phosphate dehydrogenase (G6PD) deficiency. Fatalities have occurred and this may have been related to ascorbic acid toxicity. Doses in excess of 4 g per day for three days increases the rate of erythro-lysis *ex vivo*. In experiments with rats, repeated intravenous injection of dehydroascorbic acid (8 mg/day for three or four days) can damage islet cells in the pancreas and produce permanent diabetes.[10]

Ascorbic acid has a similar chemical structure and reactivity to glucose. Both are reducing agents and both undergo "glycation" reactions with proteins. In a similar manner to glucose, ascorbic acid will irreversibly form covalent linkages with primary amino groups of proteins. These reactions have been suggested to contribute to the modification of lens protein and development of cataract.[11]

6. Ascorbic Acid — Antioxidant or Pro-Oxidant

Ascorbic acid is a reducing agent and an essential water-soluble vitamin for humans and certain mammals; it is believed to have a role in sustaining the redox state of the cell possibly maintaining sulphydryl compounds, including glutathione, in their reduced state. Ascorbic acid is also capable of scavenging free radicals and singlet oxygen and interacts with oxidants and oxidizing agents. Under physiological conditions it can also scavenge the myeloperoxidase-derived oxidant hypochlorous acid at rates sufficient to protect important targets such as α_1 antiprotease.

Ascorbic acid has a number of effects on cellular redox systems. It has long been established that it can enhance iron absorption and also stabilize iron-binding proteins. *In vitro* it can cause decomposition of lipid hydroperoxides and hydrogen peroxide through its ability to reduce Fe^{3+} to Fe^{2+}. This dual property has been reported in the literature, suggesting that low concentrations of ascorbate enhance oxygen radical activity whilst high concentrations scavenge hydroxyl radicals, singlet oxygen and lipid peroxides. Clearly, the dose of ascorbate that is protective *in vitro* may not be pertinent, *in vivo*.[12] There are few studies to date that have repeatedly observed a profound protective effect of ascorbic acid against cancer in spite of many anecdotal claims. There is some evidence that ascorbic acid affects the risk of cancer of a number of sites when used in doses well above 60 mg per day. The most convincing evidence for a relationship with decreased risk of cancer comes from data relating specifically tumours of the stomach and less convincing data for mouth, pharynx, oesophagus, lung, pancreas and cervix.[13] Below, we report on a pilot, placebo-controlled intervention study of ascorbic acid in thirty normal individuals supplemented with 500 mg per day. The aim of the study was to determine whether the use of antioxidant supplements could influence the effects of oxidative damage to DNA *in vivo*, hence providing an explanation for its reported anti-cancer effects.

7. Ascorbic Acid Intervention

The ascorbic acid intervention study we initiated was a single-blinded placebo-controlled trial approved by the Leicestershire Area Health Authority Ethics Committee. It consisted of 30 healthy volunteer subjects, 16 females and 14 males with ages ranging from 17 to 49. Written informed consent was obtained; smokers, subjects taking vitamin supplements and/or salicylates were all excluded from the study. Peripheral venous blood samples from fasting subjects along with first-void urine samples were obtained at 3 weekly intervals for 12 weeks, then at week 19. A 6-week course of placebo (500 mg Ca CO_3) preceded a 6 week course of ascorbic acid. At each collection plasma ascorbate levels were determined, PBMC — derived DNA extracted and urinary creatinine levels were measured to correct for excreted oxidative DNA damage products measured by immunoassay.

8. Measuring Oxidative Damage to DNA

An increasingly popular marker of *in vivo* oxidative damage to DNA is 8-oxoguanine (8-oxo-G). It can be measured as the base product by either gas chromatography-mass spectrometry (GC-MS) with selected ion monitoring, high performance liquid chromatography (HPLC) with electrochemical detection (EC) and guanase digestion (guanase assay), or as the deoxynucleoside by reversed-phase HPLC-EC (deoxynucleoside assay).[14, 15] The measurement of this important and potentially mutagenic lesion has been associated with some methodological difficulties. The effort that has followed such interest in measuring this lesion is underlined by the formation of a group to standardize its quantitation; ESCODD, the European Standards Committee on Oxidative DNA Damage.[16] The problem that is being specifically addressed is the baseline level of 8-oxo-G in human cells. ESCODD has defined important areas of concern, (a) inaccuracies in DNA quantitation; (b) poor optimization of extraction procedures for DNA;

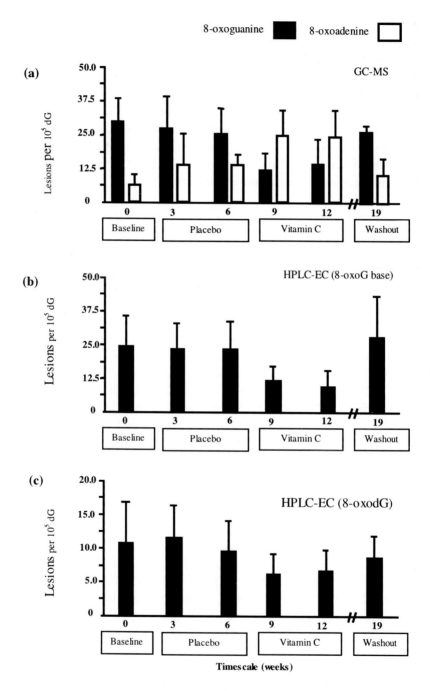

Fig. 1. The effects of supplementation of 500 mg/day vitamin C on normal subjects in terms of modulating oxidative DNA damage. The 8oxo-dG was measured by three separate techniques, because of the controversy surrounding the quantification of this lesion (reproduced with permission IOS Press).

(c) incomplete, or excess DNA hydrolysis/digestion, and/or artefactual damage induced during sample manipulation. In its first report,[16] ESCODD established the importance of appropriate standardization and quality control material particularly for GC-MS. In a later report our research group confirmed unequivocally that GC-MS was overestimating the level of 8-oxo-G by two to three times the target value within a synthetic oligomer.[17] In the ascorbic acid intervention study, 8-oxo-dG was measured by the three assays described above [Figs. 1(a)–(c)]. The level of 8-oxo-G was of the order of 2–3 times higher when measured by GC-MS and guanase assays, compared to the deoxynucleoside assay. The GC-MS and guanase assays share a methodological similarity, formic acid hydrolysis, which differs from the deoxynucleoside assay in which enzymatic hydrolysis is used. This difference in methodology may account for the higher levels of 8-oxo-G seen with the former procedures. Furthermore, the GC-MS method showed a 20% higher level of 8-oxo-G compared to the guanase assay [Figs. 1(a) and (b)] that might be accounted for by artefactual production of 8-oxo-G during derivatization.

9. Urine and Serum Levels of 8-oxo-dG

Levels of PBMC 8-oxo-G showed a significant negative correlation ($p < 0.0001$) with plasma ascorbic acid levels, which was mirrored by a significant positive correlation between 8-oxo-A and ascorbic acid. An albeit less significant correlation ($p < 0.04$), was also seen between PBMC 8-oxo-dG and serum ascorbic acid. A highly significant ($p < 0.001$) increase in serum 8-oxo-dG was observed after 6 weeks of ascorbic acid supplementation that returned approximately to baseline levels following washout (Fig. 2). Urinary 8-oxo-dG showed no such correlation with ascorbic acid. This is probably due to a lag-time between elimination of 8-oxo-dG from the cell and its appearance in serum and then subsequently urine. Comparison between our basal levels of urinary 8-oxo-dG and those quantitated by HPLC-EC reveal the former to be higher, although in strong agreement with other groups also using antibody technology. An explanation for the higher levels could be that the antibody-based method does not distinguish between free 8-oxo-dG and 8-oxo-dG containing oligonucleotides. Indeed, we have recently shown in our laboratory that the antibody recognizes 8-oxo-dG located within a random 20-mer oligonucleotide (Cooke *et al.* unpublished observations). Whatever, the level of the baseline measurement for 8-oxo-G, vitamin C at 500 mg/day reduced it and this has now been shown by several other groups. Rehman, has in fact reproduced a similar change in the levels of oxidative damage lesions 8-oxo-dG and 8-oxo-dA following 6 weeks of vitamin C and iron co-supplementation.[18] A similar initial increase in 8oxo-A was seen in PBMC following administration of the plant antioxidant lycopene to human volunteers.[19]

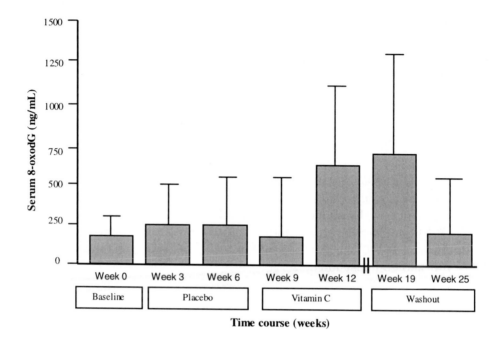

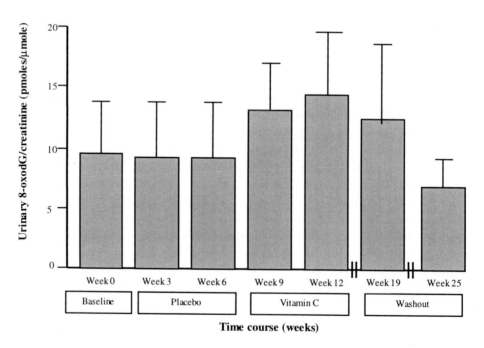

Fig. 2. The effect of supplementation of 500 mg/day vitamin C on normal subjects with respect to repair of the oxidative DNA lesion, 8-oxo-dG, measured in serum and urine. (Data reproduced from Ref. 20, with permission.)

10. Ascorbic Acid and DNA Repair

Our initial analysis of the 8-oxo-G levels in PBMC suggested that ascorbic acid might indeed be acting directly as an antioxidant *in vivo* because the reduction was clearly observed upon supplementation and returned to baseline on removal of the ascorbic acid. However, the simultaneous increase in 8-oxo-A observed also suggested a concomitant pro-oxidant effect. The observation that 8-oxo-dG was released into the serum (possibly in the form of oligonucleotide fragments) and subsequently increased in urine, indicated that the lowering action of ascorbic acid was not due to inhibition of 8-oxo-G formation, but rather promotion of its removal.[20] There are at least two possible sources of 8-oxo-dG in urine; first, through cell turnover and DNA degradation; second, through repair of DNA/the nucleotide pool. To account for the increase in serum/urinary 8-oxo-dG, ascorbic acid could influence a number of processes. Ascorbic acid may: (i) act as a pro-oxidant for guanine moieties not contained in DNA e.g. dGTP giving rise to 8-oxo-dGTP, ultimately yielding 8-oxo-dG; (ii) promote the repair and/or purging of 8-oxo-dG from the nucleotide pool/DNA.[21] The timing of the maximum increase in urine (Fig. 2) compared to serum and PBMC would suggest ascorbic acid is having a residual effect, detectable long after plasma values have returned to baseline. It could be that the processing of such lesions accounts for the delay between removal of the lesion from DNA and its appearance in the urine. Account can now, therefore, be taken of the observed simultaneous increase in 8-oxo-A. 8-oxo-G is known to be repaired by a glycosylase enzyme hOGG1.[22] This enzyme removes only the base, however, our results in serum and urine suggest removal of the deoxynucleoside since the antibody used measures only the deoxynucleoside and not the base. The increase in 8-oxo-A could only be explained if ascorbic acid initially increases both lesions, via a pro-oxidant activity priming an adaptive response which up-regulates DNA repair processes, which in turn specifically eliminates 8-oxo-dG. We are currently investigating the role ascorbic acid may have in up-regulating the genes responsible for repair of oxidative damage to DNA, our hypothesis being that ascorbic acid stimulates the synthesis of repair enzymes designed to selectively remove important mutagenic lesions, such as 8-oxo-dG.

11. Redox Regulation of AP-1

AP-1 is an inducible transcription factor containing the protein products *fos* and *jun*. The DNA binding activity of *fos* and *jun* is regulated *in vitro* by a redox-dependent, post-translational mechanism. A conserved cysteine is the focus for both oxidation and reduction leading to DNA binding. Under antioxidant conditions strong DNA binding and transactivation of genes regulated by AP-1 is observed, whereas under pro-oxidant conditions there is only weak DNA binding

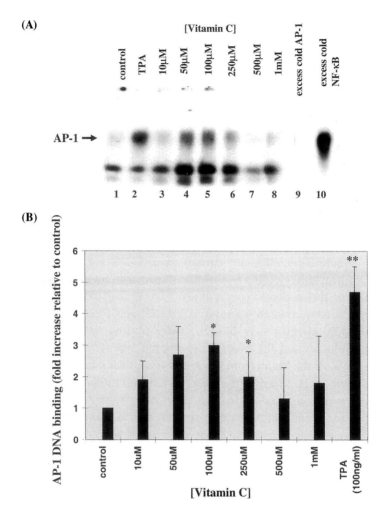

Fig. 3. *In vitro* activation of AP-1 DNA binding by addition and incubation of ascorbic acid with CCRF cells, showing a dose response up to levels present in serum or plasma following supplementation. Doses higher than 100 μM inhibit binding. No cell death was recorded up to 500 μM ascorbate. Data is reproduced in modified form from K. Holloway *et al.* (submitted).

of AP-1 and weak transactivation of genes.[23] In Fig. 3, we show the effect ascorbic acid on AP-1 DNA binding following incubation of CCRF cells with physiologically relevant concentrations of ascorbic acid. Binding is optimal after 3 h incubation (three fold increase), with 50–100 μM ascorbic acid. This coincides with the known extra-cellular concentration of ascorbic acid in individuals undergoing supplementation (500 mg/day) with the vitamin over a period of 6 weeks. Is this increased binding of AP-1 to DNA due to the pro-oxidant or anti-oxidant effect of ascorbic acid? It appears that formation of a

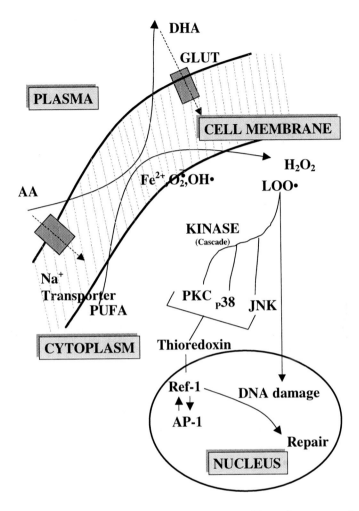

Fig. 4. Cell signalling by ascorbic acid–hypothetical view of how dietary supplementation may induce oxidative damage to DNA and subsequent activation of repair.

heterodimer of *c-fos* and *c-jun* proteins or a *c-jun* homodimer is involved in promoting AP1-DNA binding. Transcription of the *c-fos* promoter is regulated by serum response factor SRF/TCF and the cyclic AMP-responsive factor CREB, while AP-1 activates *c-jun* promoter as a positive auto-regulator.[24] Our observations on AP-1 activation and binding suggest that *c-jun* mRNA is induced relative to *c-fos* as indicated by the incubation of cells with antibodies to both *jun* and *fos* (unpublished observations) leading to weak DNA binding and weak transactivation. Under conditions of incubation with 50 μM ascorbic acid the *c-jun* mRNA is induced prior to *c-fos* while upon antioxidant stimulation (100 μM) *c-fos* mRNA is induced first followed by induction of *c-jun*. In vitamin C stimulated CCRF cells newly synthesized *c-fos* may interact with pre-existing *c-jun* molecules to form AP-1 which in turn activates the *c-jun* promoter by binding to its AP-1 site.

12. AP-1 Binding and DNA Repair

Ref-1 (alternatively referred to as human apurinic/apyrimidimic endonuclease hAPE) is a difunctional protein, which in addition to functioning as a DNA repair protein, is an important component of the signal transduction processes that regulate eukaryotic gene expression in response to cellular stress (it is itself induced by oxidative stress).[25] Thioredoxin maintains ref-1 in a reduced state thereby facilitating its interaction with *fos* and *jun* elements of AP-1 binding and promoting AP-1 binding to DNA.[26, 27] Our most recent results show that vitamin C induces *fos* and predominantly *jun*, *de novo* synthesis, presumably through the redox activation of AP-1 binding. An important sequealae of this may be the induction of p53, cell cycle arrest and repair of 8 oxo-dG, but not 8 oxo-dA, through a pro-oxidant action. The induction of nucleotide excision repair by this link would be entirely consistent with previous literature findings indicating links between AP-1 p53, ERCC-1 and nucleotide excision repair.[28, 29]

13. Conclusions

Our choice of markers for monitoring oxidative damage was 8-oxo-dG and 8-oxo-A, both lesions are products of oxidative attack on the guanine and adenine residues of DNA respectively. The trial reported here was designed so that each healthy volunteer was placed first on placebo (calcium carbonate 500 mg per day) for six weeks followed by six weeks of 500 mg ascorbic acid per day, hence each subject acted as their own control. The results illustrated a complex effect with 8-oxo-dG being decreased by intervention, while 8-oxo-A was increased, significantly, over the six weeks. This is entirely consistent with similar studies measuring oxidative DNA damage *in vivo* following supplementation with antioxidants.[18, 19] Our conclusion was that ascorbic acid had what appeared to be a profound antioxidant effect *in vivo* (exemplified by a reduction in 8-oxo-dG), which was partly balanced by a pro-oxidant effect on adenine. This result was difficult to explain in terms of a putative antioxidant effect of ascorbic acid. We have subsequently discovered that elimination and excretion of 8-oxo-dG can be tracked through DNA, serum and urine measurements and that this may be an indication of the *in vivo* repair of the lesion.[20] Results within the same study indicated that whilst ascorbic acid reduced 8-oxo-dG in the DNA of PBMC, it increased serum and urine levels. This strongly suggests that the apparent antioxidant effect of ascorbic acid may initially have been mistaken for a classical scavenging activity, whereas up-regulation, or priming of the enzyme(s) which eliminate 8-oxo-dG, may in fact have occurred. If this hypothesis is correct it could explain how levels of 8-oxo-G decreased while 8-oxo-A values increased in our supplementation study. Ascorbic acid can act as a pro-oxidant *in vitro* and may initiate a priming of DNA repair. This finding has precedence in the field of free radical biochemistry.[30] From radiation biology studies we know that healthy tissue surrounding a tumour can be protected by low dose irradiation, and that this elicits an adaptive response in the healthy tissue that helps to protect it from the subsequent cytotoxic dose directed against the tumour. An adaptive response has been implicated in previous intervention studies associated with the protective effect of antioxidant rich diets.[31] Such a response is likely to be influenced by transcription factors, some of which are in turn dependent on the redox status of the cell.[32] These novel findings concerning antioxidants and cell signalling open up a new frontier of research and present a challenging new area for exploitation of natural substances in the human diet as protective agents against genotoxcity.

Acknowledgments

J. Lunec would like to thank The Food Standards Agency and The Scottish Office for financial support of the Oxidative Stress Group.

References

1. Hornig, D. H. and Moser, U. (1981). The safety of high vitamin C intakes in man. *In* "Vitamin C (ascorbic acid)" (J. N. Counsell, and D. H. Horning, Eds.), pp. 225–248, New Jersey Applied Science Publishers.
2. Levine, M. *et al.* (1995). Determination of vitamin C requirements in humans, *Am. J. Clin. Nutri.* **62(Suppl.)**: 1347s–1356s.
3. COMA. (1991). Dietary Reference Values for Food Energy and Nutrients for the United Kingdom. Report of the Panel on Dietary Reference Values of the Committee on the Medical Aspects of Food Policy (COMA) Department of Health RHSS 41 London, HMSO.
4. Department of Health/Ministry of Agriculture Fisheries and Food. (1991). Report of the Working Group on Dietary Supplements and Health Foods.
5. Welch, R. W., Wang, Y., Crossman, A., Jr., Park, J. B., Kirk, K. L. and Levine, M. (1995). Accumulation of vitamin C (ascorbate) and its oxidized metabolite dehydroascorbic acid occurs by separate mechanisms. *J. Biol. Chem.* **270**: 12 584–12 592.
6. Kallner, A. *et al.* (1985). Formation of carbon dioxide from ascorbate in man. *Am. J. Clin. Nutri.* **41**: 609–613.

7. Frei, B., England, L. and Ames, B. N. (1989). Ascorbate is an outstanding antioxidant in human blood plasma. *Proc. Natl. Acad. Sci.* **86**: 6377–6381.
8. Halliwell, B. (1996). Vitamin C: antioxidant or pro-oxidant in vivo? *Free Radic. Res.* **25**: 439–454.
9. Herbert, V. *et al.* (1994). Vitamin C supplements are harmful to lethal for over 10% of Americans with high iron stores. *FASEB* **8**: A678.
10. Patterson, J. W. (1950). The diabetogenic effect of dehydroascorbic acids. *J. Biol. Chem.* **183**: 81–88.
11. Ortwerth, B. J., Linetsky, M. and Olesen, P. R. (1995). Ascorbic acid glycation of lens proteins produces UVA sensitizers similar to those in human lens. *Photochem. Photobiol.* **62**: 454–562.
12. Nienhuis ,A. W. (1981). Vitamin C and iron. *New England J. Med.* **304**: 170–171.
13. Bendioh, A. and Langseth, L. (1995). The health effects of vitamin C supplementation: a review. *J. Am. Coll. Nutri.* **14**: 124–136.
14. Podmore, I. D., Griffiths, H. R., Herbert, K. E., Mistry, N., Mistry, P. and Lunec, J. (1998). Vitamin C exhibits pro-oxidant properties. *Nature* **392**: 559.
15. Herbert, K. E., Evans, M. D., Finnegan, M. T. V., Farooq, S., Mistry, N., Podmore, I. D., Farmer, P. and Lunec, J. (1996). A novel HPLC procedure for the analysis of 8-oxoguanine in DNA. *Free Radic. Biol. Med.* **20**: 467–473.
16. Lunec, J. (1998). ESCODD: European Standards Committee on Oxidative DNA Damage. *Free Radic. Res.* **29**: 601–608.
17. Lunec, J., Herbert, K. E., Jones, G. D. D., Dickenson, L., Evans, M. D., Mistry, N., Chauhan, D., Capper, G. and Zheng, Q. (2000). Development of a quality control material for the measurement of 8-oxo-7,8 dihydro-2- deoxyguanosine, an in vivo marker of oxidative stress, and comparison of results from different laboratories. *Free Radic. Res.* **33**: S21–S26.
18. Rehman, A., Colli, S. C., Yang, M., Kelly, M., Diplock, A. T., Halliwell, B. and Rice-Evans, C. (1998). The effects of co-supplementation of iron and vitamin C on oxidative damage to DNA in healthy volunteers. *Biochem. Biophys. Res. Commun.* **246**: 293–298.
19. Rehman, A., Bourne, L. C., Halliwell, B. and Rice Evans, C. (1999). Tomato consumption modulates oxidative DNA damage in humans. *Biochem. Biophys. Res. Commun.* **262**: 828–831.
20. Cooke, M. S., Evans, M. D., Podmore, I. D., Herbert, K. E., Mistry, P., Hickenbotham, P. T., Husseini, A., Griffiths, H. R. and Lunec, J. (1998). Novel repair action of vitamin C upon in vivo oxidative DNA damage. *FEBS Letts.* **439**: 363–367.
21. Cooke, M. S., Evans, M. D., Herbert, K. E. and Lunec, J. (2000). Urinary 8-oxo2'-deoxyguanosine, source, significance and supplements. *Free Radic. Res.* **32**: 381–397.

22. Arai, K., Morishita, K., Shinmura, K., Kohno, T., Kim, S. R., Nohmi, T., Taniwaki, M., Ohwada, S. and Yokata, J. (1997). Cloning of human homolog of the yeast OGG1 gene that is involved in the repair of oxidative DNA damage. *Oncogene* **14**: 2857–2861.
23. Meyer, M., Pahl, H. L. and Baeuerle, P. A. (1994). Regulation of the transcription factors NF-κB and AP-1 by redox changes. *Chemico-Biological Interactions* **91**: 91–100.
24. Groseh, S. and Kaina, B. (1999). Transcription activation of Apurinic/ Apyrimidimic Endonuclease (Ape, Ref-1) by oxidative stress requires CREB. *Biochem. Biophys. Res. Commun.* **261**: 859–863.
25. Xanthoudakis, S. and Curran, T. (1996). Redox regulation of AP-1. A link between transcription factor signalling and DNA repair. *In* "Biological Reactive Intermediates V" (R. Snyder, Ed.), pp. 69–75, Plenum Press New York.
26. Parvis, G., Mustachich, D. and Coon, A. (2000). The role of the redox protein thioredoxin in cell growth and cancer. *Free Radic. Biol. Med.* **29**: 312–322.
27. Schenk, H., Klein, M., Edbrugger, W., Droge, W. and Schulze-Ostholt, K. (1994). Distinct effects of thioredoxin and antioxidants on the activation of transcription factors NF-κB and AP-1. *Proc. Natl. Acad. Sci. USA* **91**: 1672–1676.
28. Li, Q., Tsang, B., Bostick-Bruton, F. and Reed E. (1999). Modulation of excision repair cross complementation group 1 (ERCC-1) mRNA expression by pharmacological agents in human ovarian carcinoma cells. *Biochem Pharmacol.* **57**: 347–353.
29. Kirch, H. C., Flasivinkels, S., Rump, F. H., Brockmann, D. and Esche, H. (1999). Expression of human p53 requires synergistic activation of transcription from the p53 promoter by AP-1, NF-κB and *myc/max*. *Oncogene* **18**: 2728–2738.
30. Kaina, B., Kiesel, S., Grombacher, T. and Fritz, G. (1998) Inducible protective functions of mammalian cells against DNA damaging agents. *Periodicum Biologorum* **100**: 339–344.
31. Ramana, C. V., Boldogh, I., Iziumi, T. and Mitra S. (1998). Activation of apurinic apyrimidimic endonuclease in human cells by reactive oxygen species and its correlation with their adaptive response to genotoxicity. *Proc. Natl. Acad. Sci. USA* **95**: 5061–5066.
32. Adler, V., Zhimin, Y., Tew, K. D. and Ronai, Z. (1999). Role of redox potential and reactive oxygen species in stress signalling. *Oncogene* **18**: 6104–6111.

Chapter 35

Carotenoids and Oxidative Stress

Norman I. Krinsky

Norman I. Krinsky • Department of Biochemistry, School of Medicine, and the Jean Mayer USDA Human Nutrition Research Center on Aging at Tufts University, Boston, MA 02111 USA
Tel: +1-617-636-6861, E-mail: norman.krinsky@tufts.edu

1. Introduction

There have been numerous reviews that have appeared in the last few years detailing various aspects of the antioxidant,[1–3] or even pro-oxidant,[4,5] actions of carotenoids. In fact, some of these reviews question whether carotenoids have an antioxidant action *in vivo*.[6,7]

This chapter will focus on recent results indicating the evidence for an antioxidant action of carotenoids *in vitro*, *ex vivo*, and *in vivo*, and will also touch on the evidence concerning pro-oxidation. In addition, I will discuss the interesting hypothesis of Jandacek[8] concerning carotenoids levels in humans with reference to oxidative stress and disease.

2. Carotenoids as Antioxidants

There have been many reports concerning the relative antioxidant efficacy of carotenoids, with many different results. Part of the problem with attempting to evaluate efficacy is that very different systems have been used to dissolve the carotenoids, initiate oxidant stress, and then evaluate efficacy. There is probably no system that would be most useful, but some recent *in vitro*, *ex vivo* and *in vivo* investigations are detailed below.

2.1. *In vitro* Studies

It is very clear that the nature of the interaction between the carotenoids and the matrix in which they are studied dictates the effect. This is seen most clearly in the study of Liebler *et al.*[9] who reported that when β-carotene was incorporated into liposomes, it was an effective antioxidant against 2,2′-azobis (2-amidinopropane) dihydrochloride (AAPH)-induced lipid peroxidation, but its effectiveness was lost when it was added to preformed liposomes. Other studies using liposomes have indicated that both astaxanthin and canthaxanthin can protect liposomes versus Cu^+-initiated lipid peroxidation[10] and that zeaxanthin can react with peroxynitrous acid (HOONO) in liposomes, and presumably protect them.[11] On the basis of that observation, these authors suggested that zeaxanthin may protect the macular region of the retina from peroxynitrite attack. The activity of zeaxanthin in protecting liposomes against AAPH attack was reported to be equivalent to that seen with alpha-tocopherol, while β-carotene, canthaxanthin, and astaxanthin were somewhat weaker and lycopene was least effective.[12] In fact, these authors reported that after 60 mins of incubation, lycopene became a pro-oxidant.

Many of the earlier investigation used the development of thiobarbituric acid-related substances (TBARS) as an index of lipid peroxidation (reviewed in Ref. 13), but this assay is quite non-specific, and as Kikugawa *et al.* have demonstrated recently,[14] the oxidation of β-carotene by either nitrogen dioxide or oxygen itself results in measurable TBARS activity.

The order of effectiveness of carotenoids in preventing radical- or oxidative stress-initiated damage has been studied by a number of authors,[15-18] but at this time the results are so variable that it does not appear to be useful with respect to the possible effectiveness of different carotenoids in humans.

2.2. *Ex vivo* Studies

Most of the *ex vivo* studies involving carotenoids have used either low-density lipoproteins (LDL) or microsomal fractions from various tissues. The LDL investigations consisted of evaluating the antioxidant ability of carotenoids added directly to LDL or introduced into LDL by oral ingestion of supplements or fruits and vegetables. In the last few years, more and more investigators are using this latter approach, which presumably inserts the carotenoids "appropriately" in the LDL particle.

The most recent studies in which carotenoids have been added to either plasma or isolated LDL fractions demonstrate that β-carotene addition is protective,[19-22] although one study reports that β-carotene addition results in a pro-oxidative action, as demonstrated by an increase in TBARS.[23] However, in this latter study there was no change in the lag period or rate of LDL oxidation upon addition of β-carotene. In the studies that looked at carotenoids other than β-carotene, very mixed results were reported. In some cases carotenoids such as canthaxanthin and zeaxanthin were effective antioxidants[19] as were lycopene, alpha-carotene, β-cryptoxanthin, zeaxanthin and lutein.[21] However, in some studies in which β-carotene was effective, the addition of either lutein or lycopene actually increased LDL oxidation.[20] Based on the above, it would seem that the final answer to the efficacy of carotenoids added to either plasma or LDL to act as effective antioxidants remains to be answered.

In the case of LDL particles enriched through dietary intervention with fruits and vegetables or by supplementation with carotenoids, there are very clearly two different patterns observed. In one case, supplementation with either β-carotene,[24, 25] or mixed carotenoids to a population depleted of dietary carotenoids[26] resulted in protection of the isolated LDL. Green vegetable supplementation does not protect LDL in either smokers or non-smokers, whereas red vegetable supplementation was protective only in non-smokers, and not in smokers.[27] Lycopene from tomato based products was reported to be effective[28] whereas pure lycopene supplementation was ineffective.[24]

It should be mentioned that not all studies of carotenoid supplementation have resulted in a change in LDL oxidizability. A 12 week period of daily supplementation with either 13 mg lycopene or 112 mg β-carotene resulted in an increase in LDL carotenoids, but no change in LDL oxidizability.[29] Similar results were seen with lutein supplementation for 1–2 weeks that resulted in a 4–6 fold increase in serum lutein but was without effect on LDL lag time.[30]

When fruits and vegetables are added to the diet, not only do plasma carotenoids increase, but vitamin C and other potential antioxidants such as polyphenols and flavonoids may also increase. Therefore, it is very difficult to interpret whether the changes observed in LDL oxidizability are due to an increase in carotenoids or to other components of the fruit and vegetables. One recent study of added fruits and vegetables reported an increase in the resistance of LDL to oxidation[31] whereas two other studies did not find any effect of the diet on LDL resistance.[30, 32] These variable results might be attributed to different lengths of time on the diets, different degrees of changes in the plasma carotenoid levels, and certainly, different study populations.

2.3. *In vivo* Studies

Various animal species have been used for many years in attempts to evaluate the *in vivo* antioxidant effect of carotenoids. However, these studies are marred by the fact that most experimental animals are very poor absorbers of carotenoids, and only large, pharmacological doses permit absorption of carotenoids into these animals. Some animals that can absorb dietary carotenoids, such as the ferret, gerbils and pre-ruminant calves, have been used to study carotenoid absorption, but virtually nothing has been done with respect to antioxidant effectiveness in these species.[33] Thus, we are left with humans, who have an almost unlimited capacity to absorb dietary carotenoids,[34] but have a somewhat limited ability to indicate their oxidative stress status. It has been suggested that the evidence for an antioxidant role for carotenoids *in vivo* is not very strong.[7] Nevertheless, there are some studies that are worth discussing.

The key issue in determining whether dietary carotenoids alter the oxidative stress status in humans is the selection of appropriate biomarkers. For many years, determination of thiobarbituric acid-reactive substances (TBARS), such as malondialdehyde (MDA) was assumed to be a valid measure of lipid peroxidation, but we now know that this is a somewhat non-specific biomarker. Nevertheless, using TBARS or MDA, investigators have evaluated the effect of added carotenoids in several instances where an oxidative stress might arise.

Dixon and her associates[35, 36] put women on carotenoid-deficient diets, and observed an increase in plasma MDA levels. This effect could be reversed when the diets were supplemented with a mixture of carotenoids, strongly supporting the idea that dietary carotenoids can serve to decrease oxidative stress in humans.

Another group that exhibits oxidative stress are patients suffering from cystic fibrosis (CF), which by preventing pancreatic enzymes secretion prevents appropriate absorption of fat-soluble vitamins such as vitamin E. Children with CF are routinely supplemented with vitamin E, but even so, their plasma MDA levels may be above control subjects. When treated with β-carotene (0.5 mg/kg) for 3 months, not only does the elevated MDA level fall, but there was also a prolongation in the lag time of LDL oxidation.[37] Other groups have also demonstrated normalization of MDA levels in CF children treated with either 13 mg β-carotene/day for 2 months[38] or 50 mg/day for 10 weeks.[39]

Another study reported that in a group of Iranian men that had high MDA levels, supplementation with 30 mg/day β-carotene for 10 weeks could significantly reduce the MDA.[40]

In addition to lipid products, damage to DNA has also been used as a biomarker of oxidative damage. The most common product measured has been 8-hydroxy-2'-deoxyguanosine (8-OH-dG), even though there is still some question as to the relative importance of this marker in terms of evaluating DNA damage.[41] In addition, levels of this marker, either in urine or lymphocytes, have not decreased when diets were supplemented with either β-carotene,[42, 43] lutein, or lycopene.[43] In contrast to these observations have been the reports of significant decreases in 8-OH-dG levels following supplementation with carrot juice[44] or increased fruit and vegetable consumption.[45] Another marker of DNA damage is the number of strand breaks observed in lymphocytic DNA, and increased fruit and vegetable intake can decrease that biomarker.[44] These strand breaks can also be induced by treating lymphocytes with hydrogen peroxide (H_2O_2), and pre-treatment of the donors of these lymphocytes with a tomato puree supplement also significantly decreases strand breakage.[46]

Other markers of oxidative stress that have been evaluated with respect to carotenoid plasma levels, and intake from fruits and vegetables or supplements have included the ferric reducing ability assay (FRAP)[47] as well as a total antioxidant capacity assay.[48] It has been reported that supplementation with either β-carotene or spinach (whole-leaf, minced or liquefied) did not result in any change in the plasma FRAP level.[49] Lee *et al.* added tomato products (tomato soup and canned tomatoes) to the diet with either olive oil or sunflower oil, and determined FRAP activity.[50] Using either oil, the plasma lycopene level increased significantly, but only the olive oil arm resulted in an increase in FRAP levels, whereas the sunflower oil did not improve antioxidant activity. This would suggest that it is not the carotenoid component of the tomato products that was associated with FRAP activity, but something else whose absorption was modified by the type of oil in the diet. It could not be the vitamin E content, for the sunflower oil contained 14 times that found in the olive oil.

To summarize, the addition of carotenoids, either as supplements or from food sources, does not seem to play too important a role in determining antioxidant status in human plasma.

3. Carotenoids as Pro-Oxidants

The concept that carotenoids might behave as pro-oxidants is derived from the conclusions of Burton and Ingold[51] that at high, non-physiological, oxygen tensions (760 torr; 100% oxygen), relatively high concentrations (> 0.5 mM) of β-carotene behaved as a pro-oxidant. However, close inspection of the data in that important paper strongly suggests that the phenomenon observed was actually a decrease in antioxidant activity under the above conditions, and not necessarily a pro-oxidant effect. Thus, at 150 torr (20% oxygen), β-carotene was an effective antioxidant in inhibiting the oxidation of methyl linoleate initiated by the radical generator, AIBN.[51] At 760 torr, and with prolonged time, there was a marked decrease in the antioxidant effect, suggesting an autocatalytic inactivation of the β-carotene. In a subsequent paper, Burton concluded that at the low partial pressures of oxygen found in mammalian tissues, β-carotene has the potential to act as an antioxidant, complementing the role of vitamin E, which is effective at higher oxygen tensions.[52] Nevertheless, many investigators confused the high oxygen tensions (760 torr) used by Burton and Ingold[51] with the oxygen tension in the lung, which would be 150 torr for the inspired air, and then drop rapidly to 15 torr or less in the tissues.

There have been additional reports that at either high oxygen tensions or in experiments using high concentrations of carotenoids, there can be some evidence of a pro-oxidant effect. Much of this material has been covered in the excellent review by Palozza[5] so only a few new articles will be discussed here. With respect to oxygen tension, the only recent report indicates that at 150 torr, β-carotene loses only 4% of the effectiveness observed at 15 torr in protecting human serum albumin from oxidation by AAPH, but at 760 torr, 1.6 μM β-carotene increased protein oxidation by 26%.[53] Human plasma contains about 1–2 μM total carotenoids,[54] so the concentration used in this study is in the physiological range.

The effects of carotenoid concentration on antioxidant/pro-oxidant effectiveness has been studied by Lowe *et al.*,[55] who added either β-carotene or lycopene to HT29 colon carcinoma cells and used xanthine/xanthine oxidase to induce oxidative damage. They measured both the comet assay for DNA damage and changes in membrane integrity using ethidium bromide uptake. At physiological levels (1–3 μM), both β-carotene and lycopene prevented cellular damage, but at higher doses (4–10 μM), the ability to protect these cells was lost, and their data suggests that the membrane integrity was more sensitive to these high doses than was the DNA damage. Bestwick and Milne[56] treated Caco-2 cell cultures with varying levels of β-carotene (0.1–50 μM) and reported that at 50 μM β-carotene, there was a significant reduction in intracellular levels of reactive oxygen species, but at the same time, there were indications of decreased resistance of a H_2O_2 challenge with respect to enhanced Trypan blue staining, indicating increased membrane liability.

In other cell systems, antioxidant, pro-oxidant, or no effect have been observed. The antioxidant effects were observed in human lung cells pre-treated with β-carotene and exposed to tobacco-specific nitrosamines,[57] whereas a pro-oxidant effect was reported for Hep62 cells treated with 10 μM β-carotene and then exposed to H_2O_2, in which case the cells treated with β-carotene showed increased levels of DNA strand breaks.[58] Adding β-carotene directly to human plasma and then exposing the LDL to Cu^{2+}-mediated oxidation resulted in a large increase in MDA, but with only a modest increase in the oxidation rate of LDL.[23] HL-60 cells pre-loaded with β-carotene (up to 1.5 nmol/10^6 cells) and treated with 2,2'-azobis(2,4-dimethylvaleronitrile) (AMVN) did not show any effect on either cell viability or oxidation of *cis*-parinaric acid incorporation into a variety of membrane phospholipid classes.[59]

Finally, there is a report that a 4-week period of β-carotene supplementation (60 mg/day) can decrease oxidative DNA damage (8-OH-dG) in leukocytes of nonsmokers, but increased the level of DNA damage in smokers.[60] However, this treatment had no impact on the level of DNA damage, as evaluated by strand breaks (Comet assay), in either the smokers or nonsmokers, which suggests that the correlation between 8-OH-dG measurements and the Comet assay is not too strong. These results are confounded by the very marked differences in the baseline levels of 8-OH-dG in the groups prior to supplementation.

4. Carotenoids and Aging

Whether one accepts the idea that oxidants play a major role in initiating DNA damage leading to aging[61] or that proof of a causal relationship of oxidative DNA damage as an important carcinogenic factor is still lacking,[62] we still have to account for the epidemiological evidence that diets rich in fruits and vegetables are associated with a lower incidence of chronic diseases. And then, we have to try to relate those diets to carotenoid content. Most of the epidemiological evidence is associative in nature, and even when we find reports of a positive association between fruit and vegetable intake and a decreased incidence of chronic diseases, we cannot attribute that outcome to the carotenoid content of the diet. A recent review concludes that "the current evidence is insufficient to conclude that antioxidant vitamin supplementation (including β-carotene) materially reduces oxidative damage in humans".[63] Another recent review finds that "a causal role for oxidative stress in the aging process has not been clearly established".[64] Virtually nothing has been done to follow-up on Cutler's report of a positive correlation between species longevity and the levels of various antioxidants, including carotenoids.[65] This correlation probably reflects dietary differences, for we now know that carotenoids serve as one of the best markers of fruit and vegetable intake.[66] Therefore, the various reports of associations between β-carotene intake and behavioral alterations in "older" populations are probably reporting on associations between fruit and vegetable intake and the measured physiological function.

A recent report has suggested that the inverse relationship observed between plasma carotenoid levels and some chronic diseases may actually reflect the destruction of the carotenoids by the oxidant stress associated with the diseases.[8] This is a very provocative idea, and would suggest that the oxidative breakdown products of carotenoids should be investigated for any role that they may play in the disease process. This concept had been proposed earlier following the observation that lung tissue from ferrets exposed to cigarette smoke accumulated such breakdown products following incubation with β-carotene,[67] and that this observation may be related to the human intervention trials that showed an increase in lung tumors in smokers given large dose β-carotene supplementation.[68, 69]

5. Summary

In 1989, a comprehensive review appeared entitled "Antioxidant Functions of Carotenoids" that listed many reports of antioxidant actions of carotenoids in animals, cells and in *in vitro* experiments.[70] The strongest data came from experiments demonstrating the ability of carotenoid pigments to quench singlet excited oxygen. In the intervening 12 years, we have accumulated more evidence of an antioxidant action *in vitro*, somewhat less so for *ex vivo* experiments, such as those using LDL particles, and even less so for *in vivo* studies. There is also new evidence that under high oxygen tensions, i.e. those above ambient pressures, carotenoids can lose their antioxidant properties and may even behave as pro-oxidants. When we attempt to analyze the results of intervention with fruits and vegetables in humans, we find that the results are very confounded with respect to carotenoids. Although they may serve as excellent biomarkers of fruit and vegetable intake, there are many other compounds in these types of diets that may affect the level of "oxidant stress' and possibly, overall good health. Twelve years ago, when summarizing the antioxidant properties of carotenoids, the statement was made that "we still have much to learn about these molecules".[70] We seem to be in the same position today.

References

1. Krinsky, N. I. (1998). The antioxidant and biological properties of the carotenoids. *Ann. NY Acad. Sci.* **854**: 443–447.
2. Paiva, S. A. and Russell, R. M. (1999). β-Carotene and other carotenoids as antioxidants *J. Am. Coll. Nutri.* **18**: 426–433.

3. Terao, J., Oshima, S., Ojima, F., Lim, B. P. and Nagao, A. (1997). Carotenoids as antioxidants. *In* "Food and Free Radicals" (M. Hiramatsu, Ed.), pp. 21–29, Plenum Press, New York.

4. Edge, R. and Truscott, T. G. (1997). Prooxidant and antioxidant reaction mechanisms of carotene and radical interactions with vitamins E and C. *Nutrition* **13**: 992–994.

5. Palozza, P. (1998). Prooxidant actions of carotenoids in biologic systems. *Nutri. Rev.* **56**: 257–265.

6. Halliwell, B. (1999). Antioxidant defence mechanisms: from the beginning to the end (of the beginning). *Free Radic. Res.* **31**: 261–272.

7. Rice-Evans, C., Sampson, J., Bramley, P. M. and Holloway, D. E. (1997). Why do we expect carotenoids to be antioxidants in vivo? *Free Radic. Res.* **26**: 381–398.

8. Jandacek, R. J. (2000). The canary in the cell: a sentinel role for β-carotene. *J. Nutri.* **130**: 648–651.

9. Liebler, D. C., Stratton, S. P. and Kaysen, K. L. (1997). Antioxidant actions of β-carotene in liposomal and microsomal membranes: role of carotenoid-membrane incorporation and α-tocopherol. *Arch. Biochem. Biophys.* **338**: 244–250.

10. Rengel, D., Díez-Navajas, A., Serna-Rico, A., Veiga, P., Muga, A. and Milicua, J. C. G. (2000). Exogenously incorporated ketocarotenoids in large unilamellar vesicles. Protective activity against peroxidation. *Biochim. Biophys. Acta* **1463**: 179–187.

11. Scheidegger, R., Pande, A. K., Bounds, P. L. and Koppenol, W. H. (1998). The reaction of peroxynitrite with zeaxanthin. *Nitric Oxide* **2**: 8–16.

12. Woodall, A. A., Britton, G. and Jackson, M. J. (1997). Carotenoids and protection of phospholipids in solution or in liposomes against oxidation by peroxyl radicals: relationship between carotenoid structure and protective ability. *Biochim. Biophys. Acta* **1336**: 575–586.

13. Palozza, P. and Krinsky, N. I. (1994). Antioxidant properties of carotenoids. *In* "Retinoids: From Basic Science to Clinical Applications" (M. A. Livrea, and G. Vidali, Eds.), pp. 35–41, Birkhaüser, Basel.

14. Kikugawa, K., Hiramoto, K. and Hirama, A. (1999). β-carotene generates thiobarbituric acid-reactive substances by interaction with nitrogen dioxide in air. *Free Radic. Res.* **31**: 517–523.

15. Jiménez-Escrig, A., Jiménez-Jimenez, I., Sánchez-Moreno, C. and Saura-Calixto, F. (2000). Evaluation of free radical scavenging of dietary carotenoids by the stable radical 2,2-diphenyl-1-picrylhydrazyl. *J. Sci. Food Agric.* **80**: 1686–1690.

16. Naguib, Y. M. (2000). Antioxidant activities of astaxanthin and related carotenoids. *J. Agric. Food Chem.* **48**: 1150–1154.

17. Miller, N. J., Sampson, J., Candeias, L. P., Bramley, P. M. and Rice-Evans, C. A. (1996). Antioxidant activities of carotenes and xanthophylls. *FEBS Lett.* **384**: 240–242.
18. Mortensen, A. and Skibsted, L. H. (1997). Free radical transients in photobleaching of xanthophylls and carotenes. *Free Radic. Res.* **26**: 549–563.
19. Carpenter, K. L., van der Veen, C., Hird, R., Dennis, I. F., Ding, T. and Mitchinson, M. J. (1997). The carotenoids beta-carotene, canthaxanthin and zeaxanthin inhibit macrophage-mediated LDL oxidation. *FEBS Lett.* **401**: 262–266.
20. Dugas, T. R., Morel, D. W. and Harrison, E. H. (1998). Impact of LDL carotenoid and alpha-tocopherol content on LDL oxidation by endothelial cells in culture. *J. Lipid Res.* **39**: 999–1007.
21. Panasenko, O. M., Sharov, V. S., Briviba, K. and Sies, H. (2000). Interaction of peroxynitrite with carotenoids in human low density lipoproteins. *Arch. Biochem. Biophys.* **373**: 302–305.
22. Romanchik, J. E., Harrison, E. H. and Morel, D. W. (1997). Addition of lutein, lycopene, or β-carotene to LDL or serum in vitro: effects on carotenoid distribution, LDL composition, and LDL oxidation. *J. Nutri. Biochem.* **8**: 681–688.
23. Bowen, H. T. and Omaye, S. T. (1998). Oxidative changes associated with β-carotene and α-tocopherol enrichment of human low-density lipoproteins. *J. Am. Coll. Nutri.* **17**: 171–179.
24. Dugas, T. R., Morel, D. W. and Harrison, E. H. (1999). Dietary supplementation with β-carotene, but not with lycopene, inhibits endothelial cell-mediated oxidation of low-density lipoprotein. *Free Radic. Biol. Med.* **26**: 1238–1244.
25. Levy, Y., Kaplan, M., Ben-Amotz, A. and Aviram, M. (1996). Effect of dietary supplementation of β-carotene on human monocyte-macrophage-mediated oxidation of low density lipoprotein. *Isr. J. Med. Sci.* **32**: 473–478.
26. Lin, Y., Burri, B. J., Neidlinger, T. R., Muller, H. G., Dueker, S. R. and Clifford, A. J. (1998). Estimating the concentration of β-carotene required for maximal protection of low-density lipoproteins in women. *Am. J. Clin. Nutri.* **67**: 837–845.
27. Chopra, M., O'Neill, M. E., Keogh, N., Wortley, G., Southon, S. and Thurnham, D. I. (2000). Influence of increased fruit and vegetable intake on plasma and lipoprotein carotenoids and LDL oxidation in smokers and nonsmokers. *Clin. Chem.* **46**: 1818–1829.
28. Agarwal, S. and Rao, A. V. (1998). Tomato lycopene and low density lipoprotein oxidation: a human dietary intervention study. *Lipids* **33**: 981–984.
29. Carroll, Y. L., Corridan, B. M. and Morrissey, P. A. (2000). Lipoprotein carotenoid profiles and the susceptibility of low density lipoprotein to oxidative modification in healthy elderly volunteers. *Eur. J. Clin. Nutri.* **54**: 500–507.

30. Chopra, M., McLoone, U., O'Neill, M., Williams, N. and Thurnham, D. I. (1996). Fruit and vegetable supplementation — effect on ex vivo LDL oxidation in humans. *In* "Natural Antioxidants and Food Quality in Atherosclerosis and Cancer Prevention" (J. T. Kumpulainen, and J. T. Salonen, Eds.), pp. 150–155, The Royal Society of Chemistry, London.

31. Hininger, I., Chopra, M., Thurnham, D. I., Laporte, F., Richard, M. J., Favier, A. and Roussel, A. M. (1997). Effect of increased fruit and vegetable intake on the susceptibility of lipoprotein to oxidation in smokers. *Eur. J. Clin. Nutri.* **51**: 601–606.

32. van het Hof, K. H., Brouwer, I. A., West, C. E., Haddeman, E., Steegers-Theunissen, R. P., van Dusseldorp, M., Weststrate, J. A., Eskes, T. K. and Hautvast, J. G. (1999). Bioavailability of lutein from vegetables is 5 times higher than that of beta-carotene. *Am. J. Clin. Nutri.* **70**: 261–268.

33. Lee, C. M., Boileau, A. C., Boileau, T. W., Williams, A. W., Swanson, K. S., Heintz, K. A. and Erdman, J. W., Jr. (1999). Review of animal models in carotenoid research. *J. Nutri.* **129**: 2271–2277.

34. Parker, R. S., Swanson, J. E., You, C.-S., Edwards, A. J. and Huang, T. (1999). Bioavailability of carotenoids in human subjects. *Proc. Nutri. Soc.* **58**: 155–162.

35. Dixon, Z. R., Burri, B. J., Clifford, A., Frankel, E. N., Schneeman, B. O., Parks, E., Keim, N. L., Barbieri, T., Wu, M.-M., Fong, A. K. H., Kretsch, M. J., Sowell, A. L. and Erdman, J. W., Jr. (1994). Effects of a carotene-deficient diet on measures of oxidative susceptibility and superoxide dismutase activity in adult women. *Free Radic. Biol. Med.* **17**: 537–544.

36. Dixon, Z. R., Shie, F. S., Warden, B. A., Burri, B. J. and Neidlinger, T. R. (1998). The effect of a low carotenoid diet on malondialdehyde-thiobarbituric acid (MDA-TBA) concentrations in women: a placebo-controlled double- blind study. *J. Am. Coll. Nutri.* **17**: 54–58.

37. Winklhofer-Roob, B. M., Puhl, H., Khoschsorur, G., van't Hof, M. A., Esterbauer, H. and Shmerling, D. H. (1995). Enhanced resistance to oxidation of low density lipoproteins and decreased lipid peroxide formation during β-carotene supplementation in cystic fibrosis. *Free Radic. Biol. Med.* **18**: 849–859.

38. Lepage, G., Champagne, J., Ronco, N., Lamarre, A., Osberg, I., Sokol, R. J. and Roy, C. C. (1996). Supplementation with carotenoids corrects increased lipid peroxidation in children with cystic fibrosis. *Am. J. Clin. Nutri.* **64**: 87–93.

39. Rust, P., Eichler, I., Renner, S. and Elmadfa, I. (1998). Effects of long-term oral beta-carotene supplementation on lipid peroxidation in patients with cystic fibrosis. *Int. J. Vitam. Nutri. Res.* **68**: 83–87.

40. Meraji, S., Ziouzenkova, O., Resch, U., Khoschsorur, A., Tatzber, F. and Esterbauer, H. (1997). Enhanced plasma level of lipid peroxidation in Iranians

could be improved by antioxidants supplementation. *Eur. J. Clin. Nutri.* **51**: 318–325.

41. Halliwell, B. (2000). Why and how should we measure oxidative DNA damage in nutritional studies? How far have we come? *Am. J. Clin. Nutri.* **72**: 1082–1087.

42. van Poppel, G., Poulsen, H., Loft, S. and Verhagen, H. (1995). No influence of beta carotene on oxidative DNA damage in male smokers. *J. Natl. Cancer Inst.* **87**: 310–311.

43. Collins, A. R., Olmedilla, B., Southon, S., Granado, F. and Duthie, S. J. (1998). Serum carotenoids and oxidative DNA damage in human lymphocytes. *Carcinogenesis* **19**: 2159–2162.

44. Pool-Zobel, B. L., Bub, A., Muller, H., Wollowski, I. and Rechkemmer, G. (1997). Consumption of vegetables reduces genetic damage in humans: first results of a human intervention trial with carotenoid-rich foods. *Carcinogenesis* **18**: 1847–1850.

45. Haegele, A. D., Gillette, C., O'Neill, C., Wolfe, P., Heimendinger, J., Sedlacek, S. and Thompson, H. J. (2000). Plasma xanthophyll carotenoids correlate inversely with indices of oxidative DNA damage and lipid peroxidation. *Cancer Epidemiol. Biomark. Prev.* **9**: 421–425.

46. Porrini, M. and Riso, P. (2000). Lymphocyte lycopene concentration and DNA protection from oxidative damage is increased in women after a short period of tomato consumption. *J. Nutri.* **130**: 189–192.

47. Benzie, I. F. F. and Strain, J. J. (1996). The ferric reducing ability of plasma (FRAP) as a measure of "antioxidant power": the FRAP assay. *Anal. Biochem.* **239**: 70–76.

48. Miller, J., Rice-Evans, C., Davies, M. J., Gopiathan, V. and Milner, A. (1993). A novel method for measuring antioxidant capacity and its application to monitoring the antioxidant status in premature neonates. *Clin. Sci.* **84**: 407–412.

49. Castenmiller, J. J., Lauridsen, S. T., Dragsted, L. O., van het Hof, K. H., Linssen, J. P. and West, C. E. (1999). β-carotene does not change markers of enzymatic and nonenzymatic antioxidant activity in human blood. *J. Nutri.* **129**: 2162–2169.

50. Lee, A., Thurnham, D. I. and Chopra, M. (2000). Consumption of tomato products with olive oil but not sunflower oil increases the antioxidant activity of plasma. *Free Radic. Biol. Med.* **29**: 1051–1055.

51. Burton, G. W. and Ingold, K. U. (1984). β-carotene: an unusual type of lipid antioxidant. *Science* **224**: 569–573.

52. Burton, G. W. (1989). Antioxidant action of carotenoids. *J. Nutri.* **119**: 109–111.

53. Zhang, P. and Omaye, S. T. (2000). β-carotene and protein oxidation: effects of ascorbic acid and α-tocopherol. *Toxicology* **146**: 37–47.

54. Yong, L.-C., Forman, M. R., Beecher, G. R., Graubard, B. I., Campbell, W. S., Reichman, M. E., Taylor, P. R., Lanza, E., Holden, J. M. and Judd, J. T. (1994). Relationship between dietary intake and plasma concentrations of carotenoids in premenopausal women: application of the USDA-NCI carotenoid food-consumption database. *Am. J. Clin. Nutri.* **60**: 223–230.

55. Lowe, G. M., Booth, L. A., Young, A. J. and Bilton, R. F. (1999). Lycopene and β-carotene protect against oxidative damage in HT29 cells at low concentrations but rapidly lose this capacity at higher doses. *Free Radic. Res.* **30**: 141–151.

56. Bestwick, C. S. and Milne, L. (2000). Effects of β-carotene on antioxidant enzyme activity, intracellular reactive oxygen and membrane integrity within post confluent Caco-2 intestinal cells. *Biochim. Biophys. Acta* **1474**: 47–55.

57. Weitberg, A. B. and Corvese, D. (1997). Effect of vitamin E and beta-carotene on DNA strand breakage induced by tobacco-specific nitrosamines and stimulated human phagocytes. *J. Exp. Clin. Cancer Res.* **16**: 11–14.

58. Woods, J. A., Bilton, R. F. and Young, A. J. (1999). β-Carotene enhances hydrogen peroxide-induced DNA damage in human hepatocellular HepG2 cells. *FEBS Lett.* **449**: 255–258.

59. Day, B. W., Bergamini, S., Tyurina, Y. Y., Carta, G., Tyurin, V. A. and Kagan, V. E. (1998). β-carotene. An antioxidant or a target of oxidative stress in cells? *In* "Subcellular Biochemistry" (P. J. Quinn, and V. E. Kagan, Eds.), Vol. 30, pp. 209–217, Plenum, New York.

60. Welch, R. W., Turley, E., Sweetman, S. F., Kennedy, G., Collins, A. R., Dunne, A., Livingstone, M. B., McKenna, P. G., McKelvey-Martin, V. J. and Strain, J. J. (1999). Dietary antioxidant supplementation and DNA damage in smokers and nonsmokers. *Nutri. Cancer* **34**: 167–172.

61. Ames, B. N. and Shigenaga, M. K. (1992). Oxidants are a major contributor to aging. *Ann. NY Acad. Sci.* **663**.

62. Loft, S. and Poulsen, H. E. (1996). Cancer risk and oxidative DNA damage in man. *J. Mol. Med.* **74**: 297–312.

63. McCall, M. R. and Frei, B. (1999). Can antioxidant vitamins materially reduce oxidative damage in humans? *Free Radic. Biol. Med.* **26**: 1034–1053.

64. Fukagawa, N. K. (1999). Aging: is oxidative stress a marker or is it casual? *Proc. Soc. Exp. Biol. Med.* **222**: 293–298.

65. Cutler, R. G. (1991) Antioxidants and aging. *Am. J. Clin. Nutri.* **53**: 373S–379S.

66. McEligot, A. J., Rock, C. L., Flatt, S. W., Newman, V., Faerber, S. and Pierce, J. P. (1999). Plasma carotenoids are biomarkers of long-term high vegetable intake in women with breast cancer. *J. Nutri.* **129**: 2258–2263.

67. Wang, X. D., Liu, C., Bronson, R. T., Smith, D. E., Krinsky, N. I. and Russell, M. (1999). Retinoid signaling and activator protein-1 expression in ferrets given β-carotene supplements and exposed to tobacco smoke. *J. Natl. Cancer Inst.* **91**: 60–66.

68. The Alpha-Tocopherol Beta Carotene Cancer Prevention Study Group. (1994). The effect of vitamin E and beta carotene on the incidence of lung cancer and other cancers in male smokers. *New England J. Med.* **330**: 1029–1035.
69. Omenn, G. S., Goodman, G. E., Thornquist, M. D., Balmes, J., Cullen, M. R., Glass, A., Keogh, J. P., Meyskens, F. L., Jr., Valanis, B., Williams, J. H., Jr., Barnhart, S., Cherniack, M. G., Brodkin, C. A. and Hammar, S. (1996). Risk factors for lung cancer and for intervention effects in CARET, the beta-carotene and retinol efficacy trial. *J. Natl. Cancer Inst.* **88**: 1550–1559.
70. Krinsky, N. I. (1989). Antioxidant functions of carotenoids. *Free Radic. Biol. Med.* **7**: 617–635.

Chapter 36

Functional Foods

Toshihiko Osawa

Toshihiko Osawa • Laboratory of Food and Biodynamics, Nagoya University Graduate School of Bioagricultural Sciences, Chikusa, Nagoya 46408601, Japan Tel: 81-52-789-4125, E-mail: osawat@agr.nagoya-u.ac.jp

1. Introduction

"Functional Foods" research project started first in 1984, under the sponsorship of the Japanese Ministry of Education, Science and Culture, and the concept for "tertiary function" has been proposed first in the world. This paper mainly focused on foods with "tertiary function" which are expected to contribute to disease prevention by modulating physiological systems such as immune, endocrine, nervous, circulatory and digestive systems. Following the activation and development of systematic and large-scale "Functional Foods Research Project" sponsored by the Ministry of Education, Science and Culture, a national policy has approved "Functional Foods" in terms of FOSHU (Foods for Specified Health Uses) by the Japanese Ministry of Welfare. Until now, about 200 food items have been approved and introduced to Japanese market as FOSHU according to new legislation by the Japanese Ministry of Welfare. In this chapter, background and reason why "Functional Foods" started in Japan will be discussed together with the worldwide recent progress of Functional Foods Research. Moreover, the recent progress of research on functions of dietary antioxidants in oxidative stress will also be discussed as one example for "Functional Foods".

2. Definition of Functional Foods

Recently, many researches have been carried out on food components that have "tertiary" function. The primary or nutritional function is the fundamental and basic function and much public interest has been focused on "primary function" when the most Japanese suffered extreme food shortage. During the improvement in lifestyle in the 1960s, Japanese graduary put more focus on sensory satisfaction along with the rapid development of food industries. By the 1980s, much attention has been focused on the foods having "tertiary" functional activity. With the definition of functional activity, "primary" function means the role of standard nutrient components, and "secondary" function has been proposed as "sensory" functions related with flavor, taste, color and texture etc. By the 1980s, many Japanese scientists began to recognize the importance of the concept of prevention of age-related and geriatric disease through daily dietary habits. As shown in Table 1, recent research had shown that a variety of food components could be expected to disease prevention by modulating physiological systems such as immune, endocrine, nervous and circulatory and digestive systems. In 1984, Japanese Ministry of Education, Science and Culture funded basic scientific research at Universities to design and create the "Physiologically Functional Foods" (simply, "Functional Foods") on the basis of substances that have tertiary functions.

In 1992, a new academic research project (Grant-in Aid for Scientific Research on Priority Area No. 320) has been created under the sponsorship by the Japanese

Table 1. Functions of Food

Primary function	nutrition
Secondary function	sensory satisfaction
Tertiary function	modulation of physiological system (immune, endocrine, nervous, circulatory and digestive)

Table 2. "Grant-in-Aid" Research Program for "Functional Foods"[1]

I. Analysis and Design of Body-regurating Food Factors
   (a) Factors of the active form
     1. To function by inducing endogenous substances
     2. To function as if they were endogenous substances
   (b) Factors in the precursor form (Protein-derived peptides)
     1. To function at the preabsorptive stage
     2. To function at the postaborptive stage

II. Analysis and Design of Body-defending Factors of Foods
   (a) Factors involved in the immunological mechanism
     1. Immunostimulants
     2. Immunosuppressants
   (b) Factors involved in the immunological mechanism
     1. Anti-infection factors
     2. Anti-tumor factors

III. Development of a Technological Basis for the Design of Functional Foods
   (a) Design of microscopic structures
     1. Molecular breeding
     2. Molecular tailoring
   (b) Design of Macroscopic Structures
     1. Development of new methods for the structure conversion
     2. Development of new methods for structural analysis

Ministry of Education, Science and Culture. This third 3-year project entitled "Analysis and Molecular Design of Functional Foods"[1] includes total of 59 research teams from 23 universities, and details of the research program are shown in Table 2. The design and construction of functional foods would be positioned to use for targeting to enrich the concentration of the quantities of functional food components, or to remove some unfavorable toxic components. In 1990, "Designer Foods" project started in USA, and new concept, "Food Phytochemicals for Cancer Prevention", has been created.[2] In 1995, International Conference entitled "Food Factors for Cancer Prevention" was organized in Hamamatsu in December 1995, and more than 1000 participants attended this Conference. The conference brought together leading researcher from all over the world to present the most up-to-date findings on the role of diets in cancer prevention, and the proceedings of this Conference has been published in 1997.[3] In 2000, Second International Conference on Food Factors was held in Kyoto, and concept of "Food Factors" has been proposed not only for cancer prevention but also for health promotion. Two volume of Proceedings for Second International Conference on Food Factors have been published recently,[4] and new concept "Food Factors" is now acceptable not only in Japan but also USA and Europe.

3. Definition of Food for Specific Health Use (FOSHU)

In the late 1980s, the Japanese Ministry of Health and Welfare has started to make an investigation to establish a category of foods that have health promoting effects to reduce the escalating cost of health care in Japan, because of the decreased consumption of fruits and vegetables by the emerging Western trends in Japanese diets.

From these backgrounds, the Japanese Ministry of Health and Welfare decided to introduce new concept "Food for Specific Health Use" (FOSHU). The Japanese Ministry of Health and Welfare defined FOSHU as "processed foods containing ingredients that aid specific bodily functions in addition of nutritional effects" that (a) to which some ingredients to help get into shape are added, (b) from which allergens are removed, (c) the result of such addition/removal is scientifically evaluated; and (d) to which the Japanese Ministry of Health and Welfare has given permission to indicate the nature of effectiveness to the health.

In order to get an approval for FOSHU, each food product should be judged based on the comprehensive examination and permission/approval will be issued by Ministry of Health and Welfare only to those food products that have cleared the examination of product-specific documents submitted by the applicants. The experienced specialist in a broad research area conducts such comprehensive examination. Until now, more than 200 food items have been approved and introduced to market as FOSHU according to new legislation by the Ministry of Health and Welfare. In order to get permission/approval, the applicants are required to use an ingredient that is approved, then they must ask Ministry of Health and Welfare for the examination of the FOSHU product that they have developed using that ingredient.[5] Now, there are 13 categories of the ingredients for FOSHU products:

(1) Dietary fiber.
(2) Oligosaccharides.
(3) Sugar alcohols.
(4) Polyunsaturated fatty acids.

(5) Peptides and proteins.
(6) Glycosides, isoprenoids and vitamins.
(7) Alcohols and phenols.
(8) Cholines.
(9) Lactic acid bacteria.
(10) Minerals.
(11) Diacylglycerols.
(12) Polyphenols.
(13) Others.

After 15 years of development of FOSHU products, we are greatly disappointed and have to clarify the distinction between "Functional Foods" and "FOSHU", because the Japanese Ministry of Health and Welfare decided not to use the term "Functional Foods" for the legislation. Many researchers in academic mainly put the focus on prevention of life-style related diseases including cancer, heart diseases and diabetes by food components that have tertiary functions including new physiological activities. Now, the term "Functional Foods" became much popular and evaluated as the best term compared with "Designer Foods" or "Nutraceuticals". Our research group has been involved in development of novel type of dietary antioxidants to protect from oxidative stress. In this chapter, we also cover the recent progress of research on functions of dietary antioxidants as one example for "Functional Foods".

4. Case of "Functional Foods" for Prevention of Oxidative Stress

Although "tertiary function" means many different type of biological activities, much attention has been focused on prevention of oxidative stress and modification of immune systems.[6] Our research group has been involved in isolation and identification of dietary antioxidants for a long time, and now developing novel and convenient evaluation systems for oxidative stress by application of immunochemical methods to prevent the life-style related diseases such as atherosclerosis, diabetes and cancer, etc.

Recently, much attention has been focused on oxidative damages caused by degradation products or free radicals formed during lipid peroxidation of the cell membranes, although the active species were not identified. Excess production of oxygen radical species such as hydroxyl radicals can easily initiate the lipid peroxidation in the cell membranes to form the lipid peroxides. Lipid peroxidation is known to be a free radical chain reaction that takes place *in vivo* and *in vitro*, and forms lipid hydroperoxides and secondary products such as malondialdehyde (MDA) and 4-hydroxynonenal (HNE). These lipid peroxidation products are highly reactive and have been shown to interact with many biological components such as proteins, amino acids, amines, and DNA. Until now, many different types of detection and quantification methods for lipid peroxidation products have been developed by an instrumental analyses including application of HPLC. Recently we have been involved in developing novel type of evaluation systems for oxidative stress using immunochemical methods by application of monoclonal antibodies which are specific to 13-hydroperoxy linoleic acid (13-HPODE),[7,8] malondialdehyde (MDA),[9] 4-hydeoxynonenal (HNE)[10] and acrolein,[11] because immunochemical methods are specific, simple and convenient. We have also succeeded in developing a novel monoclonal antibody that are specific to dityrosine[12] as the new biomarker in the oxidative protein damage caused by reactive oxygen species (ROS). There are many indications that these oxidative damages may play an important role for the cause of many age-related diseases, and there is speculation that oxidative damages can be occurred in DNA during the peroxidative breakdown of the membrane polyunsaturated fatty acids, in particular, oxidation of 2'-deoxyguanosine to 8-hydroxydeoxy-guanosine (8-OH-dG) by hydroxyl radical. Although development of many sensitive methods for the detection of 8-OH-dG, in particular, ECD (Electrochemical Detector) equipped HPLC technique became the most popular method, despite the cost of the apparatus and requirement of many steps for sample preparation.[13] In order to develop a more sensitive and convenient method, we decided to make an evaluation for the protective role of dietary antioxidants against to oxidative damages by monitoring the amount of 8-OH-dG in biological samples using the monoclonal antibody method.[14] After obtaining specific monoclonal antibody to 8-OH-dG, we developed a new ELISA (enzyme-linked immunosorbent assay) method in quantitating 8-OH-dG by competitive inhibition.[15] These immunochemical methods for detection of oxidatively damaged proteins and DNAs are very specific and useful technique to investigate the mechanism for oxidative stress from the viewpoint of molecular level. By application of the immunochemical technique to the antioxidative assay systems, we can make the reliable, simple and convenient evaluation methods of antioxidative substances.

5. Dietary Antioxidants as Functional Foods

From our hypothesis that endogenous antioxidants in plants must play an important role for antioxidative defense systems, an intensive search for novel type of natural antioxidants has been carried out from numerous plant materials, including those used as foods. Until now, we have succeeded in isolation and identification of a number of lipid-soluble and water-soluble dietary antioxidants from bean seeds, sesame seeds and some spices. Recently, we have succeeded in isolating antioxidative color pigments present in colored pea beans (*Phaseolus vulgaris* L.), and identified the main antioxidative pigment as cyanidine-3-β-D-glucoside (C3G), which was found to exhibit the strong antioxidative activity in the acidic regions, and also antioxidative mechanism of C3G have been examined.[16] By our detailed examination on antioxidative mechanism of C3G, C3G was found to have the strong antioxidative activity and be converted to protocatechuic acid that also posses the antioxidative activity when scavenging free radicals.[17] Recently, we succeeded in analyzing protection mechanisms of C3G against oxidative stress induced by peroxinitrite.[18]

Sesaminol is the main antioxidative component present in sesame oil[19] and an unique antioxidant because it has a superior heat stability, and also is able to effectively increase the availability of tocopherols in biological systems. Recently, the protective role of sesaminol against oxidative damage of low-density lipoprotein has also been examined, and this data showed that the presence of sesaminol effectively reduced oxidative modification of apoprotein determined by application of antibodies specific to MDA and 4-HNE.[20] Recently, we examined antioxidative mechanism of sesaminol in details using 2,2'-azobis (2,4-dimethylvalero-nitrile) as a free radical inducer, it was found that sesaminol reacted with alkylperoxyradicals to form stable reaction products.[21]

We have also made a large-scale investigation on water-soluble antioxidants, in particular, sesaminol glucosides (SG), which we found in sesame seeds in a large quantity.[22] Sesaminol triglucoside, main water-soluble antioxidant present in sesame seeds together with mono and diglucosides (about 1% concentration). Sesaminol can be formed from sesamolin during the refining process of sesame oil and also from SG, present in the defatted sesame flour after (DSF) stripping off oils from sesame seeds. These background prompted us to evaluate antiatherogenic activity of sesaminol and its precursors by *in vivo* system using high cholesterol fed rabbits by feeding DSF which contains about 1% SG for 90 days. From this result, we have found that feeding DSF to rabbits does not protect cholesterol-induced hypercholesterolemia, but may decrease susceptibility to oxidative stress in rabbits fed cholesterol, perhaps due to the antioxidative activity of sesaminol.[23]

Recently, we have also examined antioxidative effect of SG in Watanabe Heritable Hyperlipidemia (WHHL) rabbits. The percentage area of aorta covered with plaque in the SG-treated rabbits was reduced compared to the control, and it was showed that lipid peroxide was decreased while an increase in the activity of glutathione peroxidase and glutathione S-transferase in tissues including liver and aorta was observed. This study indicates that the reduction of arteriosclerosis by SG-treatment relies not on its cholesterol lowering effect but more heavily on its antioxidant potential to inhibit LDL oxidative modification in WHHL rabbits.[24]

The authors focused on curcumin, main yellow pigments in *Curcuma longa* (turmeric). Curcumin has been used widely and for a long time in the treatment of sprain and inflammation in indigenous medicine. Curcumin is the main component of turmeric, and two minor components are also present. Curcuminoids possess antioxidant activity, and also are responsible for the yellow color of curry cooking. Curcumin was reported to inhibit the microsome-mediated mutagenicity of benzo(a)pyrene and 7,12-dimethyl-benz(a)anthracene, and more recently, it was also reported that curcumin acts as a strong inhibitor of tumor promotion, and this effect can be explained to roughly parallel the relative antioxidant activity[25] Recently, we reported on the preventive effect of curcumin on radiation-induced tumor initiation in rat mammary glands.[26, 27]

Recently, we succeeded in obtaining a strong lipid-soluble antioxidant, tetrahydrocurcumin (THC), by hydrogenation using Pd-C (or Raney-nickel) as the catalyst. We also found that THC was converted from curcumin during absorption in the intestines, and more potent antioxidant than curcumin. Recently, we found that THC is a promising chemopreventive agent than curcumin in the 1,2-dimethyl-hydrazine (DMH) induced mouse colon carcinogenesis model.[28]

Fermented plant foods are also good source of dietary antioxidants. It has been reported that chocolate and cocoa are stable against oxidative deterioration and believed to be because of fatty acid composition and presence of polyphenols in cacao beans although the detailed chemical investigation on antioxidative polyphenolic substances has not been carried out. From this background, our research group has started to isolate and identify antioxidative components in cacao liquor polyphenols (CLP), and we found several flavans (catechin and epicatechin) and clovamides had a potent antioxidative activity in the microsomal lipid peroxidation system.[29, 30] Furthermore, the physiological effects of CLP were examined with experimental animal models, and CLP were found to show these physiological functions:

(1) antiulceric activity induced by ethanol;[31]
(2) inhibitory effect on oxidative stress in vitamin deficient rats;[32]
(3) inhibitory effect on mutagenic activity induced by heterocyclic amines;[33]
(4) anticlastogenic activity of cacao: inhibitory effect of cacao liquor polyphenols against mitomycin C-induced DNA damage;[32]
(5) inhibitory effect on tumor promotion in two-stage carcinogenesis in mouse skin.[32]

6. Conclusion

The term of "Functional Foods" and its concept are now internationally accepted. It seems that such an international interest in "Functional Foods" reflects the current research activities in Japan,[6] however, the Japanese Ministry of Health and Welfare decided not to use the term "Functional Foods" and adopted the term FOSHU instead. Looking back on the FOSHU systems over five years after introduction to Japanese Food Industries, many companies have a frustration because of the limitation of health claim and labeling. However, research scientists in both academics and industries in many countries including Japan are now ready to develop a new age "Functional Foods" which are evaluated by clinical testing and approved to make a health labeling for prevention of life-style related diseases including arteriosclerosis, diabetes and cancer etc.

There are many indications that lipid peroxidation play an important role in carcinogenesis, although there is no definite evidence. Modification of immune systems are also very important approach to prevent diseases, however, I mainly focused on prevention of oxidative stress by dietary antioxidative components as one of the examples for "Functional Foods" in this chapter.[34] Our group started novel approach to develop evaluation and detection methods of lipid peroxidation products by application of monoclonal and polyclonal antibodies. We have also started our new project to investigate a novel biomarker of oxidative DNA damage other than 8-OH-dG. Of course, research efforts on metabolic pathways of dietary antioxidants in the digestive tracts are also required. Although these approaches have started just recently, they help us to understand the relationship between antioxidative activity and health promotion including cancer.

References

1. Arai, S. (1996). Studies on functional foods in Japan-State of the Art. *Biosci. Biotech. Biochem.* **60**: 9–15.
2. Huang, M.-T., Osawa, T., Ho, C.-T. and Rosen, R. T., Eds. (1994). Food Phytochemicals for Cancer Prevention, I and II, American Chemical Society, Washington.
3. Ohigashi, H., Osawa, T., Terao, J., Watanabe, S. and Yoshikawa, T., (1997). *Food Factors for Cancer Prevention*, Springer, Tokyo.
4. Ohigashi, H., Osawa, T., Terao, J., Watanabe, S. and Yoshikawa, T., (2000). *Food Factors for Health Promotion, J. Biofactors*, **12**(1–4) and **15**(1–4).
5. Swinbanks, D. and O'Brien, J. (1993). Japan explores the bundary between food and medicine. *Nature* **364**: 180.
6. Goldberg, I. (1994). *Functional Foods*, Chapman and Hall, New York.
7. Kato, Y., Mori, Y., Morimitsu, Y., Hiroi, S., Ishikawa, T. and Osawa, T. (1999). Formation of Ne-(hexanonyl)lysine in protein exposed to lipid hydroperoxide. *J. Biol. Chem.* **274**(29): 20 406–20 414.
8. Kato, Y., Miyake, Y., Yamamoto, K., Shimomura, Y., Ochi, H., Mori, Y. and Osawa, T. (2000). Preparation of monoclonal antibody to Ne-(hexanonyl)lysine: application to the evaluation of oxidative modification of rat skeletal muscle by exercise with flavonoid supplimentation. *Biochem. Biophys. Res. Commun.* **274**: 389–393.

9. Uchida, K., Sakai, K., Itakura, K., Osawa, T. and Toyokuni, S. (1997). Protein modification by lipid peroxidation products. Formation of malondialdehyde-derived Ne-(2-propenal)lysine in proteins. *Arch. Biochem. Biophys.* **346**: 45–52.

10. Toyokuni, S., Miyake, N., Hiai, H., Hagiwara, M., Kawakishi, S., Osawa, T. and Uchida, K. (1995). The monoclonal antibody specific for the 4-hydroxy-2-nonenal histidine adduct. *FEBS Lett.* **359**: 189–191.

11. Uchida, K., Kanematsu, M., Sakai, K., Matsuda, T., Hattori, N., Mizuno, Y., Suzuki, D., Miyama, T., Noguchi, N., Niki, E. and Osawa, T. (1998). Protein-bound acrolein: potential markers for oxidative stress. *Proc. Natl. Acad. Sci. USA* **95**: 4882–4887.

12. Kato, Y., Wu, X., Naito, M., Nomura, H., Kitamoto, N. and Osawa, T. (2000). Immunochemical detection of protein dityrosine in atherosclerotic lesion of apo-E-deficient mice using a novel monoclonal antibody. *Biochem. Biophys. Res. Commun.* **275**: 5–11.

13. Toyokuni, S., Tanaka, T., Hattori, Y., Nishiyama, Y., Yoshida, A., Uchida, K., Ochi, H. and Osawa, T. (1996). Quantitative immuno-histochemical determination of 8-hydroxy-2'-deoxyguanosine by a monoclonal antibody N45.1: its application to ferric nitrilotriacetate-induced renal carcinogenesis model. *Lab. Invest.* **76**: 365–374.

14. Hattori, Y., Nishigari, C., Tanaka, T., Uchida, K., Nikaido, O., Osawa, T., Hiai, H., Imamura, S. and Toyokuni, S. (1996). Formation of 8-hydroxy-2'-deoxy-guanosine in epidermal cells of hairless mice after chronic UVB exposure. *J. Invest. Dermatol.* **107**: 733–737.

15. Erhola, M., Toyokuni, S., Okada, K., Tanaka, T., Hiai, H., Ochi, H., Uchida, K., Osawa, T., Nieminen, M. M.,Alho, H. and K-Lehtinen, P. (1997). Biomarker evidence of DNA oxidation in lung cancer patients: association of urinary 8-hydroxy-2'-deoxyguanosine excrerin with radiotherapy, chemotherapy, and response to treatment. *FEBS Lett.* **409**: 287–291.

16. Tsuda, T., Shiga, K., Ohshima, K. and Osawa T. (1996) Inhibition of lipid peroxidation and active oxygen radical scavenging effect of anthocyanin pigments isolated from *Phaselous vulgaris* L. *Biochem. Pharmacol.* **52**: 1033–1039.

17. Tsuda, T., Horio, F. and Osawa, T. (1999). Absorption and metabolism of cyanidin 3-O-β-D-glucoside in rats. *FEBS Lett.* **449**: 179–182.

18. Tsuda, T., Kato, Y. and Osawa, T. (2000). Mechanism for the peroxynitrite scavenging activity by anthocyanins. *FEBS Lett.* **484**: 207–210.

19. Osawa, T., Nagata, M., Namiki, M. and Fukuda, Y. (1985). Sesamolinol, a novel antioxidant isolated from sesame seed. *Agric. Biol. Chem.* **49**: 3351–3352.

20. M.-H, Kang., Katsuzaki, H. and Osawa, T. *et al.* (1998). Inhibition of 2,2'-azobis(2,4-dimethyl-valeronitrile)-induced lipid peroxidation by sesaminols. *Lipids* **33**: 1031–1036.

21. Kang, M.-H., Naito, M., Sakai, K., Uchida, K. and Osawa, T. (2000). Mode of action of sesame lignans in protecting low-density lipoprotein against oxidative damage in vitro. *Life Sci.* **66**: 161–171.

22. Katsuzaki, H., Kawakishi, S. and Osawa, T. (1994). Sesaminol glucosides in sesame seeds. *Phytochem.* **35**: 773–776.

23. Kang, M.-H., Kawai, Y., Naito, M. and Osawa, T. (1999). Dietary defatted sesame flour decrease susceptibility to oxidative stress in hypercholesterolemic rabbits. *J. Nutri.* **129**: 1885–1890.

24. Kang, M.-H., Naito, M. and Osawa, T. (2001). Prevention of atherosclerosis in LDL receptor-deficient Watanabe heritable hyperlipidemia (WHHL) rabbits by sesaminol glucosides-feeding. *Life Sci.* **in press**

25. Nakamura, Y., Ohto, Y., Murakami, A., Osawa, T. and Ohigashi, H. (1998). Inhibitory effects of curcumin and tetrahydrocurcuminoids on the tumor promoter-induced reactive oxygen species generation in leukocytes in vitro and in vivo. *Japan J. Cancer Res.* **89**: 361–370.

26. Inano, H., Onoda, M., Inafuku, N., Kubota, M., Kamada, Y., Osawa, T., Kobayashi, H. and Wakabayashi, K. (1999). Chemoprevention by curcumin during the promotion stage of tumorigenesis of mammary gland in rats irradiated with gamma-rays. *Carcinogenesis* **20**: 1011–1018.

27. Inano, H., Onoda, M., Inafuku, N., Kubota, M., Kamada, Y., Osawa, T., Kobayashi, H. and Wakabayashi, K. (2000). Potent preventive action of curcumin on radiation-induced initiation of mammary tumorigenesis in rats. *Carcinogenesis* **21**: 1835–1841

28. Kim, J.-M., Araki, S., Kim, D.-J., Park, C.-B., Takasuka, N., Baba-Toriyama, H., Ota, T., Nir, Z., Khachik, F., Shimidzu, N., Tanaka, Y., Osawa, T., Uraji, T., Murakoshi, M., Nishino, H. and Tsuda, H. (1998). Chemopreventive effects of carotenoids and curcumins on mouse colon carcinogenesis after 1,2-dimethylhydrazine initiation. *Carcinogenesis* **19**: 81–85.

29. Sanbongi, C., Osakabe, N., Natsume, M., Takizawa, T., Gomi, S. and Osawa, T. (1998). Antioxidative polyphenols isolated from *Theobroma cacao*. *J. Agric. Food Chem.* **46**: 454–457.

30. Osakabe, N., Yamagishi, M., Sanbongi, C., Natsume, M., Takizawa, T. and Osawa, T. (1998). The antioxidant substances in cacao liquor. *J. Nutri. Sci. Vitam.* **44**: 312–321.

31. Osakabe, N., Yamagishi, M., Natsume, M., Takizawa, T., Nakamura, T. and Osawa, T. (1999). Antioxidative polyphenolic substances in cacao liquor. *In* "Caffeinated Beverages" (T. Parliament, C.-T. Ho, and P. Schieberle, Eds.), pp. 88–101, American Chemical Society.

32. Osakabe, N., Sanbongi, C., Yamagishi, M., Takizawa, T. and Osawa, T. (1998). Effects of polyphenol substances derived from *Theobroma cacao* on gastric mucosal lesion induced by ethanol. *Biosci. Biotechnol. Biochem.* **62**: 1535–1538.

33. Yamagishi, M., Natsume, M., Nagaki, A., Adachi, T., Osakabe, N., Takizawa, T., Kumon, H. and Osawa, T. (2000). Antimutagenic activity of cacao: inhibitory effect of cacao liquor polyphenols on the mutagenic action of heterocyclic amines. *J. Agric. Food Chem.* **48**: 5074–5078.
34. Osawa, T. (1999). Protective role of dietary polyphenols in oxidative stress. *Mech. Aging Dev.* **111**: 133–139.

Chapter 37

Nutrition and Phase II Reactions — The Role of Glutathione

Tammy M. Bray* and Jun Li

Tammy M. Bray and **Jun Li** • Department of Human Nutrition, The Ohio State University, 350E Campbell Hall, 1787 Neil Ave., Columbus, OH 43210
*Corresponding Author.
Tel: 614-292-7594, E-mail: bray.21@osu.edu

1. Introduction

Metabolism of therapeutic drugs or environmental chemicals may directly affect the therapeutic activity of the drugs as well as the toxicity of chemicals. The metabolic reactions can be classified into two categories: Phase I reactions, which are mainly oxidation, reduction and hydroxylation; and Phase II reactions, which are also referred to as conjugation reactions, and involve major conjugating agents such as glutathione (GSH), glucuronic acid, glycine, methionine and sulfate.[1] The availability of these conjugation agents and the activities of the enzymes involved are vital to the Phase II reactions, and also subject to nutritional modulation.

Of the Phase II reactions, GSH conjugation has been extensively studied. Particularly, in the past decade GSH has become one of the research foci because it is not only a crucial substrate in Phase II reaction, but also a key component in cell signal transduction. For instance, GSH may affect cellular redox status by shifting the balance of oxidized/reduced GSH (GSSG/GSH) and thiol/disulfide, which in turn may influence the activation of transcription factors. Nutritional status has been found to impact both GSH synthesis and the GSSG/GSH ratio, which may explain the interaction between nutrition and immune responses, as both GSH and the GSSG/GSH ratio appear to be associated with inflammation and immunity. In addition, phytochemicals have been shown to influence the activity of GSH S-transferase (GST), the other key component of the GSH conjugation. Many of the phytochemicals are found in natural foods. Hence, the focus of this review is to highlight the important roles of diet and nutrition in modulating tissue levels of GSH, the activity of GST, and the interrelationships among Phase II reactions, signal transduction and GSH. In addition, the implication of the interactions between diet and Phase II reactions with regard to health and disease will also be discussed.

2. Glutathione in Phase II Reaction and Cell Signal Transduction

GSH (L-γ-glutamyl-L-cysteinyl-glycine) is the most abundant non-protein thiol in mammalian cells. One of the most well recognized functions of GSH is its role as a substrate for GST [EC 2.5.1.18], an enzyme that catalyzes the GSH conjugation reaction that leads to the detoxification of xenobiotic compounds. Another important role of GSH is serving as a substrate for GSH peroxidase [EC 1.11.1.9], an enzyme that catalyzes reactions for the antioxidation of reactive oxygen species (ROS) and free radicals.

Most xenobiotics are activated by P450 enzymes (mixed function oxidases) to a reactive intermediate that can be more readily excreted after conjugation than the parent compounds. However, the reactive intermediate may also have greater potential to react with cellular macromolecules, including cell membrane lipids, structure proteins as well as RNA and DNA. The results of the interaction may lead to many deleterious effects including mutagenesis. Interaction of xenobiotics with any or all of these macromolecules is considered a possible mechanism for xenobiotic toxicity.[2]

In addition, some reactive intermediates of xenobiotics may be indirectly toxic by generating ROS. Paraquat and alloxan, which are toxic in the lung and pancreas, respectively, are two such examples. The cytotoxicity of both compounds has been associated with increased ROS generation following their activation to reactive intermediates by mixed function oxidases. Similar to reactive intermediates, ROS may induce cytotoxicity by oxidizing cellular lipids, proteins, DNA and RNA.[3, 4]

In normal situations, toxicity induced by xenobiotics and ROS are counteracted by cellular defense systems. GSH plays a pivotal role in the system due to its role in both xenobiotic detoxification and free radical metabolism. For instance, GSH serves as a substrate for the GST. This family of enzymes catalyzes the addition of the thiol group of GSH to the activated intermediate of a variety of xenobiotics, which will facilitate their excretion from the cell.[5] In addition, GSH is also utilized by GSH peroxidase to reduce hydrogen peroxide (H_2O_2) and other hydroperoxides to less damaging intermediates.[6] Therefore, it is believed that GSH plays a critical role in the cellular defense against chemical- and ROS-induced toxicity and injury.[7]

It is also acknowledged that GSH has other important physiological roles, including the storage and transport of cysteine, regulation of leukotriene and prostaglandin metabolism, as well as involvement in the deoxyribonucleotide synthesis, immune function, and cell proliferation. In addition, there has been growing interest in the role of GSH in cell signaling pathways, as GSH is the major intracellular redox buffer in almost all cell types,[8, 9] while the redox status of the cell has been shown to regulate signal transduction, gene transcription and post-translational modification of proteins.[4, 10, 11] Thus, the ability of cells to maintain appropriate GSH levels is critically important in maintaining cellular function and integrity.[12]

Two transcription factors in particular, nuclear factor kappa B (NFκB) and activator protein-1 (AP-1), are regulated by the intracellular redox status and xenobiotic electrophiles.[4] Both NFκB and AP-1 are activated by oxidative stress and xenobiotics and induce the expression of a variety of proteins that function in the immune system and/or cellular detoxification systems.[4, 13, 14]

NFκB, in its most common form, is a heterodimer of p50 and p65, two of the five-member family of NFκB DNA-binding sub-units.[15] Normally, NFκB is found in its inactive form, bound to IκB, an inhibitory sub-unit that keeps NFκB to the cytoplasm and prevents it from binding to DNA.[16] It has been proposed that NFκB activation requires the cytokine-activated protein kinase complex, IKK, which phosphorylates IκB.[17] Loss of the IκB sub-unit allows NFκB to translocate to the nucleus where it induces the transcription of genes involved in the immune response.[18]

Many different stimuli have been identified that activate NFκB, including ROS, toxins, bacterial lipopolysaccharide (LPS) and some cytokines.[13, 16] ROS mediated activation of NFκB was first proposed by Schreck *et al.*[16] This hypothesis was supported by the observation that tumor necrosis factor-α (TNF-α), interleukin-1 (IL-1), UV light and γ rays, known inducers of oxidative stress, activate NFκB.[11] Later experiments also suggest that H_2O_2, or a by-product of H_2O_2 metabolism, may be responsible for NFκB activation.[16]

Activation of NFκB leads to increased expression of cytokines and other chemotactic factors involved inflammation. This then causes infiltration of immune cells such as macrophages and leukocytes that produce ROS as a killing or defense mechanism. Hence, an initial activation of NFκB by xenobiotics or ROS leads to increased cytokine activation, which in turn could increase NFκB activity through cytokine-mediated degradation of the IκB inhibitory unit. Thus, the inflammatory and autoimmune reactions initiated by xenobiotics and ROS could be amplified via a positive feedback loop with NFκB, which may be the mechanism of certain xenobiotic and ROS induced cytotoxicity.[18]

Research on NFκB and GSH indicates that in some cell lines 12-O-tetradecanoylphorbol-13-acetate (TPA) activates NFκB and also increases GSSG levels along with the GSSG/GSH ratio. It has also been found that 1,3-bis (2-chloroethyl-1-nitrosurea (BCNU), an agent that elevates GSSG concentrations by inhibiting GSH reductase, enhances NFκB activation, suggesting a role of GSH in inhibiting NFκB activation or of GSSG in enhancing NFκB activation.[19] Although a direct link between GSH and NFκB activity has not yet been established, it appears that manipulation of GSH, the GSSG/GSH ratio, and/or ROS levels can have a dramatic effect on NFκB activation and points to the importance of GSH in maintaining cellular homeostasis.

However, findings by other researchers led to the proposal of a bi-phasic model for NFκB activation. That is, the activation of NFκB proceeds through two phases: one in the cytoplasm where an oxidizing environment is required to facilitate the translocation of NFκB into the nucleus; the other in the nucleus where the actual binding of NFκB to the corresponding DNA sequence demands a reducing environment.[20] This proposal is supported by findings that addition of not only GSH precursors, but also GSSG, could inhibit the activation of NFκB,[18] which according to previous hypothesis, should enhance the activation. These findings by no means weakens the importance of the role of GSH in NFκB activation — on the contrary, they solidify the key role of GSH in the activation of this transcription factor and hence the consequent immune responses. More intensive research will be expected in this area, particularly with regard to the interactions of GSH with nutrition, compartmentation and translocation of GSSG and GSH, as well as potential clinical applications.

Before ending this section of the review, it is interesting to notice that thioredoxin (Trx), one of the major endogenous redox-regulating molecules with thiol reducing capacity, also appears to play dual and opposing roles in NFκB activation. In the cytoplasm, Trx interferes with the IκB kinases signaling pathway, thus blocking the degradation of IκB and subsequent activation of NFκB; however, when Trx is in the nucleus, it enhances NFκB transcriptional activities by enhancing its ability to bind DNA.[21] It was further found that UVB irradiation and TNF-α markedly induce Trx expression in cells, and the induced Trx appeared to be translocated to the nucleus as early as 1 h after the stress/stimulation. Thus, the endogenously expressed Trx mediated by xenobiotics (and irradiation) seems to act as a positive regulator of NFκB activation.[21] Therefore, it would be of great interest to determine how GSH and its precursors influence the activation of NFκB by looking at the potential changes of redox status in the cytoplasm and nucleus.

3. Glutathione and Thioredoxin: A "Parallel" Redox Regulation System

Trx is a ubiquitous 12-kDa protein with a large number of biological activities. Reduced Trx is a potent protein disulfide oxido-reductase which is important in antioxidant defense, regulation of cellular proliferation and gene expression through activation of transcription factors. Oxidized Trx is converted back to its reduced form by Trx reductase (TrxR; EC 1.6.4.5), whose physiological substrates also include protein disulfide isomerase.[22, 23]

A number of factors are believed to determine the thiol-disulfide balance in cytoplasm. While the principal thiol-disulfide redox buffer is composed of GSH, Trx is another important player that helps reduce protein disulfide bonds. Since GSH does not efficiently reduce Trx, how these two systems work together is of interest to researchers.

Although the well-known structure of Trx shows no binding site for GSH,[24] it is suggested that GSSG reduction can be supported at a high rate by the Trx/TrxR system in glutathione reductase-deficient cells. This finding may be relevant not only to certain stages of the malarial parasite, but also to cells containing high concentrations of GSSG in other organisms like dormant forms of neurospora, glutathione reductase-deficient yeast mutants, or CD4+ lymphocytes of AIDS patients.[25]

It has also been found that the bovine and rat TrxR sequences revealed a close homology to glutathione reductase including the conserved active site sequence (Cys-Val-Asn-Val-Gly-Cys).[26] Thus, it appears that the GSH/GSH reductase and Trx/TrxR system could provide backup for each other. This is further supported by findings that NADPH and human TrxR with or without Trx are efficient electron donors to human plasma GSH peroxidase. It was further demonstrated that incubation of 0.05 μM TrxR with 0.25 μM GSH peroxidase, in GSH-free system, resulted in reduction of t-butyl hydroperoxide. Addition of Trx, 2.5 and 5 μM, respectively, further increased the rate of the reaction.[27]

Therefore, as the functions of Trx and GSH appear to overlap to a certain extent, the Trx/TrxR system is proposed to function as a backup to the GSH/GSH reductase system, thus solidifying the antioxidation defenses and cellular redox regulation.

4. Glutathione: Synthesis, Nutritional Modulation, and Its Clinical Implication

The rate-limiting step of *in vivo* GSH synthesis is catalyzed by γ-glutamylcysteine synthetase to form γ-glutamylcysteine from glutamate and cysteine. Although the plasma concentration of cysteine is relatively low, cysteine can be supplied via cleavage of the disulfide cystine, by cysteine synthesis from methionine through the cystathionine pathway, or through supplementation of cysteine precursors such as N-acetylcysteine (NAC) or 2-oxothiozolidine-4-carboxylate (OTC). The rate-limiting step in GSH synthesis is regulated by feedback inhibition of GSH in *in vitro* systems,[28] suggesting a regulatory mechanism that limits the maximum tissue concentration of GSH *in vivo*.

The γ-glutamyl bond of GSH can resist intracellular peptidases, while C-terminal glycine can resist γ-glutamyl cyclotransferase, both of which contribute to the intracellular stability of GSH.[29] However, γ-glutamyltranspeptidase, an enzyme located on the external surface of cell membranes of a variety of tissues, has the ability to cleave the γ-glutamyl bond. Therefore, this enzyme appears to influence the tissue specificity of GSH uptake.[30, 31] For example, the activity of γ-glutamyltranspeptidase is very low in the liver, which might explain the low hepatic uptake of plasma GSH, even though liver is the major organ for synthesis and export of GSH.[32] On the other hand, tissues such as the lung and kidney have higher γ-glutamyltranspeptidase activity, which enables extracellular degradation of GSH to cysteinyl glycine and γ-glutamyl amino acids. After being transported into the cell, these dipeptides are further metabolized to the amino acid constituents of GSH by dipeptidases, γ-glutamyl cyclotransferase and oxoprolinase. Therefore, uptake of plasma GSH via γ-glutamyl transpeptidase contributes to the intracellular GSH concentration in extrahepatic tissues like lung and kidney.

The availability of dietary sulphur amino acids influences tissue GSH concentrations. This was demonstrated in studies when hepatic GSH concentrations were decreased following consumption of diets deficient in protein or sulphur amino acids, and returned to normal when the rats were placed on normal protein diets or low protein diets supplemented with sulphur amino acids.[33, 34] Later it was shown that when rats fed a 7.5% protein diet were supplemented with cysteine or its precursors at a concentration equivalent to the sulphur amino acid content of the normal (15%) protein diet, hepatic GSH levels increased to the concentration observed in the normal protein group.[35]

Hepatic GSH levels are tightly regulated. In a study of rats fasted for 24 h or fed a diet containing 0.5% protein for 2 weeks, hepatic GSH concentrations did not fall below 3 μmol/g of tissue.[36] When rats were fed high protein (30% or 45%) diets with a sulphur amino acid content 2 or 3 times higher than normal, hepatic GSH concentrations did not exceed the normal physiological maximum of 8 μmol/g.[35] Additionally, tissue concentrations of GSH were not enhanced when dietary protein was adequate or above the requirement level.[37, 38]

Previous dietary protein status affects the diurnal response of hepatic GSH concentration to sulphur amino acid supplementation. Sulphur amino acid supplementation to rats previously fed a normal protein (15%) diet for two weeks indicated that the rate of increase or the peak concentration of hepatic GSH of the diurnal cycle did not change compared to the unsupplemented group. However, in rats that were previously fed a low protein (7.5%) diet for two weeks and then supplemented with sulphur amino acid, the hepatic GSH concentration increased more rapidly and was sustained at a higher concentration than in rats that were previously fed a normal protein (15%) diet for two weeks.[35] In rats previously fed a 0.5% protein diet for two weeks, this initial increase was even more pronounced than in rats fed a 7.5% protein diet and the peak concentration exceeded the physiological maximum.[36] The difference in response to sulphur amino acids between rats fed low protein diets and rats fed adequate protein diets is not readily explained, even though hepatic GSH concentration was similar before sulphur amino acid supplementation (i.e. the feedback inhibition mechanism is not involved) and the amount of supplementation was identical for both low and normal protein groups.

Nutritional status may also influence tissue GSH concentrations by affecting the uptake of extracellular GSH into extrahepatic tissues via γ-glutamyl trans-peptidase and by affecting the transport of plasma amino acids into tissues. The influence of nutritional status on extrahepatic tissue GSH concentrations was demonstrated in studies where lung GSH concentration was decreased by long-term severe dietary protein restriction, i.e. 0.5% protein diet for 2 weeks,[36] but there was no difference in lung GSH concentrations between rats fed 7.5 and 15% protein diets. Supplementation of a cysteine prodrug to the 0.5% protein group increased lung GSH levels but the concentration was not affected as dramatically as in liver, and it was not restored to the level seen in rats fed the 15% protein diet.[36] However, the enzyme activity of γ-glutamyltranspeptidase was not measured in any of these studies, and the role of this enzyme in the relationship between the protein status and the uptake of GSH is still unknown. In addition to the uptake of GSH via the γ-glutamyltranspeptidase mechanism, amino acid transport mechanisms for substrates of GSH synthesis such as cysteine, cystine, methionine, and glutamate may also affect the extrahepatic tissue concentration of GSH. An imbalance of the plasma amino acid profile may influence the uptake of these amino acids since they compete for some of the same transport systems.[39] It has been demonstrated that cystine uptake into cultured endothelial cells is inhibited competitively by glutamate and that the GSH concentration in these cells decreases when they are cultured in a glutamate-enriched medium.[40] However, the specific effects of the balance of dietary amino acids and nutritional status on amino acid transport for GSH synthesis are unknown.

Since the key enzyme controlling GSH synthesis is γ-glutamylcysteine synthetase, the intracellular availability of cysteine and glutamate become the critical factors in GSH synthesis. The apparent K_m values of γ-glutamylcysteine synthetase for cysteine and glutamate are 0.3 and 1.8 mM, respectively.[28] This reflects the high affinity of cysteine and low affinity of glutamate for this enzyme and the relative intracellular concentrations of these two substrates under normal metabolism. Generally, glutamate and glutamine are in large excess in the cell whereas cysteine concentrations are low.[41] Also, hepatic cellular cysteine concentrations are maintained at a level approximately 50 times lower than that of tissue GSH. In mice fed protein deficient diets (4%), cellular cysteine is maintained at a concentration of 0.05 μmol/g compared to GSH that is 2–3 μmol/g. In mice fed a protein excess diet (35%), hepatic cysteine is maintained at 0.2 μmol/g compared 8–10 μmol/g for GSH.[35] Therefore, it is probable that the rate of GSH synthesis and consequently, the GSH concentration, is significantly influenced by the availability of intracellular cysteine. This suggests that cysteine is the rate limiting substrate in GSH synthesis.

Therefore the supply of cysteine and not glutamate appears to be critical for the synthesis of GSH. Furthermore, if GSH synthesis is dependent on the availability of the cysteine, then increases in cysteine concentration should result in the enhancement of intracellular GSH. This has been demonstrated in several *in vitro* and *in vivo* models.[42] In models where hepatocytes had been stressed with oxidants, GSH concentrations were depleted. Supplementation of cysteine increased GSH concentration and protected cells from oxidant damage.[43] Other researchers also showed that in *in vitro* models, addition of exogenous cysteine alone to various culture systems resulted in an increase in intracellular GSH and protection from oxidant damage.[44] In addition, the beneficial effects of cysteine supplementation on GSH concentration can also be observed *in vivo*. Intravenous supply of cysteine to GSH-depleted rats caused marked increases in hepatic GSH *in vivo*.[45]

Other cysteine derivatives, such as NAC, have been shown to safely increase intracellular GSH and protect tissues from acetaminophen toxicity.[46] OTC, a stable derivative of cysteine, produces sustained and efficient intracellular delivery of cysteine for GSH synthesis.[35, 47] Oral administration of OTC can elevate hepatic GSH in protein-deficient rats and in addition gives protection against pulmonary oxygen toxicity during hyperoxia exposure.[36] Because of the limiting nature of cysteine, introduction of any cysteine source to cells results in GSH synthesis.

Our laboratory has found that efficacies of these cysteine derivatives (such as NAC and OTC), as well as methionine and GSH, on the enhancement of tissue GSH levels are different in mice fed a low protein diet (0.5%). In addition, the efficacy of GSH enhancement by these agents tends to be tissue specific as well. However, levels of tissue GSSG appear to be constant regardless of the amount of protein in the diet and the presence/absence of supplementation. On the other hand, tissue levels of reduced GSH increase significantly with the addition of adequate protein or supplementation in the diet. Thus, it is obvious that the GSSG/GSH ratios are substantially higher in protein-deficient group and unsupplemented group, when compared to the protein-adequate groups and supplemented groups. Thus, the difference in redox status among those groups suggests potential links between impaired immune responses observed in protein-malnourished humans/animals and redox-status mediated activation of transcription factors. Further research on this area might reveal some of the molecular mechanisms of GSH's role in cell signal transduction, immune responses, as well as conjugation of xenobiotics, and the effect of nutritional modulation, particularly in the context of protein malnutrition.

Although only a few human diseases are directly linked with defects of GSH synthesis and metabolism,[48] GSH has a wide array of implications in many diseases due to the findings that ROS are playing a important role in those disease states and GSH is a very important antioxidant as well as a key conjugation agent. For instance, it has been proposed that tumor promotion is enhanced with decreasing of GSH/GSSG ratio and hence the oxidative challenge imposed by tumor promoters is not attenuated by the antioxidant system. On the other hand, if GSH/GSSG ratio is high, GSH peroxidase would do a much better job and the oxidative challenge would be diminished, leading to a much less potent tumor promotion.[49]

Another example is diabetes, where ROS intermediate has also been proposed for and identified in the process of β-cell death.[50] It is therefore logical to hypothesize that GSH and/or its precursors may be effective in preserving those β cells when encountered xenobiotic challenges, considering its role in conjugation and antioxidation. Our laboratory has demonstrated in the alloxan model that by supplementing NAC, β cells are preserved by a large percent, with pancreas function maintained and blood glucose levels kept within a normal range, when compared to controls.[50]

5. Glutathione-S-Transferase and Phytochemicals: Implications in Health and Disease

In the past two decades, phytochemicals have been a focus of research of nutrition and Phase II reactions since epidemiological studies have suggested that a diet rich in fruits and vegetables, which contains a variety of phytochemicals, is associated with a reduced risk for a number of common cancers. Since many carcinogens go through Phases I and II reactions, modification of Phase II reactions and detoxification of carcinogens might be one of the mechanisms by which phytochemicals modify carcinogensis. While GSH and GST are components of the GSH conjugation, quinone reductase (QR) and UDP-glucuronosyltransferase (UDP-GT) are also important Phase II reaction enzymes.[51]

Organosulfur compounds are a group of phytochemicals that have been shown to possess anticarcinogenic efficacy in multiple organs such as the lung, esophagus, stomach, liver and colon.[52, 53] Of possible mechanisms involved in chemoprevention, activities of Phase I (cytochrome P450) and Phase II (including GST) enzymes have been considered to be most important.[54] Organosulfur compounds can be categorized mainly into 3 classes: ally sulfur compounds, glucosinolates and isothiocyanates, as well as other organosulfur compounds.[55]

Most of the allyl sulphur compounds have been identified in garlic or onions. Both diallyl sulfide (DAS) and diallyl disulfide (DADS) were found to increase the activity of GST and *p*-nitrophenol UDP-GT.[56] A significant reduction in the excretion of urinary mutagens induced by benzo[a]pyrene was seen in rats fed garlic (*Allium sativum*), which was associated with a stimulation in the activities of liver cytosolic GST and liver and lung QR, suggesting that the antimutagenic potential of garlic may be mediated through induction of detoxification enzymes in target tissues.[57]

Another important organosulfur group is organic isothiocyanates and glucosinolates, both of which are present in substantial amounts in cruciferous vegetables.[54, 58] A high intake of glucosinolate-containing cruciferous vegetables, such as Brussel sprouts (Brassica oleraceae), has been linked to a decreased cancer risk. Consumption of Brussel sprouts for one week resulted in increased rectal GST-alpha and -pi isozyme levels. These enhanced detoxification enzyme levels may partly explain the epidemiological association between a high intake of glucosinolates (cruciferous vegetables) and a decreased risk of colorectal cancer.[59]

Polyphenols present in tea, wine and other plants have been associated with reduced risk of a variety of illnesses, including heart disease and certain types of cancer. It has been shown that the topical application or oral administration of the polyphenol fraction of green tea inhibited carcinogen-DNA adduct formation in the epidermis,[60, 61] and significantly enhanced GSH peroxidase, catalase, NADPH-QR and GST activity in the small intestine, lung and liver.[62]

Catechin in drinking water and dietary turmeric significantly inhibited the forestomach tumor burden and incidence in hamsters, while the induction of oral tumors was delayed. In addition, a single i.p. injection of catechin to male Swiss mice induced increased forestomach and hepatic GST activity, suggesting that catechin and turmeric which are regularly consumed natural products, are effective in mice or golden hamsters as chemopreventive agents.[63]

Ellagic acid (EA), a naturally occurring plant polyphenol, is also proposed to possess broad chemoprotective properties. Dietary EA has been shown to reduce the incidence of N-2-fluorenylacetamide-induced hepatocarcinogenesis and N-nitrosomethylbenzylamine (NMBA)-induced esophageal tumors in rats. The activities of the hepatic phase II enzymes GST and NADPH-QR were all increased as well. It was therefore hypothesized that the chemoprotective effect of EA against various chemically induced cancers may involve the induction of phase II enzymes, thereby enhancing the ability of target tissues to detoxify reactive intermediates of xenobiotic metabolism.[64]

Isoflavones are another important class of phytochemicals. They have been shown to increase the activities of GST, as well as *p*-nitrophenol and *p*-hydroxybiphenyl specific UDP-GT, which are also important enzymes involved in Phase II reactions.[65] Isoflavones in soy may play a role in the prevention of cancer through their capacity to affect antioxidant or protective phase II enzyme activities. In rats consuming diets high in isoflavones, the activity of liver GSH peroxidase, GSH reductase and UDP-GT was increased, as was the activity of GST and QR in kidney and colon. After 13 weeks of isoflavone supplementation, an inverse relationship between tumor incidence and isoflavone intake was observed, supporting the proposed mechanism of soy and soy isoflavones as antioxidant and phase II enzyme inducers.[66]

6. Summary

In summary, intracellular GSH plays a critical role in chemical detoxification and antioxidant defense, two intimately related protective mechanisms involved in cellular defense against xenobiotic insults and oxidative stress. In addition to these functions, recent advances have revealed that GSH and sulphur amino acids may also play a critical role in signal transduction and gene transcription. By maintaining cellular redox balance and protecting critical sulfhydryl groups, GSH may determine interactions not only among proteins, but also between transcription factors and DNA. The involvement of ROS in the activation of oxidative stress response transcription factors, particularly NFκB and AP-1, begins to reveal another dimension in the control and regulation of cellular homeostasis. The regulation of gene expression by GSH has promising clinical and therapeutic implications. However important GSH appears to be in the redox regulation, it does not work by itself. Trx also plays an important role in the reduction of protein disulfides. Trx, along with TrxR, is believed to provide backup to the GSH/GSH reductase system, and vice versa. As tissue GSH levels are influenced by nutritional status and other dietary factors, the activity of GST, a key enzyme in GSH conjugation reaction, can also be enhanced by dietary components, particularly fruits and vegetables, which contain large amounts of phytochemicals. The enhanced activity of this conjugating enzyme is associated with reduced tumorigenesis in lab animals, which is echoed by the negative association between the reduced incidences of certain cancers and the consumption level of fruit and vegetables. Despite these exciting findings, our knowledge of the influence of nutritional status on these complicated processes remains at a rudimentary level. Strategies which utilize GSH, its precursors or Trx for disease prevention and treatment of toxicity need to be based upon a sound understanding of the interrelationships between nutrition, GSH, signal transduction and the mechanisms of chemical toxicity.

References

1. Mandel, H. G. (1971). Pathways of drug biotransformation: biochemical conjugation. *In* "Fundamentals of Drug Metabolism and Drug Disposition" (N. B. LaDu, G. H. Mandel, and W. Leong., Eds.), pp. 149–186, The Williams & Wilkins Company, Baltimore.
2. Bray, T. M., Ho, E. and Levy, M. A. (1998). Glutathione, sulphur amino acids and chemical detoxication. *In* "Nutrition and Chemical Toxicity" (I. Costas, Ed.), pp. 183–200, John Wiley & Sons Ltd., New York.
3. Pinkus, R., Weiner, L. M. and Daniel, V. (1996). Role of oxidants and anti-oxidants in the induction of AP-1, NF-κB, and glutathione S-transferase gene expression. *J. Biol. Chem.* **271**(23): 13 422–13 429.
4. Sen, C. K. and Packer, L. (1996). Antioxidant and redox regulation of gene transcription [see comments]. *FASEB J.* **10**(7): 709–720.
5. Habig, W. H. and Jakoby, W. B. (1981). Assays for differentiation of glutathione S-transferases. *Meth. Enzymol.* **77**: 398–405.
6. Eklow, L., Moldeus, P. and Orrenius, S. (1984). Oxidation of glutathione during hydroperoxide metabolism. A study using isolated hepatocytes and the glutathione reductase inhibitor 1,3-bis(2-chloroethyl)-1-nitrosourea. *Eur. J. Biochem.* **138**(3): 459–463.
7. Bellomo, G., Palladini, G. and Vairetti, M. (1997). Intranuclear distribution, function and fate of glutathione and glutathione-S-conjugate in living rat hepatocytes studied by fluorescence microscopy. *Microsc. Res. Tech.* **36**(4): 243–252.
8. Meister, A. (1989). On the biochemistry of glutathione. *In* "Glutathione Centennial: Molecular Perspectives and Clinical Implications" (T. H. N. Taniguchi, Y. Sakamoto, and A. Meister, Eds.), pp. 3–22, Academic Press, New York.

9. Staal, F. J. *et al.* (1994). Redox regulation of signal transduction: tyrosine phosphorylation and calcium influx. *Proc. Natl. Acad. Sci. USA* **91**(9): 3619–3622.

10. Anderson, M. T. *et al.* (1994). Separation of oxidant-initiated and redox-regulated steps in the NF-κB signal transduction pathway. *Proc. Natl. Acad. Sci. USA* **91**(24): 11 527–11 531.

11. Sies, H. (1991). Hydroperoxides and thiol oxidants in the study of oxidative stress in intact cells and organs. *In* "Oxidative Stress" (H. Sies, Ed.), pp. 73–90, Academic Press, London.

12. Shi, M. M. *et al.* (1994). Quinone-induced oxidative stress elevates glutathione and induces gamma-glutamylcysteine synthetase activity in rat lung epithelial L2 cells. *J. Biol. Chem.* **269**(42): 26 512–26 517.

13. Janssen, Y. M. *et al.* (1995). Asbestos induces nuclear factor kappa-B (NF-κB) DNA-binding activity and NF-κB-dependent gene expression in tracheal epithelial cells. *Proc. Natl. Acad. Sci. USA* **92**(18): 8458–8462.

14. Pinkus, R., Bergelson, S. and Daniel, V. (1993). Phenobarbital induction of AP-1 binding activity mediates activation of glutathione S-transferase and quinone reductase gene expression. *Biochem J.* **290(Part 3)**: 637–640.

15. Muller, J. M., Rupec, R. A. and Baeuerle, P. A. (1997). Study of gene regulation by NF-κB and AP-1 in response to reactive oxygen intermediates. *Methods* **11**(3): 301–312.

16. Schreck, R., Rieber, P. and Baeuerle, P. A. (1991). Reactive oxygen intermediates as apparently widely used messengers in the activation of the NF-κB transcription factor and HIV-1. *EMBO J.* **10**(8): 2247–2258.

17. DiDonato, J. A. *et al.* (1997). A cytokine-responsive IκB kinase that activates the transcription factor NF-κB [see comments]. *Nature* **388**(6642): 548–554.

18. Mihm, S., Galter, D. and Droge, W. (1995). Modulation of transcription factor NF-κB activity by intracellular glutathione levels and by variations of the extracellular cysteine supply. *FASEB J.* **9**(2): 246–252.

19. Droge, W. *et al.* (1994). Functions of glutathione and glutathione disulfide in immunology and immunopathology. *FASEB J.* **8**(14): 1131–1138.

20. Ginn-Pease, M. E. and Whisler, R. L. (1998). Redox signals and NF-κB activation in T cells. *Free Radic. Biol. Med.* **25**(3): 346–361.

21. Hirota, K. *et al.* (1999). Distinct roles of thioredoxin in the cytoplasm and in the nucleus. A two-step mechanism of redox regulation of transcription factor NF-κB. *J. Biol. Chem.* **274**(39): 27 891–27 897.

22. Holmgren, A. (1985). Thioredoxin. *Ann. Rev. Biochem.* **54**: 237–271.

23. Holmgren, A. and Bjornstedt, M. (1995). Thioredoxin and thioredoxin reductase. *Meth. Enzymol.* **252**: 199–208.

24. Holmgren, A. (1979). Thioredoxin catalyzes the reduction of insulin disulfides by dithiothreitol and dihydrolipoamide. *J. Biol. Chem.* **254**(19): 9627–9632.

25. Kanzok, S. M. *et al.* (2000). The thioredoxin system of the *Malaria Parasite Plasmodium falciparum*. Glutathione Reduction Revisited. *J. Biol. Chem.* **275**(51): 40 180–40 186.

26. Zhong, L. *et al.* (1998). Rat and calf thioredoxin reductase are homologous to glutathione reductase with a carboxyl-terminal elongation containing a conserved catalytically active penultimate selenocysteine residue. *J. Biol. Chem.* **273**(15): 8581–8591.

27. Bjornstedt, M. *et al.* (1994). The thioredoxin and glutaredoxin systems are efficient electron donors to human plasma glutathione peroxidase. *J. Biol. Chem.* **269**(47): 29 382–29 384.

28. Richman, P. G. and Meister, A. (1975). Regulation of gamma-glutamyl-cysteine synthetase by nonallosteric feedback inhibition by glutathione. *J. Biol. Chem.* **250**(4): 1422–1426.

29. DeLeve, L. D. and Kaplowitz, N. (1990). Importance and regulation of hepatic glutathione. *Semin. Liver Dis.* **10**(4): 251–266.

30. Hahn, R., Wendel, A. and Flohe, L. (1978). The fate of extracellular glutathione in the rat. *Biochim. Biophys. Acta* **539**(3): 324–337.

31. Griffith, O. W. and Meister, A. (1979). Glutathione: interorgan translocation, turnover, and metabolism. *Proc. Natl. Acad. Sci. USA* **76**(11): 5606–5610.

32. Lauterburg, B. H., Adams, J. D. and Mitchell, J. R. (1984). Hepatic glutathione homeostasis in the rat: efflux accounts for glutathione turnover. *Hepatology* **4**(4): 586–590.

33. Boebel, K. P. and Baker, D. H. (1983). Blood and liver concentrations of glutathione, and plasma concentrations of sulfur-containing amino acids in chicks fed deficient, adequate, or excess levels of dietary cysteine. *Proc. Soc. Exp. Biol. Med.* **172**(4): 498–501.

34. Cho, E. S., Johnson, N. and Snider, B. C. (1984). Tissue glutathione as a cyst(e)ine reservoir during cystine depletion in growing rats. *J. Nutri.* **114**(10): 1853–1862.

35. Bauman, P. F., Smith, T. K. and Bray, T. M. (1988). The effect of dietary protein and sulfur amino acids on hepatic glutathione concentration and glutathione-dependent enzyme activities in the rat. *Can. J. Physiol. Pharmacol.* **66**(8): 1048–1052.

36. Taylor, C. G. *et al.* (1992). Elevation of lung glutathione by oral supplementation of L-2-oxothiazolidine-4-carboxylate protects against oxygen toxicity in protein-energy malnourished rats. *FASEB J.* **6**(12): 3101–3107.

37. Williamson, J. M., Boettcher, B. and Meister, A. (1982). Intracellular cysteine delivery system that protects against toxicity by promoting glutathione synthesis. *Proc. Natl. Acad. Sci. USA* **79**(20): 6246–6249.

38. Hazelton, G. A., Hjelle, J. J. and Klaassen, C. D. (1986). Effects of cysteine pro-drugs on acetaminophen-induced hepatotoxicity. *J. Pharmacol. Exp. Ther.* **237**(1): 341–349.

39. Christensen, H. N. (1990). Role of amino acid transport and countertransport in nutrition and metabolism. *Physiol. Rev.* **70**(1): 43–77.

40. Miura, K. *et al.* (1992). Cystine uptake and glutathione level in endothelial cells exposed to oxidative stress. *Am. J. Physiol.* **262(1 Part 1)**: C50–C58.

41. Bannai, S. and Tateishi, N. (1986). Role of membrane transport in metabolism and function of glutathione in mammals. *J. Membr. Biol.* **89**(1): 1–8.

42. Beatty, P. and Reed, D. J. (1981). Influence of cysteine upon the glutathione status of isolated rat hepatocytes. *Biochem. Pharmacol.* **30**(11): 1227–1230.

43. Goss, P. M., Bray, T. M. and Nagy, L. E. (1994). Regulation of hepatocyte glutathione by amino acid precursors and cAMP in protein-energy malnourished rats. *J. Nutri.* **124**(3): 323–330.

44. Hiraishi, H. *et al.* (1994). Protection of cultured rat gastric cells against oxidant-induced damage by exogenous glutathione. *Gastroenterology* **106**(5): 1199–1207.

45. Aebi, S. and Lauterburg, B. H. (1992). Divergent effects of intravenous GSH and cysteine on renal and hepatic GSH. *Am. J. Physiol.* **263(2 Part 2)**: R348–R352.

46. Wong, B. K., Chan, H. C. and Corcoran, G. B. (1986). Selective effects of N-acetylcysteine stereoisomers on hepatic glutathione and plasma sulfate in mice. *Toxicol. Appl. Pharmacol.* **86**(3): 421–429.

47. Nishina, H., Ohta, J. and Ubuka, T. (1987). Effect of L-2-oxothiazolidine-4-carboxylate administration on glutathione and cysteine concentrations in guinea pig liver and kidney. *Physiol. Chem. Phys. Med. NMR* **19**(1): 9–13.

48. Meister, A. and Anderson, M. E. (1983). Glutathione. *Ann. Rev. Biochem.* **52**: 711–760.

49. Perchellet, J. P. *et al.* (1985). Inhibitory effects of glutathione level-raising agents and d-alpha-tocopherol on ornithine decarboxylase induction and mouse skin tumor promotion by 12-O-tetradecanoylphorbol-13-acetate. *Carcinogenesis* **6**(4): 567–573.

50. Ho, E., Chen, G. and Bray, T. M. (1999). Supplementation of N-acetylcysteine inhibits NFκB activation and protects against alloxan-induced diabetes in CD-1 mice. *FASEB J.* **13**(13): 1845–1854.

51. Wargovich, M. J. (1997). Experimental evidence for cancer preventive elements in foods. *Cancer Lett.* **114**(1–2): 11–17.

52. Wattenberg, L. W. (1992). Inhibition of carcinogenesis by minor dietary constituents. *Cancer Res.* **52(7 Suppl.)**: 2085S–2091S.

53. Wattenberg, L. (1992). Chemoprevention of cancer by naturally occurring and synthetic compounds. *In* "Cancer Chemoprevention" (L. M. Wattenberg, C. W. Boone, and G. J. Kellof, Eds.), pp. 19–39, CRC Press, Boca Raton, FL.

54. Zhang, Y. and Talalay, P. (1994). Anticarcinogenic activities of organic isothiocyanates: chemistry and mechanisms. *Cancer Res.* **54(7 Suppl.)**: 1976S–1981S.

55. Wargovich, M. (1992). Inhibition of gastrointestinal cancer by organosulfur compounds in garlic. *In* "Cancer Chemoprevention" (L. M. Wattenberg, C. W. Boone, and G. J. Kellof, Eds.), pp. 195–203, CRC Press, Boca Raton, FL.
56. Haber, D. *et al.* (1994). Modification of hepatic drug-metabolizing enzymes in rat fed naturally occurring allyl sulphides. *Xenobiotica* **24**(2): 169–182.
57. Polasa, K. and Krishnaswamy, K. (1997). Reduction of urinary mutagen excretion in rats fed garlic. *Cancer Lett.* **114**(1–2): 185–186.
58. Hecht, S. S. (1995). Chemoprevention by isothiocyanates. *J. Cell. Biochem. Suppl.* **22**: 195–209.
59. Nijhoff, W. A. *et al.* (1995). Effects of consumption of Brussels sprouts on intestinal and lymphocytic glutathione S-transferases in humans. *Carcinogenesis* **16**(9): 2125–2128.
60. Wang, Z. Y. *et al.* (1992). Protection against benzo[a]pyrene- and N-nitrosodiethylamine-induced lung and forestomach tumorigenesis in A/J mice by water extracts of green tea and licorice. *Carcinogenesis* **13**(8): 1491–1494.
61. Wang, Z. Y. *et al.* (1992). Inhibition of N-nitrosodiethylamine- and 4-(methylnitrosamino)-1-(3-pyridyl)-1-butanone-induced tumorigenesis in A/J mice by green tea and black tea. *Cancer Res.* **52**(7): 1943–1947.
62. Khan, S. G. *et al.* (1992). Enhancement of antioxidant and phase II enzymes by oral feeding of green tea polyphenols in drinking water to SKH-1 hairless mice: possible role in cancer chemoprevention. *Cancer Res.* **52**(14): 4050–4052.
63. Azuine, M. A. and Bhide, S. V. (1994). Adjuvant chemoprevention of experimental cancer: catechin and dietary turmeric in forestomach and oral cancer models. *J. Ethnopharmacol.* **44**(3): 211–217.
64. Ahn, D. *et al.* (1996). The effects of dietary ellagic acid on rat hepatic and esophageal mucosal cytochromes P450 and phase II enzymes. *Carcinogenesis* **17**(4): 821–828.
65. Siess, M. H. *et al.* (1989). Induction of monooxygenase and transferase activities in rat by dietary administration of flavonoids. *Xenobiotica* **19**(12): 1379–1386.
66. Appelt, L. C. and Reicks, M. M. (1999). Soy induces phase II enzymes but does not inhibit dimethylbenz[a]anthracene-induced carcinogenesis in female rats. *J. Nutri.* **129**(10): 1820–1826.

Chapter 38

Oligomeric Proanthocyanidins in Human Health and Disease Prevention

Debasis Bagchi* and Harry G. Preuss

Debasis Bagchi • Department of Pharmaceutical and Administrative Sciences, Creighton University School of Pharmacy & Allied Health Professions, 2500 California Plaza, Omaha, NE 68178
*Corresponding Author.
Tel: (402) 280 2950, Fax: (402) 280 1883, E-mail: debsis@creighton.edu
Harry G. Preuss • Department of Physiology and Biophysics, Georgetown University Medical Center, Med-Dent Building, Room 103 SE, 3900 Reservoir Road NW, Washington, DC 20007
Tel: (202) 687 1441, E-mail: preusshg@georgetown.edu

1. Introduction

In recent years, there has been increased interest in alternative therapies, which use a wide range of natural products. Concerning these natural products, many polyphenolic compounds such as catechin and its derivatives, oligomeric proanthocyanidins (OPC) and resveratrol, have captured considerable attention. Polyphenols, in general, are considered important phytopharmaceuticals, i.e. chemicals with potential benefits in human health and disease prevention isolated from medicinal plants in their natural form. According to the estimate of the World Health Organizations, nearly 80% of the population of developing countries (over two billion people) rely on traditional medicines, mostly plant-derived drugs or phytopharmaceuticals for their primary health care needs. In the USA alone, sales of herbal products have reached in excess of four billion annually. Unfortunately, most of these phytopharmaceuticals are not properly standardized and/or backed by peer-reviewed scientific research.

Polyphenols, integral components of the human diet, are extensively involved in the normal biochemical and physiological functions of plants. These functions include growth, reproduction, resistance against pathogens and predators, protection from plague, and preharvest seed germination. Interest in polyphenols has increased greatly due to a broad spectrum of medicinal, pharmacological and therapeutic efficacies including free radical scavenging, metal-chelating and antioxidant ability, antiinflammatory, cardioprotective and anticarcinogenic properties.[1, 2] In particular, *in vitro*, *in vivo* and human clinical studies have focused on a specific group of polyphenols called oligomeric proanthocyanidins, as well as the phytoalexin resveratrol.

2. French Paradox and Oligomeric Proanthocyanidins (OPC)

Renewed interest has been placed on the roles of wine and its nonalcoholic constituents because of the phenomenon known as the *French Paradox*.[3, 4] Epidemiological studies have shown that high dietary intakes of saturated fat, cholesterol, total fat, and total caloric intake positively correlate with mortality from heart disease in most developed countries. However the French, who consume foods high in saturated fats and cholesterol, exercise less and smoke more, have a significantly lower mortality rate from coronary heart disease compared to other western societies, nearly half that of USA. This phenomenon is called the *French Paradox*. Specifically, the French have a high intake of saturated fat (14–15% of energy) similar to that consumed in the United Kingdom and the United States. In the UK and USA, this diet is thought to lead to high serum cholesterol concentrations and elevated rates of coronary heart disease. However in France, this is not the case. The mortality rate from coronary heart disease of the French is much closer to that of Mediterranean countries, Japan and

China, where the rate is less than one third that seen in the USA and other westernized societies.[4-7]

Lifestyle may play a role in the *French Paradox*. Specifically, many believe the generous consumption of wine among the French population is responsible for this phenomenon. Epidemiological studies in 1979 by St. Leger and colleagues[5] linked a significant reduction in mortality from coronary heart disease with higher consumption of wine. Further research showed that consumption of wine in France (20–30 g per day) can reduce the risk of coronary heart disease by at least 40%.[4, 6, 7] In a study conducted in Denmark between 1976 and 1988 involving 13,285 subjects, a low-to-moderate intake of wine was associated with lower mortality from cardiovascular and cerebrovascular disease.[6, 8] The protective effect of red wine was observed in both men and women, in the elderly, and in smokers and non-smokers. Thus, moderate consumption of wine is a negative risk factor for the development of atherosclerosis and coronary artery disease, especially in France and other Mediterranean areas where wine is an integral part of the diet. In contrast, the study also showed a comparable intake of alcohol alone caused an increased risk in cardiovascular and cerebrovascular disease. Thus, the health benefits of red wine were attributed not to the alcohol, but to the non-alcoholic polyphenolic constituents of red wine, especially oligomeric proanthocyanidins and resveratrol. What possibilities are behind this phenomenon? Polyphenolic constituents, including oligomeric proanthocyanidins (OPC) and resveratrol, are potent antioxidants that appear to decrease the potential for thrombosis. Antioxidants can increase the supply of nitric oxide, and increased nitric oxide has been demonstrated to relax aortic endothelium and may overcome hyperlipidemia in rats.[3, 8]

3. Chemistry of Polyphenols and Oligomeric Proanthocyanidins

Phenolics belong to a class of compounds possessing an aromatic benzene ring bearing one or more hydroxyl groups. Plant polyphenolics are known as secondary plant metabolites and are biosynthesized either by the shikimate or acetate pathways. Their chemical structures vary from the comparatively simple phenol or phenolic acid derivatives to high molecular weight and highly complex polymeric structures such as lignins and tannins. In 1986, approximately 4,000 polyphenols were identified. The list is constantly expanding, as there are more than 8,000 phenolic structures currently known.[9]

A limited number of basic phenolic structures exist in nature, which are further hydroxylated, methoxylated and glycosylated. Depending on the basic chemical structure, polyphenols can be divided into at least ten different classes of compounds. Bioflavonoids, which constitute the most important single group of polyphenols, can be further subdivided into ten different classes chalcone, aurone, flavone, flavonol, flavanol, flavanone, flavandiol, anthocyanidin, isoflavonoid and oligomeric proanthocyanidins. The basic chemical skeleton of a flavonoid consists of two aromatic rings linked through three carbons that form an oxygenated heterocyclic ring. Bioflavonoids occur naturally both as aglycones and as glycosides.[9]

Flavones such as luteolin, apigenin, etc. flavonols such as quercetin, kaempferol etc. and flavanones such as naringenin, hesperidin etc. and their glycosides are well known examples of natural flavonoids. There are approximately 380 flavonol glycosides and more than 200 different quercetin and kaempferol glycosides. Flavonols occur as *O*-glycosides i.e. the flavonoid unit is attached to the sugar unit through an oxygen linkage, while flavones and flavanones *O*-glycosides and *C*-glycosides (carbon–carbon linkage) are very common. It is important to know that sugars in *O*-glycosides can be cleaved easily by acid hydrolysis, while sugars in *C*-glycosides cannot be cleaved by acid hydrolysis. Examples of isoflavones include genistein, diadzein etc. Anthocyanins refer to the glycosides of anthocyanidins such as malvidin, cyanidin, pelagonidin etc. which are water soluble plant pigments responsible for the color of flowers and fruits.

OPC is the acronym for "oligomeric proanthocyanidins", a class of polyphenolic bioflavonoids found in fruits and vegetables, which are highly concentrated in the seeds of grapes and the bark of maritime pine trees. In fact, predominant compounds in grape seed and pine bark are the OPCs, which are made of proanthocyanidins sub-units called monomers. The term oligomeric in simple means more than one. Thus, OPCs consist of two or more monomers linked together chemically. OPCs containing two monomers are called dimers, while three monomers are called trimers and four monomers are called tetramers. There are basically two proanthocyanidins monomers called catechin and epicatechin. Each of these bind at either the alpha or beta position on their molecular structure to form dimers, trimers, etc. In addition, these catechin and epicatechin derivatives can also react with gallic acid to form gallate ester, or with various sugar molecules to form glycosides.

Catechins and galloylated catechins including epigallocatechin, epicatechin gallate and epigallocatechin gallate, are commonly found in green teas. These catechin derivatives are potent antioxidants, effective inhibitors of phase I enzymes, and demonstrate inhibition of tumor initiation and promotion in skin and other organs.[10, 11] Catechin, epicatechin and OPC dimers B2, B3, B4 and B5 and galloylated derivatives are found in grape seeds, while OPC of the B1 series (B1, B7, epicatechin-epicatechin-catechin trimer, and epicatechin-epicatechin-epicatechin-catechin tetramer) are found in almond. Dimer A2 is found in horse chestnut shells and the glucosylated catechin catechin-3-O-glucose is found in lentils. Catechin and epicatechin derivatives, and OPC dimers B1, B2 and B5 have antibacterial, antimicrobial and angioprotective properties.[12] Proanthocyanidin trimers C1 and EEC have been shown to inhibit platelet aggregation.[13] Figure 1 illustrates the structures of procyanidin monomers, dimmers, trimers and tetramers.

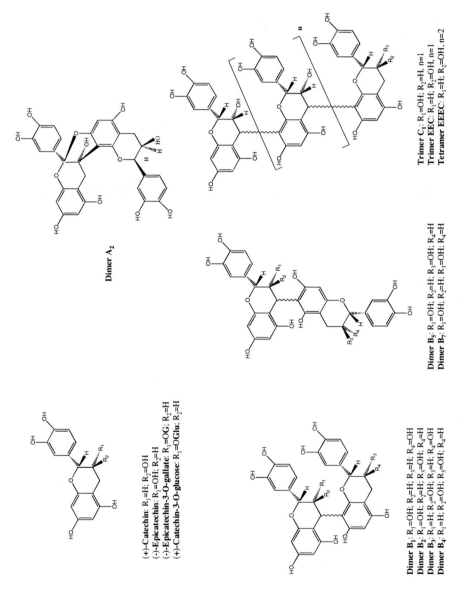

Dimer A₂

(+)-**Catechin**: R₁=H; R₂=OH
(−)-**Epicatechin**: R₁=OH; R₂=H
(−)-**Epicatechin-3-O-gallate**: R₁=OG; R₂=H
(+)-**Catechin-3-O-glucose**: R₁=OGlu; R₂=H

Dimer B₁: R₁=OH; R₂=H; R₃=H; R₄=OH
Dimer B₂: R₁=OH; R₂=H; R₃=OH; R₄=H
Dimer B₃: R₁=H; R₂=OH; R₃=H; R₄=OH
Dimer B₄: R₁=H; R₂=OH; R₃=OH; R₄=H

Dimer B₅: R₁=OH; R₂=H; R₃=OH; R₄=H
Dimer B₇: R₁=OH; R₂=H; R₃=OH; R₄=H

Trimer C₁: R₁=OH; R₂=H, n=1
Trimer EEC: R₁=H; R₂=OH, n=1
Tetramer EEEC: R₁=H; R₂=OH, n=2

Fig. 1. Structures of procyanidin monomers, dimers, trimers and tetramers.

Tannins are highly hydroxylated polyphenolic compounds of intermediate to high molecular weight. Tannins form insoluble complexes with carbohydrates, cell wall polysaccharides and proteins. Plant tannins are divided into two major categories: (a) hydrolyzable tannins and (b) condensed tannins. Hydrolyzable tannins consist of gallic acid and its dimeric condensation product glycosylated to a sugar molecule. Two or more of these units can oxidatively condense and form high-molecular weight polymers. These tannins are easily hydrolyzed with acid, alkali or enzymes. The best known example of hydrolyzable tannin is tannic acid. Autooxidative or enzymatic polymerization of flavanol and flavandiol units have been suggested as the process leading to the formation of condensed tannins. Proanthocyanidins can occur as high molecular weight polymers (> 30,000 Da) with degrees of polymerization of 50 or more to form condensed tannins. Literature reviews on condensed tannins or proanthocyanidins from plant sources primarily refer to the dimeric, trimeric and tetrameric derivatives of oligomeric proanthocyanidins. Low molecular weight oligomeric proanthocyanidins and hydrolyzable tannins are soluble in different aqueous and organic solvents, such as water, acetone and alcohol, while high-molecular weight condensed and hydrolyzable tannins are insoluble.[14]

4. Occurrence of Oligomeric Proanthocyanidins in Food and Beverage

Polyphenols are distributed in diverse plant foods including cereals, vegetables, legumes, fruits, nuts, seeds, etc. and beverages including tea, coffee, cocoa, beer, wine, etc. Polyphenols are partially responsible for the sensory and nutritional qualities, as well as astringency and bitterness of plant foods. The occurrence of polyphenols in plant foods largely depends on the genetic factors, environmental conditions, as well as conditions of storage and processing. The formation of phenol and phenol glycosides greatly depends on light, and their highest concentrations are found generally in the leaves and outer parts of plants.

Polyphenolic content in dry cereals are less than 1%. Dark varieties of legumes such as red kidney beans, black beans etc. have comparatively higher polyphenolic content including isoflavones. Vegetables in the outer parts of the plant contain primarily flavonoid glycosides, while berries have high anthocyanin content. Apples and citrus fruits are rich in phenolic acids and flavonoids. The highest concentrations of phenolic components occur in the skin of the fruits. Nuts are rich in tannins, and oil seeds primarily contain phenolic acids. Catechins and proanthocyanidins accumulate principally in the lignified portion of grape clusters, especially in the seeds.[15] Olive oil contains both phenolic acids and hydrolyzable tannins.

The polyphenol content of fruit juices in general ranges from 5–500 mg/ml. Green tea contains significant amount of catechin derivatives including epigallocatechin gallate, while black tea contains large amounts of oxidized polyphenols such as theaflavins and thearubigins. Coffee beans contain chlorogenic acid, while cocoa beans contain epicatechin, tannins and anthocyanins. The polphenols in wine include phenolic acids, anthocyanins, tannins, and other flavonoids. Generally, white wines contain 0.15 g/liter of polyphenolic components, while red wines contain 1–5 g/liter of polyphenolic components.[16] The color of red wine is formed following condensation of anthocyanins and other flavonoids. There are also significant variations of polyphenolic content between young and aged wines, as well as important differences in the nature of the polyphenols present in aged wines as compared to grape juices and young wines. Goldberg *et al.*[17, 18] has demonstrated that during wine fermentation the complex polymeric and glycosidic forms of flavonoids are broken down to monomeric forms and that these forms remain stable in wines containing 10% or more alcohol. The flavonoid content of an average North American diet could be enhanced 40% by consumption of two glasses of red wine per day.[19]

5. Dietary Intake and Bioavailability of Oligomeric Proanthocyanidins

Recommendations by various health professionals to consume five to seven servings of fresh fruits and vegetables daily, would certainly assure supplementation of a requisite quantity of flavonoids. According to a study conducted by Dr. Kuhnau[20] in 1976, humans consume an estimated 1 g per day of mixed flavonoids. Leth and Justesen[21] estimated the total intake of flavones, flavonols and flavanones in Denmark to be 28 mg/day. The total flavonoid/phenolic levels in a typical fruit serving of 200 g is in the range of 50–500 mg, with apples having over 200 mg.[22] It has been demonstrated that bioflavonoid concentration can be maintained in the plasma following regular supplementation of flavonoids in the diet. Following ingestion of red onion, the maximum plasma quercetin concentration occurs at 3.3 h and the elimination half-life is 16.8 hours.[23]

The absorption, distribution, metabolism and excretion of food polyphenolics are primarily governed by their chemical structure, degree of glycosylation and polymerization, solubility, conjugation with other phenolics and interaction with transport receptors. A classification between extractable and non-extractable polyphenols was suggested for nutritional purposes. Extractable polyphenols are low and intermediate molecular weight phenolics, some hydrolyzable tannins and proanthocyanidins that can be extracted using different solvents like water, alcohol and acetone. On the other hand, non-extractable polyphenols are high molecular weight compounds bound to dietary fiber or protein that remain insoluble in regular solvents. Studies have demonstrated that extractable polyphenols are digested and absorbed in the gut and excreted in minor amounts, while non-extractable polyphenols are extensively recovered in the feces. This confirms the resistance of these compounds to intestinal digestion and absorption. Jimenez-Ramsey *et al.*[24] demonstrated the distinct relative absorption of various extractable polyphenols, depending on their extractability with different solvents. These researchers demonstrated that monomeric and oligomeric proanthocyanidins fractions, soluble in water and ethanol are partially absorbed *in vivo* from the intestinal tract and extensively detected in all tissues and plasma. However, the proanthocyanidins fractions, insoluble in water and ethanol but soluble in aqueous acetone, are excreted in the feces and not detected in the plasma or tissues. It is interesting to note that non-extractable polyphenols that are different from highly polymerized tannins can be liberated under certain conditions and thus made available for digestion and bioavailability. Extractable polyphenols are metabolized in the gastrointestinal tract. The availability of fermentive bacteria in the digestive tract and chemical structures of the phenolic acids and flavonoids determine the absorption sites in the small and large intestines. Flavonoids such as quercetin, genistein etc. and phenolic acids are absorbed through the small intestinal mucosa. Ferulic and caffeic acids are absorbed in the intestinal tract. Hollman *et al.*[25, 26] have demonstrated that flavonoid glycosides are more bioavailable and readily absorbed *in vivo* as compared to their corresponding aglycones.

6. Nutritional Benefits of Oligomeric Proanthocyanidins (OPC)

Oligomeric proanthocyanidins (OPCs) are among the most potent plant antioxidants, which make them effective scavengers of noxious reactive oxygen species (ROS). They are chelators of metals and inhibit the Fenton and Haber–Weiss reactions, which are important sources of ROS. OPCs protect DNA, lipids and proteins from oxidative damage, inactivate procarcinogens, detoxify xenobiotics, downregulate mutant genes and interfere in several steps that lead to the development of malignant tumors. Although the major focus has been on dimeric, trimeric and tetrameric OPCs, Hagerman *et al.*[14] demonstrated that non-extractable polyphenols containing polymeric OPCs and high-molecular weight hydrolyzable tannins, are 15 to 30 times more effective at quenching peroxyl radicals compared to the simple phenols. Since non-extractable polyphenols are not absorbed *in vivo*, they theoretically could exert their antioxidant activity within the digestive gastrointestinal tract. Thus, they could protect lipids, proteins and carbohydrates from oxidative degradation during digestion and spare soluble antioxidants. The pharmacological and medicinal properties of OPCs have been extensively reviewed.[1] OPCs have potential applications as antibiotics, antidiarrheal, antiulcer, antiallergic, antibacterial, antiviral, antiinflammatory, antimutagenic, anticarcinogenic and antiproliferative agents, as well as in the treatment of hypertension and cardioprotection, vascular fragility and hypercholesterolemia.[1, 27–34] OPCs possess diverse therapeutic effects including vasodilatory, immune-stimulating and estrogenic activities, as well as potent inhibitors of LDL oxidation and platelet aggregation, and protect vitamin E in LDL from oxidative damage.[13, 27–34]

7. Recent Advances in Natural Grape Seed Proanthocyanidin Research

Over the past decade, research has continuously demonstrated that various phytopharmaceuticals offer significant protection against diseases originating from diverse etiologies. This is due, at least in part, to their powerful antioxidant and/ or chemoprotective properties. Studies utilizing these phytopharmaceuticals (popularly known as nutraceuticals) give new hope to an overwhelmingly fatigued human population who rely heavily upon allopathic medicine. Allopathic medicines captured the attention of many since it is backed by solid scientific research and approved by FDA, whereas, most nutraceutical supplements are not. However, this disparity is undergoing rapid change. Nutraceutical sales have already reached the multibillion dollar mark within the last few years, and it appears that consumer confidence is switching rapidly in their favor. Significant research studies have been conducted on nutraceuticals at the organ, cellular, subcellular, and molecular levels. Accordingly, the gap between these two arms of health care industry, i.e. modern allopathic medicine and nutraceuticals is shrinking.

A significant amount of *in vitro*, *in vivo* and human clinical studies have been conducted in many laboratories including ours to unveil the mechanistic aspects of cytoprotection by grape seed proanthocyanidins. We have examined a grape seed extract (GSPE, commercially known as ActiVin), which is a patented, standardized water-ethanol extract from red grape seeds. GCMS and HPLC studies demonstrated that GSPE contains 54% dimeric, 13% trimeric and 7% tetrameric OPC, and a small amount of monomeric bioflavonoids and high molecular weight OPC.

7.1. Safety of GSPE

A broad spectrum of toxicity and safety studies were conducted on GSPE.[35, 36] The LD_{50} of GSPE was found to be greater than 5000 mg/kg body weight (b.w.) when administered orally <u>via</u> gastric intubation to fasted male and female albino rats. No gross findings were observed at the scheduled necropsy. The LD_{50} of GSPE was found to be greater than 2,000 mg/kg b.w. when administered once for 24 h to the clipped, intact skin. In addition, 2,000 mg/kg b.w. was found to be a no observed effect level (NOEL) for systemic toxicity. GSPE also passed the primary dermal irritation and primary eye irritation studies as demonstrated by the USEPA and Toxic Substances Control Act Health Effects Test Guidelines,

40 CFR 798.4500. Dose-dependent (0, 100, 250 and 500 mg/kg b.w./day) chronic GSPE exposure (p.o.) for six months did not cause any toxicity in the eight vital organs including brain, liver, lung, spleen, heart, kidneys, duodenum and pancreas, of female B6C3F1 mice. Serum chemistries [ALT (alanine aminotransferase), BUN (blood urea nitrogen) and CK (creatine kinase)] and histopathology of the eight vital organs remained unchanged and appeared close to normal. In a separate study, long-term (twelve months) chronic *in vivo* GSPE exposure (100 mg/kg b.w./day) did not cause any toxicity in the seven vital organs including brain, heart, intestine, kidney, liver, lung, and spleen, of male B6C3F1 mice. Serum chemistry and histopathology of the seven vital organs remained unchanged and appeared close to normal. All these studies demonstrate the relative safety of GSPE.[35, 36]

7.2. Free Radical Scavenging Activity

GSPE demonstrates excellent free radical scavenging ability against bio-chemically generated superoxide anion, hydroxyl radicals and peroxyl radicals, and provides significantly better protection as compared to vitamins C, E and β-carotene.[37, 38] Using laser scanning confocal microscopy, concentration-dependent protective ability of GSPE was assessed against hydrogen peroxide-induced oxidative stress in cultured J774A.1 macrophage and adrenal pheochromocytoma PC-12 cells. At 100 μg/ml concentration, GSPE provided approximately 70% protection against hydrogen peroxide-induced modulation of intracellular oxidized states and increases in fluorescent intensities in these cells.[39]

7.3. Protection Against Drug- and Chemical-Induced Multiorgan Toxicity

Drug or chemically induced degradation of DNA in cells has serious biological consequences such as apoptotic and necrotic cell deaths, mutation, and/or carcinogenic transformation. A series of studies were conducted to determine the cytoprotective role of GSPE against structurally diverse drug- and chemical-induced multiorgan toxicity. GSPE exhibited dramatic protection against smokeless tobacco-induced oxidative stress and apoptotic cell death in a primary culture of human oral keratinocytes, and provided better protection as compared to vitamins C and E, singly and in combination.[40] Comparative *in vivo* protective abilities of GSPE, vitamins C and E, and β-carotene were assessed against 12-O-tetradecanoylphorbol-13-acetate (TPA)-induced hepatic and brain lipid peroxidation and DNA fragmentation and peritoneal macrophage activation in mice. Pretreatment of mice with GSPE (25, 50 and 100 mg/kg b.w.) resulted in a significant dose-dependent inhibition of TPA-induced production of ROS in the peritoneal macrophages, and lipid peroxidation and DNA fragmentation in brain and liver tissues compared to GSPE-untreated control animals. Furthermore, GSPE provided significantly greater protection than vitamins C, E and β-carotene, as well as combined of vitamins C and E.[41]

Short- and long-term protective effects of GSPE were examined on acetaminophen (AAP) overdose-induced lethality and hepatotoxicity in mice. GSPE dramatically reduced AAP-induced mortality, serum ALT activity (a marker of liver toxicity) and hepatic DNA fragmentation. Histopathological evaluation of liver sections showed a remarkable interference of GSPE against AAP toxicity and substantial inhibition of apoptotic and necrotic cell death. Furthermore, GSPE alone enhanced the expression of bcl-XL gene, a positive regulator of bcl_2 family of genes, and dramatically reduced AAP-induced inactivation of bcl-XL gene.[42] The protective effect of GSPE was assessed on AAP-induced nephrotoxicity, serum chemistry changes, especially changes in BUN, and genomic DNA damage in kidneys. GSPE protected against AAP-induced increases in BUN level and DNA fragmentation in kidneys. These findings were further strengthened by the histopathological evaluation of kidney sections. Thus, GSPE ameliorated kidney and renal functions in mice from AAP-induced nephrotoxicity.[43]

A series of *in vivo* studies were conducted to determine the protective ability of GSPE pre-exposure against amiodarone-induced pulmonary toxicity, dimethylnitrosamine (DMN)-induced splenotoxicity, cadmium chloride-induced nephrotoxicity, doxorubicin-induced cardiotoxicity, and O-ethyl-S,S-dipropyl phosphorodithioate (MOCAP)-induced neurotoxicity in mice. Parameters of analysis included changes in serum chemistry, histopathology and integrity of genomic DNA. Results indicate that 7-day GSPE pre-exposure prior to these drugs/ toxicants provided near complete protection in terms of serum chemistry changes (ALT, BUN and CK), genomic DNA fragmentation, histopathological changes, and abolished both forms of cell death, e.g. apoptosis and necrosis, in the lung, spleen, kidney and heart tissues. Some protection was observed against MOCAP-induced brain injury. Taken together, these studies suggest that GSPE is bioavailable to multiple target organs and may protect multiple target organs from a variety of toxic assaults including environmental pollutants, drugs and chemicals.[43-45]

7.4. Cardioprotection

Oxidative stress play a major role in the pathogenesis of myocardial ischemia-reperfusion injury. Garlick *et al.*[46] and Zweier[47] have shown that abrupt reperfusion of the ischemic myocardium led to a massive formation of oxygen free radicals and enhanced formation of oxygen free radicals could be important in the development of reperfusion-induced ventricular arrhythmias including ventricular tachycardia and ventricular fibrillation. It is important to mention that ventricular arrhythmias remains an important source of mortality in ischemic heart diseases, therefore every effort must be made to increase the recovery of cardiac function and minimize sudden cardiac death caused by ventricular fibrillation. GSPE supplementation (100 mg/kg b.w./day for 3 weeks) provided elevated cardioprotection against myocardial ischemia-reperfusion injury in rats. This was evidenced by improved post-ischemic left ventricular functions (dp, dp/dt$_{max}$), better aortic flow, and reduced creatine kinase release and malondialdehyde formation in the coronary effluent compared to the GSPE-untreated control animals. Furthermore, myocardial infarct size was reduced approximately 25% in the GSPE-fed animals.[38] Following oral supplementation of 50 and 100 mg/kg b.w./day of GSPE for three weeks, the incidence of reperfusion-induced ventricular fibrillation was reduced from its control value by 92% to 42% and 25%, respectively, as compared to control GSPE-untreated animals. The prevalence of reperfusion-induced ventricular tachycardia showed the same pattern. Rats treated with 100 mg GSPE/kg b.w./day for three weeks, the recovery of postischemic cardiac function including coronary flow, aortic flow, and the left ventricular developed pressure were improved about 25, 50, and 25%, respectively. Electron spin resonance studies indicate that GSPE significantly inhibited the formation of oxygen free radicals. These results demonstrate that GSPE possesses novel cardioprotective efficacy against reperfusion-induced arrhythmias through their abilities to reduce or remove the ROS in the ischemic/reperfused myocardium.[48] Furthermore, reperfusion of ischemic myocardium induces apoptosis in concert with the enhancement of proapoptotic factors, *c-jun* and JNK-1. GSPE reduces apoptotic cell death probably by attenuating ischemia/reperfusion-induced increased abundance of *c-jun* and JNK-1 proteins.[49] Thus, GSPE may be used as a potential therapeutic agent to cure ischemic heart disease and myocardial arrhythmias.

7.5. Protection Against Stress-Induced Oxidative Gastrointestinal Injury

GSPE supplementation provided significant protection against acute and chronic stress-induced gastrointestinal mucosal lipid peroxidation, genomic DNA fragmentation and membrane microviscosity in rats, as well as against enhanced production of oxygen free radicals in these tissues.[50]

7.6. Cancer Chemoprevention

GSPE was shown to induce selective cytotoxicity towards cultured human MCF-7 breast cancer, A-427 lung cancer and CRL-1739 gastric adenocarcinoma cells, while enhancing the growth and viability of normal human gastric mucosal cells and murine macrophage J774A.1 cells.[51] In a separate study, we assessed the protective ability of GSPE against chemotherapeutic drug-induced cytotoxicity towards normal human liver cells. Cultured Chang liver cells were treated *in vitro*, with idarubicin (Ida) (30 nM) or 4-hydroxyperoxycyclophosphamide

(4-HC) (1 μg/ml) with or without GSPE (25 μg/ml). The cells were grown *in vitro* and the growth rate of the cells were determined using MTT assay. GSPE dramatically reduced the growth inhibitory effects of Ida and 4-HC on liver cells. Since these chemotherapeutic drugs are known to induce apoptosis in the target cells, these cells were also analyzed for presence of apoptotic cells using flow cytometry. To determine the mechanisms of ameliorating effects of GSPE, the expressions of apoptosis/cell cycle/growth related genes, bcl$_2$, p53 and *c-myc* were determined in the treated and control cells using Western blotting and RT-PCR. There was an increased expression of bcl$_2$ in the cells treated with GSPE. However, there was a significant decrease in the expression of other cell cycle related genes such as p53 and *c-myc* in these cells following treatment with GSPE. These results suggest that GSPE can serve as a potential candidate to ameliorate the toxic effects associated with chemotherapeutic agents used in the treatment of cancer.[52, 53]

Although carcinogenesis is a genetically regulated multistage process, our knowledge to prevent or, at least, ameliorate its initiation, promotion and progression stages are limited. A study was conducted to determine whether pre-, post- and co-exposure of GSPE followed by dimethylnitrosamine (DMN) prevents/reduces/delays the onset of liver tumor formation in male B6C3F1 mice and, if so, whether GSPE interferes with any of the stages of cancer development. The results demonstrate that long-term GSPE exposure may negatively influence and interfere with all the three stages of DMN-induced liver carcinogenesis.[54]

7.7. Age-Related Hypertension and Glycosylated Hemoglobin Study

The pathogenesis of the aging phenomenon and the many chronic diseases associated with aging is attributed, in part, to glycosylation of biological macromolecules and augmented free radical formation causing increased tissue damage. Oral supplementation of GSPE in conjunction with niacin-bound chromium and zinc monomethionine, singly and in combination, significantly reduced systolic blood pressure, renal lipid peroxidation and glycosylated hemoglobin (HbA1c) in normotensive and hypertensive rats, and this combination was shown to affect the glucose-insulin and renin-angiotensin systems without producing any obvious toxicity.[55, 56]

7.8. Improvement of Lipid Profile in Hypercholesterolemic Human Subjects

Hypercholesterolemia, a significant cardiovascular risk factor, is prevalent in the American population. Many drugs lower circulating cholesterol levels, but they are not infrequently associated with severe side effects. We examined forty hypercholesterolemic subjects (total cholesterol 210–300 mg/dl) in a randomized, double-blind, placebo-controlled study. The four groups of ten subjects received either placebo bid, niacin-bound chromium (Cr) 200 µg elemental chromium bid, GSPE 100 mg bid, or a combination of Cr plus GSPE at the same dosage bid for eight weeks. The incidence of adverse events was low and consisted predominantly of gastrointestinal symptoms. There were no reported adverse gastrointestinal events in the GSPE group, three in the Cr group, one in the Cr plus GSPE combination group, and four in the placebo group. Over two months, the average percent change ± SEM in total cholesterol from baseline among groups was: placebo −3.5% ± 4, GSPE −2.5% ± 2, Cr −10% ± 5, and GSPE plus Cr combination group −16.5% ± 3, while the changes in LDL levels were: placebo −3.0% ± 4, GSPE −1.0% ± 2, Cr −14% ± 4, and GSPE plus Cr combination group −20% ± 6. HDL levels and triglyceride concentrations did not change significantly among the groups. The levels of autoantibodies to oxidized LDL decreased by −30.7% and −44.0% in the GSPE and GSPE plus Cr groups in contrast to −17.3% and −10.4% in the placebo and Cr groups, Thus, 50% of subjects (9 out of 18 subjects) receiving GSPE had a greater than 50% decrease in autoantibodies to oxidized LDL compared to 16% (3 out of 19 subjects) in the two groups not receiving GSPE. It is noteworthy to mention that oxidized LDL has been demonstrated as the most notorious element for cardiovascular dysfunction and death. Accordingly, a combination of GSPE and Cr can decrease total cholesterol and LDL levels significantly. Furthermore, there was a trend to decrease the circulating antibodies to oxidized LDL in the two groups receiving GSPE.[57, 58]

7.9. Inflammation, Angiogenesis, SPF and Skin Health

Cell adhesion represents a process that is centrally important in immune function, inflammation and in a variety of skin pathologic conditions. The effects of GSPE was evaluated on the expression of TNFα-induced ICAM-1 (intracellular adhesion molecule-1, CD54) and VCAM-1 (vascular cell adhesion molecule-1, CD106) expression in primary human umbilical vein endothelial cells (HUVEC). GSPE at low concentrations (1–5 µg/ml). GSPE was associated with downregulated TNFα-induced VCAM-1 expression but not ICAM-1 expression. The potent inhibitory effect of low concentrations of GSPE on agonist-induced VCAM-1 expression suggests therapeutic potential of GSPE in a variety of inflammatory and skin pathologic conditions and other pathologies involving altered expression of VCAM-1.[59]

In a more recent study, modulatory effect of GSPE was assessed on angiogenesis, which plays a central role in wound healing. Among many known growth factors, VEGF (vascular endothelial growth factor) is believed to be the most prevalent, efficacious and long-term signal that is known to stimulate angiogenesis in wounds. Using a ribonuclease protection assay (RPA), the ability of GSPE to regulate oxidant-induced changes in several angiogenesis-related genes were studied. While mRNA responses were studied using RPA, VEGF protein release from cells into the culture medium was studied using ELISA. Pretreatment of immortalized human HaCaT keratinocytes with GSPE upregulated both hydrogen peroxide and TNFα-induced VEGF expression and release. These results suggest that GSPE may have beneficial therapeutic effects in promoting dermal wound healing and other inflammatory skin disorders.[60]

Proactive application of 1% GSPE creme to human volunteers 30 min prior to UVA-UVB radiation provided a 9% increase in sun protection factor (SPF) protection. Thus, topical application of GSPE enhanced SPF through tissue conditioning and presumably by scavenging radiation-induced oxygen free radicals.[35]

7.10. Human Pancreatitis Study

The protective effect of GSPE was investigated in three patients suffering from chronic pancreatitis, two with a history of alcohol excess and one idiopathic. Treatment with narcotic analgesics, pancreatic enzyme supplements failed to control their symptoms. In contrast, GSPE supplementation significantly reduced the frequency and intensity of abdominal pain and resolved vomiting in these patients.[61]

7.11. Mechanistic Pathways of Cytoprotection by GSPE

The above studies demonstrate that GSPE is a very potent antioxidant, and provides significantly better protection when compared to vitamins C, E and β-carotene. Free radicals and oxidative stress have been demonstrated for a broad spectrum of degenerative diseases. By significantly scavenging free radicals, GSPE may prevent or delay the onset of multiple degenerative diseases and potentially improve the quality of human life. The involvement of GSPE with other biochemical or molecular events have been described and discussed in a series of publications. A series of studies demonstrate the profound antiendonucleolytic, antilipoperoxidative, antiapoptogenic, antinecrogenic and gene modulatory potential of GSPE. The series of studies unequivocally demonstrate GSPE's ability to defend multiple organs from toxicities generated under diverse conditions. Five striking observations in favor of GSPE's cytoprotective capabilitites are: (i) its' potential to detoxify multiple structurally and functionally different drugs and chemicals, and/or their metabolites in multiple target organs in the body, and (ii) its' universal bioavailability and abundant quantity in individual target organs, which can detoxify drugs/ toxicants/metabolites or their effects, (iii) its' interference with endonuclease activity, (iv) its' interference with DNA methylation, and (v) its' modulatory effect on bcl₂ and bcl-XL genes. Another noteworthy property of GSPE which might significantly contribute to drug/chemical interactions is its influence on the drug metabolizing enzymes. Acetaminophen is primarily metabolized by cytochrome P450 2E1, and in a recent study we have demonstrated that GSPE is a potent inhibitor of cytochrome P450 2E1 activity in both *in vitro* and *in vivo* models.[62] This effect of GSPE may have handled acetaminophen hepato- and nephro-toxicity efficiently. Likewise, amiodarone is metabolized by cytochrome P450 3A. Although, the effect of GSPE on cytochrome P450 3A is not known, grape fruit juice is a powerful inhibitor of cytochrome P450 3A.[63] Thus, anti-amiodarone potential of GSPE may have been due to its inhibitory influence on cytochrome P450 3A. GSPE may also function in the detoxification of other drugs/chemicals. Furthermore, at the molecular level expression of several genes, viz. bcl₂, bcl-XL, p53, *c-myc*, bax, bid, and the members of ICE family have been recognized to modify upstream or downstream events related to cell death. These genes have the power to enhance/modulate/delay the onset of the cell death process. Increased expression of bcl₂ and bcl-XL may also ameliorate the toxic effects of structurally and functionally diverse drugs and chemicals, and environmental pollutants.

8. Conclusion

Epidemiological evidence suggests that high antioxidant status can lower the incidence of many degenerative diseases including cancer. The increased consumption of fresh vegetables and fruits, with their antioxidant potential, and the decreased intake of fish, meats and fats, are associated with good health. Nevertheless, supplementation of oligomeric proanthocyanidins (OPC) and safe antioxidants are essential, because we do not get enough antioxidants from the foods and beverages consumed daily. Examining all evidence, the nutritional significance of proanthocyanidins and their potential health benefits may play a much greater role than acting merely as antioxidants. Research suggests that oligomeric proanthocyanidins can enhance chemoprevention and prolong a healthy life.

References

1. Rice-Evans, C. (2001). Flavonoid antioxidants, *Curr. Med. Chem.* **8**: 797–807, Marcel Dekker, Inc., New York.
2. Bors, W., Heller, W., Michel, C. and Stettmaier K. (1996). Flavonoids and poly-phenols: chemistry and biology. *In* "Hanbook of Antioxidants" (E. Cadenas, and L. Packer, Eds.), pp. 409–466. Marcel Dekker, Inc., New York.

3. Criqui, M. H. and Ringel, B. L. (1994). Does diet or alcohol explain the French paradox? *Lancet* **344**: 1719–1723.
4. Renaud, S. and de Lorgeril, M. (1992). Wine, alcohol, platelets and the French paradox for coronary heart disease. *Lancet* **339**: 1523–1526.
5. St Leger, A. S., Cochrane, A. L. and Moore, F. (1979). Factors associated with cardiac mortality in developed countries with particular reference to the consumption of wine. *Lancet* **1**: 1017–1020.
6. Gronbaek, M., Deis, A., Sorensen, T. I. A., Becker, U., Schnohr, P. and Jensen, G. (1995). Mortality associated with moderate intakes of wine, beer, or spirits. *Br. Med. J.* **310**: 1165–1169.
7. Marques-Vidal, P., Cambou, J.-P., Nicaud, V., Luc, G., Evans, A., Arveiler, D., Bingham, A. and Cambien, F. (1995). Cardiovascular risk factors and alcohol consumption in France and Northern Ireland. *Atherosclerosis* **115**: 225–232.
8. Anonymous. (1993). Inhibition of LDL oxidation by phenolic substances in red wine: a clue to the French paradox? *Nutri. Rev.* **51**: 185–187.
9. Bravo, L. (1998). Polyphenols: chemistry, dietary sources, metabolism, and nutritional significance. *Nutri. Rev.* **56**: 317–333.
10. Gali, H. U., Perchellet, E. M., Gao, X. M., Karchesy, J. J. and Perchellet, J. P. (1994). Comparison of the inhibitory effects of monomeric, dimeric and trimeric procyanidins on the biochemical markers of skin tumor promotion in mouse epidermis in vivo. *Planta Medica* **60**: 235–239.
11. Valcic, S., Timmermann, B. N., Alberts, D. S., Wachter, G. A., Krutzsch, M., Wymer, J. and Guillen, J. M. (1996). Inhibitory effect of six green tea catechins and caffeine on the growth of four selected human human tumor cell lines. *Anticancer Drugs* **7**: 461–468.
12. Vennat, B., Pourrat, A., Pourrat, H., Gross, D., Bastide, P. and Bastide, J. (1988). Procyanidins from the roots of *Fragaria vesca*: characterization and pharmacological approach. *Chem. Pharm. Bull.* **36**: 828–833.
13. Chang, W. C. and Hsu, F. L. (1989). Inhibition of platelet aggregation and arachidonate metabolism on platelets by procyanidins. *Prostaglandins Leukotrienes and Essential Fatty Acids* **38**: 181–188.
14. Hagerman, A. E., Riedl, K. M., Jones, G. A., Sovik, K. N., Ritchard, N. T., Hartzfeld, P. W. and Riechel, T. L. (1998). High molecular weight polyphenolics (tannins) as biological antioxidants. *J. Agric. Food. Chem.* **46**: 1887–1892.
15. Kovac, V., Alonso, E., Bourzeix, M. and Revilla, E. (1992). Effect of several ecological practices on the content of catechins and proanthocyanidins of red wine. *J. Agric. Food Chem.* **40**: 1953–1957.
16. Pierpoint, W. S. (1986). Flavonoids in the human diet. *Prog. Clin. Biol. Res.* **213**: 125–140.
17. Goldberg, D. M., Hahn, S. E. and Parkes, J. G. (1995). Beyond alcohol: beverage consumption and cardiovascular mortality. *Clin. Chim. Acta* **237**: 155–187.
18. Goldberg, D. M. (1995). Does wine work? *Clin. Chem.* **41**: 14–16.

19. German J. B. (1998). Nutritional studies of flavonoids in wine. *In* "Flavonoids in Health and Disease" (C. A. Rice-Evans, and L. Packer, Eds.), pp. 343–358, Marcel Dekker, Inc., New York.

20. Kunhau, J. (1976). The flavonoids. A class of semi-essential food components: their role in human nutrition. *World Rev. Nutri. Diet.* **24**: 117–191.

21. Leth, T. and Justesen, U. (1998). Analysis of flavonoids in fruits, vegetables and beverages by HPLC-UV and LC-MS and estimation of the total daily flavonoid intake in Denmark. *In* "Polyphenols in Food: Office for Official Publications of the European Communities" (R. Amado, H. Andersson, S. Bardocz, and F. Serra, Eds.), pp. 39–40, Luxembourg.

22. Macheix, J.-J., Fleuriet, A. and Billot, J. (1990). *Fruit Phenolics*. CRC Press, Boca Raton, FL.

23. Hollman, P. C., v d Gaag, M., Mengelers, M. J., van Trijp, J. M., de Vries, J. H. and Katan, M. B. (1996). Absorption and disposition kinetics of the dietary antioxidant quercetin in man. *Free Radic. Biol. Med.* **21**: 703–707.

24. Jimenez-Ramsey, L. M., Rogler, J. C., Housley, T. L., Butler, L. G. and Elkin, R. G. (1994). Absorption and distribution of ^{14}C-labeled condensed tannins and related sorghum phenolics in chickens. *J. Agric. Food Chem.* **42**: 963–967.

25. Hollman, P. C., Bijsman, M. N., van Gameren, Y., Cnossen, E. P., de Vries, J. H. and Katan, M. B. (1999). The sugar moiety is a major determinant of the absorption of dietary flavonoid glycosides in man. *Free Radic. Res.* **31**: 569–573.

26. Hollman, P. C. and Katan, M. B. (1998). Bioavailability and health effects of dietary flavonols in man. *Arch. Toxicol. Suppl.* **20**: 237–248.

27. Afanas'ev, I. B., Dorozhko, A. I., Brodskii, A. V., Kostyuk, V. A. and Potapovitch, A. I. (1989). Chelating and free radical mechanisms of inhibitory action of rutin and quercetin in lipid peroxidation. *Biochem. Pharmacol.* **38**: 1763–1769.

28. Bors, W. and Saran, M. (1987). Radical scavenging by flavonoid antioxidants. *Free Radic. Res. Commun.* **2**: 289–294.

29. Buening, M. K., Chang, R. L., Huang, M. T., Fortner, J. G., Wood, A. W. and Conney, A. H. (1981). Activation and inhibition of benzo(a)pyrene and aflatoxin B1 metabolism in human liver microsomes by naturally occurring flavonoids. *Cancer Res.* **41**: 67–72.

30. Chen, Z. Y., Chan, P. T., Ho, K. Y., Fung, K. P. and Wang, J. (1996). Antioxidative activity of natural flavonoids is governed by number and location of their aromatic hydroxyl groups. *Chem. Phys. Lipids* **79**: 157–163.

31. Hanefeld, M. and Herrmann, K. (1976). On the occurrence of proanthocyanidins, leucoanthocyanidins and catechins in vegetables. *Z. Lebensm. Unters. Forsch.* **161**: 243–248.

32. Kolodziej, H., Haberland, C., Woerdenbag, H. J. and Konings, A. W. T. (1995). Moderate cytotoxicity of proanthocyanidins to human tumor cell lines. *Phytotherapy Res.* **9**: 410–415.

33. Masquelier, J., Michaud, J., Laparra, J. and Durnon, M. C. (1979). Flavonoids and pycnogenols. *Int. J. Vit. Nutri. Res.* **49**: 307–311.
34. Fine, A. M. (2000). Oligomeric proanthocyanidin complexes: history, structure, and phytopharmaceutical applications. *Altern. Med. Rev.* **5**: 144–151.
35. Bagchi, D., Bagchi, M., Stohs, S. J., Das, D. K., Ray, S. D., Kuszynski, C. A., Joshi, S. S. and Preuss, H. G. (2000). Free radicals and grape seed pro-anthocyanidin extract: Importance in human health and disease prevention. *Toxicology* **148**: 187–197.
36. Ray, S., Bagchi, D., Lim, P. M., Bagchi, M., Gross, S. M., Kothari, S. C., Preuss, H. G. and Stohs, S. J. (2000). Acute and long-term safety evaluation of a novel IH636 grape seed proanthocyanidin extract. *Res. Commun. Mol. Pathol. Pharmacol.* **109**: 165–197.
37. Bagchi, D., Garg, A., Krohn, R. L., Bagchi, M., Tran, M. X. and Stohs, S. J. (1997). Oxygen free radical scavenging abilities of vitamins C and E, and a grape seed proanthocyanidin extract in vitro. *Res. Commun. Mol. Pathol. Pharmacol.* **95**: 179–190.
38. Sato, M., Maulik, G., Ray, P. S., Bagchi, D. and Das, D. K. (1999). Cardio-protective effects of grape seed proanthocyanidin against ischemic reperfusion injury. *J. Mol. Cell. Cardiol.* **31**: 1289–1297.
39. Bagchi, D., Kuszynski, C., Balmoori, J., Bagchi, M. and Stohs, S. J. (1998). Hydrogen peroxide-induced modulation of intracellular oxidized states in cultured macrophage J774A.1 and neuroactive PC-12 cells, and protection by a novel grape seed proanthocyanidin extract. *Phytotherapy Res.* **12**: 568–571.
40. Bagchi, M., Balmoori, J., Bagchi, D., Ray, S. D., Kuszynski, C. and Stohs, S. J. (1999). Smokeless tobacco, oxidative stress, apoptosis, and antioxidants in human oral keratinocytes. *Free Radic. Biol. Med.* **26**: 992–1000.
41. Bagchi, D., Garg, A., Krohn, R. L., Bagchi, M., Bagchi, D. J., Balmoori, J. and Stohs, S. J. (1998). Protective effects of grape seed proanthocyanidins and selected antioxidants against TPA-induced hepatic and brain lipid peroxidation and DNA fragmentation, and peritoneal macrophage activation in mice. *Gen. Pharmacol.* **30**: 771–776.
42. Ray, S. D., Kumar, M. A. and Bagchi, D. (1999). A novel proanthocyanidin IH636 grape seed extract increases in vivo bcl-X$_L$ expression and prevents acetaminophen-induced programmed and unprogrammed cell death in mouse liver. *Arch. Biochem. Biophys.* **369**: 42–58.
43. Ray, S. D., Patel, D., Wong, V. and Bagchi, D. (2000). In vivo protection of DNA damage associated apoptotic and necrotic cell deaths during acetaminophen-induced nephrotoxicity, amiodarone-induced lung toxicity and doxorubicin-induced cardiotoxicity by a novel IH636 grape seed proanthocyanidin extract. *Res. Commun. Mol. Pathol. Pharmacol.*, **107**: 137–166.
44. Ray, S. D., Wong, V., Rinkovsky, A., Bagchi, M., Raje, R. R. and Bagchi, D. (2000). Unique organoprotective properties of a novel IH636 grape seed proanthocyanidin extract on cadmium chloride-induced nephrotoxicity,

dimethylnitrosamine (DMN)-induced spleenotoxicity and MOCAP-induced neurotoxicity in mice. *Res. Commun. Mol. Pathol. Pharmacol.*, **107**: 105–128.

45. Bagchi, D., Ray, S. D., Patel, D. and Bagchi, M. (2001). Protection against drug- and chemical-induced multiorgan toxicity by a novel IH636 grape seed proanthocyanidin extract. *Drugs. Exptl. Clin. Res.*, **XXVII**: 3–15.

46. Garlick, P. B., Davies, M. J., Hearse, D. J. and Slater, T. F. (1987). Direct detection of free radicals in the reperfused rat heart using electron spin resonance spectroscopy. *Circ. Res.* **61**: 757–760.

47. Zweier, J. L. (1988). Measurement of superoxide derived free radical mechanism of reperfusion injury. *J. Biol. Chem.* **263**: 1353–1357.

48. Pataki, T., Bak, I., Kovacs, P., Bagchi, D., Das, D. K.. and Tosaki, A. (2002). Grape seed proanthocyanidins reduced ischemia/reperfusion-induced injury in isolated rat hearts. *Am. J. Clin. Nutr.* **in press**.

49. Sato, M., Bagchi, D., Tosaki, A. and Das, D. K. (2001). Grape seed proanthocyanidin reduces cardiomyocyte apoptosis by inhibition ischemia/ reperfusion-induced activation of JNK-1 and *c-jun*. *Free Radic. Biol. Med.* **31**: 729–737.

50. Bagchi, M., Milnes, M., Williams, C. B., Balmoori, J., Ye, X., Stohs, S. J. and Bagchi, D. (1999). Acute and chronic stress-induced oxidative gastrointestinal injury in rats, and the protective ability of a novel grape seed proanthocyanidin extract. *Nutri. Res.* **19**: 1189–1199.

51. Ye, X., Krohn, R. L., Liu, W., Joshi, S. S., Kuszynski, C. A., McGinn, T. R., Bagchi, M., Preuss, H. G., Stohs, S. J. and Bagchi, D. (1999). The cytotoxic effects of a novel IH636 grape seed proanthocyanidin extract on cultured human cancer cells. *Mol. Cell. Biochem.* **196**: 99–108.

52. Joshi, S. S., Kuszynski, C. A., Benner, E. J., Bagchi, M. and Bagchi, D. (1999). Amelioration of the cytotoxic effects of chemotherapeutic agents by grape seed proanthocyanidin extract. *Antiox. Redox Signal* **1**: 563–570.

53. Joshi, S. S., Kuszynski, C. A., Bagchi, M. and Bagchi, D. (2000). Chemo-preventive effects of grape seed proanthocyanidin extract on Chang liver cells. *Toxicology*, **155**: 83–90.

54. Ray, S. D., Parikh, H., Ali, S. and Bagchi, D. (2000). IH636 grape seed proanthocyanidin extract (GSPE) exposure significantly attenuates dimethyl-nitrosamine (DMN)-induced liver cancer and mortality in ICR mice. *Proc. Amer. Assoc. Cancer Res.* **41(Abstract 2928)**: 460.

55. Tyson, D. A., Talpur, N. A., Echard, B. W., Bagchi, D. and Preuss, H. G. (2000). Acute effects of grape seed extract and niacin-bound chromium on cardiovascular parameters of normotensive and hypertensive rats. *Res. Commun. Pharmacol. Toxicol.* **5**: 91–106.

56. Preuss, H. G., Montamarry, S., Echard, B. W., Scheckenbach, R. and Bagchi, D. (2001). Long-term effects of chromium, grape seed extract, and zinc on various metabolic parameters of rats. *Mol. Cell. Biochem.* **223**: 95–102.

57. Preuss, H. G., Wallerstedt, D., Talpur, N., Tutuncuoglu, S. O., Echard, B., Myers, A., Bui, M. and Bagchi, D. (2000). Effects of niacin-bound chromium and grape seed proanthocyanidin extract on the lipid profile of hyper-cholesterolemic subjects: A pilot study. *J. Med.* **31**: 227–246.

58. Talpur, N. A., Echard, B., Tutuncuoglu, S. O., Wallerstedt, D., Bui, M. N., Manohar, V., Bagchi, D. and Preuss, H. G. (2000). Effect of chromium and grape seed extract on lipid profile of hypercholesterolemic patients. *FASEB J.* **14(Abstract 503.8)**: A727.

59. Sen, C. K. and Bagchi, D. (2001). Regulation of inducible adhesion molrecule expression in human endothelial cells by grape seed proanthocyanidin extract. *Mol. Cell. Biochem.*, **216**: 1–7.

60. Khanna, S., Roy, S., Bagchi, D., Bagchi, M. and Sen, C. K. (2001). Upregulation of oxidant-induced VGEF expression in cultured keratinocytes by a grape seed proanthocyanidin extract. *Free Radic. Biol. Med.*, **31**: 38–42.

61. Banerjee, B. and Bagchi, D. (2001). Beneficial effects of a novel IH636 grape seed proanthocyanidin extract in the treatment of chronic pancreatitis. *Digestion*, **63**: 203–206.

62. Ray, S. D., Parikh, H., Hickey, E., Bagchi, M. and Bagchi, D. (2001). Differential effects of IH636 grape seed proanthocyanidin extract and a DNA repair modulator 4-aminobenzamide on liver microsomal cytochrome P450 2E1-dependent aniline hydroxylation. *Mol. Cell. Biochem.*, **218**: 27–33.

63. Libersa, C. C., Brique, S. A., Motte, K. B., Caron, J. F, Guedon-Moreau, L. M., Humbert, L., Vincent, A., Devos, P. and Lhermitte, M. A. (2000). Dramatic inhibition of amiodarone metabolism induced by grapefruit juice. *Br. J. Clin. Pharmacol.* **49**: 373–378.

Chapter 39

Effects of Antioxidant Vitamins C and E on Gene Expression in the Vasculature

Helen R. Griffiths* and S.J. Rayment

Helen R. Griffiths and **S.J. Rayment** • Pharmaceutical Sciences Research Institute, Aston University, Aston Triangle, Birmingham B4 7ET

*Corresponding Author.
Tel: +44 121 359 3611 × 5226, E-mail: h.r.griffiths@aston.ac.uk

1. Introduction

Cellular phenotype is governed through the cell specific expression of the genome. Superimposed on the expressed proteome is the capacity of cells to respond to extracellular stimuli or stress through modulation of gene expression. This enables cells to rapidly adapt to environmental changes and is achieved in at least two ways (1) via recognition of specific consensus sequences in the promoter regions of affected genes by activated transcription factors (e.g. haemoxygenase 1)[1] and (2) by modulation of translational efficiency (ferritin).[2]

Oxidative stress is known to elicit a change in gene expression and represent a homeostatic response offering cells a complex repertoire of defence against oxidative stress through upregulation of antioxidant enzymes (thioredoxin, Mn-SOD, glutathione-S transferase, haem oxygenase),[3] cell cycle control and DNA repair genes.[4] The expected corollary of this is that antioxidant treatment prior to oxidative stress can prevent these changes in gene expression. In part this hypothesis is supported, but increasingly antioxidants alone have been described as capable of modulating gene expression.

Oxidative stress is undoubtably linked with ageing and accelerated atherosclerosis. Oxidant induced vascular smooth muscle cell proliferation, endothelial and monocyte activation, with increased adhesion via integrin/cellular adhesion molecule expression are pivotal to this process. Benefits of antioxidant supplementation are described in the literature. Thus, this review will address the importance of antioxidant vitamins C and E on gene expression, with particular reference to the vasculature in order to examine the evidence for antioxidant benefits from altered gene expression.

2. Redox Sensitive Transcription Factors

In order for RNA polymerase II to initiate transcription of a mammalian gene, there is an absolute requirement for an active transcription factor to be bound in the consensus sequence structural motif in the 5′ region of the gene. Many mammalian transcription factors have been identified, where each transcription factor has a unique mechanism of activation facilitating selectivity of response; for example steroid receptors which are triggered through binding of steroids. The transcription factors AP-1 and NF-κB appear to be crucial in modulating the oxidant/antioxidant response.

3. Nuclear Factor Kappa Binding Protein NF-κB

First identified in the long terminal repeat of HIV-1 and believed to be restricted to B cells, NF-κB has since been shown to be induced in T cells and monocytes.[5]

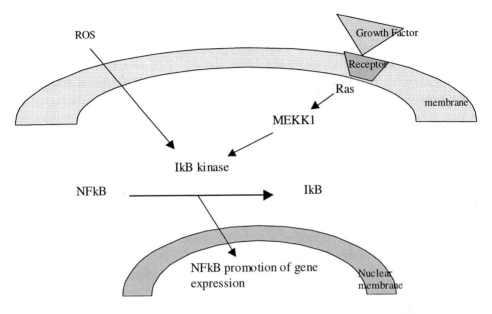

Fig. 1. A schematic diagram illustrating the oxidant signalling sequence to NF-κB activation.

It appears to play a pivotal role in the activation of immune and inflammatory genes. Functional NF-κB is a heterodimer made from any members of the Rel family. This family of proteins is broadly divided into two classes which possess high sequence homology, p50 and p65 proteins, and the resultant complexes can vary in transcriptional activity. NF-κB becomes activated in the absence of protein synthesis thus suggesting that it is maintained in a latent form. Indeed the inhibitor I-κB retains inactive NF-κB in the cytoplasm until triggered for degradation by I-κB kinase.[6] The subsequent ubiquitination marks I-κB for degradation by the cytoplasmic proteasomal apparatus, thereby allowing NF-κB to move to the nucleus and bind to the consensus sequence (see Fig. 1). The upstream signalling cascades of NF-κB activation vary from cell to cell and between stimuli, however MEKK1 and ras cascades are important players.[7]

Unusually NF-κB is activated by a large number of apparently unrelated conditions and agents (e.g. TNF alpha, lipopolysacharide and UV light). However, it appears that most of these rely on the generation of intracellular ROS as evidenced by the inhibitory action of several antioxidants including N-acetyl cysteine.[8] Further confirmation comes from the observation that overexpression of catalase inhibited TNF mediated activation of NF-κB.[9]

Some of the proteins upregulated by NF-κB in response to oxidative stress include immunoreceptors and adhesion molecules, chemokines and cytokines, and members of the Rel family, via an auto-regulatory process.

4. Activator Protein 1 (AP1)

The transcription factor AP1 is also sensitive to redox status. It recognizes and binds to the TPA (or antioxidant) response element (TRE or ARE respectively).[10] This element is present in the promoter regions of many genes including metallothionein, collagenase, stromelysin, *myc*, *fos*, *jun* and IL-2. ARE plays an important role in mediating the induction of early response genes associated with changes in extracellular growth signals/stress, where protein kinase C (PKC) plays a pivotal role in transducing upstream signals.[11] Again a heterodimeric protein, AP1 comprises the proteins from the *c-Jun* and *c-Fos* families, where dimerization is achieved through leucine zipper motifs. Several proteins form the *Jun* and *Fos* proteins including *Jun B*, *Jun D*, *Fos D*, and FRA1, where transactivational activity depends on the composition of the heterodimer. The activation and binding of latent *Fos-Jun* complexes to DNA is redox regulated, where inactivation is associated with oxidation of a cysteine thiol group on *Jun*. Further regulation is achieved through Ref-1, which facilitates binding of AP-1 to DNA under oxidizing conditions.[12] Thus a complex picture emerges whereby initial activation of latent *Jun* is mediated via pro-oxidant signals, but subsequent regulation of induced AP-1 is sensitive to antioxidants, and following an initial trigger by oxidants is later inhibited by oxidants.[13]

An important mechanism by which oxidants can modulate signalling appears to be through increases in intracellular calcium.[14] The major calcium transporters of the ER have redox sensitive thiols, which on oxidation, lead to leakage of calcium into the cytosol. The increase in calcium plays a pivotal role in inducing several protein kinases including PKC, tyrosine kinase and MAP kinases. Other redox sensitive thiol groups within the signalling cascades remain to be elucidated, hampered by the lack of an obvious stable fingerprint such as phosphorylation.

This simplistic overview introduces the two major classes of transcription factors responsible for redox mediated regulation of gene expression, however it is important to note that these TFs are subject to further regulation by other TFs e.g. Sp1.[15]

Antioxidants are increasingly recognized as having several modes of action intracellularly including antioxidant, pro-oxidant and co-factor; each of which may elicit effects on gene expression via redox sensitive transcription factors.

In order to facilitate understanding the effects of specific dietary antioxidants on gene expression it is important to address their pharmacology, including uptake, bioavailability and degradation, as variation in response may reflect variable metabolism.

5. Vitamin C Effects on Gene Expression in the Vasculature

The importance of ascorbate as an aqueous phase antioxidant has been established for many decades, and the chemistry of ROS scavenging is well described. Additionally it serves to maintain membrane alpha tocopherol and enzyme activities including hepatic mixed function oxidase activity. Many of the enzymes which utilize ascorbate as a co-factor are located in the endoplasmic reticulum, such as prolyl and lysyl hydrolases, which catalyze the production of stable crosslinked collagen through post-translational modification and folding.[16]

As ascorbate (AA) and its oxidation product dehydroascorbate (DHA) are charged water-soluble species, transporters must exist for their intracellular transfer. Bioavailability of dietary ascorbate is dependent on absorption in the small intestine by an energy driven saturable Na^+-dependent transport process. Of the cell types examined to date, it appears that both high affinity and low affinity transporters exist with respective K_m values of 5–150 µM (normal plasma range) and 1–5 mM.[17, 18] Ascorbate exists predominantly in the blood and body in its reduced form, where DHA comprises only 5–7% of total plasma vitamin C. Compared with plasma levels (50–100 µM), tissue levels are generally far higher (6–8 mM in leukocytes, 5 mM in HUVECs), where ascorbate is transported against a concentration gradient.[19] DHA transport is mediated by the several isoforms of the glucose transporter (GLUT), where the oxidized form is transported faster than the reduced form.[20] In this way, plasma DHA is maintained at low levels. In support of this, AA uptake as DHA is promoted by insulin, and inhibited by hyperglycemia, yielding scurvy-like tissue in diabetics.[21] Reduction of DHA to AA both enzymatically and non-enzymatically occurs in all tissues, most likely in the lumen of the endoplasmic reticulum.[22] Several candidates for catalysis of this reaction have been proposed; e.g. glutathione dependent protein disulphide

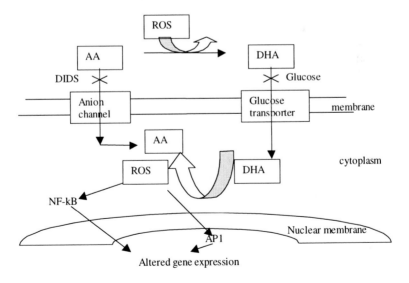

Fig. 2. The mechanism for uptake of ascorbic acid (AA) and dehydroascorbic acid (DHA), into leukocytes and endothelial cells. DIDS = selective anion channel inhibitor.

isomerase or NADPH oxidoreductase, with the concomitant generation of oxidizing equivalents within the lumen. This is shown schematically in Fig. 2. Support for the latter is presented through the observation of a reactive oxygen species burst following incubation of dihydrodichlorofluorescein loaded U937 monocytes with 150 μM AA (see Fig. 3).

AA has potent pro-oxidant activity in the presence of metal ions, where it acts as a reducing agent to provide catalysts for the Fenton reaction. Through this mechanism it is a potent inducer of apoptosis,[23] however, this mechanism of action is likely to reflect a signalling process and act merely as an activator of caspase cascade rather than in the regulation of gene expression. The physiological existence and relevance of the pro-oxidant nature of AA has been questioned, but evidence (from supplementation studies) of increased levels of DNA oxidation products in leukocytes from subjects given vitamin C, lends some credence to the hypothesis that AA may act as pro-oxidant *in vivo*.[24, 25]

There are limited reports in the literature citing the capacity of AA to inhibit redox induced changes in gene expression, where the protection of NF-κB activation appears to orchestrate these effects. This is partially due to the complex behaviour of AA as both oxidant and antioxidant. Table 1 summarizes these studies.

Whilst the role of AA in collagen biosynthesis is well documented a recent study has shown that ascorbate differentially regulates elastin and collagen biosynthesis in vascular smooth muscle cells at the pre-translational level. At doses of 10–200 μM, elastin mRNA levels were reduced in association with diminished elastin production. In contrast, collagen I and II mRNAs levels increased due to increased stability of the collagen I mRNA species, demonstrating that ascorbate

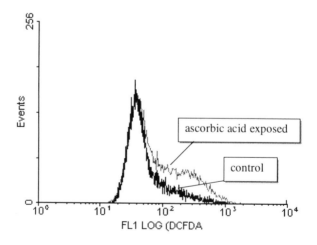

Fig. 3. The intracellular detection of ROS by dihydrodichlorofluorescein diacetate in control U937 cells (black) and U937 cells exposed to ascorbic acid for 40 min (grey), showing a population of cells with increased ROS after AA treatment.

Table 1. Summary of Studies Evaluating Effects of Ascorbic Acid on Stimulus Induced Gene Expression

Cells Studied	Antioxidant Dose	Stimulus	End Point	Reference
HUVECs	Plasma collection 2 h post oral dose of 2 g vitamin C	Smoking derived oxidants in plasma	ICAM-1 expression (no change)	27
Arterial smooth muscle	Vitamin C 100 µM/24 h	Moderately oxidized LDL	Reduced apoptosis	26

has an important role in modulating matrix production at the gene expression level.[28] The extracellular matrix can affect proliferation, differentiation and expression of smooth muscle and endothelial cells,[29] due to immobilization of growth factors.[30] Ascorbate has been shown to exert a proliferative effect on different cell lines, which depends on dose.[29, 31] In vascular smooth muscle cells, fibroblasts and endothelial cells, AA has a biphasic effect, promoting growth to 250 µM whilst acting as a potent inhibitor of cell growth from 1–5 mM, where the latter is not due to cell death. One plausible mechanism for the effect of ascorbate is that an alteration in extracellular matrix production elicits change in growth support. However, oxidants have also been demonstrated to increase VEGF, indicating that control of growth at the level of gene expression may be a mechanism by which AA can exert its effect.[32]

A recent report by Siow *et al*.[34] has examined the induction of the antioxidant like stress proteins in vascular cells (including HOX1, MSP23) by oxidized LDL, and investigated the potential for antioxidant protection by vitamin C. These workers describe the protection against ox-LDL induced gene expression of MSP23 by pre-incubation of cells for 24 h with AA. However, levels of MSP23 mRNA appear to be elevated in control cells receiving AA only. This suggests that AA alone may elicit expression of MSP23, but can also subsequently protect against subsequent oxidative insult. In a similar study, we have investigated the response of primary monocytes to hydrogen peroxide in the presence or absence of AA on expression of immune response genes IL-6, TNF-α and ICAM-1. Again AA alone for 2 h, was found to induce mRNA for TNF-α, but offered protection against ICAM-1 and IL-6 induction (unpublished observations, Rayment). Since the genes in question are all under control of NF-κB, a pro-oxidant mechanism seems likely, where a temporal difference in gene expression may explain the observed induction of TNF-α mRNA in the absence of induction of ICAM-1 or IL-6.

If the uptake of DHA with subsequent intracellular generation of ROS is a valid and an important mechanism for accumulation of intracellular reduced AA, then AA exposure may lead to a transient pro-oxidant effect, capable of eliciting changes in gene expression in cells lining the vasculature, but offering protection in the long term.

6. Vitamin E Effects on Gene Expression in the Vasculature

Vitamin E has long been defined as an essential nutrient for reproduction, but much of its pharmacological activity remains elusive. It is a potent chain breaking antioxidant, scavenging peroxyl radicals and also plays an important role in membrane stabilization. The term vitamin E refers collectively to tocopherols and tocotrienols, of which alpha tocopherol (AT) has the highest biological activity as an antioxidant.[35] Absorption of alpha tocopherol from the diet is low (20–40%) where supplements in the order of tenfold are required in order to double plasma levels. The selective utilization of RRR-alpha tocopherol has been attributed to the hepatic tocopherol transfer protein.[36] Site specific mutations or a deletion in the terminal portion of this protein can render an individual vitamin E deficient despite a vitamin E replete diet, leading to the development of serious neurological abnormalities.[37]

Transport of AT in the blood is principally via lipoproteins, initially through incorporation into VLDL in the liver, and subsequently transferred to other lipoproteins during metabolism, where delivery to tissue is achieved via high affinity lipoprotein receptors.

The absorbed tocopherols are metabolized by oxidation to quinones or conjugated with glucuronic acid, appearing in the urine as the corresponding isometric CEHC.[38]

Vitamin E has been shown to be an effective inhibitor of gene expression induced by pro-inflammatory stimuli such as cytokines *in vitro* (see Table 2). In cases of increased transcriptional activity the capacity of AT to scavenge peroxyl radicals is postulated to interfere with the signal transduction factors leading to NF-κB activation. Other studies describe interference in signalling leading to altered mobilization and surface expression e.g. CD11b. One study has reported the effect of *in vivo* supplementation with AT on PMN integrin expression, indicating physiological benefit *in vivo* through reduction in PMN adherence to the endothelium. However, a recent AT supplementation study examined the shedding of E-selectin, as a marker of increased expression, but did not demonstrate any significant reduction of soluble E-selectin.[47] Clearly, the importance of AT-mediated modulation of gene transcription *in vivo* remains to be determined.

In a study to examine effects on resting levels of chemokine expression, Wu *et al.* reported that supplementation of U937 monocytes with 60 μM vitamin E for 20 h prevents spontaneous production of the chemokine IL-8, but has no effect in preventing spontaneous production of MCP-1 and IL-6.[39] As these chemokines are under the regulation of NF-κB this may represent the scavenging of basal oxidants necessary for spontaneous IL-8 expression by AT, whereas secondary transcription factors may overide such inhibition for IL-6 and MCP-1. Similarly, the effects of AT on MnSOD expression in rat aortic smooth muscle

Table 2. Summary of Studies Evaluating the Effects of Alpha-Tocopherol/Vitamin E on Stimulus Induced Gene Expression

Cell Under Study	Antioxidant Dose	Stimulus	Endpoint mRNA/Protein	Reference
U937 monocytes	Vitamin E 60 μM	Il-1 beta	Inhibition of MCP-1, IL-8, IL-6	39
Endothelial cells	Vitamin E 60 μM	Il-1 beta	Inhibition of ICAM-1, VCAM-1 and E-selectin	39
Human aortic endothelial cells	Gamma tocopherol	Oxidized LDL	I-κB degradation/apoptosis	40
Ex vivo PMN leukocytes	Vitamin E — 600 mg/day for 10 days	Oxidized LDL or fMLP	CD11b/CD18	41
U937 monocytes	Alpha tocopherol (50 μM)	LPS or fMLP	CD11b, VLA-4, EMSA for NF-κB	42
HUVECs	Alpha tocopherol	IL-1 beta	VCAM-1, ICAM-1	43
PMNs	Alpha tocopherol	PAF	CD11b/CD18	44
THP-1 cells transfected with TNF-α promotor	Alpha tocopherol succinate	LPS	Reporter gene attached to NF-κB promotor	45
Rat kupffer cells	Dietary alpha tocopherol	Carbon tocopherol	AP-1 EMSA	46

cells demonstrated that whilst 2 days exposure to 50 μM AT promotes expression, incubation for 7 days caused a decrease in both mRNA and protein activity.[48] This is postulated to reflect an adaptive response to raised to AT. The effects of sustained supplementation *in vivo* and *in vitro* merits further investigation.

Alpha tocopherol has been shown to possess a signalling role in smooth muscle cells; an attribute not reflected by other tocopherols with the same antioxidant profile. At the post-translational level, alpha tocopherol inhibits protein kinase C and 5 lipoxygenase, whilst activating protein phosphatase 2A and diacylglycerol kinase.[49] The mechanism of PKC regulation by AT has been investigated in several cell types,[50] where dose and time-dependent inhibition of activity has been described. This is suggested to be independent of an antioxidant effect, as beta tocopherol does not mimic the effects of AT, and furthermore is not due to direct binding of AT to PKC. The inhibition of PKC is associated with its dephosphorylation, possibly by protein phosphatase Type 2A which is itself up-regulated by AT. The activation of PP2A1 and deactivation of PKC alpha affect the AP-1 transcription factor, resulting in a change in its composition and DNA binding activity. AP-1 is an important mediator in cellular proliferation and activation or on the other hand, apoptosis, where changes in AP-1 composition and phosphorylation state modulate its transactivational activity. This presents a mechanism for the modulation of apoptotic signalling and proliferation conferred by AT in vascular tissue.[51] However, other mechanisms of control PKC have been identified by Lee *et al.*[52] In vascular smooth muscle cells, the hypergylycemia induced activity of diacyl glycerol (DAG) activated PKC, is inhibited by AT. The mechanism postulated to explain the observed inhibition of PKC is through increased DAG kinase activity, converting DAG to phosphatidic acid, thus removing the trigger for PKC activation. Indeed PKC contains unique structural features that are susceptible to oxidative modification and regulation by antioxidants.[7]

At the transcriptional level, the scavenger receptor CD36, alpha-tocopherol transfer protein, alpha tropomysin and collagenase are all subject to regulation by alpha tocopherol.[53] The scavenger receptor CD36, which shows specificity for oxidized LDL, is expressed in macrophages, endothelial cells and smooth muscle cells. Preincubation of HL-60 macrophages and SMCs with 50 µM AT down regulates CD36 mRNA and subsequent protein expression by reducing its promoter activity.[54, 55] Similar observations have been made on the effect of AT on collagenase expression. The age dependent increase in collagenase expression of skin fibroblasts in association with an increase in PKC alpha protein expression can be inhibited *in vitro* by AT.[56]

Alpha tocopherol is increasingly recognized as exerting pro-oxidant effects, where tocopherol mediated peroxidation of LDL has implications for its role as an anti-atherogenic supplement.[57] Perhaps the non-antioxidant effects of AT on gene expression are a result of a pro-oxidant mechanism.

7. Conclusions

Antioxidant vitamin E is a potent inhibitor of gene expression in model systems of the inflamed/atherosclerotic vasculature. This effect appears to be via inhibition of oxidant signalling to inhibitor kappa kinase. Studies on the effects of vitamin C *in vitro* are limited. However, both vitamins C and E (via protein kinase C) appear to modulate basal signalling processes leading to altered gene expression via a non-antioxidant mechanism, possibly pro-oxidant. The effects of antioxidant supplementation *in vivo* on basal gene expression and on *ex vivo* stimulated expression needs investigation.

Acknowledgments

HRG and SJR gratefully acknowlege financial support from the Food Standards Agency, N04015.

References

1. Keyse, S. M. and Tyrell, R. M. (1989). Haem oxygenase is a major 32 kDa stress protein induced in human skin fibroblasts by UVA, hydrogen peroxide and sodium arsenite. *Proc. Natl. Acad. Sci.* **86**: 99–103.
2. Klausner, R. D., Rauault, T. A. and Harford, J. B. (1993). Regulating the fate of mRNA: the control of cellular iron metabolism. *Cell* **72**: 19–28.
3. Beg, A. A. and Baltimore, D. (1996) An essential role for NF-κB in preventing TNF-α induced cell death. *Science* **274**: 782–784.
4. Griffiths, H. R., Mistry, P., Herbert, K. E. and Lunec, J. (1998). Molecular and cellular effects of ultraviolet light-induced genotoxicity. *Crit. Rev. Clin. Lab. Sci.* **35**: 189–237.
5. Bauerle P. A. and Henkel, T. (1994). Function and activation of NF-κB in the immune system. *Ann. Rev. Immunol.* **12**: 141–179.
6. Didenata, J. A., Haykow, M., Rothwarf, D. W., Zandi, E. and Karin, M. (1997). A cytokine responsive I-κB kinase that activates transcription factor NF-κB. *Nature* **388**: 548–554.
7. Janssen-Heininger, Y. M., Macara, I. and Mossman, B. T. (1999). Cooperativity between oxidants and tumor necrosis factor in the activation of nuclear factor (NF)-kappaB: requirement of Ras/mitogen-activated protein kinases in the activation of NF-kappaB by oxidants. *Am. J. Respiratory Cell Mol. Biol.* **20**: 942–952.
8. Schreck, R. and Bauerle, P. A. (1991) A role for oxygen radicals as second messengers. *Trends in Cell Biol.* **1**: 39–42.
9. Schmidt, K. N., Amstad, P., Cerutti, P. and Bauerle, P. A. (1995). Roles of hydrogen peroxide and superoxide anion as messengers in activation of the transcription factor NF-κB. *Chem. Biol.* **2**: 13–22.
10. Angel, P., Baumann, I., Stein, B., Delius, H., Rahmsdoerf, H. J. and Henlich, P. (1987) TPA induction of human collagenase gene is mediated by an enhancer element located in the 5′ flanking region. *Mol. Cell. Biol.* **7**: 2256–2266.
11. Angel, P. and Karin, M. (1991) The role of *jun, fos* and the AP-1 complex in cell proliferation and transformation. *Biochem. Biophys. Acta* **1072**: 129–157.
12. Xanthoudakis, S. and Curran, T. (1992). Identification and characterisation of Ref-1, a nuclear factor that facilitates AP-1 DNA binding activity. *EMBO J.* **11**: 653–665.
13. Arrigo, A. P. and Kretz-Remy, C. (1998). Regulation of mammalian gene expression by free radicals. *In* "Molecular Biology of Free Radicals in Human Diseases" (O. I. Aruoma, and B. Halliwell, Eds.) pp. 183–223.
14. Chakraborti, T., Ghosh, S. K., Michael, J. R., Batabyal, S. K. and Chakraborti, S. (1998). Targets of oxidative stress in cardiovascular system. *Mol. Cell. Biochem.* **187**: 1–10.

15. Seol, D. W., Chen, Q. and Zarnegar, R. (2000). Transcriptional activation of the hepatocyte growth factor receptor (c-met) gene by its ligand (hepatocyte growth factor) is mediated through AP-1. *Oncogene* **19**: 1132–1137.

16. Suberlich, H. A. (1994). Pharmacology of vitamin C. *Ann. Rev. Nutri.* **14**: 371–391.

17. Rose, R. C. (1988). Transport of vitamin C and other water soluble vitamins. *Biochem. Biophys. Acta* **947**: 335–366.

18. Tsukaguchi, H., Tokui, T., Mackenzie, B., Berger, U. V., Chen, X.-Z., Wang, Y.-X, Bubaker, R. F. and Hediger, M. A. (1999). A family of mammalian sodium dependent L-ascorbic acid transporters. *Nature* 70–75.

19. Ek, A., Strom, K. and Colgreave, I. A. (1995). The uptake of ascorbic acid into HUVECs and its effect on oxidant insult. *Biochem. Pharmacol.* **50**: 1339–1346.

20. Vera, J. C., Rivas, C. I., Fischborg, J. and Golde, D. W. (1993). Mammalian facilitative hexose transporters mediate transport of dehydroascorbic acid. *Nature* **364**: 79–82.

21. Cunningham, J. J. (1998). The glucose/insulin system and vitamin C: implications in insulin dependent diabetes mellitus. *J. Am. Coll. Nutri.* **17**: 105–108.

22. Bode, A. (1997). Metabolism of vitamin C in health and disease. *Adv. Pharmacol.* **38**: 21–47.

23. Sakagami, H., Satoh, K., Hakeda, Y. and Kumegawa, M. (2000). Apoptosis inducing activity of vitamin C and vitamin K. *Cell Mol. Biol.* **46**: 129–143.

24. Podmore, I. D., Griffiths, H. R., Herbert, K. E., Mistry, N., Mistry, P. and Lunec, J. (1998). Vitamin C exhibits pro-oxidant properties *Nature* **392**: 559.

25. Rehman, A., Collis, C. S., Yang, M., Kelly, M., Diplock, A. T., Halliwell, B. and Rice-Evans, C. (1998). The effects of iron and vitamin C co-supplementation on oxidative damage to DNA in healthy volunteers. *Biochem. Biophys. Res. Commun.* **246**: 293–298.

26. Siow, R. C., Richards, J. P., Pedley, K. C., Leake, D. S. and Mann, G. E. (1999). Vitamin C protects human vascular smooth muscle cells against apoptosis induced by moderately oxidized LDL containing high levels of lipid hydroperoxides. *Art. Thromb. Vasc. Biol.* **19**: 2387–2394.

27. Adams, M. R., Jessup, W. and Celermajer, D. S. (1997). Cigarette smoking is associated with increased human monocyte adhesion to endothelial cells: reversibility with oral L-arginine but not vitamin C. *J. Am. Coll. Cardiol.* **29**: 491–497.

28. Davidson, J. M., LuValle, P. A., Zoia, O., Quaglino, D. and Giro, M. (1997). Ascorbate differentially regulates elastin and colagen biosynthesis in vascular smooth muscle cells and skin fibroblasts by pre-translational mechanisms. *J. Biol. Chem.* **272**: 345–352.

29. Ivanov, V. O., Ivanov, S. and Niedzwiecki, A. (1997). Ascorbate affects proliferation of guinea-pig smooth muscle cells by direct and extracellular matrix effects. *J. Mol. Cell. Cardiol.* **29**: 3293–3303.

30. Smith, J. C., Singh, J. P., Lilliquist, J. S., Goon, D. S. and Stiles, C. D. (1982). Growth factors adherent to cell substrate are mitogenically active in situ. *Nature* **296**: 154–156.
31. Murad, S., Tajima, S., Johnson, G. R., Sivarajah, A. and Pinnell, S. R. (1983). Collagen synthesis in cultured human skin fibroblasts: effects of ascorbic acid and its analogs. *J. Invest. Dermatol.* **81**: 158–162.
32. Segawa, Y., Shirao, Y., Yamagishi, S., Higashide, T., Kobayashi, M., Katsuno, K., Iyobe, A., Harada, H., Sato, F., Miyata, H., Asai, H., Nishimura, A., Takahira, M., Souno, T,. Segawa, Y., Maeda, K., Shima, K., Mizuno, A., Yamamoto, H. and Kawasaki, K. (1998). Upregulation of retinal vascular endothelial growth factor mRNAs in spontaneously diabetic rats without ophthalmoscopic retinopathy. A possible participation of advanced glycation end products in the development of the early phase of diabetic retinopathy. *Ophthalmic Res.* **30**: 333–339.
33. Gopalakrishana, R., Gundimeda, U., Anderson, W. B., Colburn, N. H. and Slaga, T. J. (1999). Tumor promoter benzoyl peroxide induces sulfhydryl oxidation in protein kinase C: its reversibility is related to the cellular resistance to peroxide-induced cytotoxicity. *Arch. Biochem. Biophys.* **363**: 246–258.
34. Siow, R. C., Satao, H., Leake, D. S., Ishii, T., Bannai, S. and Mann, G. E. (1999). Induction of antioxidant stress proteins in vascular endothelial and smooth muscle cells: protective effect of vitamin C against atherogenic lipoproteins. *Free Radic. Res.* **31**: 309–318.
35. Traber, M. G. (1997) Regulation of human plasma vitamin E. *Adv. Pharm.* **38**: 49–63.
36. Fechner, H., Schlame, M., Guthmann, F., Stevens, P. A. and Rustow, B. (1998). Alpha- and delta-tocopherol induce expression of hepatic alpha-tocopherol-transfer-protein mRNA. *Biochem. J.* **331**: 577–581.
37. Ouachi, K., Arita, M., Kayden, H., Hentati, F., Ben Hamida, M., Sokel, R., Arai, H., Inoue, K., Mandel, J. L. and Koenig, M. (1995). Ataxia with isolated vitamin E deficiency is caused by mutations in the alpha-tocopherol-transfer protein. *Nature Genetics* **9**: 141–145.
38. Flohe, B. R. and Traber, M. G. (1999). Vitamin E: function and metabolism. *FASEB J.* **13**: 1145–1155.
39. Wu, D., Koga, T., Martin, K. R. and Meydani, M. (1999). Effect of vitamin E on human aortic endothelial cell production of chemokines and adhesion to monocytes. *Atherosclerosis* **147**: 297–307.
40. Li, D., Saldeen, T. and Mehta, J. L.(1999). Gamma-tocopheol decreases ox-LDL mediated activation of NF-κB and apoptosis in human coronary artery endothelial cells. *Biochem. Biophys. Res. Commun.* **259**: 157–161.
41. Yoshida, N., Yoshikawa, T. Y., Manabe, H., Terasawa, Y., Kondon, M., Noguchi, N. and Niki, E. (1999). Vitamin E protects against polymorphonuclear leukocyte-dependent adhesion to endothelial cells. *J. Leukocye Biol.* **65**: 757–763.

42. Islam, K. N., Devaraj, S. and Jialal, I. (1998). Alpha-tocopherol enrichment of monocytes decreases agonist induced adhesion to human endothelial cells. *Circulation* **98**: 2255–2261.

43. Nakamura, T., Goto, M., Matsumoto, A. and Tanaka, I. (1998) Inhibition of NF-κB transcriptional activity by alpha-tocopheryl acetate. *Biofactors* **7**: 21–30.

44. Yoshikawa, T., Yoshida, N., Manabe, H., Terasawa, Y., Takemura, T. and Kondo, M. (1998). Alpha-tocopherol protects against expression of adhesion molecules on neutrophils and endothelial cells. *Biofactors* **7**: 15–19

45. Erl, W., Weber, C., Wardemann, C. and Weber, P. C. (1997). Alpha-tocopheryl succinate inhibits monocyte adhesion to endothelial cells by suppressing NF-κB mobilization. *Am. J. Physiol.* **273**: H634–H640.

46. Camandoloa, S., Aragno, M. and Cutrin, J. C. (1999). Liver AP-1 activation due to carbon tetrachloride is potentiated by 1,2-dibronoethane but is inhibited by alpha-tocopherol or gadolinium chloride. *Free Radic. Biol. Med.* **26**: 1108–1116.

47. Seiljeflot, I., Amensen, H., Brude, I. R., Nenseter, M. S., Drevon, C. A. and Hjermann, I. (1998). The effects of omega-3 fatty acids and or antioxidants on endothelial cell markers. *J. Clin. Invest.* **28**: 629–635.

48. Huang, W., Chan, P., Chen, Y., Chen C., Liao, S., Chin, W. and Cheng, J. (1999). Changes in superoxide dismuatase in cultured rat aortic smooth muscle cells by an incubation of vitamin E. *Pharmacol.* **59**: 275–282.

49. Azzi, A. and Stocker, A. (2000). Vitamin E: non-antioxidant roles. *Prog. Lip. Res.* **39**: 231–255.

50. Ricciarelli, R., Tasinato, A., Clement, S., Ozer, N., Boscoboinik, D. and Azzi, A. (1998). Alpha-tocopherol specifically inactivates protein kinase C alpha by changing its phosphorylation state. *Biochem. J.* **334**: 243–249.

51. De Nigris, F., Farnconi, F., Maida, I., Palumbo, G., Anania, V. and Napoli, C. (2000). Modulation by alpha- and gamma-tocopherol and oxidized LDL of apoptotic signalling in human coronary smooth muscle cells. *Biochem. Pharm.* **59**: 1477–1487.

52. Lee, I. K., Koya, D., Ishi, H., Kanoh, H. and Koing, G. L. (1999). d-alpha tocopherol prevents the hyperglycemia induced activation of DAG-protein kinase C pathway in vascular smooth cells by an increase of DAG kinase activity. *Diabetes Res. Clin. Prac.* **45**: 183–190.

53. Azzi, A., Breyer, I., Feher, M., Pastori, M., Ricciarelli, R., Spycher, S., Staffieri, M., Stocker, A., Zimmer, S. and Zingg, J. M. (2000). Specific cellular responses to alpha-tocopherol. *J. Nutri.* **130**: 1649–1652.

54. Teupser, D., Thiery, J. and Siedel, D. (1999). Alpha-tocopherol down-regulates scavenger receptor activity in macrophages. *Atherosclerosis* **144**: 109–115.

55. Ricciarelli, R., Zingg, J. M. and Azzi, A. (2000). Vitamin E reduces the uptake of oxidized LDL by inhibiting CD36 scavenger receptor expression in cultured aortic smooth muscle cells. *Circulation* **102**: 82–87.

56. Ricciarelli, R., Maroni, P., Ozer, N., Zingg, J. M. and Azzi, A (1999). Age dependent increase of collagenase expression can be reduced by alpha-tocopherol via protein kinase C inhibition. *Free Radic. Biol. Med.* **27**: 729–737.
57. Upston, J. M., Terentis, A. C. and Stocker, R. (1999). Tocopherol-mediated peroxidation of lipoproteins: implications for vitamin E as a potential anti-atherogenic supplement. *FASEB J.* **13**: 977–994.

SECTION 6

Endogenous Antioxidants
and Repair

Chapter 40

Oxidative Defense Mechanisms

Cheryl Y. Teoh and Kelvin J.A. Davies*

Cheryl Y. Teoh and **Kelvin J.A. Davies** • Ethel Percy Andrus Gerontology Center and Division of Molecular Biology, University of Southern California, 3715 McClintock Avenue, Room 306, Los Angeles, CA 900089-0191, USA

*Corresponding Author.
Tel: (213) 740-8959, E-mail: kelvin@usc.edu

1. Introduction

Reactive oxygen species cause damage to all cellular macromolecules, including DNA, lipids, and proteins. Generated during O_2 reduction, reactive oxygen species can attack DNA nucleotides causing strand breaks and/or damaged bases.[1] At the same time, reactive oxygen species can oxidize lipids or proteins, generating intermediates that react with DNA and form covalent adducts.[2] Oxidized proteins can also form large aggregates through crosslinking and hydrophobic bonding, which are detrimental to normal cell function.

The cell has various defense and repair mechanisms to protect itself from oxygen radical attack. Antioxidant enzymes are present to intercept reactive oxygen species. Specific repair enzymes repair oxidized DNA and oxidized lipids, and oxidized proteins are degraded. Cells are also able to adapt to oxidative stress by increasing the expression of various protective genes.

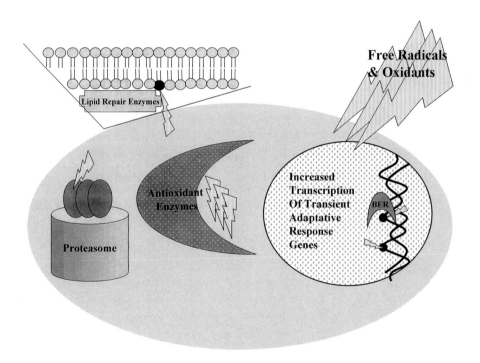

Fig. 1. **The Cell's Response To Oxidative Stress.** The figure depicts free radicals/oxidants (shown as lightning) oxidizing the cell, particularly DNA, lipids, and proteins, and the cell's defense and repair system. There are repair enzymes that repair DNA by base excision repair mechanisms (BER). Oxidized lipids are removed by phospholipase A_2, and glutathione peroxidases may be involved in repair. Oxidized proteins are recognized and selectively degraded by the proteasome. Antioxidant enzymes remove free radicals, and the cell can also increase transcription of genes involved in transient adaptive responses to oxidative stress.

2. Antioxidant Enzymes and Compounds

A variety of antioxidant enzymes, such as superoxide dismutase (SOD), catalase, glutathione peroxidase and DT diaphorase, are important in limiting the levels of reactive oxygen species within cells. The expression of these enzymes must be coordinately regulated because they need to work cooperatively in order to protect cells effectively. This is emphasized in experiments involving over-expression of Cu, Zn-SOD and catalase conducted by Sohal and Orr[3] that resulted in a 30% increase in lifespan, a longer mortality rate doubling time, a lower amount of protein oxidative damage, and a delayed loss in physical performance in *Drosophila*.

Superoxide dismutase (SOD) catalyzes the reaction $O_2^{\cdot-} + O_2^{\cdot-} \rightarrow H_2O_2 + O_2$. All members of the SOD family utilize transition metals at their active sites. The product of SOD, H_2O_2, is detoxified by catalases and glutathione peroxidases, which covert H_2O_2 to oxygen and water. Glutathione peroxidases, especially selenium glutathione peroxidase, are more effective in removing H_2O_2 than is catalase.[4] DT diaphorase, also known as NAD(P)H:(quinone acceptor) oxy-reductase, catalyzes the reduction of many (dehydro)quinones to (dehydro)quinols by direct divalent reduction.[5]

In addition to the antioxidant enzymes, of course, we have long known that ascorbic acid (vitamin C) and α-tocopherol (vitamin E) are essential components of antioxidant defenses. Vitamin C protects proteins and nucleotides in aqueous cell compartments, whereas vitamin E is a major, chain-breaking, antioxidant within biological membranes. These, and other antioxidant compounds, are discussed in much greater detail in other chapters of the book.

3. DNA Repair

Oxidation of DNA can produce different types of DNA damage: strand breaks, sister chromatid exchange, DNA-protein crosslinks, sugar damage, abasic sites, and base modifications. There are several types of DNA repair mechanisms but it is believed that base excision repair is the primary cell defense against oxidatively damaged DNA.

The base excision repair pathway involves several DNA glycosylases that recognize and release modified bases by cleaving the *N*-glycosylic bond between the modified base and the deoxyribose sugar group. This leaves an abasic site, which will be further processed by apurinic/apyrimidinic (AP) endonucleases. Oxidized pyrimidines, such as thymine glycols, are recognized by endonuclease III and endonuclease VIII in *E. coli*. Oxidized purines, parti-cularly 8-oxo-7,8-dihydro-2'-deoxyguanosine (oxo-8-dG), are recognized by formamidopyrimidine glycosylase (Fpg or MutM), MutT, and MutY. Oxo-8-dG is a major form of oxidized guanine and it is found that these three enzymes suppress guanine oxidation. The enzyme MutM removes oxo-8-dG paired with cytosine,[6, 7] while MutT hydrolyzes the oxidized nucleotide oxo-8-dGTP to the nucleoside monophosphate,[8, 9] and MutY removes adenine when paired with oxo-8-dG.[10–12] Human homologues of endonuclease III,[13, 14] MutM,[15–18] MutT,[19] MutY[20] have been cloned.

4. Lipid Degradation and Repair

Polyunsaturated fatty acid residues of phospholipids are sensitive to oxidation due to polyunsaturation and to the presence of a methylene group between two double bonds.[21] The high concentrations of polyunsaturated fatty acids in many membranes makes them easy targets for oxidation, and causes them to be involved in long free radical chain reactions. Lipid hydroperoxides are the initial products of lipid oxidation; however, they can be reduced by glutathione peroxidase to unreactive fatty acid alcohols, or they may react with metals to produce new reactive molecules, such as epoxides and aldehydes. Cholesterol has also been shown to undergo oxidation, resulting in the production of various epoxides and alcohols. The major aldehyde products are malondialdehyde (MDA) and 4-hydroxynonenal (HNE).[22] These products can react with DNA to form various DNA adducts.

Oxidized lipid bilayers are substrates for phospholipase enzymes. Phospholipase A_2 acts at the sn-2 position of the phospholipid glycerol backbone to generate a free fatty acid and a lysophospholipid. When phospholipid hydroperoxides are incubated with phospholipase A_2, the free fatty acid hydroperoxides become available substrates for Se-dependent glutathione peroxidase,[23] indicating that phospholipase A_2 and glutathione peroxidase may participate in a lipid "repair" pathway. Another kind of glutathione peroxidase called phospholipid hydroperoxide glutathione peroxidase reduces fatty acid hydroperoxides to their corresponding hydroxy-fatty acids.[24] It has also been found that glutathione transferase B and AA also removes lipid hydroperoxides.[25]

5. Protein Degradation

Extensive studies have revealed that oxidized proteins are recognized by proteases and completely degraded into amino acids; at the same time, new replacement proteins are synthsized *de novo*.[26–28] During periods of high oxidative stress, the proteolytic capacity of cells starts to decline due to oxidative inhibition. Under such circumstances, oxidized proteins may not undergo proteolytic digestion and may undergo chemical fragmentation, or may accumulate and form crosslinks or hydrophobic bonds with one another.[29–32]

Soluble, oxidized proteins in the nucleus and cytoplasm of eukaryotes are recognized and degraded by the proteasome complex.[33-37] The proteasome complex is made up of around 14 different polypeptides, each present in multiple (copies, ranging) from 20–35 kDa; however, the exact composition varies with species and cell type. Oxidation of amino acids causes some amino acid residues to become more hydrophilic; however, changes in charge relationships within a protein result in significant unfolding, leading to loss of protein function.[26, 35-39] This partial denaturation exposes hidden hydrophobic sequences, increasing surface hydrophobicity. These hydrophobic patches are recognized and bind to the proteasome; thus, ensuring degradation of the damaged proteins.[26, 34-36] During aging, a decline in proteasome activity can lead to significant accumulation of damaged, crosslinked protein aggregates.[27, 28, 30-32, 34, 38, 40]

6. Adaptive Response

Researchers have found temporary adaptation to oxidative stress in cells where cell cultures were exposed to lower levels of oxidant (pre-treatment) before being exposed to a lethal dose of the same oxidant and cells were able to survive in normally lethal conditions,[41-43] which would be probably due to the widespread gene expression alterations in the cell due to exposure to oxidative stress. This form of adaptation is temporary resistance to oxidative stress, as measured by cell proliferation capacity.

Two main regulons are responsible for bacterial adaptive response to oxidative stress: the oxyR regulon[44] and the soxRS regulon.[45, 46] Eukaryotes, on the other hand, have no "master regulation molecules" but at least 40 gene products, comprised of stress-related genes, antioxidant defense genes and, or repair enzymes, are involved in adaptive response.

Several things happen during cellular adaptive responses: transient growth arrest, and the increase in transcription of stress-related genes, antioxidant defense genes and/or repair enzymes. An early part of cell adaptive responses is transient growth arrest, and the genes *gadd153*, *gadd45*, and *adapt15* are involved in inducing transient growth-arrest.[47-49] It has been found that the gadd45 family proteins are cofactors of nuclear hormone receptors.[50] Adapt66 (a *mafG* homolog) is a transcription factor, which is probably responsible for inducing transcription of other adaptive genes.[51] A number of "adapt" genes have recently been discovered such as *adapt78* (also called *DSCR1*),[52-54] *adapt33*,[55] and *adapt73/PigHep3*.[56] Adapt78 has been found to bind to calcineurin and inhibit its function.[57-59] The function of the other "adapt" genes are still not yet clear.

During exposure to mild levels of hydrogen peroxide, *c-jun*, *egr*, and JE;[60, 61] proto-oncogenes *c-fos* and *c-myc*;[60] CL100 phosphatase;[62] interleukin-8;[63] catalase, glutathione peroxidase, mitochondrial mangano-superoxide dismutase;[64] and mitogen-activated protein[65] can be induced. The complete list of oxidant-stress inducible genes is much longer than this[66] and apologies are extended to those investigators whose studies have not been cited. Oxidants can also cause a broad spectrum of cellular responses, ranging from accelerated mitosis, to transient growth-arrest, to temporary adaptation, to permanent growth-arrest, to apoptosis or necrosis, depending on the oxidant used, the cell type, the concentration of oxidant, and the conditions of exposure.[67]

7. Conclusions

There is an overlap between the various defense and repair mechanisms. Oxidative agents produced intracellularly, or from extracellular sources, oxidize macromolecules such as lipids and proteins, which in turn, play a role in damaging DNA. The cell uses antioxidant enzymes to detoxify the oxidative agents before they do more damage and repair enzymes repair oxidized DNA. Oxidized proteins and lipids, on the other hand, are handled by a mixture of damage removal and repair enzymes. Oxidative stress resistance can also be increased by exposing cells to moderate levels of oxidants. Such treatments modify the expression of over 30 genes to increase stress resistance and improve survival capacity.

Acknowledgments

The research underlying this review paper was made possible by NIH-NIEHS grant number ES03598, and NIH-NIA grant number AG16256 to K.J.A.D.

References

1. Ward, J. F., Limoli, C. L., Calabro-Jones, P. and Evans, J. W. (1987). Radiation versus chemical damage to DNA. *In* "Anticarcinogenesis and Radiation Protection" (P. A. Cerutti, O. F. Nygaard, and M. G. Simic, Eds.), pp. 321–327, Plenum, New York.
2. Esterbauer, H., Eckl, P. and Ortner, A. (1990). Possible mutagens derived from lipids and lipid precursors. *Mutat. Res. Rev. Genet. Toxicol.* **238**: 223–233.
3. Orr, W. C. and Sohal, R. S. (1994). Extension of life-span by overexpression of superoxide dismutase and catalase in *Drosophila melanogaster*. *Science* **263**: 1128–1130.
4. Singh, A. K., Dhaunsi, G. S., Gupta, M. P., Orak, J. K., Asayama, K. and Singh, I. (1994). Demonstration of glutathione peroxidase in rat liver peroxisomes and its intraorganellar distribution. *Arch. Biochem. Biophys.* **315**: 331–338.

5. Iyanagi, T. and Yamazaki, I. (1970). One-electron-transfer reactions in biochemical systems. V. Difference in the mechanism of quinone reduction by the NADH dehydrogenase and the NAD(P)H dehydrogenase (DT-diaphorase). *Biochim. Biophys. Acta* **216**: 282–294.

6. Cabrera, M., Ngheim, Y. and Miller, J. H. (1988). MutM, a second mutator locus in *Escherichia coli* that generates G:C → T:A transversions. *J. Bacteriol.* **170**: 5405–5407.

7. Michaels, M. L., Pham, L., Cruz, C. and Miller, J. H. (1991). MutM, a protein that prevents G:C → T:A transversions, is formamidopyrimidine-DNA glycosylase. *Nucleic Acids Res.* **19**: 3629–3632.

8. Maki, H. and Sekiguchi, M. (1992). MutT protein specifically hydrolyses a potent mutagenic substrate for DNA synthesis. *Nature* **355**: 273–275.

9. Mo, J. Y., Maki, H. and Sekiguchi, M. (1992). Hydrolytic elimination of a mutagenic nucleotide, 8-oxo-dGTP, by human 18-kilodalton protein: sanitization of nucleotide pool. *Proc. Natl. Acad. Sci. USA* **89**: 11 021–11 025.

10. Ngheim, Y., Cabrera, M., Cupples, C. G. and Miller, J. H. (1988). The *mutY* gene: a mutator locus in *Escherichia coli* that generates G:C → T:A transversions. *Proc. Natl. Acad. Sci. USA* **85**: 2709–2713.

11. Au, K. G., Clark, S., Miller, J. H. and Modrich, P. (1989). *Escherichia coli mutY* gene encodes an adenine glycosylase active on G:A mispairs. *Proc. Natl. Acad. Sci. USA* **86**: 8877–8881.

12. Michaels, M. L., Cruz, C., Grollman, A. P. and Miller, J. H. (1992). Evidence that MutM and MutY combine to prevent mutations by an oxidatively damaged form of guanine in DNA. *Proc. Natl. Acad. Sci. USA* **89**: 7022–7025.

13. Aspinwall, R., Rothwell, D. G., Roldan-Arjona, T., Anselmino, C., Ward, C. J., Cheadle, J. P., Sampson, J. R., Lindahl, T., Harris, P. C. and Hickson, I. D. (1997). Cloning and characterization of a functional human homolog of *Escherichia coli* endonuclease III. *Proc. Natl. Acad. Sci. USA* **94**: 109–114.

14. Hilbert, T. P., Chaung, W., Boorstein, R. J., Cunningham, R. P. and Teebor, G. W. (1997). Cloning and expression of the cDNA encoding the human homolog of the DNA repair enzyme *Escherichia coli* endonuclease III. *J. Biol. Chem.* **272**: 6733–6740.

15. Radicella, J. P., Dherin, C., Desmaze, C., Fox, M. S. and Boiteux, S. (1997). Cloning and characterization of hOGG1, a human homolog of the OGG1 gene of *Saccharomyces cerevisiae*. *Proc. Natl. Acad. Sci. USA* **94**: 8010–8015.

16. Roldan-Arjona, T., Wei, Y. F., Carter, K. C., Klungland, A., Anselmino, C., Wang, R. P., Augustus, M. and Lindahl, T. (1997). Molecular cloning and functional expression of a human cDNA encoding the antimutator 8-hydroxyguanine-DNA glycosylase. *Proc. Natl. Acad. Sci. USA* **94**: 8016–8020.

17. Lu, R., Nash, H. M. and Verdine, G. L. (1997). A mammalian DNA repair enzyme that excises oxidatively damaged guanines maps to a locus frequently lost in lung cancer. *Curr. Biol.* **7**: 397–407.

18. Arai, K., Morishita, K., Shinmura, K., Kohno, T., Kim, S. R., Nohmi, T., Taniwaki, M., Ohwada, S. and Yokota, J. (1997). Cloning of a human homolog of the yeast *OGG1* gene that is involved in the repair of oxidative DNA damage. *Oncogene* **14**: 2857–2861.

19. Sakumi, K., Furuichi, M., Tsuzuki, T., Kakuma, T., Kawabata, S., Maki, H. and Sekiguchi, M. (1993). Cloning and expression of cDNA for a human enzyme that hydrolyzes 8-oxo-dGTP, a mutagenic substrate for DNA synthesis. *J. Biol. Chem.* **268**: 23 524–23 530.

20. Slupska, M. M., Baikalov, C., Luther, W. M., Chiang, J. H., Wei, Y. F. and Miller, J. H. (1996). Cloning and sequencing a human homolog (*hMYH*) of the *Escherichia coli mutY* gene whose function is required for the repair of oxidative DNA damage. *J. Bacteriol.* **178**: 3885–3892

21. Porter, N. A. (1986). Mechanisms for the autoxidation of polyunsaturated lipids. *Acc. Chem. Res.* **19**: 262–268.

22. Schauenstein, E. and Esterbauer, H. (1978). Formation and properties of reactive aldehydes. *Submol. Biol. Cancer Ciba Fnd.* **67**: 225–241.

23. van Kuijk, F. J., Handelman, G. J. and Dratz, E. A. (1985). Consecutive action of phospholipase A_2 and glutathione peroxidase is required for reduction of phospholipid hydroperoxides and provides a convenient method to determine peroxide values in membranes. *J. Free Radic. Biol. Med.* **1**: 421–427.

24. Ursini, F., Maiorino, M. and Gregolin, C. (1985). The selenoenzyme phospholipid hydroperoxide glutathione peroxidase. *Biochim. Biophys. Acta* **839**: 62–70.

25. Prohaska, J. R. (1980). The glutathione peroxidase activity of glutathione S-transferases. *Biochim. Biophys. Acta* **611**: 87–98.

26. Grune, T., Reinheckel, T. and Davies, K. J. A. (1997). Degradation of oxidized proteins in mammalian cells. *FASEB J.* **11**: 526–534.

27. Berlett, B. S. and Stadtman, E. R. (1997). Protein oxidation in aging, disease, and oxidative stress. *J. Biol. Chem.* **272**: 20 313–20 316.

28. Taylor, A., Shang, F. and Obin, M. (1997). Relationships between stress, protein damage, nutrition, and age-related eye diseases. *Mol. Aspects Med.* **18**: 305–414.

29. Ivy, G. O., Schottler, F., Wenzel, J., Baudry, M. and Lynch, G. (1984). Inhibitors of lysosomal enzymes: accumulation of lipofuscin-like dense bodies in the brain. *Science* **226**: 985–987.

30. Sitte, N., Huber, M., Grune, T., Ladhoff, A., Doecke, W.-D., von Zglinicki, T. and Davies, K. J. A. (2000). Proteasome inhibition by lipofuscin/ceroid during postmitotic aging of fibroblasts. *FASEB J.* **14**: 1490–1498.

31. Sitte, N., Merker, K., von Zglinicki, T., Grune, T. and Davies, K. J. A. (2000). Protein oxidation and degradation during cellular senescence of BJ-fibroblasts: Part I effects of proliferative senescence. *FASEB J.* **14**: 2495–2502.

32. Sitte, N., Merker, K., von Zglinicki, T., Davies, K. J. A. and Grune, T. (2000). Protein oxidation and degradation during cellular senescence of BJ-fibroblasts: Part II aging of non-dividing cells. *FASEB J.* **14**: 2503–2510.

33. Rivett, A. J. (1985). Preferential degradation of the oxidatively modified form of glutamine synthetase by intracellular mammalian proteases. *J. Biol. Chem.* **260**: 300–305.

34. Pacifici, R. E., Salo, D. C. and Davies, K. J. A. (1989). Macroxyproteinase (M.O.P.): a 670 kDa proteinase complex that degrades oxidatively denatured proteins in red blood cells. *Free Radic. Biol. Med.* **7**: 521–536.

35. Salo, D. C., Pacifici, R. E., Lin, S. W., Giulivi, C. and Davies, K. J. A. (1990). Superoxide dismutase undergoes proteolysis and fragmentation following oxidative modification and inactivation. *J. Biol. Chem.* **265**: 11 919–11 927.

36. Giulivi, C., Pacifici, R. E. and Davies, K. J. A. (1994). Exposure of hydrophobic moieties promotes the selective degradation of hydrogen peroxide-modified hemoglobin by the multicatalytic proteinase complex, proteasome. *Arch. Biochem. Biophys.* **311**: 329–341.

37. Ullrich, O., Reinheckel, T., Sitte, N., Haass, G., Grune, T. and Davies, K. J. A. (1999). Poly-ADP-ribose-polymerase activates nuclear proteasome to degrade oxidatively damaged histones. *Proc. Natl. Acad. Sci. USA* **96**: 6223–6228.

38. Fucci, L., Oliver, C. N., Coon, M. J. and Stadtman, E. R. (1983). Inactivation of key metabolic enzymes by mixed-function oxidation reactions: possible implication in protein turnover and ageing. *Proc. Natl. Acad. Sci. USA* **80**: 1521–1525.

39. Dean, R. T., Thomas, S. M. and Garner, A. (1986). Free-radical-mediated fragmentation of monoamine oxidase in the mitochondrial membrane. *Biochem. J.* **240**: 489–494.

40. Conconi, M., Szweda, L. I., Levine, R. L., Stadtman, E. R. and Friguet, B. (1996). Age-related decline of rat liver multicatalytic proteinase activity and protection from oxidative inactivation by heat-shock protein 90. *Arch. Biochem. Biophys.* **331**: 232–240.

41. Spitz, D. R., Dewey, W. C. and Li, G. C. (1987). Hydrogen peroxide or heat shock induces resistance to hydrogen peroxide in Chinese hamster fibroblasts. *J. Cell Physiol.* **131**: 364–373.

42. Laval, F. (1988). Pretreatment with oxygen species increases the resistance to hydrogen peroxide in Chinese hamster fibroblasts. *J. Cell Physiol.* **201**: 73–79.

43. Wiese, A. G., Pacifici, R. E. and Davies, K. J. A. (1995). Transient adaptation to oxidative stress in mammalian cells. *Arch. Biochem. Biophys.* **318**: 231–240.

44. Christman, M. F., Morgan, R. W., Jacobson, F. S. and Ames, B. N. (1985). Positive control of a regulon for defenses against oxidative stress and some heat-shock proteins in *Salmonella typhimurium*. *Cell* **41**: 753–762.

45. Wu, J. and Weiss, B. (1991). Two divergently transcribed genes, *soxR* and *soxS*, control a superoxide response regulon of *Escherichia coli*. *J. Bacteriol.* **173**: 2864–2871.

46. Amabile-Cuevas, C. F. and Demple, B. (1991). Molecular characterization of the *soxRS* genes of *Escherichia coli*: two genes control a superoxide stress regulon. *Nucleic Acids Res.* **19**: 4479–4484.

47. Fornace, A. J., Jr., Nebert, D. W., Hollander, M. C., Luethy, J. D., Papathanasiou, M., Fargnoli, J. and Holbrook, N. J. (1989). Mammalian genes coordinately regulated by growth arrest signals and DNA-damaging agents. *Mol. Cell Biol.* **9**: 4196–4203.

48. Bartlett, J. D., Luethy, J. D., Carlson, S. G., Sollott, S. J. and Holbrook, N. J. (1992). Calcium ionophore A23187 induces expression of the growth arrest and DNA damage inducible CCAAT/enhancer-binding protein (C/EBP)-related gene, *gadd153*. Ca2+ increases transcriptional activity and mRNA stability. *J. Biol. Chem.* **267**: 20 465–20 470.

49. Crawford, D. R., Schools, G. P. and Davies, K. J. A. (1996). Oxidant-inducible *adapt15* RNA is associated with growth arrest and DNA damage-inducible *gadd153* and *gadd45*. *Arch. Biochem. Biophys.* **329**: 137–144.

50. Yi, Y. W., Kim, D., Jung, N., Hong, S. S., Lee, H. S. and Bae, I. (2000). Gadd45 family proteins are coactivators of nuclear hormone receptors. *Biochem. Biophys. Res. Commun.* **272**: 193–198.

51. Crawford, D. R., Leahy, K. P., Wang, Y., Schools, G. P., Kochheiser, J. C. and Davies, K. J. A. (1996). Oxidative stress induces the levels of a *MafG* homolog in hamster HA-1 cells. *Free Radic. Biol. Med.* **21**: 521–525.

52. Fuentes, J. J., Pritchard, M. A., Planas, A. M., Bosch, A., Ferrer, I. and Estivill, X. (1995). A new human gene from the Down syndrome critical region encodes a proline-rich protein highly expressed in fetal brain and heart. *Human Mol. Genet.* **4**: 1935–1944.

53. Crawford, D. R., Leahy, K. P., Abramova, N., Lan, L., Wang, Y. and Davies, K. J. A. (1997). Hamster adapt78 mRNA is a Down syndrome critical region homologue that is inducible by oxidative stress. *Arch. Biochem. Biophys.* **342**: 6–12.

54. Leahy, K. P., Davies, K. J. A., Dull, M., Kort, J. J., Lawrence, K. W. and Crawford, D. A. (1999). *Adapt78*, a stress-inducible mRNA, is related to the glucose-regulated family of genes. *Arch. Biochem. Biophys.* **368**: 67–74.

55. Wang, Y., Crawford, D. R. and Davies, K. J. A. (1996). *Adapt33*, a novel oxidant-inducible RNA from hamster HA-1 cells. *Arch. Biochem. Biophys.* **332**: 255–260.

56. Crawford, D. R. and Davies, K. J. A. (1997). Modulation of a cardiogenic shock inducible RNA by chemical stress: *adapt73/PigHep3*. *Surgery* **121**: 581–587.

57. Rothermel, B., Vega, R. B., Yang, J., Wu, H., Bassel-Duby, R. and Williams, R. S. (2000). A protein encoded within the Down syndrome critical region is enriched in striated muscles and inhibits calcineurin signaling. *J. Biol. Chem.* **275**: 8719–8725.

58. Fuentes, J. J., Genesca, L., Kingsbury, T. J., Cunningham, K. W., Perez-Riba, M., Estivill, X. and de la Luna, S. (2000). *DSCR1*, overexpressed in Down syndrome, is an inhibitor of calcineurin-mediated signaling pathways. *Human Mol. Genet.* **9**: 1681–1690.

59. Kingsbury, T. J. and Cunningham, K. W. (2000). A conserved family of calcineurin regulators. *Genes Dev.* **14**: 1595–1604.

60. Amstad, P., Crawford, D., Muehlematter, D., Zbinden, I., Larsson, R. and Cerutti, P. (1990). Oxidants stress induces the proto-oncogenes, *c-fos* and *c-myc* in mouse epidermal cells. *Bulletin du Cancer* **77**: 501–502.

61. Nose, K., Shibanuma, M., Kikuchi, K., Kageyama, H., Sakiyama, S. and Kuroki, T. (1991). Transcriptional activation of early-response genes by hydrogen peroxide in a mouse osteoblastic cell line. *Eur. J. Biochem.* **201**: 99–106.

62. Keyse, S. M. and Enslie, E. A. (1992). Oxidative stress and heat shock induce a human gene encoding a protein-tyrosine phosphatase. *Nature* **359**: 644–647.

63. DeForge, L. E., Preston, A. M., Takeuchi, E., Kenney, J., Boxer, L. A. and Remick, D. G. (1993). Regulation of *interleukin-8* gene expression by oxidant stress. *J. Biol. Chem.* **268**: 25 568–25 576.

64. Shull, S., Heintz, N. H., Periasamy, M., Manohar, M., Janssen, Y. M., Marsh, J. P. and Mossman, B. T. (1991). Differential regulation of antioxidant enzymes in response to oxidants. *J. Biol. Chem.* **266**: 24 398–24 403.

65. Guyton, K. Z., Liu, Y., Gorospe, M., Xu, Q. and Holbrook, N. J. (1996). Activation of mitogen-activated protein kinase by H_2O_2. Role in cell survival following oxidant injury. *J. Biol. Chem.* **271**: 4138–4142.

66. Crawford, D. R., Suzuki, T. and Davies, K. J. A. (1999). Oxidant modulated gene statement. *In* "Antioxidant and Redox Regulation of Genes" (C. K. Sen, H. Sies, and P. Baeuerle, Eds.), pp. 21–49, Academic Press, San Diego.

67. Davies, K. J. A. (1999). The broad spectrum of responses to oxidants in proliferating cells: a new paradigm for oxidative stress. *IUBMB Life* **48**: 41–47.

Chapter 41

Heme Oxygenase: Its Regulation and Role

Herbert L. Bonkovsky* and Kimberly K. Elbirt[†]

Herbert L. Bonkovsky and **Kimberly K. Elbirt** • University of Massachusetts Medical School, UMass Memorial Health Care, Center for the Study of Iron and Porphyrin Metabolism, 55 Lake Avenue North, Worcester, MA 01655

*Corresponding Author.
Tel: 508-856-3068, E-mail: BonkovsH@ummhc.org
[†]Departments of Medicine; Biochemistry and Molecular Pharmacology
Tel: 508-856-3525, E-mail: Kimberly.Elbirt@umassmed.edu

1. Introduction

The enzymatic mechanism responsible for the specific cleavage of the heme macrocycle at the α methene bridge was first described just over 30 years ago.[1] It was given the name heme oxygenase (HO) and was initially thought to involve a cytochrome P450, but involvement of cytochromes P450 was soon shown not to be required.[2] In independent studies of heat shock and oxidative and other "stress" responses, a highly induced protein with a molecular mass of about 32 kDa (hsp 32) was described[3,4] and shown to be identical to the first isoform of HO (HO-1). Although there have been rare exceptions,[5] most authors studying "stress" and aging responses have concluded that the induction of HO-1 is a protective or adaptive response that functions to ameliorate cell and tissue injury and death. Thus, HO, and particularly the highly inducible HO-1, is a key component of the "stress" response.

2. General Characteristics and Isoforms of Heme Oxygenase

Heme oxygenase (HO) is a ubiquitous enzyme responsible for the physiologic cleavage of heme. In association with NADPH-ferrihemoprotein reductase (NADPH-cytochrome P450 reductase), and in the presence of oxygen and NADPH, HO catalyzes the degradation of heme into equimolar amounts of carbon monoxide (CO), biliverdin (BV), and iron. Intermediates include a peroxy compound, α-meso-hydroxyheme, and verdoheme (Fig. 1). The alpha-methene bridge of the heme macrocycle is the specific target of HO, and the sole source of the carbon atom of CO derived from the breakdown of heme. In most mammalian, but not

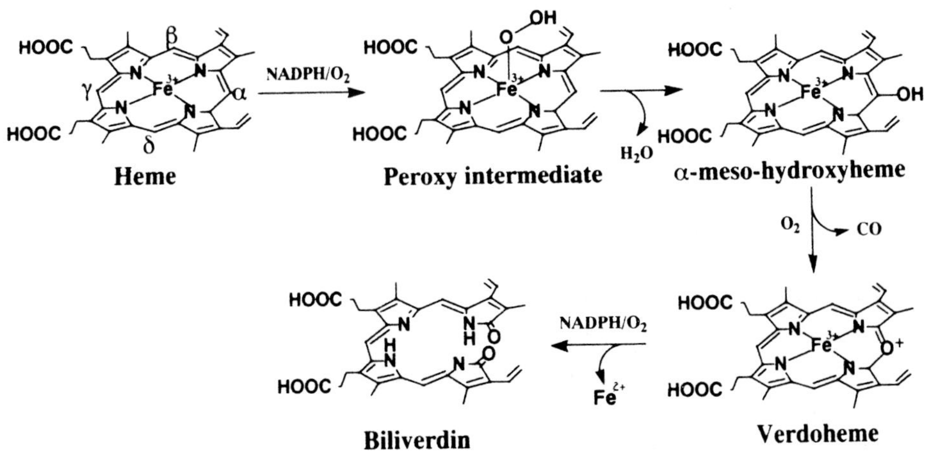

Fig. 1. Summary of the reaction catalyzed by heme oxygenase.

Table 1. Characteristics of the Isoforms of Heme Oxygenase

Characteristic	Isoforms		
	HO-1	HO-2	HO-3
Major function	Catalytic heme degradation and formation of CO, BV	Catalytic heme degradation and formation of CO, BV	Heme sensing/ binding protein
Inducibility	Highly inducible by many conditions	Constitutive	Uninducible
Tissue localization	Ubiquitous, highest in spleen, liver, kidneys	Mainly brain and testes	Ubiquitous, highest in liver, prostate, kidneys
Molecular mass (kDa)	~32–33	~36	~32–33
mRNA (Number of Transcripts/ Transcript Size)	1 of ~1.8 kb ([†]~1258 base pairs)	2 of ~1.3, ~1.9 kb	1 of ~2.4 kb; *others detected of ~1.7–3.7 kb
Vmax (nmol BV/mg prot/min)	56.7 ([†]580 ± 44)	4.0	0.4
Apparent Km_{heme} (µM)	0.24 ([†]3.8 ± 0.5)	0.67	N/D

BV, biliverdin; kDa, kilo Dalton; Km, Michaelis constant N/D, not determined; Vmax, maximal velocity.
*Several tissue-specific mRNA transcripts have been identified for HO-3 in heart and testes.[8]
[†]Values for the chick form of the enzyme.[9] Other values are for the rat form.[62]

avian, species biliverdin is subsequently reduced to bilirubin by biliverdin reductase, a cytosolic soluble enzyme found in large amount in many cells and tissues.

Three isoforms of HO, called HO-1, HO-2, and HO-3, have been described in mammals. Table 1 presents a summary of the three isoforms, which are the products of separate genes.[6-8] HO-1 is nearly ubiquitous in distribution and highly inducible. The region from amino acids 128–136 and His[25] are conserved among HO-1 proteins from rats, humans, and chickens, highlighting their importance in substrate binding and catalysis.[9]

HO-2 is constitutively active and virtually uninducible, but may be a key factor in maintaining homeostasis under basal conditions or mild physiologic stress.[6, 7, 10] It is present in highest concentrations in brains and testes of mammals, and it appears to play an important role in protecting the brain from ischemic injury.[11] HO-3 has a much lower specific activity and may function chiefly in heme sequestration or as a heme sensor.[7]

HO-1 has been identified as the isoform chiefly responsible for protecting cells from oxidative and other stress.[3, 4, 12, 15–17] In addition to cytoprotection, HO-1 is thought to contribute to regulating the amount of heme available to form hemoproteins, such as the microsomal cytochromes P450, which play key roles in a vast array of reactions, including steroid biogenesis and phase I drug metabolism.[13–15] The protective role of HO can be attributed to several functions: reduction in free heme concentrations, production of the potent antioxidants,

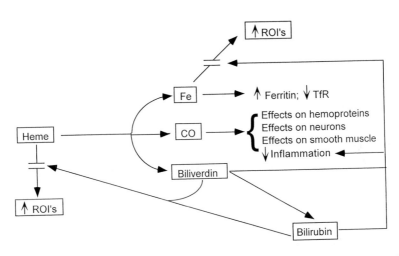

Fig. 2. The products of the heme oxygenase reaction and some of their important functions.

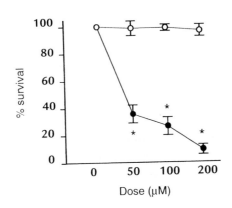

Fig. 3. Increased heme-dependent cytotoxicity in lymphocyte lines from a child with severe HO-1 deficiency (closed circles) versus normal controls (open circles). Cells were exposed to the concentrations of heme shown on the x-axis for 24 hour. Dead or apoptotic cells were identified by lowered forward light scatter or high annexin V binding using flow cytometry. Data show M ± SD, $n = 5$. *Differs from control, $p < 0.001$. (From Ref. 16.)

biliverdin and bilirubin,[16, 17] and production of CO, which has wide-ranging effects on neuromuscular, circulatory, and inflammatory functions, through its role as a signaling molecule (Fig. 2).[18, 19]

In combination with the rate-limiting enzyme of heme biosynthesis, namely, delta-aminolevulinate (ALA) synthase, HO helps to maintain low intracellular levels of loosely bound or "free" heme. This is important because the heme molecule itself is a prooxidant capable of producing cellular oxidative stress. By helping to maintain "free" cellular heme levels within a narrow range, HO controls the potentially deleterious effects of acute increases of such heme that occur during tissue injury or oxidative insult.

The most convincing evidence for this protective role of HO-1 comes from studies of cells with severely deficient activities of HO-1 due to spontaneous[16] or targeted[20, 21] disruptions of the HO-1 gene. Figure 3 shows the marked increase in cytotoxicity due to heme in transformed lymphocyte lines from a child with severe deficiency of HO-1.

3. Functions of the Products of the HO Reaction

3.1. Carbon Monoxide (CO)

Carbon monoxide (CO) has molecular and biological properties akin to those of nitric oxide (NO), which is now recognized to be a key molecule for a plethora of pathways and functions.[22] For example, CO, like NO, may sometimes increase levels of cGMP through activation of guanylate cyclase. Increased cGMP produces widespread relaxation of smooth muscle, especially in blood vessels (vasodilatation, decreases in blood pressure) and in the gastrointestinal tract (relaxation of sphincters, decreases in peristaltic activity). Other actions of CO are independent of effects on guanylate cyclase, which is activated more readily by NO than by CO. Another general mechanism of action of CO probably involves its tight binding to hemoproteins involved in oxidation-reduction reactions or in the carriage and transport of oxygen. CO typically binds to hemoproteins (e.g. cytochromes P450, hemoglobin, myoglobin) more tightly by several orders of magnitude than does O_2, the physiologic ligand or substrate.[23] Thus, for example, the inhibition of contractility of bile canaliculi by CO has been attributed to its inhibition of a cytochrome P450-dependent epoxygenase.[24] In contrast, CO decreases hepatic canalicular secretion of bile acids[25] and sinusoidal vascular pressure[26] by cGMP-dependent mechanisms. Increases in HO-1 and thus levels of CO in splanchinc organs play a role in the pathogenesis of portal hypertension.[27]

HO and its product CO have several roles in the central nervous system, including central reductions in systemic blood pressure and production of long-term neuronal adaptation[28] and retrograde potentiation,[29] suggesting a role for CO in development of memory.

3.2. Biliverdin and Bilirubin

Biliverdin and its product bilirubin, formed in most mammals, are potent antioxidants.[30, 31] They are present normally in concentrations sufficient to be consistent with their playing a role in the protection of mammalian cells and tissues from oxidative stress. Unconjugated bilirubin is hydrophobic as a consequence of its intramolecular internal hydrogen bonding. Thus, it is quite soluble in lipids (e.g. in biological membranes), and with vitamin E is believed to be a major endogenous inhibitor of lipid peroxidation. Biliverdin is water-soluble and thus more likely to play a role in inhibiting peroxidation reactions in aqueous environments.

3.3. Iron

In contrast to CO, biliverdin, or bilirubin, "free" iron, which is also released during heme degradation catalyzed by HO (Fig. 2), increases oxidative stress. The iron generated from heme breakdown enters a "regulatory" iron pool, which modulates expression of many mRNAs (e.g. the form of the divalent cation transporter-1 (DCT-1) that contains an iron regulatory element (IRE), IREG, ferritin, transferrin receptor, and the erythroid form of ALA synthase) by affecting the conformation of iron regulatory protein (IRP)-1 and its binding to IREs in the 5'- or 3'-untranslated regions (UTRs) of the mRNAs of these and other genes.[32, 33] The production of "free" iron is potentially damaging, but there exists a panoply of mechanisms to limit toxicity of such iron, including upregulation of ferritin synthesis and down-regulation of transferrin receptors.

Indeed, humans and mice with severe deficiencies of HO-1 show pathological accumulations of iron in liver and kidneys, indicating that such iron deposition is not a consequence of normal HO activity, but rather of its lack.[16, 20]

4. Regulation of Expression of HO-1

Table 2 lists some of the many chemical substances and physical factors that are known to induce HO-1. The regulation of HO-1 depends on a variety of factors that affect the cell, including the intracellular concentration of "free" or loosely bound heme.

Two general mechanisms for induction of HO-1 have been defined. Heme, the heme precursor ALA, and phenobarbital-like drugs (e.g. phenobarbital, glutethimide, or phenytoins) act via a heme-dependent mechanism, whereas

Table 2. Some Conditions That Induce Heme Oxygenase-1

Conditions	References
Aging (in brain)	Iijima *et al.*, 1999[59]
Heme	Lincoln *et al.*, 1988;[13] Gabis *et al.*, 1996;[37] Shan *et al.*, 2000[55]
Other metalloporphyrins	Llesuy *et al.*, 1994;[63] Cable *et al.*, 1994;[36] Shan *et al.*, 2000[55]
Hemoglobin, myoglobin	Nath *et al.*, 2000[61]
Transition metals	Kikuchi and Yoshida, 1983;[34] Lincoln *et al.*, 1988[13]
Arsenicals-arsenite; phenylarsine oxide	Keyse and Tyrrell 1989;[3] Gabis *et al.*, 1998;[51] Shan *et al.*, 1999[50]
Inflammatory cytokines	Fukuda and Sassa, 1993;[64] Strandell *et al.*, 1995[65]
Prostaglandins	Rossi and Santoro, 1995;[66] Koizumi *et al.*, 1995[67]
Ultraviolet light (UV)	Keyse and Tyrrell, 1989;[3] Vile *et al.*, 1994[12]
Phorbol esters	Alam *et al.*, 1995[40]
Heat shock	Shibahara *et al.*, 1987;[6] Gabis *et al.*, 1996[37]
Hydrogen peroxide	Keyse and Tyrrell, 1989[3]
Lipopolysaccharide (LPS)	Camhi *et al.*, 1995;[45] Vaccharajani *et al.*, 2000;[30] Wiesel *et al.*, 2000[44]
Thiol scavengers (GSH depletion)	Oguro *et al.*, 1996[46]
Tobacco smoke	Müller and Gebel, 1994;[68] Pinot *et al.*, 1997[69]
Ischemia/reperfusion	Tacchini *et al.*, 1993;[70] Amersi *et al.*, 1999[71]
Hypoxia, hyperoxia	Lee *et al.*, 2000;[54] Kacimi *et al.*, 2000[72]
Portal hypertension (in splanchnic organs)	Fernandez and Bonkovsky, 1999[27]
Ets-family proteins	Bertrand *et al.*, 1999[73]

Due to limitations of space, this list and its bibliography are not exhaustive.

physical and chemical inducers of oxidative stress act by heme-independent mechanism(s).[34–38] Heme exerts regulatory effects on HO-1 enzyme activity and gene expression by increasing the transcription of HO-1, leading to greater production of HO-1 protein, which catalyzes heme degradation. Distinctions among the mechanisms utilized by metals, heme, and heat shock to induce HO-1 have also been described.[13, 37–39] Despite these differences, the effects of diverse factors on HO-1 gene expression appear to be exerted chiefly at the transcriptional level, suggesting that a limited number of signal transduction pathways mediate induction of HO-1 gene transcription in response to a multitude of cellular perturbations.

4.1. Structure of the HO-1 Gene and Its Promoter and Enhancer Regions

The HO-1 genes from rodents, humans, and chicken have been cloned and characterized (see Ref. 39 for review). The general structures of all are similar, containing 4 introns and 5 exons. Computer-assisted analysis of putative (consensus) transcription factor binding sites present in the first 7.1 kb of the chick HO-1 promoter and homologous regions in mammalian HO-1 genes revealed numerous possible enhancer and/or promoter elements. The signal transduction pathways and transcription factor complexes that target these elements are beginning to be elucidated. Signaling pathways that have been identified thus far include the mitogen activated protein (MAP) kinases, protein kinase A, protein kinase C and the JAK-STAT signaling cascades. Studies using stably transfected HO-1 reporter gene constructs have delineated elements required for induction of the murine HO-1 gene by heme and heavy metals.[10, 38] Other work has identified regions of the HO-1 promoter that are important for responses to cadmium, heat shock, hypoxia, and lipopolysaccharide (LPS).[10, 39–41]

Induction of HO-1 gene expression by several cellular stressors occurs through the utilization of AP-1 binding elements (TGAGTCA) present at several sites in the promoter regions of human, murine, and avian HO-1.[39] AP-1 proteins are essential mediators of early- and late-response genes during the cellular response to external stress or insult. (For a review of the AP-1 family of transcription factors, see Karin *et al.*[42]). Thioredoxin may induce HO-1 expression by the reduction of disulfides of the AP-1 proteins *Fos* and *Jun*, which facilitates promoter binding and transactivation.[43] Studies by Alam *et al.*[10, 40] in murine hepatoma (Hepa) and fibroblast (L929) cells have identified and characterized enhancer sequences in the mouse HO-1 gene, including consensus AP-1 elements that bind *c-Fos* and *c-Jun* heterodimers and are responsive to phorbol ester, cadmium, heme, and several other inducers. Transcriptional activation of HO-1 by LPS in mouse macrophages has also been attributed to a distal 5' AP-1 binding site located approximately 4 kb upstream of the transcription start site.[44] Induction of rat liver HO-1 following depletion of glutathione (GSH) was shown to occur through AP-1 activation.[45] In human fibroblasts, GSH depletion induced the expression of both HO-1 and *c-Fos*, a transcription factor that binds consensus AP-1 binding elements.[46]

Research performed in our laboratory has investigated the role of MAP kinases, and AP-1 transcription factor complexes and promoter binding sites in the regulation of chick HO-1 gene expression in response to the arsenical compounds sodium arsenite, a prototypic oxidative stress inducer, and the phosphatase inhibitor phenylarsine oxide (PAO). Arsenite has been proposed to affect gene expression by modulating the activities of transcription factor complexes bound to AP-1 elements in the promoter regions of several genes.[39, 47]

In electrophoretic mobility shift assays, arsenite and cobalt chloride increased binding of nuclear proteins(s) to an AP-1 consensus element in the chick HO-1 promoter.[48, 49] DNA-binding and mutagenesis studies showed that both an AP-1 element at −1580 to −1573 and an MRE/*c-Myc* complex at −52 to −41 (nt's upstream of the transcription starting point) were important for full arsenite induction, whereas the induction by $CoCl_2$ involved only the AP-1 element.[49] The MRE/*c-Myc* complex shares similarities to the upstream stimulatory factor element identified in the promoter regions of mammalian HO-1 and other stress-regulated genes; it plays a major role in the basal level of expression of the chick HO-1 gene. Pretreatment with the antioxidants N-acetyl cysteine or quercetin reduced induction of the endogenous chick HO-1 gene or the chick HO-1 promoter-reporter constructs induced by arsenite or $CoCl_2$ and also reduced nuclear protein binding to the AP-1 element in electromobility shift assays.[49] We found similar effects for another arsenical, phenylarsine oxide, which is believed to act primarily by binding vicinal −SH groups in protein phosphatases thereby inhibiting phosphatase activities.[50]

MAP kinase signaling cascades mediate the arsenite induction of HO-1 through AP-1 elements.[39, 51] The roles of ERK and p38 MAP kinases in arsenite-mediated HO-1 induction were elucidated through the use of specific inhibitors of MEK or p38, which blocked most of the arsenite-mediated increase in HO-1 promoter-luciferase reporter gene activity (Fig. 4). Pretreatment with either or both of the inhibitors also resulted in decreased levels of endogenous HO-1 mRNA in response to arsenite treatment.[51] The results of these and other studies involving MAPK expression constructs indicate that ERK and p38, but not the JNK, MAP kinase pathways modulate HO-1 induction by arsenite.

Recent data from other labs further implicate tyrosine phosphorylation/dephosphorylation reactions and inhibition of serine/threonine phosphatase 1 and 2A (by okadaic acid) in stress-mediated induction of HO-1 via changes in cAMP.[52, 53] Other data point to a role of STAT proteins for optimal induction of HO-1 in murine macrophages.[54]

In contrast, heme or cobalt protoporphyrin (CoPP) induction of chick HO-1 occurs via a pathway independent of MAP kinase/stress pathways, as indicated by the differences in induction mediated by heme versus other stressors such as arsenite, cadmium, and heat shock[39, 49–51] and by the lack of effect of N-acetyl cysteine on heme- or CoPP-dependent induction.[55] In the chick HO-1 gene, the key elements required for the metalloporphyrin-mediated induction is located between −5.6 to −3.6 kb upstream of the transcription starting point and does not involve consensus AP-1 binding elements.[55]

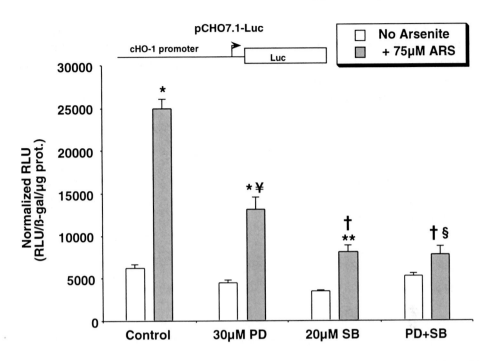

Fig. 4. Effects of specific inhibitors of the MAP kinases MEK (PD 98059) and/or p38 (SB203580) to diminish arsenite-mediated induction of a chick HO-1 promoter-reporter construct. Chick hepatoma (LMH) cells, transfected with the construct shown were treated with arsenite (ARS) alone or with the MAP kinase inhibitors plus arsenite. After six h, cells were harvested and assayed for luciferase activity. Data are presented as mean + S.E., $n = 3$. *, differs from no ARS, $p = 0.0001$. **, differs from no ARS, $p = 0.005$. +, differs from ARS control, $p < 0.01$. φ, differs from ARS control, $p = 0.05$. §, differs from PD + ARS, $p = 0.01$. RLU, relative light units. (From Ref. 51.)

5. Anti-Inflammatory and Antioxidant Properties of HO-1

Several lines of investigation have identified HO-1 as an essential component in cytoprotective responses to oxidative or inflammatory stress. Its expression is highly induced at sites of acute or chronic inflammation. Inhibitors of HO-1 exacerbate inflammatory responses, whereas chemically or genetically mediated overexpression of HO-1 alleviates the ill effects of inflammation.[30, 56] These anti-inflammatory effects of HO-1 in murine macrophages are attributable to the actions of CO, in addition to the well-established antioxidant activities of biliverdin and bilirubin. CO exerts its potent anti-inflammatory effects by suppression of proinflammatory cytokines (TNF-α, IL-1β, MIP-1β) and increase of the anti-inflammatory cytokine IL-10. These effects of CO were produced not through guanylate cyclase-cGMP or NO, but through other signaling cascades that up-regulated p38 MAP kinase.[30] This leads to cytoprotection in macrophages and monocytes, and suggests new therapeutic strategies for septic shock and the systemic inflammatory response syndrome. In LPS-treated rats, biliverdin or bilirubin was found to inhibit P- and E-selectins, which recruit leukocytes to sites of inflammation. HO-1 induction in vascular beds thus protects against disproportionate, uncontrolled inflammatory reactions.[56]

6. Heme Oxygenase and Aging

As pointed out elsewhere in this book, oxidative stress may play an important role in aging. One mechanism for the oxidative stress contribution to aging is the net increase in free radical production in aging tissues. This may lead to tissue injury *via* lipid, protein, or nucleic acid (per)oxidative damage.

Extensive studies of HO gene and protein expression and activities in normal aging have not yet been done. In rats, monkeys, and humans activities of hepatic HO are highest in fetuses and newborns. In rats they decline to adult levels by 3–4 weeks of age. In contrast, HO activities in spleens of fetuses and newborn rats are only about 50% of those of adults (for review, see Ref. 57). One study found no evidence of a change in hepatic HO activity as rats aged.[58]

HO-1 expression is increased in brain regions of aged versus young rats.[59] HO-1 is also present in neurofibrillary tangles in Alzheimer disease associated with tau protein.[60] In Alzheimer disease, neurotoxicity of amyloid deposits is associated with oxidative stress due to free radicals. In neuronal cells over-expressing transfected HO-1, levels of tau protein were markedly decreased, and inhibition of HO activity returned tau expression to high levels. These results are consistent with the concept that increased oxidative stress and activation of MAP kinases play a role in production of neurofibrillary tangles, including tau, and that HO-1 induction may function to control the damage.

Increased HO-1 expression has also been associated with renal aging in rats.[61] In this model, protective maintenance of the glomerular filtration rate and renal plasma flow may result from vasodilation mediated by CO derived from the HO-1 reaction. The importance of HO-1 in the maintenance of renal functioning is supported by the observation of acute renal failure and death in HO-1 knockout mice exposed to hemoprotein in a model of oxidative stress.[61] Renal lesions have been noted in the one human described thus far with severe HO-1 deficiency[16] and in mice with targeted deletion of the HO-1 gene.[20, 21]

Acknowledgments

We thank Jean Clark for help with typing the manuscript and the U.S.P.H.S., National Institutes of Health for grant and contract support (RO1-DK 38825 and

NO1 DK 92326 to H.L.B.). The opinions expressed in this chapter are those of the authors. They do not necessarily reflect the official views of the U.S.P.H.S., N.I.H., or the University of Massachusetts Medical School.

References

1. Tenhunen, R., Marver, H. S. and Schmid, R. (1969). Microsomal heme oxygenase: characterization of the enzyme. *J. Biol. Chem.* **244**: 6388–6394.
2. Maines, M. D. and Kappas, A. (1974). Cobalt induction of hepatic heme oxygenase, with evidence that cytochrome P450 is not essential for this enzyme activity. *Proc. Natl. Acad. Sci. USA* **71**: 4293–4297.
3. Keyse, S. M. and Tyrrell, R. M. (1989). Heme oxygenase is the major 32 kDa stress protein induced in human skin fibroblasts by UVA radiation, hydrogen peroxide, and sodium arsenite. *Proc. Natl. Acad. Sci. USA* **86**: 99–103.
4. Taketani, S., Kohno, H., Yoshinaga, T. and Tokunaga, R. (1989). The human 32 kDa stress protein induced by exposure to arsenite and cadmium ions is heme oxygenase. *FEBS Lett.* **245**: 173–176.
5. Nutter, L. M., Sierra, E. E. and Ngo, E. O. (1994). Heme oxygenase does not protect human cells against oxidant stress. *J. Lab. Clin. Med.* **123**: 506–514.
6. Shibahara, S., Yoshizawa, M., Suzuki, H., Takeda, K., Meguro, K. and Endo, K. (1993). Functional analysis of cDNAs for two types of human heme oxygenase and evidence for their separate regulation. *J. Biochem.* **113**: 214–218.
7. McCoubrey, W. K., Jr. and Maines, M. D. (1994). The structure, organization and differential expression of the gene encoding rat heme oxygenase-2. *Gene* **139**: 155–161.
8. McCoubrey, W. K., Jr., Huang, T. J. and Maines, M. D. (1997). Isolation and characterization of a cDNA from the rat brain that encodes hemoprotein heme oxygenase-3. *Eur. J. Biochem.* **247**: 725–732.
9. Bonkovsky, H. L., Healey, J. F. and Pohl, J. (1990). Purification and characterization of heme oxygenase from chick liver — comparison of the avian and mammalian enzymes. *Eur. J. Biochem.* **189**: 155–166.
10. Alam, J., Cai, J. and Smith, A. (1994). Isolation and characterization of the mouse heme oxygenase-1 gene. Distal 5′ sequences are required for induction by heme or heavy metals. *J. Biol. Chem.* **269**: 1001–1009.
11. Doré, S., Goto, S., Sampei, K., Blackshaw, S., Hester, L. D., Ingi, T., Sawa, A., Traystman, R. J., Koehler, R. C. and Snyder, S. H. (2000). Heme oxygenase-2 acts to prevent neuronal death in brian cultures and following transient cerebral ischemia. *Neuroscience* **4**: 587–592.
12. Vile, G. F., Basu-Modak, S., Waltner, C. and Tyrrell, R. M. (1994). Heme oxygenase 1 mediates an adaptive response to oxidative stress in human skin fibroblasts. *Proc. Natl. Acad. Sci. USA* **91**: 2607–2610.

13. Lincoln, B. C., Healey, J. F. and Bonkovsky, H. L. (1988). Regulation of hepatic haem metabolism. Disparate mechanisms of induction of haem oxygenase by drugs and metals. *Biochem. J.* **250**: 189–196.

14. Bhat, G. and Padmanaban, G. (1988). Heme is a positive regulator of cytochrome P450 gene transcription. *Arch. Biochem. Biophys.* **264**: 584–590.

15. Sardana, M. K., Sassa, S. and Kappas, A. (1980). Adrenalectomy enhances the induction of heme oxygenase and the degradation of cytochrome P450 in liver. *J. Biol. Chem.* **255**: 11 320–11 323.

16. Yachie, A., Niida, Y., Wada, T., Igarashi, N., Kaneda, H., Toma, T., Ohta, K., Kasahara, Y. and Koizumi, S. (1999). Oxidative stress causes enhanced endothelial cell injury in human heme oxygenase-1 deficiency. *J. Clin. Invest.* **103**: 129–135.

17. Yamaguchi, T., Terakado, M., Horio, F., Aoki, K., Tanaka, M. and Nakajima, H. (1996). Role of bilirubin as an antioxidant in an ischemia-reperfusion model of rat liver injury and induction of heme oxygenase. *Biochem. Biophys. Res. Commun.* **223**: 129–135.

18. Maines, M. D. (1997). The heme oxygenase system: a regulator of second messenger gases. *Ann. Rev. Pharmacol. Toxicol.* **37**: 517–554.

19. Zakhary, R., Poss, K. D., Jaffrey, S. R., Ferris, C. D., Tonegawa, S. and Snyder, S. H. (1997). Targeted gene deletion of heme oxygenase-2 reveals neural role for carbon monoxide. *Proc. Natl. Acad. Sci. USA* **94**: 14 848–14 853.

20. Poss, K. D. and Tonegawa, S. (1997). Heme oxygenase-1 is required for mammalian iron reutilization. *Proc. Natl. Acad. Sci. USA* **94**: 10 919–10 924.

21. Poss, K. D. and Tonegawa, S. (1997). Reduced stress defense in heme oxygenase-1 deficient cells. *Proc. Natl. Acad. Sci. USA* **94**: 10 925–10 930.

22. Monacada, S., Palmer, R. M. and Higgs, E. A. (1991). Nitric oxide: physiology, pathophysiology, and pharmacology. *Pharmacol. Rev.* **2**: 109–142.

23. Stryer, L. (1988). *Biochemistry.* Third Edition, p. 149 Freeman, New York

24. Suematsu, M. and Ishimura, Y. (2000). The heme oxygenase-carbon monoxide system: a regulator of hepatobiliary function. *Hepatology* **31**: 3–6.

25. Sano, S., Shiomi, M., Wakabayashi, Y., Shinoda, Y., Goda, N., Yamaguchi, T., Nimura, Y. and Suematsu, M. (1997). Carbon monoxide generated by heme oxygenase modulates bile-acid dependent biliary transport in perfused rat liver. *Am. J. Physiol.* **272**: G1268–G1275.

26. Suematsu, M., Goda, N., Sano, T., Kashiwagi, S., Egawa, T., Shinoda, Y. and Ishimura, Y. (1995). Carbon monoxide: an endogenous modulator of sinusoidal tone in the perfused rat liver. *J. Clin. Invest.* **5**: 2431–2437.

27. Fernandez, M. and Bonkovsky, H. L. (1999). Increased heme oxygenase-1 gene expression in liver cells and splanchnic organs from portal hypertensive rats. *Hepatology* **6**: 1672–1679.

28. Zufall, F. and Leinders-Zufall, T. (1997). Identification of a long-lasting form of odor adaptation that depends on the carbon monoxide/cGMP second-messenger system. *J. Neurosci.* **15**: 2703–2712.

29. Stevens, C. F. and Wang, Y. (1993). Reversal of long-term potentation by inhibitors of haem oxygense. *Nature* **364**: 147–149.
30. Vaccharajani, T. J., Work, J., Issekutz, A. C. and Granger, D. N. (2000). Heme oxygenase modulates selectin expression in different regional vascular beds. *Am. J. Physiol.* **278**: H1613–H1617.
31. Stocker, R., Yamamoto, Y., McDonagh, A. F., Glazer, A. N. and Ames, B. N. (1987). Bilirubin is an antioxidant of possible physiological importance. *Science* **235**: 1043–1046.
32. Eisenstein, R. S. and Blemings, K. P. (1998). Iron regulatory proteins, iron responsive elements and iron homeostasis. *J. Nutri.* **128**: 2295–2298.
33. Hentze, M. W. (1996). Iron-sulfur clusters and oxidant stress responses. *Trends Biochem. Sci.* **8**: 282–283.
34. Kikuchi, G. and Yoshida, T. (1983). Function and induction of the microsomal heme oxygenase. *Mol. Cell. Biol.* **53/54**: 163–183.
35. Cable, E., Greene, Y., Healey, J., Evans, C. O. and Bonkovsky, H. L. (1990). Mechanism of synergistic inducti of hepatic heme oxygenase by glutethimide and iron: studies in cultured chick embryo liver cells. *Biochem. Biophys. Res. Commun.* **168**: 176–181.
36. Cable, E. E., Pepe, J. A., Karamitsios, N. C., Lambrecht, R. W. and Bonkovsky, H. L. (1994). Differential effects of metalloporphyrins on messenger RNA levels of δ-aminolevulinate synthase and heme oxygenase. Studies in cultured chick embryo liver cells. *J. Clin. Invest.* **94**: 649–654.
37. Gabis, K. K., Gildemeister, O. S., Pepe, J. A., Lambrecht, R. W. and Bonkovsky, H. L. (1996). Induction of heme oxygenase-1 in LMH cells. Comparison of LMH cells to primary cultures of chick embryo liver cells. *Biochim. Biophys. Acta* **1290**: 113–120.
38. Eyssen-Hernandez, R., Ladoux, A. and Frelin, C. (1996). Differential regulation of cardiac heme oxygenase-1 and vascular endothelial growth factor mRNA expressions by hemin, heavy metals, heat shock and anoxia. *FEBS Lett.* **382**: 229–233.
39. Elbirt, K. K. and Bonkovsky, H.L. (1999). Heme oxygenase: recent advances in understanding its regulation and role. *Proc. Assoc. Am. Physicians* **111**: 438–447.
40. Alam, J., Camhi, S. and Chio, A. M. (1995). Identification of a second region upstream of the mouse heme oxygenase-1 gene that functions as a basal level and inducer-dependent transcription enhancer. *J. Biol. Chem.* **270**: 11 977–11 984.
41. Lee, P. J., Jiang, B. H., Chin, B. Y., Iyer, N. V., Alam, J., Semenza, G. L. and Choi, A. M. K. (1997). Hypoxia-inducible factor-1 mediates transcriptional activation of the heme oxygenase-1 gene in response to hypoxia. *J. Biol. Chem.* **272**: 5375–5381.
42. Karin, M., Liu and Zandi, E. (1997). AP-1 function and regulation. *Curr. Opin. Cell Biol.* **9**: 240–246

43. Wiesel, P., Foster, L., Pelloacani, A., Layne, M. D., Hsieh, C.-M., Huggins, G. S., Strauss, P., Yet, S.-F. and Perrella, M. A. (2000). Thioredoxin facilitates the induction of heme oxygenase-1 in response to inflammatory mediators. *J. Biol. Chem.* **275**: 24 840–24 846.

44. Camhi, S. L., Alam, J. and Otterbein, L. (1995). Induction of heme oxygenase-1 gene expression by lipopolysaccharide is mediated by AP-1 activation. *Am. J. Respir. Cell. Mol. Biol.* **13**: 387–398.

45. Oguro, T., Hayashi, M., Numazawa, S., Asakawa, K. and Yoshida, T. (2000). Heme oxygenase-1 gene expression by a glutathione depletor, phorone, mediated through AP-1 activation in rats. *Biochem. Biophys. Res. Commun.* **221**: 259–265.

46. Numazawa, S., Yamada, H., Furusho, A., Nakahara, T., Oguro, T. and Yoshida, T. (1997). Cooperative induction of *c-fos* and heme oxygenase gene products under oxidative stress in human fibroblastic cells. *Exp. Cell Res.* **237**: 434–444.

47. Guyton, K. Z., Xu, Q. and Holbrook, N. J. (2000). Induction of the mammalian stress response gene *gadd153* by oxidative stress: role of AP-1 element. *Biochem. J.* **314**: 547–554.

48. Lu, T. H., Lambrecht, R. W., Pepe, J., Shan, Y., Kim, T. and Bonkovsky, H. L. (1998). Molecular cloning, characterization, and expression of the chicken heme oxygenase-1 gene in transfected primary cultures of chick embryo liver cells. *Gene* **207**: 177–186.

49. Lu, T. H., Shan Y., Pepe, J., Lambrecht, R. W. and Bonkovsky, H. L. (2000). Upstream regulatory elements in chick heme oxygenase-1 promoter: a study in primary cultures of chick embryo liver cells. *Mol. Cell. Biochem.* **209**: 17–27.

50. Shan, Y., Lambrecht, R. W., Lu, T. H. and Bonkovsky, H. L. (1999). Effects of phenylarsine oxide on expression of heme oxygenase-1 reporter constructs in transiently transfected cultures of chick embryo liver cells. *Arch. Biochem. Biophys.* **372**: 224–229.

51. Elbirt, K. K., Whitmarsh, A. J., Davis, R. J. and Bonkovsky, H. L. (1998). Mechanism of sodium arsenite-mediated induction of heme oxygenase-1 in hepatoma cells — role of mitogen-activated protein kinases. *J. Biol. Chem.* **273**: 8922–8931.

52. Masuya, Y., Hioki, K., Tokunaga, R. and Taketani, S. (1998). Involvement of the tyrosine phosphorylation pathway in induction of human heme oxygenase-1 by hemin, sodium arsenite, and cadmium chloride. *J. Biochem.* **124**: 628–633.

53. Immenschuh, S., Hinke, V., Katz, N. and Kietzmann, T. (2000). Transcriptional induction of heme oxygenase-1 gene expression by okadaic acid in primary rat hepatocyte cultures. *Mol. Pharmacol.* **57**: 610–618.

54. Lee, P. J., Camhi S. L., Chin, B. Y., Alam, J. and Choi, A. M. (2000). AP-1 and STAT mediate hyperoxia-induced gene transcription of heme oxygenase-1. *Am. J. Physiol.* **279**: L175–L182.

55. Shan, Y., Pepe, J., Lu, T. H., Elbirt, K. K., Lambrecht, R. W. and Bonkovsky, H. L. (2000). Induction of the heme oxygenase-1 gene by metalloporphyrins. *Arch. Biochem. Biophys.* **380**: 219–227

56. Otterbein, L. E., Bach, F. H., Alam, J., Soares, M., Tao Lu, H., Wysk, M., Davis, R. J., Flavell, R. A. and Choi, A. M. (2000). Carbon monoxide has anti-inflammatory effects involving the mitogen-activated protein kinase pathway. *Nat. Med.* **6**: 422–428.

57. Rodgers, P. A. and Stevenson, D. K. (1990). Developmental biology of heme oxygenase. *Clin. Perinatol.* **17**: 275–291.

58. Plewka, A. and Bienioszek, M. (1994). Effects of age, phenobarbital, β-naphthoflavone and dexamethasone on rat hepatic heme oxygenase. *Arch. Toxicol.* **68**: 32–36.

59. Iijima, N., Tamada, Y., Hayashi, S., Tanaka, M., Ishihara, A., Hasegawa, M. and Ibata, Y. (1999). Expanded expression of heme oxygenase-1 (HO-1) in the hypothalamic median eminence of aged as compared with young rats: an immunocytochemical study. *Neurosci. Lett.* **271**: 113–116.

60. Takeda, A., Perry, G., Abraham, N. G., Dwyer, B. E., Kutty, R. K., Laitinen, J. T., Petersen, R. B. and Smith, M. A. (2000). Overexpression of heme oxygenase in neuronal cells, the possible interaction with tau. *J. Biol. Chem.* **275**: 5395–5399.

61. Reckelhoff, J. F., Kanji, V., Racusen, L. C., Schmidt, A. M., Yan, S. D., Marrow, J., Roberts, L. J., II and Salahudeen, A. K. (1998). Vitamin E ameliorates enhanced renal lipid peroxidation and accumulation of F2-isoprostanes in aging kidneys. *Am. J. Physiol.* **274**: 767–774.

62. Maines, M. D., Trakshel, G. M. and Kutty, R. K. (1986). Characterization of two constitutive forms of rat liver microsomal heme oxygenase. Only one moleuclar species of the enzyme is inducible. *J. Biol. Chem.* **1**: 411–419.

63. Llesuy, S. F. and Tomaro, M. L. (1994). Heme oxygenase and oxidative stress. Evidence of involvement of bilirubin as physiological protector against oxidative damage. *Biochem. Biophys. Acta* **1**: 9–14

64. Fukuda, Y. and Sassa, S. (1993). Effect of interleukin-1β on the levels of mRNAs encoding heme oxygenase and haptoglobin in human HEPG2 hepatoma cells. *Biochem. Biophys. Res. Commun.* **193**: 297-302.

65. Strandell, E., Buschard, K., Saldeen, J., and Welsh, N. (1995). Interleukin-1β induces the expression of HSP70, heme oxygenase and Mn-SOD in FACS-purified rat islet β-cells, but not in α-cells. *Immunol. Lett.* **48**: 145–148.

66. Rossi, A. and Santoro, M. G. (1995). Induction by prostaglandin A1 of haem oxygenase in myoblastic cells: an effect independent of expression of the 70 kDa heat shock protein. *Biochem. J.* **308**: 455–463.

67. Koizumi, T., Odani, N., Okuyana, T., Ichikawa, A. and Negishi, M. (1995). Identification of a *cis*-regulatory element for delta 12-prostaglandin J2-induced expression of the rat heme oxygenase gene. *J. Biol. Chem.* **21**: 21 779–21 784.

68. Müller, T. and Gebel, S. (1994). Heme oxygenase expression in Swiss 3T3 cells following exposure to aqueous cigarette smoke fractions. *Carcinogenesis* **15**: 67–72.
69. Pinot, F., Yaagoubi, A., Christie, P., Dinh-Xuan, A. T. and Polla, B. S. (1997). Induction of stress proteins by tobacco smoke in human monocytes: modulation by antioxidants. *Cell Stress Chaperones* **2**: 156–161.
70. Tacchini, L., Schiaffonati, L., Pappalardo, C., Gatti, S. and Bernelli-Zazzera, A. (1993). Expression of hsp70, immediate-early response and heme oxygenase genes in ischemic-reperfused rat liver. *Lab. Invest.* **68**: 465–471.
71. Amersi, F., Beulow, R., Kato, H., Ke, B., Coito, A. J., Shen, X. D., Zhao, D., Zaky, J., Melinek, J., Lassman, C. R., Kolls, J. K., Alam, J., Ritter T., Volk, H. D., Farmer, D. G., Ghobrian, R. M., Busuttill, R. W. and Kupiec-Weglinski, J. W. (1999). Upregulation of heme oxygenase-1 protects genetically fat Zucker rat livers from ischemia/reperfusion injury. *J. Clin. Invest.* **104**: 1631–1639.
72. Kacimi, R., Chentoufi, J., Honbo, N., Long, C. S. and Karliner, J. S. (2000). Hypoxia differentially regulates stress proteins in cultured cardiomyocytes: role of the p38 stress-activated kinase signaling cascade, and relation to cyroprotection. *Cardiovasc. Res.* **46**: 139–150.
73. Bertrand, M. J., Deramaudt, M., Remy, P. and Abraham, N. G. (1999). Upregulation of human heme oxygenase gene expression by Ets-family proteins. *J. Cell. Biochem.* **72**: 311–321.

Chapter 42

Metallothioneins

Y. James Kang

Y. James Kang • Department of Medicine, University of Louisville School of Medicine, 530 S. Jackson St, 3rd Floor, Ambulatory Care Building, Louisville, KY 40292
Tel: (502) 852-8677, E-mail: yjkang01@athena.louisville.edu

1. Introduction

Metallothioneins (MTs) are low-molecular weight, cysteine-rich, metal-binding proteins. These proteins have been studied for more than forty years since the discovery of the first cadmium-binding MT from horse kidney in 1957.[1] MTs are found in all eukaryotes as well as some prokaryotes.[2] MTs have been divided into three classes based on their structural similarities: classes I, II and III. Mammalians contain all of the three isoforms of MTs with a species-specific and tissue-specific expression of a particular isoform. For example, MT-I is highly expressed in rodents and MT-II is in humans. Both MT-I and –II are found in multiple organ systems, but MT-III is highly localized in the brain and some in the kidney.[3] It has been well known that MTs play an important role in detoxification of heavy metals.[4, 5] Other biological function of MTs is likely related to the physiologically relevant metals that these proteins bind. In mammals, MTs are found to bind zinc and copper under normal physiological conditions. Both zinc and copper are trace metals that are essential for life. MTs function as a metal chaperone for the regulation of gene expression, and for synthesis and functional activity of proteins such as metalloproteins and metal-dependent transcription factors.[6–8]

That MTs would function in protection against aging process is suggested from the pivotal role of oxidative stress in aging and the antioxidant function of MTs. Because Sec. 10 of this volume is exclusively designated to discuss aging and oxidative stress, the antioxidant action of MT will be the focus of this chapter.

2. Antioxidant Action of MTs

The hypothesis that MTs function as an antioxidant against reactive oxygen and nitrogen species has received extensive experimental support from *in vitro* studies. Studies using a cell-free system have demonstrated the ability of MTs as a free radical scavenger.[9–11] However, these results have not been demonstrated in intact animal studies and most *in vivo* experimental data provide only indirect evidence for the free radical scavenging action of MTs. Nevertheless, many *in vivo* studies indicate that MTs indeed provide protection against oxidative injury in multiple organ systems.

2.1. Evidence Obtained from *In Vitro* Studies

That MTs function as an antioxidant was first suggested in a study examining the effect of MTs on radiosensitivity of cultured human epithelial (HE) cell line and mouse fibroblast (C1 1D) cells.[12] Following this study, a detailed examination of the kinetics and mechanism of reaction of MTs with superoxide and hydroxyl radicals was undertaken.[9] An MT containing zinc and/or cadmium was shown to scavenge hydroxyl and superoxide radicals produced by the xanthine/xanthine oxidase reaction system.[9] The data suggested that all 20 cysteine sulfur atoms are involved in the radical quenching process and MTs appear to be an extraordinarily efficient hydroxyl radical scavenger.[9] The rate constant for the reaction of hydroxyl radical with MTs ($K_{\cdot OH/MT}$ = 2700) is about 340-fold higher than that with GSH ($K_{\cdot OH/GSH}$ = 8).[9] Further studies have shown that MTs are about 800-fold more potent than GSH (on a molar basis) in preventing hydroxyl radical-generated DNA degradation *in vitro*.[10]

2.2. Evidence Obtained from Studies Using Cultured Cells

Studies using cultured cells have provided further evidence to support the antioxidant function of MTs. Erythrocyte ghosts incubated with xanthine/xanthine oxidase/Fe(III) underwent hydrogen peroxide- and superoxide-dependent lipid peroxidation.[11] Both cadmium- and zinc-MT strongly inhibited lipid peroxidation when present during incubation.[11] A study using HL-60 cells has demonstrated a direct reaction of hydrogen peroxide with the sulfhydryl groups of MTs.[13] The thiolate groups in the MT were shown to be the preferred targets attacked by hydrogen peroxide compared to sulfhydryl residues from GSH and other protein fractions.[13] In another study, transfection of NIH 3T3 cells with a plasmid containing mouse MT-I gene yielded a 4-fold increase in intracellular MT-I, which was localized in the cytoplasm as determined by immunofluorescence and confocal microscopy.[14] These cells were six times more resistant to the cytotoxic effects of tert-butyl hydroperoxide, whose toxicity is mediated by free radicals, relative to control transfectants (with an inverted MT-I gene that resulted in no increase in cellular MT concentrations).[14] These MT-overexpressing NIH 3T3 cells were also 4-fold more resistant to the cytotoxicity of the NO donor S-nitroso-N-acetylpenicillamine.[15] Taken together, these studies using cultured cells have demonstrated that MT protects cells from all of the toxicologically significant reactive oxygen and nitrogen species.

2.3. Evidence Obtained from Intact Animal Studies

Many studies were undertaken to determine whether or not MT protects against oxidative injury in intact animals. Prior to the use of genetically engineered mice, the bulk of these studies involved various MT inducers. Although these inducer studies have been informative, it should be noted that all of the agents used to induce MTs were all pleiotropic, causing a panoply of biological responses. Induction of MTs in rats by cadmium, zinc, alpha-hederin or lipopolysaccharide has been shown to increase hepatic resistance to oxidative stress.[16] Pretreatment of rat with tumor necrosis factor or interleukin-6 induced MT synthesis and prevented liver damage and lipid peroxidation caused by carbon tetrachloride.[17] Zinc pretreatment significantly increased MT concentrations in the renal cortex and depressed proximal tubule necrosis and acute renal failure caused by injection of gentamicin,[18] whose nephrotoxicity has been shown to be mediated by hydroxyl radical and MTs presumably scavenge hydroxyl radicals.[19]

Rats exposed to cadmium aerosol displayed an increase in MT content in alveolar epithelial Type-II cells. When isolated from these animals, the cells were found to suffer much less from oxidant-induced cytotoxicity than controls.[20] Pulmonary MTs were also increased by bismuth or zinc in mice, and this MT induction correlated with suppression of carcinogenesis by *cis*-platinum or melphalan in the lung.[21] Induction of MTs in the lungs of mice by zinc was also found to inhibit paraquat-induced pulmonary lipid peroxidation.[22] Paraquat lethality from intratracheal installation was also significantly decreased in mice pretreated with zinc; this correlated with MT levels in the lung, but not with MT contents in the liver or kidney.[22] Recent studies have been extended to the function of MTs in the brain.[23] It was shown that MT-III is predominantly expressed in zinc-containing neurons in the brain and is particularly abundant in the hippocampus. Thus, MT-III is likely to play an important role in neuromodulation by zinc-containing neurons.[24] Injection of zinc or copper intracerebroventricularly in rats increased MT levels in some areas of the brain,[25] and 6-hydroxydopamine, a free radical generator, increased MT contents in brain cells.[26]

Several studies have shown that the inducing agents can also induce MT expression in the heart.[27–29] However, the significance of this induction in protection against oxidative stress in the heart had not been addressed until recently. The state-of-the-art molecular approaches have provided powerful tools to elucidate the role of MTs in myocardial protection against oxidative stress.

2.4. Direct Evidence Obtained from Genetically Altered Mouse Models

In cancer chemotherapy, doxorubicin (DOX) is widely used and a most effective anticancer agent. However, it causes a severe cardiotoxicity, which limits its clinical application. The mechanism for DOX-induced cardiotoxicity likely involves reactive oxygen species accumulation due to DOX metabolism. To address a possible role of MTs in cardiac protection against DOX toxicity, we have developed a cardiac-specific, MT-overexpressing, transgenic mouse model.[30] The MT was constitutively overexpressed in the heart only, and other antioxidant components including GSH, GSHpx, GSH reductase, catalase, and SOD were not altered in the MT-overexpressing heart. Using this unique experimental model we have demonstrated that MTs provide protection against DOX-induced cardiomyopathy as examined by light and electron microscopy and biochemical markers and functional alterations.[30] To further determine unequivocally the role of MTs in cardiac protection against DOX, we have established a primary neonatal mouse cardiomyocyte culturing system.[31] As compared to non-transgenic controls, transgenic cardiomyocytes displayed a marked resistance to DOX toxicity, as measured by morphological alterations, cell viability, mitochondrial malmetabolism and lactate dehydrogenase leakage.[32] This cytoprotective effect of MTs correlated with inhibition of DOX-induced lipid peroxidation.[32] It was further demonstrated that MTs prevent DOX-induced apoptosis in the myocardium through inhibition of DOX-activated p38 MAPK pathway.[33]

Myocardial damage induced by ischemia-reperfusion of the heart has been proposed to be caused, at least in part, by the generation of reactive oxygen species.[34, 35] However, direct evidence to support the role of free radicals in this myocardial injury has not been obtained. We applied a Langendorff perfusion model to examine directly the effects of MTs on ischemia-reperfusion-induced derangement of contractile activity of the heart, myocyte injury as estimated by creatine phosphokinase release, and cell death as measured by the size of infarct zone. The transgenic hearts showed significantly better post-ischemic recovery of the suppressed contractile force.[36] Upon reperfusion, a high creatine phosphokinase activity was detected in the effluent collected from non-transgenic mouse hearts. A much lower activity was detected in the effluent collected from the transgenic mouse hearts; the peak value was about 1/3 of that from controls.[36] The zone of infarction induced by a 90-min reperfusion following 50-min of ischemia was suppressed by about 40% in the transgenic hearts.[36] These observations were further confirmed in an *in vivo* open chest model of cardiac ischemia-reperfusion in mice.[37]

Based on these observations, we speculate that MTs may be useful in the prevention of oxidative heart injury. The regulation of MT expression has been well studied and several agents have been identified to selectively elevate MT levels in the heart, such as bismuth subnitrate,[22] tumor necrosis factor-alpha,[28] and isoproterenol.[29] Therefore, the basis for developing pharmaceutical agents to increase MT concentrations in the heart already exists. Exploring the potential for MT to protect against cardiac oxidative injury would likely result in novel approaches to the heart disease and could positively influence clinical outcomes.

3. Possible Mechanisms of Antioxidant Action of MTs

Studies *in vitro* have demonstrated that MTs directly react with all reactive oxygen species.[9–11] However, it is questionable whether these *in vitro* observations are applicable to the *in vivo* action of MTs. In particular, all of the reactive oxygen species, especially hydroxyl radicals, are very reactive and have an extremely short half-life. It is speculated that MTs can only be effective as a free radical scavenger *in vivo* if it is located sufficiently close to the site of production of the radicals to interact with them before their reaction with other cellular components. Depending on the local concentrations of MTs, this may predict that the direct interaction between MTs and the radicals as a major mechanism of action *in vivo* is impracticable.

It has been demonstrated that the cluster structure of Zn-MT provides a chemical basis by which the cysteine ligands can induce oxidoreductive properties.[38] This structure allows for thermodynamic stability of zinc in MT while permitting zinc to retain kinetic lability. This is demonstrated by the fast zinc exchange between MT isoforms,[8] between MTs and the zinc cluster in the Gal4 transcription factor,[8] and between MTs and the apoforms of various zinc proteins.[7, 39] Importantly, zinc release from MTs is modulated by both GSH and GSH disulfide (GSSG).[39–41] GSH inhibits zinc release in the absence of GSSG, indicating that MT is stabilized at relative high cellular GSH concentrations. The presence of GSSG, or any other oxidizing agent, results in a release of zinc that is synergistically increased by GSH. The rate of zinc release depends linearly on the amount of GSSG, i.e. the more oxidative the redox state becomes, the more efficiently zinc is released from MTs.[7]

Mobilization of zinc from MTs by an oxidative reaction may either constitute a general pathway by which zinc is distributed in the cell or it may be restricted to conditions of stress where zinc is needed in antioxidant defense systems.[40] It has been argued that the primary determinant of MT protection against oxidative stress is the release of zinc sequestered by MTs and its subsequent uptake by plasma membranes, since zinc protects against lipid peroxidation and thereby stabilizes membranes.[11, 42] In addition, released zinc may suppress lipid peroxidation by affecting many different cellular functions, such as decreasing iron uptake and inhibiting NADPH-cytochrome c-reductase.[43]

If the reaction between Zn-MT and disulfides triggers a mechanism of Zn release and the protection by MTs against oxidative injury is mediated by the released Zn, a dynamic change in the level of Zn bound to MTs during oxidative stress would occur. In conjunction with this change, MTs would become oxidized. These events have not been observed in intact animals, and experiments to test this idea are not straightforward. However, it is crucial to elucidate these mechanisms in order to understand the antioxidant action of MTs *in vivo*. With the advances in technology and recently developed state-of-the-art approaches, novel insights into the biological function of this unusual and ubiquitous protein are on the horizon.

4. MTs and Aging

There are not many studies that specifically address the relationship between MTs and aging. Because oxidative stress is critically involved in aging process and MTs are potent antioxidants, it is reasonable to believe that MTs function in preventing aging. There were evidences that decreased production of MTs in older animals was closely related to enhanced sensitivity to toxic insults. In a study examining nephrotoxicity following a single injection of *cis*-platin between rats aged 52 weeks and 9 weeks, it was found that the level of renal MTs in young animals increased 3 days after the injection, but decreased remarkably in older animals. Morphological evaluation revealed that degeneration of the proximal tubules in the older animals occurred.[44] In a study to investigate possible role of Zn-MT in human pigment epithelium with regard to age-related changes, it was found that macular retinal pigment epithelium cells in donors younger than 70 years old contained 3 folds more MTs than those in donors older than 70 years old.[45] It is worthwhile to note that it is important to measure dynamic changes in MT concentrations in tissues should a comparison be made between younger and aging subjects. For example, after prepuberal and adult male rats were chronically stressed for a month with several acute stressors in a random schedule, there were no changes in hepatic MTs in either young or adult rats. However, a significant increase in serum MT concentrations was observed in the young rats.[46]

Age-related detrimental alterations have been identified in multiple organ systems including myocardial tissues. We have used the cardiac-specific, MT-overexpressing, transgenic mice to test the hypothesis that MT inhibits the age-related myocardial degeneration. Morphological changes in myocardial tissues were compared between transgenic mice and non-transgenic controls at age of 13 months. Degenerative ultrastructual changes, in particular, myocardial apoptosis were observed in non-transgenic mice. In addition, mitochondrial cytochrome *c* release and activation of caspase-3 were also detected. These changes were all dramatically diminished in the transgenic mouse hearts, demonstrating that MT inhibits age-related myocardial degeneration through inhibition of cytochrome c-dependent apoptotic pathway.[47]

MT-III is a brain-specific member of the MT family.[3] Its absence has been implicated in the development of Alzheimer's disease.[24, 48, 49] To address the roles of MT-III in brain physiology and pathophysiology, an MT-III knock-out mouse model was generated.[50] MT-III-deficient mice had decreased concentrations of zinc in several brain regions. The age-related glial fibrillary acidic protein expression was increased, although no neuropathology or behavioral deficits were detected in 2-year-old MT-III-deficient mice. Moreover, MT-III-deficient mice were more susceptible to seizures induced by kainic acid and subsequently exhibited greater neuron injury. This study thus suggested a potential role of MT-III in regulating neural response to stimulation and thereby preventing neural degeneration.

5. Conclusion

MTs likely function in preventing aging process because oxidative stress plays a pivotal role in aging and MTs are a group of proteins that protect from oxidative injury. The antioxidant function of MTs was suggested in the early 1980s. Studies *in vitro* have indeed demonstrated a direct reaction between MTs and reactive oxygen species. However, it has been debated whether MTs function as an antioxidant *in vivo* because the *in vitro* action of MTs as a free radical scavenger has never been demonstrated in intact animal studies. In the last few years, studies using cardiac-specific, MT-overexpressing, transgenic mouse models have produced direct evidence to support the antioxidant and protective function of MTs from oxidative injury in the heart. The supporting data gathered from diverse experimental settings, including both acute and chronic oxidative stress conditions, are compelling. However, the debate concerning the antioxidant function of MTs *in vivo* cannot be settled until a comprehensive understanding of the mechanism of action of MTs is obtained. With the advances in molecular biotechnology and an understanding of the zinc cluster structure of MTs, it is foreseeable that novel insights into this problem are forthcoming.

Acknowledgments

The author is a university scholar of the University of Louisville. The research work cited in this review is currently supported in part by NIH grants CA68125 and HL59225, an Established Investigator Award from the American Heart Association (9640091N), and a grant from the Jewish Hospital Foundation, Louisville, Kentucky. The author thanks Dr. Lu Cai for helping preparation of this chapter.

References

1. Margoshes, M. and Vallee, B. L. (1957). A cadmium protein from equine kidney cortex. *J. Am. Chem. Soc.* **79:** 1813–18148.
2. Webb, M. and Cain, K. (1982). Functions of metallothionein. *Biochem. Pharmacol.* **31:** 137–142.

3. Davis, S. R. and Cousins, R. J. (2000). Metallothionein expression in animals: a physiological perspective on function. *J. Nutri.* **130:** 1085–1088.

4. Klaassen, C. D., Liu, J. and Choudhuri, S. (1999). Metallothionein: an intracellular protein to protect against cadmium toxicity. *Ann. Rev. Pharmacol. Toxicol.* **39:** 267–294.

5. Templeton, D. M. and Cherian, M. G. (1991). Toxicological significance of metallothionein. *Meth. Enzymol.* **205:** 11–24.

6. Zeng, J., Vallee, B. L. and Kagi, J. H. (1991). Zinc transfer from transcription factor IIIA to thionein clusters. *Proc. Natl. Acad. Sci. USA* **88:** 9984–9988.

7. Jacob, C., Maret, W. and Vallee, B. L. (1998). Control of zinc transfer between thionein, metallothionein, and zinc proteins. *Proc. Natl. Acad. Sci. USA* **95:** 3489–3494.

8. Maret, W., Larsen, K. S. and Vallee, B. L. (1997). Coordination dynamics of biological zinc "clusters" in metallothioneins and in the DNA-binding domain of the transcription factor Gal4. *Proc. Natl. Acad. Sci. USA* **94:** 2233–2237.

9. Thornalley, P. J. and Vasak, M. (1985). Possible role for metallothionein in protection against radiation-induced oxidative stress. Kinetics and mechanism of its reaction with superoxide and hydroxyl radicals. *Biochim. Biophys. Acta* **827:** 36–44.

10. Abel, J. and de Ruiter, N. (1989). Inhibition of hydroxyl-radical-generated DNA degradation by metallothionein. *Toxicol. Lett.* **47:** 191–196.

11. Thomas, J. P., Bachowski, G. J. and Girotti, A. W. (1986). Inhibition of cell membrane lipid peroxidation by cadmium- and zinc-metallothioneins. *Biochim. Biophys. Acta* **884:** 448–461.

12. Bakka, A., Johnsen, A. S., Endresen, L. and Rugstad, H. E. (1982). Radio-resistance in cells with high content of metallothionein. *Experientia* **38:** 381–383.

13. Quesada, A. R., Byrnes, R. W., Krezoski, S. O. and Petering, D. H. (1996). Direct reaction of H_2O_2 with sulfhydryl groups in HL-60 cells: zinc-metallothionein and other sites. *Arch. Biochem. Biophys.* **334:** 241–250.

14. Schwarz, M. A., Lazo, J. S., Yalowich, J. C., Reynolds, I., Kagan, V. E., Tyurin, V., Kim, Y. M, Watkins, S. C. and Pitt, B. R. (1994). Cytoplasmic metallothionein overexpression protects NIH 3T3 cells from tert-butyl hydroperoxide toxicity. *J. Biol. Chem.* **269:** 15 238–15 243.

15. Schwarz, M. A., Lazo, J. S., Yalowich, J. C., Allen, W. P., Whitmore, M., Bergonia, H. A., Tzeng, E., Billiar, T. R., Robbins, P. D. and Lancaster, J. R. J. (1995). Metallothionein protects against the cytotoxic and DNA-damaging effects of nitric oxide. *Proc. Natl. Acad. Sci. USA* **92:** 4452–4456.

16. Iszard, M. B., Liu, J. and Klaassen, C. D. (1995). Effect of several metallothionein inducers on oxidative stress defense mechanisms in rats. *Toxicology* **104:** 25–33.

17. Sato, M., Sasaki, M. and Hojo, H. (1995). Antioxidative roles of metallothionein and manganese superoxide dismutase induced by tumor necrosis factor-alpha and interleukin-6. *Arch. Biochem. Biophys.* **316:** 738–744.
18. Du, X. H. and Yang, C. L (1994). Mechanism of gentamicin nephrotoxicity in rats and the protective effect of zinc-induced metallothionein synthesis. *Nephrol. Dial. Transplant.* **9(Suppl. 4):** 135–140.
19. Yang, C. L., Du, X. H., Zhao, J. H., Chen, W. and Han, Y. X. (1994). Zinc-induced metallothionein synthesis could protect from gentamicin nephrotoxicity in suspended proximal tubules of rats. *Renal. Fail.* **16:** 61–69.
20. Hart, B. A., Eneman, J. D., Gong, Q. and Durieux-Lu, C. C. (1995). Increased oxidant resistance of alveolar epithelial Type-II cells. Isolated from rats following repeated exposure to cadmium aerosols. *Toxicol. Lett.* **81:** 131–139.
21. Satoh, M., Kondo, Y., Mita, M., Nakagawa, I., Naganuma, A. and Imura, N. (1993). Prevention of carcinogenicity of anticancer drugs by metallothionein induction. *Cancer Res.* **53:** 4767–4768.
22. Satoh, M., Naganuma, A. and Imura, N. (1992). Effect of preinduction of metallothionein on paraquat toxicity in mice. *Arch. Toxicol.* **66:** 145–148.
23. Ebadi, M., Iversen, P. L., Hao, R., Cerutis, D. R., Rojas, P., Happe, H. K., Murrin, L. C. and Pfeiffer, R. F. (1995). Expression and regulation of brain metallothionein. *Neurochem. Int.* **27:** 1–22.
24. Aschner, M. (1996). The functional significance of brain metallothioneins. *FASEB J.* **10:** 1129–1136.
25. Gasull, T., Giralt, M., Hernandez, J., Martinez, P., Bremner, I. and Hidalgo, J. (1994). Regulation of metallothionein concentrations in rat brain: effect of glucocorticoids, zinc, copper, and endotoxin. *Am. J. Physiol.* **266:** E760–E767.
26. Shiraga, H., Pfeiffer, R. F. and Ebadi, M. (1993). The effects of 6-hydroxydopamine and oxidative stress on the level of brain metallothionein. *Neurochem. Int.* **23:** 561–566.
27. Sharma, G., Nath, R. and Gill, K. D. (1992). Effect of ethanol on the distribution of cadmium between the cadmium metallothionein and non-metallothionein-bound cadmium pools in cadmium-exposed rats. *Toxicology* **72:** 251–263.
28. Sato, M., Sasaki, M. and Hojo, H. (1992). Tissue specific induction of metallothionein synthesis by tumor necrosis factor-alpha. *Res. Commun. Chem. Pathol. Pharmacol.* **75:** 159–172.
29. Namikawa, K., Okazaki, Y., Nishida, S., Kimoto, S., Akai, F., Tomura, T. and Hashimoto, S. (1993). Changes in myocardial metallothionein on isoproterenol-induced myocardial injury. *Yakugaku. Zasshi.* **113:** 591–595.
30. Kang, Y. J., Chen, Y., Yu, A., Voss-McCowan, M. and Epstein, P. N. (1997). Overexpression of metallothionein in the heart of transgenic mice suppresses doxorubicin cardiotoxicity. *J. Clin. Invest.* **100:** 1501–1506.

31. Wang, G.-W., Schuschke, D. A. and Kang, Y. J. (1999). Metallothionein-overexpressing neonatal mouse cardiomyocytes are resistant to H2O2 toxicity. *Am. J. Physiol.* **276:** H167–H175.

32. Wang, G.-W. and Kang, Y. J. (1999). Inhibition of doxorubicin toxicity in cultured neonatal mouse cardiomyocytes with elevated metallothionein levels. *J. Pharmacol. Exp. Ther.* **288:** 938–944.

33. Kang, Y. J., Zhou, Z.-X., Wang, G.-W., Buridi, A. and Klein, J. B. (2000). Suppression by metallothionein of doxorubicin-induced cardiomyocyte apoptosis through inhibition of p38 mitogen-activated protein kinases. *J. Biol. Chem.* **275:** 13 690–13 698.

34. Flitter, W. D. (1993). Free radicals and myocardial reperfusion injury. *Br. Med. Bull.* **49:** 545–555.

35. Steare, S. E. and Yellon, D. M. (1995). The potential for endogenous myocardial antioxidants to protect the myocardium against ischaemia-reperfusion injury: refreshing the parts exogenous antioxidants cannot reach? *J. Mol. Cell. Cardiol.* **27:** 65–74.

36. Kang, Y. J., Li, G. and Saari, J. T. (1999). Metallothionein inhibits ischemia-reperfusion injury in mouse heart. *Am. J. Physiol.* **276:** H993–H997.

37. Kang, Y. J. and Wang, J.-F. (1998). Cardiac protection by metallothionein against ischemia-reperfusion injury and its possible relation to ischemic preconditioning. *In* "Metallothionein IV" (C. D. Klaassen, Ed.), pp. 511–516, Birkhauser, Basel, Boston, Berlin.

38. Maret, W. and Vallee, B. L. (1998). Thiolate ligands in metallothionein confer redox activity on zinc clusters. *Proc. Natl. Acad. Sci. USA* **95:** 3478–3482.

39. Jiang, L. J., Maret, W. and Vallee, B. L. (1998). The glutathione redox couple modulates zinc transfer from metallothionein to zinc-depleted sorbitol dehydrogenase. *Proc. Natl. Acad. Sci. USA* **95:** 3483–3488.

40. Maret, W. (1995). Metallothionein/disulfide interactions, oxidative stress, and the mobilization of cellular zinc. *Neurochem. Int.* **27:** 111–117.

41. Maret, W. (1994). Oxidative metal release from metallothionein via zinc-thiol/disulfide interchange. *Proc. Natl. Acad. Sci. USA* **91:** 237–241.

42. Chvapil, M., Ryan, J. N. and Zukoski, C. F. (1972). Effect of zinc on lipid peroxidation in liver microsomes and mitochondria. *Proc. Soc. Exp. Biol. Med.* **141:** 150–153.

43. Coppen, D. E., Richardson, D. E. and Cousins, R. J. (1988). Zinc suppression of free radicals induced in cultures of rat hepatocytes by iron, t-butyl hydroperoxide, and 3-methylindole. *Proc. Soc. Exp. Biol. Med.* **189:** 100–109.

44. Namikawa, K., Kinsoku, A., Minami, T., Okazaki, Y., Kadota, E., Teramura, K. and Hashimoto, S. (1995). Relationship between age and nephrotoxicity following single low-dose cisplatin (CDDP) injection in rats. *Biol. Pharm. Bull.* **18:** 957–962.

45. Tate, D. J., Jr., Newsome, D. A. and Oliver, P. D. (1993). Metallothionein shows an age-related decrease in human macular retinal pigment epithelium. *Invest. Ophthalmol. Vis. Sci.* **34:** 2348–2351.
46. Armario, A., Hidalgo, J., Bas, J., Restrepo, C., Dingman, A. and Garvey, J. S. (1987). Age-dependent effects of acute and chronic intermittent stresses on serum metallothionein. *Physiol. Behav.* **39:** 277–279.
47. Zhou, Z.-X. and Kang, Y. J. (1999). Age-related myocardial apoptosis is inhibited by metallothionein in mouse. *Free Radic. Biol. Med.* **27(Suppl. 1):** S47.
48. Uchida, Y., Takio, K., Tutani, K., Ihara, Y. and Tomonaga, M. (1991). The growth inhibitory factor that is deficient in the Alzheimer's disease brain is a 68 amino acid metallothionein-like protein. *Neuron* **7:** 337–347.
49. Vallee, B. L. (1995). The function of metallothionein. *Neurochem. Int.* **27:** 23–33.
50. Erickson, J. C., Hollopeter, G., Thomas, S. A., Froelick, G. J. and Palmiter, R. D. (1997). Disruption of the metallothionein-III gene in mice: analysis of brain zinc, behavior, and neuron vulnerability to metals, aging, and seizures. *J. Neurosci.* **15:** 1271–1281.

Chapter 43

Human Genomic DNA Repair Enzymes

Timothy R. O'Connor*

Timothy R. O'Connor • Department of Biology, Beckman Research Institute of the City of Hope, Duarte, California 91010 USA
Tel: 626-301-8220, E-mail: toconnor@coh.org
*On leave from *UMR1772 Physicochimie et Pharmacologie des Macromolecules Biologiques*

1. Introduction

Cells are constantly bombarded by insults from both exogenous and endogenous agents.[1,2] As an initial defense mechanism, layers of antioxidants and enzymes protect DNA from damage (see chapters on antioxidants). Despite the enormous cellular effort expended by these systems to eliminate

Fig. 1. Theory of somatic cell mutation.

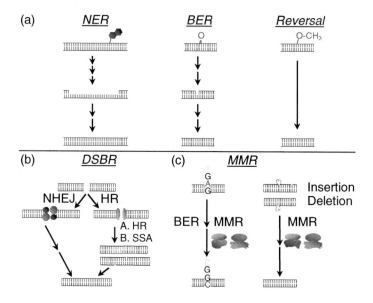

Fig. 2. General strategies used by cells to repair DNA damage. There are some differences depending on the organism, but these general strategies are ubiquitous. (a) Nucleotide, base, and reversal of DNA damage mechanisms. Nucleotide excision repair (NER) involves the removal of an oligodeoyxribonucleotide from DNA as for some bulky adducts. Base excision repair (BER) removes a base from DNA as for the 8-oxo-G base removal, and reversal of DNA damage directly repairs a damaged base without another repair intermediate as for the O6-meG base. (b) Double strand break repair mechanisms. Two mechanisms are possible for double strand break repair (DSBR). In human cells, the major pathway for DSBR is by non-homologous end-joining (NHEJ). Homologous recombination (HR) is a minor pathway in human cells, but the major pathway in *S. cerevisiae*. (c) Mechanisms for repair of mispaired bases. Cells use both BER and a system of mismatch repair (MMR) to correct mispaired bases. The MMR system repairs mismatched DNA bases differently than insertion and deletion mutations.

exposure of DNA to reactive agents, however, oxidative DNA damage occurs. The persistence of these adducts leads to the accumulation of mutations following DNA synthesis. If the adducts are formed in critical growth control genes, deleterious mutations that alter gene product function can accumulate. Subsequent selection of harmful mutations in other critical genes can trigger cellular transformation (Fig. 1). Therefore, to shield against DNA damage, cells have evolved DNA repair systems to maintain genetic stability.[3] These systems include base, nucleotide, double-strand break, and mismatch repair. This chapter will provide an introduction to the systems that insulate DNA from mutation and cell death. Probably the most striking observation about DNA repair is that the strategies for the elimination of damage from genetic material have been conserved during evolution; the same basic mechanisms found in *S. cerevisiae*, baker's yeast, are also found in human cells (Fig. 2).

Table 1. Genes Whose Products are Implicated in DNA Repair Processes

Nucleotide Excision Repair (NER)		Base Excision Repair (BER)	
hXPC	hXPB	hMBD4	hPOLβ
hHR23B	hXPD	hMPG	hPOLδ
hDDB1	hXPG	hNTH	hPOLε
hDDB2	hXPF	hOGG1	hFEN1
hXPA	hERCC1	hTDG	hPCNA
hRPAp14, 32, 70	hPOLδ	hUDG	hXRCC1
hCSA	hPOLε	hXPG	hPARP
hCSB	hPCNA	hAPE1	hLIG1
hTFIIHp34, 44, 52	hLIG1	BRCA1	hLIG3
hCDK7-B		hRPAp14, 32, 70	
hRFCp36, 37, 38, 40B, 140			

Double-Strand Break Repair (DSBR)			Mismatch Repair (MMR)		
Non-Homologous End Joining (NHEJ)	Homologous Recombination (HR)				
hKu70	hRAD51	POLβ (?)	hMSH1	Exonuclease I	hOGG1
hKu80	hRAD52	XRCC2	hMSH2	FEN1	hMYH
hDNA-PKcs	hRAD54	XRCC3	hMSH3	DNA POLβ	hMBD4
POLβ (?)	hRAD55	BRCA1	hMSH5	DNA POLε	hTDG
hXRCC4	hRAD57	BRCA2	hMSH6	hRPAp14, 32, 70	hPOLβ
hLIG4	hRAD50	hMSH2	hPMS1	hRFCp36, 37, 38, 40B, 140	hPOLδ
hRAD50	hMRE11	HMSH3	hPMS2	PCNA	hPOLε
hMRE11	hXRS2		hMLH1	hRPAp14, 32, 70	hPCNA
	(NBS1)		hMLH3		hXRCC1
hXRS2 (NBS1)			POLδ		hPARP
					hLIG1
					hLIG3

2. DNA Repair Systems

This chapter will present an overview of the various DNA repair systems and human diseases associated with DNA repair. For further reading, you are referred to recent reviews on the individual repair pathways, the other chapter in this book that addresses DNA repair in mitochondria, and several recent books dedicated to DNA repair and mutagenesis.[4-7] For repairing oxidative or ionizing radiation damage to DNA, cells use virtually all DNA repair mechanisms, with the possible of exception of reversal of DNA damage. Only after understanding how these individual systems work can we better comprehend how DNA repair systems function together in a cell to effect repair and preserve genetic stability. Figure 2 outlines the strategies for each type of repair system, and Table 1 lists the different human gene products known to function in each pathway.

3. DNA Polymerases, a Central Role in DNA Repair

Prior to discussion of the various DNA repair systems, it is useful to discuss DNA polymerases because these enzymes play central roles in DNA metabolism including DNA replication, repair, and recombination. Understanding the function of DNA polymerases leads to a comprehension of evolutionary development as well as the etiology of cancer, since all the mutations that develop in cells are the result of DNA polymerase errors.[8-26]

The DNA polymerases found in human cells and their main functions are listed in Table 2. The DNA polymerases used during replication include POLα, POLδ, and POLε. The process of replication leading to cell division is characterized by an extremely low frequency of errors and requires numerous cofactors. The fidelity of a DNA polymerase refers to the probability of the insertion of an incorrect base opposite the template base. Replication with high fidelity, however, does not eliminate all the errors. The replication process has significant proofreading capacity that limits the errors placed in DNA, but repair processes often are effected by DNA polymerases with lower fidelity. Therefore, cells use high fidelity DNA polymerases and a number of auxiliary proteins to replicate chromosomes. However, high fidelity polymerases are generally blocked by DNA adducts. In an effort to complete synthesis, cells have evolved other DNA polymerases that perform trans-lesion synthesis (TLS) to assist in the completion of replication. However, not all TLS DNA polymerases reduce the mutation frequency in cells. For example, hsREV3p part of POLζ, one of the TLS DNA polymerases, increases mutation frequency.[27] Recently, structural data on DNA polymerases have provided an intimate view of how DNA polymerases function that leads to a corresponding knowledge about how errors can arise during DNA synthesis.[28-34] This in turn will permit us to determine the contribution of misreplication due to the ensemble of DNA polymerases found in human cells (Fig. 1).

Table 2. Human DNA Polymerases

POL	Function	Subunit kDa	Fidelity	Reference
α	Initiation of replication, DSBR, Telomere length, cell cycle regulation	165 67 58 48	High	18
β	BER, DSBR, meiosis	39	Low	17, 18
γ	Mitochondrial replication	125 35	High	18
δ	Leading/lagging strand replication, DSBR, BER, NER, cell cycle regulation	125 66 50	High	10, 17, 18
ε	Lagging strand replication, DSBR, BER, NER, cell cycle regulation	261 59	High	17, 18
ζ	Error-prone TLS (Rev3p•Rev7p)	353	Low	7–9, 18
η	Error-free TLS XPV (Rad30Ap)	78	High	7–9, 18
θ	Repair of interstrand crosslinks	198	?	19, 20
ι	Error-prone TLS (Rad30Bp)	138	Low	18, 24
κ	Error-free TLS (dinB *E. coli*, Rad30p) (?)	99	High	12, 21
λ	Sister chromatid cohesion (Trf4p)	60	Low (?)	22
hREV1	Error-prone TLS (Rev1p)	138	Low	9, 18, 23

Other names for the homologous DNA polymerases in yeast are indicated in parentheses. There is some controversy about the names of DNA polymerases θ, κ, and λ that will need to be addressed by the appropriate international committees. XPV is the Xeroderma pigmentosum variant complementation group.

4. Nucleotide Excision Repair (NER)

The nucleotide excision repair (NER) system is one form of excision repair that is mainly responsible for the removal of bulky adducts from DNA. Despite its role in the excision of such adducts, this system also repairs some types of DNA damage resulting from oxidizing agents.[2, 35–42] NER is slightly different depending on the adduct and its position in chromatin. If an adduct is in a RNA polymerase II transcribed region and on the transcribed strand, repair is generally effected by transcription-coupled repair (TCR) (Fig. 3),[43–50] otherwise NER is performed by the global genome repair (GGR) mechanism (Fig. 3).[43] Some adducts are not subject to TCR even if in the appropriate position. In addition to monofunctional adducts, NER also has a role in the removal of DNA cross-links,[38, 51, 52] but most

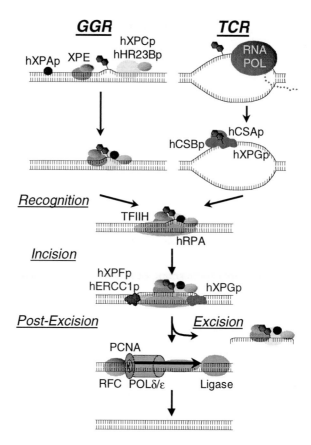

Fig. 3. Model for the excision repair of adducts by the human NER system in both non-transcribed (GGR-global genome repair) and RNA polymerase II-transcribed (TCR-transcription coupled repair) regions. The first step in NER is the recognition of the damaged base. In GGR this can involve hXPAp, hXPE (hDDB1p•hDDB2p), and hXPCp•hHR23Bp. In TCR, RNA Polymerase II, hCSAp, hCSBp, and hXPGp can all play a role in damage recognition. The TFIH complex with hRPA proteins then binds to the damaged region and opens the region. The hXPGp endonuclease incises 3' to the damaged site followed by cleavage by the hXPFp•hERCC1p endonuclease. Helicase activity removes the damaged oligodeoxyribonucleotide (16–29 nucleotides), and resynthesis is performed in a PCNAp dependent manner. Repair is completed by DNA Ligase I catalyzed sealing of the nicked DNA.

recently, cross-link repair by NER has been shown to be futile in mammalian cell extracts that could explain in part the toxicity of such lesions.[51] In NER, an oligodeoxyribonucleotide of 16–27 bases containing the damaged base is eliminated from DNA. The pathway is broken down into the recognition, incision, excision, and post-excision steps.[37, 43, 48–50, 53–56] In the recognition step, several important proteins identify the damaged base, and bind to the site. The incision step of NER is catalyzed by two endonucleases; the first cleavage at the 3' side by XPGp is followed by XPFp•ERCC1p cleavage on the 5'side. Helicase activity excises the oligodeoxyribonucleotide, and the post-excision steps are performed by POLδ or ε and DNA ligase I.

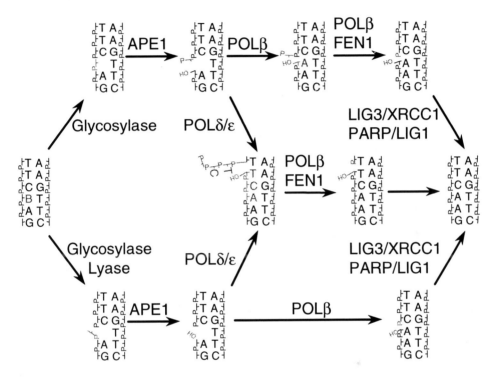

Fig. 4. Model for the excision repair of adducts by the human BER system. The upper pathway is directed by a DNA glycosylase without an AP-lyase activity. Base excision is followed by AP endonuclease incision leaving a 5'deoxyribose phosphate. A base is inserted at the site by DNA POLβ displacing the 5'deoxyribose phosphate. The 5'deoxyribose phosphate is removed by a lyase activity associated with DNA POLβ. Finally, a complex involving hXRCC1•hPARP•LIG3 or LIG1 completes repair. The lower pathway is followed when excision of the adduct is accomplished by a DNA glycosylase that also has an associated AP-lyase activity. AP lyases perform a β-elimination reaction that is chemically distinct from the hydrolysis performed by an AP endonuclease. The AP endonuclease has an associated activity that can remove the unsaturated deoxyribose moiety from the 3'hydroxyl group to permit DNA synthesis to proceed. Once the AP endonuclease has acted, DNA POLβ fills in the gap and permits synthesis to occur. The rest of the steps following synthesis are identical to that of BER followed along the upper pathway. The central pathway is considered a salvage pathway in human cells, but the major pathway in yeast. The role of this pathway in mammalian cells is not yet understood. The central pathway is PCNAp-dependent and uses DNA POLδ and POLε. This pathway could be used in mammalian cells if lesions altering the sugar structure did not permit the other two pathways to proceed. Alternatively, NER could also play a role in the repair of such damage.

5. Base Excision Repair (BER)

The other excision repair pathway that is principally responsible for the repair of structurally small DNA adducts is base excision repair (BER).[1, 5, 12, 57–62] Adducts eliminated by BER include a number of oxidized bases, fragmented deoxyribose structures, and abasic sites. Single-strand breaks formed by oxidative damage or radiation that leave 5'phosphoryl groups are also processed by BER.

The first step in BER is the recognition and departure of the damaged base from DNA that is catalyzed by DNA glycosylases (Fig. 4). These versatile enzymes recognize many DNA adducts and cleave the glycosylic bond, the bond that links the base to the deoxyribose. Some DNA glycosylases also cleave the phosphodiester backbone in a concerted β-elimination reaction. Following the removal of the adducted base, an endonuclease specific for abasic (AP) sites incises the DNA. Alternatively, if the phosphodiester bond has already been cleaved by the DNA glycosylase, the AP endonuclease has other activities (phosphatase and endonuclease) that leave a 3' hydroxyl group that is necessary for DNA polymerase to conduct the following steps. DNA POLβ inserts a base into the gap or at the 3' hydroxyl group. DNA POLβ has another activity that removes the deoxyribose 5' phosphate following DNA synthesis. The final step in the BER pathway is the ligation catalyzed by DNA ligase I or III.

Another BER pathway is a major pathway in *S. cerevisiae*, but considered minor in human cells. That pathway branches at the AP-endonuclease position and is PCNA-dependent sometimes called long patch BER, appears to involve strand displacement synthesis, and a different DNA polymerase for repair synthesis. The exact role that this BER mechanism has in mammalian cells, however, is still under investigation.[63, 64]

Originally the BER pathway was believed to function without complex formation between proteins. However, data at present indicate that protein complexes have an integral role in the BER pathway.[65–68] In the future, the interplay of the BER complexes with other DNA repair systems must be explored.

6. Reversal of DNA Damage

Mechanisms of repair based on reversal of DNA damage for eliminating oxidized bases have not yet been described. However, these mechanisms exist for the repair of O6-meG and for cyclobutane pyrimidine dimer sequences formed by methylating agents and UV radiation, respectively.[69–72]

7. Single-Strand Break Repair

Exposure to oxidizing agents or ionizing radiation induces single-strand breaks in DNA. The breaks generated leave 5'phosphoryl groups and modified deoxyribose groups. The most effective repair mechanism for elimination of single-stranded breaks is BER (Fig. 4). AP-endonuclease cleaves the unsaturated deoxyribose moiety and the repair follows that of BER. If normal BER is not possible due to the presence of deoxyribose structures that do not permit cleavage by AP-endonuclease, it is possible that the long patch BER mechanism is used for repair. Alternatively, the use of NER or recombination repair can also provide backup.

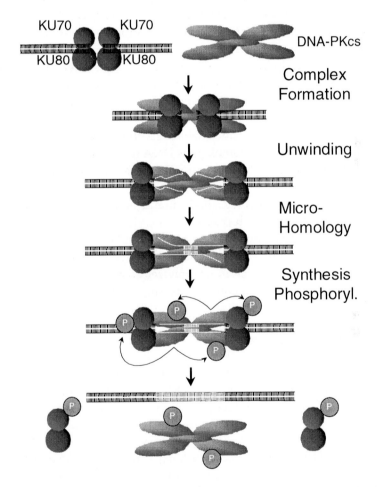

Fig. 5. Model for the major DSBR pathway in human cells via NHEJ that also has a role in V(D)J recombination. The KU70p/80p heterodimers bind to the end of the DSB and DNA-PKcsp binds to the complex. The ends are unwound by DNA helicase activity of the KU heterodimer or perhaps the hRAD50p•hMRE11p•NBS1p heterodimer. Microhomology aligns part of the DNA, and nuclease activity of hRAD50p•hMRE11p may be required to trim the ends. DNA polymerase, possibly POLβ based on *S. cerevisiae* studies,[136] fills in gaps that are sealed by DNA Ligase IV.

8. Double-Strand Break Repair (DSBR)

DSBs are generated during exposure of cells to oxidizing agents, ionizing radiation, or blocked replication forks. Accumulation of a single DSB in cells is a lethal event, and defects in DSBR are associated with human pathologies.[73–78] Two pathways can be used to repair these double-stranded breaks, but in human cells the major mechanism is by non-homologous end joining (NHEJ).[75–88]

NHEJ is initiated by the binding of the DNA PKCsp and Ku70p•Ku80p complex to the ends of the DSB (Fig. 5). The KU heterodimer unwinds the ends using a helicase activity that permits the annealing of small regions of homology. Some nuclease activity of the hRAD50p•hMRE11p complex may also be required. A DNA polymerase, probably POLβ,[89] then fills in any gaps. Finally, both ends are connected using XRCC4p and DNA ligase IV to complete repair.

DSB repair is also effected by homologous recombination (HR).[76, 82, 90] Many gene products involved in HR repair are now known. There are two sub pathways for HR. For both pathways, repair is initiated by hRAD52p binding to both ends of the DNA. The hRAD51p product then coats the DNA near the strand break and a strand is transferred to a homologous DNA region requiring a DNA helicase to unwind the DNA strand. The DNA strand invasion is followed by a resolution of the structure.

Another type of HR is termed single-strand annealing (SSA). Single-strand annealing is a more simple process than HR. The initial step again requires hRAD52p, but there is no requirement for hRAD51p in this repair. The homologous regions pair, excluding ends that are non-homologous. The exclusion of these ends is followed by cleavage that is also referred to as trimming. After trimming, the DNA strands are ligated to complete repair.

9. Mismatch Repair (MMR)

DNA repair is the last resort following the failure of anti-oxidants to avoid DNA damage, and mismatch repair (MMR) is the last step in repair. MMR can recognize in some cases the adduct paired with an incoming base and can also protect against DNA damage that occurs following an error in DNA synthesis. Defects in the MMR system are closely associated with human cancers.[91–94]

One type of MMR involves a system similar to that of the GO system described in *E. coli*.[95, 96] This system uses the BER pathway, and there are indications that a homologous pathway also functions in human cells.[97–101] The elimination of G → T mutations is accomplished in this system by the specificity of the hOGG1p and hMYHp (Fig. 6). DNA polymerases can insert an A opposite an 8-oxo-G that would lead to a mispair, and eventually to a G → T mutation. But the 8-oxo-G excision by hOGG1p is not favored thermodynamically in the 8-oxo-G•A base pair. The hMYHp, a DNA glycosylase, excises the A from the 8-oxo-G•A pair and

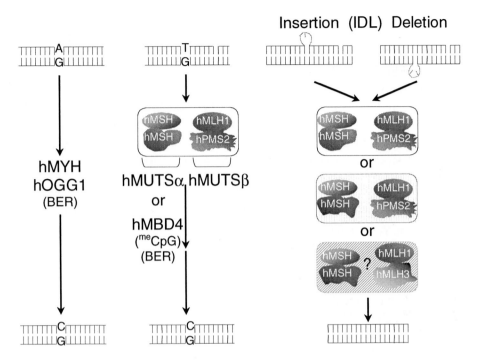

Fig. 6. Models for correction of mispaired bases by BER and MMR in human cells. From left, the mammalian equivalent of the GO system that functions in repair of A•G mispairs and does not require a strand nick to function, and follows BER after the A is removed. The repair of T•G mispairs does not require a strand break if the hMBD4p acts and the remainder of the repair is effected by the BER system. If the hMUTSα•hMUTSβ complex is used however, the repair requires a nick present in a newly synthesized strand. Insertion and deletion (IDL) mutations can be eliminated by the hMUTSα•hMUTSβ complex, but also by the (hMSH2•hMSH3)•(hMLH1•hPMS2) complex. There is also a possibility that an hMLH1•hMLH3 complex also has a role in elimination of IDLs. The different backgrounds indicate the formation of different complexes involved in recognition.

subsequent insertion of a C opposite the 8-oxo-G permits the excision of the position by hOGG1p to initiate BER.

Mispairs that are G•T are special in mammalian cells at CpG sequences. A fraction of C residues found in CpG sequences in mammalian cells are modified by DNA methyltransferases to form 5-methylcytosine (5-meC). The 5-meCs are important in the epigenetic control of gene expression.[102] Deamination of 5-meC by enzymatic or other means generates a T in DNA. Since no nick exists in the DNA and T is a normal base, cells have evolved a means to eliminate the T specifically from CpG sequences. To accomplish that without knowledge of the strand, the hMBD4p has a domain that specifically recognizes CpG sequences and another domain with DNA glycosylase activity that removes the T from the

G•T mispair.[103] Another DNA glycosylase, hTDGp, also removes the T from the same mispair, but the activity is minor in comparison to that of the hMBD4p. The actual role of BER in correcting such mispairs is not yet established.

Other types of mismatched bases are repaired using the protein complexes formed by the series of proteins shown in Fig. 6. It is critical for the complex to distinguish between the recently synthesized strand and the template strand. MMR effected by BER uses an intact DNA strand, whereas MMR performed by complexes recognizing mispairs uses a nick in the recently synthesized strand as a signal to target the strand needing repair. A complex of hMutSα (hMSH2p•hMSH6p) and hMutSβ (hMLH1p•hPMS2p) forms to recognize the mispair.

The correction of insertion and deletion mutations (IDL) is also tackled by the MMR system. Although the hMutSα•hMutSα complex can initiate IDL repair, two other complexes, (hMSH2p•hMSH3p)•hMutSβ and (hMSH2p•hMSH3p)•(hMLH1p•hMLH3p), also have a role in the repair of IDLs. As in the repair of mismatched bases, the signal that defines the synthesized strand is a nick. Some of the proteins involved in the later steps of human MMR repair following recognition include hEXO1p, hFEN1, POLδ/ε, hRPA, hRFC, and hPCNA.[104] The requirement of a helicase found in both NER and DSBR, however, has not yet been demonstrated. Clearly more work must be performed to understand mammalian MMR in as much depth as the system *E. coli* MMR system.[105, 106]

10. Repair of DNA–DNA and Protein-DNA Crosslinks

Genomic DNA is constantly in contact with proteins in cells. Exposure to drugs or radiation can result in DNA–DNA and protein-DNA crosslinks. Crosslinks present lethal consequences with respect to both transcription and replication. Repair of DNA–DNA crosslinks was discussed above. Another type of DNA protein crosslink that is observed by treatment with ionizing radiation or by topoisomerase I acting drugs is a tyrosine coupled to the 3-hydroxy position of a deoxyribose of DNA to form a phosphodiester bond. A tyrosine phospho-diesterase from yeast has been isolated that cleaves the tyrosine-DNA linkage.[107, 108] A homologous coding sequence has also been identified in human cells, but the function of the protein remains unknown.[109]

11. DNA Repair Heterogeneity

Genomic DNA repair is a finely tuned process that operates at several different levels. These different levels have already been presented in Fig. 1 of Chap. 16. Most of the levels of heterogeneity have been established using NER.[43] NER as shown in Fig. 3 involves a series of complexes, some of which engage proteins from other biological processes (e.g. transcription) to complete repair. Therefore, in addition to the complex systems often required for repair, it is also necessary to superimpose the constraints of different types of DNA repair heterogeneity.

Genomic DNA is organized into chromatin. Chromatin is composed of histones and other associated proteins. Regions of chromatin that are poised for transcription undergo transitions that render the chromatin structure more accessible to the cellular transcription apparatus. Differences in NER repair rates in open versus closed chromatin are associated with chromatin structure, but such differences have not been observed for BER. Heterogeneities of either MMR or DSBR have not been reported.

Transcription-coupled repair (TCR) is observed for NER. TCR is linked to the TFIIH complex that is also involved in transcription elongation of genes that are transcribed by the RNA polymerase II complex (Fig. 3). The signal for TCR is associated with the presence of adducts in DNA that block progression of the transcription complex. The exact mechanism for TCR in human cells, however, is not known. Moreover, even adducts that block transcription are not necessarily subject to TCR. Despite the coupling noted for NER in RNA polymerase II transcribed genes, NER in RNA polymerase III transcribed genes is not transcription-coupled.

TCR in BER is very different from that observed for NER. The repair of thymine glycol is transcription-coupled in both yeast and human cells.[65, 66, 110–113]. TCR of thymine glycol is associated with an acceleration of excision that is linked to a complex that forms between XPG and hNTHp, a DNA glycosylase that excises thymine glycol. A type of TCR has also been described for repair of 8-oxoguanine. The repair of 8-oxoguanine by the hOGG1p, another DNA glycosylase, is specific for the non-transcribed sequences.[66, 114] Yet other adducts removed by BER, e.g. 3-methyadenine and 7-methylguanine, do not manifest TCR.

Repair of bases is also dependent on the position with respect to genes. The presence of the proteins in the promoter regions can alter the repair rate of adducts. One interesting feature is that the changes in adduct repair rates is sometimes slower, but also can be faster. Faster repair rates for both NER and BER are generally observed near the transcription start sites.

The last type of DNA repair heterogeneity is sequence context dependent repair (SCD). SCD repair is observed for both NER and BER. SCD repair rates for NER of cyclobutane pyrimidine adducts vary over 15-fold in active genes. The differences in repair rate are coupled to the mutations observed in the tumor suppressor P53 gene. SCD is also evident in BER of 3-methyladenine and 7-methylguanine adducts whose repair rates vary up to 30-fold.[115] The rate of repair of 7-methylguanine adducts moreover is correlated with the rate of excision found in a simple *in vitro* system. SCD repair is most probably also found for oxidative base damage eliminated by NER or BER, but the determination of the repair rates at nucleotide resolution for these damaged bases requires more work. These large position-to-position differences in SCD repair may actually have a greater role in shaping mutational maps in tumors than strand specificity.[115]

For DSB repair, no heterogeneity with respect to strand would be anticipated, but for MMR, SSB, and crosslink repair, some strand specificity would not be surprising. Nonetheless, no heterogeneity of these DNA repair systems has been reported.

12. Regulation of DNA Repair Genes

One example of the role of MMR proteins emphasizes the necessity to maintain equilibrium levels of repair proteins. In cells selected with methotrexate, there is an amplification of the DHFR gene that also increases the amount of hMSH3p.[116] The increase in hMSH3p complexes the hMSH2p present in cells and increases both transition and transversion mutations, but not IDL mutations (Fig. 6). This increased mutation rate arises since the hMSH2p must form a complex with hMSH6p to repair mismatched bases, but the excess hMSH3p reduces this interaction. This illustrates two important points about the levels of repair proteins in cells. The first point is that it is possible to decrease the genetic stability of cells without introducing a mutation in a critical gene. The second point is that repair processes require a delicate balance to adjust repair so that genetic stability is maintained.

Such delicate balance has also been observed for the expression of different BER enzymes.[117] Expressing certain DNA glycosylases can either reduce survival or mutation frequency. These changes are often deleterious, but can be put to therapeutic use. Expressing DNA repair genes to render cells more sensitive to damaging agents can be used to sensitize target cells to chemotherapeutic agents.[118]

Therefore, in addition to understanding individual gene function, it is critical to shift to systems that are more complex to study DNA repair in larger contexts. This is facilitated by techniques that follow repair using DNA microarray chips, serial analysis of gene expression, or differential display. All these techniques have advantages, and are addressed in other chapters of this book. In the future, understanding how each repair system functions in concert to reduce mutation rates will depend on such an integrated approach. Initial results for the damage response of *S. cerevisiae* suggest that examination of global RNA expression levels is useful.[119] In addition to the examination of RNA expression levels, it will also be important to study the protein levels in response to DNA damage. All these areas should help researchers understand the equilibrium necessary to provide a person with genetic stabilty.

13. Human DNA Repair Syndromes

In light of the complex nature of DNA repair it is not startling that defects in DNA repair result in human syndromes. Some of these will be discussed in other chapters, but it is important to note human diseases that result from DNA repair defects.[9, 45, 46, 65, 73–76, 78, 83, 120–133] Most of these syndromes are genetic and do not afflict a large percentage of the population. Some of these genetic diseases have several complementation groups identified by biochemical analysis or genetic analysis. A list of human syndromes associated with DNA repair defects along with their associated genes is found in Table 3. One group of genetic diseases that has been identified in the last five years is that associated with RecQ related helicases.[134, 135] These are often involved in repair mechanisms.

It is critical to note, however, that even changes in the equilibrium of repair genes can render individuals more prone to developing a neoplasm, as suggested in the previous section. Moreover, subtle mutations in genes or single nucleotide polymorphisms (SNPs) could have similar effects that are not easily detectable. In the future, population screening and identification of such mutations may help to further demystify the process of tumorigenesis. Therefore, although we have

Table 3. Selected Human Syndromes Associated with DNA Repair Defects

Syndrome	Defective Repair	References
Xeroderma Pigmentosum		
Groups A-G	NER	2, 49, 120, 121
XPV	POLη	9, 18
Hereditary non-polyposis colon cancer	MMR	91, 93, 104
Nijmegen breakage syndrome	DSBR, HR	73–75, 77, 78
Cockayne's Syndrome		
A-B	NER	50, 121, 122
Fanconi's Anemia	?	132, 133
Trichothiodystrophy	NER	50, 121
Bloom's Syndrome	Recombination	125, 134, 135
DNA Ligase I deficiency	NER	124
Werner's Syndrome	Recombination	126, 134, 135
Rothmund-Thomson	Recombination	127, 134, 135
Ataxia telangectasia	DSBR	128, 129, 134, 135
Li-Fraumeni	NER (P53p)	130, 131

learned much about the systems of DNA repair that protect cells, there are yet many secrets to discover.

14. Conclusion

Human genomic repair enzymes represent a critical last defense against oxidative DNA damage. Failure of these systems to function properly directs cells toward cellular transformation and contributes to aging.

Acknowledgments

The author would like to thank the Beckman Research Institute and the National Institutes of Health for funding that made writing this chapter possible and Dr. Gerald P. Holmquist for his comments on the manuscript.

References

1. Lindahl, T. (2000). Suppression of spontaneous mutagenesis in human cells by DNA base excision-repair. *Mutat. Res.* **462:** 129–135.
2. Marnett, L. J. (2000). Oxyradicals and DNA damage. *Carcinogenesis* **21:** 361–370.
3. Schmutte, C. and Fishel, R. (1999). Genomic instability: first step to carcinogenesis. *Anticancer Res.* **19:** 4665–4696.
4. Friedberg, E. C., Walker, G. C. and Siede, W. (1995). *DNA Repair and Mutagenesis*, ASM Press, Washington, D.C.
5. Hickson, I. D. (1997). *Base Excision Repair of DNA Damage*, Landes Bioscience, Chapman & Hall, New York, Austin, Texas, USA.
6. Nickoloff, J. A. and Hoekstra, M. F. (1998). *DNA Damage and Repair*, Humana Press, Totowa, New Jersey.
7. Naegeli, H. (1997). *Mechanisms of DNA Damage Recognition in Mammalian Cells*, Landes Bioscience, Chapman & Hall, New York, Austin, Texas, USA.
8. Burgers, P. M. (1998). Eukaryotic DNA polymerases in DNA replication and DNA repair. *Chromosoma* **107:** 218–227.
9. Cordonnier, A. M. and Fuchs, R. P. (1999). Replication of damaged DNA: molecular defect in xeroderma pigmentosum variant cells. *Mutat. Res.* **435:** 111–119.
10. Goodman, M. F. and Fygenson, K. D. (1998). DNA polymerase fidelity: from genetics toward a biochemical understanding. *Genetics* **148:** 1475–1482.
11. Goodman, M. F. and Tippin, B. (2000). Sloppier copier DNA polymerases involved in genome repair. *Curr. Opin. Genet. Dev.* **10:** 162–168.

12. Hindges, R. and Hubscher, U. (1997). DNA polymerase delta, an essential enzyme for DNA transactions. *Biol. Chem.* **378**: 345–362.

13. Jager, J. and Pata, J. D. (1999). Getting a grip: polymerases and their substrate complexes. *Curr. Opin. Struct. Biol.* **9**: 21–28.

14. Johnson, R. E., Prakash, S. and Prakash, L. (2000). The human DINB1 gene encodes the DNA polymerase (POLθ). *Proc. Natl. Acad. Sci. USA* **97**: 3838–3843.

15. Johnson, R. E., Washington, M. T., Prakash, S. and Prakash, L. (2000). Fidelity of human DNA polymerase-eta. *J. Biol. Chem.* **275**: 7447–7450.

16. Masutani, C., Kusumoto, R., Iwai, S. and Hanaoka, F. (2000). Mechanisms of accurate translesion synthesis by human DNA polymerase-eta. *EMBO J.* **19**: 3100–3109.

17. Matsuda, T., Bebenek, K., Masutani, C., Hanaoka, F. and Kunkel, T. A. (2000). Low fidelity DNA synthesis by human DNA polymerase-eta. *Nature* **404**: 1011–1013.

18. Osheroff, W. P., Jung, H. K., Beard, W. A., Wilson, S. H. and Kunkel, T. A. (1999). The fidelity of DNA polymerase-beta during distributive and processive DNA synthesis. *J. Biol. Chem.* **274**: 3642–3650.

19. Wood, R. D. and Shivji, M. K. (1997). Which DNA polymerases are used for DNA-repair in eukaryotes? *Carcinogenesis* **18**: 605–610.

20. Hubscher, U., Nasheuer, H. P. and Syvaoja, J. E. (2000). Eukaryotic DNA polymerases, a growing family. *Trends Biochem. Sci.* **25**: 143–147.

21. Burtis, K. C. and Harris, P. V. (1997). A possible functional role for a new class of eukaryotic DNA polymerases [letter; comment]. *Curr. Biol.* **7**: R743–R744.

22. Sharief, F. S., Vojta, P. J., Ropp, P. A. and Copeland, W. C. (1999). Cloning and chromosomal mapping of the human DNA polymerase-theta (POLθ), the eighth human DNA polymerase. *Genomics* **59**: 90–96.

23. Ohashi, E., Ogi, T., Kusumoto, R., Iwai, S., Masutani, C., Hanaoka, F. and Ohmori, H. (2000). Error-prone bypass of certain DNA lesions by the human DNA polymerase-kappa. *Genes Dev.* **14**: 1589–1594.

24. Wang, Z., Castano, I. B., de Las Penas, A., Adams, C. and Christman, M. F. (2000). Polymerase-kappa: a DNA polymerase required for sister chromatid cohesion [In Process Citation]. *Science* **289**: 774–779.

25. Nelson, J. R., Lawrence, C. W. and Hinkle, D. C. (1996). Deoxycytidyl transferase activity of yeast REV1 protein. *Nature* **382**: 729–731.

26. McDonald, J. P., Rapic-Otrin, V., Epstein, J. A., Broughton, B. C., Wang, X., Lehmann, A. R., Wolgemuth, D. J. and Woodgate, R. (1999). Novel human and mouse homologs of *Saccharomyces cerevisiae* DNA polymerase-eta. *Genomics* **60**: 20–30.

27. Gibbs, P. E., McGregor, W. G., Maher, V. M., Nisson, P. and Lawrence, C. W. (1998). A human homolog of the *Saccharomyces cerevisiae* REV3 gene,

which encodes the catalytic subunit of DNA polymerase-zeta. *Proc. Natl. Acad. Sci. USA* **95**: 6876–6880.

28. Wilson, S. H. and Kunkel, T. A. (2000). Passing the baton in base excision repair [news]. *Nat. Struct. Biol.* **7**: 176–178.

29. Beard, W. A., Bebenek, K., Darden, T. A., Li, L., Prasad, R., Kunkel, T. A. and Wilson, S. H. (1998). Vertical-scanning mutagenesis of a critical tryptophan in the minor groove binding track of HIV-1 reverse transcriptase. Molecular nature of polymerase-nucleic acid interactions. *J. Biol. Chem.* **273**: 30 435–30 442.

30. Beard, W. A. and Wilson, S. H. (1998). Structural insights into DNA polymerase-beta fidelity: hold tight if you want it right. *Chem. Biol.* **5**: R7–R13.

31. Sawaya, M. R., Prasad, R., Wilson, S. H., Kraut, J. and Pelletier, H. (1997). Crystal structures of human DNA polymerase-beta complexed with gapped and nicked DNA: evidence for an induced fit mechanism. *Biochemistry* **36**: 11 205–11 215.

32. Pelletier, H. and Sawaya, M. R. (1996). Characterization of the metal ion binding helix-hairpin-helix motifs in human DNA polymerase-beta by X-ray structural analysis. *Biochemistry* **35**: 12 778–12 787.

33. Pelletier, H., Sawaya, M. R., Wolfle, W., Wilson, S. H. and Kraut, J. (1996). A structural basis for metal ion mutagenicity and nucleotide selectivity in human DNA polymerase-beta. *Biochemistry* **35**: 12 762–12 777.

34. Pelletier, H., Sawaya, M. R., Wolfle, W., Wilson, S. H. and Kraut, J. (1996). Crystal structures of human DNA polymerase-beta complexed with DNA: implications for catalytic mechanism, processivity, and fidelity. *Biochemistry* **35**: 12 742–12 761.

35. Bohr, V. A. and Dianov, G. L. (1999). Oxidative DNA damage processing in nuclear and mitochondrial DNA. *Biochimie* **81**: 155–160.

36. Le Page, F., Gentil, A. and Sarasin, A. (1999). Repair and mutagenesis survey of 8-hydroxyguanine in bacteria and human cells. *Biochimie* **81**: 147–153.

37. Petit, C. and Sancar, A. (1999). Nucleotide excision repair: from *Escherichia coli* to man. *Biochimie* **81**: 15–25.

38. Reardon, J. T., Bessho, T., Kung, H. C., Bolton, P. H. and Sancar, A. (1997). In vitro repair of oxidative DNA damage by human nucleotide excision repair system: possible explanation for neurodegeneration in xeroderma pigmentosum patients. *Proc. Natl. Acad. Sci. USA* **94**: 9463–9468.

39. Satoh, M. S., Jones, C. J., Wood, R. D. and Lindahl, T. (1993). DNA excision-repair defect of xeroderma pigmentosum prevents removal of a class of oxygen free radical-induced base lesions. *Proc. Natl. Acad. Sci. USA* **90**: 6335–6339.

40. Scott, A. D., Neishabury, M., Jones, D. H., Reed, S. H., Boiteux, S. and Waters, R. (1999). Spontaneous mutation, oxidative DNA damage, and the

roles of base and nucleotide excision repair in the yeast *Saccharomyces cerevisiae*. *Yeast* **15**: 205–218.

41. Swanson, R. L., Morey, N. J., Doetsch, P. W. and Jinks-Robertson, S. (1999). Overlapping specificities of base excision repair, nucleotide excision repair, recombination, and translesion synthesis pathways for DNA base damage in *Saccharomyces cerevisiae*. *Mol. Cell Biol.* **19**: 2929–2935.

42. Kuraoka, I., Bender, C., Romieu, A., Cadet, J., Wood, R. D. and Lindahl, T. (2000). Removal of oxygen free-radical-induced 5',8-purine cyclodeoxy-nucleosides from DNA by the nucleotide excision-repair pathway in human cells. *Proc. Natl. Acad. Sci. USA* **97**: 3832–3837.

43. Balajee, A. S. and Bohr, V. A. (2000). Genomic heterogeneity of nucleotide excision repair. *Gene* **250**: 15–30.

44. Bhatia, P. K., Wang, Z. and Friedberg, E. C. (1996). DNA repair and transcription. *Curr. Opin. Genet. Dev.* **6**: 146–150.

45. Friedberg, E. C. (1996). Relationships between DNA repair and transcription. *Ann. Rev. Biochem.* **65**: 15–42.

46. Hanawalt, P. C. (1996). Role of transcription-coupled DNA repair in suscepti-bility to environmental carcinogenesis. *Env. Health Perspect.* **104(Suppl. 3)**: 547–551.

47. Marinoni, J. C., Rossignol, M. and Egly, J. M. (1997). Purification of the transcription/repair factor TFIIH and evaluation of its associated activities in vitro. *Methods* **12**: 235–253.

48. Sancar, A. (1996). DNA excision repair [published erratum appears in *Ann. Rev. Biochem.* 1997; **66**: VII]. *Ann. Rev. Biochem.* **65**: 43–81.

49. Svejstrup, J. Q., Vichi, P. and Egly, J. M. (1996). The multiple roles of transcription/repair factor TFIIH. *Trends Biochem. Sci.* **21**: 346–350.

50. Benhamou, S. and Sarasin, A. (2000). Variability in nucleotide excision repair and cancer risk: a review. *Mutat. Res.* **462**: 149–158.

51. Mu, D., Bessho, T., Nechev, L. V., Chen, D. J., Harris, T. M., Hearst, J. E. and Sancar, A. (2000). DNA interstrand cross-links induce futile repair synthesis in mammalian cell extracts. *Mol. Cell Biol.* **20**: 2446–2454.

52. Bessho, T., Mu, D. and Sancar, A. (1997). Initiation of DNA interstrand cross-link repair in humans: the nucleotide excision repair system makes dual incisions 5' to the cross-linked base and removes a 22- to 28-nucleotide-long damage-free strand. *Mol. Cell Biol.* **17**: 6822–6830.

53. Araujo, S. J. and Wood, R. D. (1999). Protein complexes in nucleotide excision repair [published erratum appears in *Mutat. Res.* 2000 March 20; **459**(2): 171–172]. *Mutat. Res.* **435**: 23–33.

54. Batty, D. P. and Wood, R. D. (2000). Damage recognition in nucleotide excision repair of DNA. *Gene* **241**: 193–204.

55. Wood, R. D. (1996). DNA repair in eukaryotes. *Ann. Rev. Biochem.* **65**: 135–167.

56. Taylor, E. M. and Lehmann, A. R. (1998). Conservation of eukaryotic DNA repair mechanisms. *Int. J. Rad. Biol.* **74**: 277–286.

57. Memisoglu, A. and Samson, L. (2000). Base excision repair in yeast and mammals. *Mutat. Res.* **451**: 39–51.

58. Moller, P. and Wallin, H. (1998). Adduct formation, mutagenesis and nucleotide excision repair of DNA damage produced by reactive oxygen species and lipid peroxidation product. *Mutat. Res.* **410**: 271–290.

59. Beard, W. A. and Wilson, S. H. (2000). Structural design of a eukaryotic DNA repair polymerase: DNA polymerase-beta. *Mutat. Res.* **460**: 231–244.

60. McCullough, A. K., Dodson, M. L. and Lloyd, R. S. (1999). Initiation of base excision repair: glycosylase mechanisms and structures. *Ann. Rev. Biochem.* **68**: 255–285.

61. Cadet, J., Bourdat, A. G., D'Ham, C., Duarte, V., Gasparutto, D., Romieu, A. and Ravanat, J. L. (2000). Oxidative base damage to DNA: specificity of base excision repair enzymes. *Mutat. Res.* **462**: 121–128.

62. Lindahl, T., Karran, P. and Wood, R. D. (1997). DNA excision repair pathways. *Curr. Opin. Genet. Dev.* **7**: 158–169.

63. Fortini, P., Parlanti, E., Sidorkina, O. M., Laval, J. and Dogliotti, E. (1999). The type of DNA glycosylase determines the base excision repair pathway in mammalian cells. *J. Biol. Chem.* **274**: 15 230–15 236.

64. Fortini, P., Pascucci, B., Parlanti, E., Sobol, R. W., Wilson, S. H. and Dogliotti, E. (1998). Different DNA polymerases are involved in the short- and long-patch base excision repair in mammalian cells. *Biochemistry* **37**: 3575–3580.

65. Cooper, P. K., Nouspikel, T., Clarkson, S. G. and Leadon, S. A. (1997). Defective transcription-coupled repair of oxidative base damage in Cockayne syndrome patients from XP group G. *Science* **275**: 990–993.

66. Le Page, F., Kwoh, E. E., Avrutskaya, A., Gentil, A., Leadon, S. A., Sarasin, A. and Cooper, P. K. (2000). Transcription-coupled repair of 8-oxoguanine: requirement for XPG, TFIIH, and CSB and implications for Cockayne syndrome. *Cell* **101**: 159–171.

67. Klungland, A., Hoss, M., Gunz, D., Constantinou, A., Clarkson, S. G., Doetsch, P. W., Bolton, P. H., Wood, R. D. and Lindahl, T. (1999). Base excision repair of oxidative DNA damage activated by XPG protein. *Mol. Cell* **3**: 33–42.

68. Bessho, T. (1999). Nucleotide excision repair 3' endonuclease XPG stimulates the activity of base excision repairenzyme thymine glycol DNA glycosylase. *Nucleic Acids Res.* **27**: 979–983.

69. Deisenhofer, J. (2000). DNA photolyases and cryptochromes. *Mutat. Res.* **460**: 143–149.

70. Todo, T. (1999). Functional diversity of the DNA photolyase/blue light receptor family. *Mutat. Res.* **434**: 89–97.

71. Pegg, A. E. (2000). Repair of O(6)-alkylguanine by alkyltransferases. *Mutat. Res.* **462**: 83–100.

72. Bignami, M., O'Driscoll, M., Aquilina, G. and Karran, P. (2000). Unmasking a killer: DNA O(6)-methylguanine and the cytotoxicity of methylating agents. *Mutat. Res.* **462:** 71–82.
73. Carney, J. P., Maser, R. S., Olivares, H., Davis, E. M., Le Beau, M., Yates, J. R., III, Hays, L., Morgan, W. F. and Petrini, J. H. (1998). The hMre11/ hRad50 protein complex and Nijmegen breakage syndrome: linkage of double-strand break repair to the cellular DNA damage response. *Cell* **93:** 477–486.
74. Carney, J. P. (1999). Chromosomal breakage syndromes. *Curr. Opin. Immunol.* **11:** 443–447.
75. Digweed, M., Reis, A. and Sperling, K. (1999). Nijmegen breakage syndrome: consequences of defective DNA double strand break repair. *Bioessays* **21:** 649–656.
76. Karran, P. (2000). DNA double strand break repair in mammalian cells. *Curr. Opin. Genet. Dev.* **10:** 144–150.
77. Petrini, J. H., Bressan, D. A. and Yao, M. S. (1997). The RAD52 epistasis group in mammalian double strand break repair. *Semin. Immunol.* **9:** 181–188.
78. Petrini, J. H. (2000). The Mre11 complex and ATM: collaborating to navigate S phase. *Curr. Opin. Cell. Biol.* **12:** 293–296.
79. Featherstone, C. and Jackson, S. P. (1999). DNA double-strand break repair. *Curr. Biol.* **9:** R759–R761.
80. Featherstone, C. and Jackson, S. P. (1999). Ku, a DNA repair protein with multiple cellular functions? *Mutat. Res.* **434:** 3–15.
81. Flores-Rozas, H. and Kolodner, R. D. (2000). Links between replication, recombination and genome instability in eukaryotes. *Trends Biochem. Sci.* **25:** 196–200.
82. Haber, J. E. (2000). Partners and pathways repairing a double-strand break. *Trends Genet.* **16:** 259–264.
83. Hall, J. and Angele, S. (1999). Radiation, DNA damage and cancer. *Mol. Med. Today* **5:** 157–164.
84. Jeggo, P. A. (1998). Identification of genes involved in repair of DNA double-strand breaks in mammalian cells. *Rad. Res.* **150:** S80–S91.
85. Jin, S., Inoue, S. and Weaver, D. T. (1997). Functions of the DNA dependent protein kinase. *Cancer Surv.* **29:** 221–261.
86. Labhart, P. (1999). Nonhomologous DNA end joining in cell-free systems. *Eur. J. Biochem.* **265:** 849–861.
87. Olive, P. L. (1998). The role of DNA single- and double-strand breaks in cell killing by ionizing radiation. *Rad. Res.* **150:** S42–S51.
88. Pfeiffer, P. (1998). The mutagenic potential of DNA double-strand break repair. *Toxicol. Lett.* **96–97:** 119–129.
89. Wilson, T. E. and Lieber, M. R. (1999). Efficient processing of DNA ends during yeast nonhomologous end joining. Evidence for a DNA polymerase (POLβ) beta-dependent pathway. *J. Biol. Chem.* **274:** 23 599–23 609.

90. Haber, J. E. (1999). DNA recombination: the replication connection. *Trends Biochem. Sci.* **24:** 271–275.

91. Jiricny, J. and Nystrom-Lahti, M. (2000). Mismatch repair defects in cancer. *Curr. Opin. Genet. Dev.* **10:** 157–161.

92. Bocker, T., Ruschoff, J. and Fishel, R. (1999). Molecular diagnostics of cancer predisposition: hereditary non-polyposis colorectal carcinoma and mismatch repair defects. *Biochim. Biophys. Acta* **1423:** O1–O10.

93. Umar, A. and Kunkel, T. A. (1996). DNA-replication fidelity, mismatch repair and genome instability in cancer cells. *Eur. J. Biochem.* **238:** 297–307.

94. Minnick, D. T. and Kunkel, T. A. (1996). DNA synthesis errors, mutators and cancer. *Cancer Surv.* **28:** 3–20.

95. Michaels, M. L., Tchou, J., Grollman, A. P. and Miller, J. H. (1992). A repair system for 8-oxo-7,8-dihydrodeoxyguanine. *Biochemistry* **31:** 10 964–10 968.

96. Michaels, M. L., Cruz, C., Grollman, A. P. and Miller, J. H. (1992). Evidence that *MutY* and *MutM* combine to prevent mutations by an oxidatively damaged form of guanine in DNA. *Proc. Natl. Acad. Sci. USA* **89:** 7022–7025.

97. Lu, R., Nash, H. M. and Verdine, G. L. (1997). A mammalian DNA repair enzyme that excises oxidatively damaged guanines maps to a locus frequently lost in lung cancer. *Curr. Biol.* **7:** 397–407.

98. Roldan-Arjona, T., Wei, Y. F., Carter, K. C., Klungland, A., Anselmino, C., Wang, R. P., Augustus, M. and Lindahl, T. (1997). Molecular cloning and functional expression of a human cDNA encoding the antimutator enzyme 8-hydroxyguanine-DNA glycosylase. *Proc. Natl. Acad. Sci. USA* **94:** 8016–8020.

99. Radicella, J. P., Dherin, C., Desmaze, C., Fox, M. S. and Boiteux, S. (1997). Cloning and characterization of hOGG1, a human homolog of the OGG1 gene of *Saccharomyces cerevisiae*. *Proc. Natl. Acad. Sci. USA* **94:** 8010–8015.

100. Slupska, M. M., Luther, W. M., Chiang, J. H., Yang, H. and Miller, J. H. (1999). Functional expression of hMYH, a human homolog of the *Escherichia coli MutY* protein. *J. Bacteriol.* **181:** 6210–6213.

101. Slupska, M. M., Baikalov, C., Luther, W. M., Chiang, J. H., Wei, Y. F. and Miller, J. H. (1996). Cloning and sequencing a human homolog (hMYH) of the *Escherichia coli mutY* gene whose function is required for the repair of oxidative DNA damage. *J. Bacteriol.* **178:** 3885–3892.

102. Jones, P. A. and Laird, P. W. (1999). Cancer epigenetics comes of age. *Nature Genet.* **21:** 163–167.

103. Hendrich, B., Hardeland, U., Ng, H. H., Jiricny, J. and Bird, A. (1999). The thymine glycosylase MBD4 can bind to the product of deamination at methylated CpG sites [published erratum appears in *Nature* 2000 March 30; **404**(6777): 525]. *Nature* **401:** 301–304.

104. Kolodner, R. D. and Marsischky, G. T. (1999). Eukaryotic DNA mismatch repair. *Curr. Opin. Genet. Dev.* **9**: 89–96.

105. Modrich, P. (1987). DNA mismatch correction. *Ann. Rev. Biochem.* **56**: 435–466.

106. Modrich, P. (1989). Methyl-directed DNA mismatch correction. *J. Biol. Chem.* **264**: 6597–6600.

107. Yang, S. W., Burgin, A. B., Jr., Huizenga, B. N., Robertson, C. A., Yao, K. C. and Nash, H. A. (1996). A eukaryotic enzyme that can disjoin dead-end covalent complexes between DNA and Type-I topoisomerases. *Proc. Natl. Acad. Sci. USA* **93**: 11 534–11 539.

108. Burgin, A. B., Jr., Huizenga, B. N. and Nash, H. A. (1995). A novel suicide substrate for DNA topoisomerases and site-specific recombinases. *Nucleic Acids Res.* **23**: 2973–2979.

109. Pouliot, J. J., Yao, K. C., Robertson, C. A. and Nash, H. A. (1999). Yeast gene for a Tyr-DNA phosphodiesterase that repairs topoisomerase I complexes. *Science* **286**: 552–555.

110. Gowen, L. C., Avrutskaya, A. V., Latour, A. M., Koller, B. H. and Leadon, S. A. (1998). BRCA1 required for transcription-coupled repair of oxidative DNA damage. *Science* **281**: 1009–1012.

111. Leadon, S. A., Dunn, A. B. and Ross, C. E. (1996). A novel DNA repair response is induced in human cells exposed to ionizing radiation at the G1/S-phase border. *Rad. Res.* **146**: 123–130.

112. Leadon, S. A., Barbee, S. L. and Dunn, A. B. (1995). The yeast RAD2, but not RAD1, gene is involved in the transcription-coupled repair of thymine glycols. *Mutat. Res.* **337**: 169–178.

113. Leadon, S. A. and Lawrence, D. A. (1992). Strand-selective repair of DNA damage in the yeast GAL7 gene requires RNA polymerase II. *J. Biol. Chem.* **267**: 23 175–23 182.

114. Le Page, F., Klungland, A., Barnes, D. E., Sarasin, A. and Boiteux, S. (2000). Transcription coupled repair of 8-oxoguanine in murine cells: the OGG1 protein is required for repair in nontranscribed sequences but not in transcribed sequences. *Proc. Natl. Acad. Sci. USA* **97**: 8397–8402.

115. Ye, N., Holmquist, G. P. and O'Connor, T. R. (1998). Heterogeneous repair of N-methylpurines in normal human cells at the nucleotide level. *J. Mol. Biol.* **284**: 269–285.

116. Marra, G., Iaccarino, I., Lettieri, T., Roscilli, G., Delmastro, P. and Jiricny, J. (1998). Mismatch repair deficiency associated with overexpression of the MSH3 gene. *Proc. Natl. Acad. Sci. USA* **95**: 8568–8573.

117. Frosina, G. (2000). Overexpression of enzymes that repair endogenous damage to DNA. *Eur. J. Biochem.* **267**: 2135–2149.

118. Limp-Foster, M. and Kelley, M. R. (2000). DNA repair and gene therapy: implications for translational uses. *Env. Mol. Mutagen.* **35**: 71–81.

119. Jelinsky, S. A. and Samson, L. D. (1999). Global response of *Saccharomyces cerevisiae* to an alkylating agent. *Proc. Natl. Acad. Sci. USA* **96**: 1486–1491.

120. Cleaver, J. E. (2000). Common pathways for ultraviolet skin carcinogenesis in the repair and replication defective groups of xeroderma pigmentosum. *J. Dermatol. Sci.* **23**: 1–11.

121. De Boer, J. and Hoeijmakers, J. H. (2000). Nucleotide excision repair and human syndromes. *Carcinogenesis* **21**: 453–460.

122. Hanawalt, P. C. (2000). DNA repair. The bases for Cockayne syndrome [news]. *Nature* **405**: 415–416.

123. Brash, D. E. and Ponten, J. (1998). Skin precancer. *Cancer Surv.* **32**: 69–113.

124. Prigent, C., Satoh, M. S., Daly, G., Barnes, D. E. and Lindahl, T. (1994). Aberrant DNA repair and DNA replication due to an inherited enzymatic defect in human DNA ligase I. *Mol. Cell Biol.* **14**: 310–317.

125. Ellis, N. A., Groden, J., Ye, T. Z., Straughen, J., Lennon, D. J., Ciocci, S., Proytcheva, M. and German, J. (1995). The Bloom's syndrome gene product is homologous to RecQ helicases. *Cell* **83**: 655–666.

126. Lombard, D. B. and Guarente, L. (1996). Cloning the gene for Werner syndrome: a disease with many symptoms of premature aging. *Trends Genet.* **12**: 283–286.

127. Kitao, S., Shimamoto, A., Goto, M., Miller, R. W., Smithson, W. A., Lindor, N. M. and Furuichi, Y. (1999). Mutations in RECQL4 cause a subset of cases of Rothmund-Thomson syndrome. *Nature Genet.* **22**: 82–84.

128. Savitsky, K., Sfez, S., Tagle, D. A., Ziv, Y., Sartiel, A., Collins, F. S., Shiloh, Y. and Rotman, G. (1995). The complete sequence of the coding region of the ATM gene reveals similarity to cell cycle regulators in different species. *Human Mol. Genet.* **4**: 2025–2032.

129. Shiloh, Y. and Rotman, G. (1996). Ataxia-telangiectasia and the ATM gene: linking neurodegeneration, immunodeficiency, and cancer to cell cycle checkpoints. *J. Clin. Immunol.* **16**: 254–260.

130. Ford, J. M. and Hanawalt, P. C. (1995). Li-Fraumeni syndrome fibroblasts homozygous for p53 mutations are deficient in global DNA repair but exhibit normal transcription-coupled repair and enhanced UV resistance. *Proc. Natl. Acad. Sci. USA* **92**: 8876–8880.

131. Little, J. B., Nove, J., Dahlberg, W. K., Troilo, P., Nichols, W. W. and Strong, L. C. (1987). Normal cytotoxic response of skin fibroblasts from patients with Li-Fraumeni familial cancer syndrome to DNA-damaging agents in vitro. *Cancer Res.* **47**: 4229–4234.

132. Carreau, M. and Buchwald, M. (1998). The Fanconi anemia genes. *Curr. Opin. Pediatr.* **10**: 65–69.

133. (1996). Positional cloning of the Fanconi anaemia group A gene. The Fanconi anaemia/breast cancer consortium. *Nature Genet.* **14**: 324–328.

134. Shen, J. C. and Loeb, L. A. (2000). The Werner syndrome gene: the molecular basis of RecQ helicase-deficiency diseases. *Trends Genet.* **16:** 213–220.
135. Karow, J. K., Wu, L. and Hickson, I. D. (2000). RecQ family helicases: roles in cancer and aging. *Curr. Opin. Genet. Dev.* **10:** 32–38.
136. Wilson, S. H. (1998). Mammalian base excision repair and DNA polymerase-beta. *Mutat. Res.* **407:** 203–215.

Chapter 44

DNA Repair in Mammalian Mitochondria

Nadja C. Souza-Pinto and Vilhelm A. Bohr*

Nadja C. Souza-Pinto and **Vilhelm A. Bohr** • Laboratory of Molecular Genetics, National Institute on Aging, NIH, Baltimore, MD

*Corresponding Author. 5600 Nathan Shock Dr., Baltimore, MD 21224, USA
Tel: 410-558-8162, Fax: 410-558-8157, E-mail: BohrV@grc.nia.nih.gov

1. Introduction

The free radical theory of aging states that free radicals are produced as a by-product of the cellular utilization of molecular oxygen (reactive oxygen species, ROS), and that mutations and cancer arise as a consequence of free radical reactions with cellular components.[1] This theory was later modified to propose that mitochondria might be the "clock" that determines life span.[2] Since the mitochondrial DNA (mtDNA) lies at the inner mitochondrial membrane,[3] in close proximity to the electron transport chain, mtDNA is likely to be the most critical target for ROS. As a result of the accumulation of damage and mutations to the mtDNA over time, loss of mitochondrial function would be an early event in a cascade that leads to aging.

The reaction of ROS with DNA generates a variety of DNA lesions, such as base oxidation products, sugar oxidation products and stand breaks.[4] More than 30 different base oxidation products have been detected upon reaction of DNA with different oxidants,[5, 6] and it is likely that there are over 100 different oxidative DNA base lesions in mammalian DNA. Among these, 8-oxo-dG is one of the most abundant.[7] It is considered to be mutagenic because it can mispair with adenine during DNA replication, what results in G-T transversion mutations.[8] Many laboratories have demonstrated that the level of 8-oxo-dG in the mtDNA increases with age,[9-11] and this provided additional support for the mitochondrial theory of aging. Other aspects of the relationship between mtDNA damage and aging will be discussed elsewhere in this book. This review will focus on the mechanisms by which mitochondria repair damage to their DNA.

2. DNA Repair Mechanisms in Mammalian Mitochondria

The initial observation[12] that mitochondria are unable to remove UV-induced damage from their genome led to the notion that mitochondria lack DNA repair capacity. Although it is true that mitochondria do not remove UV induced DNA damage, subsequent studies clearly showed that mitochondria repair some types of damage from their DNA.[13-15] One of the earliest reports on mitochondrial DNA repair showed that alkyation damage was removed from rat liver mtDNA, and that the kinetics was similar to that of a nuclear sequence.[15] Our knowledge about the DNA repair mechanisms in mitochondria are based on the identification of specific enzymes and on studies on the repair of various lesions. Different DNA lesions are repaired via different repair pathways that have been identified in studies on nuclear or total DNA. Figure 1 shows DNA repair pathways in nuclear and mtDNA, as currently understood. These will be discussed in the following.

Studies on mitochondrial DNA damage and repair have traditionally required the purification of mtDNA. This imposes a methodological problem, since most

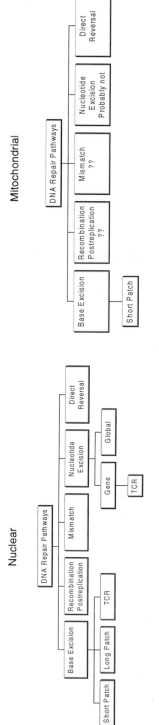

Fig. 1. DNA repair pathways in nuclei and mitochondria.

purification schemes may introduce oxidative damage into the DNA. As an alternative approach this laboratory have modified the gene specific repair assay developed by Bohr *et al.*[16] to detect various DNA lesions other than UV induced dimers. This method allows detection of repair in both the mitochondrial and nuclear DNA from the same biological sample, without the need for isolation of the mtDNA. In addition, with this technique, oxidative lesions can be detected in parts or in the entire mitochondrial genome, and strand bias (or transcription coupled repair) can be analyzed.

Using the gene specific repair assay and a variety of damaging agents, efficient repair of strand breaks and alkali sensitive sites has been demonstrated in rodent and human mitochondrial DNA. Repair of fapy glycosylase (*Fpg*)–sensitive sites in mtDNA has been reported for rat,[13] CHO[14] and human cells.[17] In the latter study, human cultured fibroblasts were exposed to the photo activated dye methylene blue, which generates mainly 8-oxo-dG. The removal of 8-oxo-dG from the mitochondrial genome was very efficient, after 9 hours 47% of the initial lesions had been repaired. In addition, analysis of repair on both strands of the highly transcribed ribosomal region and non-ribosomal regions showed that repair of 8-oxo-dG in mitochondria is without bias for the transcribed strand, as seen in nuclear DNA repair, and thus is not coupled to transcription.[17]

The isolation of enzymes that can carry out base excision repair (BER) suggested that this repair pathway exists in mitochondria. In this regard, apurinic/apyrimidinic (AP) endonucleases classes I and II, glycosylases, DNA ligase and DNA polymerase have been identified in mammalian mitochondria (Ref. 18 and references therein).

3. Base Excision Repair Pathways in Mitochondria

Base excision repair is one of the main pathways for the removal of small base modification from DNA. BER is initiated by a glycosylase, which will specifically recognize a damaged base and cleave the N-glycosyl bond between the base and the sugar, generating an abasic (AP) site. The AP site is further processed by an AP endonuclease, which introduces a strand break 5' to the baseless sugar and generates a 5'-deoxyribose phosphate (dRP) terminus. This intermediate is a blocking end for ligation and removal of the dRP moiety is often the rate-limiting step during BER. A DNA polymerase, POLβ in nuclei and POLγ in mitochondria, removes the dRP motif and introduces a new nucleotide. A DNA ligase then seals the gap (Ref. 19 and references therein).

Whereas *in vitro* DNA repair assays have proven very useful in the investigation of the nucleotide excision repair (NER) mechanisms,[20] no such assays had been adapted to study mitochondrial DNA repair. Using protein extracts obtained from *Xenopus* mitochondria, Ryoji *et al.*[21] used a repair incorporation assay to

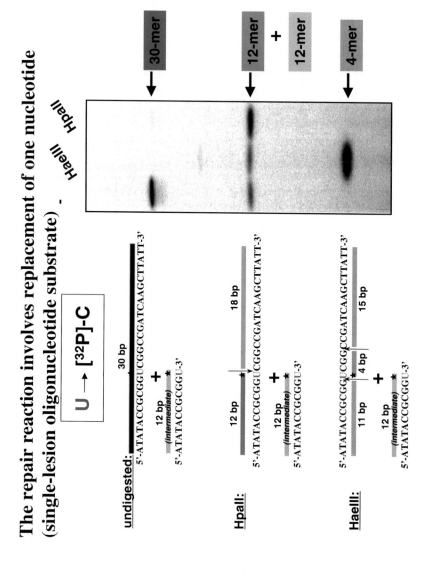

Fig. 2. Measurement of DNA repair patch in mitochondrial DNA using cell extracts. The figure shows repair incorporation in a DNA construct with a single uracil lesion (U). Using restriction mapping we can detect where the incorporated radioactive label is located. Cutting with HpaII, there is label only in the 12 bp fragment, and cutting with HaeIII, there is label only in the 4 bp fragment, indicating a pathway of single nucleotide exchange. For details, see Ref. 24.

demonstrate that plasmid DNA treated with H_2O_2 was repaired. In another approach, Pinz and Bogenhagen[22] reconstituted *in vitro* the repair of abasic sites in an oligomer substrate using solely proteins isolated from *Xenopus* mitochondria, an AP endonuclease, DNA POLγ and a mtDNA ligase, which was suspected to be related to the nuclear DNA ligase III. In fact, recently it was confirmed that the human DNA ligase III gene encodes both a nuclear and mitochondrial isoforms.[23]

In mammalian systems, we recently demonstrated that protein extracts from rat liver mitochondria support the repair of uracil containing substrates.[24] Using a repair incorporation assay and single lesion constructs, we showed incorporation of radioactive dCTP into a double strand oligonucleotide containing uracil opposite deoxyguanine and also in plasmid DNA.

In the nucleus of mammalian cells there are two pathways for BER, the short patch or one nucleotide replacement, and the long patch, which results in the incorporation of 2–6 nucleotides. These two pathways involve different subsets of proteins and operate independently.[25, 26] The utilization of a single lesion construct in the repair incorporation assay allowed us to study the patch size for the repair of uracil by mitochondrial extracts. This experiment is shown in Fig. 2. We found this repair event proceeds solely via the short patch, or single nucleotide replacement.[24] We also found that there is no incorporation of radioactive dCTP 5′ to the lesion, suggesting that only BER is operative. Whether or not mitochondrial extracts can support long patch BER remains to be determined.

4. DNA Glycosylases in Mammalian Mitochondria

DNA glycosylases are a class of enzymes that recognize specific base modifications in DNA and attack the N-glycosyl bond, releasing the free damaged base. There are two classes of glycosylases: (a) monofunctional glycosylases, that only release the modified base leaving an abasic site as product, and (b) bifunctional glycosylases, or glycosylases/AP lyases. The latter release the damage base and attack the phosphodiester bond generating a single strand break. The lyase activity usually proceeds through a β, δ elimination reaction[27] Below we summarize the characteristics of the three best-known mitochondrial glycosylases.

4.1. UDG1

Uracil in DNA is one of the most common types of DNA damage. It is formed by deamination of cytosine or by misincorporation of dUTP during DNA synthesis.[28] Uracil DNA glycosylase was one of the first repair activities detected in mitochondrial extracts.[29] The enzyme, a 30 kD protein, was later purified from human cells by affinity chromatography.[30] This activity was later named UDG1 to differentiate from the nuclear uracil DNA glycosylase, UDG2. The same gene, *ung*, encodes both enzymes. The two different isoforms are generated via two different promoters and alternative splicing.[31] Although both isoforms show a peak in expression during the S phase of the cell cycle, the two enzymes are expressed in a different manner, and show different tissue abundance in humans.[32]

UDG is a monofunctional glycosylase, in other words, it has no associated AP lyase activity. This enzymes recognizes the U:G mismatch and releases the uracil residue generating an abasic site. For the completion of the repair process, however, it requires the coupled activity of an AP endonuclease.[19]

4.2. mtODE (Mitochondrial Oxidative Damage Endonuclease)

The observation that mammalian mitochondria efficiently removed *Fpg*-sensitive sites from their genomes[13, 14, 17] led to the speculation that these organelles possessed a glycosylase that specifically recognizes oxidative modified purines. Using rat liver mitochodria, an activity that recognizes and incises at 8-oxo-dG and abasic sites in duplex DNA was identified. This activity was purified and characterized, and called mitochondrial oxidative damage endonuclease (mtODE).[33] Based on gel filtration chromatography, mtODE's molecular mass is predicted to be between 25 and 30 kDa. mtODE has a monovalent cation optimum between 50 and 100 mM KCl and a pH optimum between 7.5 and 8.0. mtODE does not require any co-factors and is active in the presence of 5 mM EDTA. It is specific for 8-oxo-G and preferentially incises at 8-oxo-G:C base pairs. mtODE was found to be a 8-oxo-dG glycosylase/lyase enzyme, because it can be covalently linked to the 8-oxo-G oligonucleotide by sodium borohydride reduction. In addition, we confirmed its glycosylase activity by measuring the release of 8-oxo-dG from oxidized DNA using HPLC-EC (Hudson *et al.*, unpublished results).

Analysis of the *ogg1* gene showed a mitochondrial localization signal upstream from the coding sequence, suggesting that this enzyme could be transported into the mitochondria. Further experiments using CHO cells transfected with an *ogg1*-containing construct identified 5 different isoforms, from which 3 were localized to the mitochondria.[34] We recently demonstrated that the major 8-oxo-dG glycosylase/AP lyase activity in mouse liver mitochondria is, in fact, encoded by the *ogg1* gene, since liver mitochondria from *ogg1⁻/⁻* animals showed no detectable incision activity with a construct containing a single 8-oxo-dG.[55]

In vitro extension assays with POLγ and defined substrates containing an abasic site or an 8-oxo-dG revealed that POLγ inserts dA opposite to both substrates at significant rates.[35] To minimize such mismatches cells possess an adenine DNA glycosylase, *hMutY*, which excises the incorporated adenine in the context of an 8-oxo-dG:dA mispair. Using an epitope-tagged *hMutY*, this protein was also localized to mitochondria.[34]

In addition to the two glycosylases mentioned above, mitochondria also posses an 8-oxo-dGTPase to minimize the incorporation of the mutagenic nucleotide in the newly replicated DNA.[36]

4.3. mtTGEndo (Mitochondrial Thymine Glycol Endonuclease)

This laboratory has also recently identified and purified a thymine glycol endonuclease (mtTGendo) from rat liver mitochondria.[37] This activity was identified following the incision of a duplex oligonucleotide containing a single thymine glycol lesion. After purification using different chromatography columns, the most pure active fractions contained a single band of approximately 37 kDa. mtTGendo is active within a broad KCl concentration range and is EDTA-resistant. Furthermore, it has an associated apurinic/apyrimidinic-lyase activity, as demonstrated by sodium borohydrate trapping of the Schiff base intermidiate.[37] mtTGendo does not incise 8-oxo-dG or uracil-containing duplexes or thymine glycol in single-stranded DNA. Based upon functional similarity, it was concluded that mtTGendo might be a rat mitochondrial homologue of the *Escherichia coli* endonuclease III protein. A mitochondrial isoform of the NTH1 glycosylase has also been identified.[34] Whether mtTGendo is the mitochondrial isoform of the NTH1 protein remains to be determined.

5. Other Mitochondrial Enzymes Involved in DNA Repair

5.1. DNA Polymerase Gamma

DNA polymerase-gamma is the major DNA polymerase in mitochondria. Although a POLβ like activity has recently been isolated from bovine heart mitochondria, its function remains undetermined. POLγ is the replicative polymerase and is also implicated in DNA repair. Using *in vitro* repair systems with mitochondrial extracts[24] or with isolated proteins,[38] it was observed that POLγ was responsible for filling the single nucleotide gap generated during repair of a uracil containing substrate. In addition a dRP lyase activity was identified in the catalytic subunit of human POLγ, indicating that it can process the dRP intermediate to make it a substrate for ligation by DNA ligase.

5.2. AP Endonucleases

AP endonucleases are classified according to their cleavage pattern. Class I endonucleases cleave on the 3' side of the abasic site, and the products are an unsaturated baseless sugar and a 5' phosphate. Class II enzymes cleave 5' to the abasic site leaving 3′ OH and 5′ dRP ends.[39] Two mitochondrial class II AP endonucleases have been purified from mouse plasmacytoma cells.[40] Although both activities cross-reacted with antibodies raised against the major AP endonuclease of HeLa cells, it is unlikely that the mitochondrial endonucleases were derived from the gene for the major nuclear AP endonuclease. Further support for that came from experiments in which Takao *et al.*[34] failed to see mitochondrial localization of hAPE.

5.3. Damage Reversal

The simplest form of DNA repair is the direct chemical reversal of the damaged adducts. A major class of enzymes that perform damage reversal are photolyases. These enzymes use light to monomerize UV induced pyrimidine dimmers, and although it is unknown whether photolyases exist in mammalian mitochondria, they have been localized to both the nucleus and mitochondria in yeast.[41]

Another form of damage reversal is mediated by DNA alkyltransferases. These "suicide" enzymes transfer the chemical adduct directly to a cysteine residue within the active site thereby inactivating themselves. An O6-methylguanine DNA transferase has been partially purified from rat liver mitochondria.[42] In this study, the authors found that the kinetics of removal for the O6-methyl-2′-deoxyguanine were similar between the mtDNA and nuclear DNA. However, there was no removal of the bulkier adduct O6-buthyl-2′-deoxyguanine. Further support for mitochondrial methyltransferases came from the work of Satoh *et al.*[15] They reported the removal of O6-ethyl-2′-deoxyguanine from rat liver mtDNA after the exposure to N-ethyl-N-nitrosurea *in vivo*. It is, however, unknown what are the relative contributions of BER and methyltransferases in the repair of small alkylation damage in mitochondria.

5.4. Mismatch Repair

In bacteria, three proteins, *mut L*, *mut H* and *mut S* mediate mismatch repair. Homologues for those proteins have been found in higher organisms and shown to also be involved in the mismatch repair pathway. While mismatch repair has not been directly measured in mitochondria, *mut S* homologues (MSH1) have been identified in yeast[43] and coral[44] mitochondria. In *S. cerevisiae*, disruption of MSH1 caused gross mtDNA rearrangements and an increased rate of mtDNA mutations. So far, no mismatch repair proteins have been isolated from mammalian mitochondria. However, given the severity of the phenotype of the yeast with a defective *mut S* homologue, it is unlikely that such mutations in humans would be compatible with life.

5.5. Recombinational Repair

Interstrand cross-links are thought to be repaired via recombinational repair. In mammalian mitochondria, interstrand cross-links induced by *cis*-platin are efficiently removed from mtDNA by CHO cells,[45] while cross-links induced by psoralen are not.[46] This suggests that the existence of a mitochondrial recombinational repair pathway needs further study. However, several lines of evidence suggest that mitochondria possess a homologous recombination pathway. These observations come from an *in vitro* study[47] and from the analysis of recombination intermediates in patients with Kearns-Sayre syndrome and external ophthalomoplegia.[48] In a recent study by Tang *et al.*,[49] using transmitochondrial cell lines harboring various proportions of wt-, delta-, and duplicated-mtDNAs, the authors found that cells that contained initially 100% dup-mtDNAs became heteroplasmic, containing both wild type and rearranged mtDNAs. These were likely to have been generated via intramolecular recombination events. Therefore it is likely that mitochondria may be able to perform recombinational repair. However, the definitive identification of such pathway and the proteins involved awaits further research.

6. Changes in mtDNA Repair with Age

There is clear evidence that mtDNA oxidative damage does, in fact, increase with age, despite great controversy about the actual levels of oxidative damage and about the different methods for its measurement. For a comprehensive review of this topic, please refer to Anson and Bohr.[50] It has been a common notion that DNA repair declines with age, and this leads to the accumulation of DNA damage and genomic instability. Over the past years, various attempts have been made to measure DNA repair capacity changes with age. An initial correlation between DNA repair capacity and aging was provided by Hart and Setlow,[51] who demonstrated a linear correlation between the logarithm of lifespan and the DNA repair capacity in cells from different mammalian species, suggesting that higher DNA repair activity is associated with longer life span. Further studies showed limitations to this correlation, and there have since been numerous studies on the changes in repair with age. The results have varied and there is no clear consensus. In addition, studies have measured only global genome repair, and may not have detected potential differences in the repair at actively transcribed and non-transcribed regions of the genome. Guo *et al.*[52] have recently addressed this question. They reported that the removal of UV induced damage in actively transcribed regions is lower in hepatocytes isolated from old than from young rats, suggesting a more relevant decrease in DNA repair with age. All these studies, however, measured only repair in nuclear DNA.

To address the question of whether DNA repair declines with age in mitochondria, we investigated changes in mtDNA repair activities with age in both rat[53] and mouse[56] mitochondria. We found that the 8-oxo-dG incision activity increases with age in both rat heart and liver mitochondria. The incision activity towards a oligomer substrate containing a single 8-oxo-dG activity was significantly higher at 14 and 23 months than at 6 months of age. The highest activity was found at 14 months, suggesting a peak in middle age. We also measured mtUDG activity in those same extracts and found no changes with age. These results suggested a specific up-regulation of the repair of oxidative damage with age in liver and heart mitochondria, and that the 8-oxo-dG incision activity may be induced with age in rat mitochondria.

Using mouse liver, we investigated changes in 8-oxo-dG incision activity in both nucleus and mitochondria. Interestingly, we found that activity to increase with age in mitochondria, from 6 to 14 months of age, but not in the nucleus.[56] It is not clear whether this increase in activity is due to higher protein levels or other mechanisms, such as protein import into the mitochondria. The differential response to age for the mitochondrial and the nuclear 8-oxo-dG glycosylase activities suggests that the expression of the two isoforms is differentially regulated. The mitochondrial genome is more prone to the attack by reactive oxygen species (ROS) than nuclear DNA. The mitochondrial DNA is in close proximity to the inner mitochondrial membrane, which is the major cellular site for ROS generation, and lacks protective structural proteins. The nuclear DNA, in contrast, is protected by the chromatin structure, and possesses backup mechanisms for the repair of oxidative damage. Since BER seems to be the only line of defense against oxidative DNA damage in the mitochondria, it seems reasonable that cells would evolve mechanisms to protect the mitochondrial DNA against massive accumulation of oxidative damage with age through up-regulation of BER enzymes.

Together, these results suggest that the mitochondrial capacity to repair 8-oxo-dG does not decrease, but rather increases with age. It has been proposed that the rate of free radical production is the determinant factor in the aging process.[54] Thus, it is possible that the rate of damage formation exceeds the mitochondrial DNA repair capacity leading to damage accumulation. In that case, the induction of the glycosylase activity may represent a cellular response in an attempt to counteract increased damage formation.

7. Perspectives

Great advances have been made in the last few years in our understanding of how DNA repair takes place in the nuclear or total DNA. In contrast, little is still known about these processes in mammalian mitochondria. The field is now moving rapidly and it has become clear that mitochondrial repair is more important than previously considered and that accumulation of DNA damage in these organelles may be an important elementy in the aging phenotype. Mitochondria clearly possess base excision repair capacity, but it is a great future challenge to determine which other DNA repair pathways operate here.

Acknowledgment

We thank R. Stierum for help with Fig. 2

References

1. Harman, D. (1956). Aging: A theory based on free radical and radiation chemistry. *J Gerontol.* **11**: 298–300.
2. Harman, D. (1972). The biologic clock: the mitochondria. *J. Am. Geriatr. Soc.* **20**: 145–147.
3. Shearman, C. W. and Kalf, G. F. (1977). DNA replication by a membrane-DNA complex from rat liver mitochondria. *Arch. Biochem. Biophys.* **182**: 573–586.
4. Aust, A. E. and Eveleigh, J. F. (1999). Mechanisms of DNA oxidation. *Proc. Soc. Exp. Biol. Med.* **222**: 246–252.
5. Kasai, H. and Nishimura, S. (1983). Hydroxylation of the C-8 position of deoxyguanosine by reducing agents in the presence of oxygen. *Nucleic Acids Symp. Ser.* **12**: 165–167.
6. Dizdaroglu, M. (1994). Chemical determination of oxidative DNA damage by gas chromatography-mass spectrometry. *Meth. Enzymol.* **234**: 3–16.
7. Park, J. W., Cundy, K. C. and Ames, B. N. (1989). Detection of DNA adducts by high-performance liquid chromatography with electrochemical detection. *Carcinogenesis* **10**: 827–832.
8. Grollman, A. P. and Moriya, M. (1993). Mutagenesis by 8-oxoguanine: an enemy within. *Trends Genet.* **9**: 246–249.
9. Hudson, E. K., Hogue, B. A., Souza-Pinto, N. C., Croteau, D. L., Anson, R. M. *et al.* (1998). Age-associated change in mitochondrial DNA damage. *Free. Radic. Res.* **29**: 573–579.
10. Beckman, K. B. and Ames, B. N. (1997). Oxidative decay of DNA. *J. Biol. Chem.* **272**: 19 633–19 636.
11. Takasawa, M., Hayakawa, M., Sugiyama, S., Hattori, K., Ito, T. *et al.* (1993). Age-associated damage in mitochondrial function in rat hearts. *Exp. Gerontol.* **28**: 269–280.
12. Clayton, D. A., Doda, J. N. and Friedberg, E. C. (1974). The absence of a pyrimidine dimer repair mechanism in mammalian mitochondria. *Proc. Natl. Acad. Sci. USA* **71**: 2777–2781.

13. Driggers, W. J., LeDoux, S. P. and Wilson, G. L. (1993). Repair of oxidative damage within the mitochondrial DNA of RINr 38 cells. *J. Biol. Chem.* **268**: 22 042–22 045.

14. LeDoux, S. P., Wilson, G. L., Beecham, E. J., Stevnsner, T. and Wassermann, K. *et al.* (1992). Repair of mitochondrial DNA after various types of DNA damage in Chinese hamster ovary cells. *Carcinogenesis* **13**: 1967–1973.

15. Satoh, M. S., Huh, N., Rajewsky, M. F. and Kuroki, T. (1988). Enzymatic removal of O6-ethylguanine from mitochondrial DNA in rat tissues exposed to N-ethyl-N-nitrosourea in vivo. *J. Biol. Chem.* **263**: 6854–6856.

16. Bohr, V. A., Smith, C. A., Okumoto, D. S. and Hanawalt, P. C. (1985). DNA repair in an active gene: removal of pyrimidine dimers from the DHFR gene of CHO cells is much more efficient than in the genome overall. *Cell* **40**: 359–369.

17. Anson, R. M., Croteau, D. L., Stierum, R. H., Filburn, F. and Parsell, R. *et al.* (1998). Homogenous repair of singlet oxygen-induced DNA damage in differentially transcribed regions and strands of human mitochondrial DNA. *Nucleic Acids Res.* **26**: 662–668.

18. Croteau, D. L., Stierum, R. H. and Bohr, V. A. (1999). Mitochondrial DNA repair pathways. *Mutat. Res.* **434**: 137–148.

19. Demple, B. and Harrison, L. (1994). Repair of oxidative damage to DNA: enzymology and biology. *Ann. Rev. Biochem.* **63**: 915–948.

20. Wood, R. D. (1997). Nucleotide excision repair in mammalian cells. *J. Biol. Chem.* **272**: 23 465–23 468

21. Ryoji, M., Katayama, H., Fusamae, H., Matsuda, A. and Sakai, F. *et al.* (1996). Repair of DNA damage in a mitochondrial lysate of *Xenopus laevis* oocytes. *Nucleic Acids Res.* **24**: 4057–4062.

22. Pinz, K. G. and Bogenhagen, D. F. (1998). Efficient repair of abasic sites in DNA by mitochondrial enzymes. *Mol. Cell Biol.* **18**: 1257–1265.

23. Lakshmipathy, U. and Campbell, C. (1999). The human DNA ligase III gene encodes nuclear and mitochondrial proteins. *Mol. Cell Biol.* **19**: 3869–3876.

24. Stierum, R. H., Dianov, G. L. and Bohr, V. A. (1999). Single-nucleotide patch base excision repair of uracil in DNA by mitochondrial protein extracts. *Nucleic Acids Res.* **27**: 3712–3719.

25. Dianov, G. and Lindahl, T. (1994). Reconstitution of the DNA base excision-repair pathway. *Curr. Biol.* **4**: 1069–1076.

26. Frosina, G., Fortini, P., Rossi, O., Carrozzino, F., Raspaglio, G., Cox, L. S., Lane, D. P., Abbondandolo, A. and Dogliotti, E. (1996). Two pathways for base excision repair in mammalian cells. *J. Biol. Chem.* **271**: 9573–9578.

27. Krokan, H. E., Standal, R. and Slupphaug, G. (1997). DNA glycosylases in base excision repair of DNA. *Biochem. J.* **325**: 1–16.

28. Lindahl T. (2000). Suppression of spontaneous mutagenesis in human cells by DNA base excision-repair. *Mutat Res.* **462**: 129–135.

29. Anderson, C. T. and Friedberg, E. C. (1980). The presence of nuclear and mitochondrial uracil-DNA glycosylase in extracts of human KB cells. *Nucleic Acids Res.* **8**: 875–888.
30. Domena, J. D. and Mosbaugh, D. W. (1985). Purification of nuclear and mitochondrial uracil-DNA glycosylase from rat liver. Identification of two distinct subcellular forms. *Biochemistry* **24**: 7320–7328.
31. Nilsen, H., Otterlei, M., Haug, T., Solum, K. and Nagelhus, T. A. *et al.* (1997). Nuclear and mitochondrial uracil-DNA glycosylases are generated by alternative splicing and transcription from different positions in the *ung* gene. *Nucleic Acids Res.* **25**: 750–755
32. Haug, T., Skorpen, F., Aas, P. A., Malm, V., Skjelbred, C. and Krokan, H. E. (1998). Regulation of expression of nuclear and mitochondrial forms of human uracil-DNA glycosylase. *Nucleic Acids Res.* **26**: 1449–1457.
33. Croteau, D. L., ap Rhys, C. M., Hudson, E. K., Dianov, G. L. and Hansford, R. G. *et al.* (1997). An oxidative damage-specific endonuclease from rat liver mitochondria. *J. Biol. Chem.* **272**: 27 338–27 344.
34. Takao, M., Aburatani, H., Kobayashi, K. and Yasui, A. (1998). Mitochondrial targeting of human DNA glycosylases for repair of oxidative DNA damage. *Nucleic Acids Res.* **26**: 2917–2922.
35. Pinz, K. G., Shibutani, S. and Bogenhagen, D. F. (1995). Action of mitochondrial polymerase gamma at sites of base loss or oxidative damage. *J. Biol. Chem.* **270**: 9202–9206.
36. Kang, D., Nishida, J., Iyama, A., Nakabeppu, Y., Furuichi, M., Fujiwara, T., Sekiguchi, M. and Takageshi, K. (1995). Intracellular localization of 8-oxo-dGTPase I human cells, with special reference to the role of the enzyme in mitochondria. *J. Biol. Chem.* **270**: 14 659–14 665.
37. Stierum, R. H., Croteau, D. L. and Bohr, V. A. (1999). Purification and characterization of a mitochondrial thymine glycol endonuclease from rat liver. *J. Biol. Chem.* **274**: 7128–7136.
38. Longley, M. J., Prasad, R., Srivastava, D. K., Wilson, S. H. and Copeland, W. C. (1998). Identification of 5′-deoxyribose phosphate lyase activity in human DNA polymerase gamma and its role in mitochondrial base excision repair in vitro. *Proc. Natl. Acad. Sci. USA* **95**: 12 244–12 248.
39. Doetsch, P. W. and Cunningham, R. P. (1990). The enzymology of apurinic/apyrimidinic endonucleases. *Mutat. Res.* **236**: 173–201.
40. Tomkinson, A. E., Bonk, R. T. and Linn, S. (1988). Mitochondrial endonuclease activities specific for apurinic/apyrimidinic sites in DNA from mouse cells. *J. Biol. Chem.* **263**: 12 532–12 537.
41. Yasui, A., Yajima, H., Kobayashi, T., Eker, A. P. and Oikawa, A. (1992). Mitochondrial DNA repair by photolyases. *Mutat. Res.* **273**: 231–236.
42. Myers, K. A., Saffhill, R. and O'Connor, P. J. (1988). Repair of alkylated purines in the hepatic DNA of mitochondria and nuclei in the rat. *Carcinogenesis* **9**: 285–292.

43. Reenan, R. A. and Kolodner, R. D. (1992). Isolation and characterization of two *Saccharomyces cerevisiae* genes encoding homologs of the bacterial HexA and MutS mismatch repair proteins. *Genetics* **132**: 963–973.

44. Pont-Kingdon, G. A., Okada, N. A., Macfarlane, J. L., Beagley, C. T., Wolstenholme, D. R., Cavalier-Smith, T. and Clark-Walker, G. D. (1995). A coral mitochondrial mutS gene. *Nature* **375**: 109–111.

45. Pirsel, M. and Bohr, V. A. (1993). Methyl methanesulfonate adduct formation and repair in the DHFR gene and in mitochondrial DNA in hamster cells. *Carcinogenesis* **14**: 2105–2108.

46. Cullinane, C. and Bohr, V. A. (1998). DNA interstrand cross-links induced by psoralen are not repaired in mammalian mitochondria. *Cancer Res.* **58**: 1400–1404.

47. Thyagarajan, B., Padua, R. A. and Campbell, C. (1996). Mammaliam mitochondria possess homologous DNA recombination activity. *J. Biol. Chem.* **271**: 27 536–27 543.

48. Poulton, J., Deadman, M. E., Bindoff, L., Morten, K., Land, J. and Brown, G. (1993). Families of mtDNA re-arrangements can be detected in patients with mtDNA deletions: duplications may be a transient intermediate form. *Human Mol. Genet.* **2**: 23–30.

49. Tang, T., Manfredi, G., Hirano, M. and Schon, E. A. (2000). Maintenance of human rearranged mitochondrial DNAs in long-term cultured trans-mitochondrial cell lines. *Mol. Biol. Cell* **11**: 2349–2358.

50. Anson, R. M. and Bohr, V. A. (2000). Mitochondria, oxidative DNA damage and aging. *J. Amer. Aging Assoc.* **23**: 199–218.

51. Hart, R. W. and Setlow, R. B. (1974). Correlation between deoxyribonucleic acid excision repair and life span in a number of mammalian species. *Proc. Natl. Acad. Sci. USA* **71**: 2169–2173

52. Guo, Z., Heydari, A. and Richardson, A. (1998). Nucleotide excision repair of actively transcribed versus nontranscribed DNA in rat hepatocytes: effect of age and dietary restriction. *Exp. Cell Res.* **245**: 228–238.

53. Souza-Pinto, N. C., Croteau, D. L., Hudson, E. K., Hansford, R. G. and Bohr, V. A. (1999). Age-associated increase in 8-oxo-deoxyguanosine glycosylase/AP lyase activity in rat mitochondria. *Nucleic Acids. Res.* **27**: 1935–1942.

54. Perez-Campo, R., Lopez-Torres, M., Cadenas, S., Rojas, C. and Barja, G. (1998). The rate of free radical production as a determinant of the rate of aging: evidence from the comparative approach. *J. Comp. Physiol.* **B168**: 149–158.

55. Souza-Pinto, N. C., Hogue, B. A., Eide, L., Stevnsner, T., Seeberg, E., Klungland, A. and Bohr, V. A. (2001). Repair of 8-oxo-dG in mitochondrial DNA in mice depends on the OGG1 gene. *Cancer Res.* **61**: 5378–5381.

56. Souza-Pinto, N. C., Hogue, B. A. and Bohr, V. A. (2001). DNA repair and aging in mouse liver: 8-oxo-dG glycosylase activity increases in mitochondrial but not in nucleus. *Free Radic. Biol. Med.* **30**(8): 916–923.

Chapter 45

α-Lipoic Acid

Chandan K. Sen*, Savita Khanna and Sashwati Roy

Chandan K. Sen, Savita Khanna and **Sashwati Roy** • Laboratory of Molecular Medicine, Departments of Surgery and Molecular & Cellular Biochemistry, 512 Heart & Lung Research Institute, The Ohio State University Medical Center, 473 W. 12th Avenue, Columbus, OH 43210

*Corresponding Author.
Tel: 1 614 247 7786, E-mail: sen-1@medctr.osu.edu

1. Introduction

α-Lipoic acid also known as thioctic acid, 1,2-dithiolane-3-pentanoic acid, 1,2-dithiolane-3-valeric acid or 6,8-thioctic acid has generated considerable clinical interest as a thiol-replenishing and redox-modulating agent.[1-5] At physiological pH, lipoic acid is anionic and commonly referred to as *lipoate*. Oxidant-antioxidant reactions are essentially redox reactions in which reactive forms of oxygen are reduced by electrons from antioxidants; in the process, the antioxidant is oxidized often to its functionally inert form. Effective functioning of redox-active antioxidants requires the recycling of the oxidized form of antioxidant to its potent reduced form. Based on the current literature it may be argued that α-lipoic is not endogenously generated in the free form at levels that could support its antioxidant property.[6] Uniquely, lipoic acid is the only thiol-replenishing safe human nutrient known so far that is readily taken up by cells and promptly reduced by cellular enzymes to dihydrolipoic acid. To this end, lipoic acid has a clear advantage over N-acetyl-L-cysteine as a thiol-replenishing antioxidant nutrient.[7]

Thiols, ubiquitously distributed in the aerobic cells, have been linked with a variety of key physiological functions.[1, 4, 5, 8] Disulfide linkages are critical determinants of protein structure. The cellular glutathione pool not only serves antioxidant functions but also serves as a reservoir of cysteine for protein synthesis. The oxidation-reduction or redox state of protein-thiols is suggested to regulate a number of key signal transduction processes.[1, 5] Much of our current interest in thiols instigate from consistent observations that several pathophysiological conditions such as some forms of cancer and AIDS are associated with lowered thiol status in cells such as lymphocytes.[9, 10] Corrections of such perturbations in cellular thiol homeostasis have resulted in beneficial clinical measures of outcome. Lipoic acid presents significant therapeutic opportunity to treat disorders involving perturbed redox status of cellular thiols. Lipoic acid has been shown to beneficially influence age-related changes in the heart and liver[11, 12] as well as exercise-induced oxidative stress.[13] This chapter presents a concise overview of the current developments in lipoic acid research.

2. Reduction of Lipoate to Dihydrolipoate: The Metabolic Antioxidant

Cellular reducing equivalents such as NADH or NADPH produced as a result of cellular metabolism, serve as cofactors of enzymes such as reductases or dehydrogenases in bio-reduction processes. For physiological antioxidants like glutathione (GSH), glutathione disulfide (GSSG) reductase utilizes NADPH to reduce GSSG. The problem, however, with GSH as a nutritional supplement[14] is that it is poorly bio-available to mammalian cells/tissues. Among the several agents that have been tested for their efficacy as thiol replenishing drugs N-acetylcysteine and lipoic acid have been proven to be safe and effective.[7] Lipoic acid is often referred to as a *metabolic antioxidant*. This terminology is used to reflect the fact that enzymes in the human cells and tissues recognize lipoic acid as a substrate and reduce it to dihydrolipoic acid at the expense of cellular NAD(P)H.[2] As cells treated with lipoic acid utilize more of these reducing equivalents, more substrate is taken up and metabolized by the cell. Indeed, lipoic acid is known to stimulate glucose uptake by cells.[15-18] Thus, a unique property of lipoic acid as a supplement is that it can harness the power of the cell's own metabolic processes for its recycling and potency.

Lipoic acid contains a disulfide-bond that forms a dithiolane ring. Reduction of this disulfide bond results in ring-opening forming the vicinal dithiol, dihydrolipoic acid. The redox potential of the lipoate-dihydrolipoate couple is −320 mV. Thus, dihydrolipoic acid is a strong reductant capable of chemically reducing GSSG, the redox potential of the GSSG-GSH couple being −240 mV.[19] Exogenously supplemented lipoic acid gets reduced to dihydrolipoic acid in several biological systems including mitochondria, bacteria, perfused rat liver, isolated hepatocytes, erythrocytes, keratinocytes and lymphocytes.[20] The mitochondrial E3 enzyme, dihydrolipoyl dehydrogenase reduces lipoic acid to dihydrolipoic acid at the expense of NADH. The enzyme shows a marked preference for the naturally occurring R-enantiomer of lipoic acid.[21] Lipoic acid is also a substrate for the NADPH dependent enzyme GSSG reductase.[22] GSSG reductase shares a high degree of structural homology with lipoamide dehydrogenase. Both are homo-dimeric enzymes with 50 kDa subunits conserved between all species. In contrast to dihydrolipoyl dehydrogenase, however, GSSG reductase exhibits a preference for the S-enantiomer of lipoic acid. Although lipoic acid is recognized as a substrate by GSSG reductase, the rate of reduction is much slower than that of the natural substrate GSSG. Whether lipoic acid will be reduced in a NADH or NADPH dependent mechanism is largely tissue-specific. Thioredoxin reductase catalyzes the NADPH dependent reduction of oxidized thioredoxin. Thioredoxin reductase from calf thymus and liver, human placenta and rat liver efficiently reduces both lipoic acid as well as lipoamide with Michaelis-Menton type of kinetics in NADPH dependent reactions.[23] Under similar conditions at 20°C, pH 8.0 mammalian thioredoxin reductase reduced lipoic acid fifteen times more efficiently than the corresponding NADH dependent lipoamide dehydrogenase action. The relative contribution of the three different enzymes found to reduce lipoic acid in mammalian cells is tissue and cell specific depending on the presence or absence of mitochondrial activity as well as natural substrates such as oxidized thioredoxin and GSSG. Studies with Jurkat T cells have shown that when added to the culture medium, lipoic acid readily enters the cell where it is reduced to dihydrolipoic acid. This reduced product first accumulates in the cell and is then released to the culture medium.[20, 24]

3. Lipoyl Residues in Oxidative Metabolism

In 1951 α-lipoic acid was identified to be an essential cofactor in oxidative metabolism.[25] Biologically, lipoic acid exists as lipoamide in at least five proteins where it is covalently linked to a lysyl residue. Four of these proteins are found in α-ketoacid dehydrogenase complexes, the pyruvate dehydrogenase complex, the branched chain keto-acid dehydrogenase complex and the α-ketoglutarate dehydrogenase complex. Three of the lipoamide containing proteins are present in the E2 enzyme dihydrolipoyl acyltransferase, which is different in each of the complexes and specific for the substrate of the complex. One lipoyl residue is found in protein X, which is the same in each complex. The fifth lipoamide residue is present in the glycine cleavage system.[26] Lipoamide containing enzymes of the α-keto acid dehydrogenase complexes oxidatively decarboxylate their substrates producing NADH. Human pyruvate dehydrogenase complex contains two lipoyl domains. Protein X is suggested to play a structural role to allow proper functioning of the pyruvate dehydrogenase complex.[27] Each complex comprises of about six copies of protein X that contains a lipoate-bound domain. The glycine cleavage system catalyzes the oxidation of glycine to CO_2 and NH_3 generating NADH and 5,10-methylenetetrahydrofolate as a result.[2, 28]

Dihydrolipoamide dehydrogenase, an integral component of α-ketoacid dehydrogenase complex and the glycine cleavage system, oxidizes dihydro-lipoamide to lipoamide. In the presence of lipoic acid, this reaction is also driven in the reverse direction at the expenses of NADH.[26] It seems likely that dihydrolipoamide dehydrogenase may also act independent of the α-ketoacid dehydrogenase complexes and the glycine cleavage system. Plasma membrane associated dihydrolipoamide dehydrogenase has been found in *Trypanosoma brucei* and rat adipocytes.[29–31] The exact function of this membrane-associated enzyme is yet unclear. In eukaryotes, the enzyme has been connected with insulin-stimulated hexose transport. Although the possible role of dihydrolipamide dehydrogenase in hexose transport is not yet clear, several studies have reported enhanced glucose uptake following exogenous lipoic acid treatment.[15–18, 32]

Lipoic acid is required as lipoyl residues for the proper functioning of the mitochondrial enzymes discussed above. Therefore, lipoic acid is often projected as a mitochondrial coenzyme.[11, 33] In this context, it is important to note that so far there is no convincing evidence to prove that lipoyl residues covalently bound to protein may have antioxidant properties. Evidence supporting that lipoic acid is a potent antioxidant is derived from the study of exogenously supplemented free lipoic acid. In addition, free lipoic acid has not been detected in any biological sample that has not been supplemented with exogenous lipoic acid.[6, 34] At the same time, it has been observed that lipoic acid supplementation does not alter the lipoyllysine (tissue-bound) form of lipoic acid. While nutritional lipoic acid supplementation increased tissue levels of free lipoic acid in the tissue, the lipoyllysine pools of the same tissue was not affected.[6] These results suggest that the activity of tissue lipoyltransferase[35-37] is not limited by the availability of free lipoic acid and that other regulatory mechanisms tightly control tissue lipoylation. In summary, it is true that lipoyl residues participate in mitochondrial oxidative metabolism and that free lipoic acid is a biological antioxidant. However, it is unlikely that supplemented lipoic acid will influence the function of lipoyl-residue containing mitochondrial enzymes or that lipoyl-residues in mitochondrial enzymes would function as an antioxidant.

4. Redox Properties: Biological Implications

In 1959, Rosenberg and Culik[38] suggested an antioxidant function of lipoic acid. They observed that administration of lipoic acid prevented symptoms of scurvy in vitamin C deficient guinea-pigs. They also observed that lipoic acid prevented vitamin E deficiency symptoms in rats fed with a tocopherol-deficient diet. Lipoic acid scavenges several reactive species including hydroxyl radicals, hydrogen peroxide, hypochlorous acid and singlet oxygen. In addition, lipoic acid has transition metal chelation properties by virtue of which it may avert the transformation of superoxide anion to the deleterious hydroxyl radical. The chemical and biochemical aspects of the antioxidant and redox properties of lipoic acid and dihydrolipoic acid have been extensively reviewed.[2, 26, 39] The following section will focus on the various biological systems that are influenced by the redox-active properties of lipoic and dihydrolipoic acid.

4.1. NF-κB

NF-κB is a redox sensitive transcription factor the function of which has been related to a number of clinical disorders. The activity of NF-κB is inducible in response to a wide range of stimuli including peroxide, cytokines, phosphatase inhibitors and viral products.[1, 40, 41] It is suggested that reactive oxygen species may serve as a common intracellular messenger for NF-κB activation in response to the diverse range of stimuli. A member of the Rel family of transcription factors, NF-κB in its dormant form is localized in the cytosol as hetero- or homodimeric proteins associated with a inhibitory protein IκB. Following stimulation of cells by appropriate stimuli, IκB is phosphorylated, dissociated from the dimer and degraded. As a result, NF-κB proteins are able to migrate to the nucleus under the guidance of a nuclear localization signal. In the nucleus, NF-κB binds to a consensus κB site under reducing conditions. Depending on the nature of the NF-κB protein bound to the DNA and the local microenvironment, transcription is switched on. A large number of genes including those of cytokines, adhesion molecules, growth factors, immunoreceptors, NO synthase, some viral genes and IκB are regulated by NF-κB activity.[42]

Several antioxidants and reducing agents including ebselen, β-mercaptoethanol, pyrrolidinedithiocarbamate, desferrioxamine, dithiolthione, N-acetylcysteine, vitamin E derivatives and butylated hydroxyanisole have proven effective for their ability to inhibit NF-κB activation.[1, 10] Treatment of Jurkat T cells with high micromolar concentrations of lipoic acid suppressed phorbol ester or tumor necrosis factor TNFα induced activation of NF-κB in a dose-dependent manner.[43, 44] This NF-κB inhibitory effect was also seen with dihydrolipoic acid. Direct addition of dihydrolipoic acid to the cell culture medium suppressed TNFα induced NF-κB activation.[45] Both R- and S-enantiomers of lipoic acid were effective with respect to the NF-κB inhibitory function.[43, 44] Later it was observed that the ability of lipoic acid to inhibit NF-κB activation is not dependent on its ability to increase cellular glutathione. Simultaneous treatment of cells with a glutathione synthesis inhibitor buthionine sulfoximine and lipoic acid for 18 h resulted in 95% cellular GSH depletion. Lipoic acid treatment could not increase glutathione levels in these cells, however, lipoic acid was able to inhibit NF-κB activation induced by phorbol ester, TNFα or hydrogen peroxide.[1] Thus, the effects of lipoic acid on inducible NF-κB signaling were not simply mediated by enhanced cellular glutathione.

Although reactive oxygen species have been suggested to function as a common intracellular messenger in the NF-κB activation cascade in response to a variety of stimuli, the underlying mechanisms remain to be characterized. Jurkat T cells are not responsive to hydrogen peroxide with respect to NF-κB activation, however, in a subclone of these cells developed by Dr. Patrick Baeuerle (Frieburg, Germany) and named Wurzburg cells hydrogen peroxide treatment results is marked activation of NF-κB.[46, 47] These two related cell lines with contrasting peroxide sensitivity were used to reveal the possible factors that are responsible for oxidant-sensitivity of Wurzburg cells.

Determination of intracellular Ca^{2+} concentration ($[Ca^{2+}]i$) revealed that 0.25 mM hydrogen peroxide treatment increased $[Ca^{2+}]i$. In the presence of extracellular calcium chelators, it was observed that the Ca^{2+} flux was mainly contributed by calcium released from intracellular stores. Although Wurzburg cells are derived from Jurkat T cells, a marked difference in the nature of this oxidant induced calcium flux was noted in these two cell types. In Jurkat, the calcium flux was rapid and transient. Within 10–15 min after oxidant treatment, intracellular calcium concentration was restored to pre-treatment levels. In contrast, the calcium response in Wurzburg cells was slower in kinetics and sustained for a longer time.[47]

Two major steps in the activation of NF-κB are the phosphorylation and degradation of IκB. Both are likely to be supported by high $[Ca^{2+}]i$. IκB contains a PEST sequence of amino acid and is thus highly susceptible to proteolytic cleavage. It has been shown that degradation of such PEST containing sequence may be catalyzed by proteases such as m-calpain,[48] the activity of which is calcium dependent. Because high $[Ca^{2+}]i$ could influence both phosphorylation and degradation of IκB, the hypothesis that the peroxide induced differential calcium response in Jurkat and Wurzburg cells is linked to their respective NF-κB responses was tested. In Wurzburg cells loaded with the lipophilic esterified calcium chelator EGTA-AM, hydrogen peroxide failed to activate NF-κB. This observation provided the first hint that intracellular calcium flux in response to hydrogen peroxide treatment may be involved in the NF-κB activation process. Slow and sustained flux of calcium within the cell was observed to be a significant factor in oxidant-induced NF-κB activation.[47] To substantiate this contention, experiments were designed to test whether hydrogen peroxide would be able to trigger NF-κB activation under conditions where intracellular free calcium levels were maintained high on a sustained basis. Such manipulation of the intracellular calcium level was made possible by treating the cells with thapsigargin, a sarco-endoplasmic reticulum calcium pump inhibitor. The sarco-endoplasmic reticulum serves as a major storehouse of intracellular calcium. Calcium is sequestered from the cytosol and retained in this organelle against a high concentration gradient by the active function of the sarco-endoplasmic reticular calcium pumps. Inhibition of these pumps resulted in a release of stored calcium to the cytosol resulting in a high level of intracellular free calcium for at least 1 hour. Thapsigargin treatment alone resulted in a weak NF-κB activation. This activation was markedly potentiated by hydrogen peroxide co-treatment of the Jurkat cells. Thus, NF-κB activation in Jurkat T cells did respond to hydrogen peroxide under conditions of elevated intracellular free calcium levels. This activation could be completely inhibited by the intracellular calcium chelator EGTA-AM demonstrating the involvement of intracellular calcium in oxidant-induced NF-κB activation.[47]

Both lipoic acid and N-acetyl-L-cysteine are known to suppress NF-κB activation in response to a wide variety of activation stimuli. Pretreatment of cells with these antioxidants altered oxidant-induced calcium response. Studies with indo-1 loaded cells revealed that such pretreatment certainly decreases oxidant-induced perturbation of intracellular calcium homeostasis.[1, 47] In addition to suppression of cytosolic activation of NF-κB, lipoic acid also influences the binding of activated NF-κB proteins to the consensus κB site. *In vitro* DNA binding studies showed that dihydrolipoic acid enhances such DNA binding.[45] This effect of dihydrolipoic acid may be attributed to its strong reducing properties because reductants are well known to enhance NF-κB DNA binding. NF-κB dependent transactivation requires an oxidizing environment in the cytosol for NF-κB activation, and a reducing environment in the nucleus for DNA binding. During electrophoretic mobility shift assay of NF-κB this reducing atmosphere of the DNA binding mixture is achieved by the use of mM concentrations of the potent reductant dithiothreitol.

4.2. Acquired Immunodeficiency Syndrome (AIDS)

Human immunodeficiency virus infection eventually leads to a substantial fall of helper T cell (CD4$^+$) count in peripheral circulation. As a result, patients yield to opportunistic infections, widespread immune dys-regulation and certain neoplasms. The long terminal repeat region of HIV-1 proviral DNA contains two NF-κB binding sites.[49, 50] DNA binding of NF-κB can activate HIV transcription. Thus, strategies to suppress NF-κB dependent transactivation may be of therapeutic importance to delay the onset and progression of AIDS. Previously we have discussed that N-acetylcysteine is able to suppress NF-κB activation in response to a wide variety of stimuli. Consistently, N-acetylcysteine also inhibited HIV LTR-directed expression of β-galactosidase gene in response to TNFα and phorbol ester.[51] N-acetylcysteine facilitates cellular GSH synthesis by improving the supply of the rate limiting substrate cysteine.[7] However, whether transcriptional regulation of N-acetylcysteine is linked to the pro-GSH effect of N-acetylcysteine is yet unclear. Certain transcription dependent effects of N-acetylcysteine such as the enhanced survival of PC12 cells is GSH independent.[52]

α-Lipoic acid inhibited the replication of HIV-1 in cultured lymphoid T cells.[53] Jurkat, SupT1 and Molt-4 cells were infected with HTLV IIB and HIV-1 Wal, and these cells were treated with lipoic acid 16 h after infection. A dose dependent effect of lipoic acid on the inhibition of reverse transcriptase activity and plaque formation was observed. At 70 mg/ml or 350 μM of lipoic acid, reverse transcriptase activity was decreased by 90% and at 35 μg/ml plaque forming units were completely eliminated. In a previous section of this chapter, it has been discussed that high intracellular calcium is involved in oxidant-induced NF-κB activation. Using thapsigargin, cytosolic levels of free calcium can be elevated. Such condition resulted in oxidant-induced NF-κB activation in Jurkat T cells that are otherwise known to be insensitive to such activation.[46] Independent studies have shown that thapsigargin treatment of T-lymphocyte cells results in a marked activation of HIV production.[54] Viral activation was manifested by increases in soluble viral core p24 production, increases in cellular levels of viral antigens, and increased viral transcription. Such calcium dependent activation of the transcription of pro-viral HIV may be mediated by Ca^{2+}-dependent NF-κB activation as suggested previously.[1, 47] Pretreatment of cultured T cells with lipoic acid or N-acetylcysteine diminished oxidant induced perturbation of intracellular calcium homeostasis.[1, 47] This mechanism could partly explain some of the beneficial properties of both lipoic acid as well as N-acetylcysteine against HIV infection.

Characteristically, HIV+ individuals have decreased levels of acid-soluble thiols, cysteine and GSH in particular, in their plasma and leukocytes.[55] Because decreased thiol status is a typical marker of oxidative stress, this observation was the first to associate HIV infection with oxidative stress. One other critical observation made by the same group was that the plasma level of glutamate of HIV infected patients was significantly higher.[56] Availability of cysteine is the rate-limiting step in cellular glutathione synthesis.[7] Because of its marked instability in the reduced form more than 90% of extracellular cysteine is present as cystine.[57] Glutamate competitively inhibits cystine uptake by cells.[4, 7, 58] In this way, substrate for GSH synthesis within the cell is limited in the presence of high concentrations of extracellular glutamate. In Jurkat T cells, high levels of extracellular glutamate results in a 50% reduction in cellular GSH level.[59]

HIV infected individuals have compromised T cell thiol status. The decrease in T cell GSH did not correlate with the absolute CD4 number and was only slightly correlated with the stage of the disease indicating that the perturbation of cellular GSH homeostasis occurs very early after HIV infection. An interesting trend in these observations was that a group of thiol-rich T cells were the most affected by infection.[60] This sub-population of T cells was decreased or absent in infected patients. N-acetylcysteine supplementation corrected T cell thiol status in HIV infected subjects and prolonged survival.[60] Using flow cytometry[61] to detect cellular thiols it has been observed that α-lipoic acid increases cellular GSH levels on a dose-dependent manner from 10 to 100 μM. α-Lipoic acid supports cells to circumvent inhibition of glutathione synthesis by glutamate. Based on results obtained in our laboratory we have developed the following mechanistic explanation. When treated to cells, α-lipoic acid rapidly enters the cells and is reduced to dihydrolipoic acid. Dihydrolipoic acid, expelled from the cell to the culture medium, reduces extracellular cystine to cysteine. Cysteine can bypass glutamate inhibition of cystine uptake, and can provide sufficient substrate for cellular GSH synthesis. Because these effects of lipoic acid are observed at concentrations 100 μM or below, they should be considered to be clinically relevant.[7]

Among the clinically relevant thiol-replenishing agents tested so far undoubtedly N-acetylcysteine and lipoic acid hold the most promise. Although in many respects the effect of N-acetylcysteine are quite similar to that of lipoic acid, much higher concentrations of N-acetylcysteine are required to produce comparable effects. Usually under experimental conditions 10–30 mM N-acetylcysteine is used to obtain NF-κB inhibition or similar other effects. A 10–40 fold less concentration of lipoic acid is clearly effective in suppressing NF-κB activation in response to a large number of stimuli. In Jurkat T cells although 100 μM N-acetylcysteine fail to enhance cellular GSH level, lipoic acid does so in a dose dependent manner from 10 to 100 μM.[59] Much of these concentration differences may be explained by the mechanism by which lipoic acid functions. As discussed previously, lipoic acid is able to harness the metabolic power of the cell to continuously regenerate its potent reduced form. In this way, the cellular dihydrolipoic acid pool may be continuously renewed at the expense of the cell's metabolic power. In contrast, hydrolysis of N-acetylcysteine serves as a donor of a single cysteine residue.[4, 7] N-acetylcysteine and lipoic acid were directly compared in an *in vitro* model of HIV-directed gene expression. At 200 μM lipoic acid treatment resulted in 40% decrease in HIV-1 p24 antigen expression in TNFα stimulated OM 10.1 cells latently transfected with HIV1. In contrast, 10 mM N-acetylcysteine was required to produce comparable effect.[62]

4.3. Caspase-Dependent Cell Death

Caspases are critical mediators of apoptotic cell death. The caspase family is diversified by the presence of over a dozen members. Caspase-1 was discovered first. The finding was based on the sequence similarity to the *C. elegans* death gene, ced-3. Caspase-1 was originally labeled ICE (for interleukin-1-beta-converting enzyme). Subsequently, the entire gene family was unveiled. The term *caspase* was created to denote the Cysteine requiring ASPartate proteASE activity of these enzymes. The protease activity of the caspase family is unique in that they cleave following (C-terminal to) aspartate residues (Asp-X), a property shared only by the cytotoxic lymphocyte serine protease, granzyme B. Among members of the caspase family, caspase 3 was shown to have the highest similarity to ced-3. All members of the caspase family contain the sequence QACXG that contains the active site cysteine. The putative active site of caspase 3 (CPP32, apopain, YAMA) contains a cysteine residue[63] that is subject to redox control.[64] Both thioredoxin and glutathione have been shown to be required for caspase 3 activity to induce apoptosis.[65]

Programmed cell death, or apoptosis, represents a highly controlled form of cell death in which single cells are selectively eliminated without release of cellular debris and perturbation of neighboring tissues.[66-68] Apoptosis regulates several key physiological functions. For example, developing lymphocytes undergo extensive cell death during selection of the immune repertoire. Understanding the fundamental mechanism of apoptosis is crucial to developing therapeutic strategies for controlling apoptosis in diseased tissues. Such information may be used to protect healthy cells against apoptosis, and also to selectively kill diseased cells. For example, inhibitors of apoptosis may be utilized to induce resistance to chemotherapeutic drugs and irradiation and inducers of apoptosis may be used to control neoplastic events that result from uncontrolled cell proliferation. Caspase 3 has been investigated as a therapeutic target to induce death of cancer cells.[69-71] Under certain conditions, reactive oxygen and nitrogen species have been shown to intercept caspase 3 mediated death. A physiological example of this is the NADPH oxidase-derived oxidants generated by stimulated neutrophils that prevent caspase activation in these cells.[72] Impairment of caspase-mediated death is also observed under conditions of GSH deficiency.[73] Other thiol blocking agents such as N-ethylmaleimide, or iodoacetamide have been shown to impair caspase function as well.[74]

The transmembrane Fas Ag is a member of the TNF/nerve growth factor receptor family, which can trigger apoptosis. Interaction between Fas–Fas ligand (FasL) transduces apoptotic signals in sensitive target cells. This pathway to induce programmed cell death has been suggested to be of potential use in cancer treatment. Treatment with anti-Fas Ab has been shown to suppress the growth of Fas bearing (Fas+) tumor cells.[75] Also, malignant glioma cells are susceptible to Fas mediated apoptosis triggered by agonistic Ab.[76] The killing of myelogenous leukemia cells by the Fas-FasL pathway has the remarkable potential of serving as a novel and effective approach for leukemia immunotherapy.[77] Proliferation of vascular smooth muscle cells in response to injury plays a central role in the pathogenesis of vascular disorders. FasL gene transfer to the wall of blood vessel induced apoptosis of Fas$^+$ vascular smooth muscle cells and inhibited neointima formation in injured rat carotid artery.[78] Thus, Fas-mediated apoptosis is expected to have therapeutic potential in certain disorders, especially cancer treatment. Reactive oxygen and nitrogen species have been observed to confer resistance to Fas-mediated apoptosis by inhibiting inducible caspase 3 activity. While excessive oxidative stress is likely to trigger necrosis and levels of reactive oxygen species much lower than that required to inflict oxidative damage may trigger death-signaling,[79–83] intermediary levels of reactive oxygen species in the cell interrupt programmed cell death by intercepting caspase function.[84–86]

α-Lipoic acid remarkably potentiated Fas mediated cell death in leukemic Jurkat cells, but not in healthy peripheral blood lymphocytes.[64] Previously chemotherapeutic agents such as doxorubicin, vincristine, and the alkaloid taxol have been shown to facilitate Fas-mediated cell death.[76, 87] Doxorubicin, vincristine and taxol are anti-tumor drugs that are used for cancer therapy. At high concentrations, these agents *per se* are toxic to cells. In contrast, lipoic acid is a safe nutrient, mostly known for its ability to bolster cellular glutathione levels, alter intracellular redox state and help protect against diabetic complications.[17, 88, 89] This work presented first evidence showing that a redox active agent may potentiate Fas-mediated death in leukemic Jurkat cells. Because the potentiating effect of lipoic acid treatment on Fas-mediated apoptosis was observed in one of the earliest markers of apoptosis (externalization of membrane phosphatidyl serine), it was suspected that lipoic acid regulates one or more early intracellular events. Expression of the Fas receptor was not influenced by lipoic acid treatment, suggesting that intracellular events signaling for apoptosis may have been influenced. An early event in Fas-mediated apoptosis that was strikingly influenced by lipoic acid treatment was the activation of the caspase 3. The potentiating effect of lipoic acid treatment on Fas-mediated apoptosis of Jurkat cells was markedly decreased by a caspase 3 inhibitor, indicating that indeed increased caspase 3 activity in lipoic acid treated Fas-activated cells played a significant role in potentiating cell death. As described in a previous section, low concentration of hydrogen peroxide inhibits caspase activity in Jurkat cells.[90] Consistently, it was observed that the activity of purified caspase 3 protein was inhibited by hydrogen peroxide. This report provided first evidence showing that indeed caspase 3 activity may be also potentiated by intracellular reducing agents such as dihydrolipoic acid demonstrating that inducible caspase 3 activity may be up-regulated pharmacologically.[64] In a later study, Baker *et al.* provided further support to the contention that reductants may facilitate caspase 3 activity.[91] They examined the ability of various recombinant human thioredoxins to activate caspase 3. The EC_{50} for caspase 3 activation by reduced thioredoxin-1 was 2.5 µM, by reduced glutathione 1.0 mM and by the bench-top reductant dithiothreitol 3.5 mM. A catalytic site redox-inactive mutant thioredoxin-1 was almost as active as thioredoxin-1 in activating caspase 3. Caspase activation was shown to correlate with the number of reduced cysteine residues in the thioredoxins. Reduced insulin and serum albumin were as effective on a molar basis as thioredoxin-1 in activating caspase 3.[91] Unwanted harmful cells may be sensitized to inducible death by the use of redox-active nutrients such as α-lipoic acid.

4.4. Cell Adhesion

Adhesion of leukocytes to endothelial cells in post capillary venules is one of the earliest as well as a critical step in acute and chronic inflammatory responses and is dependent upon the expression of a variety of cell surface receptors also known as cell adhesion molecules.[92] Apart from immune functions these cell adhesion molecules participate in orchestrating other vital biological processes, such as embryogenesis, cell growth, differentiation and wound repair.[93] The expression of cell adhesion molecules is induced in response to several stimuli such as cytokines, PMA, and lipopolysaccharide.[92] Thiol antioxidants down-regulate cytokine- or oxidant-induced expression of adhesion molecules.[94–102] At a concentration as high as 20 mM, N-acetylcysteine strongly inhibits ICAM-1 expression induced by H_2O_2 or cytokines in keratinocytes, whereas under the same conditions pyrrolidine dithiocarbamate was less effective in preventing inducible ICAM-1 expression.[99] However, both N-acetylcysteine and pyrrolidine dithiocarbamate have been reported to modulate the induction of cell adhesion molecule expression by IL-1α in astrocytoma cells.[97] In HUVEC, interleukin-1β activated VCAM-1 gene expression has been observed to be repressed approximately 90% by the antioxidants PDTC (50 µM) and N-acetylcysteine (30 mM).[94] N-acetylcysteine (5–10 mM) and exogenous GSH (10-1000 µg/ml) have been shown to decrease ICAM-1 expression also in hyperoxia-exposed human pulmonary artery endothelial cells and HUVEC.[96] In contrast to the effect of N-acetylcysteine at high concentration, lower concentrations of the antioxidant (0.2 and 1 mM) potentiate IL-1α-induced expression of ICAM-1 and VCAM-1.[97]

Among the two thiol antioxidants, N-acetylcysteine and pyrrolidine dithio-carbamate, that have been widely studied for their ability to regulate the expression of molecules and cell–cell adhesion, N-acetylcysteine is a clinically safe drug.[7] However, in all studies showing the inhibitory effects of N-acetylcysteine on inducible ICAM-1 or VCAM-1 expression, high millimolar range (5–30 mM) of the thiol antioxidant was required. Pharmacokinetic study in humans show that only up to 25 μM of N-acetylcysteine is available in human plasma following oral intake.[103] Pyrrolidine dithiocarbamate has never been tested for safety in humans. Pharmaco-kinetic studies of α-lipoic acid show that following a single orally administered dose (10 mg/kg body wt), the plasma concentration may reach up to 60–70 μM. Higher concentrations of α-lipoic acid may be achieved if the drug is administered intravenously.[104] At clinically relevant doses (50–100 μM), lipoic acid down-regulated agonist induced adhesion of T-cells to endothelial cells, and agonist-induced ICAM-1 and VCAM-1 expression on endothelial cells.[101, 105]

4.5. Reductive Stress

In pathologies such as diabetes and ischemia-reperfusion, reductive (high NAD(P)H/NAD(P)$^+$ ratio) stress has been considered as one of the major factors contributing to these metabolic disorders.[106] α-Lipoic acid has been suggested to lower reductive stress in pathologies such as diabetes and ischemic injury by utilizing cellular NAD(P)H for its reduction to dihydrolipoic acid and thus lowering the cellular NAD(P)H/NAD(P)$^+$ ratio.[32] Consistently, beneficial effects of lipoic acid supplementation have been observed in diabetes and ischemia-reperfusion injury.[89, 107]

4.6. Glucose Uptake

α-Lipoic acid has been widely used in Germany for the treatment of diabetes related complications such as neuropathies. The therapeutic potential of lipoic acid in treating diabetic polyneuropathies has been reviewed.[108] The effects of lipoic acid were studied in two multicenter, randomized, double-blind placebo-controlled trials in Germany. Intravenous treatment with lipoic acid (600 mg/day) over 3 weeks was observed to be safe and effective in reducing symptoms of diabetic peripheral neuropathy. Oral treatment with 800 mg/day of lipoic acid for 4 months tended to improve cardiac autonomic dysfunction in non-insulin dependent diabetes mellitus.[109–113] In Type-II diabetic patients, lipoic acid also improved glucose disposal.[114] Several studies on experimental animals support that lipoic acid may have potent implication in the treatment of diabetes and related complications.[16, 88, 115–120] One of the major mechanisms by which lipoic acid may have beneficial effects in diabetes is stimulation of skeletal muscle glucose uptake. It has been shown that some of the elements of the insulin-signaling pathway participate in mediating the skeletal muscle glucose uptake stimulatory effect of lipoic acid.[15] Treatment of L6 myotubes with lipoic acid resulted in intracellular redistribution of GLUT1 and GLUT4 glucose transporters, similar to that caused by insulin, with minimal effects on GLUT3 transporters. Such effect of lipoic acid on the intracellular redistribution of glucose transporters was dependent on phosphatidylinositol 3-kinase activity.[15] More recent studies with rats show that the signaling pathway involved in mediating the skeletal muscle glucose uptake stimulatory effect of lipoic acid is not identical to that for insulin. It was observed that although a portion of lipoic acid action on glucose transport in mammalian skeletal muscle is mediated via the insulin signal transduction pathway, the majority of the direct effect of lipoic acid on skeletal muscle glucose transport is independent insulin signaling pathway.[117] Khanna *et al.* demonstrated that under conditions of acute infection that is accompanied with insulin resistance, lipoic acid might have therapeutic implications in restoring glucose availability in tissues such as the skeletal muscle.[17]

4.7. A Novel Form of Lipoic Acid

α-Lipoic acid has been detected in the form of lipoyllysine in various natural sources including edible plant and animal products.[6, 34] Following enzymatic reduction of lipoic acid to dihydrolipoic acid, most of the reduced form is rapidly effluxed from the cell to the culture medium.[20, 24] To improve the retention of dihydrolipoic acid in cells, the lipoic acid molecule has been chemically modified (Fig. 1) to confer a positive charge at physiological pH. As a result, N,N-dimethyl,N'-2-amidoethyl, 6,8-dimercapto octanoic acid was synthesized (US patent number 6090842, July 2000). Because this derivative of lipoic acid is positively charged as physiological pH it is commonly referred to as lipoic acid plus or LA-plus.[121] The chemical structure of LA-plus bears close resemblance with lipoyllysine. At physiological pH, LA-plus bears a positive charge and is expected to be taken up by cytoplasmic organelles such as mitochondria. The uptake of LA-plus by and retention of DHLA-Plus (corresponding dihydrolipoic acid form) in Wurzburg T cells was markedly better compared to those following lipoic acid treatment. Thus, LA-plus may be expected to be effective at much lower concentration than that required of lipoic acid. This hypothesis has been validated in studies considering inducible NF-κB activation as well as prevention of apoptotic death of thymocytes.[121]

The thymus is an organ that is larger in embryos and that gradually involutes throughout life. Apoptosis plays a major role in this age-related involution process. Thus, thymocytes are in a continuous state of spontaneous apoptosis. LA-plus, but not lipoic acid, at a concentration of 5 μM significantly retarded the rate of spontaneous apoptosis in isolated thymocytes. Previously, it has been shown that 4 mM dihydrolipoic acid or 2 mM lipoamide is necessary to inhibit etoposide-induced apoptosis in rat thymocytes, and that lipoic acid does not protect even

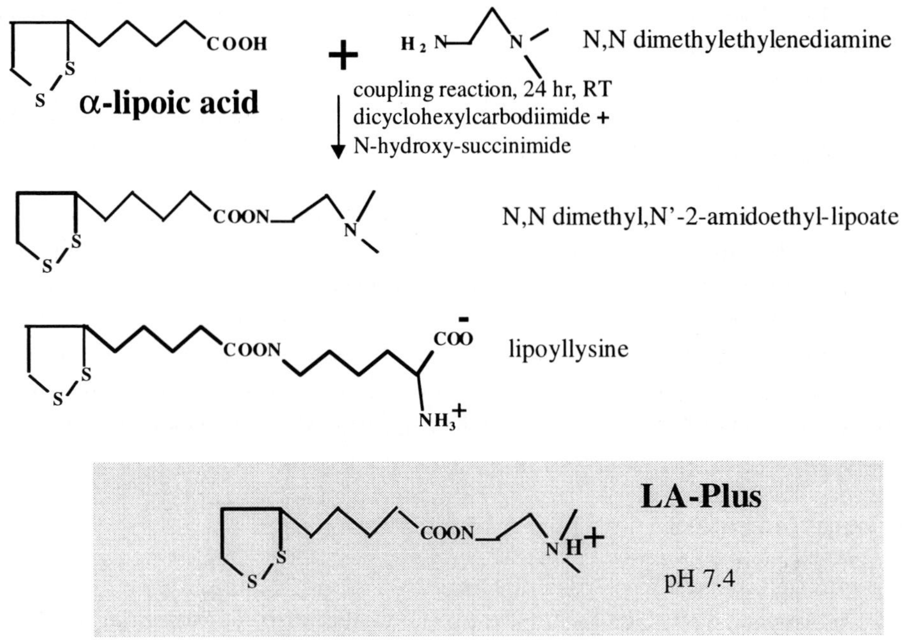

Fig. 1. **Overview of LA-plus synthesis and structural analogy of LA-plus with the naturally occurring form of α-lipoic acid, lipoyllysine.** LA-plus is synthesized by the coupling of lipoic acid with dimethylethylenediamine. For details see Sen *et al.*, 1998[121] and Tirosh *et al.*, 1999.[123]

at such high concentrations.[122] Consistent with the previous report 10 μM lipoic acid failed to afford any protection against etoposide-induced apoptosis. In contrast, only 5 μM of LA-plus was able to completely inhibit etoposide-induced apoptosis in isolated rat thymocytes. Thus at a concentration almost three orders of magnitude less than the concentration at which lipoic acid has been previously reported to be ineffective, LA-plus treatment provided complete protection.[121] In a later study, the relative efficacy of LA-plus and lipoic acid was compared in a model of glutamate-induced death of hippocampal neuronal cells.[123] Elevated levels of extracellular glutamate have been linked to reactive oxygen species mediated neuronal damage and brain disorders. Glutamate treatment for 12 h resulted in cytotoxicity. Measurement of intracellular peroxides showed marked (up to 200%) increase after 6 h of glutamate treatment. Compared to lipoic acid, LA-plus was more effective in (a) protecting cells against glutamate induced cytotoxicity, (b) preventing glutamate induced loss of intracellular GSH, and (c) disallowing increase of intracellular peroxide level following the glutamate challenge. The protective effect of LA-plus was found to be independent of its stereochemistry and was synergistically enhanced by selenium.[123] These results demonstrate that LA-plus is an exciting research tool that needs to be explored for its therapeutic potential.

5. Conclusion

α-Lipoic acid is safe nutrient that has been used by humans for decades. Given that oxidants play a key role in triggering oxidative stress as well as in driving redox-sensitive signal transduction processes that are closely linked to health and disease, the significance of thiol antioxidants has become even more prominent. Among the thiol-based antioxidant nutrients known so far lipoic acid clearly has several advantages over other alternatives. The development of LA-plus has opened up yet another exciting facet of lipoic acid research.

Acknowledgment

Supported by NIH GM 27345, the Surgery Wound Healing Research Program and US Surgical, Tyco Healthcare Group. The Laboratory of Molecular Medicine is the research division of the Center for Minimally Invasive Surgery.

References

1. Sen, C. K. and Packer, L. (1996). Antioxidant and redox regulation of gene transcription. *FASEB J.* **10**: 709–720.
2. Packer, L., Roy, S. and Sen, C. K. (1997). Alpha-lipoic acid: a metabolic antioxidant and potential redox modulator of transcription. *Adv. Pharmacol.* **38**: 79–101.
3. Sen, C. K. (1998). Redox signaling and the emerging therapeutic potential of thiol antioxidants. *Biochem. Pharmacol.* **55**: 1747–1758.
4. Sen, C. K. and Packer, L. (2000). Thiol homeostasis and supplements in physical exercise. *Am. J. Clin. Nutri.* **72**: 653S–669S.
5. Sen, C. K. (2000). Cellular thiols and redox-regulated signal transduction. *Curr. Topic Cell Regul.* **36**: 1–30.
6. Khanna, S., Atalay, M., Lodge, J. K., Laaksonen, D. E., Roy, S., Hanninen, O., Packer, L. and Sen, C. K. (1998). Skeletal muscle and liver lipoyllysine content in response to exercise, training and dietary alpha-lipoic acid supplementation. *Biochem. Mol. Biol. Int.* **46**: 297–306.
7. Sen, C. K. (1997). Nutritional biochemistry of cellular glutathione. *J. Nutri. Biochem.* **8**: 660–672.
8. Deneke, S. M. (2000). Thiol-based antioxidants. *Curr. Topic Cell Regul.* **36**: 151–180.

9. Sen, C. K., Packer, L. and Hanninen, O. (2001). *Handbook of Oxidants and Antioxidants in Exercise*, p. 1207, Elsevier, Amsterdam.

10. Sen, C. K., Sies, H. and Baeuerle, P. A. (2000). *Antioxidant and Redox Regulation of Genes*, p. 556, Acdemic Press, San Diego.

11. Lykkesfeldt, J., Hagen, T. M., Vinarsky, V. and Ames, B. N. (1998). Age-associated decline in ascorbic acid concentration, recycling, and biosynthesis in rat hepatocytes — reversal with (R)-alpha-lipoic acid supplementation. *FASEB J.* **12**: 1183–1189.

12. Hagen, T. M., Ingersoll, R. T., Lykkesfeldt, J., Liu, J., Wehr, C. M., Vinarsky, V., Bartholomew, J. C. and Ames, A. B. (1999). (R)-alpha-lipoic acid-supplemented old rats have improved mitochondrial function, decreased oxidative damage, and increased metabolic rate. *FASEB J.* **13**: 411–418.

13. Khanna, S., Atalay, M., Laaksonen, D. E., Gul, M., Roy, S. and Sen, C. K. (1999). Alpha-lipoic acid supplementation: tissue glutathione homeostasis at rest and after exercise. *J. Appl. Physiol.* **86**: 1191–1196.

14. Sen, C. K., Atalay, M. and Hanninen, O. (1994). Exercise-induced oxidative stress: glutathione supplementation and deficiency. *J. Appl. Physiol.* **77**: 2177–2187.

15. Estrada, D. E., Ewart, H. S., Tsakiridis, T., Volchuk, A., Ramlal, T., Tritschler, H. and Klip, A. (1996). Stimulation of glucose uptake by the natural coenzyme alpha-lipoic acid/thioctic acid: participation of elements of the insulin signaling pathway. *Diabetes* **45**: 1798–1804.

16. Khamaisi, M., Potashnik, R., Tirosh, A., Demshchak, E., Rudich, A., Tritschler, H., Wessel, K. and Bashan, N. (1997). Lipoic acid reduces glycemia and increases muscle GLUT4 content in streptozotocin-diabetic rats. *Metabolism* **46**: 763–768.

17. Khanna, S., Roy, S., Packer, L. and Sen, C. K. (1999). Cytokine-induced glucose uptake in skeletal muscle: redox regulation and the role of alpha-lipoic acid. *Am. J. Physiol.* **276**: R1327–R1333.

18. Yaworsky, K., Somwar, R., Ramlal, T., Tritschler, H. J. and Klip, A. (2000). Engagement of the insulin-sensitive pathway in the stimulation of glucose transport by alpha-lipoic acid in 3T3-L1 adipocytes. *Diabetologia* **43**: 294–303.

19. Jocelyn, P. C. (1967). The standard redox potential of cysteine-cytine from the thiol-disulfide exchange reaction with glutathione and lipoic acid. *Eur. J. Biochem.* **2**: 327–331.

20. Handelman, G. J., Han, D., Tritschler, H. and Packer, L. (1994). Alpha-lipoic acid reduction by mammalian cells to the dithiol form, and release into the culture medium. *Biochem. Pharmacol.* **47**: 1725–1730.

21. Haramaki, N., Han, D., Handelman, G. J., Tritschler, H. J. and Packer, L. (1997). Cytosolic and mitochondrial systems for NADH- and NADPH-dependent reduction of alpha-lipoic acid. *Free Radic. Biol. Med.* **22**: 535–542.

22. Pick, U., Haramaki, N., Constantinescu, A., Handelman, G. J., Tritschler, H. J. and Packer, L. (1995). Glutathione reductase and lipoamide dehydrogenase have opposite stereospecificities for alpha-lipoic acid enantiomers. *Biochem. Biophys. Res. Commun.* **206**: 724–730.

23. Arner, E. S., Nordberg, J. and Holmgren, A. (1996). Efficient reduction of lipoamide and lipoic acid by mammalian thioredoxin reductase. *Biochem. Biophys. Res. Commun.* **225**: 268–274.

24. Han, D., Handelman, G., Marcocci, L., Sen, C. K., Roy, S., Kobuchi, H., Tritschler, H. J., Flohe, L. and Packer, L. (1997). Lipoic acid increases de novo synthesis of cellular glutathione by improving cystine utilization. *Biofactors* **6**: 321–338.

25. Reed, L. J., DeBusk, B. G., Gunsalus, I. C. and Hornbeger, J. (1951). Crystalline-α-lipoic acid: a catalytic agent associated with pyruvate dehydrogenase. *Science* **114**: 93–94.

26. Packer, L., Witt, E. H. and Tritschler, H. J. (1995). Alpha-lipoic acid as a biological antioxidant. *Free Radic. Biol. Med.* **19**: 227–250.

27. Lawson, J. E., Behal, R. H. and Reed, L. J. (1991). Disruption and mutagenesis of the *Saccharomyces cerevisiae* PDX1 gene encoding the protein X component of the pyruvate dehydrogenase complex. *Biochemistry* **30**: 2834 2839.

28. Packer, L. and Tritschler, H. J. (1996). Alpha-lipoic acid: the metabolic antioxidant. *Free Radic. Biol. Med.* **20**: 625–626.

29. Danson, M. J. (1988). Dihydrolipoamide dehydrogenase: a "new" function for an old enzyme? *Biochem. Soc. Trans.* **16**: 87–89.

30. Danson, M. J., Conroy, K., McQuattie, A. and Stevenson, K. J. (1987). Dihydrolipoamide dehydrogenase from *Trypanosoma brucei*. Characterization and cellular location. *Biochem. J.* **243**: 661–665.

31. Danson, M. J., Eisenthal, R., Hall, S., Kessell, S. R. and Williams, D. L. (1984). Dihydrolipoamide dehydrogenase from halophilic archaebacteria. *Biochem. J.* **218**: 811–818.

32. Roy, S., Sen, C. K., Tritschler, H. J. and Packer, L. (1997). Modulation of cellular reducing equivalent homeostasis by alpha-lipoic acid. Mechanisms and implications for diabetes and ischemic injury. *Biochem. Pharmacol.* **53**: 393–399.

33. Femiano, F., Gombos, F., Scully, C., Busciolano, M. and Luca, P. D. (2000). Burning mouth syndrome (BMS): controlled open trial of the efficacy of alpha-lipoic acid (thioctic acid) on symptomatology. *Oral. Dis.* **6**: 274–277.

34. Lodge, L., Handelman, G. J., Konishi, T., Matsugo, S., Mathur, V. V. and Packer, L. (1997). Natural sources of lipoic acid: determination of lipoyllysine released from protease-digested tissues by high performane liquid chromatography incorporating electrochemical detection. *J. Appl. Nutri.* **in press**.

35. Fujiwara, K., Okamura-Ikeda, K. and Motokawa, Y. (1990). cDNA sequence, in vitro synthesis, and intramitochondrial lipoylation of H-protein of the glycine cleavage system. *J. Biol. Chem.* **265**: 17 463–17 467.

36. Fujiwara, K., Okamura-Ikeda, K. and Motokawa, Y. (1997). Cloning and expression of a cDNA encoding bovine lipoyltransferase. *J. Biol. Chem.* **272**: 31 974–31 978.
37. Fujiwara, K., Okamura-Ikeda, K. and Motokawa, Y. (1996). Lipoylation of acyltransferase components of alpha-ketoacid dehydrogenase complexes. *J. Biol. Chem.* **271**: 12 932–12 936.
38. Rosenberg, H. R. and Culik, R. (1959). Effect of α-lipoic acid on vitamin C and vitamin E deficiencies. *Arch. Biochem. Biophys.* **80**: 86–93.
39. Packer, L., Tritschler, H. J. and Wessel, K. (1997). Neuroprotection by the metabolic antioxidant alpha-lipoic acid. *Free Radic. Biol. Med.* **22**: 359–378.
40. Ginn-Pease, M. E. and Whisler, R. L. (1998). Redox signals and NF-kappaB activation in T cells. *Free Radic. Biol. Med.* **25**: 346–361.
41. Mercurio, F. and Manning, A. M. (1999). Multiple signals converging on NF-kappaB. *Curr. Opin. Cell. Biol.* **11**: 226–232.
42. Baeuerle, P. A. and Henkel, T. (1994). Function and activation of NF-kappa B in the immune system. *Ann. Rev. Immunol.* **12**: 141–179.
43. Suzuki, Y. J., Agarwal, B. B. and Packer, L. (1992). Alpha-lipoic acid is a potent inhibitor of NF-κB activation in human T cells. *Biochem. Biophys. Res. Commun.* **189**: 1709–1715.
44. Suzuki, Y. J. and Packer, L. (1994). Alpha-lipoic acid is a potent inhibitor of NF-kappa B activation in human T-cells: does the mechanism involve antioxidant activities. *In* "Biological Oxidants and Antioxidants: New Strategies in Prevention and Therapy" (L. Packer, and E. Cadenas, Eds.), Hippocrates Verlag, Stuttgart.
45. Suzuki, Y. J., Mizuno, M., Tritschler, H. J. and Packer, L. (1995). Redox regulation of NF-kappa B DNA binding activity by dihydrolipoate. *Biochem. Mol. Biol. Int.* **36**: 241–246.
46. Staal, F. J., Roederer, M. and Herzenberg, L. A. (1990). Intracellular thiols regulate activation of nuclear factor kappa B and transcription of human immunodeficiency virus. *Proc. Natl. Acad. Sci. USA* **87**: 9943–9947.
47. Sen, C. K., Roy, S. and Packer, L. (1996). Involvement of intracellular Ca^{2+} in oxidant-induced NF-kappa B activation. *FEBS Lett.* **385**: 58–62.
48. Watt, F. and Molloy, P. L. (1993). Specific cleavage of transcription factors by the thiol protease, m-calpain. *Nucleic Acids Res.* **21**: 5092–5100.
49. Nabel, G. and Baltimore, D. (1987). An inducible transcription factor activates expression of human immunodeficiency virus in T cells [published erratum appears in Nature 1990 March 8; 344(6262): 178]. *Nature* **326**: 711–713.
50. Nabel, G. J., Rice, S. A., Knipe, D. M. and Baltimore, D. (1988). Alternative mechanisms for activation of human immunodeficiency virus enhancer in T cells. *Science* **239**: 1299–1302.
51. Roederer, M., Staal, F. J., Raju, P. A., Ela, S. W. and Herzenberg, L. A. (1990). Cytokine-stimulated human immunodeficiency virus replication is inhibited by N-acetyl-L-cysteine. *Proc. Natl. Acad. Sci. USA* **87**: 4884–4888.

52. Yan, C. Y., Ferrari, G. and Greene, L. A. (1995). N-acetylcysteine-promoted survival of PC12 cells is glutathione-independent but transcription-dependent. *J. Biol. Chem.* **270**: 26 827–26 832.

53. Baur, A., Harrer, T., Peukert, M., Jahn, G., Kalden, J. R. and Fleckenstein, B. (1991). Alpha-lipoic acid is an effective inhibitor of human immuno-deficiency virus (HIV-1) replication. *Klin Wochenschr* **69**: 722–724.

54. Papp, B. and Byrn, R. A. (1995). Stimulation of HIV expression by intracellular calcium pump inhibition. *J. Biol. Chem.* **270**: 10 278–10 283.

55. Eck, H. P., Gmunder, H., Hartmann, M., Petzoldt, D., Daniel, V. and Droge, W. (1989). Low concentrations of acid-soluble thiol (cysteine) in the blood plasma of HIV-1-infected patients. *Biol. Chem. Hoppe Seyler* **370**: 101–108.

56. Droge, W., Eck, H. P., Naher, H., Pekar, U. and Daniel, V. (1988). Abnormal amino-acid concentrations in the blood of patients with acquired immuno-deficiency syndrome (AIDS) may contribute to the immunological defect. *Biol. Chem. Hoppe Seyler* **369**: 143–148.

57. Droge, W., Eck, H. P. and Mihm, S. (1992). HIV-induced cysteine deficiency and T-cell dysfunction — a rationale for treatment with N-acetylcysteine. *Immunol. Today* **13**: 211–214.

58. Sen, C. K., Khanna, S., Roy, S. and Packer, L. (2000). Molecular basis of vitamin E action. Tocotrienol potently inhibits glutamate-induced pp60(c-Src) kinase activation and death of HT4 neuronal cells. *J. Biol. Chem.* **275**: 13 049–13 055.

59. Sen, C. K., Roy, S., Han, D. and Packer, L. (1997). Regulation of cellular thiols in human lymphocytes by alpha-lipoic acid: a flow cytometric analysis. *Free Radic. Biol. Med.* **22**: 1241–1257.

60. Herzenberg, L. A., De Rosa, S. C., Dubs, J. G., Roederer, M., Anderson, M. T., Ela, S. W. and Deresinski, S. C. (1997). Glutathione deficiency is associated with impaired survival in HIV disease. *Proc. Natl. Acad. Sci. USA* **94**: 1967–1972.

61. Sen, C. K., Roy, S. and Packer, L. (1999). Flow cytometric determination of cellular thiols. *Meth. Enzymol.* **299**: 247–258.

62. Merin, J. P., Matsuyama, M., Kira, T., Baba, M. and Okamoto, T. (1996). Alpha-lipoic acid blocks HIV-1 LTR-dependent expression of hygromycin resistance in THP-1 stable transformants. *FEBS Lett.* **394**: 9–13.

63. Fernandes-Alnemri, T., Litwack, G. and Alnemri, E. S. (1994). CPP32, a novel human apoptotic protein with homology to *Caenorhabditis elegans* cell death protein Ced-3 and mammalian interleukin-1 beta-converting enzyme. *J. Biol. Chem.* **269**: 30 761–30 764.

64. Sen, C. K., Sashwati, R. and Packer, L. (1999). Fas mediated apoptosis of human Jurkat T-cells: intracellular events and potentiation by redox-active alpha-lipoic acid. *Cell Death Differ.* **6**: 481–491.

65. Ueda, S., Nakamura, H., Masutani, H., Sasada, T., Yonehara, S., Takabayashi, A., Yamaoka, Y. and Yodoi, J. (1998). Redox regulation of

caspase 3(-like) protease activity: regulatory roles of thioredoxin and cytochrome c. *J. Immunol.* **161**: 6689–6695.

66. Granville, D. J., Carthy, C. M., Hunt, D. W. and McManus, B. M. (1998). Apoptosis: molecular aspects of cell death and disease. *Lab. Invest.* **78**: 893–913.

67. Green, D. R. (1998). Apoptotic pathways: the roads to ruin. *Cell* **94**: 695–698.

68. Ashkenazi, A. and Dixit, V. M. (1998). Death receptors: signaling and modulation. *Science* **281**: 1305–1308.

69. Martinez-Lorenzo, M. J., Gamen, S., Etxeberria, J., Lasierra, P., Larrad, L., Pineiro, A., Anel, A., Naval, J. and Alava, M. A. (1998). Resistance to apoptosis correlates with a highly proliferative phenotype and loss of Fas and CPP32 (caspase 3) expression in human leukemia cells. *Int. J. Cancer* **75**: 473–481.

70. Shinoura, N., Muramatsu, Y., Yoshida, Y., Asai, A., Kirino, T. and Hamada, H. (2000). Adenovirus-mediated transfer of caspase 3 with Fas ligand induces drastic apoptosis in U-373MG glioma cells. *Exp. Cell Res.* **256**: 423–433.

71. Yamabe, K., Shimizu, S., Ito, T., Yoshioka, Y., Nomura, M., Narita, M., Saito, I., Kanegae, Y. and Matsuda, H. (1999). Cancer gene therapy using a pro-apoptotic gene, caspase 3. *Gene Ther.* **6**: 1952–1959.

72. Fadeel, B., Ahlin, A., Henter, J. I., Orrenius, S. and Hampton, M. B. (1998). Involvement of caspases in neutrophil apoptosis: regulation by reactive oxygen species. *Blood* **92**: 4808–4818.

73. Boggs, S. E., McCormick, T. S. and Lapetina, E. G. (1998). Glutathione levels determine apoptosis in macrophages. *Biochem. Biophys. Res. Commun.* **247**: 229–233.

74. Mohr, S., Zech, B., Lapetina, E. G. and Brune, B. (1997). Inhibition of caspase 3 by S-nitrosation and oxidation caused by nitric oxide. *Biochem. Biophys. Res. Commun.* **238**: 387–391.

75. Shimizu, M., Yoshimoto, T., Nagata, S. and Matsuzawa, A. (1996). A trial to kill tumor cells through Fas (CD95)-mediated apoptosis in vivo. *Biochem. Biophys. Res. Commun.* **228**: 375–379.

76. Roth, W., Fontana, A., Trepel, M., Reed, J. C., Dichgans, J. and Weller, M. (1997). Immunochemotherapy of malignant glioma: synergistic activity of CD95 ligand and chemotherapeutics. *Cancer Immunol. Immunother.* **44**: 55–63.

77. Komada, Y. and Sakurai, M. (1997). Fas receptor (CD95)-mediated apoptosis in leukemic cells. *Leuk. Lymphoma* **25**: 9–21.

78. Sata, M., Perlman, H., Muruve, D. A., Silver, M., Ikebe, M., Libermann, T. A., Oettgen, P. and Walsh, K. (1998). Fas ligand gene transfer to the vessel wall inhibits neointima formation and overrides the adenovirus-mediated T cell response [In Process Citation]. *Proc. Natl. Acad. Sci. USA* **95**: 1213–1217.

79. Kim, D. K., Cho, E. S. and Um, H. D. (2000). Caspase-dependent and -independent events in apoptosis induced by hydrogen peroxide. *Exp. Cell. Res.* **257**: 82–88.

80. Li, A. E., Ito, H., Rovira, II, Kim, K. S., Takeda, K., Yu, Z. Y., Ferrans, V. J. and Finkel, T. (1999). A role for reactive oxygen species in endothelial cell anoikis. *Circ. Res.* **85**: 304–310.

81. Lieberthal, W., Triaca, V., Koh, J. S., Pagano, P. J. and Levine, J. S. (1998). Role of superoxide in apoptosis induced by growth factor withdrawal. *Am. J. Physiol.* **275**: F691–F702.

82. Olejnicka, B. T., Dalen, H. and Brunk, U. T. (1999). Minute oxidative stress is sufficient to induce apoptotic death of NIT-1 insulinoma cells. *APMIS* **107**: 747–761.

83. Turner, N. A., Xia, F., Azhar, G., Zhang, X., Liu, L. and Wei, J. Y. (1998). Oxidative stress induces DNA fragmentation and caspase activation via the *c-Jun* NH2-terminal kinase pathway in H9c2 cardiac muscle cells. *J. Mol. Cell. Cardiol.* **30**: 1789–1801.

84. Clement, M. V. and Stamenkovic, I. (1996). Superoxide anion is a natural inhibitor of FAS-mediated cell death. *EMBO J.* **15**: 216–225.

85. Lin, K. I., Pasinelli, P., Brown, R. H., Hardwick, J. M. and Ratan, R. R. (1999). Decreased intracellular superoxide levels activate Sindbis virus-induced apoptosis. *J. Biol. Chem.* **274**: 13 650–13 655.

86. Hampton, M. B. and Orrenius, S. (1998). Redox regulation of apoptotic cell death. *Biofactors* **8**: 1–5.

87. Roth, W., Wagenknecht, B., Grimmel, C., Dichgans, J. and Weller, M. (1998). Taxol-mediated augmentation of CD95 ligand-induced apoptosis of human malignant glioma cells: association with bcl-2 phosphorylation but neither activation of p53 nor G2/M cell cycle arrest. *Br. J. Cancer* **77**: 404–411.

88. Jacob, S., Streeper, R. S., Fogt, D. L., Hokama, J. Y., Tritschler, H. J., Dietze, G. J. and Henriksen, E. J. (1996). The antioxidant alpha-lipoic acid enhances insulin-stimulated glucose metabolism in insulin-resistant rat skeletal muscle. *Diabetes* **45**: 1024–1029.

89. Jacob, S., Ruus, P., Hermann, R., Tritschler, H. J., Maerker, E., Renn, W., Augustin, H. J., Dietze, G. J. and Rett, K. (1999). Oral administration of RAC-alpha-lipoic acid modulates insulin sensitivity in patients with Type-2 diabetes mellitus: a placebo-controlled pilot trial. *Free Radic. Biol. Med.* **27**: 309–314.

90. Hampton, M. B. and Orrenius, S. (1997). Dual regulation of caspase activity by hydrogen peroxide: implications for apoptosis. *FEBS Lett.* **414**: 552–556.

91. Baker, A., Santos, B. D. and Powis, G. (2000). Redox control of caspase 3 activity by thioredoxin and other reduced proteins. *Biochem. Biophys. Res. Commun.* **268**: 78–81.

92. Albelda, S. M., Smith, C. W. and Ward, P. A. (1994). Adhesion molecules and inflammatory injury. *FASEB J.* **8**: 504–512.

93. Frenette, P. S. and Wagner, D. D. (1996). Adhesion molecules — Part I. *New England J. Med.* **334**: 1526–1529.

94. Marui, N., Offermann, M. K., Swerlick, R., Kunsch, C., Rosen, C. A., Ahmad, M., Alexander, R. W. and Medford, R. M. (1993). Vascular cell adhesion molecule-1 (VCAM-1) gene transcription and expression are regulated through an antioxidant-sensitive mechanism in human vascular endothelial cells. *J. Clin. Invest.* **92**: 1866–1874.

95. Offermann, M. K., Lin, J. C., Mar, E. C., Shaw, R., Yang, J. and Medford, R. M. (1996). Antioxidant-sensitive regulation of inflammatory-response genes in Kaposi's sarcoma cells. *J. Acquir. Immune Defic. Syndr. Human Retrovirol.* **13**: 1–11.

96. Aoki, T., Suzuki, Y., Suzuki, K., Miyata, A., Oyamada, Y., Takasugi, T., Mori, M., Fujita, H. and Yamaguchi, K. (1996). Modulation of ICAM-1 expression by extracellular glutathione in hyperoxia-exposed human pulmonary artery endothelial cells. *Am. J. Respir. Cell Mol. Biol.* **15**: 319–327.

97. Moynagh, P. N., Williams, D. C. and La, O. N. (1994). Activation of NF-kappa B and induction of vascular cell adhesion molecule-1 and intracellular adhesion molecule-1 expression in human glial cells by IL-1. Modulation by antioxidants. *J. Immunol.* **153**: 2681–2690.

98. Weber, C., Erl, W., Pietsch, A., Strobel, M., Ziegler-Heitbrock, H. W. and Weber, P. C. (1994). Antioxidants inhibit monocyte adhesion by suppressing nuclear factor-kappa B mobilization and induction of vascular cell adhesion molecule-1 in endothelial cells stimulated to generate radicals. *Arterioscler. Thromb.* **14**: 1665–1673.

99. Ikeda, M., Schroeder, K. K., Mosher, L. B., Woods, C. W. and Akeson, A. L. (1994). Suppressive effect of antioxidants on intercellular adhesion molecule-1 (ICAM-1) expression in human epidermal keratinocytes. *J. Invest. Dermatol.* **103**: 791–796.

100. Kawai, M., Nishikomori, R., Jung, E. Y., Tai, G., Yamanaka, C., Mayumi, M. and Heike, T. (1995). Pyrrolidine dithiocarbamate inhibits intercellular adhesion molecule-1 biosynthesis induced by cytokines in human fibroblasts. *J. Immunol.* **154**: 2333–2341.

101. Roy, S., Sen, C. K., Kobuchi, H. and Packer, L. (1998). Antioxidant regulation of phorbol ester-induced adhesion of human Jurkat T-cells to endothelial cells. *Free Radic. Biol. Med.* **25**: 229–241.

102. Roy, S., Sen, C. K. and Packer, L. (1999). Determination of cell–cell adhesion in response to oxidants and antioxidants. *Meth. Enzymol.* **300**: 395–401.

103. Holdiness, M. R. (1991). Clinical pharmacokinetics of N-acetylcysteine. *Clin. Pharmacokinet* **20**: 123–134.

104. Peter, G. and Borbe, H. O. (1995). Absorption of 7,8-^{14}C-rac-a-lipoic acid from in situ ligated segments of gastroinstinal tract of the rat. *Arzneimittel-forschung/Drug Res.* **45**: 293–229.

105. Roy, S. and Packer, L. (1998). Redox regulation of cell functions by alpha-lipoate: biochemical and molecular aspects. *Biofactors* **8**: 17–21.
106. Williamson, J. R., Kilo, C. and Ido, Y. (1999). The role of cytosolic reductive stress in oxidant formation and diabetic complications. *Diabetes Res. Clin. Pract.* **45**: 81–82.
107. Panigrahi, M., Sadguna, Y., Shivakumar, B. R., Kolluri, S. V., Roy, S., Packer, L. and Ravindranath, V. (1996). Alpha-lipoic acid protects against reperfusion injury following cerebral ischemia in rats. *Brain Res.* **717**: 184–188.
108. Biewenga, G., Haenen, G. R. and Bast, A. (1997). The role of lipoic acid in the treatment of diabetic polyneuropathy. *Drug Metab. Rev.* **29**: 1025–1054.
109. Ziegler, D., Schatz, H., Conrad, F., Gries, F. A., Ulrich, H. and Reichel, G. (1997). Effects of treatment with the antioxidant alpha-lipoic acid on cardiac autonomic neuropathy in NIDDM patients. A 4-month randomized controlled multicenter trial (DEKAN Study). Deutsche Kardiale Autonome Neuropathie. *Diabetes Care* **20**: 369–373.
110. Ziegler, D. and Gries, F. A. (1997). Alpha-lipoic acid in the treatment of diabetic peripheral and cardiac autonomic neuropathy. *Diabetes* **46(Suppl. 2)**: S62–S66.
111. Ziegler, D., Hanefeld, M., Ruhnau, K. J., Hasche, H., Lobisch, M., Schutte, K., Kerum, G. and Malessa, R. (1999). Treatment of symptomatic diabetic polyneuropathy with the antioxidant alpha-lipoic acid: a 7-month multicenter randomized controlled trial (ALADIN III Study). ALADIN III Study Group. Alpha-Lipoic Acid in Diabetic Neuropathy. *Diabetes Care* **22**: 1296–1301.
112. Ziegler, D., Hanefeld, M., Ruhnau, K. J., Meissner, H. P., Lobisch, M., Schutte, K. and Gries, F. A. (1995). Treatment of symptomatic diabetic peripheral neuropathy with the anti-oxidant alpha-lipoic acid. A 3-week multicentre randomized controlled trial (ALADIN Study). *Diabetologia* **38**: 1425–1433.
113. Ziegler, D., Reljanovic, M., Mehnert, H. and Gries, F. A. (1999). Alpha-lipoic acid in the treatment of diabetic polyneuropathy in Germany: current evidence from clinical trials. *Exp. Clin. Endocrinol. Diabetes* **107**: 421–430.
114. Jacob, S., Henriksen, E. J., Schiemann, A. L., Simon, I., Clancy, D. E., Tritschler, H. J., Jung, W. I., Augstin, H. J. and Dietze, G. J. (1995). Enhancement of glucose disposal in patients with Type-2 diabetes by alpha-lipoic acid. *Drug Res.* **45**: 872–874.
115. Cameron, N. E. and Cotter, M. A. (1999). Effects of antioxidants on nerve and vascular dysfunction in experimental diabetes. *Diabetes Res. Clin. Pract.* **45**: 137–146.
116. Cameron, N. E., Cotter, M. A., Horrobin, D. H. and Tritschler, H. J. (1998). Effects of alpha-lipoic acid on neurovascular function in diabetic rats: interaction with essential fatty acids. *Diabetologia* **41**: 390–399.
117. Henriksen, E. J., Jacob, S., Streeper, R. S., Fogt, D. L., Hokama, J. Y. and Tritschler, H. J. (1997). Stimulation by alpha-lipoic acid of glucose transport

activity in skeletal muscle of lean and obese Zucker rats. *Life Sci.* **61**: 805–812.

118. Khamaisi, M., Rudich, A., Potashnik, R., Tritschler, H. J., Gutman, A. and Bashan, N. (1999). Lipoic acid acutely induces hypoglycemia in fasting nondiabetic and diabetic rats. *Metabolism* **48**: 504–510.

119. Low, P. A., Nickander, K. K. and Tritschler, H. J. (1997). The roles of oxidative stress and antioxidant treatment in experimental diabetic neuropathy. *Diabetes* **46(Suppl. 2)**: S38–S42.

120. Streeper, R. S., Henriksen, E. J., Jacob, S., Hokama, J. Y., Fogt, D. L. and Tritschler, H. J. (1997). Differential effects of lipoic acid stereoisomers on glucose metabolism in insulin-resistant skeletal muscle. *Am. J. Physiol.* **273**: E185–E191.

121. Sen, C. K., Tirosh, O., Roy, S., Kobayashi, M. S. and Packer, L. (1998). A positively charged alpha-lipoic acid analogue with increased cellular uptake and more potent immunomodulatory activity. *Biochem. Biophys. Res. Commun.* **247**: 223–228.

122. Bustamante, J., Slater, A. F. and Orrenius, S. (1995). Antioxidant inhibition of thymocyte apoptosis by dihydrolipoic acid. *Free Radic. Biol. Med.* **19**: 339–347.

123. Tirosh, O., Sen, C. K., Roy, S., Kobayashi, M. S. and Packer, L. (1999). Neuroprotective effects of alpha-lipoic acid and its positively charged amide analogue. *Free Radic. Biol. Med.* **26**: 1418–1426.

Chapter 46

Coenzyme Q_{10} Prevents DNA Oxidation and Enhances Recovery from Oxidative Damage

Renata Alleva

Renata Alleva • Institute of Biochemistry-School of Medicine-University of Ancona, Corso Mazzini 27, San Benedetto del Tr. 63039-(AP) Italy
*Corresponding Author.
Tel: 0039 0735 592755, E-mail: renalle@libero.it

1. Introduction

Coenzyme Q_{10} (CoQ_{10}) is an essential lipid component of the mitochondrial respiratory chain, participating as a redox active molecule in electron transfer reactions of diverse cellular functions.[1] Besides its redox activity in the electron transport chain, ubiquinol-10, the reduced form of CoQ_{10}, acts as a highly efficient antioxidant contributing to the antioxidant defences against radical-mediated oxidative damage in cell membranes,[2,3] blood,[4] lipoproteins,[5-7] seminal fluids.[8] The available information on the antioxidant role of ubiquinol-10 is consistent with the assumption that it acts primarily by preventing the formation of lipid peroxyl radicals (LOO•) and by reducing the perferryl radical.[9] Ubiquinol-10 may act by removing LOO• either directly or through the regeneration of α-tocopherol from the α-tocopheroxyl radical.[10] Thus, ubiquinol-10 may prevent both the initiation and the propagation of lipid peroxidation in contrast to α-tocopherol which exerts its antioxidant activity through a chain-breaking mechanism, thus acting only by inhibiting propagation.[11] The ability of ubiquinol-10 in preventing both of these events, is feasible due to its favourable location in the inner hydrophobic region of the membrane phospholipid bilayer, where such reactions occur and where the proton-motive Q cycle is also located, which is capable of regenerating ubiquinol-10 from ubisemiquinone radical.[12] Till recently a plethora of studies demonstrating the antioxidant power of ubiquinol-10 have been carried out with lipid peroxidation as the parameter investigated. However, it is widely demonstrated that oxidative damage in biological systems occur in many other molecular species, including proteins[13] and DNA.[14-16] In particular, oxidative DNA damage has gained increasing attention by researchers, owing to its implication in the aetiology of many degenerative diseases, ageing and cancer.[17-19] Although DNA damage inflicted by reactive oxygen species (ROS) is not a rare event and it is estimated that a human cell sustains an average of 10^5 "oxidative hits" per day due to cellular oxidative metabolism,[20] DNA is functionally very stable, so that the incidence of cancer is much lower than one would expect, taking into account the high frequency of oxidative insults.

To counteract the hazardous insult of ROS, cells have developed a variety of non-enzymatic and enzymatic defence systems responsible for DNA protection.

In particular, cells are endowed with a broad array of enzymes involved in DNA repair and the importance of these repair systems is indicated by the existence in cells of a variety of enzymes capable of rejoining single lesions and the high degree of homology of DNA repair genes in various species.[21] Moreover, the critical role of repair in the prevention of cancer development is pointed out by several syndromes in which repair systems are blocked.[22] Thus, owing to the role that DNA damage seems to play in the pathogenesis of cancer, much attention has been recently paid to the mechanism(s) modulating such a repair activity and to the factors that may modulate and/or enhance DNA recovery from oxidative damage. The hypothesis has been recently formulated that the activity of DNA repair enzymes might be modulated by the redox status of cells, which in turn is affected by antioxidants.[23-25] The cancer-preventing potential of natural antioxidants, arises from several epidemiological studies indicating that fruit and vegetable intake is inversely related with the risk of cancer.[26, 27] Moreover, different authors observed a higher resistance against oxidative damage in DNA of lymphocytes isolated from subjects who had undergone antioxidant supplementation.[28, 29] The potential protective activity against oxidative DNA damage has been documented for some of the quantitatively most important antioxidants present in human blood, such as ascorbic acid,[30] α-tocopherol[31] and β-carotene,[32] as well as for other antioxidants naturally occurring in food, as lycopenes, lutein, flavonoids, that despite their low concentrations in human blood, may also provide protection against cancer, by preventing oxidative damage inflicted to DNA.[33-35] Antioxidants may act by enhancing the endogenous defence systems, such as cytosolic proteins that protect against DNA damaging factors.[36] However, it should be also reported that other studies investigating the role of antioxidants in preventing DNA damage have found conflicting results.[37, 38] This disagreement may arise from the different techniques used to detect oxidative lesions in DNA. Among the methods available to assay oxidative DNA damage and repair, the single cell gel-electrophoresis, Comet assay, offers many advantages, being a highly sensitive and versatile method.[39] In fact, the assay requires only a few hundred cells that are placed on a mini gel electrophoresis and lysed for electrophoresis under alkaline conditions. Oxidative modification and strand breaks in DNA release its supercoiling and allow the uncoiled DNA to extend towards the anode. Visualized by fluorescent dye and fluorescent microscopy the relaxed DNA extension looks like a comet. The assay can be expanded further by the use of specific DNA repair enzymes, that will introduce strand breaks at sites of damage-endonuclease III specific for oxidized pyrimidines and formamido-pyrimidine glycosylase (fpg) which recognizes ring-opened purines as well as 8-oxo-dG.[40] By means of the comet assay, CoQ_{10} has been recently investigated to assess its ability to prevent induced DNA damage in human lymphocytes.[41] This article reports on data concerning the ability of both ubiquinone-10 and ubiquinol-10 in preventing DNA damage and modulating its repair and highlights a novel aspect of CoQ_{10} antioxidant power.

2. DNA Damage and Repair Systems

Reactive oxygen species are continuously generated during normal cell metabolism from molecular oxygen, where their harmful formation is strictly controlled by a variety of enzymatic and non-enzymatic antioxidant systems. Despite such antioxidant defences, ROS cause constant damage to oxidizable molecules, which are repaired and replaced in a dynamic equilibrium. Either an excessive production of ROS or impaired functioning of antioxidant defences can result in oxidative stress and cell injury. Oxidation of DNA induced by ROS, leads to 100 different modifications[42] which include DNA single-strand and double strand-breaks, base modifications, sites of base loss (apurinic/apyrimidinic sites, AP sites) and cross-links.[43] Among these, double-strand breaks are highly toxic and mutagenic and are assumed to be responsible for some of the chromosomal aberrations and large deletion observed with ionizing radiation and other kind of oxidative damage. In contrast, single strand breaks are readily rejoined; however since the repair kinetics proceeds following a biphasic mechanism,[44] it cannot be excluded that also a sub-type of single strand breaks can contribute to the mutagenicity of some oxidants. DNA damage profiles formed are dependent on the oxidant inducing the damage and the determination of DNA damage profiles may help to understand the consequences of a normal or enhanced generation of ROS in cells and tissues. However, the identification of the species directly responsible for the oxidative stress is difficult in cellular environment, where the various ROS can be converted into one another by a number of reactions and formation of one species leads to generating many others.[43] Among these, hydroxyl radical is considered the most harmful species due to its ability to directly inflict DNA damage, by reacting with all DNA components: the deoxyribose backbone, purine and pyrimidine bases. Moreover, hydroxyl radical mediates the mutagenic effect of oxidative stress *in vivo*, as suggested by the evidence that, ROS-specific DNA modifications found in several cancer tissues, are reproducible *in vitro* in ·OH generating systems. The most frequent DNA damage takes place at base levels, the ubiquitous one is 8-deoxyguanosine (8-OH-Gua) occurring approximately one in 100 000 guanidine residues in normal human cells. Oxidized bases can be repaired by following two general pathways, which consist of base excision repair and nucleotide excision repair. Glycosilases and apurinic/apirimidinic (AP) endonucleases act in synergy respectively by removing oxidized bases and cleaving the phosphodiester backbone resulting in the loss of the abasic sugar. The consequent nucleotide gap is then filled in and sealed by the action of DNA polymerase and DNA ligase. Several other repair enzymes are induced in mammalian cells in response to DNA damage. DNA strand breaks induced by a variety of oxidants activate a nuclear enzyme (poly-ADP-ribose polymerase) involved in repair of DNA lesions[45] Antioxidants may act by enhancing the enzyme repair systems through a post-transcriptional gene regulation of transcription factors.[23-25] At least two well-known transcription factors, NF-κB and activator protein-1 (AP-1), involved in the inducible expression of a variety of genes implicated in oxidative stress and cellular response, are known to be regulated by antioxidants and other factors influencing the cellular redox status.[24, 25] Thus, it is might be argued that the efficacy of different antioxidants in stimulating TFs is feasibly dependent on the redox power of the antioxidant molecule.

3. Oxidative DNA Damage and Recovery in DNA of Lymphocytes *In Vitro* Enriched with Ubiquinol-10 or Ubiquinone-10

In certain cell types, particularly lymphocytes, DNA damage leads to a rapid suicide,[46] for this reason, lymphocytes represent a useful model to investigate the effect of antioxidant supplementation against oxidative DNA damage. The concentration of CoQ_{10} in lymphocytes isolated from human blood, is ranging from 15 to 20 pmol per 10^6 cells of which 50–70% is present as ubiquinol-10. When control lymphocytes (C-lymphocytes) are exposed to hydrogen peroxide (H_2O_2), in a system generating hydroxyl radical, the proportion of ubiquinol-10 rapidly decreases from 47.0 ± 2.2 to $22.0 \pm 3.0\%$, where as CoQ_{10} concentration remains unchanged, indicating that ubiquinol-10 is stoichiometrically oxidized into ubiquinone-10 (Fig. 1). The decrease of cells' ubiquinol-10 content leads to an accumulation of DNA strand breaks, which markedly rise after 30 min of incubation with H_2O_2 and coincided with a significant cell death (Figs. 2 and 3). Strand breaks are a major species of oxidative DNA damage, so they are a valid indicator of oxidative damage induced by H_2O_2 *in vitro*. By treating ubiquinol-10 enriched lymphocytes ($CoQ_{10}H_2$-lymphocytes) as control ones, a comparable loss of ubiquinol-10 is seen at the same extent as that observed in control cells; however, despite this loss a relatively larger proportion of CoQ_{10} still remains in the reduced antioxidant active form (Fig. 1). The enrichment of lymphocytes with ubiquinol-10 also affects the viability and the amount of DNA strand breaks, being significantly lower than that occurring in control lymphocytes. Thus, in this system, ubiquinol-10 acts as an early target for H_2O_2 derived oxidants, where as other cellular antioxidants, such as α-tocopherol and β-carotene, are either not oxidized, or their oxidized forms are reduced back effectively. This finding is reminiscent of the situation occurring in lipoproteins, where ubiquinol-10 is the first antioxidant to be consumed in the early stages of lipid peroxidation.[5] Surprisingly, also the enrichment of lymphocytes with ubiquinone-10 (CoQ_{10}-lymphocytes) protects against oxidative damage occurring in the DNA of cells exposed to H_2O_2, as observed in $CoQ_{10}H_2$- lymphocytes (Fig. 4). To test further the effect of ubiquinol-10 and ubiquinone-10 enrichment on different DNA damage profiles, we exposed control and $CoQ_{10}H_2$ or CoQ_{10} enriched cells to other oxidizing agents, such as atmospheric oxygen and (R)-1-[(10-chloro-4-oxo-3-phenyl-4H-benzo[a]quinolizin-1-yl) carbonyl]-2-pyrrolidinemethanol (Ro 19–8022). Exposure of lymphocytes to atmospheric oxygen produces specifically DNA strand breaks detectable within the first hour, and repair of existing, endogenous oxidized bases is seen throughout the incubation [Figs. 5(a)–(c)]. When oxidation is carried out with oxygen, DNA of CoQ_{10}- or $CoQ_{10}H_2$ enriched cells is less susceptible to oxidation than DNA of control lymphocytes, exhibiting a lower amount of strand breaks and showing a faster recovery from oxidative damage by leading to a

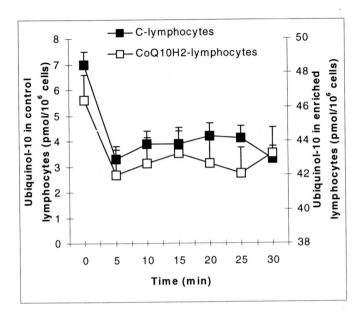

Fig. 1. Kinetic of ubiquinol-10 loss in ubiquinol-10-enriched and control lymphocytes exposed to 100 μM H₂O₂. Aliquots of the reaction mixture (corresponding to 3×10^6 cells) were removed at the time point indicated and the cellular content of ubiquinol-10 analyzed by HPLC-ED as described in the methods section. Note the differences in scale in the two y-axes.

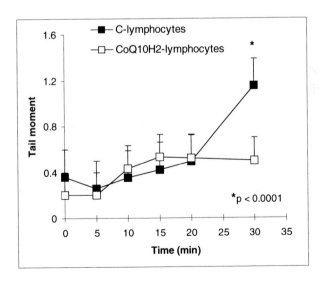

Fig. 2. Kinetic of tail moment values in ubiquinol-10-enriched and control lymphocytes during exposure to H₂O₂. Ubiquinol-10-enriched-lymphocytes and control cells were exposed to 100 μM H₂O₂, aliquots (corresponding to 3×10^6 cells) removed at the time point indicated, and DNA damage evaluated as described in the methods section.

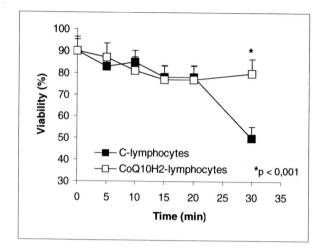

Fig. 3. Time-dependent change in viability of ubiquinol-10-enriched and control lymphocytes during exposure to H_2O_2. The experimental condition were as described in the legend to Fig. 2. Cell viability was evaluated by trypan blue exclusion.

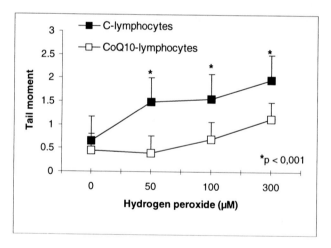

Fig. 4. H_2O_2 concentration-dependent DNA damage in ubiquinone-10-enriched and control lymphocytes. Ubiquinone-10-enriched and control cells were oxidized with 0, 50, 100, 300 μM of H_2O_2 on ice for 30 min before the extent of DNA strand breaks was evaluated by comet assay as described in the method section.

faster rejoining of strand breaks (Table 1). These findings confirm the ability of the antioxidants to counteract strand break formation as observed in lymphocytes exposed to H_2O_2 (Figs. 2 and 4). Conversely, ubiquinol-10 and ubiquinone-10 are ineffective when the damage is induced with Ro 19-8022, which oxidizes purines (Fpg sensitive sites) accumulated as the main oxidation product, although some

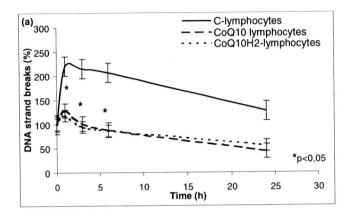

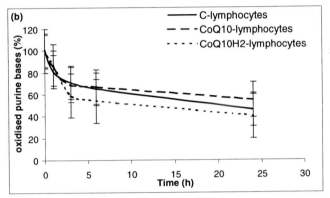

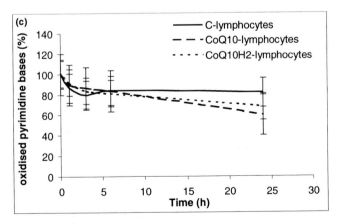

Fig. 5. Kinetics of DNA strand break (a), oxidized purine DNA base (b) and oxidized pyrimidine DNA base (c) repair in ubiquinone-10 or ubiquinol-10 *in vitro* enriched and control lymphocytes exposed to oxygen atmospheric.

Table 1. Kinetic Parameters of DNA Strand Break (SBs) Repair in *in Vitro* Enriched CoQ_{10} and $CoQ_{10}H_2$-Lymphocytes Exposed to Oxygen or to Ro 19-8022

	Oxygen Oxidation				Ro 19-8022 Oxidation			
	ΔSBs (a.u.)	K (min⁻¹ × 10⁻⁴)	V (a.u. × 10⁻³/min)	T½ (h)	ΔSBs (a.u.)	K (min⁻¹ × 10⁻⁴)	V (a.u. × 10⁻³/min)	T½ (h)
C-lymphocytes	66 ± 15	0.8 ± 0.2	10 ± 3	144 ± 45	24 ± 10	6.3 ± 2.3	93 ± 21	18 ± 5
CoQ_{10}-lymph.	24 ± 10*	5.5 ± 2.1*	59 ± 8*	21 ± 8*	15 ± 5	9.0 ± 4.0	133 ± 52	13 ± 4
$CoQ_{10}H_2$-lymph.	13 ± 5*	4.7 ± 2.3*	39 ± 6*	24 ± 10*	7 ± 3	9.6 ± 5.0	136 ± 48	12 ± 6

*P < 0.05

a.u. = arbitrary units

ΔSBs = SBs after oxidation − SBs basal

The first order rate constants (K), velocity of repair (V) and half-time (T½) were determined using logarithmic plots of data obtained from DNA damage to time point where damage was completely repaired.

oxidized pyrimidines and strand breaks are also detectable (Fig. 6). In these experimental conditions, neither the extent of the damage induced on Fpg sites nor the recovery are affected by the enrichment with the antioxidant, showing both control and enriched cells a similar DNA damage profile and a kinetic of DNA recovery [Figs. 7(a)–(c); Table 1]. Together, the above data indicate that ubiquinol-10 or ubiquinone-10 are able to efficiently counteract the formation of strand breaks. These data concerning the protection of ubiquinol-10 against DNA oxidative damage add new information about the versatility of ubiquinol-10 in preventing oxidative damage in different systems, however they are not striking since ubiquinol-10 antioxidant power is widely documented,[4–10] therefore reasonably expected. Conversely, the antioxidant activity of ubiquinone-10 was unexpected since the antioxidant property of CoQ_{10} is commonly ascribed to the quinolic group of ubiquinol-10 and to its ability of directly scavenging free lipid radicals.[4] The mechanism proposed for such antioxidant activity could be related to the localization of ubiquinone-10 within the lipid bilayer among phospholipid molecules, where it can exert a strong ordering and condensing effect restricting the number of hydroxyl radicals capable of reaching the DNA of lymphocytes. Interestingly, our data indicate that even ubiquinol-10 seems to act through a different mechanism than the conventional ones. In fact, even though the increased DNA resistance of cells loaded with ubiquinol-10 supports the notion that the

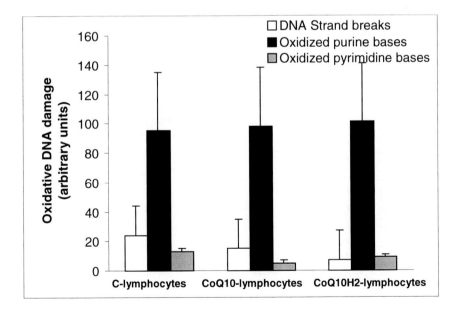

Fig. 6. DNA strand break, oxidized purine and pyrimidine DNA base formation in ubiquinone-10 or ubiquinol-10 *in vitro* enriched and control lymphocytes exposed to Ro 19-8022.

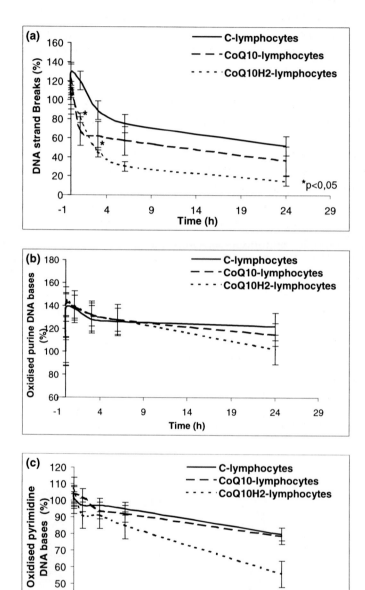

Fig. 7. Kinetics of DNA strand break (a), oxidized purine DNA base (b) and oxidized pyrimidine DNA base (c) repair in ubiquinone-10 or ubiquinol-10 *in vitro* enriched and control lymphocytes pre-oxidized by Ro 19-8022.

reduced form of CoQ_{10} inhibits the formation of H_2O_2 or oxygen induced-strand breaks, strand breaks and the loss of viability are preceded by both the consumption of $CoQ_{10}H_2$ and a time during which further $CoQ_{10}H_2$ consumption is not observed, indicating that ubiquinol-10 does not act directly inhibiting DNA oxidation.

4. Effect of CoQ_{10} *In Vivo* Supplementation on DNA Damage and Its Repair in Primary Cultured Lymphocytes

The effect of dietary supplementation of CoQ_{10} on oxidative DNA damage has not yet been widely investigated, therefore not many data are available on this issue. In order to assess the effect of CoQ_{10} dietary intake on oxidative DNA damage and its repair, we carried out a supplementation study in which a group of healthy subjects were given two consecutive doses of CoQ_{10}, respectively 100 and 300 mg/day, and lymphocytes taken before and after the supplement were exposed to oxygen or Ro 19-8022. The intake of a single oral dose of 100 or 300 mg affected the CoQ_{10} content in lymphocytes with increases of 65% and 144% respectively accompanied by a slight increase of $CoQ_{10}H_2/CoQ_{10}$ ratio (Table 2). CoQ_{10} enriched lymphocytes are more resistant to oxidative DNA damage compared to native non-supplemented lymphocytes and lymphocytes isolated from wash-out plasma. In fact, when exposed to oxygen, we observed that strand breaks markedly increase both in DNA of native lymphocytes and wash-out lymphocytes after 1 h of oxygen exposure, whereas no additional formation of strand breaks occur in CoQ_{10} enriched lymphocytes (Fig. 8). More interestingly, we found an association between oxidative damage and plasma or cellular CoQ_{10} concentration. In fact the highest levels of damage is seen in subjects with low

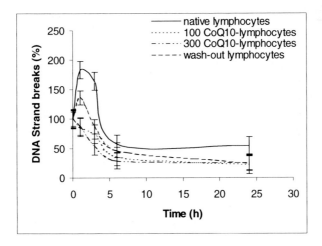

Fig. 8. Kinetics of DNA strand break repair in native and CoQ_{10} *in vivo* enriched lymphocytes exposed to oxygen atmospheric.

Table. 2. Concentrations of Lipid-Soluble Antioxidants in Lymphocytes Before and After Different Doses of CoQ$_{10}$ Oral Supplementation

Antioxidants (pmol/10^6 cells)	Native Lymphocytes	100 CoQ$_{10}$ Lymphocytes	300-CoQ$_{10}$ Lymphocytes	Wash-Out Lymphocytes
α-tocopherol	236 ± 166	324 ± 171	295 ± 123	314 ± 184
β-carotene	4.7 ± 2.7	6.0 ± 5.2	6.3 ± 4.0	5.8 ± 2.3
CoQ$_{10}$	4.4 ± 1.9	6.9 ± 3.5	5.0 ± 1.1	4.6.0 ± 1.1
CoQ$_{10}$H$_2$	9.6 ± 5.2	15.5 ± 8.6*	26.7 ± 9.6*	15.1 ± 6.0
CoQ$_{10}$H$_2$ (%)	69 ± 5	69 ± 8	84 ± 10	72 ± 6

*$p < 0.05$ native versus 100 e 300 mg/day supplementation

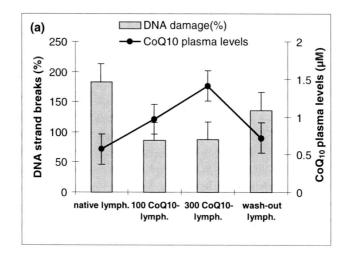

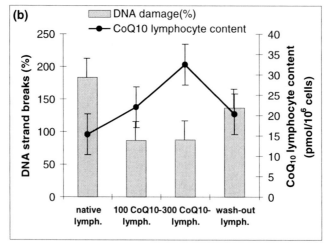

Fig. 9. DNA strand break formation after oxygen exposure in relation to CoQ$_{10}$ content in plasma (a) and in lymphocytes (b).

Table 3. Kinetic Parameters of DNA Strand Break (SBs) Repair *In Vivo* CoQ_{10} and $CoQ_{10}H_2$-Enriched Lymphocytes Exposed to Oxygen or to Ro 19-8022

	Oxygen Oxidation				Ro 19-8022 Oxidation			
	ΔSBs (a.u.)	K ($min^{-1} \times 10^{-4}$)	V (a.u. $\times 10^{-3}$/min)	$T_{\frac{1}{2}}$ (h)	ΔSBs (a.u.)	K ($min^{-1} \times 10^{-4}$)	V (a.u. $\times 10^{-3}$/min)	$T_{\frac{1}{2}}$ (h)
Native lymph.	49 ± 15	16.7 ± 6.2	181 ± 52	7 ± 4	18 ± 7	5.9 ± 2.3	88 ± 21	20 ± 5
100-CoQ₁₀lymph.	13 ± 6*	12.6 ± 8.3	102 ± 35	9 ± 5	22 ± 10	11.7 ± 4.0	175 ± 52	14 ± 4
300-CoQ₁₀lymph.	12 ± 4*	15.5 ± 7.0	123 ± 42	7 ± 3	26 ± 11	13.5 ± 5	239 ± 48	9 ± 5
Wash-out lymph.	20 ± 8	15.3 ± 7.5	115 ± 48	7 ± 4	20 ± 9	11.0 ± 5.0	158 ± 48	11 ± 6

*$P < 0.05$

a.u. = arbitrary units

ΔDSBs = SBs after oxidation − SBs basal

The first order rate constants (K), velocity of repair (V) and half-time ($T_{\frac{1}{2}}$) were determined using logarithmic plots of data obtained from DNA damage to time point where damage was completely repaired.

levels of CoQ_{10} in plasma or cells, which inversely correlates with the extent of strand breaks induced by oxygen [Figs. 9(a) and (b)].

The kinetics of strand break repair proceeds faster in native and wash-out lymphocytes than in CoQ_{10} enriched lymphocytes, suggesting that strand breaks accumulated after oxygen exposure are repaired faster than the endogenous ones of CoQ_{10} enriched cells (Table 3).

Conversely, *in vivo* supplementation with CoQ_{10} does not prevent endogenous formation of oxidized purines and pyrimidines or affect their repair.

Reminiscent of the situation occurring *in vitro*, CoQ_{10} supplementation does not affect time-dependent oxidation of purines and pyrimidines bases or their repair when treated with Ro19-8022. The ability of CoQ_{10} to increase the DNA repair rate in lymphocytes exposed to oxygen, is likely due to an inhibition of additional damage by protecting cells against further oxidation. Repair of DNA lesions is a critical event in the prevention of cancer since unrepaired bases may undergo mutagenic processes likely leading to carcinogenesis. However, DNA repair rates in normal cells are quite difficult to estimate.[47] An alternative way of measuring repair enzyme activity is achieved by a modification of the Comet assay, which allows an assessment of repair ability in a simple extract, provided with a DNA substrate (gel-embedded nucleoids) carrying specific kinds of damage. In detail, the assay consist of a lymphocyte extract containing DNA enzymes incubated with a substrate consisting of DNA oxidized by exposing lymphocytes to oxidising agents. Enzymes contained in cellular extract introduce breaks at the level of DNA oxidized bases, and DNA containing breaks is drawn out to form a comet tail, the intensity of which represents the break frequency.

The time course of break production on the specifically damage DNA substrate is measure of repair capability.

5. Effect of *In Vivo* CoQ_{10} Supplementation on DNA Repair Enzyme Expression

Using the modified Comet assay, we further investigated the role of CoQ_{10} supplementation in modulating the repair enzyme activity in lymphocytes exposed to Ro 19-8022. As shown in Fig. 10, cellular extracts from CoQ_{10} enriched lymphocytes exhibit a markedly higher DNA repair activity compared to native and wash-out lymphocytes, indicating that supplementation with CoQ_{10} improves DNA repair.

It might be argued that repair processes, maintain oxidative damage at a level that is tolerable in terms of maintenance of DNA stability.[47] However deficiency in such repair would be expected to give rise to a significant mutational burden-as seen in yeast deficient in the OGG1 gene.[48] CoQ_{10} supplementation clearly enhances enzyme repair activity, but the mechanism of such stimulatory effect

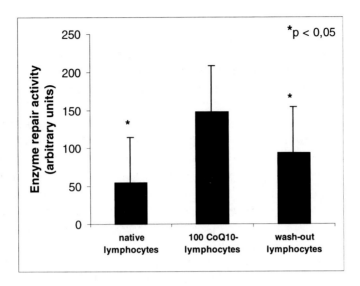

Fig. 10. Enzyme repair activity in native and CoQ_{10} *in vivo* enriched lymphocytes.

has not been clarified. The study of the mechanisms and factors implicated in the modulation of DNA recovery and repair is an intriguing matter and that is why it has recently gained increasing interest by researchers.[49-51] Increasing evidence exists that cellular redox status is an important regulator of various aspects of cellular events, including proliferation, apoptosis and gene expression.[24] Furthermore, a link between the regulation of transcription factors (TFs), oxidative signalling and DNA binding activity has, in fact, been recently proposed.[24, 25]

6. Conclusion

The "antioxidant hypothesis" attempts to explain the cancer-protective effect of some dietary antioxidants. To clarify whether endogenous DNA damage is influenced by dietary antioxidants, it is first important to establish a correlation between DNA damage and levels of antioxidants derived from diet, second to assess whether the manipulation of these levels by antioxidant supplementation has significant effect.

We found that the ability of CoQ_{10} in inhibiting oxidative DNA damage is dependent on the DNA damage profile occurring on DNA molecules and on the oxidizing species inducing the damage. In the study carried out by supplementing healthy volunteers with CoQ_{10}, the improved antioxidant status of $CoQ_{10}H_2$ or CoQ_{10} enriched lymphocytes was confirmed by treating them *in vitro* with H_2O_2 or oxygen: less damage, in terms of DNA strand break accumulation, occurs in supplemented lymphocytes. However, even though there is evidence of a link

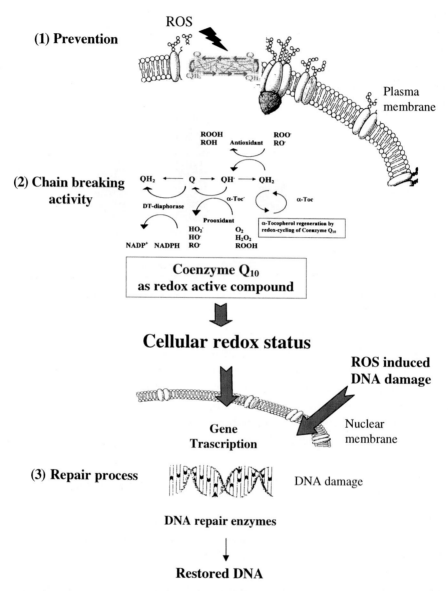

Scheme 1 — Hypothetical scheme illustrating the role of CoQ$_{10}$ in the prevention of DNA oxidative damage. (1) CoQ$_{10}$ may prevent DNA damage either by directly scavenging ROS or by restricting the number of ROS capable to reach DNA due to its ordering and condensing effect within the lipid bilayer of cell membrane; (2) CoQ$_{10}$ inhibits lipid peroxidation by both a chain-breaking mechanism preventing the propagation of peroxidation processes which can generate harmful radicals able to reach DNA and by recycling α-tocopherol; (3) CoQ$_{10}$ may play a role in the transactivation of DNA repair enzymes by affecting the cellular redox status involved in the regulation of transcriptional factors, such as NF-κB and of regulatory proteins governing the transcriptional regulation of formamidopyrimidine (Fpg) DNA glycosilase, a base excision repair proteins. ROS, reactive oxygen species; Q, ubiquinone-10; QH$_2$, ubiquinol-10; QH, ubisemiquinone-10.

between CoQ$_{10}$ content in cells and levels of base oxidation in cellular DNA, the protective activity of CoQ$_{10}$ against oxidative DNA damage is exerted through different mechanisms than the "conventional" one. The enhancement of the endogenous repair enzyme systems leading to a faster recovery from oxidative insult in CoQ$_{10}$ enriched lymphocytes than in native ones, suggests that the antioxidant might influence the intracellular redox status, which in turn stimulates transcription factors involved in gene expression of DNA repair enzymes (see Scheme 1). The ability of both ubiquinol-10 and ubiquinone-10 in enhancing DNA recovery from oxidative damage is a novel aspect of CoQ$_{10}$ antioxidant power, therefore it is worthwhile to further explore and clarify the mechanisms at the basis of DNA protection.

Acknowledgments

Special thanks; to Dr. Collins for giving us the opportunity of working in his lab and for his help in realizing the CoQ$_{10}$ supplementation study; to Dr. M. Tomasetti for his valuable work in carrying out the experiments and his technical assistance in the preparation of the article; to Mr. Keith Smith for revising the English. CoQ$_{10}$ tablets was a generous gift from Pharmanord (Denmark).

References

1. Crane, F., Sun, L. and Sun, E. (1993). The essential functions of Coenzyme Q. *Clin. Invest.* **71**: S55–S59.
2. Navarro, F., Arroyo, A., Martin, S. F., Bello, R. I., de Cabo, R., Burgess, J. R., Navas, P. and Villalba, J. M. (1999). Protective role of ubiquinone-10 in vitamin E and selenium-deficient plasma membranes. *Biofactors* **9**: 163–170.
3. Battino, M., Ferri, E., Gattavecchia, E., Sassi, S. and Lenaz G. (1991). Coenzyme Q$_{10}$ as a possible membrane protectin agent against γ-irradiation damages. *In* "Biomedical and Clinical Aspects of Coenzyme Q" (K. Folkers, Y. Yamamura, and G. P. Littarru, Eds.), Vol. 6, pp. 181–190, Elsevier, Amsterdam.
4. Frei B., Kim, M. C. and Ames, B. (1990). Ubiquinol-10 is an effective lipid-soluble antioxidant at physiological concentration *Proc. Natl. Acad. Sci. USA* **87**: 4879–4883.
5. Stocker, R., Bowry, V. W. and Frei, B. (1991). Ubiquinol-10 protects human low density lipoprotein more efficiently against lipid peroxidation than does alpha-tocopherol. *Proc. Natl. Acad. Sci. USA* **88**: 1646–1650.
6. Kontush, A., Hubner, C., Finckh, B., Kohlschutter, A. and Beisiegel, U. (1995). Antioxidant activity of ubiquinol-10 at physiologic concentrations in human low density lipoproteins. *Biochim. Biophys. Acta* **1258**: 177–187.

7. Alleva, R., Tomasetti, M., Battino, M., Curatola, G. and Littarru, G. P. (1995). The roles of Coenzyme Q_{10} and vitamin E on the preoxidation of human low density subfractions. *Proc. Natl. Acad. Sci. USA* **92**: 9388–9391.

8. Alleva, R., Scaramucci, A., Mantero, F., Bompadre, S., Leoni, L. and Littarru, G. P. (1997). The protective role of ubiquinol-10 against formation of lipid hydroperoxides in human seminal fluid. *Mol. Aspects Med.* **18(Suppl)**: S221–S228.

9. Ernster, L. and Forsmark-Andrèe, P. (1993). Ubiquinol-10: an endogenous antioxidant in aerobic organisms. *Clin. Invest.* **71**: S60–S65.

10. Kagan, V., Serbinova, E. and Packer, L. (1990). Antioxidant effects of ubiquinone-10s in microsomes and mitocondria are mediated by tocoperol recycling. *Biochem. Biophys. Res. Commun.* **169**: 851–857.

11. Ernster, L. and Forsmark-Andrée, P. (1993). Ubiquinol-10: an endogenous antioxidant in aerobic organisms. *Clin. Invest.* **71**: S60–S65.

12. Ernster, L. and Dallner, G. (1995). Biochemical, physiological and medical aspects of ubiquinone-10 function. *Biochim. Biophys. Acta* **1271**: 195–204.

13. Zwizinski, C. W. and Smid, H. H. O. (1992). Peroxidative damage to cardiac mitochondria: identification and purification of modified adenine nucleotide translocase *Arch. Biochem. Biophys.* **2294**: 178–183.

14. Hruszewycz, A. M. and Bergold, D. S. (1990). The 8-hydroxyguanine content of isolated mitochondria increases with lipid peroxidation *Mutat. Res.* **244**: 123–128.

15. Wiseman, H. and Halliwell, B. (1996). Damage to DNA by reactive oxygen and nitrogen species: role in inflammatory disease and progression of cancer. *Biochem. J.* **313**: 17–29.

16. Ames, B. N., Shigenaga, M. K. and Gold, L. S. (1993). DNA lesions, inducibile DNA repair, and cell division: three key factors in mutagenesis and carcinogenesis. *Env. Health Perspect.* **101**: S35–S44.

17. Loft, S. and Poulsen, H. E. (1996). Cancer risk and oxidative DNA damage in man. *J. Mol. Med.* **74**: 297–312.

18. Poulsen, H. E., Prieme, H. and Loft, S. (1998). Role of oxidative DNA damage in cancer initiation and promotion. *Eur. J. Cancer Prev.* **7**: 9–16.

19. Ames, B. N. (1989). Endogenous oxidative DNA damage, ageing, and cancer. *Free Radic. Res. Commun.* **71**: 121–128.

20. Fraga, C. G., Motchnik, P. A., Shigenaga, M. K., Helbock, H. J., Jacob, R. A. and Ames, B. N. (1991). Ascorbic acid protects against endogenous oxidative DNA damage in human sperm. *Proc. Natl. Acad. Sci.* **88**: 11 003–11 006.

21. Janssen, Y. M. V., van Houten, B., Borm, P. J. A. and Mossman. B.T. (1993). Biology of disease: cell and tissue responses to oxidative damage. *Lab. Invest.* **69**(3): 261–271.

22. Lehamann, A. R. (1982). Xeroderma pigmentosum, cockaine syndrome and ataxia-telagiectasia: disorders relating DNA repair to carcinogenesis. *Cancer Surv.* **1**: 93–118.

23. Xanthoudakis, S., Miao, G., Wang, F., Pan, Y. C. and Curran, T. (1992). Redox activation of *Fos-Jun* DNA binding activity is mediated by DNA repair enzyme. *EMBO J.* **9**: 3323–3335.
24. Hirota, K., Matsui, M., Iwata, S., Nishiyama, A., Mori, K. and Yodoi, J. (1997). AP-1 transcriptional activity is regulated by a direct association between thioredoxin and Ref-1 *Proc. Natl. Acad. Sci. USA* **94**: 3633–3638.
25. Schenk, H., Klein, M., Erdbrugger, W., Droge, W. and Schulze-Osthoff, K. (1994). Distinct effects of thioredoxin and antioxidants on the activation of transcription factors NF-kappa B and AP-1 *Proc. Natl. Acad. Sci. USA* **91**: 1672–1676.
26. Van't Veer, P., Jansen, M. C., Klerk, M. and Kok, F. J. (2000). Fruits and vegetables in the prevention of cancer and cardiovascular disease. *Public Health Nutri.* **3**(1): 103–107.
27. Strandhagen, E., Hansson, P. O., Bosaeus, I., Isaksson, B. and Eriksson, H. (2000). High fruit intake may reduce mortality among middle-aged and elderly men. The study of men born in 1913. *Eur. J. Clin. Nutri.* **54**(4): 337–341.
28. Fillion, L., Collins, A. R. and Southon, S. (1998). β-carotene enhances the recovery of lymphocytes from oxidative damage. *Acta Biochim. Pol.* **45**(1): 183–190.
29. Green, M. H. L., Lowe, J. E., Waugh, A. P. W., Aldrige, K. E., Cole, J. and Arlett, C. F. (1994). Effect of diet and vitamin C on DNA strand breakage in freshly-isolated human white blood cells. *Mutat. Res.* **316**: 91–102.
30. Cooke, M. S., Evans, M. D., Podmore, I. D., Herbert, K. E., Mistry, N., Mistry, P., Hickenbothman, P. T., Husseine, A., Griffiths, H. R. and Lunec, J. (1998). Novel repair action of vitamin C upon in vivo oxidative damage. *FEBS Lett.* **363**: 363–367.
31. Royack, G. A,, Nguyen, M. P., Tong, D. C., Poot, M. and Oda, D. (2000). Response of human oral epithelial cells to oxidative damage and the effect of vitamin E. *Oral Oncol.* **36**(1): 37–41.
32. Collins, A. R., Olmedila, B., Southon, S., Granado, F. and Duthie, S. J. (1998). Serum carotenoids and oxidative DNA damage in human lymphocytes. *Carcinogenesis* **19**: 2159–2162.
33. Porrini, M. and Riso, P. (2000). Lymphocyte lycopene concentration and DNA protection from oxidative damage is increased in women after a short period of tomato consumption. *J. Nutri.* **130**(2): 189–192.
34. Torbergsen, A. C. and Collins, A. R. (2000). Recovery of human lymphocytes from oxidative DNA damage; the apparent enhancement of DNA repair by carotenoids is probably simply an antioxidant effect. *Eur. J. Nutri.* **39**(2): 80–85.
35. Anderson, R. F., Amarasinghe, C, Fisher, L. J., Mak ,W. B. and Packer, J. E. (2000). Reduction in free-radical-induced DNA strand breaks and base damage through fast chemical repair by flavonoids. *Free Radic. Res.* **33**(1): 91–103.

36. Pool-Zobel, B. L., Bub, A., Liegibel, U. M., Treptow-van Lishaut, S. and Rechkemmer, G. (1998). Mechanisms by which vegetable consumption reduces genetic damage in humans. *Cancer Epidemiol. Biomarkers Prev.* **7**(10): 891–899.

37. Prieme, H., Loft, S., Nyyssonen, K., Salonen, J. T. and Poulsen, H. E. (1997). No effect of supplementation with vitamin E, ascorbic acid, or coenzyme Q_{10} on oxidative damage estimated by 8-oxo-7,8-dihydro-2′-deoxyguanosine excretion in smokers. *Am. J. Clin. Nutri.* **65**(2): 503–507.

38. Garcia, R., Gonzalez, C. A., Agudo, A. and Riboli, E. (1999). High intake of specific carotenoids and flavonoids does not reduce the risk of bladder cancer. *Nutri. Cancer* **35**(2): 212–214.

39. Singh, N. P., McCoy, M. T., Tice, R. R. and Schneider, E. L. (1988). A simple technique for quantification of low levels of DNA damage in individual cells. *Exp. Cell. Res.* **175**: 184–191.

40. Collins, A. R., Duthie, S. J. and Dobson, V. L. (1993). Direct enzyme detection of endogenous oxidative base damage in human lymphocyte DNA. *Carcinogenesis* **14**: 1733–1735.

41. Tomasetti, M., Littarru, G. P., Stocker, R. and Alleva, R. (1999). Coenzyme Q_{10} enrichment decreases oxidative DNA damage in human lymphocytes. *Free Radic. Biol. Med.* **27**(9/10): 1027–1032.

42. Von Sonnentag, C. (1987). *The Chemical Basis for Radiation Biology*. Taylor and Francis Eds, London.

43. Epe, B. (1995). DNA damage profiles induced by oxidizing agents. *Rev. Physiol. Biochem. Pharmacol.* **127**: 223–249.

44. Churchill, M. E., Peak, J. G. and Peak, M. J. (1991). Repair of near-visible and blue-light-induced DNA single-strand breaks by the CHO cell lines AA8 and EM9. *Photochem. Photobiol.* **54**: 639–644.

45. Carlson, D. A., Seto, S., Wasson, D. B. and Carrera, C. J. (1996). DNA strand breaks, NAD metabolism, and programmed cell death. *Exp. Cell Res.* **164**: 273–281.

46. Cohen, J. J., Duke, R. C., Fadok, V. A. and Sellins, K. S. (1992). Apoptosis and programmed cell death in immunity. *Ann. Rev. Immunol.* **10**: 267–293.

47. Collins, A. R. (1999). Oxidative DNA damage, antioxidants, and cancer. *Bioessay* **21**: 238–246.

48. Lu, R., Nash, H. M. and Verdine, G. L. (1997). A mammalian DNA repair enzyme that excises oxidatively damaged guanines maps to a locus frequently lost in lung cancer. *Curr. Biol.* **7**(6): 397–407.

49. Pflaum, M., Will, O., Mahaler, H.-C. and Epe, B. (1998). DNA oxidation products determined with repair endonucleases in mammalian cells: types, basal levels and influence of cell proliferation. *Free Radic. Res.* **29**(6): 585–594.

50. Parshad, R., Price, F. M., Bohr, V. A., Cowans, K. H., Zujewski, J. A. and Sanford, K. K. (1996). Deficient DNA repair capacity, a predisposing factor in breast cancer. *Br. J. Cancer* **74**(1): 1–5.

51. Yang, L. Y., Li, L., Jiang, H., Shen, Y. and Plunkett, W. (2000). Expression of ERCC1 antisense RNA abrogates gemicitabine-mediated cytotoxic synergism with cisplatin in human colon tumour cells defective in mismatch repair but proficient in nucleotide excision repair. *Clin Cancer Res.* **6**(3): 773–781.

Chapter 47

In Vivo Assessment of Antioxidant Status

Mohsen Meydani* and Ligia Zubik

Mohsen Meydani and **Ligia Zubik** • Vascular Biology Laboratory, Jean Mayer, USDA Human Nutrition Research Center on Aging at Tufts University, Boston, MA; 711 Washington Street, Boston, MA 02111

*Corresponding Author.
Tel: (617) 556-3126, E-mail: MMeydani@HNRC.TUFTS.EDU

1. Summary

Oxidative stress and antioxidants have been recognized to be important factors in the modulation of aging and age-associated diseases. Measuring the *in vivo* antioxidant status would assist in the evaluation of health status and the planning of strategies to reduce the burden of disease, slow the aging process, and reduce the risk of diseases where oxidative stress is believed to be the important player. Measuring the dietary intake of antioxidants by various methods is one approach whereby a rough estimate of antioxidant status can be achieved. However, it does not account for factors such as bioavailability, contribution of non-nutritive antioxidants and endogenous antioxidants to the total antioxidant status. Many sensitive and specific analytical methods are available to measure the level of individual antioxidants in biological samples. The analytical procedures are difficult and tedious; thus most of the studies have limited their analysis to well-known antioxidants such as vitamins C and E, carotenoids, and selenium. Thus, several methods to measure total antioxidant capacity have been developed. The methods differ in the choice of oxidation source as well as target and type of measurement used to detect the oxidized products. Some modifications have improved the utility and application of methods such as oxygen radical absorbance capacity (ORAC) or enhanced chemiluminesence measurements. However, most of the methods mainly measure the contribution of the water-soluble antioxidants and do not recognize the contribution of fat-soluble antioxidants to the total antioxidant capacity in the biological samples. Further, most of the methods overestimate the total antioxidant capacity when the level of uric acid or albumin is elevated, the pathological conditions associated with disease. Thus, the application of methods and an interpretation of the results for the *in vivo* assessment of antioxidant status should be validated with a complementary measurement of specific antioxidants including both water- and fat-soluble antioxidants.

2. Introduction

According to the free radical theory of aging, the constant generation of endogenous reactive oxygen species and exogenous oxidative stress together with an antioxidant defense system in part defines the survival, longevity, and death of an aerobic organism. Thus, many defense mechanisms within the organisms have evolved to limit the levels of reactive oxygen species and to protect the organism from their deleterious effects. The complex antioxidant defense mechanism is an important one, which includes enzymatic and non-enzymatic systems. The enzymatic system is comprised of antioxidant enzymes such as superoxide dismutase (SOD), catalase, glutathione peroxidase/reductase, peroxidases, and DT diaphorases. These enzymes function together in a cooperative manner with other antioxidant defense systems. The non-enzymatic systems are comprised of macromolecules such as albumin and several hydrophilic and hydrophobic small molecule antioxidants such as vitamin C, glutathione, uric acid, ubiquinone, vitamin E, and carotenoids. Several structural defense systems are also in place, which help to reduce oxidative stress such as sequestration of H_2O_2 generating enzymes in peroxisomes and chelation of divalent cations such as iron by ferritin and copper by ceruloplasmin. In addition, the function of other enzymatic systems contributes to the elimination and correction of damaged protein or oxidized DNA. The presence, composition, and activity of these defense systems is dependent on the cell type, tissue, the nature of extra-cellular fluid, and the extent of local and global oxidative stress within an organism. A decrease in the antioxidant defense capacity resulting from a reduced level of endogenous or exogenous antioxidants and due to environmental factors such as radiation, environmental oxidants, infections, and drugs may significantly contribute to the development of chronic and acute conditions as well as the aging process. It has been proposed that the changes in the balance between the antioxidant defense mechanism and oxidative stress is important in the health and disease status, which in turn defines the survival and longevity of an organism.[1–4]

The association of antioxidant capacity with the maximum lifespan potential of animals has been investigated.[5] Maximum lifespan potential in the animal kingdom has been demonstrated to be directly associated with the level of several antioxidant defense systems including SOD, catalase, vitamins C and E, and inversely with the level of oxidative stress. For example, short-lived animals such as rats or mice produce relatively more 8-hydroxy deoxyguanosine (8OH-DG), a damaged product of DNA, compared to long-lived mammals such as monkeys and humans.[5, 6] Additional evidence regarding the association of antioxidant capacity and oxidative stress to the aging process and longevity has emerged from the investigation of increased longevity by dietary restrictions in animal models. In the past two decades, several studies have demonstrated that an increase in longevity by dietary restrictions (i.e. underfeeding without malnutrition), which results in the alteration of the aging process, is believed to be in part mediated through changing the oxidative stress and antioxidant defense status. mRNA expression of Cu–Zn SOD and catalase is reported to decrease with age, while dietary restrictions increased SOD, catalase, and glutathione peroxidase activities and glutathione levels in the livers of aged animals.[7–9] In this paradigm the oxidative stress level was noted to be lower in diet-restricted rats than in rats fed *ad libitum*.[7, 10, 11] Along with the increase of longevity with food restrictions, the capacity of several bodily functions such as immune function[12] and kidney function[13] were improved, and the onset of several age-associated degenerative diseases including spontaneous tumor and cataract formation were delayed.[14, 15] In support of the contribution of increasing the antioxidant defense system and decreasing the oxidative stress to longevity, Sohal *et al.*[16] noted that in genetically modified fruit flies overexpression of the antioxidant enzymes SOD and catalase prolonged the metabolic life of fruit flies while it ameliorated the age-related oxidative damage and susceptibility to molecular damage in response to acute oxidative stress. There is compelling evidence from experimental animals and human epidemiological and clinical studies that dietary antioxidants significantly contribute to health status and may play an important role in the prevention of several age-associated degenerative diseases such as cancer,[6] cardiovascular diseases,[17] and Alzheimer's disease.[18] There is also experimental evidence from rodent studies indicating that increasing the dietary intake of antioxidants may contribute to longevity.[19]

Therefore, considering the potential roles that oxidative stress may play in both the aging process *per se* and the development of age-associated diseases, the determination of specific markers of oxidative stress and the measurement of antioxidant defense capacity along with specific surrogate markers of chronic conditions can be used to assess the current health status of individuals and plan strategies to improve and/or maintain the status as a potential means to improve well being and probable longevity.

The measurement of *in vivo* oxidative stress and antioxidant defense has been a major challenge for scientists in the past decade. Several specific and sensitive methods for the measurement of *in vivo* markers of oxidative stress have been developed, which are discussed in other chapters in this volume. *In vivo* assessment of the antioxidant status in humans is limited by the availability of biological samples for analysis. Biological fluids such as plasma, serum, saliva, and urine are commonly used; blood cells, sperm, epithelial and mucosal cells also can be collected non-invasively. Here, the methodologies dealing with *in vivo* assessment of antioxidants status are reviewed in brief.

3. Dietary Intake of Antioxidants

The measurement of dietary intake of antioxidants is one of the means to assess the antioxidant status of an individual. Dietary recall, dietary record, and a food frequency questionnaire (FFQ) are the common tools to measure dietary intake of energy as well as macro- and micro-nutrients including antioxidant nutrients. The FFQ is the most commonly used tool, which calculates the long-term intake of antioxidants by taking into account the frequency of the foods consumed over a certain period of time and calculating the intake of specific dietary antioxidants using USDA food composition tables. Dietary intake only provides information on the level of antioxidants from exogenous sources, and in certain modified forms of the FFQ, it includes antioxidants from supplemental sources.[20-22] This method has been used to evaluate the association of specific dietary antioxidants with health and diseases status in which antioxidants and oxidative stress are believed to be important modulators of risk factors. It is now well recognized that food composition tables need to include phytochemicals of which many contain antioxidant activity. It is also important to note that dietary assessment methods provide the level of antioxidants present in food but they do not take into account the destruction of antioxidants during the processing of food and the bioavailability of antioxidants, which is affected by food matrix, intestinal microflora, metabolism, and elimination following lymphatic or hepatic absorption.[23] Therefore, the dietary intake measurement provides an approximation of antioxidant status and requires additional complementary measurements.

4. Measurement of Individual Antioxidants

The measurement of plasma levels of antioxidants has been used to assess the antioxidant status of individuals and its relationship to specific conditions. Table 1 shows the level of compounds in plasma with antioxidant activity. Several specific and sensitive methods have been developed to measure major antioxidants in plasma and other biological samples. Although the plasma levels of antioxidants are usually in equilibrium with tissue levels, some antioxidants such as tocopherols and carotenoids, which can accumulate in specific tissues, may not reflect the target tissue level. Further, the transient increases in plasma antioxidant concentrations following the consumption of antioxidant-rich foods may provide

Table 1. Nonenzymatic Antioxidants in Human Plasma[a]

Antioxidant	Plasma Content (μmoles/liter)
Water soluble	
Ascorbate	30–150
Glutathione	1–2
Urate	160–450
Bilirubin	5–20
Lipid soluble (lipoprotein associated)	
α-tocopherol	15–40
γ-tocopherol	3–5
α-carotene	0.05–0.1
β-carotene	0.3–0.6
Lycopene	0.5–1.0
Lutein	0.1–0.3
Zeaxanthin	0.1–0.2
Ubiquinol-10	0.4–1.0

[a]From Sies *et al.*[70]

inaccurate information about the habitual antioxidant status of the individual. Therefore, plasma measurements may reflect the recent, rather than the long-term status, whereas measurement of blood cell's antioxidants may better reflect the long-term antioxidant status of the individual. For example, the determination of white blood cell vitamin C concentration provides a better index of the long-term vitamin C status than measuring the plasma concentration.[24] The total antioxidant capacity of a sample can be calculated by the summation of a measured concentration of individual antioxidants multiplied by the stoichiometric factor available from different sources. This calculation method does not include the antioxidant contribution of unknown compounds with antioxidant activity present in the biological samples such as phytochemicals and those of antioxidant enzymes. Further, measurement of individual antioxidants in the biological samples does not include the contribution to total antioxidant activity from the potential interactions between the antioxidants, which occurs during oxidative stress. For example, tocopherol is believed to be regenerated from its tocopherol radical by the antioxidant action of ascorbate.[25, 26] Ascorbate has been shown to protect the reduced glutathione from oxidation and to recycle flavonoid radicals;[27] also it may synergistically interact with the flavonoid quercetin against oxidative damage.[28]

5. Total Antioxidant Capacity

In order to overcome the above problems and eliminate tedious determination of individual antioxidants, methods capable of measuring antioxidant activity of all the compounds present in a sample with one simple determination is desirable. Several methods have been developed including ORAC (oxygen radical absorbance capacity), TRAP (total radical-trapping antioxidant potential), TEAC (Trolox equivalent antioxidant capacity), FRAP (ferric reducing antioxidant power), luminol-based assay, and DCFH-DA (dichlorofluorescein-diacetate) assay.

These methods claim to be capable of measuring total antioxidant capacity of foods, body fluids, and tissues but each has certain capabilities with some shortcomings, which will be discussed below. These and other methods were devised based on the antioxidants that are present in the biological samples to inhibit oxidative stress, which is artificially introduced into the assay system. Methods differ in the choice of oxidation source as well as the target and type of measurements used to detect the oxidized products. Below are brief descriptions of the most common methods, which have been used to demonstrate and to validate the association of *in vivo* antioxidant status with dietary intake of antioxidants and with certain pathological and physiological conditions.

The **TEAC assay** was originally reported by Miller *et al.*[29] and then modified and is currently available as a commercialized kit from Randox Laboratories

(San Francisco, CA). The assay is based on the generation of ABTS$^{\cdot+}$ radical monitored spectrophotemetrically at 734 nm. The assay uses ABTS [2,2'-azino bis(3-ethylbenzothiazoline-6-sulfonic acid)] and H_2O_2 to generate ABTS$^{\cdot+}$ radical cations in the presence of metmyoglubolin as a peroxidase. Thus the antioxidant suppression of absorbance is directly related to the antioxidant capacity of testing material in solution. In this assay, H_2O_2 is reacted with metmyoglubolin to generate ferrylmyoglubolin. The inhibition of the absorbance over a fixed time (3 min in the Randox kit) is used to calculate the TEAC as it is defined based on the equivalent of mM concentration of Trolox (6-hydroxyl-2,5,7,8,-tetramethylchroman-2-carboxylic acid) inhibition of ABTS$^{\cdot+}$ radical.

The **ORAC assay** is also an inhibition assay, which is based on the original assay developed by Glazer[30] using phycoerythrin, an oxidizable protein as the target substrate for free radicals generated by 2,2'-azobis(2-amidinopropane) dihydrochloride (AAPH). AAPH produces peroxyl radicals at a constant rate in a set temperature. The generated peroxyl radicals react with phycoerythrin and decrease its fluorescence intensity. The fluorescence decay curve of phycoerythrin can be monitored over the period of time by spectrofluorometer set at excitation wavelength of 540 nm and an emission wavelength of 565 nm. The presence of antioxidants in the test material in the solution quenches peroxyl radicals generated from AAPH and delays the onset of phycoerythrin fluorescence decay. Thus, the antioxidant capacity of testing material is proportional to the extent of inhibition (lag time) of fluorescence decay. Since the kinetics of phycoerythrin fluorescence quenching is not linear in the presence of peroxyl radicals and the lag time is difficult to determine, the modified method has been adopted by Cao *et al.*,[31] which uses the area under the curve, taking into account the lag time and the slope of the completed decay curve in the measurement. Differences of area under the curve between the blank and sample is calculated and converted to ORAC value using the Trolox as a control standard for total antioxidant capacity.

The **TRAP assay**, which is the most widely used assay developed by Wayner *et al.*,[32] is based on the measurement of oxygen consumption using oxygen electrodes during the oxidation of oxidizable materials in the test compound. The induction of oxidation and oxygen consumption is initiated by the addition of AAPH to generate peroxyl radicals in the test solution and the presence of antioxidants in the test compound delays the induction of oxidation. In the modified method,[33] linoleic acid is added prior to initiation to warrant the oxidation process to proceed. The delay time in the induction of oxidation is proportional to the presence of antioxidants in solution and is compared to the delay time resulting from the addition of Trolox as an internal standard.

The **FRAP assay** is claimed to be a direct measure of the total antioxidant activity of biological fluids reported by Benzi and Strain.[34] The assay system is based upon the reduction of ferric tripyridyl triazine (Fe^{3+}-TPTZ) complex to ferrous tripyridyl triazine (Fe^{2+}-TPTZ) by antioxidants in an acidic condition. The generation of ferrous tripyridyl triazine (Fe^{2+}-TPTZ) is monitored by spectro-photometer at 593 nm. Trolox is used as a standard to calculate the mM equivalent of total antioxidant capacity.

The *Luminol-based assay* is an inhibition assay, which is based on the prevention of luminol oxidation by the antioxidants present in the test sample. The luminol-based assay[35] uses AAPH to generate peroxyl radicals to oxidize luminol producing luminol radicals that emit light and is measured by a luminometer. The duration of the chemiluminescence inhibition by the sample is directly proportional to its total antioxidant potential. The enhanced chemiluminescent assay is a modified version reported by Whitehead *et al.*,[36] which uses the enzyme horseradish peroxidase to catalyze the reaction between a hydrogen accepter such as H_2O_2 or perborate, and a hydrogen doner, luminol. The kinetics of reaction, which is fast with low intensity light emission, can be influenced favorably by the addition of an enhancer such as *para*-idophenol, which produces a constant source of free radical intermediates and more intense, stable and prolonged chemiluminescent light emission. The time period of light suppression is directly related to the amount of antioxidant present in the test sample and total capacity is determined relative to Trolox.

The *dichlorofluorescin-diacetate (DCFH-DA) assay* is also an inhibition assay, which uses AAPH as a peroxyl radical generator to oxidize DCFH-DA to dichlorofluorescein and which can be monitored either fluorometrically (Excitation: 480 nm, Emission: 526 nm) or spectrophotometrically at 504 nm.[37] The presence of antioxidants in the test sample results in a delay (lag-time) in the induction of oxidation of DCFH-DA, which is compared to lag-time resulting from the addition of standard Trolox during the propagation phase following the consumption of antioxidants of the sample.

Recently, other new methods have been published including TOSC (total oxyradical scavenging capacity) assay, crosin-based assays, and cyclic voltammetry. The TOSC assay[38] is based on the oxidation of 2-keto-4-methyiolbutyric acid (KMBA) to ethylene by free radicals generated from AAPH and ethylene formation is inhibited by the antioxidant present in the sample. The crosin-based assay[39] is based on the sample's antioxidant inhibition of bleaching of the crosin by peroxyl radicals produced from AAPH. The cyclic voltammetry[40] is based on the evaluation of the tracing of the biological oxidation potential and reducing the power of low molecular weight antioxidants present in plasma. These assays however have been less commonly used in the assessment of antioxidant status as related to dietary intake or health status.

Using the above-mentioned methods, several studies have reported an association between the measurement of total antioxidant capacity with oxidative stress conditions or with the intake of foods with known antioxidant activities. For example, depletion of the total antioxidant capacity is reported to be associated with a higher incidence of diabetic complication. Opara *et al.*[41] reported that in the Type-2 diabetic patient's serum, TEAC was lower than control. Further, in diabetic patients with proteinuria, the TEAC value was even lower than those without proteinuria. Ventura *et al.*,[42] using the Randox kit, reported that with the induction of hyperhomocysteinemia by methionine oral loading, the serum TEAC value decreased. An increase of serum ORAC value following the consumption of strawberries, spinach, red wine, or vitamin C in elderly women has been demonstrated.[43] Day *et al.*[44] observed an increase of the total antioxidant capacity as measured by enhanced chemiluminescence TRAP assay following the consumption of concentrated red grape juice.

6. Pitfalls in the Methods of Assessment

There are increasing number of methods being developed to determine the antioxidant profile of individuals or compounds and extracts. Measuring dietary intake is frequently used to assess the antioxidant status of individuals in epidemiological and clinical studies and it is rarely coupled with actual *in vivo* measurements of serum antioxidants profile. As pointed out, the calculation of antioxidant intake from diet only accounts for the compounds that have been identified and known to have antioxidant activity and included in the USDA food composition tables. Currently, food composition tables do not include the non-nutritive components of foods with antioxidant activity such as polyphenoles. Sporadic data are now available on the total antioxidant capacity of a limited numbers of fruit and vegetables using some of the currently available methods.[45, 46] The compilation of these data after validation can be potentially used to calculate total dietary antioxidant intake of individuals from food sources. However, the bioavailability of the total antioxidants from food sources, which is affected by food composition, matrix, and metabolic characteristics of individuals, remains the major obstacle in the actual assessment of *in vivo* antioxidant status using the dietary intake data.

Many sensitive and specific analytical procedures are being developed to measure individual antioxidants in biological samples such as plasma, serum, and whole blood and saliva. However, most of the reports include data for those of the well known antioxidants such as vitamins C, E, total carotenoids or β-carotene and selenium, and most recently other carotenoids such as lycopene, lutein, β-carotene, and zeaxanthin have been included.[47] Rarely, reported studies contain a complete profile of plasma antioxidants as listed in Table 1. It is also important to note that the antioxidant potential of phytochemicals from fruit and vegetables[43, 48] can contribute substantially to the total antioxidant capacity of plasma in individuals or the population whose dietary habits consist of a high consumption of fruits and vegetables. The analytical limitations have been the major obstacle in the determination and measurement of these compounds in the biological samples.

Therefore, for a complete assessment of antioxidants' status and an understanding of their role in health and disease, well-validated assays for measuring total antioxidant capacity are needed. Most of the assays listed for the measurement of total antioxidant capacity could be used on plasma or serum samples. Prior and Cao[49] have recently reviewed different analytical methods for measuring total antioxidant capacity and have pointed out the problems associated with each one of the assays. Here, we highlight some of them in brief.

In the TEAC assay, the scavenger effect of antioxidants and the decrease in the rate of ABTS[.+] formation are not recognized because absorbance is measured at fixed time points regardless of ABTS[.+] formation.[50, 51] This time dependency was also observed with the TEAC value measured for quercetin at 6 min, which was lower than when it was measured after 10 minutes.[50] Strube *et al.*[50] used the post addition assay to circumvent the effect of the compounds on radical formation. Therefore, they added the test compound when ABTS[.+] had already formed. Using this assay, plasma antioxidant capacity was found to be surprisingly higher in myocardial infarction patients.[52] This assay may be better used for the *in vitro* measurement of the antioxidant capacity of food products rather than for *in vivo* situations. Valuable information has been generated on the antioxidant activities of phytochemicals using this assay.[53–56]

The measurement of area under the curve as indicated by Cao *et al.*[31] proves to be more useful and better than measuring the lag time in the phycoerythrine decay assay or ORAC. The ORAC assay is the only method that takes free radical reaction to completion and combines both inhibition percentage and the length of inhibition time of the free radical reaction into a single value. However, the assay requires a longer time than the FRAP and TEAC assays, and like most of the above-listed assays, it does not detect the contribution of lipid soluble antioxidants such as vitamin E or carotenoids present in the lipoprotein particles into the total antioxidant capacity. For example, despite a high level of serum vitamin E compared to controls in non-insulin-dependent diabetes mellitus (NIDDM) patients, the TRAP assay (measured by phycoerythrine fluorescence lag time) showed a lower serum total antioxidant capacity than controls.[57] As shown by Cao and Prior,[58] there is a weak linear correlation between serum ORAC and FRAP measurements. However, the FRAP assay, even though it is simple and inexpensive, does not measure the SH-group-containing antioxidants such as lipoic acid and glutathione, which are important antioxidants *in vivo*. Further, it does not measure the antioxidant capacity of serum albumin, which is a major protein in serum. Furthermore, Cao and Prior[58] observed no correlation between serum ORAC and TEAC or between serum TEAC and FRAP values.

The shortcoming of the TRAP assay using oxygen consumption is related to the instability of oxygen electrodes, which are needed for the duration of the measurement.[58] Using the enhanced chemiluminscence in the TRAP assay has produced inconsistent results in relation to the extent of oxidative stress *in vivo*.

For example, vascular patients who developed systemic inflammatory reaction following revascularization had a significantly reduced level of serum TRAP compared to controls.[59] Whereas in septic shock patients, the serum TRAP value was reported to be higher as opposed to lower than those of the control subjects.[60] The high level of bilirubin and uric acid were the major contributors to the increased serum TRAP value in this circumstance. Also following a half-marathon run, which resulted in significant increase in oxidative stress, the runner's TRAP value was increased due to increase of serum uric acid level.[61] Interestingly, in patients with active atherosclerosis, serum TEAC was elevated.[62] Further, Nieto *et al.*[63] reported that in atherosclerotic patients, the ORAC value was higher than control mainly due to a high serum level of uric acid. It is important to note that a high level of serum uric acid — contrary to the suggestion[63] that it is an adoptive and compensatory mechanism of a pathological condition — can have a deleterious effect. For example uric acid may increase platelet adhesiveness,[64-66] which can lead to thrombosis. Also high serum uric acid is reported to be associated with an increased risk of coronary heart disease.[63] The contribution of uric acid to a high level of total antioxidant value as measured in saliva and plasma was also observed in patients who were undergoing kidney dialysis, which decreased serum uric acid and therefore dropped the TEAC value.[67] Thus, it is important to note that a false high serum TRAP value can result from a pathological dysfunction of the liver or the kidneys in which the serum albumin, bilirubin, or uric acid is increased significantly and contributes to a higher serum TRAP value.

In HIV infected patients, no correlation was found between serum hydroperoxides level and TEAC value.[68] In experimental diabetes induced by streptozotocin in rat models while lipid peroxidation was increased, the total plasma antioxidant as measured by TRAP method of enhanced chemiluminescence method tended to increase due to an increase in the vitamin E level. When measured by TEAC (using Randox Kit), however, the total antioxidant capacity decreased due to decrease of albumin level.[69] Therefore, depending on the choice of the methods being used, the total antioxidant capacity might be over- or under-estimated or it might not correlate with the oxidative stress condition.

7. Conclusion

Several approaches exist to assess the *in vivo* status of antioxidants, each of which has certain advantages and shortcomings. The dietary assessment methods are relatively easy and practical, but provide only rough estimates whereas the measurements of individual antioxidants are limited by the difficulties of the analytical techniques. At present, there is no single perfect method available to measure total antioxidant capacity, which would include antioxidant activity in both aqueous and lipid compartment of plasma or other biological samples. The modified ORAC assay and enhanced chemiluminesence procedure relative to currently available methods appear to be more useable. Recently, a method has been proposed by Russell's group at the Jean Mayer USDA-Human Nutrition Research Center on Aging at Tufts University (personal communication), which treats the plasma sample with both a hydrophilic radical generator AAPH and a lipophilic radical generator MeO-AMVN (2.2'-azobis(4-methoxy-2,4-dimethylvaleronitrile) and monitors the oxidation of aqueous compartment with DCFH-AC and lipid compartment with BODYP (4,4-difuoro-5-(4-phenyl-1,3-butadenyl)-4-bora-3a,4a-diaza-s-indacene-3-undecanic acid. This approach appears to be promising since it would detect the contribution of both fat-soluble and water soluble antioxidants to the total antioxidant capacity. The proposed method and its utility needs to be validated with other analytical procedures and by different states of oxidative stress and antioxidant supplementation.

Acknowledgments

This project was funded in part from Federal funds from the US Department of Agriculture, Agriculture Research Service under contract number 58-1950-9-001. The contents of this publication do not necessarily reflect the views or policies of the US Department of Agriculture, nor does mention of trade names, commercial products or organizations imply endorsement by the US Government. The authors thank Stephanie Marco for the preparation of this manuscript.

References

1. Yu, B. P. (1993). Oxidative damage by free radicals and lipid peroxidation in aging. *In* "Free Radicals in Aging" (B. P. Yu, Ed.), pp. 57–88, CRC Press, Boca Raton, FL.
2. Halliwell, B. (1987). A radical approach to human disease. *In* "Oxygen Radicals and Tissue Injury" (B. Halliwell, Ed.), pp. 139–143, FASEB J, Bethesda.
3. Halliwell, B. (1994). Free radicals and antioxidants: a personal view. *Nutri Rev.* **52**: 253–265.
4. Harman, D. (1994). Aging: prospect for further increases in the functional life span. *Age* **17**: 119–146.
5. Cutler, R. G. (1991). Antioxidants and aging. *Am. J. Clin. Nutri.* **53**: S373–S379.
6. Ames, B. N., Shigenaga, M. K. and Hagen, T. M. (1993). Oxidants, antioxidants, and the degenerative diseases of aging. *Proc. Natl. Acad. Sci. USA* **90**: 7915–7922.
7. Rao, G., Xia, E., Nadakavukaren, M. J. and Richardson, A. (1990). Effect of dietary restriction on the age-dependent changes in the expression of antioxidant enzymes in rat liver. *J. Nutri.* **120**: 602–609.

8. Richardson, A., Simsei, I., Rutherford, M. S. and Butler, J. A. (1987). Effect of dietary restriction of the expression of specific genes. *Fed. Proc.* **47**: 568.

9. Mune, M., Meydani, M., Jahngen-Hodge, J., Martin, A., Smith, D., Palmer, V., Blumberg, J. B. and Taylor, A. (1995). Effect of calorie restriction on liver and kidney gluthathione in aging emory mice. *Age* **18**: 43–49.

10. Sohal, R. S., Agarwal, S., Candas, M., Forster, M. J. and Lal, H. (1994). Effect of age and caloric restriction on DNA oxidative damage in different tissues of C57BL/6 mice. *Mech. Aging Dev.* **76**: 215–224.

11. Koizumi, A., Weindruch, R. and Walford, R. L. (1987). Influences of dietary restriction and age on liver enzyme activities and lipid peroxidation in mice. *J. Nutri.* **117**: 361–367.

12. Fernandes, G., Venkatraman, J. T., Turturro, A., Attwood, V. G. and Hart, R. W. (1997). Effect of food restriction on life span and immune functions in long-lived Fischer-344 x Brown Norway F1 rats. *J. Clin. Imnunol.* **17**: 85–95.

13. Teillet, L., Verbeke, P., Gouraud, S., Bakala, H., Borot-Laloi, C., Heudes, D., Bruneval, P. and Corman, B. (2000). Food restriction prevents advanced glycation end product accumulation and retards kidney aging in lean rats. *J. Am. Soc. Nephrol.* **11**: 1488–1497.

14. Taylor, A., Lipman, R. D., Jahngen-Hodge, J., Palmer, V., Smith, D., Padhye, N., Dallal, G. E., Cyr, D. E., Laxman, E., Shepard, D., Marrow, F., Salmon, R., Perrone, G., Asmundsson, G., Meydani, M., Blumberg, J., Mune, M., Harrison, D., Archer, J. R. and Shigenaga, M. (1995). Dietary calorie restriction in the Emory mouse: effect on lifespan, eye lens cataract prevalence and progression, level of ascorbate, glutathione, glucose, and glycohemoglubin, tail collagen breaktime, DNA and RNA oxidation, skin integrity, fecundity, and cancer. *Mech. Ageing Dev.* **79**: 33–57.

15. Taylor, A., Jahngen-Hodge, J., Smith, D. E., Palmer, V. J., Dallal, G. E., Lipman, R. D., Padhye, N. and Frei, B. (1995). Dietary restriction delays catract and reduces ascorbate levels in Emory mice. *Exp. Eye Res.* **61**: 55–62.

16. Sohal, R. S., Agarwal, A., Agarwal, S. and Orr, W. C. (1995). Simultanous overexpression of copper- and zinc-containing superoxide dismutase and catalase retard age-related oxidative damage and increases metabolic potential on *Drosophila melanogaster*. *J. Biol. Chem.* **270**: 15 671–15 674.

17. Meydani, M. (1995). Vitamin E. *Lancet* **345**: 170–175.

18. Sano, M., Ernesto, M. S., Thomas, R. G., Klauber, M. R., Schafer, K., Grundmer, M., Woodbury, P., Growder, J., Cotman, C. W., Pfeiffer, E., Schneider, L. S. and Thal, L. J. (1997). A controlled trial of selegiline, alpha-tocopherol, or both as treatment for Alzheimer's disease. *New England J. Med.* **336**: 1216–1222.

19. Bezlepkin, V. G., Siroat, N. P. and Gaziev, A. I. (1996). The prolongation of survival in mice by dietary antioxidants depends on their age by the start of feeding this diet. *Mech. Ageing Dev.* **92**: 227–234.

20. Block, G. (1991). Vitamin C and cancer prevention: the epidemiologic evidence. *Am. J. Clin. Nutri.* **53**: S270–S282.

21. Block, G., Patterson, B. and Subar, A. (1992). Fruit, vegetables, and cancer prevention: a review of the epidemiological evidence. *Nutri. Cancer* **18**: 1–29.

22. Stampfer, M. J., Hennekens, C. H., Manson, J. E., Colditz, G. A., Rosner, B. and Willett, W. C. (1993). Vitamin E consumption and the risk of coronary disease in women. *New England J. Med.* **328**: 1444–1449.

23. Papas, A. M. (1999). Diet and antioxidant status. *In* "Antioxidant Status, Diet, Nutrition, and Health" (A. M. Papas, Ed.), pp. 89–106, CRC Press, New York.

24. Omaye, S. T., Schaus, E. E., Kutnink, M. A. and Hawkes, W. C. (1987). Measurement of vitamin C in blood components by high-performance liquid chromatography. Implication in assessing vitamin C status. *Ann. NY Acad. Sci.* **498**: 389–401.

25. Niki, E. (1987). Interaction of ascorbate and α-tocopherol. *Ann. NY Acad. Sci.* **498**: 186–199.

26. Igarashi, O., Yonekawa, Y. and Fujiyama-Fujihara, Y. (1991). Synergestic action of vitamin E and vitamin C in vivo using a new mutant of Wistar-strain rats, ODS, unable to synthesize vitamin C. *J. Nutri. Sci. Vitaminol.* **37**: 359–369.

27. Bors, W., Michel, C. and Schikora, S. (1995). Interaction of flavonoids with ascorbate and determination of their univalent redox potentials: a pulse radiolysis study. *Free Radic. Biol. Med.* **19**: 45–52.

28. Skaper, S. D., Fabris, M., Ferrari, V., Dalle Carbonare, M. and Leon, A. (1997). Quercetin protects cutaneous tissue-associated cell types including sensory neurons from oxidative stress induced by glutathione depletion: cooperative effects of ascorbic acid. *Free Radic. Biol. Med.* **22**: 669–678.

29. Miller, N. J., Rice-Evans, C., Davis, M. J., Gopinathan, V. and Milner, A. (1993). A novel method for measuring antioxidant capacity and its application to monitoring the antioxidant status in premature neonates. *Clin. Sci.* **84**: 407–412.

30. Glazer, A. N. (1990). Phycoerythrin fluorescence-based assay for reactive oxygen species. *Meth. Enzymol.* **186**: 161–168.

31. Cao, G., Alessio, H. M. and Cutler, R. G. (1993). Oxygen-radical absorbance capacity assay for antioxidants. *Free Radic. Biol. Med.* **14**: 303–311.

32. Wayner, D. D., Burton, G. W., Ingold, K. U. and Locke, S. (1985). Quantitative measurement of the total, peroxyl radical-trapping antioxidnt capacity of human blood plasma by controlled peroxidation. *FEBS Lett.* **187**: 33–37.

33. Wayner, D. D., Burton, G. W. and Ingold, K. U. (1986). The antioxidant efficiency of vitamin C is concentration-dependent. *Biochem. Biophys. Acta* **884**: 119–123.

34. Benzie, I. F. F. and Strain, J. J. (1999). Ferric reducing/antioxidant power assay: direct measure of total antioxidant activity of biological fluids and modified version for simultanous measurement of total antioxidant power and ascorbic acid concentration. *Meth. Enzymol.* **299**: 15–27.

35. Alho, H. and Leinonen, J. (1999). Total antioxidant activity measured by chemiluminescence methods. *Meth. Enzymol.* **299**: 3–14.
36. Whitehead, T. P., Robinson, D., Allaway, S., Syms, J. and Hale, A. (1995). Effect of red wine ingestion on the antioxidant capacity of serum. *Clin. Chem.* **41**: 32–35.
37. Valkonen, M. and Kuusi, T. (1997). Spectrophotometric assay for total peroxyl radical-trapping antioxidant potential in human serum. *J. Lipid Res.* **38**: 823–833.
38. Winston, G. W., Regoli, F., Dugas, A. J. J., Fong, J. H. and Blanchard, K. A. (1998). A rapid gas chromatographic assay for determining oxyradical scavenging capacity of antioxidants and biological fluids. *Free Radic. Biol. Med.* **24**: 480–493.
39. Tubaro, F., Ghiselli, A., Papuzzi, P., Maiorino, M. and Urcini, F. (1998). Analysis of plasma antioxidant capacity by competition kinetics. *Free Radic. Biol. Med.* **24**: 1228–1234.
40. Kohen, R., Beit-Yannai, E., Berry, E. M. and Tirosh, O. (1999). Overall low molecular weight antioxidant capacity of biological fluids and tissues by cyclic voltammetry. *Meth. Enzymol.* **300**: 285–296.
41. Opara, E. C., Abdel-Rahman, E., Soliman, S., Kamel, W. A., Souka, S., Lowe, J. E. and Abdel-Aleem, S. (1999). Depletion of total antioxidant capacity in Type-2 diabetes. *Metabolism* **48**: 1414–1417.
42. Ventura, P., Panini, R., Verlato, C., Scarpetta, G. and Salvioli, G. (2000). Peroxidation indices and total antioxidant capacity in plasma during hyperhomocysteinemia induced by methionine oral loading. *Metabolism* **49**: 225–228.
43. Cao, G., Russell, R. M., Lischner, N. and Prior, R. L. (1998). Serum antioxidant capacity is increased by consumption of strawberries, spinach, red wine or vitamin C in elderly women. *J. Nutri.* **128**: 2383–2390.
44. Day, P. A., Kemp, H. J., Bolton, C., Hartog, M. and Stansbie, D. (1997). Effect of concentrated red grape juice consumption on serum antioxidant capacity and low-density lipoprotein oxidation. *Ann. Nutri. Metab.* **41**: 35–357.
45. Wang, H., Cao, G. and Prior, R. L. (1996). Total antioxidant capacity of fruits. *J. Agric. Food Chem.* **44**: 701–705.
46. Cao, G., Sofic, E. and Prior, R. L. (1996). Antioxidant capacity of tea and common vegetables. *J. Agric. Food Chem.* **44**: 3426–3431.
47. Fotouhi, N., Meydani, M., Santos, M. S., Meydani, S. N., Hennekens, C. H. and Gaziano, J. M. (1996). Carotenoids and tocopherol concentration in plasma, peripheral blood mononuclear cells and red blood cells following long term β-carotene supplementation in men. *Am. J. Clin. Nutri.* **63**: 553–558.
48. Cao, G., Booth, S. L., Sadowski, J. A. and Prior, R. L. (1998). Increases in human plasma antioxidant capacity following consumption of controlled diet high in fruits and vegetables. *Am. J. Clin. Nutri.* **68**: 1081–1087.

49. Prior, R. L. and Cao, G. (1999). *In vivo* total antioxidant capacity: comparison of different analytical methods. *Free Radic. Biol. Med.* **27**: 1173–1183.

50. Strube, M., Haenen, G. R. M. M., van Den Berg, H. and Bast, A. (1997). Pitfalls in a method for assessment of total antioxidant capacity. *Free Radic. Res.* **26**: 515–521.

51. Arnao, M. B., Cano, A., Hernabdezruiz, J., Garciacanovas, F. and Costa, M. (1996). Inhibitions by L-ascorbic acid and other antioxidants of 2,2'-azino-bis(3-ethylbenzthiazoline-6-sulfonic acid) oxidation catalyzed by peroxidase: a new approach for determining total antioxidant status of foods. *Anal. Biochem.* **236**: 255–261.

52. Guler, K., Palanduz, S., Ademoglu, E., Sahnayenli, N., Gokkusu, C. and Vatansever, S. (1998). Total antioxidant status, lipid parameters, lipid peroxidation and glutathione levels in patients with acute myocardial infarction. *Med. Sci. Res.* **26**: 105–106.

53. Rice-Evens, C., Miller, N. J., Bolwell, P. G., Bramley, P. M. and Pridham, J. B. (1995). The relative antioxidant activities of plant-derived polyphenolic flavonoids. *Free Radic. Res.* **22**: 375–383.

54. Saleh, N., Miller, N. J., Paganga, G., Tijburg, L., Boewell, G. P. and Rice-Evans, C. (1995). Polyphenolic flavonoids as scavengers of aqueous phase radicals and as chain-breaking antioxidants. *Arch. Biochem. Biophys.* **322**: 339–346.

55. Miller, N. J., Sampson, J., Candeias, L. P., Bramley, P. M. and Rice-Evans, C. A. (1996). Antioxidant activities of carotenes and xanthophylls. *FEBS Lett.* **384**: 240–242.

56. Miller, N. J., Castelluccio, C., Tijburg, L. and Rice-Evans, C. (1996). The antioxidant properties of theaflavins and their gallate esters-radical scavengers or metal chelators? *FEBS Lett.* **392**: 40–44.

57. Ceriello, A., Bortolotti, N., Pirisi, M., Crescentini, A., Tonutti, L., Motz, E., Russo, A., Giacomello, R., Stel, G. and Taboga, C. (1997). Total plasma antioxidant capacity predicts thrombosis-prone status in NIDDM patients. *Diabetes Care* **20**: 1589–1593.

58. Cao, G. and Prior, R. L. (1998). Comparison of different analytical methods for assessing total antioxidant capacity of human serum. *Clin. Chem.* **44**: 1309–1315.

59. Spark, J. I., Chetter, I. C., Gallavin, L., Kester, R. C., Guillou, P. J. and Scott, D. J. A. (1998). Reduced total antioxidant capacity predicts ischaemia-reperfusion injury after femorodisal bypass. *Br. J. Surgery* **85**: 221–225.

60. Pascual, C., Karzai, W., Meier-Hellmann, A., Oberhoffer, M., Horn, A., Bredle, D. and Reinhart, K. (1998). Total antioxidant capcity is not always decreased in sepsis. *Crit. Care Med.* **26**: 705–709.

61. Child, R. B., Wilkinson, D. M., Fallowfield, J. L. and Donnelly, A. E. (1998). Elevated serum antioxidant capacity and plasma malondialdehyde

concentration in response to a simulated half-marthon run. *Med. Sci. Sports Exer.* **30**: 1603–1607.

62. Reinsch, N., Kiechl, S., Mayr, C., Schratzeberger, P., Dunzendrofer, S., Kahler, C. M., Buratti, T., Willeit, J. and Wiedermann, C. J. (1998). Association of high plasma antioxidant capacity with new lesion formation in carotid atherosclerosis: a prospective study. *Eur. J. Clin. Invest.* **28**: 787–792.

63. Nieto, F. J., Iribarren, C., Gross, M. D., Comstock, G. W. and Cutler, R. G. (2000). Uric acid and serum antioxidant capacity: a reaction to atherosclerosis? *Atherosclerosis* **148**: 131–139.

64. Emmerson, B. T. (1979). Atherosclerosis and urate metabolism. *Aust. NZ J. Med.* **9**: 451–454.

65. Mustard, J. F., Murphy, E. A., Ogryzlo, M. A., Smyth, H. A. *et al.* (1963). Blood coagulation and platelet economy in subjects with primary goat. *Can. Med. Assoc. J.* **89**: 1207–1211.

66. Newland, H. (1975). Hyperuricemia in coronary, cerebral, and peripheral arterial disease: an explanation. *Med. Hypotheses* **1**: 152–155.

67. Meucci, E., Littarru, C., Deli, G., Luciani, G., Tazza, L. and Littarru, G. P. (1998). Antioxidant status and dialysis: plasma and saliva antioxidant activity in patients with fluctuating urate levels. *Free Radic. Res.* **29**: 367–376.

68. McLemore, J. L., Beeley, P., Thorton, K., Morrisroe, K., Blackwell, W. and Dasgupta, A. (1998). Rapid automated determination of lipid hydroperoxide concentrations and total antioxidant status of serum samples from patients infected with HIV. *Am. J. Clin. Pathol.* **109**: 268–273.

69. Feillet-Coudray, C., Rock, E., Coudray, C., Grzelkowska, K., Azaia-Braesco, V., Darvet, D. and Mazur, A. (1999). Lipid peroxidation and antioxidant status in experimental diabetes. *Clin. Chim. Acta* **31**: 31–43.

70. Sies, H., Stahl, W. and Sundquist, A. R. (1992). Antioxidant functions of vitamins. *Ann. NY Acad. Sci.* **669**: 7–20.

Index